Theory of Nucleus

Fundamental Theories of Physics

An International Book Series on The Fundamental Theories of Physics: Their Clarification, Development and Application

Volume 84

Theory of Nucleus

Nuclear Structure and Nuclear Interaction

by

Alex Sitenko

Bogolyubov Institute for Theoretical Physics,
Ukrainian Academy of Sciences,
Kiev, Ukraine

and

Victor Tartakovskii

Department of Physics,
Taras Shevchenko University,
Kiev, Ukraine

KLUWER ACADEMIC PUBLISHERS

DORDRECHT / BOSTON / LONDON

Library of Congress Cataloging-in-Publication Data

Sitenko, A. G. (Alekseĭ Grigor'evich)
Theory of nucleus : nuclear structure and nuclear interaction / by Alex Sitenko and Victor Tartakovskii.
p. cm. -- (Fundamental theories of physics ; v. 84)
Includes bibliographical references and index.
ISBN 0-7923-4423-5 (alk. paper)
1. Nuclear structure. 2. Nuclear reactions. 3. Particles (Nuclear physics) I. Tartakovskiĭ, V. K. (Viktor Konstantinovich) II. Title. III. Series.
QC793.3.S8S58 1997
539.7'4--dc21 96-52452

ISBN 0-7923-4423-5

Published by Kluwer Academic Publishers,
P.O. Box 17, 3300 AA Dordrecht, The Netherlands.

Sold and distributed in the U.S.A. and Canada
by Kluwer Academic Publishers,
101 Philip Drive, Norwell, MA 02061, U.S.A.

In all other countries, sold and distributed
by Kluwer Academic Publishers,
P.O. Box 322, 3300 AH Dordrecht, The Netherlands.

The manuscript was translated from the
Russian by Dr. V. Malnev

Printed on acid-free paper

Printed in the Netherlands

CONTENTS

PREFACE

Modern nuclear physics is a well developed branch of physical science, with wide-ranging applications of its results in engineering and industry. At the same time, the development of a consistent theory of nuclei and nuclear processes presents certain problems.

It is well known that the most important aim of nuclear physics is the study of nuclear structure and the explanation of properties on the basis of the interaction between nucleons which constitute nuclei. Difficulties of a modern theory of the nucleus are caused by both an insufficient knowledge of nuclear interactions and the multiparticle character of nuclear systems. Experimental data on nuclear interactions do not contradict the hypothesis of the pair character of nuclear forces. However, the absence of rigorous methods of calculations of many particle nuclear systems with strong interaction makes it necessary to use macroscopic nuclear models to describe particular nuclear properties. Nuclear models have been developed in different ways, and the models themselves have been modified and complicated. In spite of the visible discrepancy, different models of the nucleus significantly supplement one another. The development of nuclear models has led to considerable progress in the understanding of atomic nuclei. The current results of theoretical nuclear physics are reported in numerous scientific papers. The most important and relevant experimental and theoretical results can be found in many monographs, the best of which are written by well-known experts in the field.

This book represents the expanded versions of the lectures on the theory of the nucleus which have been taught by the authors for many years in Kiev University. It considerably differs from a previous book (A. Sitenko and V. Tartakovskii, Lectures on the theory of nuclei, *Pergamon Press, Oxford, 1975) in content and in the scope of the material included. The book presents the fundamentals of modern ideas of the structure of the atomic nucleus. Different nuclear models and their correlation are considered in detail. Emphasis is made on the fundamentals of the theory of the nucleus but recent results are also presented.*

The first chapter deals with nucleon properties and the main peculiarities of nuclear interactions. The analysis of different phenomenological potentials and calculations of particular properties of two-nucleon systems can be found here.

In the second chapter, the properties of three-nucleon nuclei and unbound three-nucleon systems, obtained from experiments, and relevant theoretical methods of the analysis of three-nucleon systems are described. The latter are harmonic polynomials, operators and Green functions in the coordinate and momentum representations, equations of motion of three-nucleon systems in bound and unbound states known as the Lippmann-Shwinger, and Faddeev, equations.

The third chapter is devoted to a detailed analysis of the boundless hypothetical substance – nuclear matter – the properties of which are determined by nucleon-nucleon interaction, while surface effects may be ignored. The hydrodynamical nuclei model in the classical and quantum versions allows this problem to be tackled. The Hartree and Hartree-Fock methods, the nuclear density matrix technique, the time-dependent Hartree-Fock approximation, used in the hydrodynamical nuclear model, are described in detail, along with applications of these methods to many-nucleon nuclei.

The shells in nuclei are the subject of the fourth chapter. The simplest shell models, having rectangular and oscillator potentials, are considered. The nulti-particle shell model is studied in detail. In particular, the problem of large magic numbers of protons and neutrons in related nuclear shells of two unstable superheavy nuclei is clarified, the completely anti-symmetric nuclear wave function in the many-particle shell model based on additional information concerning the symmetric group, which is included in the book, is obtained and the motion of nucleons in a spherically-symmetric potential well is analyzed in detail.

In chapter five the properties of deformed nuclei are studied. The rotation of nuclei and vibrations of their surface are considered. The results of theoretical calculations and experimental data are compared and their correlation is discussed. Information on the rotation-vibration excitations of nonspherical nuclei is given, and a separate section focuses on chaotic motion in nuclei. The Wigner and Poisson distributions of nuclear energy levels are cited.

The pairing of nucleons in nuclei is considered in the sixth chapter. Within the framework of the superfluid model of nuclei, the existence of a gap in the energy spectra of the even-even nuclei is explained.

The final chapter, presents a description of the quark model of hadrons and their classification, necessary information on continuous groups, in particular, the SU_n *group, additional information on strong interactions and the fundamentals of quantum chromodynamics, discussion of the correlation between nuclear forces and fundamental quark-quark interactions, detailed analysis of a few nucleon systems, in particular, a deuteron and three-nucleon nuclei, as well as many-quark systems.*

All chapters, except the last one, contain problems which illustrate the basic material of the book. Each chapter has a bibliography listing monographs and original papers relating to the theory of the nucleus and related fields where more detailed information is given on the problems considered in the book.

The book is devoted mostly to basic principles of theoretical nuclear physics; however, some recent achievements in the field are presented, too. Some problems discussed in the lectures were discussed in original papers only and have not been described in textbooks before. Since the book has been designed as a textbook, it contains a limited number of available references (with a few exceptions). The list of references given at the end of the Preface includes books on nuclear theory in which the reader can find more detailed accounts of a number of issues as well as references to the original papers.

The authors acknowledge the financial support of the International Soros Science Education Program of International Renaissance Foundation, grants SPU042060 and SPU062065.

The authors would like to thank V. Khryapa, Z. Vakhnenko, and G. Bugriy for typesetting the book.

A. Sitenko *Kiev, 1997*
V. Tartakovskii

References

De Benedetti, S. (1964) Nuclear Interactions, Wiley, N.Y.

Bethe, H.A. and Morrison, P. (1956) Elementary Nuclear Theory, Chapman & Hall, London.

Blatt, J.M. and Weisskopf, V.E. (1952) Theoretical Nuclear Physics, Wiley, N.Y.

Bohr, A. and Mottelson B.R. (1959) Nuclear Structure I, Single-Particle Motion, Benjamin, N.Y.

Bohr, A. and Mottelson B.R. (1974) Nuclear Structure II, Nuclear Deformations, Benjamin, N.Y.

Eisenberg, J.M. and Greiner W. (1970) Nuclear Theory I, Nuclear Models, North-Holland, Amsterdam.

Eisenberg, J.M. and Greiner W. (1970) Nuclear Theory II, Exitation Mechanisms of the Nucleus, North-Holland, Amsterdam.

Eisenberg, J.M. and Greiner W. (1972) Nuclear Theory III, Microscopic Theory of the Nucleus, North-Holland, Amsterdam.

Eisenbud,L. and Wigner, E.P. (1958) Nuclear Structure, Oxford University Press.

Landau, L.D. and Lifshitz, E.M. (1965) Quantum Mechanics, Pergamon Press, Oxford.

Landau, L.D. and Smorodinskii, Ya. A. (1959) Lectures on Nuclear Theory, Plenum Press, N.Y.

Sitenko, A.G., (1991) Scattering Theory, Springer-Verlag, Berlin.

Sitenko, A.G. (1990) Theory of Nuclear Reactions, World Scientific, Singapore.

CHAPTER 1

NUCLEONS AND NUCLEAR FORCES

1.1. Constitution of Nuclei and Nucleon Properties

Main characteristics of nuclei. Atomic nuclei are characterized by definite masses and electric charges. The charge of an atomic nucleus q is equal to minus the multiple of the absolute value of the electron charge $-e$, i.e.,

$$q = Ze.$$

The integer Z determines the position of the atom in the periodic table, it is called the *atomic number*. Nuclei with the same charge but different masses are called *isotopes*. Taking 12 to be the mass of the most abundant carbon isotope, one finds all the nuclear masses to be nearly integers. The integer A that is the closest to the mass value is referred to as the *mass number* of the nucleus.

According to the present-day conceptions, the atomic nucleus consists of protons and neutrons – elementary particles with approximately equal masses. These particles interact via nuclear forces. The proton possesses a positive electric charge whose value is equal in magnitude to that of the electron; the neutron is electrically neutral. In spite of differing electrical properties, protons and neutrons manifest similar behaviour in nuclear interactions and therefore are often unified by the term *nucleons*.

An atomic nucleus with the charge $q = Ze$ and mass number A consists of Z protons and $N = A - Z$ neutrons. The mass number A determines the total number of nucleons contained in the nucleus. *Isotopes* are nuclei with equal proton numbers Z and different neutron numbers N. Nuclei with equal neutron numbers N and different proton numbers Z are called *isotones*. Nuclei with equal nucleon numbers (equal mass numbers A) are called *isobars*.

The most important peculiarity of all nuclei, with exception the lightest nuclei (nuclei of heavy hydrogen isotopes and ^{3}He) and nuclei with an extra neutron numbers is the existence of their sharp border. Experiment shows that the nuclear density is almost constant inside the nucleus, and falls sharply to zero over a distance that is small compared with the dimensions of the nucleus. If we assume that the nucleus possesses spherical symmetry, the dimensions of the nucleus will be characterized by a single parameter, namely, the *nuclear radius* R.

The nuclear radii R are proportional to the cube root of the mass number A,

$$R = R_0 A^{1/3},$$

where the quantity R_0 decreases slowly with increasing A, from a value $R_0 = 1.3 \times 10^{-13}\,\text{cm}$ for light nuclei to a value $R_0 = 1.2 \times 10^{-13}\,\text{cm}$ for heavier nuclei. For most nuclei we can put, approximately,

$$R_0 = 1.2 \times 10^{-13}\,\text{cm}.$$

The proportionality coefficient R_0 does not depend on the mass number A that means that the nuclear density is almost the same for all nuclei.

Neutron and proton properties. Individual nucleon properties have been well studied. The proton mass is equal to 1836.15 electron masses or 1.007276 atomic mass units (*amu*); the neutron mass is greater by approximately 2.5 electron masses, it is equal to 1.008665*amu*. The proton and the neutron have spins of the same magnitude, equal to 1/2, and are governed by the *Fermi-Dirac statistics.* Their *magnetic moments* (in terms of nuclear magnetons) are equal respectively to

$$\mu_p = 2.7928, \quad \mu_n = -1.9130. \tag{1.1}$$

These numbers considerably differ from 1 and 0 which follow from the Dirac equation for particles with spin 1/2. Anomalous values of both proton and neutron magnetic moments indicate that Dirac's description of nucleons is incomplete.

A free neutron is unstable and by virtue of weak interaction decays into a proton, an electron, and an electronic antineutrino: $n \to p + e^- + \bar{\nu}_e$. The excess mass, approximately equivalent 1.3 MeV, is consumed to the electron production (0.5 MeV) and kinetic energy of decay products (0.8 MeV). The *mean neutron life-time* is about 15 min. Nevertheless, in spite of being unstable in the free state, neutrons can form stable nuclei together with protons. At the same time, a free proton is stable, whereas being bound in the nucleus it can, due to the weak interaction, decay into a neutron, a positron, and an electronic neutrino: $p \to n + e^+ + \nu_e$. The nuclei are stable only for certain proportions of neutron and proton numbers. If these proportions are violated, either neutron or proton decay occurs in the nucleus until the latter becomes stable.

Small difference of masses, equal spins, and similarity of some other properties, as well as the possibility of mutual transformations, make it reasonable to treat the neutron and the proton as two different states of the same particle – the *nucleon.* In such approaches, an additional intrinsic degree of freedom, which is referred to as *charge of isospin* coordinate, is attributed to the nucleon along with the space and spin coordinates.

For nonrelativistic energies, both neutrons and protons may be treated as point particles. For high energies, however, one must involve into consideration their space structure which is manifested, e.g., in the high energy scattering experiments. An adequate quantitative description of nucleon space structure and anomalous magnetic moments is provided only by quantum chromodynamics.

1.2. Nuclear Interaction between Nucleons

Short-range nature of nuclear forces. The nucleus constituents, neutrons and protons, interact via specific *nuclear forces.* These forces cannot be reduced to electromagnetic interactions since they are manifested irrespective of whether the nucleon possesses an electric charge or is electrically neutral.

A detailed information on the nature of nuclear forces can be obtained by the study of two-nucleon systems since the two-particle problem is rather simple and can be solved exactly. The two-nucleon nuclear forces are responsible for the nucleon-nucleon scattering and the formation of a composite particle – *deutron* which is the simplest nucleus consisting of a neutron and a proton. The existence of the neutron-proton bound state indicates that the neutron-proton nuclear interaction is associated with the particle attraction.

It is highly peculiar of the nuclear interaction that its range is very short whereas the interaction is very strong. Wigner was the first who assumed the nuclear interaction to be short-range. He analyzed the values of the deutron binding energy (which is small, 2.23 MeV) and binding energy of the ^{4}He nucleus (which is large, 28 MeV), and thus found the nuclear-force range to be of the order of magnitude of 2×10^{-13} cm. Assuming the neutron-proton interaction to be produced by a square well of width 2×10^{-13} cm and taking into account the small value of the deutron binding energy, one finds the potential well depth to be about 30 MeV. For comparison we note that the Coulomb interaction energy for two protons spaced to 2×10^{-13} cm is as small as 0.7 MeV.

Other peculiar features of nuclear forces. In the nonrelativistic limiting case, with particle velocities being much smaller than the light velocity, the nuclear interaction does not depend on the velocities of interacting particles and can be described by a potential. It is peculiar of nuclear forces that nuclear interaction potential depends both on the interparticle distance and on the mutual orientation of particle spins. The particle- spin-dependence of nuclear interaction immediately follows from the experiments on slow neutron scattering by molecular hydrogen. The existence of the deutron electric quadrupole moment indicates that the nuclear interaction is non-central, i.e., depends on the mutual orientation of the total spin and the relative radius-vector of the interacting particles. At last, the observation that particle scattering produces particle polarization shows that spin-orbit forces considerably contribute to the nuclear interaction.

Insofar as the stable nuclei exist, the nucleon-nucleon nuclear interaction must be mainly attractive. The experimental data on the high-energy nucleon-nucleon scattering show, however, that nuclear forces must be repulsive at very small distances (shorter that 0.4×10^{-13} cm). High energy nucleon scattering experiments also suggest the partially exchange nature of the nuclear interaction, i.e., the interacting nucleons can exchange some parameters (space coordinates, spins, charges).

It follows from the properties of *mirror nuclei* (nuclei in which the neutrons are replaced by protons and the protons by neutrons) that the forces acting between two protons are equivalent to the forces acting between two neutrons (if we exclude the *Coulomb interaction* from consideration). This property was given the name of the *charge symmetry of nuclear forces.* Later studies have shown that the charge symmetry is a manifestation of a deeper symmetry of the nuclear interaction which is called the *isotopic invariance*, or *charge independence of nuclear forces.* The isotopic invariance of nuclear forces means that the interaction in any two pairs of nucleons is the same if these pairs of nucleons are in the same states. All the presently available experimental data on the interaction of nucleons (at both low and high energies) are in agreement with the postulate of the isotopic invariance of nuclear forces.

In 1935 Yukawa employed the short-range nature of the nuclear interaction to propose a field theory of nuclear forces which predicted the existence of particles with medium masses – *mesons.* Later on, these particles have been observed experimentally. The early meson theory, however, could not explain some peculiarities of the nuclear interaction. Though the recent theory – quantum chromodynamics – provides a more consistent treatment of strong interactions, at the moment it is unsufficiently advanced to develop a systematic description of nuclei and nuclear processes. That is why we give here a phenomenological study of the problem which is essentially based on the peculiarities of nuclear interactions which follow from the experimental observations.

1.3. Ground State of the Deuteron

Main characteristics of the deuteron. As we have already mentioned, a system consisting of a neutron and a proton can be found in a bound state. The bound state of the neutron-proton system is called the *deuteron* and is the simplest compound atomic nucleus. Despite the fact that one of the particles (the neutron) constituting the deuteron is unstable, the deuteron is characterized by an infinite lifetime, that is, is a stable system.

The most important characteristic of the deuteron is its *binding energy*, which is determined by the difference in mass between the deuteron and the particles forming it. The deuteron binding energy can be measured from the threshold of the photodisintegration reaction. The value of the *deuteron binding energy* found by this method is

$$\varepsilon = 2.22457 \pm 0.00001 \text{ MeV}. \tag{1.2}$$

Other important characteristics of the deuteron are its *spin*, *magnetic moment* and *electric quadrupole moment*. The *spin of the deuteron* is equal to 1. The *magnitude of the deuteron* spin can be determined from the band intensities of the spectrum of molecular deuterium.

The magnetic moment of the deuteron is close to the sum of the magnetic moments of the neutron and proton. The magnitude of the deuteron magnetic moment, expressed in nuclear magnetons, is

$$\mu_d = 0.8574. \tag{1.3}$$

The magnetic moment of the deuteron is measured most accurately in experiments on the deflection of a molecular beam in an inhomogeneous magnetic field.

The *electric quadrupole moment of the deuteron* is

$$Q = 2.860 \times 10^{-27}\,\text{cm}^2. \tag{1.4}$$

This quantity was found from the fine structure in the radio frequency magnetic resonance spectrum of deuterium. Note the relative smallness of the deuteron electric quadrupole moment. In fact, the magnitude of the electric quadrupole moment must be compared with the spatial dimensions of the system. If we assume that the dimensions of the deuteron coincide with the range of the nuclear forces, then even in this case the quadrupole moment is approximately 50 times smaller than the estimated cross-sectional area of the deuteron. The small magnitude of the quadrupole moment of the deuteron and the approximate additivity of the magnetic moments of the neutron and proton in the deuteron imply that the ground state of the deuteron is almost spherically symmetric. But the ground state of a system is characterized by spherical symmetry in cases where the interaction is central. Therefore, it follows from the spherical symmetry of the deuteron ground state that the nuclear interaction between the neutron and the proton is also almost central.

Ground state of the deuteron in the case of central forces.[1] We shall study the properties of the *ground state of the deuteron*, assuming that the nuclear interaction between the neutron and proton is described by a potential $V(r)$, where r is the magnitude of the distance between the particles. The Schrödinger equation for the wave function $\psi(\mathbf{r})$ describing the relative motion of the neutron and proton has the form

$$\left\{-\frac{\hbar^2}{2\mu}\Delta + V(r) - E\right\}\psi(\mathbf{r}) = 0, \tag{1.5}$$

where $\mu = M/2$ is the reduced mass (for simplicity, we assume that the masses of the neutron and proton are the same and are equal to M) and E is the energy of the relative motion. For the ground state of the system, the energy E is negative and equal in magnitude to the binding energy, $E = -\varepsilon$. The wave function $\psi(\mathbf{r})$ should vanish at infinity and be finite at $r = 0$.

In the case of a central interaction, the ground state of the system is characterized by a zero value of the relative orbital angular momentum, $l = 0$ (S-state). The wave function for this state is spherically symmetric and can be written in the form

$$\psi(\mathbf{r}) = u(r)/r, \tag{1.6}$$

where $u(r)$ depends only on the magnitude of the distance between the particles. Substituting (1.6) into (1.5), we obtain the following equation for the function $u(r)$:

$$u'' - (M/\hbar^2)[V(r) + \varepsilon]u = 0, \tag{1.7}$$

where the function $u(r)$ must satisfy the boundary conditions $u(0) = u(\infty) = 0$. The nuclear interaction is characterized by a finite range r_0, so that outside the range of the nuclear forces, eqn. (1.7) is simplified:

$$u'' - \alpha^2 u = 0, \quad r > r_0, \tag{1.8}$$

Here we have introduced the notation $\alpha^2 = M\varepsilon/\hbar^2$.

The solution of (1.8) that vanishes at infinity has the form

$$u = C\mathrm{e}^{-\alpha r}, \quad r > r_0, \tag{1.9}$$

[1]The interaction of two nucleons at low energies is treated in detail in the review by Hulthen and Sugawara (1957), which contains a detailed bibliography.

where C is a normalization constant. Since the wave function and its derivative must be continuous, the solution of eqn. (1.8) in the outer region $(r \geqslant r_0)$ must match with the solution of eqn. (1.7) in the inner region $r \leqslant r_0)$, which means that the logarithmic derivatives of the functions $u(r)$ for the outer and inner regions must be equal at $r = r_0$. It follows from the expression (1.9) that the logarithmic derivative of the function $u(r)$ in the outer region $(r \geqslant r_0)$ does not depend on r and is equal to $-\alpha$:

$$u'/u = -\alpha.$$

The logarithmic derivative of the function $u(r)$ in the inner region $(r \leqslant r_0)$ must take the same value at $r = r_0$:

$$\left. \frac{u'_{\text{inner}}}{u_{\text{inner}}} \right|_{r=r_0} = -\alpha. \tag{1.10}$$

The magnitude of the logarithmic derivative of the inner function at $r = r_0$ is determined by the potential energy $V(r)$ and is a fundamental parameter characterizing the properties of the neutron-proton system. (Such an approach to the description of the neutron-proton system was first proposed by Bethe and Peierls (1935).)

According to the relation (1.9), outside the range of the nuclear forces the wave function of the deuteron decreases exponentially with increasing distance, so that the quantity $1/\alpha$ can be considered as a parameter characterizing the spatial dimensions of the deuteron (the *deuteron radius* R_d). Making use of the value (1.2) for the deuteron binding energy, we have

$$R_d \equiv \frac{1}{\alpha} = 4.3177 \times 10^{-13}\,\text{cm}.$$

The magnitude of R_d is appreciably greater that the range r_0 of the nuclear forces, and this means that the deuteron is a weakly bound system in which the neutron and proton spend a significant part of their time outside the range of the nuclear forces.

We note that the radial dependence of the wave function of the deuteron ground state outside the range of the nuclear forces is completely determined by specifying the binding energy. This circumstance permits us, in the limit of zero range of the nuclear forces, to construct an approximate wave function for the deuteron, completely independently of the form of the nuclear potential:

$$\psi_0(\mathbf{r}) = \sqrt{\alpha/2\pi}(\mathrm{e}^{-\alpha r}/r), \quad r_0 \to 0, \tag{1.11}$$

where the normalization constant is determined from the condition

$$\int d^3r\, \psi_0^2(\mathbf{r}) = 1. \tag{1.12}$$

The formula (1.11) can be used in those cases when the behavior of the wave function at distances greater than the range of the nuclear forces is important.

Relation between the width and depth of the potential well. If we assume that the nuclear potential has the form of a square well of radius r_0 and depth V_0, then the inner function $u(r)$ that is a solution of (1.7) will have the form

$$u = C' \sin K_0 r, \quad r < r_0, \tag{1.13}$$

where C' is a normalization constant and

$$K_0 = (1/\hbar)\sqrt{M(V_0 - \varepsilon)}. \tag{1.14}$$

Substituting (1.13) into the expression (1.10) for the logarithmic derivative, we can obtain a relation connecting the quantities r_0 and V_0. The relation between the width and depth of the potential well can be obtained more intuitively, however, from the following considerations. Since in the ground state the function $u(r)$ has no zeros (except for the point $r = 0$), not more than half of a wavelength can be fitted into the inner region ($r < r_0$); $r_0 < \frac{1}{2}(2\pi/K_0)$. However, to make it possible to match the solution (1.13) with the exponentially decreasing solution (1.9), it is necessary that there be not less than a quarter of a wavelength in the inner region. Consequently, the inequalities

$$\frac{1}{4}(2\pi/K_0) \leqslant r_0 < \frac{1}{2}(2\pi/K_0). \tag{1.15}$$

must be fulfilled.

Using (1.14) and neglecting the binding energy of the deuteron in comparison with the depth of the well, from (1.15) it is not difficult to determine limits within which the depth of the potential well must lie:

$$(\pi^2/4)(\hbar^2/M) \leqslant V_0 r_0^2 < \pi^2(\hbar^2/M). \tag{1.16}$$

Putting $r_0 = 1.7\times10^{-13}$ cm, we obtain a quantity of the order of 35 MeV for the lower bound on the depth V_0 of the potential well; this appreciably exceeds the binding energy of the deuteron.

Absence of excited states of the deuteron. It is not difficult to show that, if the nuclear forces are assumed to be centrally symmetric, the deuteron has no excited bound states.

For the existence of an excited bound state with zero orbital angular momentum to be possible, it is necessary that the radial function $u(r)$ in the inner region have an additional zero, that is, that more than three-quarters of a wavelength fit into the inner region. Neglecting the binding energy of the excited state, we can write this condition in the form

$$r_0 \geqslant \frac{3}{4}(2\pi\hbar/\sqrt{MV_0}), \quad \text{whence} \quad V_0 r_0^2 \geqslant \frac{9}{4}\pi^2(\hbar^2/M).$$

But this condition contradicts the inequality (1.16).

The fact that the deuteron can also have no excited bound states with non-zero values of the relative orbital angular momentum is connected with the fact that levels with $l \neq 0$ lie appreciably above the level with $l = 0$, since in a state with non-zero orbital angular momentum there is additional (centrifugal) energy. The mean value of this energy, equal to $\hbar^2 l(l+1)/Mr^2$, can be estimated if we substitute in place of r a quantity of the order of the effective dimensions of the deuteron, that is, $1/\alpha$. For $l = 1$, we obtain for the centrifugal energy a quantity equal to 2ε; that is, even a P-state should lie in the region of the continuous spectrum.

We note that in our proof of the absence of excited bound S-states we neglected the spin dependence of the nuclear forces. In reality, even if the nuclear forces are central they can depend on the mutual orientation of the spins of the neutron and proton. Therefore, in principle, two bound S-states of the neutron-proton system can exist: one with parallel orientation of the spins of the particles (a triplet state, with the total spin of the particles equal to unity) and one with the anti-parallel orientation of the spins of the particles (a singlet state, with the total spin of the particles equal to zero). However, it follows from the experimental data on the scattering of slow neutrons in molecular hydrogen that the singlet state in the neutron-proton system is not realized.

1.4. Scattering of Slow Neutrons by Protons

Scattering amplitude and phase shifts. Along with the study of the bound state of the neutron-proton system, the study of the scattering of neutrons by protons is a second very important source of information about the nuclear interaction between neutrons and protons. Unlike the case of the bound state of the neutron-proton system, when the energy is negative, in the case of scattering of a neutron by a proton the energy of the relative motion of the particles is positive. The wave function $\psi(r)$ describing the scattering of a neutron by a proton is given by the solution of the Schrödinger equation (1.5) for a positive value of the energy E. In this case, the wave function $\psi(r)$ at large distances must have the form of a sum of the incident plane wave and a spherical outgoing scattered wave:

$$\psi(\mathbf{r}) \to e^{ikz} + f(\vartheta)(e^{ikr}/r), \tag{1.17}$$

where $k = \sqrt{ME}/\hbar$ is the magnitude of the wave-vector of the relative motion of the particles, and the z-axis is chosen to lie along the direction of the incident beam of neutrons. The coefficient $f(\vartheta)$ of the outgoing wave depends on the angle ϑ between the directions of motion of the incident and scattered particles and is called the *scattering amplitude*; angle ϑ is the *scattering angle.*

In the case of a central interaction, the scattering amplitude can be expressed directly in terms of the *phase shifts* δ_l:

$$f(\vartheta) = (i/2k)\sum_{l=0}^{\infty}(2l+1)(1-e^{2i\delta_l})P_l(\cos\vartheta). \tag{1.18}$$

The phases δ_l characterize the interaction of a particle with different values of the orbital angular momentum l and depend on the energy of the particle and on the form of the potential $V(r)$. In general, the phase shifts are complex. If only elastic scattering occurs, then the phase shifts δ_l are real functions of energy.

The *scattering cross-section* is connected with the scattering amplitude by the relation

$$d\sigma = \sigma(\vartheta)do = |f(\vartheta)|^2 do, \tag{1.19}$$

where do is an element of solid angle, defining the direction of motion of the scattered particle. Substituting (1.18) into (1.19) and integrating (1.19) over the whole solid angle, we easily find the integral scattering cross-section:

$$\sigma = (4\pi/k^2)\sum_l (2l+1)\sin^2\delta_l. \tag{1.20}$$

Energy dependence of the S-scattering phase shift. We shall consider the scattering of slow neutrons by protons in the case when the wavelength $\lambda\!\!\!^{-} = k^{-1}$ of the relative motion is greater than the range r_0 of the nuclear forces. The condition $\lambda\!\!\!^{-} > r_0$ is fulfilled if the energy of the relative motion of the neutron and proton does not exceed 10 MeV. (In the laboratory frame, in which the proton is at rest before the scattering, the energy of the incident neutron is then less than 20 MeV.) Since the effective impact parameter for the collision of two particles with relative angular momentum l is equal to $l\lambda\!\!\!^{-}$, if $\lambda\!\!\!^{-} > r_0$ interaction of the particles is possible only in an S-state ($l = 0$). Then all the phase shifts δ_l (except the phase shift in the S-state, $\delta_0 \equiv \delta$) are equal to zero, and the scattering amplitude takes the form

$$f = (1/k)e^{i\delta}\sin\delta. \tag{1.21}$$

In this case, the scattering is spherically symmetric. The integral scattering cross-section is equal to

$$\sigma = (4\pi/k^2)\sin^2\delta. \tag{1.22}$$

We shall find the dependence of the phase shift δ on the energy in the limiting case of infinitesimally small range of the nuclear forces ($r_0 \to 0$). For $l = 0$, the radial function u outside the range of the nuclear forces is determined by the equation

$$u'' + k^2 u = 0, \quad r > r_0,$$

the solution of which has the form

$$u = C \sin(kr + \delta), \quad r > r_0. \tag{1.23}$$

The phase δ can be found from the conditions that the logarithmic derivatives of the wave function (1.23) in the outer region and of the wave function in the inner region be equal at $r = r_0$:

$$k \cot(kr_0 + \delta) = \left. \frac{u'_{\text{inner}}}{u_{\text{inner}}} \right|_{r=r_0} \tag{1.24}$$

If the energy of the relative motion of the particles is small compared with the depth of the potential well the wave function in the inner region for the scattering problem will not differ from the wave function in the inner region for the bound-state problem. Therefore, we replace the right-hand side of the equality (1.24) by the expression (1.10) and thus obtain the following relation

$$k \cot(kr_0 + \delta) = -\alpha, \tag{1.25}$$

or, in the limiting case of *zero range of the nuclear forces*,

$$k \cot \delta = -\alpha, \tag{1.26}$$

We emphasize that the possibility of introducing the same parameter α to describe both the scattering and the asymptotic form of the bound-state wave function is a direct consequence of the smallness of the deuteron binding energy and of the energy of the relative motion of the neutron and proton compared with the effective potential energy of the interaction of the neutron and proton.

Assuming zero range of the nuclear forces and using (1.26), we can write the cross-section for the scattering of a neutron by a proton in the form

$$\sigma = 4\pi/(\alpha^2 + k^2), \quad r_0 \to 0. \tag{1.27}$$

Spin dependence of the nuclear interaction. According to (1.27), the scattering cross-section in the limit of zero energy of relative motion is completely determined by the magnitude of the deuteron binding energy:

$$\sigma = 4\pi/\alpha^2, \quad k \to 0. \tag{1.28}$$

If we make use of the numerical value (1.2) of the deuteron binding energy, we obtain, according to (1.28), the value $2.33 \times 10^{-24}\,\mathrm{cm}^2$ for the scattering cross-section. The experimental value of the cross-section for the scattering of a neutron by a proton at zero energy is found to be considerably greater and is equal to

$$\sigma_{\mathrm{exp}} = (20.49 \pm 0.01) \times 10^{-24}\,\mathrm{cm}^2. \tag{1.29}$$

Such a significant difference between the experimental value of the scattering cross-section and the expression (1.28) points directly to the important dependence of the nuclear interaction on the spin state of the system.

The neutron-proton system can be found in two different spin states: the *triplet state* (with total spin equal to unity) and the *singlet state* (with total spin equal to zero). Since the total angular momentum (spin) of the deuteron is equal to unity and the ground state of the deuteron is even, the triplet spin state is realized in the deuteron. In the case of scattering of an unpolarized neutron by an unpolarized proton, the system can be found either in a triplet state (with weight $\frac{3}{4}$) or in a singlet state (with weight $\frac{1}{4}$). If the nuclear interaction depends substantially on the spin state, then to describe the interaction between a neutron and proton it is necessary to introduce two different parameters δ_t and δ_s, for the triplet and singlet spin states. The phase shifts δ_t and δ_s corresponding to the triplet and singlet states are connected with α_t and α_s for $r_0 \to 0$ by the relations

$$k \cot \delta_t = -\alpha_t, \qquad k \cot \delta_s = -\alpha_s. \tag{1.30}$$

The quantity α_t is related directly to the deuteron binding energy

$$\varepsilon = \hbar^2 \alpha_t^2 / M. \tag{1.31}$$

The cross-sections for scattering of neutrons by protons in the triplet and singlet states, according to the relations (1.30) and (1.22), are

$$\sigma_t = 4\pi/(\alpha_t^2 + k^2), \quad \sigma_s = 4\pi/(\alpha_s^2 + k^2), \quad r_0 \to 0. \tag{1.32}$$

In the case of scattering of unpolarized neutrons by unpolarized protons, the scattering cross-section is given by the formula

$$\bar{\sigma} = \frac{3}{4}\sigma_t + \frac{1}{4}\sigma_s, \tag{1.33}$$

where the coefficients $\frac{3}{4}$ and $\frac{1}{4}$ are the statistical weights of the triplet and singlet spin states respectively. Using the experimental value (1.29) of the averaged cross-section for the scattering of a neutron by a proton at zero energy and knowing the binding energy of the deuteron, we can determine the quantity α_s^2, or the energy formally associated with it:

$$\varepsilon' = \hbar^2 \alpha_s^2 / M = 0.069\,\mathrm{MeV}. \tag{1.34}$$

This energy is sometimes called the *virtual level of the deuteron.* (But it must be kept in mind that a bound state of the neutron-proton system with energy $-\varepsilon'$ does not exist.)

Scattering lengths. We now consider the scattering of neutrons by protons in the limiting case of zero energy of the neutrons, assuming the range of the nuclear forces to be non-zero. The equation for the radial function outside the range of the nuclear forces in the case of zero energy has the form

$$u'' = 0, \quad r > r_0, \tag{1.35}$$

whence

$$u = C'(r - a), \quad r > r_0, \tag{1.36}$$

where C' and a are constants. The solution (1.36) is the equation of a straight line intersecting the r-axis at the point a. The quantity a is called the *scattering length.*

The scattering length a is easily related to the phase shift δ. In fact, the solution (1.36) must coincide with the function u given by (1.23), for $k \to 0$. Comparison of the two formulae shows that

$$a = -\lim_{k\to 0}\left(\frac{\tan\delta}{k}\right) \tag{1.37}$$

that is, for small k, when the inequality $kr_0 \ll 1$ is fulfilled, the phase shift δ can be chosen in the form

$$\delta = n\pi - ak, \quad kr_0 \ll 1, \tag{1.38}$$

where n is an arbitrary integer[2]. Hence it follows, on the basis of formula (1.22), that the scattering cross-section in the limiting case of zero energy of the neutrons is

$$\sigma = 4\pi a^2, \quad k \to 0. \tag{1.39}$$

The scattering length depends on the spin state of the system and, therefore, we must distinguish scattering lengths a_t and a_s in the triplet and singlet states.

Measurement of the scattering cross-section at zero energy enables us to determine only the magnitude of the scattering length, and not its sign. The scattering length can be taken to be either positive or negative. The dependence of the wave function on r at zero energy for positive and negative scattering lengths is depicted schematically in Fig. 1.1. The behaviour of the wave function within the range of the nuclear forces ($r < r_0$) is almost independent of the energy and, therefore, is the same as for the bound

[2] If we demand that the phase shift goes to zero as $k \to \infty$, then it follows from Levinson's theorem that $n = 1$.

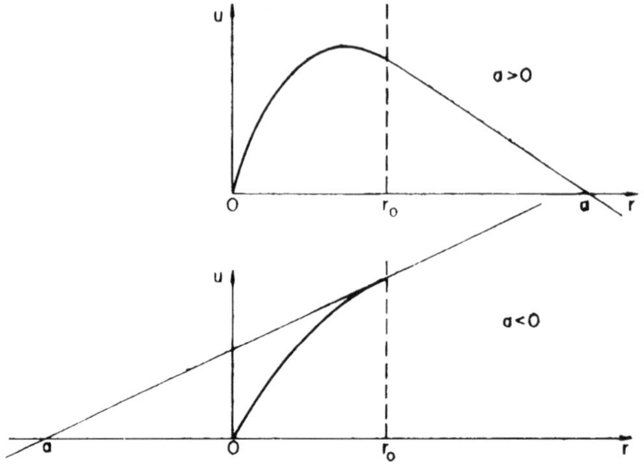

Figure 1.1. Dependence on r of the radial wave function u at zero energy, in the cases of positive and negative scattering lengths.

state, if the binding energy is not very large. From Fig. 1.1 we see that, for positive scattering lengths, the wave function for the inner region can be matched to the exponentially decreasing solution corresponding to the bound state in the outer region. For negative scattering lengths, the wave function in the inner region behaves in such a way that it is impossible to match it with the exponentially decreasing solution in the outer region, that is, a bound state of the system is impossible in this case. Thus, if $a > 0$, a bound state of the system can exist; but if $a < 0$, a bound state is impossible.

Effective range of the nuclear interaction. The formulae (1.32) determining the cross-section for the scattering of slow neutrons by protons for $k \neq 0$ were derived under the assumption that the range of the nuclear forces is equal to zero ($r_0 \to 0$). We now show how the scattering problem can be formulated in such a way that the fact that the range of the forces is finite can be taken into account consistently, irrespective of the form of the potential (Bethe, 1949). We write the exact equation for the radial function $u(r)$ that describes the motion in a field $V(r)$ and corresponds to the value E of the energy of the relative motion:

$$u'' - (M/\hbar^2)V(r)u = -k^2 u; \quad k^2 = ME/\hbar^2. \tag{1.40}$$

We denote the radial function corresponding to zero energy by $u_0(r)$. This function satisfies the equation

$$u_0'' - (M/\hbar^2)V(r)u_0 = 0. \tag{1.41}$$

Multiplying (1.40) by u_0 and (1.41) by u and subtracting the first equality from the second, we obtain the following relation:

$$\frac{d}{dr}(u_0'u - u_0u') = k^2u_0u. \tag{1.42}$$

We introduce the auxiliary functions $v(r)$ and $v_0(r)$ satisfying the equations for free motion ($v'' = -k^2v$, $v_0'' = 0$), these functions being defined in such a way that they coincide with u and u_0 for $r > r_0$. Obviously, a relation analogous to (1.42) holds for the functions v and v_0:

$$\frac{d}{dr}(v_0'v - v_0v') = k^2v_0v. \tag{1.43}$$

Subtracting the relation (1.42) from (1.43), integrating the resulting equality over r from 0 to infinity and noting that $u(0) = u_0(0) = 0$, we obtain

$$v_0(0)v'(0) - v_0'(0)v(0) = k^2\int_0^\infty dr(v_0v - u_0u). \tag{1.44}$$

We now choose the normalization of the functions u and u_0 in such a way that the functions v and v_0 go to unity at $r = 0$. Obviously, with this normalization, the functions v and v_0 have the following form:

$$v(r) = \sin(kr + \delta)/\sin\delta, \quad v_0(r) = 1 - r/a, \tag{1.45}$$

where δ and a have been defined previously.

Substituting the relations (1.45) into (1.44), we obtain the general expression

$$k\cot\delta = -(1/a) + \frac{1}{2}k^2r(0,E), \tag{1.46}$$

relating the phase shift δ to the scattering length a and the effective range

$$r(0,E) \equiv 2\int_0^\infty dr(v_0v - u_0u). \tag{1.47}$$

For the bound state, we obtain a relation analogous to (1.46) by replacing k^2 by $-\alpha^2$ and $k\cot\delta$ by $-\alpha$ in (1.46). As a result of the replacement, we have

$$\alpha = (1/a) + \frac{1}{2}\alpha^2r(0,-\varepsilon), \tag{1.48}$$

where

$$r(0,-\varepsilon) \equiv 2\int_0^\infty dr(v_0v_{-\varepsilon} - u_0u_{-\varepsilon}). \tag{1.49}$$

Since the functions v_0 and v coincide with u_0 and u outside the range of the nuclear forces, in the formulae for $r(0, E)$ and $r(0, -\varepsilon)$ the integration is performed only over the region of the nuclear interaction. We note that the relations (1.46) and (1.48) are completely independent of the form of the nuclear potential, if by the effective ranges we mean the quantities (1.47) and (1.49). (The concept of the effective range was introduced by Schwinger (1947).)

For energies E that are small compared with the effective depth of the potential well, the functions u and v within the region of the nuclear interaction are practically the same as the functions u_0 and v_0, and the quantities $r(0, E)$ and $r(0, -\varepsilon)$ can be considered as independent of the energy to a first approximation. These approximate values of $r(0, E)$ and $r(0, -\varepsilon)$ are usually denoted by r_0:

$$r_0 = 2 \int_0^\infty dr (v_0^2 - u_0^2). \tag{1.50}$$

The quantity r_0, which has the dimensions of length, is determined by the character of the nuclear interaction and may be called the *effective range of the nuclear forces.* Indeed, the integrand in (1.50) is non-zero only within the range of the nuclear forces and, in order of magnitude, is equal to unity. Therefore, the integral (1.50) characterizes the distance over which the nuclear interaction occurs.

Lengths and effective ranges for triplet and singlet states. In the low-energy region, the relation (1.46) can be represented in the form

$$k \cot \delta = -(1/a) + \frac{1}{2} k^2 r_0. \tag{1.51}$$

The effective range of the nuclear forces, like the phase shift and scattering lengths, depends on the spin state of the neutron-proton system. Denoting the effective ranges in the triplet and singlet states by r_{0t} and r_{0s}, we can write the following relations:

$$k \cot \delta_t = -(1/a_t) + \frac{1}{2} k^2 r_{0t}, \quad k \cot \delta_s = -(1/a_s) + \frac{1}{2} k^2 r_{0s}. \tag{1.52}$$

We note that for the triplet state an additional relation holds, connecting the scattering length a_t and the effective range r_{0t}:

$$1/a_t = \alpha_t - \frac{1}{2} \alpha_t^2 r_{0t}. \tag{1.53}$$

The right-hand sides in the formulae (1.52) are the first two terms of the expansion of $k \cot \delta$ in a series in powers of k^2. It can be seen that, to within terms of order k^2, the quantity $k \cot \delta$ is determined by only two parameters (for each spin state) and does not depend on the actual form of the nuclear potential. Using the formulae (1.52), it is not difficult to obtain the following expressions for the cross-sections for the scattering of slow neutrons by protons in the triplet and singlet states:

$$\sigma_t = \frac{4\pi}{\left(\frac{1}{a_t} - \frac{1}{2}k^2 r_{0t}\right)^2 + k^2}, \quad \sigma_s = \frac{4\pi}{\left(\frac{1}{a_s} - \frac{1}{2}k^2 r_{0s}\right)^2 + k^2}. \tag{1.54}$$

The averaged cross-section for the scattering of unpolarized slow neutrons by free unpolarized protons is given by the formula

$$\sigma_t = \frac{3\pi}{\left(\frac{1}{a_t} - \frac{1}{2}k^2 r_{0t}\right)^2 + k^2} + \frac{\pi}{\left(\frac{1}{a_s} - \frac{1}{2}k^2 r_{0s}\right)^2 + k^2}. \tag{1.55}$$

As can be seen, the scattering of neutrons by protons in the region of low energies of the incident neutrons is characterized by four parameters a_t, a_s, r_{0t} and r_{0s} and does not depend on the form of the potential; that is, the scattering cross-section is the same for different potentials $V(r)$ that lead to the same values of the scattering lengths and effective ranges.

Using the experimental value for the averaged cross-section for the scattering of neutrons by free protons at zero energy, and also the data on the scattering of neutrons by para-hydrogen at zero energy, we find the scattering lengths a_t and a_s:

$$a_t = (5.424 \pm 0.003) \times 10^{-13}\,\text{cm}; \quad a_s = -(23.749 \pm 0.008) \times 10^{-13}\,\text{cm}. \tag{1.56}$$

Note that the negative sign of the scattering length a_s indicates the absence of a bound singlet state in the neutron-proton system. Knowing the scattering length a_t in the triplet state and the deuteron binding energy and taking the equality (1.53) into account, we can determine the *effective range in the triplet state* with a high degree of accuracy:

$$r_{0t} = (1.760 \pm 0.005) \times 10^{-13}\,\text{cm}. \tag{1.57}$$

Using the experimental dependence on the neutron energy of the averaged cross-section for scattering of slow neutrons by free protons, we find the *effective range in the singlet state*:

$$r_{0s} = (2.77 \pm 0.05) \times 10^{-13}\,\text{cm}. \tag{1.58}$$

The error in the determination of r_{0s} turns out to be much greater that the error in the determination of r_{0t}.

The shape parameter. As we have already observed, the relation (1.51) represents the first two terms of the expansion of $k\cot\delta$ in a series in powers of k^2, the coefficients of this expansion being independent of the detailed form of the nuclear potential. The next terms of the expansion of $k\cot\delta$ in powers of k^2 depend on the form of the potential; therefore, by comparing the experimental data on the scattering of slow neutrons by protons with the theoretical formula for the cross-section in which the subsequent terms of the expansion of (1.46) are taken into account, we can obtain information about the form of the potential. With the inclusion of the terms proportional to k^4, the expansion of $k\cot\delta$ has the form

$$k\cot\delta = -(1/a) + \frac{1}{2}k^2 r_0 - Pk^4 r_0^3 + \cdots, \tag{1.59}$$

where

$$P \equiv -\frac{1}{r_0^3}\int_0^\infty dr\left\{v_0 \frac{\partial v}{\partial k^2}\bigg|_{k=0} - u_0\frac{\partial u}{\partial k^2}\bigg|_{k=0}\right\}. \tag{1.60}$$

The dimensionless quantity P is called the *shape parameter*, since it is connected with the shape of the nuclear potential $V(r)$.

To characterize potentials of different shapes, we usually introduce the potential-depth parameter s and the characteristic width b; by means of these parameters, potentials of different shapes can be compared. The *potential-depth parameter* s is defined as the number by which the potential $V(r)$ must be divided in order to reduce the binding energy of the ground state of the system to zero. If the depth parameter $s > 1$, the potential $V(r)$ admits a bound state; but if $s < 1$, the potential $V(r)$ does not lead to a bound state. The *characteristic width b of the potential* is defined as the effective range of the reduced potential $V(r)/s$ (corresponding to zero binding energy), that is,

$$b = \lim_{s\to 1} r_0, \tag{1.61}$$

where r_0 is given by the expression (1.50). For a square well, the characteristic width coincides with the ordinary width of the well, $b = r_0$, and the depth parameter is given by the expression $s = 4MV_0r_0^2/\pi\hbar^2$.

1.5. Scattering of Slow Neutrons in Molecular Hydrogen

The pseudo-potential. The scattering amplitude for scattering of a slow neutron by a proton depends on the spin state of the neutron-proton system. Since in the cross-section for elastic scattering of a neutron by a free proton the squares of the scattering lengths a_t and a_s appear, we can find only the absolute values of the quantities a_t and a_s from experiments on the scattering of neutrons by free protons. In practice one is usually concerned with the scattering of neutrons in substances containing hydrogen. Obviously, the protons can be regarded as free only if the energy of the neutrons appreciably exceeds the binding energy of the protons in the molecules of the substance. But if the energy of the neutrons is comparable with, or less than, the binding energy of the protons, the chemical binding of the protons will have an appreciable influence on the scattering.

The sign of the scattering length a_s in the singlet state can be determined by studying the scattering of slow neutrons in molecular hydrogen (Schwinger and Teller, 1937). If the wavelength of the neutron is comparable with the distance between the protons forming the molecule, or is large compared with this distance, then in the scattering of the neutron by both protons interference effects occur; a study of these effects makes it possible to determine both the absolute values and the signs of the scattering lengths.

The scattering of slow neutrons by free protons is spherically symmetric and, for $k \to 0$, is determined by a constant scattering length a. These properties of the scattering of slow neutrons by protons can be described formally by means of a special form of perturbation theory, if we introduce a certain fictitious potential that leads to the correct value of the scattering amplitude when the perturbation theory is used.

In the case od sufficiently slow neutrons (when $\lambda\!\!\!^{-} \gg r_0$) we can make use of *the zero-range approximation*, that is, assume that interaction between the neutron and protons exists only when their coordinates coincide (contact interaction). This approximation corresponds to a choice of the potential in the form of a delta-function (Breit, 1947):

$$V(\mathbf{r}) = (2\pi\hbar^2/\mu)a\delta(\mathbf{r}), \quad \mathbf{r} = \mathbf{r}_n - \mathbf{r}_p, \tag{1.62}$$

where $\mathbf{r}_n$ and $\mathbf{r}_p$ are the position vectors of the neutron and proton, and the coefficient of the delta-function is found from the requirement that the scattering amplitude in the first order of the perturbation theory in (1.62) coincide in magnitude with the scattering length, $f = -a$ (μ is the reduced mass of the neutron and proton). When we take the effects of the chemical binding into account, different powers of the ratio of the scattering length a to the molecular dimensions d or to the neutron wavelength $\lambda\!\!\!^{-}$ correspond to different orders of perturbation theory in (1.62). Therefore, we write the condition for the applicability of (1.62) in the form $a \ll d$ and $a \ll \lambda\!\!\!^{-}$. Moreover, in order that we can disregard the finite range of the nuclear forces, the following conditions must be fulfilled: $r_0 \ll d$ and $r_0 \ll \lambda\!\!\!^{-}$.

The scattering of slow neutrons in molecular hydrogen can be studied by means of the pseudo-potential (1.62), if we previously include in it the dependence of a on the spin state of the interacting particles. We introduce the Pauli matrices $\boldsymbol{\sigma}_n$ and $\boldsymbol{\sigma}_p$ to describe the spins of the neutron and proton. The total spin of the neutron-proton system is equal to

$$\boldsymbol{S} = \tfrac{1}{2}(\boldsymbol{\sigma}_n + \boldsymbol{\sigma}_p).$$

As is easily verified, the eigenvalues of the operator $\boldsymbol{\sigma}_n \cdot \boldsymbol{\sigma}_p$ in the triplet and singlet states are equal to 1 and -3 respectively. Therefore, the quantities

$$\pi_t = \tfrac{1}{4}(3 + \boldsymbol{\sigma}_n \cdot \boldsymbol{\sigma}_p), \quad \pi_s = \tfrac{1}{4}(1 - \boldsymbol{\sigma}_n \cdot \boldsymbol{\sigma}_p), \tag{1.63}$$

can be regarded as projection operators, projecting out the triplet and singlet states. Indeed, the eigenvalues of these operators are:

$$\pi_t = \begin{cases} 1, & S = 1, \\ 0, & S = 0; \end{cases} \qquad \pi_s = \begin{cases} 0, & S = 1, \\ 1, & S = 0. \end{cases}$$

The scattering length with the spin dependence of the nuclear forces taken into account can be represented by means of the projection operators in the form

$$a = a_t\pi_t + a_s\pi_s = \tfrac{1}{4}(3a_t + a_s) + \tfrac{1}{4}(a_t - a_s)\boldsymbol{\sigma}_n \cdot \boldsymbol{\sigma}_p. \tag{1.64}$$

If we are considering the scattering of slow neutrons by a system of protons, then for the potential we must take a sum of expressions of the type (1.62)

$$V = \sum_i (2\pi\hbar^2/\mu)a^{(i)}\delta(\mathbf{r}_n - \mathbf{r}_i), \tag{1.65}$$

where $\mathbf{r}_i$ is the position vector of the ith proton and $a^{(i)}$ is the scattering length for the scattering of a neutron by the ith proton. This quantity depends on the spin state of the scattered neutron and ith proton. In the case of scattering of slow neutrons in molecular hydrogen, the energy of the interaction of the neutron with a molecule can be written in the form

$$\begin{aligned} V &= (\pi\hbar^2/2\mu)[3a_t + a_s + (a_t - a_s)\boldsymbol{\sigma}_n \cdot \boldsymbol{\sigma}_1]\delta(\mathbf{r}_n - \mathbf{r}_1) \\ &+ (\pi\hbar^2/2\mu)[3a_t + a_s + (a_t - a_s)\boldsymbol{\sigma}_n \cdot \boldsymbol{\sigma}_2]\delta(\mathbf{r}_n - \mathbf{r}_2), \end{aligned} \tag{1.66}$$

where the indices 1 and 2 correspond to the two protons of the molecule.

Elastic scattering of slow neutrons in para- and ortho-hydrogen. We introduce the total spin of the molecule, $\boldsymbol{S} = \frac{1}{2}(\boldsymbol{\sigma}_1 + \boldsymbol{\sigma}_2)$. The quantum number S of the square of the spin angular momentum of the molecule can take two values: $S = 0$ and $S = 1$. If $S = 0$ the corresponding state is called *para-hydrogen*; if $S = 1$, the state is called *ortho-hydrogen*. In the interaction energy (1.66) of the neutron with the molecule, we separate the symmetric and anti-symmetric parts: $V = V_s + V_a$, where

$$V_s = (\pi\hbar^2/2\mu)[3a_t + a_s + (a_t - a_s)\boldsymbol{\sigma}_n\cdot\boldsymbol{S}][\delta(\boldsymbol{r}_n - \boldsymbol{r}_1) + \delta(\boldsymbol{r}_n - \boldsymbol{r}_2)],$$

$$V_a = (\pi\hbar^2/4\mu)(a_t - a_s)\boldsymbol{\sigma}_n\cdot(\boldsymbol{\sigma}_1 - \boldsymbol{\sigma}_2)[\delta(\boldsymbol{r}_n - \boldsymbol{r}_1) - \delta(\boldsymbol{r}_n - \boldsymbol{r}_2)]. \quad (1.67)$$

Clearly, the symmetric part V_s of the interaction energy is responsible for transitions in which the spin symmetry of the molecule does not change; in other words, V_s causes transitions from para-hydrogen to para-hydrogen, and from ortho-hydrogen to ortho-hydrogen. The anti-symmetric part V_a induces transitions in which the symmetry of the molecule changes, that is, transitions from ortho-hydrogen to para-hydrogen and vice versa.

We shall consider the scattering of slow neutrons in the limiting case when the wavelength $\lambdabar$ of the neutron is appreciably greater than the mean distance d between the protons in the hydrogen molecule. (In the equilibrium state, the mean distance between the protons in the hydrogen molecule is of the order of 0.75×10^{-8}cm.) In this case, we can neglect the difference in the positions of the two protons and assume that they are at the same point. Therefore, for $\lambdabar \gg d$, the anti-symmetric part V_a of the interaction energy vanishes and only the symmetric part V_s remains. In this case, clearly, only elastic scattering of neutrons is possible, since V_s does not depend on the relative coordinates of the protons, and the matrix elements of V_s corresponding to inelastic processes will be equal to zero by virtue of the orthogonality of the wave functions.

Comparing the relation (1.67) with (1.62), we can easily establish that the amplitude for elastic scattering of a slow neutron by a hydrogen molecule is, for $\lambdabar \gg d$,

$$A = -\tfrac{1}{2}[3a_t + a_s + (a_t - a_s)\boldsymbol{\sigma}_n\cdot\boldsymbol{S}]. \quad (1.68)$$

In this case, the cross-section for scattering of slow neutrons by a hydrogen molecule is given by the formula

$$\sigma = 4\pi(\tilde{\mu}/\mu)^2\overline{|A|^2}, \quad (1.69)$$

where $\tilde{\mu} = 2M/3$ is the reduced mass of the neutron and hydrogen molecule, $\mu = M/2$ is the reduced mass of the neutron and proton, and the line over the square of the modulus of the scattering amplitude denotes averaging over the spin orientations of the scattered neutron.

Noting that $\overline{\boldsymbol{\sigma}_n\cdot\boldsymbol{S}} = 0$ and $\overline{(\boldsymbol{\sigma}_n\cdot\boldsymbol{S})^2} = S(S+1)$, we find

$$\overline{|A|^2} = \tfrac{1}{4}\{(3a_t + a_s)^2 + S(S+1)(a_t - a_s)^2\}.$$

Consequently, the cross-section for elastic scattering of a slow neutron is

$$\sigma = \tfrac{16}{9}\pi\{(3a_t + a_s)^2 + S(S+1)(a_t - a_s)^2\}. \quad (1.70)$$

Putting $S = 0$ and $S = 1$ in this formula, we obtain values for the cross-sections for elastic scattering of slow neutrons in para- and ortho-hydrogen:

$$\left.\begin{aligned} \sigma_{\text{para}} &= \tfrac{16}{9}\pi(3a_t + a_s)^2, \\ \sigma_{\text{ortho}} &= \tfrac{16}{9}\pi\{(3a_t + a_s)^2 + 2(a_t - a_s)^2\}. \end{aligned}\right\} \tag{1.71}$$

The elastic scattering cross-section σ_{para} in para-hydrogen contains the combination $(3a_t + a_s)^2$; in the cross-section σ_{ortho}, along with this quantity the quantity $(a_t - a_s)^2$ also appears. Consequently, from experiments on the scattering of slow neutrons in para- and ortho-hydrogen, it is possible to determine the quantities $|3a_t + a_s|$ and $|a_t - a_s|$, that is, to determine the absolute values of a_t and a_s and the sign of the ratio a_t/a_s.

The ground state of the hydrogen molecule is the para-state. By studying the scattering of zero-energy neutrons by para-hydrogen molecules, we can determine, according to the relation (1.71), the quantity $|3a_t + a_s|$. Using the experimental value of the averaged cross-section for scattering of zero-energy neutrons by free protons and the fact that the bound state of the neutron-proton system is the triplet state, it is not difficult to find values for a_t and a_s (cf. §1.4.).

1.6. Proton-proton Scattering

Coulomb scattering. Important data on the nature of nuclear forces are given by the study of proton proton scattering. Since protons possess electric charge, in studying proton proton scattering we must take into account not only the nuclear interaction but also the *Coulomb interaction between the protons* .

The wave function $\psi(\boldsymbol{r})$ describing the relative motion of two protons satisfies the Scrödinger equation

$$\{-(\hbar^2/M)\Delta + (e^2/r) + V(r) - E\}\psi(\boldsymbol{r}) = 0, \tag{1.72}$$

where e^2/r is the Coulomb interaction energy, $V(r)$ is the potential of the nuclear interaction, and E is the energy of the relative motion of the protons ($E > 0$). In the case of scattering, the problem is to determine the asymptotic form of the function $\psi(\boldsymbol{r})$ for large values of $\boldsymbol{r}$. Expanding the wave function $\psi(\boldsymbol{r})$ in Legendre polynomials $P_l(\cos\vartheta)$ (ϑ is the angle between the wave vector $\mathbf{k}$ of the incident proton and the position vector $\boldsymbol{r}$)

$$\psi(\boldsymbol{r}) = \sum_l \frac{u_l(r)}{r} P_l(\cos\vartheta), \tag{1.73}$$

we obtain the following equation for the radial function $u_l(\boldsymbol{r})$ corresponding to angular momentum l of the relative motion:

$$\left\{\frac{d^2}{dr^2} + k^2 - \frac{l(l+1)}{r^2} - \frac{2\xi k}{r} - \frac{M}{\hbar^2}V(r)\right\}u_l = 0, \tag{1.74}$$

where $\xi = e^2/\hbar v$ is the so-called *Coulomb parameter* (v is the relative speed of the protons). If the nuclear interaction of the proton were absent, that is, $V(r) = 0$, the asymptotic form of the solution of eqn. (1.74) would be

$$u_l^C \to i^l(2l+1)\frac{\mathrm{e}^{i\eta_l}}{k}\sin\left(kr - \xi\ln 2kr - \frac{l\pi}{2} + \eta_l\right), \quad r \to \infty, \tag{1.75}$$

where η_l is the *Coulomb phase shift*, given by the equality

$$\mathrm{e}^{2i\eta_l} = \frac{\Gamma(l+1+i\xi)}{\Gamma(l+1-i\xi)}.$$

Using the relation (1.75), one can show that, with neglect of the nuclear interaction, the wave function $\psi(\boldsymbol{r})$ for $r \to \infty$ has the form

$$\psi^C(\boldsymbol{r}) \to \mathrm{e}^{ikz + i\xi\ln[k(r-z)]} + f^C(\vartheta)\frac{\mathrm{e}^{ikr - i\xi\ln kr}}{r}, \quad r \to \infty, \tag{1.76}$$

where $f^C(\vartheta)$ is the *Coulomb scattering amplitude*

$$\begin{aligned} f^C(\vartheta) &= \frac{i}{2k}\sum_l (2l+1)(1 - \mathrm{e}^{2i\eta_l})P_l(\cos\vartheta) \\ &= \frac{e^2}{Mv^2\sin^2(\vartheta/2)}\mathrm{e}^{-i\xi\ln\sin^2(\vartheta/2) + 2i\eta_0 + i\pi} \end{aligned} \tag{1.77}$$

(ϑ is the scattering angle in the centre-of-mass frame of the protons).

The expression (1.77) for the scattering amplitude does not take into account the indistinguishability of the protons. Since the protons have a spin equal to $\frac{1}{2}$, by virtue of the Pauli principle the complete wave function of a system of two protons must be anti-symmetric. When the total spin of the protons $S = 0$, the spatial part of the complete wave function is connected with the function $\psi(\boldsymbol{r})$, given by the expansion (1.73), by the relation $\psi_s(\boldsymbol{r}) = \psi(\boldsymbol{r}) + \psi(-\boldsymbol{r})$, and when the total spin $S = 1$, by the relation $\psi_a(\boldsymbol{r}) = \psi(\boldsymbol{r}) - \psi(-\boldsymbol{r})$. Since the replacement $\boldsymbol{r} \to -\boldsymbol{r}$ corresponds to the replacement $\vartheta \to \pi - \vartheta$, the scattering amplitudes in the singlet ($S = 0$) and triplet ($S = 1$) states are connected with the amplitude $f(\vartheta)$ calculated without taking account of the *Pauli principle* by the relations

$$f_s(\vartheta) = f(\vartheta) + f(\pi - \vartheta), \quad f_t(\vartheta) = f(\vartheta) - f(\pi - \vartheta). \tag{1.78}$$

The averaged differential cross-section for proton-proton scattering is in this case given by the formula

$$\sigma(\vartheta) = \tfrac{1}{4}|f_s(\vartheta)|^2 + \tfrac{3}{4}|f_t(\vartheta)|^2, \tag{1.79}$$

where the coefficients $\frac{1}{4}$ and $\frac{3}{4}$ characterize the weights of the singlet and triplet states.

Using the formula (1.77) for the amplitude $f^C(\vartheta)$ and taking (1.78) and (1.79) into account, we obtain the following expression for the differential cross-section for proton-proton scattering with neglect of the nuclear interaction:

$$\sigma_C(\vartheta) = \left(\frac{e^2}{Mv^2}\right)^2 \left\{ \frac{1}{\sin^4(\vartheta/2)} + \frac{1}{\cos^4(\vartheta/2)} - \frac{\cos[\xi \ln \tan^2(\vartheta/2)]}{\sin^2(\vartheta/2)\cos^2(\vartheta/2)} \right\}. \tag{1.80}$$

The first term in (1.80) gives the number of protons scattered though an angle ϑ; the second determines the number of recoiling protons moving at an angle ϑ to the initial beam; the last term describes the quantum effect of exchange of the colliding protons.

Experiments on proton-proton scattering have shown that the formula (1.80) gives an incorrect description of the scattering of protons. The experimentally observed number of protons scattered through an angle of 90° at energies of the order of 1 MeV is several times greater than the number given by the formula (1.80). Consequently, in addition to the Coulomb forces between protons, nuclear forces act, and their magnitude appreciably exceeds that of the electromagnetic interaction forces.

Proton-proton scattering with allowance for the nuclear interaction. We shall assume that the nuclear forces acting between protons, like the forces acting between neutrons and protons, are characterized by very short range. In this case, the nuclear interaction is manifested only in states with very small values of the relative angular momentum l. If the wavelength exceeds the range of the nuclear forces, then the nuclear interaction occurs only in an S-state ($l = 0$).

The proton-proton scattering amplitude in the presence of the nuclear interaction is given by the formula (if the Pauli principle is not taken into account)

$$f(\vartheta) = (i/2k)\sum_l (2l+1)\{1 - e^{2i(\eta_l+\delta_l)}\}P_l(\cos\vartheta), \tag{1.81}$$

where δ_l is the additional phase shift due to the nuclear interaction of the protons in a state with angular momentum l. Using the expression (1.77) for the *Coulomb scattering* amplitude $f(\vartheta)$, we rewrite the formula (1.81) in the form

$$f(\vartheta) = f^C(\vartheta) + (i/2k)\sum_l (2l+1)\mathrm{e}^{2i\eta_l}(1-\mathrm{e}^{2i\delta_l})P_l(\cos\vartheta). \tag{1.82}$$

If the energy of the incident proton is less than 10 MeV, all the phase shifts δ_l, apart from δ_0, can be assumed equal to zero. In this case, the scattering amplitude in which the nuclear interaction is taken into account only in the S-state is given by the formula

$$f(\vartheta) = f^C(\vartheta) + (1/k)(sin\delta)\mathrm{e}^{i\delta+2i\eta_0} \tag{1.83}$$

(in the following we shall drop the index 0 from δ_0).

Clearly, the relations (1.78) are valid both for purely Coulomb scattering and for scattering in which the nuclear interaction is taken into account. In accordance with (1.78), it is not difficult to obtain the following expressions for the proton scattering amplitudes in the singlet and triplet states:

$$\left.\begin{aligned} f_s(\vartheta) &= \frac{e^2}{Mv^2}\left\{\frac{\mathrm{e}^{-i\xi\ln\sin^2(\vartheta/2)}}{\sin^2(\vartheta/2)} + \frac{\mathrm{e}^{-i\xi\ln\cos^2(\vartheta/2)}}{\cos^2(\vartheta/2)} - \frac{4}{\xi}(\sin\delta)\mathrm{e}^{i\delta}\right\}\mathrm{e}^{2i\eta_0+i\pi}, \\ f_t(\vartheta) &= \frac{e^2}{Mv^2}\left\{\frac{\mathrm{e}^{-i\xi\ln\sin^2(\vartheta/2)}}{\sin^2(\vartheta/2)} - \frac{\mathrm{e}^{-i\xi\ln\cos^2(\vartheta/2)}}{\cos^2(\vartheta/2)}\right\}\mathrm{e}^{2i\eta_0+i\pi}. \end{aligned}\right\} \tag{1.84}$$

It can be seen that, for low proton energies, the nuclear interaction is manifested only in the singlet state.

With the nuclear interaction in the S-state taken into account, the proton-proton scattering cross-section, averaged over the spin states, has the form

$$\begin{aligned} \sigma(\vartheta) &= \left(\frac{e^2}{Mv^2}\right)^2\left\{\frac{1}{\sin^4(\vartheta/2)} + \frac{1}{\cos^4(\vartheta/2)} - \frac{\cos[\xi\ln\tan^2(\vartheta/2)]}{\sin^2(\vartheta/2)\cos^2(\vartheta/2)}\right. \\ &- \frac{2}{\xi}\sin\delta\left[\frac{\cos[\delta+\xi\ln\sin^2(\vartheta/2)]}{\sin^2(\vartheta/2)} + \frac{\cos[\delta+\xi\ln\cos^2(\vartheta/2)]}{\cos^2(\vartheta/2)}\right] \\ &+ \left.\frac{4}{\xi^2}\sin^2\delta\right\}. \end{aligned} \tag{1.85}$$

In this expression, the interference term proportional to $\sin\delta$ appears. Therefore, a comparison of the experimental data on proton-proton scattering with the theoretical formula (1.85) makes it possible to determine not only the magnitude but also the sign of the phase shift δ. A positive sign of the phase shift corresponds to forces of attraction, and a negative sign to forces of repulsion. The experimental results show that the phase shift δ is positive, that is, attractive nuclear forces act between the protons in the S-state ($l = 0$).

The proton-proton differential scattering cross-section (1.85) is symmetric about the angle 90° in the centre-of-mass frame. In the small-angle region, the Coulomb scattering is dominant. The interference between the nuclear and Coulomb scattering has a negative character, since the nuclear forces are attractive and the Coulomb forces repulsive. Therefore, in the range of angles in which the nuclear scattering begins to manifest itself, a characteristic interference minimum appears. At large angles, the nuclear scattering is dominant.

Scattering length and effective range. As in the case of scattering of a neutron by a proton, in the case of proton-proton scattering we can define a scattering length and an effective range. We write the equation for the radial function $u \equiv u_0$ of two protons with the nuclear and Coulomb interactions taken into account:

$$u'' - (2\xi k/r)u - (M/\hbar^2)V(r)u = -k^2 u. \tag{1.86}$$

Along with the function u, we introduce into the treatment a function v which coincides asymptotically with u for large r and is a solution of the equation

$$v'' - (2\xi k/r)v = -k^2 v. \tag{1.87}$$

The equation (1.87) has two linearly independent solutions: one regular (vanishing at $r = 0$) and the other irregular at the point $r = 0$. We shall denote these solutions by F and G; at large distances $r \to \infty$ they have the following form:

$$\left.\begin{aligned} F(r) &\sim \sin(kr - \xi \ln 2kr + \eta_0), \\ G(r) &\sim -\cos(kr - \xi \ln 2kr + \eta_0). \end{aligned}\right\} \tag{1.88}$$

For small values of r, the expansions

$$\begin{aligned} &F(r) \sim Ckr, \\ &G(r) \sim -(1/C)\left\{1 + \frac{r}{\rho}[\ln(r/\rho) + h(\xi) + 2\gamma - 1]\right\}, \end{aligned} \tag{1.89}$$

are valid, where

$$C^2 = \frac{2\pi\xi}{e^{2\pi\xi} - 1}, \quad \rho = \frac{\hbar^2}{Me^2}, \quad h(\xi) = \xi^2 \sum_{\nu=1}^{\infty} \frac{1}{\nu(\nu^2 + \xi^2)} - \ln\xi - \gamma,$$

and $\gamma = 0.57722\ldots$ is Euler's constant.

We shall find a solution of eqn. (1.87) having the same asymptotic form at large r as the function u which is the solution of eqn. (1.86), that is, $v \sim \sin(kr - \xi \ln 2kr + \eta_0 + \delta)$, where δ is the phase shift due to the nuclear interaction. We normalize the functions u and v in such a way that $v(0) = 1$. It is not difficult to show that the required solution v can be expressed in terms of the functions F and G in the following way:

$$v(r) = C\{F(r)\cot\delta - G(r)\}. \tag{1.90}$$

Proceeding then as in the derivation of (1.44), we obtain the analogous relation

$$[v'(r) - v_0'(r)]_{r\to 0} = k^2 \int_0^\infty dr(v_0 v - u_0 u), \tag{1.91}$$

where the function u_0 and v_0 correspond to zero energy.

According to the relations (1.90) and (1.89), at small values of r we have

$$v'(r) = (1/\rho)\{\ln(r/\rho) + 2\gamma + K(k)\}, \tag{1.92}$$

where we have introduced the notation

$$K(k) \equiv \pi\cot\delta/(\mathrm{e}^{2\pi\xi} - 1) + h(\xi). \tag{1.93}$$

The quantity $(1/\rho)K(k)$ plays the same role in the case of the interaction of two protons as $k\cot\delta$ in the case of the interaction of a neutron with a proton. This quantity has a simple physical meaning. The first term appearing in the function $(1/\rho)K(k)$ is the quantity $k\cot\delta$ multiplied by the penetration factor C^2 of the Coulomb barrier. The second term in $(1/\rho)K(k)$ takes into account the infinite range of the Coulomb interaction.

Using (1.92), we represent the relation (1.91) in the form

$$(1/\rho)K(k) = -(1/a_p) + \frac{1}{2}k^2 r(0,E), \tag{1.94}$$

where

$$-1/a_p \equiv (1/\rho)\lim_{k\to 0} K(k), \tag{1.95}$$

$$r(0,E) \equiv 2\int_0^\infty dr(v_0 v - u_0 u). \tag{1.96}$$

The quantity a_p, which does not depend on the proton energy, may be called *the proton-proton scattering length*. The quantity $r(0,E)$ takes account of the fact that the range of the nuclear forces between two protons is finite and is introduced in exactly the same way as in the case of the interaction of a neutron and a proton. The value of $r(0, E)$ at zero energy of the protons is usually called the *effective range of the nuclear forces* and is denoted by r_{0p}:

$$r_{0p} = 2 \int_0^\infty dr (v_0^2 - u_0^2). \tag{1.97}$$

Replacing $r(0, E)$ by r_{0p} in the formula (1.94) in the case of low energies, we obtain the relation (Landau and Smorodinskii, 1944)

$$(1/\rho) K(k) = -(1/a_p) + \frac{1}{2} k^2 r_{0p}, \tag{1.98}$$

which gives the dependence of the phase shift δ on the proton energy in an approximation which is independent of the shape of the potential. (The function $K(k)$ is connected with the phase shift δ by the relation (1.93).) It follows from the formula (1.98) that if we plot the dependence of the function $K(k)$ on the energy or k^2, we obtain a straight line defining two parameters: the scattering length a_p and the effective range r_{0p}. These parameters, which characterize the nuclear interaction of the protons in the low-energy region, can be found from the experimental data on the phase shift δ as a function of energy.

The cross-section for scattering of protons by protons has been measured over a broad range of energies. Experimentally determined values of the quantity $(1/\rho)K(k)$ as a function of the proton energy are shown in Fig. 1.2. We see that in the low-energy region there is indeed a linear dependence of $K(k)$ on the energy, in accordance with the formula (1.98). Comparison of the experimental data with the formula (1.98) leads to the following values for the scattering length a_p and the effective range r_{0p}:

$$a_p = -(7.806 \pm 0.003) \times 10^{-13} \text{cm}, \quad r_{0p} = (2.79 \pm 0.01) \times 10^{-13} \text{cm}. \tag{1.99}$$

The negative sign of the proton-proton scattering length a_p implies that a bound state of a system of two protons is impossible.

It is interesting to compare the parameters r_{0p} and a_p characterizing the nuclear interaction of two protons with the parameters r_{0s} and a_s characterizing the nuclear interaction between a neutron and a proton in the singlet state. The values of the effective ranges of the nuclear forces between two protons r_{0p} and between a neutron and a proton in the singlet state r_{0s} coincide within the limits of the experimental errors.

A direct comparison of the proton-proton scattering length a_p and the singlet-state neutron-proton scattering length a_s is not possible. To compare them, it is necessary to exclude the effect of the Coulomb interaction on the scattering length a_p. Since, within the range of the nuclear forces, the

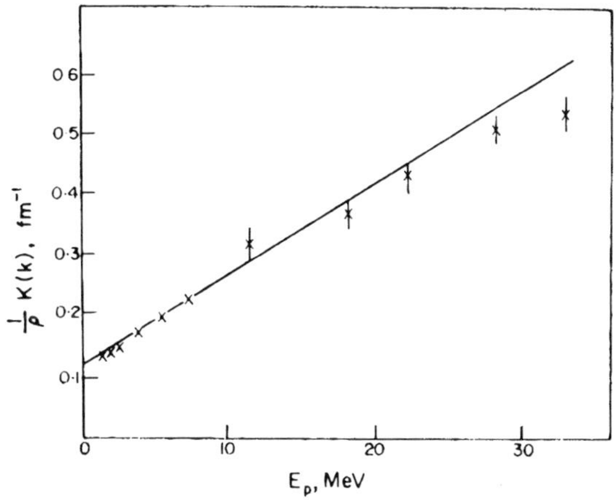

Figure 1.2. Experimental dependence of the values of $(1/\rho)K(k)$ on the proton energy. The straight line corresponds to the approximation independent of the shape of the potential.

Coulomb forces are a small perturbation, it is possible to obtain (Jackson and Blatt, 1950) an approximate relation between the proton proton scattering length a_p and the equivalent length a'_p for proton proton scattering in the absence of the Coulomb force:

$$1/a'_p = (1/a_p) + (1/\rho)[\ln(\rho/r_{0p}) - \tfrac{1}{3}].$$

By substituting the experimental values of the quantities a_p and r_{0p}, we find

$$a'_p = -(17.1 \pm 0.1) \times 10^{-13}\,\text{cm}. \tag{1.99'}$$

In the case of neutron-neutron scattering, the experimental values of the scattering length a_n and the effective radius r_{on} are given by

$$a_n = -(18.5 \pm 0.4) \times 10^{-13}\,\text{cm}, \quad r_{on} = (2.8 \pm 0.1) \times 10^{-13}\,\text{cm}. \tag{1.100}$$

Within the limits of the experimental error, the quantities a'_p and a_n correspond to the singlet-state neutron-proton scattering length (1.56). It can be assumed, therefore, that the nuclear interaction between a neutron and a proton in the singlet state is the same as that between two protons and two neutrons.

1.7. Non-central Nuclear Forces

The potential of non-central forces. Our preceding treatment has been based on the assumption that the nuclear forces are central in character;

this enabled us to explain certain properties of a two-nucleon system, in particular, the binding energy of the deuteron and the scattering of neutrons and protons at low energies. The existence of an electric quadrupole moment and the non-additivity of the magnetic moments of the neutron and proton in the deuteron, however, point to the approximate nature of this assumption. Indeed, in the case of central forces, the ground state of the deuteron should be an S-state, described by a spherically symmetric wave function. The presence of the electric quadrupole moment and the non-additivity of the magnetic moments of the particles in the deuteron mean that the ground state of the deuteron is not a pure S-state, but is a superposition of an S-state and states with non-zero angular momenta that lead to a violation of the spherical symmetry in the charge distribution and to the appearance of an additional magnetic moment. Such a complex ground state is possible only if the interaction between the particles is not purely central. This means that the nuclear interaction between a neutron and a proton must be described by a potential which depends not only on the distance between the particles but also on the orientation of the spins of the particles with respect to the relative position vector. Allowance for these *non-central or tensor* forces enables us to explain the existence of the electric quadrupole moment and the extra magnetic moment in the deuteron.

First we shall elucidate the form that the nuclear interaction potential has in the presence of non-central forces. In the case of low nucleon velocities, to which we shall confine ourselves here, the nuclear interaction potential of two nucleons depends, clearly, on the relative position vector $\boldsymbol{r}$ between the nucleons and the two spin vectors $\boldsymbol{\sigma}_1$ and $\boldsymbol{\sigma}_2$ of the nucleons. The nuclear interaction potential V is invariant with respect to spatial rotations and reflections; therefore, only the invariant combinations that can be formed from the quantities $\boldsymbol{r}$, $\boldsymbol{\sigma}_1$ and $\boldsymbol{\sigma}_2$ can appear in the potential V. Since the Pauli matrices $\boldsymbol{\sigma}$ satisfy the relations

$$\sigma_i \sigma_j = \delta_{ij} + i\varepsilon_{ijk}\sigma_k, \tag{1.101}$$

where $i, j, k = x, y, z$ and ε_{ijk} is the completely antisymmetric unit tensor, these invariant combinations are linear in $\boldsymbol{\sigma}_1$ and $\boldsymbol{\sigma}_2$. However,the quantities $\boldsymbol{n}\cdot\boldsymbol{\sigma}_1$ and $\boldsymbol{n}\cdot\boldsymbol{\sigma}_2$ where $\boldsymbol{n}$ is the unit vector in the direction of $\boldsymbol{r}$, cannot occur separately in the expression for the potential, since $\boldsymbol{\sigma}$ is a pseudo-vector and the products $\boldsymbol{n}\cdot\boldsymbol{\sigma}_1$ and $\boldsymbol{n}\cdot\boldsymbol{\sigma}_2$ are therefore not invariants and change their sign on the reflection of he coordinates. From this it follows that only the quantities $\boldsymbol{\sigma}_1 \cdot \boldsymbol{\sigma}_2$ and $(\boldsymbol{n}\cdot\boldsymbol{\sigma}_1)(\boldsymbol{n}\cdot\boldsymbol{\sigma}_2)$ can occur in the potential V, and these must occur linearly. In place of $(\boldsymbol{n}\cdot\boldsymbol{\sigma}_1)(\boldsymbol{n}\cdot\boldsymbol{\sigma}_2)$, it is convenient to introduce the quantity

$$S_{12} = 3(\boldsymbol{n}\cdot\boldsymbol{\sigma}_1)(\boldsymbol{n}\cdot\boldsymbol{\sigma}_2) - \boldsymbol{\sigma}_1\cdot\boldsymbol{\sigma}_2, \tag{1.102}$$

whose value averaged over the directions of the vector $\boldsymbol{n}$ is equal to zero.

Thus, the nuclear interaction potential between two nucleons can be represented in the form (Eisenbud and Wigner, 1941)

$$V = V_1(r) + V_2(r)\boldsymbol{\sigma}_1 \cdot \boldsymbol{\sigma}_2 + V_3(r)S_{12}, \tag{1.103}$$

where $V_1(r)$, $V_2(r)$ and $V_3(r)$ are certain functions depending on the distance between the particles. The expression (1.103) is the most general expression if we disregard the exchange character of the nuclear forces. In the formula (1.103), the first term describes the central forces that do not depend on the spins, the second term describes the central forces depending on the spins, and the third term describes the tensor forces.

Classification of the states of the neutron-proton system. We shall consider the question of the classification of the states of the neutron-proton system when the interaction between the neutron and proton is described by the potential (1.103). We note that the potential energy of the neutron-proton interaction, given by the formula (1.103), is invariant with respect to transformations of spatial rotation and reflection, and also with respect to interchange of the spins of the two particles.

The invariance of the interaction energy with respect to rotations implies the conservation of the total angular momentum, composed of the orbital and spin angular momenta, of the whole system. Therefore, the neutron-proton system can be characterized by the quantum numbers J and M, defining the *square of the total angular momentum* and the *projection of the total angular momentum* respectively. We note that since, in the general case, the orbital and spin angular momenta are not separately conserved, it is no longer possible to classify the states of the system by the values of the orbital angular momentum.

The invariance of the interaction energy with respect to reflection implies the conservation of the parity w of the neutron-proton system.

The invariance of the interaction energy with respect to interchange of the spins of the particles makes it possible to divide the states of the neutron-proton system into states that are symmetric and states that are anti-symmetric in the spin variables. The symmetric states are triplet states and correspond to the value $S = 1$ of the spin of the system. The anti-symmetric states are singlet states and correspond to the spin value $S = 0$. Consequently, in a neutron-proton system described by the potential (1.103), the square of the total spin of the particles is conserved. One can also convince oneself of this by verifying directly that the operator S^2 of the square of the spin commutes with the operators $\boldsymbol{\sigma}_1 \cdot \boldsymbol{\sigma}_1$ and S_{12}. For this, we need only observe that the operators $\boldsymbol{\sigma}_1 \cdot \boldsymbol{\sigma}_1$ and S_{12} can be represented in the form

TABLE 1.1.

J	Singlets		Triplets	
	Even	Odd	Even	Odd
0	$^1S_0$			$^3P_0$
1		$^1P_1$	$^3S_1+{}^3D_1$	$^3P_1$
2	$^1D_2$		$^3D_2$	$^3P_2+{}^3F_2$

$$\left.\begin{aligned} \boldsymbol{\sigma}_1 \cdot \boldsymbol{\sigma}_1 &= 2\boldsymbol{S}^2 - 3; \\ S_{12} &= 6(\boldsymbol{n}\cdot\boldsymbol{S})^2 - 2\boldsymbol{S}^2, \end{aligned}\right\} \tag{1.104}$$

where $\boldsymbol{S}$ is the total spin of the system: $\boldsymbol{S} = \frac{1}{2}(\boldsymbol{\sigma}_1 + \boldsymbol{\sigma}_2)$. Note, however, that the projection of the total spin on the z-axis is not a constant of the motion.

Thus, we see that a system consisting of a neutron and a proton can be characterized by specifying the quantum numbers J, S, w and M.

It follows from (1.104) that $S_{12} = 0$ in a singlet state; this means that for $S = 0$ the nuclear forces are central. Therefore, in a singlet state the orbital angular momentum l is conserved and states with $S = 0$ can be classified by the quantity l. The triplet state, which can be characterized by determining the value of the total angular momentum J, is a superposition of states with values of l determined by the rules for the addition of angular momenta. For example, the triplet state with $J = 1$ is a superposition of the states $^3S_1$, $^3P_1$ and $^3D_1$. The *parity of the state* is connected with the value of the orbital angular momentum by the relation $w = (-1)^l$. Therefore, the states $^3S_1$ and $^3D_1$ with even values of l and the state $^3P_1$ with an odd value of l are characterized by opposite parity.

In Table 1.1. we give the classification of the first few states of the neutron proton system

Ground state of the deuteron with inclusion of non-central forces. We shall consider the question of the ground state of the deuteron when the non-central character of the nuclear forces is taken into account. It follows from the experimental data that the total angular momentum of the deuteron $J = 1$. On the other hand, in the ground state an S-state is undoubtedly present, since the quadrupole moment is comparatively small and, consequently, the asymmetry in the charge distribution is slight. Therefore, the deuteron ground state is the triplet state $^3S_1 + {}^3D_1$ (see Table 1.1). The parity of this state is equal to unity. Thus, the ground state of the deuteron is characterized by the quantum numbers $J = 1$, $S = 1$ and $w = 1$, and is a superposition of S- and D-states.

The ground-state wave function of the deuteron can be presented in the form

$$\Psi_0 = \Psi_S + \Psi_D, \tag{1.105}$$

where Ψ_S and Ψ_D describe the S- and D-states and are products of radial functions and functions depending on the angles and spin variables.

The wave function $\mathcal{Y}$, which depends on the angles and spin variables and describes the state of the system with total angular momentum J and angular momentum component M, can be expressed in terms of the eigenfunctions Y_{lm} and $\chi_{s\mu}$ of the orbital and spin angular momenta:

$$\mathcal{Y}_{lSJM} = \sum_{m+\mu=M} (lmS\mu|JM) Y_{lm} \chi_{s_\mu}, \tag{1.106}$$

where l and S are the quantum numbers of the squares of the orbital and spin angular momenta, m and μ are their components, and $(lmS\mu|JM)$ are the *Clebsch-Gordan coefficients*. Therefore, the functions Ψ_S and Ψ_D can be presented in the form

$$\Psi_S = \frac{u(r)}{r}\mathcal{Y}_{011M}, \quad \Psi_D = \frac{w(r)}{r}\mathcal{Y}_{211M}, \tag{1.107}$$

where u and w are the radial functions of the S- and D-states. The functions $\mathcal{Y}_{011M}$ and $\mathcal{Y}_{211M}$ have the form

$$\left.\begin{aligned} \mathcal{Y}_{011M} &= Y_{00}\chi_{1M}; \\ \mathcal{Y}_{211M} &= \sqrt{\frac{(3+M)(2+M)}{4\times 5}} Y_{2,M+1}\chi_{1-1} \\ &\quad - \sqrt{\frac{(2-M)(2+M)}{2\times 5}} Y_{2M}\chi_{10} \\ &\quad + \sqrt{\frac{(2-M)(3-M)}{4\times 5}} Y_{2,M-1}\chi_{11} \end{aligned}\right\}. \tag{1.108}$$

Since the $\mathcal{Y}_{lSJM}$ are orthonormal functions, it follows from the normalization condition for the *wave function of the deuteron ground state*

$$\int d^3r |\Psi_0|^2 = 1 \tag{1.109}$$

that the relation $\int_0^\infty dr(u^2 + w^2) = 1$ holds. Clearly, the quantities $p_S = \int_0^\infty dr\, u^2$ and $p_D = \int_0^\infty dr\, w^2$ can be interpreted as the probabilities of finding the deuteron in the S- and D-states ($p_S + p_D = 1$). It can be verified directly without difficulty that the relations

$$\left.\begin{aligned} S_{12}\mathcal{Y}_{011M} &= \sqrt{8}\mathcal{Y}_{211M}; \\ S_{12}\mathcal{Y}_{211M} &= \sqrt{8}\mathcal{Y}_{011M} - 2\mathcal{Y}_{211M}, \end{aligned}\right\} \tag{1.110}$$

hold; therefore, the wave function of the deuteron can be written in the form (*the function of Rarita and Schwinger* (1941))

$$\Psi_0 = \frac{1}{\sqrt{4\pi}}\left\{\frac{u(r)}{r} + \frac{1}{\sqrt{8}}\frac{w(r)}{r}S_{12}\right\}\chi_{1M}, \tag{1.111}$$

where χ_{1M} is the spin function of the system with the spin component coinciding with the component M of the total angular momentum of the deuteron.

The Schrödinger equation for a neutron-proton system in a triplet state with energy E has the form

$$\{-(\hbar^2/M)\Delta + V_C + V_T S_{12} - E\}\Psi = 0, \tag{1.112}$$

where the central part V_C and the tensor part V_T of the potential, according to (1.103), are $V_C = V_1 + V_2$ and $V_T = V_3$. Substituting the function Ψ in the form (1.105) into the Schrödinger equation (1.112) and using the relations (1.110), it is not difficult to obtain the following system of coupled differential equations for the radial functions u and w:

$$\left.\begin{aligned} &\frac{\hbar^2}{M}\frac{d^2u}{dr^2} - V_C u + E u = \sqrt{8}V_T w; \\ &\frac{\hbar^2}{M}\left(\frac{d^2w}{dr^2} - \frac{6}{r^2}w\right) - V_C w + 2V_T w + E w = \sqrt{8}V_T u. \end{aligned}\right\} \tag{1.113}$$

In eqns. (1.113), two potentials $V_C(r)$ and $V_T(r)$ appear; to characterize these in the low-energy region, it is necessary to specify four parameters (the effective depth of the central potential V_C, the effective depth of the tensor potential V_T, and the ranges r_0 and r_T of these potentials). However, even in the simplest case when the potentials are chosen in the form of square wells, it is not possible to find a solution of the system (1.113) in explicit form.

Outside the range of the nuclear forces, the equations for the radial functions u and w are uncoupled. In this case, for the neutron proton bound state ($E = -\varepsilon$) the solutions have the form

$$\left.\begin{aligned} u &= A_S e^{-\alpha r}; \\ w &= A_D\left\{1 + \frac{3}{\alpha r} + \frac{3}{(\alpha r)^2}\right\}e^{-\alpha r}, \end{aligned}\right\} \tag{1.114}$$

where $\alpha = \sqrt{M\varepsilon}/\hbar$, and A_S and A_D are normalization coefficients.

In order to find the *ratio of D- to S-wave normalization coefficients* of the radial wave functions,

$$\eta = \frac{A_D}{A_S},$$

we regard the S-matrix as an analytic function of the complex wave number and find the value of this function near the pole (to be more exact, its residue at the pole), associated with the $n-p$ bound state, by analyzing the angular dependence of the proton-neutron scattering cross-section. According to (Stoks, van Campen, Spit and de Swart, 1988), we have

$$\eta = 0.0271 \pm 0.0002.$$

The relevant numerical value of the S-wave normalization coefficient is given by

$$A_S = (0.8838 \pm 0.0004) \times 10^{-13}\,\text{cm}^{-1/2}.$$

The asymptotic expression for the radial function u of the S-state is valid at distances r greater than the ranges of both the central and the tensor forces; the asymptotic expression for the radial function w of the D-state is valid for $r > r_T$. In fact, in the first equation we can discard the terms $V_C u$ and $V_T w$ only when $r > r_0$ and $r > r_T$. In the second equation, the term $V_C w$ can be neglected as soon as $r > r_T$, since it is small compared with the centrifugal energy. At small r, the radial functions u and w behave as $u \sim r$ and $w \sim r^3$ $(r \to 0)$. The function w has a sharp maximum at $r \approx r_T$, since the expression (1.114) for w falls off like $1/r^2$ for $r_T < r < 1/\alpha$. Therefore, the principal contribution to the integral $p_D = \int_0^\infty dr\, w^2$ determining the weight of the D-state is made by distances $r \approx r_T$.

Magnetic moment of the deuteron. The presence of a D-wave in the ground state of the deuteron leads to the existence of the deuteron quadrupole moment and to the deviation of the deuteron magnetic moment from the sum of the magnetic moments of the neutron and proton. In fact, because of the presence of the D-wave in the ground state, the spherical symmetry in the charge distribution is disturbed and this leads to the appearance of the quadrupole moment. There is a magnetic moment associated with the orbital angular momentum, and it is this magnetic moment which is responsible for the deviation of the deuteron magnetic moment from the sum of the neutron and proton magnetic moments.

The *deuteron magnetic moment operator* $\boldsymbol{\mu}$ (in nuclear magnetons) can be represented in the form

$$\boldsymbol{\mu} = \mu_n \boldsymbol{\sigma}_n + \mu_p \boldsymbol{\sigma}_p + \frac{1}{2}\boldsymbol{l}, \tag{1.115}$$

where μ_n and μ_p are the magnetic moments of the neutron and proton, and $\boldsymbol{l}$ is the angular momentum of the relative motion. (The factor $\frac{1}{2}$ multiplying $\boldsymbol{l}$ takes into account the fact that only the proton has charge.) Introducing the total spin $\boldsymbol{S}$ of the system, we rewrite the magnetic moment $\boldsymbol{\mu}$ of the deuteron in the form

$$\boldsymbol{\mu} = (\mu_n + \mu_p)\boldsymbol{S} + \tfrac{1}{2}(\mu_n - \mu_p)(\boldsymbol{\sigma}_n - \boldsymbol{\sigma}_p) + \tfrac{1}{2}\boldsymbol{l}.$$

Noting that the mean value of the operator $\boldsymbol{\sigma}_n - \boldsymbol{\sigma}_p$ in the ground state is equal to zero, we obtain the following expression for the effective value of $\boldsymbol{\mu}$:

$$\boldsymbol{\mu} = (\mu_n + \mu_p)\boldsymbol{J} - (\mu_n + \mu_p - \tfrac{1}{2})\boldsymbol{l}, \tag{1.116}$$

where $\boldsymbol{J} = \boldsymbol{l} + \boldsymbol{S}$ is the operator of the total angular momentum of the system.

Since the orbital angular momentum $\boldsymbol{l}$ is not conserved, to determine the magnetic moment of a deuteron in a state with a definite value of $\boldsymbol{J}$ we must replace the vector $\boldsymbol{l}$ in the formula (1.116) by its component along the direction of the *total angular momentum* $\boldsymbol{J}$:

$$\boldsymbol{l} \to \frac{\boldsymbol{l}\cdot\boldsymbol{J}}{\boldsymbol{J}^2}\boldsymbol{J} = \frac{\boldsymbol{J}^2 + \boldsymbol{l}^2 - \boldsymbol{S}^2}{2\boldsymbol{J}^2}\boldsymbol{J}.$$

As a result, we obtain the following expression for the component of the magnetic moment of the deuteron in the direction of the total angular momentum $\boldsymbol{J}$:

$$\boldsymbol{\mu} = \left\{\mu_n + \mu_p - \left(\mu_n + \mu_p - \frac{1}{2}\right)\frac{\boldsymbol{J}^2 + \boldsymbol{l}^2 - \boldsymbol{S}^2}{2\boldsymbol{J}^2}\right\}\boldsymbol{J}. \tag{1.117}$$

We shall average this expression over the ground state of the deuteron. Since $\boldsymbol{J}^2$ and $\boldsymbol{S}^2$ are constants of the motion, we have

$$\langle \boldsymbol{J}^2 \rangle = J(J+1), \quad \langle \boldsymbol{S}^2 \rangle = S(S+1).$$

Taking into account that the ground state of the deuteron is a superposition of S- and *D-waves* with *weights* p_S and p_D we find for the mean value of the square of the orbital angular momentum $\boldsymbol{l}^2$

$$\langle \boldsymbol{l}^2 \rangle = 0 \times p_S + 2 \times 3p_D = 6p_D.$$

Projecting the vector (1.117) on to the z-axis and determining the mean value of this quantity for the ground state of the deuteron ($J = S = 1$) with $M = J = 1$, we obtain the *magnetic moment of the deuteron*

$$\mu_d = \mu_n + \mu_p - \tfrac{3}{2}(\mu_n - \mu_p - \tfrac{1}{2})p_D. \tag{1.118}$$

The formula (1.118) enables us to determine the weight of the D-state in the ground state of the deuteron from the experimental values of the magnetic moments μ_n, μ_p and μ_d. Using the values (1.1) and (1.3), we obtain for the *weight of the D-wave in the ground state of the deuteron* the value

$$p_D = 0.04. \tag{1.119}$$

In deriving the formula (1.118), we have neglected relativistic effects and also effects associated with exchange currents. Allowance for these corrections can in crease the weight p_D of the D-wave by an amount of the order of 0.03. We note that the above mentioned effects can be allowed for only under certain assumptions concerning the interaction nature. Therefore, the quantity p_D turns out to be dependent on the assumption. Unlike the parameter η, the quantity p_D cannot be obtain immediately from the experiment.

Quadrupole moment of the deuteron. To conclude this section, we consider the *electric quadrupole moment of the deuteron* . The quadrupole-moment tensor of a system of equal charges is given by the formula

$$Q_{ij} = \sum(3x_i x_j - \boldsymbol{r}^2 \delta_{ij}), \tag{1.120}$$

where $\boldsymbol{r}$ is the position vector of a charge; the summation is taken over all the charges. The quadrupole moment Q is defined as the mean value of the component Q_{zz} of the tensor in the state with $J_z = J$, where J is the total angular momentum of the system. In the case of the deuteron, this component of the quadrupole moment is

$$Q_{zz} = \frac{1}{4}(3z^2 - \boldsymbol{r}^2) = \sqrt{\pi/5}\boldsymbol{r}^2 Y_{20}(\vartheta). \tag{1.121}$$

The factor $\frac{1}{4}$ being connected with the fact that the position vector $\boldsymbol{r}_p$ of the proton is equal to half the relative position vector $\boldsymbol{r}$. The quadrupole moment Q of the deuteron is defined as the mean value of (1.121) in the ground state (1.105) of the deuteron:

$$Q = (\Psi_S, Q_{zz}\Psi_S) + 2(\Psi_S, Q_{zz}\Psi_D) + (\Psi_D, Q_{zz}\Psi_D).$$

Noting that the first term in this expression vanishes because of the spherical symmetry of the function Ψ_S and neglecting the third term because of the small weight of the D-wave, we obtain for the quadrupole moment of the deuteron the expression

$$Q \approx (\sqrt{2}/10) \int_0^\infty dr\, r^2 uw. \tag{1.122}$$

Since the integrand in (1.122) contains a factor r^2, the region of small values of r makes a small contribution to the integral; therefore, in estimating the magnitude of Q, we can make use of the asymptotic expressions (1.114), valid outside the range of the nuclear forces, for the radial functions.

Consequently,

$$Q \approx \frac{\sqrt{2}}{10} A_S A_D \int_0^\infty dr\, r^2 \left(1 + \frac{3}{\alpha r} + \frac{3}{(\alpha r)^2}\right) \mathrm{e}^{-2\alpha r} = \frac{1}{\sqrt{8}} \frac{A_S A_D}{\alpha^3}. \qquad (1.123)$$

We note that $A_D = \eta A_S$ and obtain

$$Q = \frac{1}{\sqrt{2}} \eta R_d^2.$$

We substitute the numerical value for η and thus obtain for Q the number that is in agreement with (1.4). If we take into account the small weight of the D-wave, the normalization factor of the S-wave can be assumed to be the same as in the case of central forces, $A_S = \sqrt{4\pi} C \approx \sqrt{2\alpha}$. By using this value of A_S, we can relate the normalization factor of the D-wave to the magnitude of the quadrupole moment of the deuteron:

$$A_D \approx 2\alpha^{5/2} Q. \qquad (1.124)$$

Knowing the weight p_D of the D-wave and the magnitude of the quadrupole moment Q, we can estimate the magnitude of the range of the tensor forces. Indeed, the principal contribution to the integral $p_D = \int_0^\infty dr\, w^2$ determining the weight of the D-wave is made by values of r close to r_T. Assuming that the contributions of the inner and outer regions are the same and using the asymptotic expression (1.114) for w, it is not difficult to estimate the magnitude of p_D for $r_T \alpha < 1$:

$$p_D \approx 2A_D^2 \int_{r_T}^\infty dr \left(1 + \frac{3}{\alpha r} + \frac{3}{(\alpha r)^2}\right)^2 \mathrm{e}^{-2\alpha r} \approx 24 \frac{\alpha Q^2}{r_T^3}. \qquad (1.125)$$

Putting $p_D = 0.04$ and $Q = 2.86 \times 10^{-27}\,\mathrm{cm}^2$ in this relation, we obtain for the range of the tensor nuclear forces

$$r_T \approx 3 \times 10^{-13}\,\mathrm{cm}. \qquad (1.126)$$

According to (1.125), for a given value of the quadrupole moment Q, a small weight of the D-wave is incompatible with a very small value of the range r_T of the tensor forces.

1.8. Exchange Forces

Scattering of neutrons by protons at high energies and the exchange nature of nuclear forces. In treating the scattering of slow neutrons by protons, we assumed that the nuclear forces are described by a potential depending on the relative position vector between the particles and on the spins of the particles. Starting from such an assumption, we can completely explain the principal features of the scattering of neutrons by protons in the low-energy region. It has been found that , if the wavelength of the particles is of the order of or greater than the range of the nuclear forces, the scattering is entirely insensitive to the dependence of the interaction on the distance. We might expect that by increasing the energy of the interacting particles we should be able to obtain more detailed information on the radial dependence of the nuclear potential. But in fact, using a potential of the form of (1.103) it is completely impossible to explain the angular dependence of the scattering of neutrons by protons at high energies.

In fact, if the only forces acting between a neutron and a proton could be described by the potential (1.103), then, at sufficiently high energies, a large number of neutrons would be scattered through small angles in the centre-of-mass frame, and the recoiling protons would move in the opposite direction (in the laboratory frame, the protons would move principally at right angles to the incident neutron beam). But experiments on the scattering of neutrons by protons at energies greater that a few tens of MeV show that the differential cross-section for scattering of neutrons by protons has, in the centre-of-mass frame, a maximum both in the region of small scattering angles and in the region of angles close to 180°. In order to explain this property of the scattering, it is necessary to assume that, along with forces describable by the potential (1.103), additional forces, associated with exchange of the particles, act between the neutron and the proton. These forces may be called *exchange forces*, to distinguish them from the ordinary non-exchange forces. We shall show that, by introducing the exchange forces, we can explain the increase in the scattering cross-section at high energies in the large-angle region. If the energy of the neutron exceeds the effective energy of the nuclear interaction between the proton and neutron, the differential cross-section for scattering of a neutron by proton in the centre-of-mass frame can be determined by means of perturbation theory:

$$d\sigma = \frac{M^2}{16\pi^2\hbar^4} \left| \int d^3r \psi'^*(\boldsymbol{r}) V \psi(\boldsymbol{r}) \right|^2 do, \tag{1.127}$$

where $\psi(\boldsymbol{r}) = e^{i\mathbf{k}\cdot\mathbf{r}}$ and $\psi'(\boldsymbol{r}) = e^{i\mathbf{k}'\cdot\mathbf{r}}$ are the wave functions of the relative motion before and after the scattering ($\mathbf{k}$ and $\mathbf{k}'$ are the wave-vectors),

V is the nuclear interaction potential between the particles, and do is the element of solid angle containing the wave-vector after the scattering.

In the case of ordinary forces, the potential V is a function of the position vector $\boldsymbol{r}$, so that the *differential scattering cross-section has the form*

$$d\sigma = \frac{M^2}{16\pi^2\hbar^4}\left|\int d^3\boldsymbol{r}\, e^{i(\mathbf{k}-\mathbf{k}')\cdot\mathbf{r}}V(\boldsymbol{r})\right|^2 do. \tag{1.128}$$

The integral occurring in this expression differs appreciably from zero if $|\mathbf{k}-\mathbf{k}|r_0 \equiv 2k\sin(\vartheta/2)r_0 \lesssim 2\pi$, where ϑ is the scattering angle and r_0 is the range of the nuclear forces. Thus, in the case of ordinary forces, the scattering is mainly through angles $\vartheta \lesssim \lambda/r_0$, where $\lambda = 2\pi/k$ is the wavelength of the relative motion. For sufficiently high energies, $\lambda \ll r_0$ so that $\vartheta_{\text{eff}} \ll 1$.

We now consider scattering in the case of exchange forces. We represent the *potential V for exchange forces* in the form

$$V = V(\boldsymbol{r})P_r, \tag{1.129}$$

where P_r is the particle-exchange operator. The action of this operator on the function $\psi(\boldsymbol{r})$ reduces to replacing $\mathbf{r}$ by $-\mathbf{r}$:

$$P_r\psi(\boldsymbol{r}) = \psi(-\boldsymbol{r}). \tag{1.130}$$

Therefore, the differential cross-section for scattering due to exchange forces characterized by the potential (1.129) is given by the formula

$$d\sigma = \frac{M^2}{16\pi^2\hbar^4}\left|\int d^3r e^{-i(\mathbf{k}+\mathbf{k}')\cdot\mathbf{r}}V(\boldsymbol{r})\right|^2 do. \tag{1.131}$$

This expression differs appreciably from zero when $|\mathbf{k}+\mathbf{k}'|\boldsymbol{r}_0 \lesssim 2\pi$. Consequently, in the case of scattering under the influence of exchange forces, the main role is played by scattering angles for which the quantity $|\mathbf{k}+\mathbf{k}'|$ is small, that is, by scattering angles close to 180°. This result is easily understood if we assume that the incident particle is deflected through a small angle in the case of the exchange forces too, but that the nature of the colliding particles changes in the scattering process, that is, the neutron is transformed into a proton, and the proton into a neutron. Consequently, the distribution of neutrons in exchange scattering should be the same as the distribution of recoiling protons in ordinary scattering, and vice versa.

In the general case of scattering under the influence of ordinary and exchange forces, the differential cross-section has two maxima, corresponding to the forward and backward directions; the relative magnitudes of these maxima are determined by the weights of the ordinary and exchange forces in the interaction between the particles. The experimental data on the scattering of neutrons by protons at energies of a few tens of MeV show that the ordinary and exchange forces have approximately equal weight.

According to the formulae (1.128) and (1.131), which were obtained using perturbation theory, the total cross-section for scattering of a neutron by a proton at high energies should be inversely proportional to the energy, irrespective of the nature of the nuclear forces.

Exchange operators. In the general case, the potential characterizing the exchange interaction of two particles can be represented in the form of a product of some function depending on the coordinates of the particles with the exchange operator, which interchanges the coordinates of the interacting particles. Since the neutrons and protons are characterized not only by their spatial coordinates but also by spin coordinates, there exist different possibilities of exchange, corresponding to the different forms of exchange forces. Along with forces that lead to a complete exchange of the coordinates of the interacting particles, forces may exist which are associated with the exchange of the spatial or spin coordinates separately. Usually, the different exchange forces are known by the names of the investigators who first studied them (Heisenberg, 1932; Majorana, 1933; Bartlett, 1936).

Let $\psi(\boldsymbol{r}_1, \sigma_1; \boldsymbol{r}_2, \sigma_2)$ be the wave function of a two-particle system, depending on the spatial and spin coordinates $\boldsymbol{r}_1$, σ_1 and $\boldsymbol{r}_2$, σ_2 of the particles. Clearly, we can introduce three different exchange operators, $\boldsymbol{P}_r$, $\boldsymbol{P}_\sigma$ and $\boldsymbol{P}_{r\sigma}$, the action of which on the wave function reduces respectively to the interchange of the spatial coordinates, the interchange of the spin coordinates and the simultaneous interchange of the spatial and spin coordinates:

$$\left.\begin{aligned} \boldsymbol{P}_r\psi(\boldsymbol{r}_1, \sigma_1; \boldsymbol{r}_2, \sigma_2) &= \psi(\boldsymbol{r}_2, \sigma_1; \boldsymbol{r}_1, \sigma_2); \\ \boldsymbol{P}_\sigma\psi(\boldsymbol{r}_1, \sigma_1; \boldsymbol{r}_2, \sigma_2) &= \psi(\boldsymbol{r}_1, \sigma_2; \boldsymbol{r}_2, \sigma_1); \\ \boldsymbol{P}_{r\sigma}\psi(\boldsymbol{r}_1, \sigma_1; \boldsymbol{r}_2, \sigma_2) &= \psi(\boldsymbol{r}_2, \sigma_2; \boldsymbol{r}_1, \sigma_1). \end{aligned}\right\} \tag{1.132}$$

The operator $\boldsymbol{P}_r$ interchanging the spatial coordinates of the particles is usually called the *Majorana exchange operator* , the operator $\boldsymbol{P}_\sigma$ interchanging the spin coordinates of the particles is called the *Bartlett exchange operator* , and the operator $\boldsymbol{P}_{r\sigma}$ interchanging both the spatial and the spin coordinates of the particles is called the *Heisenberg exchange operator*. Obviously,

$$\boldsymbol{P}_{r\sigma} = \boldsymbol{P}_r\boldsymbol{P}_\sigma; \tag{1.133}$$

$$\boldsymbol{P}_r^2 = \boldsymbol{P}_\sigma^2 = \boldsymbol{P}_{r\sigma}^2 = 1. \tag{1.134}$$

According to (1.134), the exchange operators $\boldsymbol{P}_r$, $\boldsymbol{P}_\sigma$, and $\boldsymbol{P}_{r\sigma}$ have the eigenvalues ± 1 only. Interchange of the coordinates $\mathbf{r}_1$ and $\mathbf{r}_2$ is equivalent to a change of sign of the relative coordinate $\mathbf{r} = \mathbf{r}_1 - \mathbf{r}_2$. Therefore, for a system consisting of two particles, the operator $\boldsymbol{P}_r$ coincides with the spatial inversion operator and, consequently, the eigenvalues of the operator $\boldsymbol{P}_r$ are determined by the magnitude l of the relative angular momentum:

$$\boldsymbol{P}_r = (-1)^l. \tag{1.135}$$

The eigenvalues of the operator $\boldsymbol{P}_\sigma$ are determined by the symmetry of the spin part of the wave function. Since a system of two nucleons can be in either a triplet state ($S = 1$), with a corresponding symmetric spin function, or in a singlet state ($S = 0$), with a corresponding anti-symmetric spin function, the eigenvalues of $\boldsymbol{P}_\sigma$ can be related to the magnitude S of the total spin of the system:

$$\boldsymbol{P}_\sigma = (-1)^{S+1}. \tag{1.136}$$

The eigenvalues of the operator $\boldsymbol{P}_{r\sigma}$ are determined by the quantities l and S:

$$\boldsymbol{P}_{r\sigma} = (-1)^{l+S+1}. \tag{1.137}$$

Nuclear interaction potential with inclusion of exchange forces. Using the exchange operators $\boldsymbol{P}_r$, $\boldsymbol{P}_\sigma$ and $\boldsymbol{P}_{r\sigma}$ introduced above, it is not difficult to find a general expression for the nuclear interaction potential with the exchange forces taken into account. According to (1.103), the nuclear interaction potential in the general case consists of two parts, describing the central and non-central forces. The potentials for the central exchange forces can be represented in the form of products of the operators $\boldsymbol{P}_r$, $\boldsymbol{P}_\sigma$ and $\boldsymbol{P}_{r\sigma}$ with certain functions depending on the distance between the particles. The corresponding exchange forces are usually called the Majorana, Bartlett and Heisenberg forces. The non-exchange nuclear forces are often called Wigner forces.

The non-central nuclear forces can be of the exchange or non-exchange type. Since the tensor operator S_{12} is equal to zero in the singlet state, and the operator $\boldsymbol{P}_\sigma$ is equal to $+1$ in the triplet state, we have

$$S_{12}\boldsymbol{P}_\sigma = \boldsymbol{P}_\sigma S_{12} = S_{12}. \tag{1.138}$$

Therefore, only one form of exchange tensor force is possible – the Majorana tensor force. (The Heisenberg tensor forces, by virtue of (1.138), reduce to Majorana tensor forces.)

Thus, in the most general case, the nuclear interaction potential between two nucleons, with the exchange character and non-central nature of the nuclear forces taken into account, can be written in the form

$$\begin{aligned} V &= V_W(\boldsymbol{r}) + V_M(\boldsymbol{r})\boldsymbol{P}_r + V_B(\boldsymbol{r})\boldsymbol{P}_\sigma + V_H(\boldsymbol{r})\boldsymbol{P}_{r\sigma} \\ &+ V_{TW}(\boldsymbol{r})S_{12} + V_{TM}(\boldsymbol{r})S_{12}\boldsymbol{P}_r, \end{aligned} \tag{1.139}$$

where $V_W(\boldsymbol{r})$, $V_M(\boldsymbol{r})$, $V_B(\boldsymbol{r})$, $V_H(\boldsymbol{r})$, $V_{TW}(\boldsymbol{r})$ and $V_{TM}(\boldsymbol{r})$ are functions depending on the distance between the particles. These functions can differ both in magnitude and in the nature of their dependence on the distance. Note that there is no need to introduce separately into the relation (1.139) a term taking the spin dependence of the central forces into account, since this dependence is taken into account by the exchange terms.

The expression (1.139) for the potential, which was obtained in accordance with the requirements of invariance under spatial rotations and reflections and of the symmetry with respect to interchange of particles, contains, in the general case, six functions $V(r)$, the form of which can be determined only from experimental data. In practice, however, the question of the determination of the form of the functions $V(r)$ remains open at the present time. To a considerable extent, this is connected with the fact that nuclear phenomena in the low-energy region depend very weakly on the shape of the nuclear potential. The original hopes that it might be possible to determine the form of $V(r)$ from experiments on the scattering of two nucleons at high energies have also proved unfounded, inasmuch as the expression (1.139) was obtained in the static limit.

The Serber potential. One of the simplest potentials, and one which is often used in calculations, is the so-called *Serber potential*, incorporating a mixture of the Wigner and Majorana forces with equal weights:

$$V_S = V_S(r)(1 + P_r)/2. \tag{1.140}$$

Since the value of P_r is -1 in states with odd values of l, the interaction described by the potential (1.140) is manifested only in even states. Because of this, the Serber potential leads to scattering that is symmetric about the angle 90°. In reality, the angular distribution in the scattering of neutrons by protons at energies of the order of 100 MeV has an asymmetric character, indicating interaction in the odd states too. However, this interaction is considerably weaker than the interaction in the even states.

1.9. Isotopic Spin and Charge Independence of the Nuclear Interaction

Isotopic spin of the nucleon. As we have already observed, despite the fact that the proton has an electric charge and the neutron does not, neutrons and protons possess very similar physical properties. This similarity is manifested in the closeness of the neutron and proton masses; moreover, neutrons and protons have the same spin $\frac{1}{2}$ and obey the same statistics (Fermi-Dirac statistics); the neutron and proton can be transformed into each other, by absorbing a meson; neutrons and protons are transformed into each other in β-decay, and so on. However, the most striking manifestation of the physical similarity of the neutron and proton is the *charge independence* of their nuclear interaction. The charge independence, or, as it is more often called, the *isotopic invariance*, appears in the fact that the nuclear interaction between any two particles (neutrons or protons) does not depend on the nature of the particles and is determined only by the state in which the system is found.

The similarity of the physical properties of the neutron and proton enables us to regard them as two different states of one and the same particle – the nucleon. (These states differ in the magnitude of the charge.) The nucleon is characterized not only by its spatial and spin coordinates, but also by an additional degree of freedom – the *charge coordinate*. Since the charge coordinate (which distinguishes the neutron and proton states of the nucleon) assumes only two values in all, it is convenient to associate with it a vector $\boldsymbol{t}$, similar in its properties to the vector of an ordinary spin $\frac{1}{2}$, in some fictitious space. The vector $\boldsymbol{t}$ is called the *isotopic spin*, or *isospin* (Heisenberg, 1932). The introduction of the isotopic spin enables us to describe the isotopic invariance of the nuclear interaction very simply.

The absolute value of the isotopic-spin vector $\boldsymbol{t}$ for a nucleon is equal to $\frac{1}{2}$; therefore, the components of the vector $\boldsymbol{t}$ can be expressed directly in terms of the Pauli matrices $\boldsymbol{\tau}$:

$$\tau_x = \begin{pmatrix} 0 & 1 \\ 1 & 0 \end{pmatrix}, \quad \tau_y = \begin{pmatrix} 0 & -i \\ i & 0 \end{pmatrix}, \quad \tau_z = \begin{pmatrix} 1 & 0 \\ 0 & -1 \end{pmatrix}, \tag{1.141}$$

namely, by

$$\boldsymbol{t} = \tfrac{1}{2}\boldsymbol{\tau}. \tag{1.142}$$

The projection of the isotopic spin of a nucleon on to the z-axis can take only two values: $\nu = \pm\frac{1}{2}$. The value $+\frac{1}{2}$ is conventionally associated with the proton and the value $-\frac{1}{2}$ with the neutron. We denote the charge or isospin function of the nucleon by $\zeta(\tau)$, where τ is the charge coordinate or isospin coordinate. For the charge coordinate τ, we can choose the value of the z-component of the isospin. Wave functions for states with well-defined values of the isospin component are diagonal, $\zeta_\nu(\tau) = \delta_{\nu\tau}$, in this representation. Usually, the isospin functions corresponding to the proton and neutron states are represented in the following form:

$$\zeta_{1/2} = \begin{pmatrix} 1 \\ 0 \end{pmatrix}, \quad \zeta_{-1/2} = \begin{pmatrix} 0 \\ 1 \end{pmatrix}. \tag{1.143}$$

Since the spatial, spin and isospin coordinates are independent, the complete wave function ψ for a free nucleon can be represented in the form of a product of a spatial function φ, a spin function χ and an isospin function ζ:

$$\psi(\boldsymbol{r}, \sigma, \tau) = \varphi(\boldsymbol{r})\chi(\sigma)\zeta(\tau). \tag{1.144}$$

We introduce projection operators for the neutron and proton states:

$$\Lambda_n = \tfrac{1}{2}(1 - \tau_z), \quad \Lambda_p = \tfrac{1}{2}(1 + \tau_z). \tag{1.145}$$

The operator Λ_n leaves the neutron function unchanged and makes the proton function vanish; the operator Λ_p has the opposite effect. Using the projection operators (1.145), it is not difficult to construct operators for the charge, mass and magnetic moment of the nucleon:

$$\left.\begin{aligned} q &= \tfrac{1}{2}(1 + \tau_z)e; \\ M &= \tfrac{1}{2}(1 - \tau_z)M_n + \tfrac{1}{2}(1 + \tau_z)M_p; \\ \mu &= \tfrac{1}{2}(1 - \tau_z)\mu_n + \tfrac{1}{2}(1 + \tau_z)\mu_p. \end{aligned}\right\} \tag{1.146}$$

The charge creation and destruction operators τ_+ and τ_- have the form

$$\tau_+ = \tfrac{1}{2}(\tau_x + i\tau_y) = \begin{pmatrix} 0 & 1 \\ 0 & 0 \end{pmatrix}, \quad \tau_- = \tfrac{1}{2}(\tau_x - i\tau_y) = \begin{pmatrix} 0 & 0 \\ 1 & 0 \end{pmatrix}. \tag{1.147}$$

The operator τ_+ converts a neutron into a proton, and the operator τ_- converts a proton into a neutron.

Isotopic spin of a system of nucleons. The isotopic spin of a system of nucleons is composed of the isotopic spins of the separate particles, in accordance with the rules for vector addition of angular momenta. The isotopic spin operator of a system of A nucleons is equal to

$$\boldsymbol{T} = \sum_{i=1}^{A} \boldsymbol{t}_i, \tag{1.148}$$

where $\boldsymbol{t}_i$ is the isotopic spin operator of an individual nucleon. The eigenvalues of the square of the isotopic spin (1.148) are equal to $T(T+1)$, where T is the *total isotopic spin of system.*

If A is even, then T can take values in the range from 0 to $A/2$; if A is odd, then T takes values in the range from $\frac{1}{2}$ to $A/2$. For each T, there exist $2T+1$ eigenvalues of T_z which are integers or odd multiples of half-integers in the interval $-T \leqslant T_z \leqslant T$.

For a nucleus with atomic number Z and mass number A, containing Z protons and $N = A - Z$ neutrons, the projection of the total isotopic spin is

$$T_z = \tfrac{1}{2}(Z - N) = Z - A/2. \tag{1.149}$$

In this case, the absolute value T of the total isotopic spin of the system can be equal to one of the values in the range $|T_z| \leqslant T \leqslant A/2$. For a given number A of nucleons, according to (1.149), the magnitude of T_z determines the total charge of the nucleus. Since the charge is conserved, a conservation law holds for the projection T_z of the isotopic spin of the system.

States of a system of A nucleons with the same value T of the isotopic spin form an *isotopic multiplet.* For a given value of T, the isotopic multiplet contains $2T+1$ components. These components differ in their values of the isotopic-spin projection T_z, that is, they correspond to different charges Z of the system. Clearly, the limiting values of the charge for the components of the multiplet are equal to $A/2 \pm T$. When the electromagnetic interaction is neglected, the components of the multiplet have the same structure and are characterized by the same binding energy.

Since the nucleons possess spin equal to $\frac{1}{2}$, they satisfy the Pauli principle. We write the complete wave function of a system A nucleons in the form

$$\Psi(\boldsymbol{r}_1, \sigma_1, \tau_1; \boldsymbol{r}_2, \sigma_2, \tau_2; \ldots; \boldsymbol{r}_A, \sigma_A, \tau_A),$$

where $\boldsymbol{r}_i$, σ_i and τ_i are, respectively, the spatial, spin and charge coordinates of the ith nucleon. Then the Pauli principle for nucleus can be formulated as follows: the wave function of a system of nucleons must be anti-symmetric with respect to interchange of the spatial, spin and isospin coordinates of any pair of nucleons. Because the isospin coordinates of the nucleons are independent, the *complete wave function* Ψ of a system of nucleons can always be represented in the form of a product of a *space-spin function* Φ with an *isospin function* Z:

$$\Psi = \Phi(\boldsymbol{r}_1, \sigma_1; \boldsymbol{r}_2, \sigma_2; \ldots; \boldsymbol{r}_A, \sigma_A) Z(\tau_1, \tau_2, \ldots, \tau_A). \tag{1.150}$$

The symmetry of the isospin function Z is uniquely determined by the magnitude T of the isotopic spin of the system. By virtue of the Pauli principle, the complete wave function Ψ is anti-symmetric with respect to interchanges of the coordinates of any pair of nucleons; therefore, the functions Z and Φ must possess opposite (dual) symmetry.

Classification of the states of a two-nucleon system. We shall consider a system consisting of two nucleons. Since the isotopic spin of each nucleon is equal to $\frac{1}{2}$, the total isotopic spin of the system can take two values: $T = 1$ and $T = 0$. Therefore, all the states of the system can be divided into two groups – one corresponding to the isotopic spin $T = 1$ and the other to $T = 0$.

In the case $T = 1$, three different values of the isospin projection are possible: $T_z = 1$, 0 and -1. The corresponding states are characterized by charges $2e$, e and 0 and correspond to a system consisting of two protons, a neutron and a proton, and two neutrons. The wave functions describing these states are symmetric and can be expressed in terms of the one-nucleon isospin functions (1.143):

$$\left.\begin{aligned} Z_{1,1}(\tau_1,\tau_2) &= \zeta_{1/2}(\tau_1)\zeta_{1/2}(\tau_2); \\ Z_{1,0}(\tau_1,\tau_2) &= (1/\sqrt{2})\{\zeta_{1/2}(\tau_1)\zeta_{-1/2}(\tau_2)+\zeta_{-1/2}(\tau_1)\zeta_{1/2}(\tau_2)\}; \\ Z_{1,-1}(\tau_1,\tau_2) &= \zeta_{-1/2}(\tau_1)\zeta_{-1/2}(\tau_2). \end{aligned}\right\} \tag{1.151}$$

The functions (1.151) are normalized to unity.

In the case $T = 0$, the isospin projection $T_z = 0$. The corresponding singlet state is described by an anti-symmetric isospin function:

$$Z_{00}(\tau_1,\tau_2) = (1/\sqrt{2})\{\zeta_{1/2}(\tau_1)\zeta_{-1/2}(\tau_2)-\zeta_{-1/2}(\tau_1)\zeta_{1/2}(\tau_2)\}. \tag{1.152}$$

The anti-symmetric state (1.152) can be realized only in a system consisting of a neutron and a proton. The function (1.152) is orthogonal to the functions (1.151).

The complete wave function of a system of two nucleons can be represented in the form of a product of a space-spin function with an isospin function:

$$\Psi_{wSJMTM_T} = \Phi_{wSJM}(\boldsymbol{r}_1,\sigma_1;\boldsymbol{r}_2,\sigma_2)Z_{TM_T}(\tau_1,\tau_2). \tag{1.153}$$

Thus, the state of the system is determined if the following complete set of quantum numbers is specified: *the parity* w , *the total spin* S, the *total angular momentum* J, the *projection* M *of the total angular momentum* , the *total isospin* T and the *projection* M_T *of the isospin* . By virtue of the Pauli principle, the complete wave function (1.153) must be anti-symmetric with respect to an interchange of the nucleons, that is, with respect to a simultaneous interchange of the spatial, spin and isospin coordinates of the nucleons. The symmetry of the wave function Φ with respect to interchange of the spatial coordinates is determined by the parity w of the state, and the symmetry with respect to interchange of the spin coordinates is determined by the spin magnitude S; the symmetry of the isospin function Z with respect to interchange of the isospin coordinates is determined by the magnitude T of the isotopic spin of the system. Therefore, in a system of two nucleons, even singlet states and odd triplet states are always symmetric in the isospin variables ($T = 1$), and odd singlet states and even triplet states are always anti-symmetric in the isospin variables ($T = 0$). Hence it follows that only even singlet states and odd triplet states can be realized in a system of two identical nucleons (since the isospin function must be symmetric), whereas states of any parity and spin can be realized in systems of two different nucleons.

TABLE 1.2.

J	S = 0		S = 1	
	l–even	l–odd	l–even	l–odd
	$T = 1$	$T = 0$	$T = 0$	$T = 1$
0	$^1S_0$			$^3P_0$
1		$^1P_1$	$^3S_1 + {}^3D_1$	$^3P_1$
2	$^1D_2$		$^3D_2$	$^3P_2 + {}^3F_2$
3		$^1F_3$	$^3D_3 + {}^3G_3$	$^3F_3$
4	$^1G_4$		$^3G_4$	$^3F_4 + {}^3H_4$
5		$^1H_5$	$^3G_5 + {}^2I_5$	$^3H_5$

In Table 1.2 we give a classification of the states of a system of two nucleons for $J = 0$, 1, 2, 3, 4 and 5. In particular, it follows from the Table that the deuteron isotopic spin is equal to zero, $T = 0$.

Explicit form of the exchange operators. The Majorana, Bartlett and Heisenberg exchange operators $\boldsymbol{P}_r$, $\boldsymbol{P}_\sigma$ and $\boldsymbol{P}_{r\sigma}$ introduced in the study of nuclear exchange forces can be expressed simply in terms of the operators of the spins and isospins of the particles. The operator $\boldsymbol{\sigma}_1 \cdot \boldsymbol{\sigma}_2$ is characterized by the eigenvalues $+1$ and -3 for the triplet ($S = 1$) and singlet ($S = 0$) states respectively. Therefore, the operator $\boldsymbol{P}_\sigma$ interchanging the spin coordinates of two nucleons can be represented in the form

$$\boldsymbol{P}_\sigma = \tfrac{1}{2}(\mathbf{1} + \boldsymbol{\sigma}_1 \cdot \boldsymbol{\sigma}_2). \tag{1.154}$$

In fact, the operator (1.154) is characterized by the eigenvalues $+1$ and -1 for the triplet and singlet states respectively; therefore, $\boldsymbol{P}_\sigma \chi(\sigma_1, \sigma_2) = \chi(\sigma_2, \sigma_1)$. The formula (1.154) gives a convenient representation for the Bartlett exchange operator.

Analogously, we can introduce the operator $\boldsymbol{P}_\tau$ interchanging the isospin coordinates of two nucleons:

$$\boldsymbol{P}_\tau = \tfrac{1}{2}(\mathbf{1} + \boldsymbol{\tau}_1 \cdot \boldsymbol{\tau}_2). \tag{1.155}$$

The operator $\boldsymbol{P}_\tau$ acts on the isospin function of the nucleons: $\boldsymbol{P}_\tau \boldsymbol{Z}(\tau_1, \tau_2) = \boldsymbol{Z}(\tau_2, \tau_1)$. Since the complete wave function of a system of two nucleons is anti-symmetric with respect to interchange of the nucleons, which reduces to interchange of the spatial, spin and isospin coordinates of the nucleons, we have

$$\boldsymbol{P}_r \boldsymbol{P}_\sigma \boldsymbol{P}_\tau = -1. \tag{1.156}$$

Using this equality and noting that $\boldsymbol{P}_\tau^2 = 1$, it is not difficult to obtain the following representation for the Majorana exchange operator:

$$\boldsymbol{P}_r = -\tfrac{1}{4}(1 + \boldsymbol{\sigma}_1 \cdot \boldsymbol{\sigma}_2)(1 + \boldsymbol{\tau}_1 \cdot \boldsymbol{\tau}_2). \qquad (1.157)$$

Recalling the definition (1.133) of the Heisenberg exchange operator, we easily find that

$$\boldsymbol{P}_{r\sigma} = -\tfrac{1}{2}(1 + \boldsymbol{\tau}_1 \cdot \boldsymbol{\tau}_2), \qquad (1.158)$$

that is, the Heisenberg exchange operator $\boldsymbol{P}_{r\sigma}$ coincides in magnitude with the operator $\boldsymbol{P}_\tau$ interchanging the charge coordinates of the particles.

Isotopic invariance of the nuclear interaction. The *isotopic-spin formalism* is very convenient for the description of the charge independence or, as it is some times called, the isotopic invariance of the nuclear interaction (Breit, Condon and Present, 1936; Cassen and Condon, 1936). We shall formulate the concept of the isotopic invariance of the interaction by using the concept of the isotopic spin of the particles.

According to the definition, the isotopic spin is a vector in a certain fictitious space. The absolute value of the isotopic spin of a nucleon is equal to $\frac{1}{2}$. The state with isospin projection $+\frac{1}{2}$ corresponds to a proton, and the state with isospin projection $-\frac{1}{2}$ corresponds to a neutron. A transition form a proton to a neutron implies a change of the isospin projection from $+\frac{1}{2}$ to $-\frac{1}{2}$. Therefore, by analogy with rotations in ordinary space, such a transition can be regarded as a rotation in the isotopic space through an angle π about an axis perpendicular to the axis of quantization. If the system consists of several nucleons, their isotopic spins are combined into the total isotopic spin of the system in accordance with the rules of vector addition. In this case, states of the system corresponding to a definite value of the total isotopic spin but to different values of the isospin projection can be associated with rotations in the isotopic space. For example, a system consisting of two protons differs from a system consisting of a proton and a neutron with $T = 1$ by a rotation in isotopic space through an angle $\pi/2$; a system consisting of two protons differs from a system consisting of two neutrons by a rotation through an angle π in the isotopic space, and so on. Clearly, with this approach, the isotopic invariance of the nuclear interaction can be regarded as an invariance property with respect to rotations of the system in the isotopic space.

We note that the charge symmetry of the nuclear interaction (the invariance with respect to replacement of all the neutrons by protons and vice versa), which is a particular case of isotopic invariance, can be regarded as invariance with respect to a rotation in isotopic space through an angle π about an axis perpendicular to the quantization axis.

Thus, the isotopic invariance of the nuclear interaction means that the interaction should not depend on the orientation of the isotopic-spin vector T of the system. Consequently, the isotopic space should be isotopic as far as the nuclear interaction is concerned.

The isotopic invariance does not hold for the electromagnetic interaction. In fact, the electric charge and magnetic moment of the system are determined by the isotopic-spin projection; therefore, the electromagnetic interaction also depends on the isotopic-spin projection, which is not conserved in rotations in the isotopic space.

Since the isotopic spin possesses properties similar to those of an ordinary angular momentum, it follows from the isotropy of the isotopic space with respect to nuclear interactions that the isotopic spin T is conserved. Unlike the conservation law for the isotopic-spin projection T_z which expresses the law of conservation of the charge of the system, the conservation law for the isotopic spin T of the system is approximate, since it is valid only when the electromagnetic interaction is neglected.

We show now that the potential found earlier from considerations of the invariance with respect to spatial rotations and spatial inversion, which takes into account the non-central and exchange character of the nuclear interaction, possesses the property of isotopic invariance. In fact, using the expressions (1.154), (1.157) and (1.158) for the exchange operators, we can represent the nuclear interaction potential (1.139) of two nucleons in the form

$$\begin{aligned} V &= V_C(\boldsymbol{r}) + V_\sigma(\boldsymbol{r})(\boldsymbol{\sigma}_1 \cdot \boldsymbol{\sigma}_2) + V_\tau(\boldsymbol{r})(\boldsymbol{\tau}_1 \cdot \boldsymbol{\tau}_2) + V_{\sigma\tau}(\boldsymbol{r})(\boldsymbol{\sigma}_1 \cdot \boldsymbol{\sigma}_2)(\boldsymbol{\tau}_1 \cdot \boldsymbol{\tau}_2) \\ &+ V_T(\boldsymbol{r})S_{12} + V_{T\tau}(\boldsymbol{r})S_{12}(\boldsymbol{\tau}_1 \cdot \boldsymbol{\tau}_2), \end{aligned} \tag{1.159}$$

where

$$\left.\begin{aligned} V_C &= V_W + \tfrac{1}{2}V_B - \tfrac{1}{2}V_H - \tfrac{1}{4}V_M; \\ V_\sigma &= \tfrac{1}{2}V_B - \tfrac{1}{4}V_M; \\ V_\tau &= -\tfrac{1}{2}V_H - \tfrac{1}{4}V_M; \\ V_{\sigma\tau} &= -\tfrac{1}{4}V_M; \\ V_T &= V_{TW} - \tfrac{1}{2}V_{TM}; \\ V_{T\tau} &= -\tfrac{1}{2}V_{TM}. \end{aligned}\right\} \tag{1.160}$$

The expression (1.159) is invariant with respect to rotations in the isotopic space. Consequently, a nuclear interaction describable by the potential (1.159) is characterized by the property of isotopic invariance, and this means that the magnitude T of the isotopic spin of the system is conserved. According to (1.159), the nuclear interaction depends on the quantum numbers S and T, that is, on the state in which the system is found. The fact that (1.159) is independent of the projection of the isospin of the system implies that the interaction is independent of the nature of the nucleons (that is, of whether they are protons or neutrons).

The *hypothesis of the isotopic invariance of the nuclear interaction* agrees with the results of experiments on nucleon-nucleon scattering at low energies. The fact that the scattering lengths and effective ranges are of the same order of magnitude for proton-proton scattering, neutron-neutron scattering and singlet-state neutron-proton scattering is evidence that the interaction between two protons and the interaction between a neutron and a proton in the singlet state are identical.

1.10. Nucleon-nucleon Scattering at High Energies

Principal features of nucleon-nucleon scattering at high energies. As we have seen, the scattering of neutrons by protons and of protons by protons at low energies is completely insensitive to the shape of the nuclear interaction potential. This is associated with the fact that, at low energies, the wavelength of the particles exceeds the range of the nuclear forces and the interaction is manifested only in the S-state. To discover the dependence of the potential on the distance between the particles, it is necessary that the wavelength be less than the range of the forces. In this case, however, the interaction will appear not only in the S-state, but also in states with $l \neq 0$. It is obvious that the greater the number of partial waves participating in the scattering, the more detailed will be the information that can be obtained on the radial dependence of the potential.

It is necessary to take into account, however, that at sufficiently high nucleon energies, alongside the elastic scattering, inelastic processes can occur. Thus, if the energy of the incident nucleon exceeds 500 MeV, it is necessary to take into account a scattering process with the formation of π-mesons. In the following, we shall consider the region of energies in which only elastic scattering of nucleons is important.

At the present time, a sufficient amount of experimental data on proton-proton and neutron-neutron scattering in this energy range has been accumulated[3]. Figures 1.3 and 1.4 show the angular dependences of the experimental differential cross-sections and the polarization for proton-proton scattering (the curves are drawn through the experimental points).

As can be seen from Fig. 1.3, the differential cross-section for proton-proton elastic scattering in the (approximate) energy range from 150 to 500 MeV is practically constant and depends neither on the scattering angle nor

[3] The phenomenological theory of nucleon-nucleon scattering is described in the monograph by Wilson (1965).

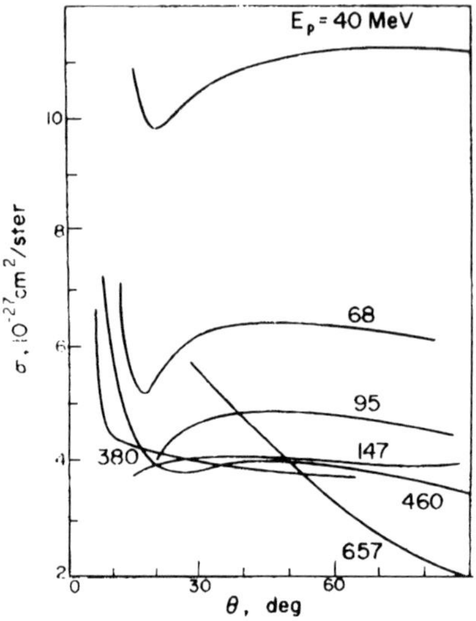

Figure 1.3. Angular dependence of the experimental differential cross-sections for scattering of protons of different energies by protons.

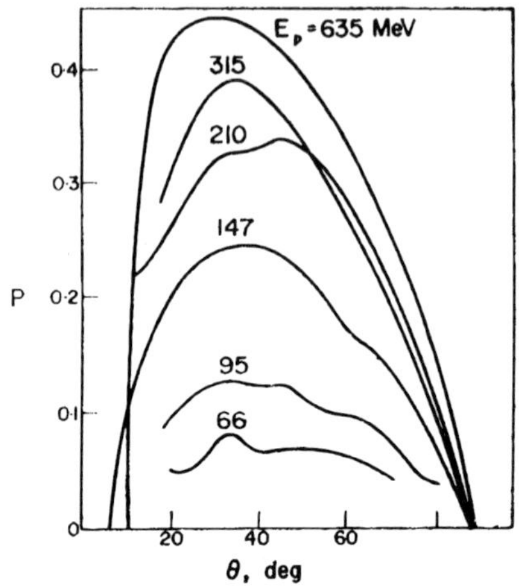

Figure 1.4. Angular dependence of the polarization in the case of scattering of protons of different energies by protons.

on the energy. The magnitude of the cross-section per unit solid angle is

$$\sigma(\vartheta) \approx 4 \times 10^{-27} \mathrm{cm}^2/\mathrm{ster} \tag{1.161}$$

within the experimental error bars. Only in the region of very small angles does the cross- section increase, because of the Coulomb scattering. At energies above 500 MeV, the isotropy is violated: a maximum appears in the angular distribution in the forward direction. This behaviour of the scattering cross-section is in sharp contradiction with the predictions of perturbation theory. This means that even at high nucleon energies we cannot assume that the nuclear interaction is weak, that is, we cannot use perturbation theory to describe it.

The experimental study of the scattering of neutrons by protons at high energies has led to a picture which is substantially different from that for the proton-proton case. Figures 1.5 and 1.6 show the angular dependences of the experimental differential cross-sections and polarization in the case of scattering of neutrons by protons.

It can be seen from Fig. 1.5 that the differential cross-section for scattering of neutrons by protons depends strongly on the scattering angle, increasing to a value at an angle 180° that is several times greater than its value at 90°. Thus, for an energy of 300 MeV, the cross-section at 90° is equal to $2 \times 10^{-27}\,\mathrm{cm}^2/\mathrm{ster}$, and at 180° is $9.2 \times 10^{-27}\,\mathrm{cm}^2/\mathrm{ster}$.

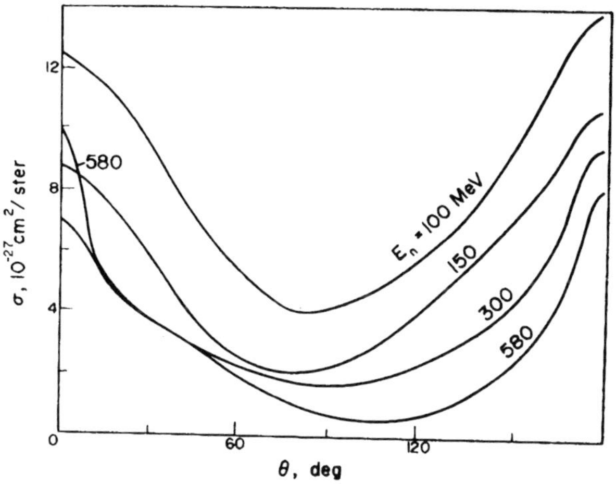

Figure 1.5. Angular dependence of the experimental differential cross-sections for scattering of neutrons of different energies by protons.

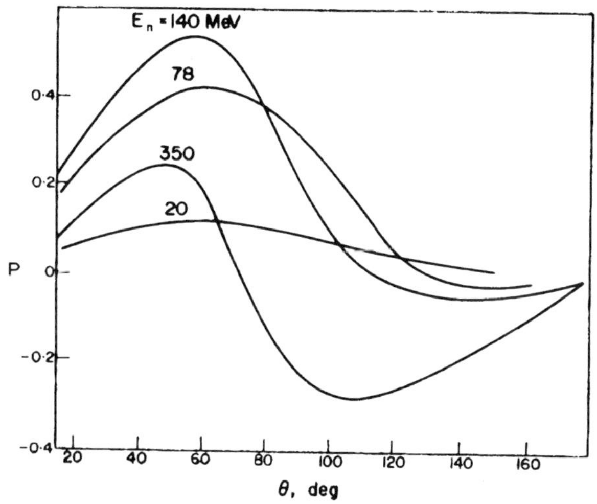

Figure 1.6. Angular dependence of the polarization in the case of scattering of neutrons of various energies by protons.

With increasing energy, the cross-section at 90° falls, reaching the value $1 \times 10^{-27}\,cm^2/ster$ at 580 MeV, while the cross-section at 180° remains almost constant. As we move from 90° towards smaller angles the cross-section does not increase so rapidly and, therefore, the angular distribution has an asymmetric character.

The total cross-sections for proton-proton and neutron-proton scattering also differ appreciably. Whereas the total cross-section for elastic proton-proton scattering in the range of energies from 150 to 500 MeV is practically constant and equal to $25 \times 10^{-27}\,\mathrm{cm}^2$, the total cross-section for elastic scattering of neutrons by protons decreases like $1/E$ up to the energies at which inelastic processes are found to be important. The small deviations from the $1/E$ law are due to the energy dependence of the phase shifts.

Inelastic scattering of nucleons. In collisions of nucleons with energies greater than 500 MeV an important role is played by *inelastic scattering*, accompanied by the creation of π-mesons. Allowance for this effect leads to an increase in the total cross-section. The *total interaction cross-section* for two protons (the sum of the cross-section for elastic scattering due to the nuclear interaction and the cross-section for creation of π-mesons) is approximately equal to 50 mbarn at energy 800 MeV and then remains almost constant up to energy 3000 MeV. The elastic scattering cross-section is practically independent of the energy and equal to 25 mbarn.

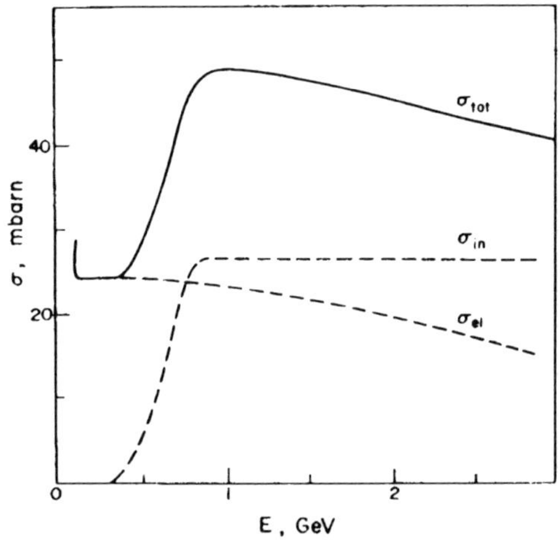

Figure 1.7. Energy dependence of the total interaction cross-section σ_{tot} for two protons, the proton-proton elastic scattering cross-section σ_{el} and the proton-proton inelastic scattering cross-section σ_{in}.

Thus, in the range of energies from 500 to 3000 MeV, the cross-sections for elastic and inelastic scattering of two protons are found to be practically equal to each other, and constant (Fig. 1.7). The angular dependence of the elastic scattering cross-section is characterized by a sharp tendency to the forward direction. Such properties characterize the diffractional scattering of particles in the presence of a perfectly absorbing sphere. In this case, the absorption cross-section and elastic scattering cross-section are equal to πR^2, where R is the radius of the sphere. If we describe the interaction of two protons with the aid of this diffraction model, we find, by putting R equal to the sum of the radii of the two protons, the magnitude of the *proton radius* R_p;

$$R_p = 0.45 \times 10^{-13}\text{cm}. \tag{1.162}$$

We note that the results of experiments on the scattering of electrons by protons are in agreement with this estimate.

High-energy nucleon scattering and isotopic invariance. Despite the fact that the scattering cross-section in the proton-proton and neutron-proton systems behave so differently, the data on *high-energy scattering* do not contradict the hypothesis of isotopic invariance of the nuclear interaction. If the assumption of isotopic invariance is correct, the forces acting between two protons also act between a neutron and a proton in the state $T = 1$. The more complicated behaviour of the neutron-proton scattering cross-section is due to the interaction in the state $T = 0$, which is not present for a system of two protons.

It is not difficult to establish a simple criterion which the cross-sections must satisfy if the interaction between the nucleons is isotopically invariant. In fact, for scattering through an angle of 90°, states with odd values of l make no contribution to the cross-section, since for odd values of l $P_l(\cos 90°) = 0$. For even values of l, a system of two protons ($T = 1$) can only be in a singlet spin state, while a neutron-proton system can be in either a singlet spin state (with $T = 1$) or in a triplet spin state (with $T = 0$). Therefore, if the interaction is isotopically invariant, the *differential scattering cross-sections* in the above cases can be written in the form

$$\left.\begin{aligned} \sigma_{pp}(90°) &= \sigma_{T=1}(90°); \\ \sigma_{np}(90°) &= \tfrac{1}{4}\sigma_{T=1}(90°) + \tfrac{3}{4}\sigma_{T=0}(90°), \end{aligned}\right\} \tag{1.163}$$

where $\frac{1}{4}$ and $\frac{3}{4}$ are the weights of the singlet and triplet spin states respectively. It follows directly from the expressions (1.163) that, if the interaction is isotopically invariant, the inequality

$$\sigma_{np}(90°) \geqslant \tfrac{1}{4}\sigma_{pp}(90°) \tag{1.164}$$

must be fulfilled for all energies.

As we have already observed the cross-section for scattering of a neutron by a proton decreases rapidly with increasing energy, whereas the proton-proton scattering cross-section remains practically unchanged. Therefore, the condition (1.164) is found to be more critical in the high-energy region. All the available experimental data on the scattering cross-sections are in agreement with the condition (1.164).

It follows from a comparison of the formulae (1.163) that, in the case of sufficiently high energies, the cross-section for scattering through an angle of 90° is considerably smaller for $T = 0$ than for $T = 1$. At the same time, the differential cross-section for scattering of a neutron by a proton through an angle of 180° is considerably greater than the proton-proton scattering cross-section. This difference is obviously connected with the interaction for $T = 0$. Thus, in the case $T = 0$, in contrast to the case $T = 1$, the scattering depends very strongly on the angle.

Using the experimental values for the cross-sections $\sigma_{pp}(\vartheta)$ and $\sigma_{np}(\vartheta)$, it is not difficult to find the scattering cross-sections in the states with $T = 1$ and $T = 0$. Clearly, the cross-section for scattering in the state $T = 1$ is the proton-proton scattering cross-section:

$$\sigma_{T=1}(\vartheta) = \sigma_{pp}(\vartheta). \tag{1.165}$$

To determine the cross-section for scattering in the state $T = 0$, we note that the isospin state of the neutron-proton system is a superposition of the states $T = 1$ and $T = 0$. According to the relations (1.151) and (1.152), we have

$$\zeta_{1/2}(1)\zeta_{-1/2}(2) = (1/\sqrt{2})\{Z_{10}(1,2) + Z_{00}(1,2)\}.$$

Therefore, the neutron proton scattering amplitude can be expressed in an analogous manner in terms of the *scattering amplitudes* in the cases $T = 1$ and $T = 0$:

$$f_{np}(\vartheta) = (1/\sqrt{2})\{f_{T=1}(\vartheta) + f_{T=0}(\vartheta)\}. \tag{1.166}$$

The neutron-proton scattering cross-section is defined as the square of the modulus of (1.166):

$$\sigma_{np}(\vartheta) = \tfrac{1}{2}\{\sigma_{T=1}(\vartheta) + \sigma_{T=0}(\vartheta) + 2\mathrm{Re}[f_{T=1}(\vartheta)f^*_{T=0}(\vartheta)]\}. \tag{1.167}$$

By virtue of the Pauli principle, the amplitudes $f_{T=1}(\vartheta)$ and $f_{T=0}(\vartheta)$ are characterized by opposite parity for a given spin value S. Therefore, in the expression $\sigma_{np}(\pi - \vartheta)$ for the cross-section for scattering through an angle $\pi - \vartheta$, the interference term occurs with the opposite sign (cf. formula (1.167)), and the sum of the cross-sections $\sigma_{np}(\vartheta)$ and $\sigma_{np}(\pi - \vartheta)$ will be equal to the sum of the scattering cross-sections in the states with $T = 1$ and $T = 0$:

$$\sigma_{np}(\vartheta) + \sigma_{np}(\pi - \vartheta) = \sigma_{T=1}(\vartheta) + \sigma_{T=0}(\vartheta) \tag{1.168}$$

The expression obtained has a simple physical meaning: it gives the total number of neutrons and protons scattered through a given angle ϑ.

Using the relations (1.168) and (1.165), we find the cross-section for scattering of nucleons in the state $T = 0$:

$$\sigma_{T=0}(\vartheta) = \sigma_{np}(\vartheta) + \sigma_{np}(\pi - \vartheta) - \sigma_{pp}(\vartheta). \qquad (1.169)$$

We note that this cross-section $\sigma_{T=0}(\vartheta)$ is always found to be positive, and this gives experimental confirmation of the hypothesis of the isotopic invariance of the nuclear interaction.

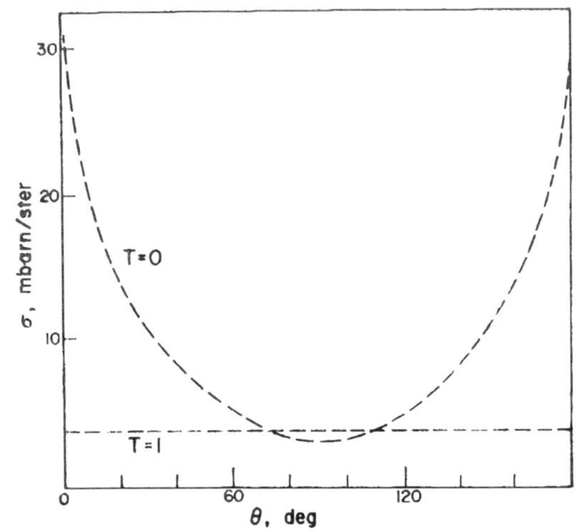

Figure 1.8. Angular dependence of the cross-sections for scattering of nucleons in states with different isospins.

The cross-sections $\sigma_{T=1}(\vartheta)$ and $\sigma_{T=0}(\vartheta)$ for a nucleon energy of the order of 400 MeV are shown schematically in Fig. 1.8. As can be seen from the figure, the angular dependence of the scattering is completely different in the cases $T = 1$ and $T = 0$. Whereas for $T = 1$ the scattering is isotopic, in the case $T = 0$ the angular distribution of the scattering is characterized by maxima in the forward and backward directions. The character of this angular dependence of the scattering points to the possibility of describing the interaction between nucleons in the state $T = 0$ by means of perturbation theory.

Determination of nucleon-nucleon scattering amplitude from experimental data. If the nuclear forces had no spin dependence, experiments on nucleon-nucleon scattering would enable us to determine, for a fixed value of the energy, only one quantity – the differential scattering cross-section $\sigma(\vartheta)$. The differential cross-section $\sigma(\vartheta)$ and the scattering amplitude $f(\vartheta)$ are connected by the relation $\sigma(\vartheta) = |f(\vartheta)|^2$. Since, in the absence of spin dependence, the scattering amplitude $f(\vartheta)$ is a scalar quantity, it follows that, if we know the differential scattering cross-section at all angles and use the optical theorem, we could in principle completely determine the scattering amplitude.

Because of the spin dependence of the nuclear interaction between the nucleons, scattering experiments enable us to determine not only the differential scattering cross-section but also the polarization of the scattered nucleon, the polarization of the recoiling nucleon, and also the correlation between the directions of the polarizations of the nucleons. If the interaction is invariant with respect to rotations in space and inversion in space and time, the scattering amplitude for scattering of two particles, each having spin $\frac{1}{2}$, can be represented in the most general case in the form (Wolfenstein, 1956)

$$\begin{aligned} f(\mathbf{k}, \mathbf{k}') &= \alpha + \beta(\boldsymbol{n} \cdot \boldsymbol{\sigma}_1)(\boldsymbol{n} \cdot \boldsymbol{\sigma}_2) + i\gamma \boldsymbol{n} \cdot (\boldsymbol{\sigma}_1 + \boldsymbol{\sigma}_2) + i\gamma' \boldsymbol{n} \cdot (\boldsymbol{\sigma}_1 - \boldsymbol{\sigma}_2) \\ &+ \delta(\boldsymbol{m} \cdot \boldsymbol{\sigma}_1)(\boldsymbol{m} \cdot \boldsymbol{\sigma}_2) + \varepsilon(\boldsymbol{l} \cdot \boldsymbol{\sigma}_1)(\boldsymbol{l} \cdot \boldsymbol{\sigma}_2), \end{aligned} \tag{1.170}$$

where $\mathbf{k}$ and $\mathbf{k}'$ are the relative of the particles before and after the scattering, $\boldsymbol{\sigma}_1$ and $\boldsymbol{\sigma}_2$ are the Pauli matrices corresponding to the spin of the interacting particles, and $\boldsymbol{n}$, $\boldsymbol{l}$ and $\boldsymbol{m}$ are a set of three mutually orthogonal unit vectors:

$$\boldsymbol{n} = \frac{\mathbf{k} \times \mathbf{k}'}{|\mathbf{k} \times \mathbf{k}'|}, \quad \boldsymbol{l} = \frac{\mathbf{k} + \mathbf{k}'}{|\mathbf{k} + \mathbf{k}'|}, \quad \boldsymbol{m} = \frac{\mathbf{k} - \mathbf{k}'}{|\mathbf{k} - \mathbf{k}'|}. \tag{1.171}$$

The coefficients α, β, γ, γ', δ and ε are complex functions depending on the angle and the energy. If we take the charge invariance of the nuclear interaction between the nucleons into account, the coefficient γ' in (1.170) must be put equal to zero:

$$\gamma' = 0. \tag{1.172}$$

Thus, to describe the elastic scattering of two nucleons it is necessary, for a given energy, to find five complex functions (α, β, γ, δ and ε) of the scattering angle. The problem is simplified somewhat if we take into account the unitarity condition on the scattering amplitude (1.170) (the generalized optical theorem). From this condition, five different integral relations between the complex functions α, β, γ, δ and ε follow. Therefore, for a complete determination of the scattering amplitude (1.170), it is necessary to measure the angular dependence, at a given energy, of five independent quantities. Obviously, the required information can be obtained only from polarization experiments.

If the nucleons are initially unpolarized, scattering experiments enable us to determine only the angular dependence of the differential cross-section $\sigma(\vartheta)$. Experiments on *double scattering* enable us to measure the angular dependence of the *polarization* $\boldsymbol{P}(\vartheta)$. (In the case of coplanar double scattering, the polarization $P(\vartheta)$ is measured by the *asymmetry of the scattering* at the second scatterer.) Experiments on *triple scattering* make it possible to measure the angular dependence of three other quantities. These are usually chosen to be the *depolarization* $D(\vartheta)$ caused by the second scatterer in the case of triple coplanar scattering, *the rotation* $R(\vartheta)$ *of the polarization vector* at the second scattering in the case of triple scattering in planes successively at right angles to each other, and the *longitudinal polarization* $A(\vartheta)$ resulting when the spin of the particle being scattered is flipped with the help of the magnetic field due to the second scatterer.

The measurable quantities $\sigma(\vartheta)$, $\boldsymbol{P}(\vartheta)$, $\mathcal{D}(\vartheta)$, $R(\vartheta)$ and $A(\vartheta)$ can be expressed in the following way in terms of the coefficients α, β, γ, δ and ε determining the scattering amplitude (Puzikov, Ryndin and Smorodinskii, 1957):

$$\left.\begin{aligned}
&\sigma(\vartheta) = |\alpha|^2 + |\beta|^2 + 2|\gamma|^2 + |\delta|^2 + |\varepsilon|^2;\\
&\sigma(\vartheta)P(\vartheta) = 2\mathrm{Im}\,[(\alpha+\beta)^*\gamma];\\
&\sigma(\vartheta)\{1 - D(\vartheta)\} = 2|\delta|^2 + 2|\varepsilon|^2;\\
&\sigma(\vartheta)R(\vartheta) = \{|\alpha|^2 - |\beta|^2 + \mathrm{Re}\,[(\delta+\varepsilon)^*(\delta-\varepsilon)]\}\cos(\vartheta/2)\\
&\qquad\qquad - 2\mathrm{Re}\,[(\alpha-\beta)^*\gamma]\sin(\vartheta/2);\\
&\sigma(\vartheta)A(\vartheta) = -\{|\alpha|^2 - |\beta|^2 + \mathrm{Re}\,[(\delta+\varepsilon)^*(\delta-\varepsilon)]\}\sin(\vartheta/2)\\
&\qquad\qquad - 2\mathrm{Re}\,[(\alpha-\beta)^*\gamma]\cos(\vartheta/2).
\end{aligned}\right\} \quad (1.173)$$

The relations (1.173), along with the unitarity relations, enable us in principle to find the quantities α, β, γ, δ and ε, that is, to determine completely the scattering amplitude (1.170). The realization of this, however, is extremely difficult in practice and, therefore, the experimental data are usually analyzed on the basis of the representation of the scattering amplitude in terms of the phase shifts.

Isotropic character of proton-proton scattering and interaction in states with $l \neq 0$. It would be possible to explain the principal feature of the nuclear scattering of protons by protons, namely, the fact that the cross-section is independent of the scattering angle, by assuming that, up to energies of the order of 500 MeV, the protons interact only in the S-state. However, the maximum possible magnitude of the scattering cross-section per unit solid angle in the S-state (equal to $\lambda\!\!\!^{-2}$) at energy of 500 MeV is found to be approximately half of the observed cross-section (1.161). Therefore, it is necessary to take into account the interaction of the protons in states with $l \neq 0$, as is also confirmed by the large magnitude of the experimentally observed polarization arising in proton-proton scattering. We remark that polarization is possible only when the spin of the system has the value $S = 1$, which is associated with states with odd l. Therefore, the large magnitude of the polarization in the scattering implies that there is strong interaction corresponding to states with odd l, and in particular to P-states.

The interaction in the ${}^3P_0$ state, like that in the ${}^1S_0$ state, leads to isotropic scattering, and the maximum possible magnitude of the cross-section per unit solid angle is also $\lambda\!\!\!^{-2}$. Therefore, allowance for the interaction of the protons in the ${}^1S_0$ and ${}^3P_0$ states makes it possible in principle to obtain for the isotropic cross-section a value in agreement with the experimental value (1.161). In doing this, however, it is necessary to assume that the phase shifts in the other states with $l = 1$ (${}^3P_1$ and ${}^3P_2$) are small, since otherwise the cross-section will depend on the angle. The marked difference between the phase shifts for ${}^3P_0$, ${}^3P_1$ and ${}^3P_2$ indicates the important role of the non-central or *spin-orbit interaction* .

The isotropic nature of the scattering can also be explained if we assume that the interaction occurs in S- and D-states, but that the corresponding phase shifts have opposite signs. In fact, if the interaction occurs only in states with $l = 0$ and $l = 2$, the scattering amplitude can be written:

$$f(\vartheta) = (1/k)\{\mathrm{e}^{i\delta_0}\sin\delta_0 + 5\mathrm{e}^{i\delta_2}(\sin\delta_2)P_2(\cos\vartheta)\}, \tag{1.174}$$

where δ_0 and δ_2 are the phase shifts in the states ${}^1S_0$ and ${}^1D_2$. (Since $T = 1$ and the parity is positive, we have $S = 0$ for the states considered.)

The differential scattering cross-section is equal to

$$\begin{aligned}\sigma(\vartheta) &= (1/k^2)\{\sin^2\delta_0 + 25(\sin^2\delta_2)P_2^2(\cos\vartheta) \\ &+ 10\sin\delta_0\sin\delta_2\cos(\delta_0 - \delta_2)P_2(\cos\vartheta)\}.\end{aligned} \tag{1.175}$$

The angular dependence of the cross-section is determined by the second and third terms in the relation (1.175). The second term is smaller by a factor of 4 at a scattering angle of 90° than at angle 0°. The sign of the third (interference) term is determined by the relative signs of the phase shifts δ_0 and δ_2 and by the value of the angle ϑ. Since the quantity $P_2(\cos\vartheta)$ is negative for $\vartheta = 90°$, allowance for the interference can lead to an increase of the cross- section at 90°, and consequently to a smoothing of the angular dependence of the cross-section, only when the phase shifts δ_0 and δ_2 have opposite signs.

Repulsion between nucleons at short distances. The change of sign of the phase shift on going from the S- to the D-state can be obtained by assuming that strong repulsive forces act between nucleons at small distances, in contrast to the attractive forces which appear at large distances (Jastrow, 1951). At sufficiently high energies, the phase shift in the S-state is determined by this repulsion, whereas the phase shifts in states with

$l \neq 0$, because of the effect of the centrifugal forces, are determined by the longer-range attractive forces.

It is not difficult to find the phase shift δ_0 in the S-state in the limiting case of infinite repulsive forces (the hard-sphere model or repulsive-core model) which appear at distances less than a certain distance $\mathbf{r}_c$, called the *repulsive-core radius*:

$$\delta_0 = -k\mathbf{r}_c. \tag{1.176}$$

The nuclear forces of attraction appearing at distances greater than $\mathbf{r}_c$ lead to a positive correction to the right-hand side of eqn. (1.176). If the nuclear potential is chosen to be of the form of repulsive core surrounded by a square potential well,

$$V(r) = \begin{cases} +\infty, & r < r_c; \\ -V_0, & r_c < r < r_0; \\ 0, & r_0 < r, \end{cases} \tag{1.177}$$

the phase shift δ_0 will be given by the expression

$$\delta_0 = -kr_c + \Delta, \tag{1.178}$$

where

$$\Delta \equiv \arctan\{(k/k_0)\tan[k_0(r_0 - r_c)]\} - k(r_0 - r_c),$$
$$k_0 = \sqrt{k^2 + (2\mu V_0/\hbar^2)}. \tag{1.179}$$

It is easily verified that the phase shift Δ given by the equality (1.179) is always positive. For this, we need use only the facts that $k < k_0$ and that $\alpha \tan x > \tan \alpha x$ for any value of x $(0 < x < \pi/2)$ if $\alpha < 1$.

At low energies $(kr_c \to 0)$, the main role is played by the attractive forces, and the negative term in (1.178) can be neglected. With increasing energy, the positive term in (1.178) decreases rapidly, while the phase shift due to the repulsive forces increases in magnitude, so that the phase shift δ_0 will change in sign from a positive value at low energies to a negative value at energies greater than a certain critical energy, the magnitude of which is determined by the parameters of the potential. In order that the change of sign of the phase shift occur at an energy in the region of 250 MeV we must choose $r_c \approx 0.4 \times 10^{-13}$ cm. Generally speaking, the magnitude of r_c depends on the shape of the attractive potential.

The conclusion that strong repulsive forces are present at short distances also follows from the energy dependence of the total scattering cross-section. In fact, if the dependence of the potential on the distance between the particles were monotonic, the phase shifts should increase monotonically with the energy. Therefore, the product

$$k^2\sigma = 4\pi \sum_l (2l+1)\sin^2\delta_l \tag{1.180}$$

should also increase monotonically with increasing energy. Figure 1.9 shows the dependence of the quantity $k^2 d\sigma/do$ at $\vartheta = 90^\circ$ on the energy in the laboratory coordinate frame in the case of proton-proton scattering, and Fig. 1.10 shows the energy dependence of $k^2\sigma$ in the case of neutron-proton scattering. According to the experimental dependences (see Figs. 1.9 and 1.10) the quantity (1.180) decreases in the energy range from 10 to 100 MeV, in which the interaction in the S-state is the most important. This means that, in this energy range, the phase shift in the S-state decreases, which is possible only in the case of strong repulsion between the nucleons at short distances.

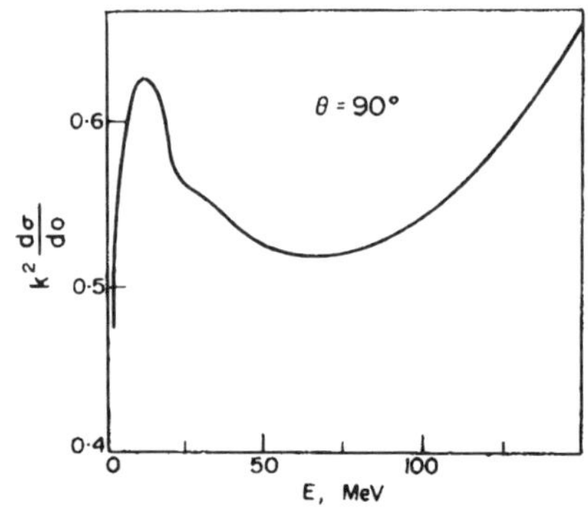

Figure 1.9. Energy dependence of the quantity $k^2 d\sigma/do$ at $\vartheta = 90^\circ$ in the case of proton-proton scattering.

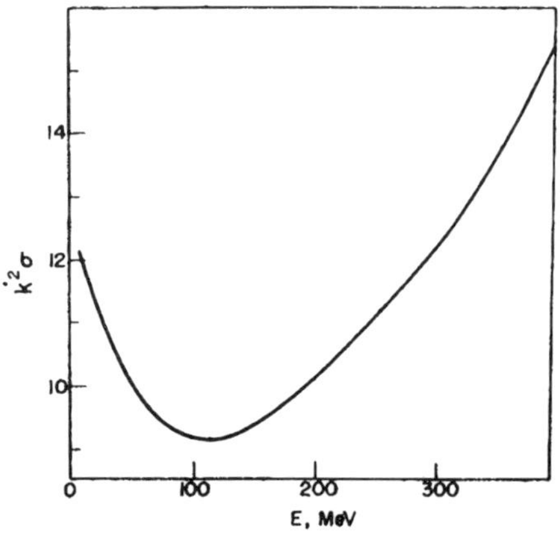

Figure 1.10. Energy dependence of the quantity $k^2\sigma$ in the case of neutron-proton scattering.

Phase-shift analysis. The most important problem in the analysis of experimental data on scattering is the determination of the phase shifts. Specifying the phase shift completely characterizes the interaction between the particles in a given state. Therefore, the extraction of all the phase shifts from the experimental data is equivalent to obtaining complete information on the nuclear interaction between the nucleons.

A knowledge of the phase shifts makes it possible in principle to find the dependence of the forces on distance, that is, to determine the nuclear interaction potential. It is true that this problem (the inverse problem of scattering theory) requires not only information on the energy dependence of the phase shifts but also additional information on the discrete levels in the system and on the asymptotic forms of the wave functions corresponding to these levels. Because of this, at the present time it has still not been possible to solve such a problem completely. We remark that the determination of the phase shifts from the experimental data on the angular dependences of the cross-section, the polarization, etc., is a laborious task which requires the use of a computer to solve it.

The energy dependence of the phase shifts can be found by using the experimental data on scattering at different energies. Clearly, the phase shifts are real only when inelastic processes can be neglected. Therefore, phase-shift analysis is meaningful up to energies of 500 MeV.

The *phase-shift analysis* of the scattering of neutrons by protons is more complicated than that of the scattering of protons by protons, since the neutron-proton system can be found in the isospin states $T=1$ and $T=0$, whereas a system of two protons can be found only in the state $T=1$. With the assumption of isotopic invariance of the nuclear interaction, the phase shifts corresponding to $T=1$ should be the same for a system of two protons and for a neutron-proton system. The phase shifts corresponding to $T=1$ can be found, therefore, from the proton-proton scattering data, and from the analysis of the scattering of neutrons by protons we need to determine only the phase shifts corresponding to $T=0$.

We shall make use of the classification of the states of a system of two nucleons (see §1.9) and characterize the different states by the quantum numbers of the isospin T, the total angular momentum J, the spin S and the parity w. If $S=0$ (singlet spin state), the total angular momentum J coincides with the orbital angular momentum l, and the scattering amplitude has the form

$$f^{S=0}(\vartheta) = (i/2k)\sum_{l}(2l+1)(1-S_l)P_l(\cos\vartheta), \tag{1.181}$$

where S_l is the unitary scattering matrix, connected with the phase shift ${}^{2S+1}\delta_l^J$ by the usual relation

$$S_l = \mathrm{e}^{2i\,{}^1\delta^J_{l=J}}. \tag{1.182}$$

If $S=1$ (the triplet spin state), the orbital angular momentum is not conserved, and the scattering amplitude can be represented in the form of a three-row matrix

$$\begin{aligned} f^{S=1}_{\mu'\mu}(\vartheta,\varphi) \quad &= \quad (i/2k)\sum_{ll'J}\sqrt{4\pi(2l+1)}Y_{l',\mu-\mu'}(\vartheta,\varphi)(l01\mu|J\mu) \\ &\times(l'\mu-\mu'1\mu'|J\mu)(\delta_{l'l}-S^J_{l'l}), \end{aligned} \tag{1.183}$$

where μ and μ' are the components of the spin of the system along a certain direction in the initial and final states, φ is the angle between the plane defined by this direction and by the direction of motion of the incident particle and the scattering plane, and $S^J_{l'l}$ is the scattering matrix. Since l is not conserved, the matrix $S^J_{l'l}$, generally speaking, is not diagonal. For a fixed value of J, the quantities l and l' can take the values $J-1$, J and $J+1$. If $l = J$, then $l' = J$ because of conservation of parity. Taking into account the unitarity of the scattering matrix, in this case we can write

$$S^J_{JJ} = e^{2i\,^3\delta^J_J}, \quad l = l' = J. \tag{1.184}$$

If l is equal to $J+1$ or $J-1$, the possible values of l' are also equal to $J+1$ or $J-1$. Therefore, $S^J_{l'l}$ is a two-row matrix. But since the scattering matrix is a unitary symmetric matrix, it can be diagonalized by means of the orthogonal matrix

$$U^J = \begin{pmatrix} \cos\varepsilon^J & \sin\varepsilon^J \\ -\sin\varepsilon^J & \cos\varepsilon^J \end{pmatrix}, \tag{1.185}$$

where ε^J is some real parameter. Thus, in the case $l, l' = J\pm1$, the scattering matrix can be represented in the form

$$S^J_{l'l} = \begin{pmatrix} \cos\varepsilon^J & -\sin\varepsilon^J \\ \sin\varepsilon^J & \cos\varepsilon^J \end{pmatrix} \begin{pmatrix} e^{2i\,^3\delta^J_{J+1}} & 0 \\ 0 & e^{2i\,^3\delta^J_{J-1}} \end{pmatrix} \begin{pmatrix} \cos\varepsilon^J & \sin\varepsilon^J \\ -\sin\varepsilon^J & \cos\varepsilon^J \end{pmatrix} \tag{1.186}$$

(the *Blatt-Biedenharn representation* (1952)). According to (1.186), the scattering matrix $S^J_{l'l}$ is determined by three real parameters: the two scattering eigenphases $^3\delta^J_{J+1}$ and $^3\delta^J_{J-1}$ in the states with $l = J+1$ and $l = J-1$, and the *mixture parameter* ε^J.

Sometimes another representation for the scattering matrix $S^J_{l'l}$ is used (the Stapp representation (1957)):

$$S^J_{l'l} = \begin{pmatrix} e^{i\,^3\bar\delta^J_{J+1}} & 0 \\ 0 & e^{i\,^3\bar\delta^J_{J-1}} \end{pmatrix} \begin{pmatrix} \cos 2\bar\varepsilon^J & i\sin 2\bar\varepsilon^J \\ i\sin 2\bar\varepsilon^J & \cos 2\bar\varepsilon^J \end{pmatrix} \begin{pmatrix} e^{i\,^3\bar\delta^J_{J+1}} & 0 \\ 0 & e^{i\,^3\bar\delta^J_{J-1}} \end{pmatrix}. \tag{1.187}$$

We give the relations connecting the phase shifts and the mixture parameters for the above representations:

$$\left.\begin{aligned} \sin(\delta_{J+1} - \delta_{J-1}) &= \sin 2\bar\varepsilon^J / \sin 2\varepsilon^J; \\ \sin(\bar\delta_{J+1} - \bar\delta_{J-1}) &= \tan 2\bar\varepsilon^J / \tan 2\varepsilon^J; \\ \delta_{J+1} + \delta_{J-1} &= \bar\delta_{J+1} + \bar\delta_{J-1}. \end{aligned}\right\} \tag{1.188}$$

In place of the parameter $\bar{\varepsilon}^J$, one often uses the quantity ϱ^J:

$$\varrho^J \equiv \sin 2\bar{\varepsilon}^J. \tag{1.189}$$

Because of the small range of the nuclear forces, the elastic scattering of nucleons can be described by means of a small number of phase shifts. In the case $T=1$, the phase-shift analysis of the scattering is performed with inclusion of the phase shifts in the states ${}^1S_0$, ${}^3P_{0,1,2}$, ${}^1D_2$, ${}^3F_{2,3,4}$, ${}^1G_4$, ${}^3H_{4,5,6}$, and the mixture parameters ε^2 and ε^4. In the case $T=0$, the phase-shift analysis is performed with inclusion of the phase shifts in the states ${}^3S_1$, ${}^1P_1$, ${}^3D_{1,2,3}$, ${}^1F_3$, ${}^3G_{3,4,5}$, and the mixture parameters ε^1 and ε^3. Figures 1.11 and 1.12 show the *energy dependences of the phase shifts* and the mixture parameters (MacGregor, Arndt and Wright, 1968a,b).

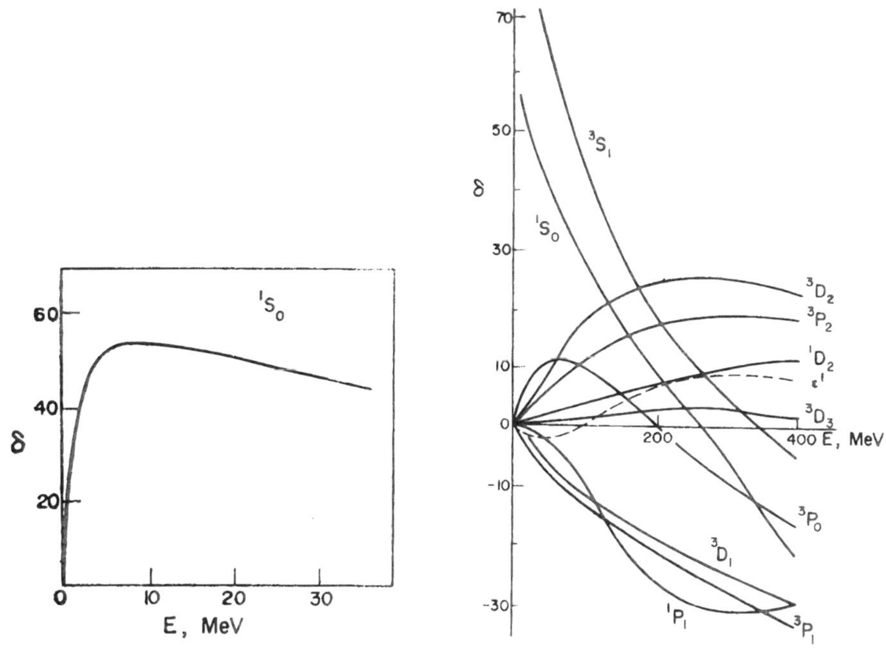

Figure 1.11. Energy dependence of phase shifts and of the mixture parameter ε^1.

The ${}^1S_0$ phase shift vanishes at low energies, indicating the absence of a singlet bound state in a system of two nucleons. With increasing energy, the ${}^1S_0$ phase shift increases, reaching values in the region of 60° at 3 MeV, and then decreases and becomes negative at energies greater than 250 MeV. As we have already observed, the change of sign of the ${}^1S_0$ phase shift indicates the existence of strong repulsion between the nucleons at short distances. The ${}^3S_1$ phase shift at zero energy is equal to 180°, and this is directly connected with the existence of a triplet bound state in a system of two nucleons, namely, the deuteron. At an energy of 325 MeV, the $^3S_1$ phase shift goes to zero, and then takes negative values. The phase shifts in the states $^3P_{0,1,2}$ differ markedly, indicating the presence of strong non-central or spin-orbit interaction between the nucleons. We note that the phase shifts in the S-, P- and D-states are quantities of the same order. The phase shifts in the F-, G- and H-states do not exceed a few degrees.

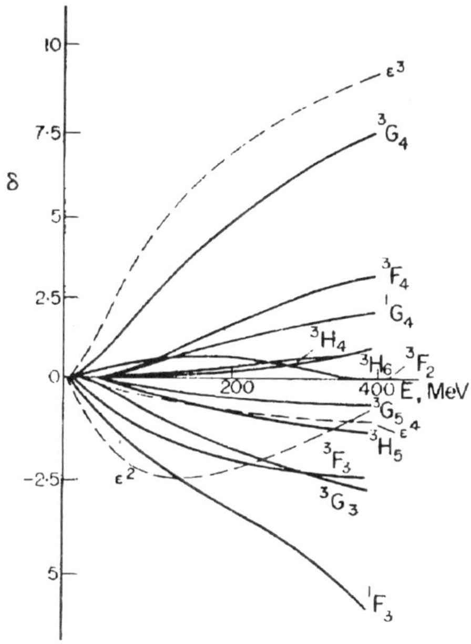

Figure 1.12. Energy dependence of phase shifts and of the mixture parameters.

Phenomenological potentials. So long as inelastic processes do not play an important role, the interaction between nucleons can be described by means of a potential. The determination of the form of the nuclear interaction potential between nucleons from scattering data is the most important problem of the non-relativistic theory. It follows from the data on nucleon-nucleon scattering at high energies that the nuclear potential should describe the exchange interaction, the strong repulsion at short distances, and the important tensor and spin-orbit interactions (in addition to the attractive forces). The numerous attempts to derive the potential of the nucleon-nucleon interaction from meson field theory have not led to the desired result. Therefore, at the present time, to describe the interaction between nucleons there is wide-spread use of potentials that are chosen phenomenologically in such a way as to give the best explanation of the accumulated experimental scattering data. Several variants of potentials giving a good description of the scattering data in the energy range from 0 to 400 MeV have been proposed. As illustrations, we shall give the expressions for the Gammel-Thaler and Hamada-Johnston potentials, which are those most often used in practice.

The *Gammel-Thaler potential* contains central, tensor and spin-orbit interactions:

$$V_{GT} = V_C(\boldsymbol{r}) + V_T(\boldsymbol{r})\boldsymbol{S}_{12} + V_{LS}\boldsymbol{L}\cdot\boldsymbol{S}. \tag{1.190}$$

The radial dependence of all the terms in (1.190) is chosen in the form of a perfectly hard sphere of radius r_c and a *potential of the Yukawa type*:

$$V(r) = \begin{cases} \infty, & r < r_c; \\ -V\mathrm{e}^{-\mu r}/\mu r, & r > r_c. \end{cases} \tag{1.191}$$

The parameters V, μ and r_c are chosen independently for each state, that is, definite values are chosen for the even singlet states and the even triplet states (for the central, tensor and spin-orbit forces separately), and analogously for the odd states. In the general case, the potential (1.190) is characterized by 24 parameters. Since, in practice, the same value of r_c has been used for the central, tensor and spin-orbit forces, the number of independent parameters has been reduced to 20. The parameter values ensuring the best agreement with the experimental data are given in Table 1.3 (Gammel and Thaler, 1957, 1959).

The Hamada-Johnston potential takes into account not only the central, tensor and spin-orbit interactions but also the squared spin-orbital interaction:

$$V_{HJ} = V_C(\boldsymbol{r}) + V_T(\boldsymbol{r})S_{12} + V_{LS}(\boldsymbol{r})\boldsymbol{L}\cdot\boldsymbol{S} + V_{LL}(\boldsymbol{r})L_{12}, \tag{1.192}$$

where

$$L_{12} = [\delta_{LJ} + 2S(S+1) - 3]\boldsymbol{L}^2 - (\boldsymbol{L}\cdot\boldsymbol{S})^2. \tag{1.193}$$

The radial dependence in (1.192) corresponds at short distances to the presence of a perfectly hard sphere of radius r_c; at large distances, the dependences of the individual terms in (1.192) are chosen in the form

$$\left.\begin{aligned} V_C(r) &= V_C Y(x)\{1 + a_C Y(x) + b_C Y^2(x)\}; \\ V_T(r) &= V_T Z(x)\{1 + a_T Y(x) + b_T Y^2(x)\}; \\ V_{LS}(r) &= V_{LS} Y^2(x)\{1 + a_{LS} Y(x)\}; \\ V_{LL}(r) &= V_{LL}(Z(x)/x^2)\{1 + a_{LL} Y(x) + b_{LL} Y^2(x)\}, \end{aligned}\right\} \tag{1.194}$$

where $Y(x) = \mathrm{e}^{-x}/x$, $Z(x) = (1 + 3/x + 3/x^2)Y(x)$, $x \equiv \mu c r/\hbar$ and μ is the meson mass ($\hbar/\mu c = 1.415 \times 10^{-13}$ cm, $\mu c^2 = 139.4$ MeV and $x_c \equiv \mu c r_c/\hbar = 0.343$).

TABLE 1.3.

Potential parameters	$S=0$		$S=1$	
	$T=1$ (l–even)	$T=0$ (l–odd)	$T=0$ (l–even)	$T=1$ (l–odd)
V_C MeV	425.5	-100 for $E_{np} \leqslant 90$ and $E_{np} = 156.$ -150 for $E_{np} = 310$	100.7 for $E_{np} \leqslant 90$ and $E_{np} = 156.60$ for $E_{np} = 310$	0
μ_C, cm^{-1}	1.45×10^{13}	1.0×10^{13}	1.23×10^{13}	–
V_r, MeV	–	–	257 for $E_{np} \leqslant 90$ 230 for $E_{np} = 156$ 175 for $E_{np} = 310$	-22.5
μ_T, cm^{-1}	–	–	1.203×10^{13}	0.8×10^{13}
V_{LS}, MeV	–	–	5000	7317.5
μ_{LS}, cm^{-1}	–	–	3.7×10^{13}	3.7×10^{13}
r_c, cm	0.4×10^{-13}	0.5×10^{-13}	0.4×10^{-13}	0.4125×10^{-13}

The potentials $V_C(r)$ and $V_T(r)$ are chosen in such a way that at large distances they coincide with the meson-theory potentials taking one-pion exchange into account:

$$V_C = 0.08(\mu c^2/3)[2S(S+1) - 3][2T(T+1) - 3];$$

$$V_T = 0.08(\mu c^2/3)[2T(T+1) - 3].$$

The values of the parameters V_C, V_T, V_{LS}, V_{LL}, a and b are given in Table 1.4 (Hamada and Johnston, 1962).

The tensor spin-orbit interaction is most important in the triplet and singlet spin states with $J = L$. The *Hamada-Johnston potential* (1.192) agrees well with the data on the interaction of two nucleons at low energies (the binding energy of the deuteron, the scattering lengths, the effective ranges, etc.) and reproduces well the energy dependence of the phase shifts up to energies of the order of 400 MeV.

We note the following features of the interaction, which are characteristic of both the *Gammel-Thaler potential* and the Hamada-Johnston potential. In all states, strong repulsive forces act between the nucleons at short distances, and these are described by a core radius of 0.4×10^{-13} cm. Outside the core region, the potential for even states corresponds to attractive forces. In the odd states the interaction is considerably weaker than

TABLE 1.4.

Potential parameters	$S=0$		$S=1$	
	$T=1$ (l–even)	$T=0$ (l–odd)	$T=0$ (l–even)	$T=1$ (l–odd)
V_C MeV	−11.2	33.5	−11.2	3.71
a_C	8.7	−8.0	6.0	−9.07
b_C	10.6	12.0	−1.0	3.48
V_T, MeV	–	–	−11.2	3.71
a_T	–	–	−0.5	−1.29
b_T	–	–	0 .2	0.55
V_{LS}, MeV	–	–	10.4	27.4
a_{LS}	–	–	−0.1	−7.12
V_{LL}, MeV	−0.124	−0.371	0.371	−0.124
a_{LL}	0.2	2.0	1.8	−7.26
b_{LL}	−0.2	6.0	−0.4	6.92

in the even states. In the odd singlet states, the interaction is repulsive. At large distances the tensor forces make a substantial contribution to the interaction. The tensor forces are attractive in the $^3S_1$ and $^3P_0$ states.

Along with the Gammel-Thaler and Hamada-Johnston potentials, other phenomenological potentials are employed. Almost all these potentials are constructed in a similar manner, so we shall restrict the consideration to brief descriptions. The first phenomenological nucleon-nucleon potentials taking into account short-distance repulsion were introduced by Gartenhaus (Gartenhaus, 1955) and Signell and Marshak (Signell and Marshak, 1957). Both these were derived in terms of the approximate statical meson theory of nuclear forces. These potentials contain both central forces with repulsive cores and tensor forces. The *Gartenhaus potential*, however, is proposed for nucleons at rest and therefore ignores the spin-orbit interaction. The Signell-Marshak potential is equal to the sum of the Gartenhaus potential and a short-range spin-orbit potential which is employed to explain the polarization phenomena and thus to improve the agreement of the theory and experiment. However, similarly to the Gammel-Thaler potential, the Signell-Marshak potential overestimates the contribution of the spin-orbit interaction. Gartenhaus and Signell-Marshak potentials are very close at long distances between nucleons, $\boldsymbol{r} > 10^{13}\mathrm{cm} \equiv 1\,\mathrm{fm}$.

When calculating the nuclear structure, many authors make use of the phenomenological nucleon-nucleon potential proposed by the Yale group (Lassila e.a., 1962). It contains terms responsible for central, tensor, spin-orbit, and quadratic spin-orbit interaction (similarly to the Hamada-Johnston potential) and, besides that, includes the one-pion exchange potential that was obtained using the simplest version of the meson field theory. The contribution of the last term is assumed to be correct and considerable for sufficiently large ($> 2\,\mathrm{fm}$) distances between the nucleons. The parameters of the *Yale-group potential* were found by comparing calculated (using this potential) and experimental values of nucleon-nucleon scattering phases. The potential contains a *hard repulsive core*; it was parametrized with the use of sums of simple functions $a_n x^{-n} \mathrm{e}^{-2x}$ (n – positive integers, $x = \frac{\mu c}{\hbar} r$).

Many authors employ the phenomenological *Reid potential* (Reid, 1968) in spite of its complexity. This potential is described by different analytical expressions for different nucleon-nucleon scattering partial waves for central, tensor, and spin-orbit interaction; their dependences on the internucleon distances are described by superpositions of terms of the form $x^{-m} \mathrm{e}^{-nx}$, where m and n are positive integers. Theoretical and experimental results were compared to find the sets of parameters of the Reid potential for the S, P, D and F scattering phases for both $T = 0$ and $T = 1$ values of the total two-nucleon isotopic spin. Both hard repulsive core, when the potential tends to infinity for $r < r_c$, and soft core, when the repulsive potential is finite though large for $r < r_c$, were considered. The sets of parameters were estimated numerically involving the data obtained in terms of the above mentioned theory of one-pion exchange potential.

Lately, many authors employ as well the Paris (Lacombe e.a., 1980) and Bonn (Macleidt e.a., 1987) phenomenological potentials. We consider the first one in brief, since modern approaches to the construction of phenomenological potentials have much in common. Usually, nucleon-nucleon potentials are constructed by fitting np data for $T = 0$ states and either np or pp data for $T = 1$ states. The *Bonn potential* was fit to np data in all states. In contrast, the *Paris potential* was fit to pp data for $T = 1$ channels. The Paris potential contains central, tensor, spin-spin, spin-orbit, and quadratic spin-orbit interactions for two isotopic states with $T = 0$ and $T = 1$. At medium and long internucleon distances ($> 0.8\,\mathrm{fm}$), the behavior of this potential is well described by the meson theory of nuclear forces which allows for the exchange of both light π-mesons and heavy ω-mesons. At short distances ($< 0.8\,\mathrm{fm}$), including the repulsive core range where the intrinsic quark-gluon structure of the nucleon makes its effect, the potential is described by a model (phenomenological) dependence. (We shall consider dependence of nuclear forces on the internucleon distances and other factors in Chapter 7). The analytic form of the Paris potential is rather simple, it is described by a finite sum of Yukawa-type terms $g_j \frac{\mathrm{e}^{-\gamma_j r}}{r}$ and their derivatives. The parameters g_j and γ_j, where $j = 1, 2, 3, \ldots$ (in the case $j = 1$ data of the one-pion-exchange theory are employed), satisfy conditions which provide that the potential is finite for $r \to 0$ (i.e., contains a soft core). The potential provides a good description of the phase shift for

$J \leqslant 7$, the deutron ground state, and is successfully applied in the theory of nuclear matter (see Chapter 5) to obtain a satisfactory value of binding energy per nucleon ($\sim 10\,\text{MeV}$) for the majority of stable nuclei.

A new high-quality nucleon-nucleon potential with explicit charge dependence and charge asymmetry (an updated version of the *Argonne potential*) was presented by R. Wiringa, V. Stoks and R. Schiavilla (1995). That potential fits both *pp* and *nn* data, as well as low-energy *nn* scattering parameters and deuteron properties.

Deuteron ground state parameters in terms of a phenomenological potential. To conclude this section, we consider the deuteron-ground-state wave function and its main characteristics calculated in terms of a modern phenomenological nucleon-nucleon potential which has been constructed using the data on nucleon-nucleon scattering in a wide energy range from very low energies ($\leqslant 1$ MeV) up to energies about 400 MeV. For higher energies, inelastic processes become very important. To find the deuteron wave function and to describe its properties by solving the Schrödinger equation with the above potential is an interesting problem since main deuteron characteristics have been obtained with high accuracy by various experiments and it seems to be promising to describe them theoretically making use of independent data on nucleon-nucleon scattering.

The description of both deuteron and low-energy nucleon-nucleon parameters is the best in the case of the Paris phenomenological potential which includes noncentral (tensor) forces. Figure 1.13 gives the calculated deuteron radial wave functions $u(r)$ and $w(r)$ in the S- and D-states, respectively (in terms of $\text{Fm}^{\frac{1}{2}}$ units). These functions attain their maximum values 0.52394 and 0.17366 approximately for $r = 1.6$ Fm and $r = 1.4$ Fm, respectively. The calculations with the Paris potential yield the following values of the deutron binding energy $\varepsilon = 2.225$ MeV, probability to find the deuteron in the D-state $p_D = 0.058$, and deuteron magnetic dipole and electric quadrupole moments: $\mu_d = 0.853\frac{e\hbar}{2M_pc}$ and $Q = 2.79 \times 10^{-13}$ cm. The relevant nucleon-nucleon scattering lengths are given by $a_t \equiv a_{np} = 5.427 \times 10^{-13}$ cm, $a_p \equiv a_{pp} = -7.810 \times 10^{-13}$ cm, $a_n \equiv a_{nn} = -17.612 \times 10^{-13}$ cm and the effective radii are equal to $r_{ot} = 1.766 \times 10^{-13}$ cm, $r(0, -\varepsilon) = 1.765 \times 10^{-13}$ cm, $r_{op} \equiv r_{pp} = 2.797 \times 10^{-13}$ cm, $r_{on} = 2.881 \times 10^{-13}$ cm.

We note that the cross-section of elastic electron-deuteron scattering (*deuteron formfactor* $A(q^2)$), calculated with the use of the deuteron wave function that corresponds to the Paris potential, is in a good accordance with experimental data for a wide range of momentum transfer ($0 \leqslant q \leqslant 5\,\text{fm}^{-1}$). The D-wave contribution to scattering is important only for sufficiently large values of the momentum transfer (for $q \geqslant 4\,\text{fm}^{-1}$).

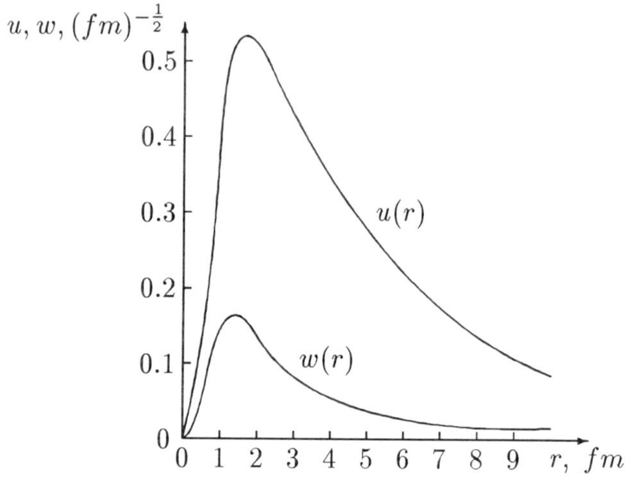

Figure 1.13. The deutron radial functions of the S- and D-states in the case of the Paris potential.

Non-relativistic theory of the nucleus. The principal problem of the *non-relativistic theory of the nucleus* is to explain the properties of nuclei, which are complex systems consisting of nucleons, on the basis of information on the properties of nucleons and on the nature of the interaction between them. At the present time, the properties of free nucleons and the character of the interaction between them in the non-relativistic energy region have been sufficiently well studied, both experimentally and theoretically.

The non-relativistic theory of the nucleus is based on the assumption that only two-body forces act between the simplest constituent parts of nuclei, that is, between the nucleons. In the non-relativistic theory of the nucleus, it is also assumed that the properties of the nucleons themselves (the mass, charge, magnetic moment, etc.) are not changed inside the nuclei, despite the presence of the strong interaction between the nucleons. Therefore, the Hamiltonian of a nucleus consisting of A nucleons can be represented in the form

$$H = \sum_i [-(\hbar^2/2M)\nabla_i^2] + \sum_{i<j} V_{ij}, \tag{1.195}$$

where $-(\hbar^2/2M)\nabla_i^2$ is the kinetic energy operator of an individual nucleon and V_{ij} is the two-body interaction potential between the nucleons i and j.

In contrast to a system of charged particles with Coulomb interaction, for which the total energy is made up of the interaction energies of the separate pairs of particles (this additivity being a consequence of the linearity of the equations of electrodynamics), in the case of nuclear systems there is no direct proof that the nuclear interaction is also additive, that is, that the interaction of two nucleons does not depend on the presence of a third nucleon. Therefore, at the present time, the only justification for the hypothesis of the *two-body nature of the nuclear forces* between the nucleons in nuclei is the absence of effects which would indicate the violation of this hypothesis.

According to the Hamiltonian (1.195), the interaction between nucleons is described by a potential in the non-relativistic theory. However, at sufficiently high energies inelastic processes associated with the formation of mesons become important and, therefore, the possibility of describing the interaction between nucleons by means of a potential is limited in the high-energy region. This restriction on the energy means that an unambiguous determination of the potential at short distances is completely impossible. However, since in nuclei the maximum possible values of the momenta of the nucleons correspond to energies less than 150 MeV, it may be hoped that relativistic effects can be neglected in the study of the structure of nuclei.

The study of the properties of nuclei containing more than two nucleons can serve as an additional source of information on the interaction between nucleons. For example, the properties of a system of two nucleons with energies close to zero can be explained by forces with zero range. However, in the case of forces of zero range, the binding energy of a system of three nucleons is found to be infinitely large. Therefore, the very fact of the existence of the nuclei ^3H and ^3He indicates that the nuclear forces are characterized by finite range. The data on nucleon-nucleon scattering at low energies (≤ 10 MeV) make it possible to determine only the effective parameters characterizing the magnitude and range of the interaction potential. As we have seen, these data are insufficient to determine the shape of the potential. Unlike the two-nucleon problem, the problem of the motion of a system consisting of three or more nucleons turns out to be very sensitive to the shape of the two-nucleon potential. This is connected with the fact that two-nucleon scattering is completely determined by the scattering matrix on the energy shell, whereas the properties of systems consisting of a larger number of nucleons depend also on the scattering matrix off the energy shell. Therefore, the indeterminacy of a potential obtained from experiments on two-nucleon scattering can in principle be eliminated if we take into account the properties of more complex nuclei.

1.11. Problems

1.1. Show that the wave function $\varphi_0(r)$ of the ground state of the neutron-proton system and the functions $\varphi_{\mathbf{k}}(r)$ the continuous spectrum form a complete set.

The energy of the neutron-proton system consists of a discrete level, corresponding to the bound state of a system, and the continuous spectrum of positive energy values, corresponding to the unbound states. In the approximation that the nuclear forces have zero range, the functions $\varphi_0(r)$ and $\varphi_{\mathbf{k}}(r)$ have the form

$$\varphi_0(\boldsymbol{r}) = \sqrt{\frac{\alpha}{2\pi}}\frac{\mathrm{e}^{-\alpha \boldsymbol{r}}}{\boldsymbol{r}}, \quad \varphi_{\mathbf{k}}(\boldsymbol{r}) = \mathrm{e}^{i\mathbf{k}\cdot\mathbf{r}} - \frac{1}{\alpha + ik}\frac{\mathrm{e}^{ikr}}{r}. \tag{1.196}$$

The functions $\varphi_0(r)$ and $\varphi_{\mathbf{k}}(r)$ correspond to the values $E_0 = -\hbar^2\alpha^2/M$ and $E_k = \hbar^2k^2/M$ of the energy of the system. In the approximation that the nuclear forces have zero range, the interaction is characterized by a single parameter α, which determines both the energy of the bound state of the system and the scattering amplitude in the unbound states.

Functions $\varphi_{\mathbf{k}}(r)$ and $\varphi'_{\mathbf{k}}(r)$ corresponding to different values $\mathbf{k}$ and $\mathbf{k}'$ of the wave vector are orthogonal:

$$\int d^3\boldsymbol{r}\varphi_{\mathbf{k}}(\boldsymbol{r})\varphi^*_{\mathbf{k}'}(\boldsymbol{r}) = (2\pi)^3\delta(\mathbf{k} - \mathbf{k}').$$

The functions $\varphi_{\mathbf{k}}(r)$ are also orthogonal to the function $\varphi_0(r)$ describing the bound state of the system:

$$\int d^3\boldsymbol{r}\varphi_{\mathbf{k}}(\boldsymbol{r})\varphi_0(\boldsymbol{r}) = 0.$$

It can easily be verified directly that the set of functions (1.196) satisfies the completeness condition

$$\varphi_0(\boldsymbol{r})\varphi_0(\boldsymbol{r}') + \int \frac{d^3\mathbf{k}}{(2\pi)^3}\varphi_{\mathbf{k}}(\boldsymbol{r})\varphi^*_{\mathbf{k}}(\boldsymbol{r}') = \delta(\boldsymbol{r} - \boldsymbol{r}'). \tag{1.197}$$

As an example of a system of functions that take into account the finite range of the nuclear forces, we can choose the bound state wave function for the *Hulthén potential*

$$\varphi_0(\boldsymbol{r}) = \sqrt{\frac{\alpha(\alpha+\beta)\beta}{2\pi(\beta-\alpha)^2}}\frac{\mathrm{e}^{-\alpha r} - \mathrm{e}^{-\beta r}}{r} \tag{1.198}$$

and complete it by the scattering functions which are orthogonal to the bound state function (1.198)

$$\varphi_{\mathbf{k}}(\boldsymbol{r}) = \mathrm{e}^{i\mathbf{k}\cdot\mathbf{r}} + f(k)\frac{\mathrm{e}^{ikr} - \mathrm{e}^{-\beta r}}{r}, \tag{1.199}$$

where

$$f(k) = -\left\{\beta - \frac{\beta^2 + k^2}{2\beta} - \frac{(\beta^2 + k^2)^2}{2(\alpha+\beta)^2\beta} + ik\right\}^{-1}, \tag{1.200}$$

and β is a parameter usually chosen to have the value $\beta = 7\alpha$. The functions (1.198) and (1.199), as well as functions (1.196), satisfy the completeness condition (1.197).

1.2. Find the normalization factor for the wave function of the ground state of the deuteron with allowance for the finite range of the nuclear forces (Smorodinskii, 1957).

The formula (1.11) for the wave function of the ground state of the deuteron gives the correct dependence on r outside the range of the nuclear forces. For many problems, it is necessary to know the wave function of the deuteron inside this region. But the normalization coefficient in (1.11) was obtained with the assumption of zero range of the nuclear forces. We shall show how the normalization coefficient in (1.11) must be changed to take account of the finite range of the nuclear forces.

We shall denote the exact wave function of the deuteron ground state by $\tilde{\varphi}_0(r) \equiv u(r)/r$. This function should be normalized in accordance with the condition

$$\int d^3\boldsymbol{r}\,\tilde{\varphi}_0^2(\boldsymbol{r}) = 1. \tag{1.201}$$

The radial function u satisfies the equation

$$u'' - \frac{M}{\hbar^2}V(r)u = \alpha^2 u. \tag{1.202}$$

We shall regard α^2 in this equation as some parameter. Differentiating eqn. (1.120) with respect to this parameter, we find

$$\frac{\partial}{\partial\alpha^2}u'' - \frac{M}{\hbar^2}V(r)\frac{\partial}{\partial\alpha^2}u = u + \alpha^2\frac{\partial}{\partial\alpha^2}u. \tag{1.203}$$

Multiplying eqn. (1.202) by $\partial u/\partial\alpha^2$ and the equality (1.203) by u and subtracting the first equality from the second, we obtain

$$u^2 = u\frac{\partial u''}{\partial\alpha^2} - u''\frac{\partial u}{\partial\alpha^2} \equiv \frac{\partial}{\partial r}\left(u^2\frac{\partial}{\partial\alpha^2}\frac{u'}{u}\right),$$

whence

$$\int_0^\infty dr\,u^2 = u^2\frac{\partial}{\partial\alpha^2}\left(\frac{u'}{u}\right)\Big|_0^\infty. \tag{1.204}$$

We introduce the approximate deuteron wave function $\varphi_0(r)$:

$$\varphi_o(r) = \frac{v(r)}{r},$$

where $v(r)$ is the approximate radial function, which coincides with the exact function $u(r)$ outside the range of the nuclear forces ($r > r_0$):

$$v(r) = u(r), \quad r > r_0.$$

For the approximate function $v(r)$, we can write a relation analogous to (1.204):

$$\int_0^\infty dr\, v^2 = v^2 \frac{\partial}{\partial \alpha^2}\left(\frac{v'}{v}\right)\Big|_0^\infty . \tag{1.205}$$

Subtracting the relation (1.205) from the relation (1.204) and taking into account that $u(0) = 0$ and that the functions u and v coincide outside the range of the nuclear forces, we find

$$\int_0^\infty dr u^2 - \int_0^\infty dr v^2 = v^2(0) \frac{\partial}{\partial \alpha^2}\left(\frac{v'(0)}{v(0)}\right). \tag{1.206}$$

We now make use of the relation (1.48) from the theory of the effective range, from which it follows that the ratio $v'(0)/v(0)$ is equal to

$$\frac{v'(0)}{v(0)} = -\frac{1}{a} - \frac{1}{2}\alpha^2 r_0,$$

and, therefore,

$$\frac{\partial}{\partial \alpha^2}\left(\frac{v'(0)}{v(0)}\right) = -\frac{1}{2} r_0 . \tag{1.207}$$

The approximate radial function $v(r)$ has the form

$$v(r) = C e^{-\alpha r}. \tag{1.208}$$

Substituting $v(r)$ into (1.206) and using the relation (1.207), we obtain

$$\int_0^\infty dr\, u^2 - C^2 \int_0^\infty dr\, e^{-2\alpha r} = -\frac{1}{2} C^2 r_0 .$$

Using the normalization condition (1.201) for the exact wave function, we find the following value for the normalization constant in the relation (1.208):

$$C = \sqrt{\frac{\alpha}{2\pi(1 - \alpha r_0)}}. \tag{1.209}$$

Thus, the approximate wave function of the deuteron ground state, corresponding to the normalization condition (1.201) for the exact wave function, has the form

$$\varphi_0(r) = \sqrt{\frac{\alpha}{2\pi(1 - \alpha r_0)}} \frac{e^{-\alpha r}}{r}. \tag{1.210}$$

For $r_0 \to 0$, this formula goes over into (1.11).

1.3. Find the probability of a reversal of spins in the scattering of a slow neutron by a proton (Fermi, 1951).

Since the nuclear interaction between the neutron and proton depends on their spins, a change in the orientation of the spins is possible in the scattering, although the total spin of the system is conserved.

We denote the spin wave functions of the neutron and proton by $\chi_\mu(n)$ and $\chi_\mu(p)$, where μ is the z-component of the spin ($\mu = \pm\frac{1}{2}$). The spin wave function of the neutron-proton system is a product of the wave functions $\chi_\mu(n)$ and $\chi_\mu(p)$. Before the scattering, the neutron-proton system can be in one of the following spin states:

$$\chi(n,p) = \begin{cases} \chi_{1/2}(n)\chi_{1/2}(p); \\ \chi_{-1/2}(n)\chi_{-1/2}(p); \\ \chi_{1/2}(n)\chi_{-1/2}(p); \\ \chi_{-1/2}(n)\chi_{1/2}(p). \end{cases} \tag{1.211}$$

We shall consider the eigenfunctions of the operator of the square of the total spin of the system. The spin functions corresponding to a total spin equal to unity (the triplet state) are

$$\left.\begin{aligned} \chi_{1,1}(n,p) &= \chi_{1/2}(n)\chi_{1/2}(p); \\ \chi_{1,0}(n,p) &= (1/\sqrt{2})\{\chi_{1/2}(n)\chi_{-1/2}(p) + \chi_{-1/2}(n)\chi_{1/2}(p)\}; \\ \chi_{1,-1}(n,p) &= \chi_{-1/2}(n)\chi_{-1/2}(p). \end{aligned}\right\} \tag{1.212}$$

The spin function corresponding to a total spin equal to zero (the singlet state) is

$$\chi_{0,0}(n,p) = (1/\sqrt{2})\{\chi_{1/2}(n)\chi_{-1/2}(p) - \chi_{-1/2}(n)\chi_{1/2}(p)\}. \tag{1.213}$$

The first two functions of (1.211), $\chi_{1/2}(n)\chi_{1/2}(p)$ and $\chi_{-1/2}(n)\chi_{-1/2}(p)$ coincide with the triplet state eigenfunctions $\chi_{1,1}(n,p)$ and $\chi_{1,-1}(n,p)$, respectively, while the third and fourth functions of (1.211) are superpositions of the triplet and singlet states $\chi_{1,0}(n,p)$ and $\chi_{0,0}(n,p)$:

$$\left.\begin{aligned} \chi_{1/2}(n)\chi_{-1/2}(p) &= (1/\sqrt{2})\{\chi_{1,0}(n,p) + \chi_{0,0}(n,p)\}; \\ \chi_{-1/2}(n)\chi_{1/2}(p) &= (1/\sqrt{2})\{\chi_{1,0}(n,p) - \chi_{0,0}(n,p)\}. \end{aligned}\right\} \tag{1.214}$$

If the neutron and proton are in the spin state $\chi_{1/2}(n)\chi_{1/2}(p)$ or $\chi_{-1/2}(n)\chi_{-1/2}(p)$, a change of orientation of the spins obviously does not occur. Then the scattering cross-section is equal to

$$\sigma_{1,2} = 4\pi a_t^2. \tag{1.215}$$

We now consider neutron-proton scattering in the spin state $\chi_{1/2}(n)\chi_{-1/2}(p)$. Since the scattering lengths are different in the triplet and singlet states, the coefficient of the outgoing spherical wave in the wave function of the neutron-proton system in the case under consideration has the form

$$\begin{aligned} (1/\sqrt{2})\left(a_t\chi_{1,0}(n,p) + a_s\chi_{0,0}(n,p)\right) &= \left((a_t + a_s)/2\right)\chi_{1/2}(n)\chi_{-1/2}(p) \\ &+ \left((a_t - a_s)/2\right)\chi_{-1/2}(n)\chi_{1/2}(p). \end{aligned}$$

The first term on the right in this expression corresponds to scattering without change of the spin orientation, and the second corresponds to scattering with change of the spin orientation. We see that the scattering lengths $(a_t + a_s)/2$ and $(a_t - a_s)/2$ correspond to these two processes respectively. Therefore, the cross-section for scattering without change of the spin orientation in the case of the spin state $\chi_{1/2}(n)\chi_{-1/2}(p)$ is equal to

$$\sigma_3 = \pi(a_t + a_s)^2. \tag{1.216}$$

If the system was in the state $\chi_{1/2}(n)\chi_{-1/2}(p)$ before the scattering, the cross-section for scattering of a neutron with change of the spin orientation is equal to

$$\sigma_3' = \pi(a_t - a_s)^2. \tag{1.217}$$

Scattering in the spin state $\chi_{-1/2}(n)\chi_{1/2}(p)$, like that in the state $\chi_{1/2}(n)\chi_{-1/2}(p)$, can occur either with or without change of the spin orientations. The corresponding cross-sections are equal to

$$\left.\begin{aligned} \sigma_4 &= \pi(a_t + a_s)^2; \\ \sigma_4' &= \pi(a_t - a_s)^2. \end{aligned}\right\} \tag{1.218}$$

In the case of scattering of unpolarized neutrons by protons, all four spin states (1.211) possess equal weight. Therefore, in this case, the probability of a change of the neutron spin orientation is equal to

$$w = \frac{(a_t - a_s)^2}{4a_t^2 + (a_t + a_s)^2 + (a_t - a_s)^2}. \tag{1.219}$$

Substituting the value of the scattering lengths into eqn.(1.219), we obtain $w = 0.65$.

1.4. Show that the cross-section for scattering of a very slow neutron by a proton contained in a heavy molecule is four times greater than the cross-section for scattering of a neutron by a free proton (Fermi, 1936).

In studying the scattering of slow neutrons in a substance containing hydrogen, we can only neglect the chemical bonding of the protons, that is, assume the protons to be free, if the neutron energy substantially exceeds the bonding energy of the protons in the molecules of the substance. The bonding energy is equal to $\hbar\omega$ in order of magnitude, where ω is the frequency of the proton vibrations in the molecule of the substance. For example, for a paraffin the magnitude of the bonding energy is about 0.4 eV. For neutron energies less than the proton bonding energy $\hbar\omega$ or of the same order, it proves to be extremely important to take account of the chemical bonding of the protons.

Using the definition of the pseudo-potential (1.62), it is not difficult to obtain a general formula, taking into account the bonding of the protons in the molecules, for the cross-section for scattering of the neutrons. We denote the wave functions of the proton in the initial and final states by $\varphi_0(\boldsymbol{r}_p)$ and $\varphi_f(\boldsymbol{r}_p)$. We choose the neutron wave functions in the form of plane waves. Thus, the initial and final functions of the neutron-proton system can be represented in the form

$$\left.\begin{aligned} \Phi_0 &= e^{i\mathbf{k}\cdot\mathbf{r}_n}\varphi_0(\boldsymbol{r}_p); \\ \Phi_f &= e^{i\mathbf{k}'\cdot\mathbf{r}_n}\varphi_f(\boldsymbol{r}_p), \end{aligned}\right\} \tag{1.220}$$

where $\mathbf{k}$ and $\mathbf{k}'$ are the initial and final wave vectors of the relative motion of the neutron and the molecule. According to perturbation theory, the scattering cross-section is given by the formula

$$d\sigma_{0f} = \frac{\tilde{\mu}^2}{4\pi^2\hbar^4}\frac{k'}{k}|V_{0f}|^2 do, \qquad (1.221)$$

where $\tilde{\mu} = MM_{\text{mol}}/(M + M_{\text{mol}})$ is the reduced mass for the relative motion of the neutron and the molecule; V_{0f} is the matrix element of the pseudo-potential (1.62). In this case, the energy conservation law

$$E = E' + E_f \quad (E = \hbar^2 k^2/2\tilde{\mu}, \quad E' = \hbar^2 k'^2/2\tilde{\mu}) \qquad (1.222)$$

holds. Here, E_f is the excitation energy of the molecule. (It is assumed that the molecule is in the ground state before the scattering.) Integrating the matrix element V_{0f} over $\mathbf{r}_n$ with the aid of the delta-function, we obtain

$$d\sigma_{0f} = \left(\frac{\tilde{\mu}}{\mu}\right)^2 \frac{k'}{k} a^2 |\mathcal{V}_{0f}|^2 do, \qquad (1.223)$$

where $\mathcal{V}_{0f}$ is the form factor defined by the integral

$$\mathcal{V}_{0f} = \int d^3\boldsymbol{r}\, e^{i(\mathbf{k}-\mathbf{k}')\cdot\boldsymbol{r}} \varphi_f^*(\boldsymbol{r})\varphi_0(\boldsymbol{r}) \qquad (1.224)$$

(μ is the reduced mass of the neutron-proton system). If the wavelength of the neutron is large compared with the dimensions of the molecule, then the factor $e^{i(k-k')\cdot r} \approx 1$. In this case, only elastic scattering of the neutrons occurs and the state of the proton in the molecule is unchanged. The elastic scattering cross-section is equal to

$$d\sigma_{00} = (\tilde{\mu}/\mu)^2 a^2 do. \qquad (1.225)$$

The scattering is characterized by spherical symmetry in the centre-of-mass system of the neutron and the molecule. The total cross-section for elastic scattering of the neutron by a bound proton is equal to

$$\sigma_{00} = (\tilde{\mu}/\mu)^2 \sigma_0, \qquad (1.226)$$

where σ_0 is the total cross-section for scattering of neutrons by free protons. In the limiting case of heavy molecules ($M_{\text{mol}} \gg M$), the ratio $\tilde{\mu}/\mu = 2$ and, consequently,

$$\sigma_{00} = 4\sigma_0, \qquad (1.227)$$

that is, because of the effect of the chemical bonding of the proton the cross-section for scattering of a very slow neutron is four times greater than the cross-section for scattering for a neutron by a free proton.

1.5. Calculate the cross-section for inelastic scattering of a neutron with excitation of a vibrational level of molecule (Bethe, 1947).

If the neutron energy exceeds an excitation energy of the molecule, the scattering of a neutron by a bound proton may be accompanied by excitation of the molecule. To determine the cross-section for such a process, it is necessary to know the wave functions of the proton in the excited states of the molecule. We shall make use of the harmonic-oscillator model to describe the vibrations of the proton in the molecule. In this case, the proton wave function satisfies the equation

$$\{-(\hbar^2/2M)\nabla_p^2 + V(r_p) - E_f\}\varphi(\boldsymbol{r}_p) = 0, \tag{1.228}$$

in which the potential energy of the proton is equal to

$$V(r_p) = \frac{1}{2}M\omega^2 r_p^2$$

(ω is the vibrational frequency of the proton in the molecule).

The proton wave function can be represented in the form of a product of three linear-oscillator wave functions:

$$\varphi(\boldsymbol{r}_p) = u_{n_x}(\xi)u_{n_y}(\eta)u_{n_z}(\zeta). \tag{1.229}$$

Here $\xi = \sqrt{\frac{2M\omega}{\hbar}}x$, $\eta = \sqrt{\frac{2M\omega}{\hbar}}y$ and $\zeta = \sqrt{\frac{2M\omega}{\hbar}}z$ are dimensionless coordinates, and n_x, n_y, and n_z are the vibrational quantum numbers of the oscillator. The functions $u_{n_x}(\xi)$ satisfy the equation

$$\left\{-\frac{d^2}{d\xi^2} + \frac{1}{4}\xi^2 - n_x\right\}u_{n_x}(\xi) = 0 \tag{1.230}$$

and are expressed in terms of Hermite polynomials

$$H_{n_x}(\xi) = \mathrm{e}^{(1/2)\xi^2}\frac{d^{n_x}}{d\alpha^{n_x}}\mathrm{e}^{(-1/2)(\xi-\alpha)^2}\bigg|_{\alpha=0}$$

as follows:

$$u_{n_x}(\xi) = (2\pi)^{-1/4}(n_x!)^{-1/2}\mathrm{e}^{(-1/4)\xi^2}H_{n_x}(\xi). \tag{1.231}$$

These functions are normalized in accordance with the condition

$$\int_{-\infty}^{\infty} d\xi u_{n_x}^2(\xi) = 1.$$

The matrix element appearing in the cross-section (1.223) breaks down into a product of three integrals of the type

$$\mathcal{V}_{0n_x} = (2\pi)^{-1/2}(n_x!)^{-1/2}\int_{-\infty}^{\infty} d\xi \mathrm{e}^{iq_x\xi}\frac{d^{n_x}}{d\alpha^{n_x}}\mathrm{e}^{(-1/2)(\xi-\alpha)^2}\bigg|_{\alpha=0}. \tag{1.232}$$

where

$$q = \sqrt{\frac{\hbar}{2M\omega}}(k-k').$$

(It is assumed that the proton was in the ground state before the scattering.) Integrating (1.232) by parts, we find

$$\mathcal{V}_{0n_x} = (iq_x)^{n_x}(n_x!)^{-1/2}\mathrm{e}^{(-1/2)q_x^2}.$$

Therefore, the cross-section for scattering of a neutron with excitation of the n-th vibrational level of the molecule with energy

$$E_f = E_n = n\hbar\omega \quad (n = n_x + n_y + n_z) \tag{1.233}$$

is given by the formula

$$d\sigma_{0n} = 4a^2 \frac{k'}{k} \sum \frac{q_x^{2n_x} q_y^{2n_y} q_z^{2n_z}}{n_x! n_y! n_z!} c^{-q^2} do,$$

where the summation is performed over all values of the quantum numbers n_x, n_y, and n_z satisfying the condition $n_x + n_y + n_z = n$. As a result of the summation, we obtain

$$d\sigma_{0n} = 4a^2 \frac{k'}{k} \cdot \frac{q^{2n}}{n!} \mathrm{e}^{-q^2} do. \tag{1.234}$$

Noting that $do = 2\pi \frac{\hbar\omega}{2\sqrt{EE'}} dq^2$ and introducing the dimensionless energy $\varepsilon = E/\hbar\omega$, we rewrite this formula in the form

$$d\sigma_{0n} = \frac{\sigma_0}{\varepsilon} \frac{q^{2n}}{n!} \mathrm{e}^{-q^2} dq^2. \tag{1.235}$$

Using the energy conservation law $E = E' + n\hbar\omega$, we see that q is confined within the limits

$$q_{\min} = \sqrt{\varepsilon} - \sqrt{\varepsilon - n}, \quad q_{\max} = \sqrt{\varepsilon} + \sqrt{\varepsilon - n}.$$

Integrating the expression (1.235) within these limits, we obtain the following formula for the total cross-section for scattering of a neutron with excitation of the n-th vibrational level of the molecule:

$$\sigma_{0n} = \frac{\sigma_0}{\varepsilon} \{ f_n(q_{\min}^2) - f_n(q_{\max}^2) \}, \tag{1.236}$$

where

$$f_n(x) = \left(1 + \frac{x}{1!} + \cdots + \frac{x^n}{n!} \right) \mathrm{e}^{-x}.$$

In the case of elastic scattering, the formula (1.236) acquires the form

$$\sigma_{00} = (\sigma_0/\varepsilon)(1 - \mathrm{e}^{-4\varepsilon}). \tag{1.237}$$

In the limiting case $\varepsilon \ll 1$, we obtain from the relation (1.237) the result (1.227) found previously:

$$\sigma_{00} = 4\sigma_0, \quad \varepsilon \ll 1. \tag{1.238}$$

For $\varepsilon = 1$, the elastic scattering cross-section is equal to $\sigma_{00} \approx 0.98\sigma_0$. For $\varepsilon \gg 1$, the elastic scattering cross-section is given by the expression

$$\sigma_{00} = \sigma_0/\varepsilon, \quad \varepsilon \gg 1. \tag{1.239}$$

For large values of ε and for any n that is not too close to ε, the second term in the relation (1.236) is very small, and the first differs little from unity. Therefore, the cross-section for scattering with excitation of any allowed level will

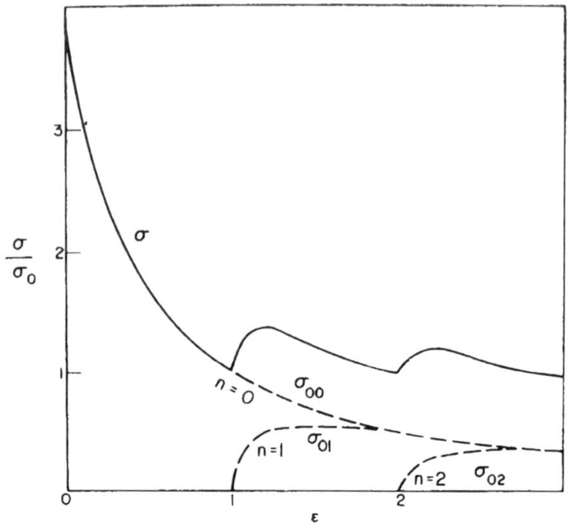

Figure 1.14. Dependence of the cross-sections σ_{00}, σ_{01}, σ_{02} and the total scattering cross-section σ on the dimensionless neutron energy $\varepsilon = E/\hbar\omega$.

be determined in this case by the same expression as that for the elastic scattering cross-section:

$$\sigma_{0n} = \sigma_0/\varepsilon, \quad \varepsilon \gg 1. \tag{1.240}$$

According to the energy conservation law, the maximum value of n is equal to ε; therefore, the total cross-section for large ε will be equal to σ_0, that is, for high neutron energies the bonding of the proton turns out to be unimportant.

Figure 1.14 shows the dependence of the cross-section σ_{00}, σ_{01} and σ_{02} and the total scattering cross-section $\sigma = \sum\limits_n \sigma_{0n}$ on the neutron energy ε. If the neutron energy is less than the energy $\hbar\omega$ of a vibrational quantum, it cannot excite vibrations in the molecule; in this case, only elastic scattering is possible. For neutron energy values that are multiples of $\hbar\omega$, there occurs a sharp increase in the scattering cross-section, due to the possibility of transfer of the neutron energy to the vibrational motion of the protons. If the neutron energy is large compared with the chemical-bond energy, the protons can be regarded as free. A small fraction of the neutron energy, spent on breaking the chemical bond of the proton in the molecule, is transferred to the molecule. The scattering in this energy region is practically elastic and is characterized by spherical symmetry in the centre-of-mass frame of the neutron and proton.

1.6. If the wavelength $\lambda\!\!\!^{-}$ of an incident neutron is comparable with the distance between the protons in the hydrogen molecule, then both elastic scattering of the neutron and inelastic scattering accompanied by a change of state of the molecule are possible. Determine the cross-section for elastic scattering of a neutron in a hydrogen molecule with allowance for the effects of retardation, and also the cross-sections of the inelastic scattering processes accompanied by excitation of the hydrogen molecule (Schwinger and Teller, 1937).

We shall consider the collision of a neutron of wave vector $\boldsymbol{k}$ with a hydrogen molecule in a state characterized by the vibrational quantum number v, rotational quantum number J and spin S. We shall determine the probability of a process in which the neutron acquires the wave vector $\boldsymbol{k}'$ and the molecule undergoes a transition to a state characterized by the quantum numbers v', J' and S'.

We choose the wave function of the initial state of the neutron and the molecule in the form

$$\Phi_i = e^{i\mathbf{k}\cdot\mathbf{r}_n}\Phi_{vJM_J}(\boldsymbol{r})\chi_\mu\chi_{S\mu_S}, \tag{1.241}$$

where $e^{i\mathbf{k}\cdot\mathbf{r}_n}$ is the wave function of the relative motion of the neutron and the molecule, $\Phi_{vJM_J}(\boldsymbol{r})$ is the wave function describing the motion of the protons in a molecule in a state characterized by the quantum numbers v, J and M_J, χ_μ is the spin function of the neutron (μ is the z-component of the neutron spin), and, finally, $\chi_{S\mu_S}$ is the spin function of the protons in a state with resultant spin S (the z-component of the spin is equal to μ_S). In an analogous way, we choose the final-state wave function in the form

$$\Phi_f = e^{i\mathbf{k}'\cdot\mathbf{r}_n}\Phi_{v'J'M'_J}(\boldsymbol{r})\chi_{\mu'}\chi_{S'\mu'_S}. \tag{1.242}$$

The differential scattering cross-section, averaged over the initial and summed over the final magnetic quantum numbers, is given by the general formula

$$d\sigma_{if} = \frac{16}{9}\frac{k'}{k}\frac{1}{2(2S+1)(2J+1)}\sum_{\mu\mu_S M_J}\sum_{\mu'\mu'_S M'_J}\left|\left(\Phi_f, \frac{\mu}{2\pi\hbar^2}V\Phi_i\right)\right|^2 do, \tag{1.243}$$

where V is the interaction potential of the neutron with the molecule, given by the expression (1.66), and do is the solid-angle element in which the wave vector of the scattered neutron lies. In the scattering, the energy conservation law

$$\frac{\hbar^2k^2}{2\tilde{\mu}} + E_{vJ} = \frac{\hbar^2k'^2}{2\tilde{\mu}} + E_{v'J'}, \tag{1.244}$$

is fulfilled, where E_{vJ} is the energy of the molecule in the state Φ_{vJM_J}; $\hbar^2k^2/2\tilde{\mu}$ is the energy of the relative motion of the neutron and the molecule ($\tilde{\mu} = 2M/3$ is the reduced mass).

We must distinguish scattering of neutrons without change of the spin state of the hydrogen molecule ($S \to S' = S$), and transitions in which the spin state of the molecule does change ($S \to S' = 1 - S$). Transitions with unchanged spin are due to the interaction energy V_s, and transitions with a change of spin are due to the interaction energy V_a (cf.(1.67)). Noting that

$$\overline{|3a_t + a_s + (a_t - a_s)\sigma_n\cdot S|^2} = (3a_t + a_s)^2 + S(S+1)(a_t - a_s)^2,$$

it is not difficult to obtain the following expression for the cross-section for scattering of neutrons without change of the spin of the molecule:

$$d\sigma_{SvJ;Sv'J'} = \frac{4}{9}\frac{k'}{k}[(3a_t + a_s)^2 + S(S+1)(a_t - a_s)^2]$$

$$\times\frac{1}{2J+1}\sum_{M_JM'_J}\left|\int d^3r\cos\left[\frac{1}{2}(\boldsymbol{k}-\boldsymbol{k}')\cdot\boldsymbol{r}\right]\Phi^*_{v'J'M'_J}(\boldsymbol{r})\Phi_{vJM_J}(\boldsymbol{r})\right|^2 do. \tag{1.245}$$

In this expression, J and J' should be numbers of the same parity. Actually, since protons obey Fermi-Dirac statistics, the total wave function of the hydrogen molecule should be anti-symmetric. Therefore, in the case of para-hydrogen, when the spin function of the molecule is anti-symmetric, the spatial function should be symmetric and, consequently, the rotational quantum number J is even. In the case of orthogydrogen the rotational quantum number J is odd.

In order to find the cross-section for scattering of a neutron with change of spin of the molecule, we calculate the mean value of the quantity $|\boldsymbol{\sigma}_n \cdot (\boldsymbol{\sigma}_1 - \boldsymbol{\sigma}_2)|^2$ over the spin states:

$$\overline{|\boldsymbol{\sigma}_n \cdot (\boldsymbol{\sigma}_1 - \boldsymbol{\sigma}_2)|^2} = (\boldsymbol{\sigma}_1 - \boldsymbol{\sigma}_2)^2 = 4[3 - S(S+1)].$$

Therefore, the cross-section for scattering of neutrons with change of the spin of the molecule is given by the formula

$$d\sigma_{SvJ;(1-S)v'J'} = \frac{4}{9}\frac{k'}{k}[3 - S(S+1)](a_t - a_s)^2$$

$$\times \frac{1}{2J+1} \sum_{M_J M'_J} \left| \int d^3r \sin\left[\frac{1}{2}(\mathbf{k} - \mathbf{k}')\cdot\boldsymbol{r}\right] \Phi^*_{v'J'M'_J}(\boldsymbol{r}) \Phi_{vJM_J}(\boldsymbol{r}) \right|^2 do. \qquad (1.246)$$

Para-ortho and ortho-para transitions are associated with a change of parity of the rotational quantum number J.

The minimum neutron energy necessary for a transition of the molecule from the state v, J to the state v', J' to be possible is equal to

$$E_{\text{min}} \equiv \frac{3\hbar^2}{4M}(k^2 - k'^2) = E_{v'J'} - E_{vJ}.$$

The initial momentum of the neutron in the laboratory frame, $\tilde{\boldsymbol{k}}$, is equal to $3\boldsymbol{k}/2$; therefore, the minimum neutron energy in the laboratory frame is equal to

$$\widetilde{E}_{\text{min}} = \frac{3}{2}(E_{v'J'} - E_{vJ}). \qquad (1.247)$$

For small values of the quantum numbers v and J, the excitation energy of the hydrogen molecule is given by the expression ($\mathcal{J}$ is the moment of inertia):

$$E_{vJ} = v\hbar\omega + \frac{\hbar^2}{2\mathcal{J}} J(J+1), \qquad (1.248)$$

where $\hbar\omega = 0.533$ eV, and $\hbar^2/\mathcal{J} = 0.015$ eV. Direct calculation shows that, in order to convert the molecule from the state $v = 0$, $J = 1$ to the state $v' = 0$, $J' = 2$, the neutron must possess an energy of 0.045 eV. For a transition from the state $v = 0$, $J = 0$ to the state $v' = 0$, $J' = 1$, an energy of 0.023 eV is necessary. Therefore, if the neutron energy is less than 0.045 eV, only the following rotational transitions are possible: $0 \to 0$, $0 \to 1$, $1 \to 0$ and $1 \to 1$.

In order to find the transition probabilities for the hydrogen molecule, we make use of the rigid-rotator model. In this case, the wave function of the molecule can be written in the form

$$\Phi_{0JM_J}(r) = (1/r)[\delta(r-d)]^{1/2} Y_{JM_J}(\vartheta_r, \varphi_r), \qquad (1.249)$$

where d is the equilibrium distance between the protons in the molecule ($d \approx 0.75 \times 10^{-8}$ cm) and ϑ_r and φ_r are the angles defining the orientation of the vector $\boldsymbol{r}$. Using (1.245) and (1.249), it is not difficult to obtain the following expression for the differential elastic cross-section ($0 \to 0$):

$$d\sigma_{0\to 0} = \frac{4}{9}(3a_t + a_s)^2 \left[\frac{\sin[\xi \sin(\vartheta/2)]}{\xi \sin(\vartheta/2)}\right]^2 2\pi \sin\vartheta \, d\vartheta, \qquad (1.250)$$

where $\xi = kd$ and ϑ is the scattering angle in the centre-of-mass frame. The integral cross-section is equal to

$$\begin{aligned} \sigma_{0\to 0} &= \frac{32}{9}\pi(3a_t + a_s)^2 \xi^{-2} \int_0^{\xi} dx \frac{\sin^2 x}{x} \\ &= \frac{16}{9}\pi(3a_t + a_s)^2 \xi^{-2} \{\ln 2\xi - \mathrm{ci}\, 2\xi + 0.5772\}, \qquad (1.251) \end{aligned}$$

where

$$\mathrm{ci}\, x \equiv -\int_x^{\infty} dt \frac{\cos t}{t}.$$

In an analogous way, it is not difficult to obtain the following expressions for the cross-sections of the transitions $0 \to 1$, $1 \to 0$ and $1 \to 1$:

$$\begin{aligned} \sigma_{0\to 1} &= 32\pi(a_t - a_s)^2 \xi^{-2} \\ &\times \left\{\frac{\sin x}{x} + \frac{\cos x - 1}{x^2} + \frac{1}{2}\ln x - \frac{1}{2}\mathrm{ci}\, x\right\} \Bigg|_{\xi - \xi^-}^{\xi + \xi^-}; \qquad (1.252) \\ \sigma_{1\to 0} &= \frac{32}{3}\pi(a_t - a_s)^2 \xi^{-2} \\ &\times \left\{\frac{\sin x}{x} + \frac{\cos x - 1}{x^2} + \frac{1}{2}\ln x - \frac{1}{2}\mathrm{ci}\, x\right\} \Bigg|_{\xi - \xi^+}^{\xi + \xi^+}; \qquad (1.253) \\ \sigma_{1\to 1} &= \frac{16}{3}\pi[(3a_t + a_s)^2 + 2(a_t - a_s)^2]\xi^{-2} \\ &\times \left\{\frac{\sin^2 \xi}{\xi^2} - \frac{3}{\xi^2}\left(\frac{\sin \xi}{\xi} - \cos \xi\right)^2 + \ln 2\xi - \mathrm{ci}\, 2\xi - 0.4228\right\}, \qquad (1.254) \end{aligned}$$

where

$$\xi^{\pm} = \sqrt{\xi^2 \pm \frac{2\tilde{\mu} d^2}{\hbar^2} E_{01}}.$$

In the scattering of very slow neutrons in molecular hydrogen, it turns out to be important to take account of the thermal motion of the molecules. If $\boldsymbol{v}$ is the velocity of the incident neutron and $\boldsymbol{u}$ is the thermal velocity of the molecule, then the relative wave vector $\mathbf{k}$ appearing in the cross-section must be assumed equal to $\mathbf{k} = (\tilde{\mu}/\hbar)(\boldsymbol{v} - \boldsymbol{u})$. The effective values of the scattering cross-sections can be found by averaging the expressions (1.251)–(1.254) over the thermal motion of the molecules

$$\bar{\sigma} = \frac{1}{v} \int d^3u |\boldsymbol{v} - \boldsymbol{u}| \sigma(\boldsymbol{v} - \boldsymbol{u}) N(u), \tag{1.255}$$

where $N(u)$ is the velocity distribution function of the molecules.

As an example, we give the results of calculations of the cross-sections in the case of scattering of a neutron with energy 0.001723 eV (this energy corresponds to a temperature of 20°K):

$$\left.\begin{aligned} \sigma_{0\to0} &= 6.444(3a_t + a_s)^2; \\ \sigma_{1\to0} &= 1.753(a_t - a_s)^2; \\ \sigma_{1\to1} &= 6.450\{(3a_t + a_s)^2 + 2(a_t - a_s)^2\}. \end{aligned}\right\} \tag{1.256}$$

Experimentally, the following quantities can be measured:

$$\sigma_{\text{para}} = \sigma_{0\to0}, \qquad \sigma_{\text{ortho}} = \sigma_{1\to0} + \sigma_{1\to1}. \tag{1.257}$$

The experimental values of these quantities at a temperature of 20°K are equal to

$$\sigma_{\text{para}} = 3.97 \times 10^{-24}\,\text{cm}^2, \quad \sigma_{\text{ortho}} = 124 \times 10^{-24}\,\text{cm}^2. \tag{1.258}$$

Using the formulae (1.256) and (1.257), one can determine the scattering lengths a_t and a_s with a high degree of accuracy from the values (1.258).

1.7. Verify that the coefficients α, β, γ, δ and ε occurring in the nucleon-nucleon scattering amplitude (1.70) are expressed in terms of the phase shifts ${}^1\delta_1^J$

and ${}^3\delta_1^J$ and mixture parameters ε^J in the following way:

$$\left.\begin{aligned}
\alpha &= \frac{1}{4k}\sum_{l=0}^{\infty}[(l+2)a_l^{l+1} + (2l+1)a_l^l + (l-1)a_l^{l-1} + (l+1)b_l^{l+1} \\
&\quad + lb_l^{l-1} + (2l+1)c_l^l]P_l(\cos\vartheta); \\
\beta &= \frac{1}{4k}\left\{\sum_{l=0}^{\infty}[(l+1)b_l^{l+1} + lb_l^{l-1} - (2l+1)c_l^l]P_l(\cos\vartheta)\right. \\
&\quad \left. - \sum_{l=2}^{\infty}\left[\frac{1}{l+1}a_l^{l+1} - \frac{2l+1}{l(l+1)}a_l^l + \frac{1}{l}a_l^{l-1}\right]P_l^2(\cos\vartheta)\right\}; \\
\gamma &= \frac{1}{4k}\sum_{l=1}^{\infty}\left[\frac{l+2}{l+1}a_l^{l+1} - \frac{2l+1}{l(l+1)}a_l^l - \frac{l-1}{l}a_l^{l-1} + b_l^{l+1} - b_l^{l-1}\right]P_l^1(\cos\vartheta); \\
\delta &= \frac{1}{4k\cos\vartheta}\left\{\sum_{l=0}^{\infty}\left[\frac{1}{2}\{(l+2)a_l^{l+1} + (2l+1)a_l^l + (l-1)a_l^{l-1}\}(\cos\vartheta - 1)\right.\right. \\
&\quad \left. + (l+1)b_l^{l+1} + lb_l^{l-1} - (2l+1)c_l^l\cos\vartheta\right]P_l(\cos\vartheta) \\
&\quad \left. + \frac{1}{2}\sum_{l=2}^{\infty}\left[\frac{1}{l+1}a_l^{l+1} - \frac{2l+1}{l(l+1)}a_l^l + \frac{1}{l}a_l^{l-1}\right](1+\cos\vartheta)P_l^2(\cos\vartheta)\right\}; \\
\varepsilon &= \frac{1}{4k\cos\vartheta}\left\{\sum_{l=0}^{\infty}\left[\frac{1}{2}\{(l+2)a_l^{l+1} + (2l+1)a_l^l + (l-1)a_l^{l-1}\}(1 + \cos\vartheta)\right.\right. \\
&\quad \left. - (l+1)b_l^{l+1} - lb_l^{l-1} - (2l+1)c_l^l\cos\vartheta\right]P_l(\cos\vartheta) \\
&\quad \left. + \frac{1}{2}\sum_{l=2}^{\infty}\left[\frac{1}{l+1}a_l^{l+1} - \frac{2l+1}{l(l+1)}a_l^l + \frac{1}{l}a_l^{l-1}\right](\cos\vartheta - 1)P_l^2(\cos\vartheta)\right\},
\end{aligned}\right\} \tag{1.259}$$

where we have introduced the notation:

$$\left.\begin{aligned}
&a_l^{J=1} \equiv \sin {}^3\delta_J^J \mathrm{e}^{i\,{}^3\delta_J^J};\\
&a_l^{J=l+1} \equiv \alpha^J \cos^2 \varepsilon^J + \beta^J \sin^2 \varepsilon^J + \tfrac{1}{2}\sqrt{\frac{J}{J+1}}(\alpha^J - \beta^J)\sin 2\varepsilon^J;\\
&b_l^{J=l+1} \equiv \alpha^J \cos^2 \varepsilon^J + \beta^J \sin^2 \varepsilon^J - \frac{1}{2}\sqrt{\frac{J+1}{J}}(\alpha^J - \beta^J)\sin 2\varepsilon^J;\\
&a_l^{J=l-1} \equiv \alpha^J \sin^2 \varepsilon^J + \beta^J \cos^2 \varepsilon^J + \frac{1}{2}\sqrt{\frac{J+1}{J}}(\alpha^J - \beta^J)\sin 2\varepsilon^J;\\
&b_l^{J=l-1} \equiv \alpha^J \sin^2 \varepsilon^J + \beta^J \cos^2 \varepsilon^J - \frac{1}{2}\sqrt{\frac{J}{J+1}}(\alpha^J - \beta^J)\sin 2\varepsilon^J;\\
&\alpha^J \equiv \sin {}^3\delta_{J-1}^J \mathrm{e}^{i\,{}^3\delta_{J-1}^J};\\
&\beta^J \equiv \sin {}^3\delta_{J+1}^J \mathrm{e}^{i\,{}^3\delta_{J+1}^J};\\
&c^{J=l} \equiv \sin {}^1\delta_J^J \mathrm{e}^{i\,{}^1\delta_J^J}.
\end{aligned}\right\} \tag{1.260}$$

1.8. Derive the integral unitarity relations for the two-nucleon elastic scattering amplitude (Puzikov, Ryndin and Smorodinskii, 1957).

If we assume that only elastic scattering is possible in a two-nucleon system, from the unitarity of the S-matrix it is not difficult to obtain the following integral relation for the matrix scattering amplitude $f(\mathbf{k}, \mathbf{k}')$:

$$f(\mathbf{k}, \mathbf{k}') - f^+(\mathbf{k}'\mathbf{k}) = \frac{i\mathbf{k}}{2\pi}\int do'' f(\mathbf{k}, \mathbf{k}'') f^+(\mathbf{k}'\mathbf{k}''). \tag{1.261}$$

where the integration is performed over the directions of the vector $\mathbf{k}''$.

For the two-nucleon elastic scattering amplitude, we make use of the Wolfenstein formula (1.170). Multiplying the left- and right-hand sides of the equality (1.261) successively by the matrices $\mathbf{1}$, $(\boldsymbol{n}\cdot\boldsymbol{\sigma}_1)(\boldsymbol{n}\cdot\boldsymbol{\sigma}_2)$, $\boldsymbol{n}\cdot(\boldsymbol{\sigma}_1+\boldsymbol{\sigma}_2)$, $(\boldsymbol{m}\cdot\boldsymbol{\sigma}_1)(\boldsymbol{m}\cdot\boldsymbol{\sigma}_2)$ and $(\boldsymbol{l}\cdot\boldsymbol{\sigma}_1)(\boldsymbol{l}\cdot\boldsymbol{\sigma}_2)$ and then calculating the traces in the left-hand sides of the equalities, we obtain five integral unitarity relations for the coefficients α, β, γ, δ and ε:

$$\left.\begin{aligned}
&(4\pi/\mathbf{k})\mathrm{Im}\,\alpha(\vartheta) = (\tfrac{1}{4})\textstyle\int do''\mathrm{Tr}\,\{f(\mathbf{k}, \mathbf{k}'')f^+(\mathbf{k}', \mathbf{k}'')\};\\
&(4\pi/\mathbf{k})\mathrm{Im}\,\beta(\vartheta) = (\tfrac{1}{4})\textstyle\int do''\mathrm{Tr}\,\{(\boldsymbol{n}\cdot\boldsymbol{\sigma}_1)(\boldsymbol{n}\cdot\boldsymbol{\sigma}_2)f(\mathbf{k}, \mathbf{k}'')f^+(\mathbf{k}', \mathbf{k}'')\};\\
&(4\pi/\mathbf{k})\mathrm{Im}\,\gamma(\vartheta) = (\tfrac{1}{8}i)\textstyle\int do''\mathrm{Tr}\,\{\boldsymbol{n}\cdot(\boldsymbol{\sigma}_1+\boldsymbol{\sigma}_2)f(\mathbf{k}, \mathbf{k}'')f^+(\mathbf{k}', \mathbf{k}'')\};\\
&(4\pi/\mathbf{k})\mathrm{Im}\,\delta(\vartheta) = (\tfrac{1}{4})\textstyle\int do''\mathrm{Tr}\,\{(\boldsymbol{m}\cdot\boldsymbol{\sigma}_1)(\boldsymbol{m}\cdot\boldsymbol{\sigma}_2)f(\mathbf{k}, \mathbf{k}'')f^+(\mathbf{k}', \mathbf{k}'')\};\\
&(4\pi/\mathbf{k})\mathrm{Im}\,\varepsilon(\vartheta) = (\tfrac{1}{4})\textstyle\int do''\mathrm{Tr}\,\{(\boldsymbol{l}\cdot\boldsymbol{\sigma}_1)(\boldsymbol{l}\cdot\boldsymbol{\sigma}_2)f(\mathbf{k}, \mathbf{k}'')f^+(\mathbf{k}', \mathbf{k}'')\},
\end{aligned}\right\} \tag{1.262}$$

where ϑ is the angle between the vectors $\mathbf{k}'$ and $\mathbf{k}$. The calculation of the traces in the right-hand sides of the equalities (1.262) is elementary but leads to cumbersome expressions, which we shall not give.

1.12. References

Bartlett, J.H. (1936) *Phys. Rev.*, **49**, p. 102.

Bethe, H.A. (1947) *Elementary Nuclear Theory* (Chap.2), Wiley, N.Y.

Bethe, H.A. (1949) *Phys. Rev.*, **76**, p. 38.

Bethe, H. and Peierls, R. (1935) *Proc. Roy. Soc.*, **A148**, p. 146.

Blatt, J.M. and Biedenharn, L.C. (1952) *Phys. Rev.*, **86**, p. 399.

Breit, G. (1947) *Phys. Rev.*, **71**, p. 215.

Breit, G., Condon, E.U. and Present, R.D. (1936) *Phys. Rev.*, **50**, p. 825.

Cassen, B. and Condon, E.U. (1936) *Phys. Rev.*, **50**, p. 846.

Eisenbud, L. and Wigner, E.P. (1941) *Proc. Nat. Acad. Sci. (USA)*, **27**, p. 281.

Fermi, E. (1936) *Ricerca Sci.*, **7**, p. 13 (in *Collected Papers of Enrico Fermi*, Chicago University Press, 1962).

Fermi, E. (1950) *Nuclear Physics*, Chicago University Press.

Gammel, J.L. and Thaler, R.M. (1957) *Phys. Rev.*, **107**, p. 291.

Gammel, J.L. and Thaler, R.M. (1959) *Phys. Rev.*, **107**, p. 1337.

Gartenhaus, S. (1955) *Phys. Rev.*, **100**, p. 900.

Hamada, T. and Johnston, I.D. (1962) *Nucl. Phys.*, **34**, p. 382.

Heisenberg, W. (1932) *Z. Phys.*, **77**, p. 1.

Hulthén, L. and Sugawara, M. (1957) "The two-nucleon problem", *Encyclopaedia of Physics* (ed. S.Flügge), Vol.39, Springer-Verlag, Berlin.

Jackson, J.D. and Blatt, J.M. (1950) *Rev. Mod. Phys.*, **22**, p. 77.

Jastrow, R. (1951) *Phys.Rev.*, **81**, p. 165.

Lacombe, M. e.a. (1980) *Phys.Rev.*, **C21**, p. 861.

Landau, L.D. and Smorodinskii, Ya.A. (1944) *Zh. Eksp. Teor. Fiz.*, **14**, p. 269, (translation in the *Collected Papers of L.D.Landau*, Pergamon Press, Oxford, 1965).

Landau, L.D. and Smorodinskii, Ya.A. (1955) *Lectures on Nuclear Theory*, Gostekhteorizdat, Moscow (translation published by Plenum Press, N.Y., 1959).

Lassila, K.E. e.a. (1962) *Phys. Rev.*, **126**, p. 881.

MacGregor, M.H., Arndt, R.A. and Wright, R.M. (1968a) *Phys. Rev.*, **169**, p. 1128.

MacGregor, M.H., Arndt, R.A. and Wright, R.M. (1968b) *Phys. Rev.*, **173**, p. 1272.

Macleidt, R. e.a. (1987) *Physics Reports*, **149**, p. 1; (1989) *Nucl. Phys.*, **19**, p. 189.

Majorana, E. (1933) *Z. Phys.*, **82**, p. 137.

Puzikov, L., Ryndin, R. and Smorodinskii, Ya. (1957) *Zh. Eksp. Teor. Fiz.*, **32**, p. 592, [(1957) *Sov. Phys. - JETP*, **5**, p. 489].

Rarita, W. and Schwinger, J. (1941) *Phys. Rev.*, **59**, p. 436.

Reid, R.V.Jr. (1968) *Ann. Phys., (N.Y.)*, **50**, p. 411.

Schwinger, J. (1947) *Phys. Rev.*, **72**, p. 742.

Schwinger, J. and Teller, E. (1937) *Phys. Rev.*, **52**, p. 286.

Signell, P.S. and Marshak, R.E. (1957) *Phys. Rev.*, **106**, p. 832; (1958) *Phys. Rev.*, **109**, p. 1229; (1957) *Phys. Rev. Lett.*, **1**, p. 416.

Smorodinskii, Ya.A. (1952) Lectures at the Conference of Yugoslav Physicists, Lošinj Island.

Stapp, H., Ypsilantis, T.J. and Metropolis, N. (1957) *Phys. Rev.*, **105**, p. 302.
Stoks, V.G., van Campen, P.C., Spit, W. and de Swart, J.J. (1988) *Phys. Rev. Lett.*,**60**, p. 1932.
Wilson, R. (1963) *The Nucleon-nucleon Interaction*, Wiley, N.Y.
Wiringa, R.W., Stoks, V.G. and Schiavilla, R. (1955) *Phys. Rev.*, **C51**, p. 38.
Wolfenstein, L. (1956) *Ann. Rev. Nucl. Sci.*, **6**, p. 43.

CHAPTER 2

FEW-NUCLEON SYSTEMS

2.1. Bound States and Scattering in a System of Two Particles

The scattering amplitude. In the previous Chapter, we considered the main peculiarities of the nuclear interaction between nucleons. It was shown that such can cause either formation of a two-nucleon bound state or *scattering of nucleons*. At the same time, in the center-of-inertia coordinate system, the energy spectrum is nondegenerate and consists of one discrete level and a continuous part. In the case of greater number of particles, when the system contains three or more interacting nucleons, the description in terms of quantum mechanics becomes more complicated. The reason is that, unlike the two-particle system, the energy spectrum of a system of three or more particles is always degenerate, i.e., discrete levels overlap with the continuous part of the spectrum. This results in the possible redistribution of particles in the interacting subsystems. This must be taken into account when one formulates the boundary conditions in the complex configuration space of the system.

Before passing to the analysis of systems consisting of three or more particles, let us consider the description of a system of two interacting particles in terms of quantum mechanics.

First of all, we consider the elastic scattering which does not change intrinsic states of the particles involved in the collision. In the center-of-mass system, the two-particle scattering problem is treated as scattering of a particle with *reduced mass* μ in the field $V(\mathbf{r})$ of a stationary force center. The Hamiltonian of the system is

$$H = H_0 + V,$$

where $H_0 = (-\hbar^2/2\mu)\Delta$ is the kinetic energy of relative motion. The scattering wave function is governed by the Schrödinger equation

$$H\psi = E\psi \tag{2.1}$$

for positive energies ($E > 0$), with the boundary conditions which require that the solution at large distances from the scatterer must be a sum of the incident and scattered waves.

In order to find the solution with the required asymptotics, we rewrite the Schrödinger equation in the form

$$(E - H_0)\psi = V\psi \tag{2.2}$$

and treat the right-hand part as a given function. Then the general solution of (2.2) may be written as

$$\psi(\mathbf{r}) = \varphi(\mathbf{r}) + \int d\mathbf{r}' G_0(E; \mathbf{r} - \mathbf{r}') V(\mathbf{r}') \psi(\mathbf{r}'), \tag{2.3}$$

where φ is the general solution of (2.2) with vanishing right-hand part, i.e.,

$$(E - H_0)\varphi = 0, \tag{2.4}$$

and $G_0(E; \mathbf{r} - \mathbf{r}')$ is the *Green function* that satisfies the inhomogeneous equation with a point source

$$(E - H_0) G_0(E; \mathbf{r} - \mathbf{r}') = \delta(\mathbf{r} - \mathbf{r}'). \tag{2.5}$$

The first term of (2.3) may be regarded as the incident wave that describes the free motion of the particle. The second term is associated with the scattered wave which can be either converging or diverging (it depends on the choice of the Green function).

If the eigenfunctions of the equation (2.4) are plane waves

$$\varphi_{\mathbf{k}'}(\mathbf{r}) = e^{i\mathbf{k}'\mathbf{r}}, \qquad E' = \frac{\hbar^2 k'^2}{2\mu}, \tag{2.6}$$

then the Green function is described by the integral

$$G_0(E; \mathbf{r} - \mathbf{r}') = \int \frac{d\mathbf{k}'}{(2\pi)^3} \frac{\exp[i\mathbf{k}(\mathbf{r} - \mathbf{r}')]}{E - E'}. \tag{2.7}$$

Integrating (2.7) over the angle variables yields

$$G_0(E; \mathbf{r} - \mathbf{r}') = \frac{\mu}{2\pi^2 i\hbar^2} \frac{1}{|\mathbf{r} - \mathbf{r}'|} \int_{-\infty}^{\infty} dk' k' \frac{\exp(ik'|\mathbf{r} - \mathbf{r}'|)}{k^2 - k'^2}, \tag{2.8}$$

where $E = \hbar^2 k^2 / 2\mu$. Integration over k' may be carried out by means of the residue method; the boundary conditions determine the procedure of encircling the poles $k' = \pm k$ of the integrand. In order to obtain a diverging wave at large distances, we take the integration contour in the plane of the complex variable k' as shown in Fig. 2.1a. Since the integral over the circle of infinitely large radius vanishes, the integral (2.8) is equal to $2\pi i$ times

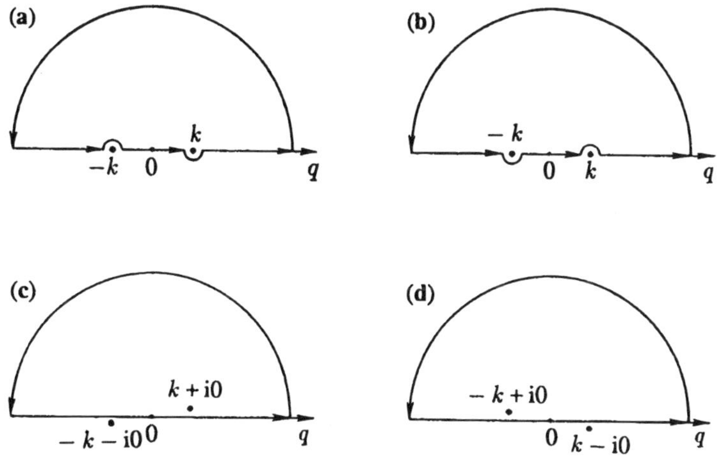

Figure 2.1. Integration contours for the Green function $G_0^{(+)}(E;\mathbf{r}-\mathbf{r}')$ *and* $G_0^{(-)}(E;\mathbf{r}-\mathbf{r}')$

to residue at the only pole enclosed by the integration contour. Thus we obtain

$$G_0^{(+)}(E;\mathbf{r}-\mathbf{r}') = -\frac{\mu}{2\pi\hbar^2}\frac{\exp(ik|\mathbf{r}-\mathbf{r}'|)}{|\mathbf{r}-\mathbf{r}'|}. \tag{2.9}$$

In order to calculate the Green function associated with the converging wave, we take the integration contour as given in Fig. 2.1b. Then

$$G_0^{(-)}(E;\mathbf{r}-\mathbf{r}') = -\frac{\mu}{2\pi\hbar^2}\frac{\exp(-ik|\mathbf{r}-\mathbf{r}'|)}{|\mathbf{r}-\mathbf{r}'|}. \tag{2.10}$$

Another way to find how to encircle the poles when calculating the Green functions, is substitute $E+i\mathcal{O}$ and $E-i\mathcal{O}$ for E in the denominator of (2.7) for the Green function associated with diverging and converging waves, respectively (Fig. 2.1c and Fig. 2.1d). Then we have

$$G_0^{(\pm)}(E;\mathbf{r}-\mathbf{r}') = \int \frac{d\mathbf{k}}{(2\pi)^3}\frac{\exp[i\mathbf{k}(\mathbf{r}-\mathbf{r}')]}{E-E'\pm i\mathcal{O}}, \tag{2.11}$$

and the integration contour which consists of the real axis and the semicircle of infinite radius in the upper half-plane, contains either the pole $k'=k+i\mathcal{O}$ in the case of $G_0^{(+)}$, or the pole $k'=-k+i\mathcal{O}$ in the case of $G_0^{(-)}$.

Taking φ in (2.3) to be the plane wave corresponding to the incident particle with momentum $\mathbf{k}$ and the Green function to be the converging wave (2.9), we rewrite (2.3) as

$$\psi_{\mathbf{k}}^{(+)}(\mathbf{r}) = \varphi_{\mathbf{k}}(\mathbf{r}) - \frac{\mu}{2\pi\hbar^2}\int d\mathbf{r}'\frac{\exp(ik|\mathbf{r}-\mathbf{r}'|)}{|\mathbf{r}-\mathbf{r}'|}V(\mathbf{r}')\psi_{\mathbf{k}}^{(+)}(\mathbf{r}'). \tag{2.12}$$

This *integral equation* determines the wave function of the scattering problem. It is equivalent to the Schrödinger equation with the boundary conditions.

At large distances, the Green function (2.9) may be approximated by the expression

$$G_0^{(+)}(E;\mathbf{r}-\mathbf{r}') \to -\frac{\mu}{2\pi\hbar^2}\frac{\exp(ikr)}{r}\mathrm{e}^{-i\mathbf{k}\mathbf{r}'} \qquad (r\to\infty), \tag{2.13}$$

where $\mathbf{k}' = (\mathbf{r}/r)k$. Therefore, if the potential is nonzero within a finite region, then the *wave function asymptotic* is given by

$$\psi_{\mathbf{k}}^{(+)}(\mathbf{r}) \to \varphi_{\mathbf{k}}(\mathbf{r}) + f(\mathbf{k},\mathbf{k}')\frac{\exp(ikr)}{r} \qquad (r\to\infty), \tag{2.14}$$

where

$$f(\mathbf{k},\mathbf{k}') = -\frac{\mu}{2\pi\hbar^2}\int d\mathbf{r}\mathrm{e}^{-i\mathbf{k}'\mathbf{r}}V(\mathbf{r})\psi_{\mathbf{k}}^{(+)}(\mathbf{r}). \tag{2.15}$$

The coefficient $f(\mathbf{k},\mathbf{k}')$ before the diverging wave is usually referred to as the scattering amplitude. Thus, $\mathbf{k}'$ must be regarded as the final particle momentum. According to (2.15), the *scattering amplitude* $f(\mathbf{k},\mathbf{k}')$ depends on the energy of relative motion, the angle between the vectors $\mathbf{k}$ and $\mathbf{k}'$, and the scattering potential.

Partial wave expansion. Let us consider in detail the problem of particle scattering in the central field, assuming that the potential V depends only on the modulus of the distance r. Since $H_0 = -\frac{\hbar^2}{2\mu}\Delta$, (2.1) may be rewritten as

$$\{\Delta - v(r) + k^2\}\psi_{\mathbf{k}} = 0, \tag{2.16}$$

where $k^2 = \frac{2\mu}{\hbar^2}E$ and $v(r) = \frac{2\mu}{\hbar^2}V(r)$ will be referred to as the *reduced potential.* The Laplace operator in the spherical coordinate system is

$$\Delta = \frac{1}{r^2}\left(\frac{\partial}{\partial r}r^2\frac{\partial}{\partial r}\right) - \frac{\mathbf{L}^2}{r^2},$$

where $\mathbf{L}^2$ is the operator of the squared angular momentum,

$$\mathbf{L}^2 = -\left\{\frac{1}{\sin\vartheta}\frac{\partial}{\partial\vartheta}\left(\sin\vartheta\frac{\partial}{\partial\vartheta}\right) + \frac{1}{\sin^2\vartheta}\frac{\partial^2}{\partial\varphi^2}\right\}.$$

Because central potentials are spherically symmetric, the angular momentum is an integral of motion. The states associated with different angular momenta take part in the scattering process independently. Therefore, it is convenient to write the solution $\psi_{\mathbf{k}}(\mathbf{r})$ of (2.16) as a superposition of partial waves associated with definite values of the angular momentum. Taking the z axis to be directed along the incident particle momentum $\mathbf{k}$, we can write the *partial wave expansion of the wave function* $\psi_{\mathbf{k}}(\mathbf{r})$ as

$$\psi_{\mathbf{k}}(\mathbf{r}) = \frac{1}{4\pi}\sum_{l}(2l+1)i^l\psi_{l,k}(r)P_l(\cos\vartheta), \tag{2.17}$$

where $P_l(\cos\vartheta)$ is the eigenfunction of the operator of squared angular momentum $\mathbf{L}^2$, i.e.,

$$\mathbf{L}^2 P_l(\cos\vartheta) = l(l+1)P_l(\cos\vartheta). \tag{2.18}$$

The coefficient in expansion (2.17) is chosen so that the function $\psi_{\mathbf{k}}(\mathbf{r})$ transforms into the plane wave $\psi^0_{\mathbf{k}}(\mathbf{r}) = \mathrm{e}^{i\mathbf{kr}}$, when there is no interaction. By using the expansion of the plane wave in terms of the partial waves, we obtain

$$\psi^0_{l,k}(r) = 4\pi j_l(kr). \tag{2.19}$$

Substituting (2.17) into (2.16) and making use of (2.18), we obtain a differential equation for the radial functions $\psi_{l,k}(r)$,

$$\left\{\frac{1}{r^2}\frac{d}{dr}(r^2\frac{d}{dr}) - \frac{l(l+1)}{r^2} - v(r) + k^2\right\}\psi_{l,k}(r) = 0 \tag{2.20}$$

This equation must be supplemented with the boundary condition for $\psi_{l,k}(r)$ which follows from (2.14). In order to derive it, we make use of the asymptotics of the plane wave radial part expansion

$$\psi^0_{l,k}(r) \equiv 4\pi j_l(kr) \underset{r\to\infty}{\longrightarrow} 4\pi\frac{\sin(kr - \frac{l\pi}{2})}{kr},$$

or

$$\psi^0_{l,k}(r) \underset{r\to\infty}{\longrightarrow} \frac{2\pi i}{kr}\left\{\mathrm{e}^{-i(kr-\frac{l\pi}{2})} - \mathrm{e}^{i(kr-\frac{l\pi}{2})}\right\}. \tag{2.21}$$

The first term within the curly braces describes the spherical wave that converges to the center, the second term is the diverging spherical wave. Far from the center the radial wave function $\psi_{l,k}$ similarly to $\psi^0_{l,k}$, is a superposition of spherical waves converging to and diverging from the center. The *particle-potential interaction* modifies only the diverging wave amplitude, whereas the converging wave amplitude is governed by the incident flux density. The amplitude variation of the wave diverging from the scattering center is described by the diagonal element of the *scattering matrix* $S_l(k)$ corresponding to the angular momentum quantum number l. Therefore, the required asymptotics of the solution of (2.20) must be

$$\psi_{l,k}(r) \xrightarrow[r\to\infty]{} \frac{2\pi i}{kr}\left\{e^{-i(kr-\frac{l\pi}{2})} - S_l(k)e^{i(kr-\frac{l\pi}{2})}\right\}. \tag{2.22}$$

A difference between amplitudes of the divergent waves in (2.21) and (2.22) determines the so-called *partial scattering amplitude*

$$f_l(k) = \frac{i}{2k}\{1 - S_l(k)\}. \tag{2.23}$$

The *total scattering amplitude* $f(\mathbf{k}, \mathbf{k}')$ can be presented as a sum:

$$f(\mathbf{k}, \mathbf{k}') = \sum_l (2l+1) f_l(k) P_l(\cos\vartheta). \tag{2.24}$$

In general, the matrix elements S_l are complex. If only elastic scattering occurs, then S_l is a function of the real *scattering phase shift* $\delta_l(k)$, i.e.,

$$S_l(k) = e^{2i\delta_l(k)}. \tag{2.25}$$

Usually one describes the scattering phase shift by an expression that vanishes as $k \to \infty$.

Substituting (2.25) into (2.22) yields the radial function asymptotic in the form

$$\psi_{l,k}(r) \xrightarrow[r\to\infty]{} 4\pi \exp(i\delta_l)\,\frac{\sin(kr - \frac{l\pi}{2} + \delta_l)}{kr}. \tag{2.26}$$

Comparing (2.26) to (2.21), we find that interaction is responsible for the phase shift δ_l in the asymptotics of the radial particle wave, which does not occur in the free wave asymptotics.

Bound states of a two-particle system. Depending on the interaction type, equation (2.2) may have solutions also for definite negative energies $E_n < 0$, (n=0,1,2,...), which form a discrete set. These solutions correspond to the *bound states of the system.* The bound state wave functions must satisfy the square integrability condition, which, in turn, determines the negative energy spectrum.

Making use of the Green function (2.9) or (2.10), we reduce the Schrödinger equation for the bound states of the system to a *homogeneous integral equation*

$$\psi^{(n)} = G_0^{(\pm)}(E_n) V \psi^{(n)}. \tag{2.27}$$

Let us consider the solutions of equation (2.2) in the general case and introduce the following notation for the energy

$$E = \frac{\hbar^2 k^2}{2\mu}.$$

In what follows we present the solutions as expansions in terms of partial waves. Let the radial function of the state with the angular momentum l be given by

$$\psi_{l,k}(r) = \frac{u_l(k,r)}{r}. \tag{2.28}$$

The equation for the radial function can be written as

$$\frac{d^2 u_l}{dr^2} + \left[k^2 - \frac{l(l+1)}{r^2} - v(r)\right] u_l = 0. \tag{2.29}$$

Different solutions of this equation are specified by the boundary conditions at the point $r = 0$ or at infinity $r \to \infty$. Suppose the quantity k is complex. Let us consider the dependence of different solutions on k. Moreover, we assume that the singularity of the potential $V(r)$ at $r = 0$ is weaker than r^{-2}, and $V(r)$ decreases at infinity faster than r^{-3}.

The *regular solution* of equation (2.29), $\Phi_l(k,r)$, which satisfies the boundary conditions at the point $r = 0$,

$$\Phi_l(k,r) \underset{r\to 0}{\longrightarrow} r^{l+1}, \tag{2.30}$$

is an analytic function without singularities in the open plane of the complex variable k for any value of r. Besides that, we consider the solutions of equation (2.29) $\mathrm{f}_l(k,r)$ and $\mathrm{f}_l(-k,r)$, which have the asymptotic properties

$$\lim_{r\to\infty} \mathrm{e}^{ikr} \mathrm{f}_l(k,r) = 1. \tag{2.31}$$

These solutions are usually called the *Jost solutions* . It is not difficult to prove that the solutions $\mathrm{f}_l(k,r)$ and $\mathrm{f}_l(-k,r)$ are independent since the relevant Wronskian is not equal to zero, i.e.,

$$W[\mathrm{f}_l(k,r), \mathrm{f}_l(-k,r)] = 2ik. \tag{2.32}$$

The solution $\mathrm{f}_l(k,r)$ is an analytic function with no singularities in the lower half-plane of the complex variable k ($Imk < 0$); it is continuous along the real axis k.

We define now the *Jost function* $\mathrm{f}_l(k)$ by the equality

$$\mathrm{f}_l(k) \equiv W[\mathrm{f}_l(k,r), \Phi_l(k,r)]. \tag{2.33}$$

It is clear that the Jost function $\mathrm{f}_l(k)$, similarly to the Jost solution $\mathrm{f}_l(k,r)$, is an analytic function with no singularities in the lower half-plane of the complex variable k. The regular solution $\Phi_l(k,r)$ can be expressed in terms of the Jost solutions $\mathrm{f}_l(k,r)$ and $\mathrm{f}_l(-k,r)$ as given by

$$\Phi_l(k,r) = -\frac{1}{2ik}\left\{\mathrm{f}_l(-k)\mathrm{f}_l(k,r) - \mathrm{f}_l(k)\mathrm{f}_l(-k,r)\right\}. \tag{2.34}$$

The asymptotics of the expression (2.34) at infinity takes the form

$$\Phi_l(k,r)\underset{r\to\infty}{\longrightarrow} -\frac{1}{2ik}\left\{\mathrm{f}_l(-k)\mathrm{e}^{-ikr} - \mathrm{f}_l(k)\mathrm{e}^{ikr}\right\}. \tag{2.35}$$

At the same time, in the case of scattering the asymptotics of the radial function ψ, normalized to the unit amplitude of the incident wave, is given by

$$\psi_{l,k}(r)\underset{r\to\infty}{\longrightarrow}\frac{2\pi i}{kr}\left\{\exp[-i(kr-\frac{l\pi}{2}) - S_l(k)\exp[i(kr-\frac{l\pi}{2})]\right\}. \tag{2.36}$$

Comparing this expression to (2.35), we obtain the relation between *the scattering matrix* $S_l(k)$ and the Jost functions $\mathrm{f}_l(k)$ and $\mathrm{f}_l(-k)$, i.e.,

$$S_l(k) = (-1)^l\frac{\mathrm{f}_l(k)}{\mathrm{f}_l(-k)}. \tag{2.37}$$

The relation between the normalized wave function $\psi_{l,k}(r)$ and the regular solution $\Phi_l(k,r)$ is given by the formula

$$\psi_{l,k}(r) = (-1)^l\frac{4\pi}{\mathrm{f}_l(-k)}\cdot\frac{\Phi_l(k,r)}{r}. \tag{2.38}$$

Zeros of the function $\mathrm{f}_l(k)$ in the lower half-plane of the complex variable k correspond to the bound states of the system. These zeros lie in the negative part of the imaginary axis, i.e.,

$$\mathrm{f}_l(k_n) = 0, \qquad k_n = -i\kappa_n, \qquad \kappa_n > 0. \tag{2.39}$$

Zeros of the function $\mathrm{f}_l(k)$ correspond to a *discrete set of energy* levels

$$E_n = -\frac{\hbar^2\kappa_n^2}{2\mu}. \tag{2.40}$$

Each level, in turn, is associated with the wave function $\Phi_l(-i\kappa_n,r)$ which is a solution of (2.29) and satisfies the square integrability condition. Therefore, it describes a bound state of the system. The functions $\Phi_l(-i\kappa_n,r)$ vanish at $r=0$ and have asymptotics

$$\Phi_l(-i\kappa_n, r) \underset{r\to\infty}{\longrightarrow} \frac{1}{\gamma_n} e^{-\kappa_n r}. \tag{2.41}$$

Using (2.35) the coefficient γ_n can easily be found to be

$$\gamma_n = -\frac{2\kappa_n}{f_l(i\kappa_n)}.$$

Note that $\Phi_l(-i\kappa_n, r)$ are real functions.

The zeros of the function $f_l(k)$ are simple since the derivative $df(k_n)/dk_n$ does not vanish

$$\left.\frac{df_l(k)}{dk}\right|_{k_n} = -2k_n\gamma_n \int\limits_0^\infty dr \Phi_l^2(k_n, r) \neq 0. \tag{2.42}$$

It is clear that $f_l(-k)$ is analytic in the upper half-plane of complex k and possesses simple zeros located on the positive imaginary semi-axis. These zeros correspond to the bound states of the system. According to the definition (2.37) the zeros of the function $f_l(-k)$ correspond to the poles of the scattering matrix $S_l(k)$. Thus, each bound state of the system is associated with a zero of the scattering matrix located on the negative imaginary semi-axis, and a symmetric (with respect to the real axis) pole of the scattering matrix on the positive imaginary semi-axis.

We denote the *normalized wave function of the bound state* by $u_l^{(n)}(r)$,

$$u_l^{(n)}(r) = C_n f_l(-i\kappa_n, r). \tag{2.43}$$

Applying (2.42), we have

$$C_n^2 = -\frac{f_l(i\kappa_n)}{\frac{df_l(-i\kappa_n)}{d\kappa_n}}. \tag{2.44}$$

Making use of the scattering matrix definition (2.37), we obtain the normalization constant C_n in the integral form,

$$C_n^2 = \frac{(-1)^l}{2\pi} \oint dk S_l(k), \tag{2.45}$$

the integration contour being a small circle that goes clockwise around the pole $k = i\kappa_n$. The *scattering matrix* in the vicinity of the pole may be written as

$$S_l(k) \simeq \frac{b_n}{k - i\kappa_n} \qquad (k \to i\kappa_n).$$

According to (2.45), the residue b_n is determined by the normalization constant of the bound state wave function, i.e.,

$$b_n = (-1)^{l+1} i C_n^2. \tag{2.46}$$

According to (2.28) and (2.43), a radial function of the bound states can be presented as

$$\psi_l^{(n)} = C_n \left\{ \frac{\mathrm{e}^{-\kappa_n r}}{r} + q_n(r) \right\}, \tag{2.47}$$

where

$$q_n(r) = \frac{1}{r}(\mathrm{f}_l(-i\kappa_n, r) - e^{-\kappa_n r})$$

is the short range or interior part of $\psi_l^{(n)}(r)$ that for large r falls off more rapidly than $\mathrm{e}^{-\kappa_n r/r}$. At the same time making use of (2.27), the *wave function of the bound state* in the momentum representation can be presented as

$$\langle \mathbf{k}|nl\rangle = -\frac{\langle \mathbf{k}|v(r)|nl\rangle}{k^2 + \kappa_n^2}, \tag{2.48}$$

where $\mathbf{k}$ is the appropriate relative momentum. The energy dominator of (2.48) has a zero at the on-shell point $k^2 = -\kappa_n^2$, while the numerator, which we call the vertex function, depends on the potential. This form for the bound state of interaction dependent vertex divided by an energy dominator that vanishes at the on-shell point is very general. The pole of (2.48) at $k^2 = -\kappa_n^2$ is related to the large r behavior of the coordinate space wave function. The asymptotic part of (2.47) comes only from the pole of (2.48). In terms of the momentum space wave function, the shorter range parts of (2.47), $q_n(r)$, translate into singularities of (2.48) that are further from the physical region ($k^2 > 0$), than the pole (Amado, 1979). From (2.48) we see that these come from the vertex function and therefore, must carry the potential range. Knowing the wave function completely is therefore equivalent to knowing C_n and $q_n(r)$. The normalization constant C_n is easy to determine empirically while information on the interior region, that is knowledge of $q_n(r)$ is far more difficult to get without knowing a good deal about the dynamics.

The Lippmann-Schwinger equation. It is convenient to rewrite the results obtained in a symbolic form that is suitable for extensions to many-particle systems. Let us consider a quantum mechanical system of several interacting particles. We assume the Hamiltonian to be a sum of two parts

$$H = H_0 + V,$$

where H_0 describes the nonperturbed motion of the system and V is the interaction which vanishes if the distance between the interacting parts of the system is sufficiently large.

In the stationary formulation, the scattering problem reduces to the Schrödinger equation

$$(E - H)\psi = 0 \tag{2.49}$$

with fixed *boundary conditions*, the energy E of the system being positive. At infinity, the solution ψ must be a sum of the incident wave φ, governed by the nonperturbed equation

$$(E - H_0)\varphi = 0, \tag{2.50}$$

and the diverging scattered wave.

The solution of (2.49) with the above mentioned boundary condition formally may be written as

$$\psi^{(+)} = \varphi + G_0^{(+)}(E)V\psi^{(+)}, \tag{2.51}$$

where $G_0^{(+)}(E)$ is the *Green function* of the nonperturbed equation (2.50),

$$G_0^{(+)}(E) = \frac{1}{E - H_0 + i\mathcal{O}}. \tag{2.52}$$

In fact, $G(E)$ is the integral operator whose kernel is a Green function for the nonperturbed equation (2.50). The choice of the integration contour enclosing the pole of (2.52) follows from the requirement that the asymptotics of $\psi^{(+)}$ must contain a diverging scattered wave.

The formal solution (2.51) of the Schrödinger equation (2.49) is an integral equation

$$\psi^{(+)} = \varphi + \frac{1}{E - H_0 + i\mathcal{O}}V\psi^{(+)}, \tag{2.53}$$

which is referred to as the *Lippmann-Schwinger equation* . Thus, the Lippmann-Schwinger equation itself consists of the boundary condition (at infinity) for the wave function, which is the most important property of this equation. It is clear, that the divergent wave in (2.51) is connected uniquely with the choice of the positive sign of the infinitesimally small imaginary addition to the energy E. The equation

$$\psi_\varepsilon = \varphi + \frac{1}{E - H_0 - i\varepsilon}V\psi_\varepsilon \tag{2.54}$$

has a unique solution for any finite ε. An ambiguity could occur if the homogeneous equation obtained from (2.54) for $\varphi = 0$ had a solution. However, the eigenvalues corresponding to the decreasing solutions of the homogeneous equation (in the asymptotic region for which $V \to 0$) must be real. Therefore, the equation

$$\psi_\varepsilon = \frac{1}{E - H_0 + i\varepsilon} V \psi_\varepsilon \tag{2.55}$$

has no solutions for any finite ε.

The scattering wave function whose asymptotics at infinity is a sum of the incident and converging scattered waves, satisfies the equation in which the integration contour must be taken in another way, i.e.,

$$\psi^{(-)} = \varphi + \frac{1}{E - H_0 - iO} V \psi^{(-)}, \tag{2.56}$$

because it is determined by the Green function

$$G_0^{(-)}(E) = \frac{1}{E - H_0 - iO}.$$

Equations (2.53) and (2.56) are written in the operator from. To write them explicitly, one has to expand the function $V\psi$ in terms of the eigenfunctions of the nonperturbed Hamiltonian H_0. This results in

$$\psi_\alpha^{(\pm)} = \varphi_\alpha + \sum_\beta \frac{\varphi_\beta(\varphi_\beta, V\psi_\alpha^{(\pm)})}{E_\alpha - E_\beta \pm iO}. \tag{2.57}$$

Thus, the *Green function* is given by

$$G_0^{(\pm)}(E) \ldots = \sum_\beta \frac{\varphi_\beta(\varphi_\beta, \ldots)}{E - E_\beta \pm iO}. \tag{2.58}$$

This relation determines the expansion of the Green function $G_0^{(\pm)}$ in terms of the complete set of eigenfunctions of the operator H_0. In the coordinate representation, we have

$$G_0^{(\pm)}(E; \mathbf{r}, \mathbf{r}') \equiv \langle \mathbf{r}' | G_0^{(\pm)}(E) | \mathbf{r} \rangle = \int \frac{d\mathbf{k}'}{(2\pi)^3} \frac{\varphi_{\mathbf{k}'}(\mathbf{r}) \varphi_{\mathbf{k}'}^*(\mathbf{r}')}{E - E' \pm iO}, \tag{2.59}$$

which is in agreement with (2.11). It is not difficult to verify that (2.57) in the coordinate representation reproduces (2.12).

Green functions $G_0^{(+)}$ and $G_0^{(-)}$ take the especially simple form in the momentum representation

$$\langle \mathbf{k}'|G_0^{(\pm)}(E)|\mathbf{k}\rangle = \frac{\delta(\mathbf{k}-\mathbf{k}')}{E - \frac{\hbar^2 k^2}{2\mu} \pm i\mathcal{O}}. \tag{2.60}$$

If the energies of relative motion are negative, then the solutions of the Schrödinger equation (2.49) are associated with the bound states of the system. In this case the Schrödinger equation reduces to the *homogeneous integral equation*

$$\psi = G_0^{(\pm)}(E)V\psi, \qquad E < 0. \tag{2.61}$$

In the general case, the *Lippmann-Schwinger equation* (2.51) has no ambiguous solutions if the system possesses bound states. Indeed, the energy levels

$$\mathcal{E} = \mathbf{P}^2/2M + E$$

($\mathbf{P}$ is the total momentum, M is the total mass, E is the energy of relative motion) are multiply degenerate because each state is associated with a certain energy distribution between the inner motion and motion of the system as a whole. Since existence of a bound state infers that the homogeneous equation (2.61) has a solution for fixed $\mathcal{E}$, the solution of the nonhomogeneous equation (2.51) becomes ambiguous. For a two-particle system, the Lippmann-Schwinger equation has an unique solution only in the c.m.s., for which $\mathbf{P} = 0$ and energy degeneration does not occur ($\mathcal{E} = E$).

The Green function G and the transition operator t. The solutions of (2.53) may be expressed in terms of the asymptotic function φ by means of the *Green function* of the equation (2.49)

$$G^{(+)}(E) = \frac{1}{E - H + i\mathcal{O}}. \tag{2.62}$$

We define the Green functions $G_0(z)$ and $G(z)$ of the complex argument z as

$$G_0(z) = \frac{1}{z - H_0}, \qquad G(z) = \frac{1}{z - H}. \tag{2.63}$$

(These functions are usually called resolvents of the operators H_0 and H.) Evidently,

$$G_0^{(\pm)}(E) = G_0(E \pm i\mathcal{O}), \qquad G^{(\pm)}(E) = G(E \pm i\mathcal{O}).$$

The Green functions $G(z)$ and $G_0(z)$ are related by the equation

$$G(z) = G_0(z) + G_0(z)VG(z). \tag{2.64}$$

It may be easily verified that the solution (2.53) may be written in the form

$$\psi^{(+)} = \lim_{\varepsilon\to 0} i\varepsilon G(E+i\varepsilon)\varphi. \tag{2.65}$$

Indeed, multiplying (2.64) for $z = E + i\varepsilon$ by $i\varepsilon$, applying it to the function φ, and making use of relation $\lim_{\varepsilon\to 0} i\varepsilon G_0(E+i\varepsilon)\varphi = \varphi$, one obtains (2.51).

The *Green functions* $G^{(\pm)}(E)$ may be expanded in terms of the complete set of eigenfunctions of the operator H in the same way as $G_0^{(\pm)}(E)$ were expanded in terms of eigenfunctions of the operator H_0. We denote the eigenfunctions of H associated with bound and continuum states as ψ_n and ψ_β, respectively. Suppose these functions form a complete set, so that

$$\sum_n \psi_n(\psi_n,\ldots) + \sum_\beta \psi_\beta(\psi_\beta,\ldots) = 1\ldots, \tag{2.66}$$

then, having applied the operator $G^{(\pm)}$ to this equality, one obtains the expansion

$$G^{(\pm)}(E)\ldots = \sum_n \frac{\psi_n(\psi_n,\ldots)}{E-E_n} + \sum_\beta \frac{\psi_\beta(\psi_\beta,\ldots)}{E-E_\beta \pm iO}, \tag{2.67}$$

with the eigenvalues E_n and E_β being negative and positive, respectively.

We rewrite (2.66, 67) in the coordinate representation for the simplest case of a particle in the external field. It is convenient to take $\psi_{\mathbf{k}}^{(+)}(\mathbf{r})$ or $\psi_{\mathbf{k}}^{(-)}(\mathbf{r})$ to be the continuum spectrum wave functions. Then the *completeness condition* takes the form

$$\sum_n \psi_n(\mathbf{r})\psi_n^*(\mathbf{r}') + \int \frac{d\mathbf{q}}{(2\pi)^3}\psi_{\mathbf{q}}(\mathbf{r})\psi_{\mathbf{q}}^*(\mathbf{r}') = \delta(\mathbf{r}-\mathbf{r}'), \tag{2.68}$$

and the Green function expansion is given by

$$\begin{aligned} G^{(\pm)}(E;\mathbf{r},\mathbf{r}') &\equiv \langle \mathbf{r}'|G^{(\pm)}(E)|\mathbf{r}\rangle \\ = \sum_n \frac{\psi_n(\mathbf{r})\psi_n^*(\mathbf{r}')}{E-E_n} &+ \int \frac{d\mathbf{q}}{(2\pi)^3}\frac{\psi_{\mathbf{q}}(\mathbf{r})\psi_{\mathbf{q}}^*(\mathbf{r}')}{E-E_q \pm iO}. \end{aligned} \tag{2.69}$$

In contrast to the free motion Green function $G_0^{(\pm)}(E;\mathbf{r},\mathbf{r}')$, the complete Green function $G^{(\pm)}(E;\mathbf{r},\mathbf{r}')$ describing the motion in the field $V(\mathbf{r})$ depends on the coordinates $\mathbf{r}$ and $\mathbf{r}'$ rather than on their difference $\mathbf{r}-\mathbf{r}'$.

It is not difficult to verify within the context of the explicit expression (2.69) that the Green function $G^{(\pm)}(E;\mathbf{r},\mathbf{r}')$ indeed satisfies the Schrödinger equation (2.49) with a point source in the right-hand part, i.e.,

$$(E-H)G^{(\pm)}(E;\mathbf{r},\mathbf{r}') = \delta(\mathbf{r}-\mathbf{r}'). \tag{2.70}$$

When considering the scattering process, it is convenient to use the *transition operator*. We introduce this operator t with the help of the equality

$$VG(z) = t(z)G_0(z). \tag{2.71}$$

Using (2.64) we obtain the following equation of the transition operator

$$t(z) = V + VG_0(z)t(z). \tag{2.72}$$

This equation is usually called the *Lippmann-Schwinger operator equation*

Acting by the both parts of the operator equation (2.72) on the free motion wave function of particles φ and comparing the result with the *Lippmann-Schwinger equation* (2.53) multiplied by the potential V from the left, we obtain

$$t(E)\varphi_{\mathbf{k}} = V\psi_{\mathbf{k}}. \tag{2.73}$$

Using this relation the solution of the Lippmann-Schwinger equation may be written in the form

$$\psi^{(+)} = \varphi + \frac{1}{E-H_0+iO}t\varphi. \tag{2.74}$$

In the coordinate representation, the second term in the right-hand part of (2.74) at large distances is a diverging scattered wave. The coefficient before the latter is just the scattering amplitude

$$f = -\frac{\mu}{2\pi\hbar^2}(\varphi', t\varphi),$$

or

$$f(\mathbf{k},\mathbf{k}') = -\frac{\mu}{2\pi\hbar^2}\langle\mathbf{k}|t|\mathbf{k}'\rangle. \tag{2.75}$$

Thus, the scattering amplitude is specified directly by a matrix element of the transition operator.

The scattering amplitude off the energy shell. Let us calculate the *scattering amplitude off the energy shell* by using the general relation between the scattering amplitude and the transition matrix. The transition operator t is governed by the Lippmann-Schwinger equation (2.72). In the momentum representation, this equation may be written as ($\hbar = 1$)

$$\langle \mathbf{k}'|t(z)|\mathbf{k}\rangle = \langle \mathbf{k}'|V|\mathbf{k}\rangle + \int \frac{d\mathbf{q}}{(2\pi)^3} \frac{\langle \mathbf{k}'|V|\mathbf{q}\rangle\langle \mathbf{q}|t(z)|\mathbf{k}\rangle}{z-(q^2/2\mu)}, \qquad (2.76)$$
$$z = E + iO,$$

where E is the energy of two-particle relative motion which is conserved under elastic scattering, μ is the reduced mass. Equation (2.76) determines the t matrix both on the energy shell, when the initial and final momenta are related to energy by

$$\frac{k^2}{2\mu} = \frac{k'^2}{2\mu} = E, \qquad (2.77)$$

and off the energy shell, when

$$\frac{k^2}{2\mu} \neq \frac{k'^2}{2\mu} \neq E. \qquad (2.78)$$

According to (2.75), the elastic scattering amplitude $f(\mathbf{k}, \mathbf{k}')$ can be expressed in terms of the t matrix on the energy shell, so that

$$f(\mathbf{k}, \mathbf{k}') = -\frac{\mu}{2\pi}\langle \mathbf{k}'|t(E+iO|\mathbf{k}\rangle. \qquad (2.79)$$

The *scattering amplitude off the energy shell*, $f(\mathbf{k}, \mathbf{k}'; z)$, can be found under the assumption that (2.79) holds when the condition (2.77) is violated.

Both the scattering amplitude $f(\mathbf{k}, \mathbf{k}'; z)$ and the transition matrix $\langle \mathbf{k}'|t(z)|\mathbf{k}\rangle$ possess singularities in the plane of complex energy z. These are poles associated with the discrete spectrum and the cut along the positive real semi-axis given rise to by the continuum spectrum of the system. The singularities can be explicitly derived from the so-called *spectral representation of the t matrix* ,

$$\langle \mathbf{k}'|t(z)|\mathbf{k}\rangle = \langle \mathbf{k}'|V|\mathbf{k}\rangle + \sum_N \frac{g_N(\mathbf{k}')g_N(\mathbf{k})}{z+(\kappa_N^2/2\mu)}$$
$$+\int \frac{d\mathbf{q}}{(2\pi)^3} \frac{\langle \mathbf{k}'|t[(q^2/2\mu)+iO]|\mathbf{q}\rangle\,\langle \mathbf{q}|t[(q^2/2\mu)-iO]|\mathbf{k}\rangle}{z-(q^2/2\mu}, \qquad (2.80)$$

where $g_N(\mathbf{k}) = [(k^2+\kappa_N^2)/2]\varphi_N(\mathbf{k})$, $\kappa_N^2/2\mu$ is the binding energy, $\varphi_N(\mathbf{k})$ is the bound state wave function in the momentum representation.

If the interaction is spherically symmetric, then the partial component expansion of the potential is given by

$$\langle \mathbf{k}'|V|\mathbf{k}\rangle = \sum_l (2l+1)V_l(k,k')P_l(\cos\vartheta)$$
$$= 4\pi\sum_{lm} V_l(k',k)Y^*_{lm}(\mathbf{n}')Y_{lm}(\mathbf{n}), \quad (2.81)$$

where $Y_{lm}(\mathbf{n})$ and $Y_{lm}(\mathbf{n}')$ are the spherical functions of the angles describing the directions of the vectors $\mathbf{k}$ and $\mathbf{k}'$; ϑ is the angle between $\mathbf{k}$ and $\mathbf{k}'$. Each term in (2.81) describes the interaction in the state with the angular momentum quantum number l. The partial component of the potential, $V_l(k',k)$, is given by

$$V_l(k',k) = 4\pi\int_0^\infty dr r^2 j_l(k'r)V(r)j_l(kr).$$

The t matrix and the *scattering amplitude* can be expanded in terms of partial components, too, so that

$$\langle \mathbf{k}'|t(z)|\mathbf{k}\rangle = \sum_l (2l+1)t_l(k',k;z)P_l(\cos\vartheta), \quad (2.82)$$

$$f(\mathbf{k},\mathbf{k}';z) = \sum_l (2l+1)f_l(k,k';z)P_l(\cos\vartheta). \quad (2.83)$$

Here $t_l(k',k;z)$ is the partial transition matrix and $f_l(k,k';z)$ is the partial *scattering amplitude off the energy shell* . The equation for the partial transition matrix readily follows from (2.72), i.e.,

$$t_l(z) = V_l + V_l\frac{1}{z-H_0}t_l(z), \quad (2.84)$$

or

$$t_l(k',k;z) = V_l(k',k)$$
$$+ \frac{1}{2\pi^2}\int_0^\infty dq q^2 V_l(k',q)\frac{1}{z-(q^2/2\mu)}t_l(q,k;z). \quad (2.85)$$

The partial scattering amplitude and the partial transition matrix satisfy a relation analogous to (2.79):

$$f_l(k,k';z) = -\frac{\mu}{2\pi}t_l(k',k;z). \quad (2.86)$$

The *partial scattering amplitude on the energy shell* is a function of the scattering phase shift $\delta_l(k)$, i.e.,

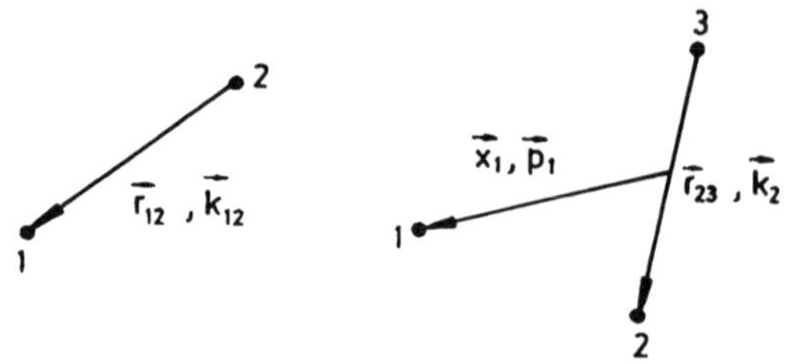

Figure 2.2. Relative coordinates and momenta in the two- and three-particle systems.

$$f_l\left(k, k'; \frac{k^2}{2\mu}\right) \equiv f_l(k) = \frac{1}{k} e^{i\delta_l} \sin \delta_l. \tag{2.87}$$

Two-particle scattering is completely described by the given scattering amplitude on the energy shell because, according (1.20), the scattering cross-section is directly expressed in terms of the phase shifts.

2.2. Quantum Mechanical Description of Three-particle Systems

Pecularities of the three-particle energy spectrum. When passing from two-particle to three-particle systems, some difficulties arise, which are associated both with the kinematics and dynamics of systems containing more than two particles. One can always pass to the center-of-mass system and thus reduce the two-particle problem to the one-particle one by considering the particle motion in the field of some potential. If the nuclear range is finite, the asymptotic scattering phase determines the wave function behavior in the whole space (except the finite interaction region). All observable characteristics of the process may be expressed directly in terms of the phases. To describe bound states and scattering in a two-particle system, it is convenient to use the Lippmann-Schwinger equation. Unfortunately, this approach is inappropriate in the case of three-particle systems. In the general case, this is due to the ambiguity of solutions of the Lippmann-Schwinger equation by virtue of peculiarities of the many-particle energy spectrum. In order to reveal the difference, we consider the simplest two- and three-particle systems. Naturally, the two systems differ in the number of independent parameters required for their description. By virtue of the translation invariance, it is convenient to describe both systems in the c.m.s. (Fig. 2.2).

The two-particle system may be treated in terms of a single relative-distance vector or the momentum:

$$\mathbf{r}_{12} = \mathbf{r}_1 - \mathbf{r}_2, \qquad \mathbf{k}_{12} = \frac{m_2\mathbf{k}_1 - m_1\mathbf{k}_2}{m_1 + m_2} \tag{2.88}$$

($\mathbf{r}_1$ and $\mathbf{r}_2$ are coordinates, $\mathbf{k}_1$ and $\mathbf{k}_2$ are momenta of individual particles having masses m_1 and m_2 respectively); the three-particle system is described by two relative-distance vectors

$$\mathbf{x}_1 = \mathbf{r}_1 - \frac{m_2\mathbf{r}_2 + m_3\mathbf{r}_3}{m_2 + m_3}, \qquad \mathbf{r}_{23} = \mathbf{r}_2 - \mathbf{r}_3, \tag{2.89}$$

or two momentum vectors

$$\mathbf{p}_1 = \frac{(m_2 + m_3)\mathbf{k}_1 - m_1(\mathbf{k}_2 + \mathbf{k}_3)}{m_1 + m_2 + m_3}, \qquad \mathbf{k}_{23} = \frac{m_3\mathbf{k}_2 - m_2\mathbf{k}_3}{m_2 + m_3} \tag{2.90}$$

($\mathbf{r}_1$, $\mathbf{r}_2$ and $\mathbf{r}_3$ are coordinates, $\mathbf{k}_1$, $\mathbf{k}_2$ and $\mathbf{k}_3$ are momenta of individual particles with masses m_1, m_2 and m_3 respectively). This difference is, however, purely qualitative and not too important. The essential difference between the two systems under consideration is manifested in their energy spectra (Fig. 2.3).

The *two-particle energy spectrum* is very simple. It consists of two clearly distinguished parts: the discrete levels associated with bound states of the system, and the continuum range corresponding to the scattering states.

The *three-particle spectrum* is much more complicate. It consists of:

a) discrete levels corresponding to the bound states of the three-particle system;

b) continuous level sequences associated with particle scattering by a bound state of two other particles (each sequence begins from the energy determined by the binding energy of the two-particle subsystem);

c) continuum spectral range given rise to by the three-particle scattering states (this range begins from energies above the three-particle break up threshold $E = 0$).

An important peculiarity of the three-particle energy spectrum is its degeneracy. The same total energy E can be associated with different physical states of the system. This makes the Lippmann-Schwinger equation inapplicable for the description of three-particle systems. Indeed, the two-particle bound-states correspond to the solutions of the homogeneous equation $\psi = G_0 V\psi$ (G_0 is the Green function of the system without interaction), while the *nonhomogeneous Lippmann-Schwinger equation*

$$\psi = \varphi + G_0 V\psi$$

allows, for positive energies, an unique solution that determines the scattering states. The system of three or more particles possesses, along with the

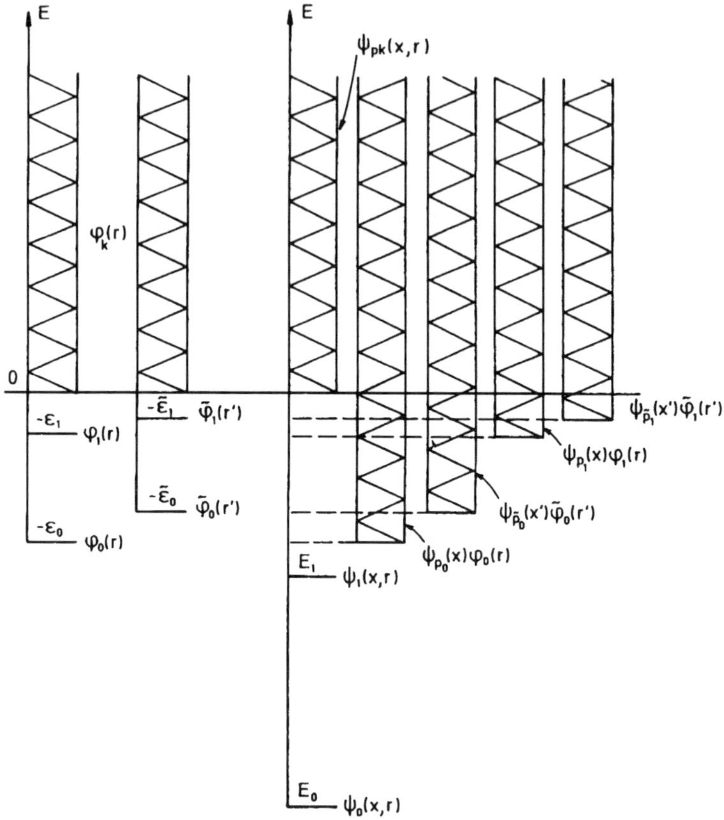

Figure 2.3. Energy spectra of two- and three-particle systems.

bound states of the whole system, also bound states of fewer-particle subsystems and thus the solutions of the inhomogeneous Lippmann-Schwinger equations are always ambiguous, and only the homogeneous equation for the bound state of the whole system in the c.m.s. has unique solutions. Ambiguity can be eliminated by appropriate rearrangement of the Lippmann-Schwinger equations. The resultant equations are usually referred to as the *Faddeev equations* (Faddeev, 1960).

The Faddeev equations. Let us derive the integral Faddeev equations for a system of three nonrelativistic spinless particles. The total Hamiltonian of the system may be written as

$$H = H_0 + V, \tag{2.91}$$

where H_0 is the particle kinetic energy operator, and V is the interaction potential. Taking into account only two-particle forces, we can write V as a sum of three terms

$$V = V_{12} + V_{23} + V_{31}, \tag{2.92}$$

where V_{ij} describes interaction between the particles i and j which decreases with increasing distance between the particles.

We consider three particles with masses m_1, m_2 and m_3. To describe three-particle relative motion, we introduce the coordinates $\mathbf{x}_1$ and $\mathbf{r}_{23}$. Instead of relative coordinates $\mathbf{x}_1$ and $\mathbf{r}_{23}$, the pairs $\mathbf{x}_2$ and $\mathbf{r}_{31}$ or $\mathbf{x}_3$ and $\mathbf{r}_{12}$ can be used:

$$\begin{aligned} \mathbf{x}_2 &= -\frac{m_1}{m_1+m_3}\mathbf{x}_1 + \frac{m_3(m_1+m_2+m_3)}{(m_1+m_3)(m_2+m_3)}\mathbf{r}_{23}, \\ \mathbf{r}_{31} &= -\mathbf{x}_1 - \frac{m_2}{m_2+m_3}\mathbf{r}_{23}; \end{aligned} \tag{2.93}$$

or

$$\begin{aligned} \mathbf{x}_3 &= -\frac{m_1}{m_1+m_2}\mathbf{x}_1 - \frac{m_2(m_1+m_2+m_3)}{(m_1+m_2)(m_2+m_3)}\mathbf{r}_{23}, \\ \mathbf{r}_{12} &= \mathbf{x}_1 - \frac{m_3}{m_2+m_3}\mathbf{r}_{23}. \end{aligned} \tag{2.94}$$

In the c.m.s., the kinetic energy operator of the system can be written in terms of the coordinates $\mathbf{x}_1$ and $\mathbf{r}_{23}$, so that

$$H_0 = -\frac{1}{2\mu_1}\Delta_1 - \frac{1}{2\mu_{23}}\Delta_{23}, \tag{2.95}$$

with the reduced masses being

$$\mu_1 = \frac{m_1(m_2+m_3)}{m_1+m_2+m_3}, \qquad \mu_{23} = \frac{m_2 m_3}{m_2+m_3}.$$

In the momentum representation, H_0 takes the form

$$H_0 = \frac{\mathbf{p}_1^2}{2\mu_1} + \frac{\mathbf{k}_{23}^2}{2\mu_{23}}. \tag{2.96}$$

The *transition operator* T is determined by the Lippmann-Schwinger equation

$$T(Z) = V + VG_0(Z)T(Z), \tag{2.97}$$

where

$$G_0(Z) = (Z - H_0)^{-1}, \qquad Z = E + iO.$$

Note that the kernel of the integral equation (2.97) is singular by virtue of delta functions which appear due to momentum conservation for the particle that is not involved in interaction with the other particles forming the pair.

In the case of two-particle interaction given by (2.92), the transition operator T can also be written as a sum

$$T(Z) = T^{(1)}(Z) + T^{(2)}(Z) + T^{(3)}(Z), \tag{2.98}$$

the individual terms of which are defined by

$$T^{(k)}(Z) = V_{ij} + V_{ij}G_0(Z)T(Z), \qquad ijl = 123, 231, 312. \tag{2.99}$$

Having substituted (2.98) for $T(Z)$ in the right-hand part of (2.99), we can treat these relations as a system of coupled operator equations for the individual terms of (2.98). Note that the iteration series $T^{(k)}(Z)$ in (2.99) contains both singular terms (of the type $V_{23}G_0(Z)V_{23}$, $V_{23}G_0(Z)V_{23}G_0(Z)V_{23}$, so on) and the terms in which the delta functions are eliminated by the intermediate integration (of the type $V_{23}G_0(Z)V_{31}$ etc.). The resultant system (2.99) is equivalent to (2.97) and therefore its solutions are ambiguous, similarly to those of the Lippmann-Schwinger equation.

In order to eliminate ambiguity, we rearrange the system (2.99) regarding the *two-particle transition operators* T_{ij} as being known. Suppose the operators T_{ij} are determined by

$$T_{ij}(Z) = V_{ij} + V_{ij}G_0(Z)T_{ij}(Z), \tag{2.100}$$

which follows from (2.97) if interaction between the particles i, j and the third particle is neglected. Note that the right-hand part of (2.100) contains all the singularities of (2.99) and hence can serve to exclude them. Separating in (2.99) the diagonal part

$$[1 - V_{ij}G_0(Z)]T^{(k)}(Z) = V_{ij} + V_{ij}G_0(Z)[T^{(i)}(Z) + T^{(j)}(Z)],$$
$$(ijk = 123, 231, 312)$$

inverting the two-particle operator $[1 - V_{ij}G_0(Z)]$, and then making use of (2.100), we rewrite the system of coupled equations for individual terms of the transition operator as

$$\begin{aligned} T^{(k)}(Z) = T_{ij}(Z) \quad &+ \quad T_{ij}(Z)G_0(Z)[T^{(i)}(Z) + T^{(j)}(Z)], \\ (ijk \quad &= \quad 123, 231, 312). \end{aligned} \tag{2.101}$$

In contrast to (2.99), this system of integral equations has a single-valued solution. The iteration series of the second term in the right-hand part of (2.101) contains no singularities. Therefore, the system of integral equations (2.101) can be solved by the Fredholm methods.

Using (2.71) for the relation between the transition operator T and the *Green function* G we obtain a system of equations for G from (2.101). Within the context of (2.98), we have

$$G(Z) = G_0(Z) + G^{(1)}(Z) + G^{(2)}(Z) + G^{(3)}(Z), \tag{2.102}$$
$$G^{(i)}(Z) = G_0(Z)T^{(i)}(Z)G_0(Z), \qquad i = 1, 2, 3.$$

The functions $G^{(i)}$ satisfy the equations

$$G^{(k)}(Z) = G_{ij}(Z) - G_0(Z) + G_0(Z)T_{ij}(Z)[G^{(i)}(Z) + G^{(j)}(Z)], \tag{2.103}$$
$$ijk = 123, 231, 312,$$

where

$$G_{ij}(Z) = G_0(Z) + G_0(Z)T_{ij}(Z)G_0(Z). \tag{2.104}$$

The equations for the corresponding wave function of the system can be derived from (2.103) by using (2.65).

In a three-particle system, the range of motion of all three particles as well as of one particle relative to the bound state of the other two can be infinite. We denote the *asymptotic wave functions* as Φ_{123} and Φ_i, where i=1,2, or 3. (The subscripts label unbound particles. All three particles are assumed to be different.)

If particle 1 is scattered by a bound state of particles 2 and 3, then the asymptotic wave function is Φ_1. We apply (2.102), multiplied by $i\varepsilon$, to the function Φ_1 and use (2.65). Then we obtain an equation for the wave function of the system Ψ_1, i.e.,

$$\Psi_1 = \Psi_1^{(1)} + \Psi_1^{(2)} + \Psi_1^{(3)},$$

$$\Psi_1^{(1)} = \Phi_1 + G_0(Z)T_{23}(Z)[\Psi_1^{(2)} + \Psi_1^{(3)}],$$
$$\Psi_1^{(2)} = G_0(Z)T_{31}(Z)[\Psi_1^{(3)} + \Psi_1^{(1)}], \tag{2.105}$$
$$\Psi_1^{(3)} = G_0(Z)T_{12}(Z)[\Psi_1^{(1)} + \Psi_1^{(2)}],$$

$$Z = E + iO.$$

The functions Ψ_2 and Ψ_3 are governed by analogous systems of equations.

We apply (2.102), multiplied by $i\varepsilon$, to the function Φ_{123} and use (2.65). Then we obtain an equation for the wave function of the system of three unbound particles Ψ_{123}, i.e.,

$$\Psi_{123} = \Phi_{123} + \Psi^{(1)}_{123} + \Psi^{(2)}_{123} + \Psi^{(3)}_{123},$$

$$\begin{aligned} \Psi^{(1)}_{123} &= \Phi_{1(23)} - \Phi_{123} + G_0(Z)T_{23}(Z)[\Psi^{(2)}_{123} + \Psi^{(3)}_{123}], \\ \Psi^{(2)}_{123} &= \Phi_{2(31)} - \Phi_{123} + G_0(Z)T_{31}(Z)[\Psi^{(3)}_{123} + \Psi^{(1)}_{123}], \\ \Psi^{(3)}_{123} &= \Phi_{3(12)} - \Phi_{123} + G_0(Z)T_{12}(Z)[\Psi^{(1)}_{123} + \Psi^{(2)}_{123}], \end{aligned} \tag{2.106}$$

$$Z = E + iO,$$

where

$$\Phi_{1(23)} = \lim_{\varepsilon \to 0} i\varepsilon G_{23}(E + i\varepsilon)\Phi_{123}, \ldots.$$

The *function* $\Phi_{1(23)}$, in contrast to Φ_{123}, takes into account interaction between the particles 2 and 3. It is not difficult to verify that at large distances the difference $\Phi_{1(23)} - \Phi_{123}$ is a divergent wave with respect to the distance between particles 2 and 3.

Note that both the infinite motion of all three particles and scattering of one particle by the bound state of the other two are described by nonuniform systems of integral equations which have unique solutions.

The wave function Ψ_0 of the bound state of the whole system is determined by a uniform set of integral equations which can be derived in a manner similar to the previous analysis. We obtain

$$\Psi_0 = \Psi^{(1)}_0 + \Psi^{(2)}_0 + \Psi^{(3)}_0,$$

$$\begin{aligned} \Psi^{(1)}_0 &= G_0(Z)T_{23}(Z)[\Psi^{(2)}_0 + \Psi^{(3)}_0], \\ \Psi^{(2)}_0 &= G_0(Z)T_{31}(Z)[\Psi^{(3)}_0 + \Psi^{(1)}_0], \\ \Psi^{(3)}_0 &= G_0(Z)T_{12}(Z)[\Psi^{(1)}_0 + \Psi^{(2)}_0]. \end{aligned} \tag{2.107}$$

The uniform set of (2.107) has solutions only for energies associated with bound states of the system. In the c.m.s., these energies are negative ($E < 0$). Note that set (2.107) can be formally derived from (2.105) if we put $\Phi_1 = 0$.

Momentum representation. The *Faddeev equations* take the simplest form in the momentum representation. Each component $\Psi^{(i)}$ of the total wave function can be written as a function of the relevant set of coordinates, i.e.,

$$\Psi^{(i)} \equiv \Psi^{(i)}(\mathbf{r}_{jk}, \mathbf{x}_i, \mathbf{R}), \qquad ijk = 123, 231, 312 \tag{2.108}$$

($\mathbf{R}$ is the center-of-mass radius).

In the *momentum representation*, $\Psi^{(i)}$ is defined by

$$\Psi^{(i)}(\mathbf{k}_{jk}, \mathbf{p}_i, \mathbf{K}) = \int d\mathbf{R} d\mathbf{x}_i d\mathbf{r}_{jk} \exp[-i\mathbf{k}_{jk}\mathbf{r}_{jk} - i\mathbf{p}_i\mathbf{x}_i - i\mathbf{K}\mathbf{R}] \times \Psi^{(i)}(\mathbf{r}_{jk}, \mathbf{x}_i, \mathbf{R}) \tag{2.109}$$

($\mathbf{K}$ is the center-of-mass momentum).

Now let us consider explicit expressions for the asymptotic functions. We label the initial momenta by the superscript zero. The free motion of all particles is described by the asymptotic function Φ_{123}

$$\Phi_{123} = \exp[i\mathbf{p}_1^0\mathbf{x}_1 + i\mathbf{k}_{23}^0\mathbf{r}_{23} + i\mathbf{K}^0\mathbf{R}] \tag{2.110}$$

$$E_{123} = \frac{p_1^{0^2}}{2\mu_1} + \frac{k_{23}^{0^2}}{2\mu_{23}} + \frac{K^{0^2}}{2M}, \qquad M = m_1 + m_2 + m_3.$$

Note that Φ_{123} is invariant with respect to interchanges of the particle coordinates.

The *asymptotic function* Φ_1 is given by

$$\Phi_1 = \exp[i\mathbf{p}_1^0\mathbf{x}_1 + i\mathbf{K}^0\mathbf{R}]\varphi_{\kappa_{23}}(\mathbf{r}_{23}), \tag{2.111}$$

$$E_{123} = \frac{p_1^{0^2}}{2\mu_1} - \frac{\kappa_{23}^2}{2\mu_{23}} + \frac{K^{0^2}}{2M},$$

where $\varphi_{\kappa_{23}}(\mathbf{r}_{23})$ is the solution of the equation

$$\left(-\frac{1}{2\mu_{23}}\Delta_{23} + V_{23} - E_{23}\right)\varphi_{\kappa_{23}}(\mathbf{r}_{23}) = 0 \tag{2.112}$$

for negative energies of relative motion $E_{23} = -\kappa_{23}^2/2\mu_{23} < 0$, i.e., the *bound state wave function* of the particles 2 and 3.

The *asymptotic function* $\Phi_{1(23)}$ can be written as

$$\Phi_{1(23)} = \exp[i\mathbf{p}_1^0\mathbf{x}_1 + i\mathbf{K}^0\mathbf{R}]\varphi_{\mathbf{k}_{23}^0}(\mathbf{r}_{23}), \qquad E_{1(23)} = E_{123}, \tag{2.113}$$

where $\varphi_{\mathbf{k}_{23}^0}(\mathbf{r}_{23})$ is the solution of (2.112) with $E_{23} = (\mathbf{k}_{23}^0)^2/2\mu_{23} > 0$, which reduces at infinity to the sum of a plane wave and a diverging spherical wave.

The kinetic energy operator in the momentum representation reduces to the multiplication operator, hence the *Green function* $G_0(Z)$ is diagonal:

$$\langle\mathbf{k}'_{jk}\mathbf{p}'_i\mathbf{K}'|G_0(Z)|\mathbf{K}\mathbf{p}_i\mathbf{k}_{jk}\rangle = (2\pi)^9\left(Z - \frac{k_{jk}^2}{2\mu_{jk}} - \frac{p_i^2}{2\mu_i} - \frac{K^2}{2M}\right)^{-1}$$
$$\times\delta(\mathbf{k}_{jk} - \mathbf{k}'_{jk})\delta(\mathbf{p}_i - \mathbf{p}'_i)\delta(\mathbf{K} - \mathbf{K}'). \tag{2.114}$$

The *two-particle scattering operator* $T_{jk}(Z)$ is diagonal in the representation of the total momentum of the system and the free particle momentum $\mathbf{p}_i$, i.e.,

$$\begin{aligned}&\langle\mathbf{k}'_{jk}\mathbf{p}'_i\mathbf{K}'|T_{jk}(Z)|\mathbf{K}\mathbf{p}_i\mathbf{k}_{jk}\rangle\\ =\;& (2\pi)^6\left\langle\mathbf{k}'_{jk}\left|\,t_{jk}\left(Z - \frac{p_i^2}{2\mu_i} - \frac{K^2}{2M}\right)\right|\mathbf{k}_{jk}\right\rangle\\ &\times\delta(\mathbf{p}_i - \mathbf{p}'_i)\delta(\mathbf{K} - \mathbf{K}'),\end{aligned} \tag{2.115}$$

where $t_{jk}(Z)$ is the two-particle t matrix determined by the integral equation (2.76).

Since all the operators entering the Faddeev equations are diagonal in the $\mathbf{K}$-representation, the wave function Ψ must contain the factor $\delta(\mathbf{K} - \mathbf{K}^0)$ which reflects conservation of the total momentum of the system. Therefore, all $\mathbf{K}$-dependencies can be excluded from consideration as soon as we convert to the center-of-mass system.

For example, we write the system of integral *Faddeev equations* (2.105) *in the momentum representation* for the case of particle scattering by the bound state of the other two particles:

$$\Psi_1 = \Psi_1^{(1)}(\mathbf{k}_{23}, \mathbf{p}_1) + \Psi_1^{(2)}(\mathbf{k}_{31}, \mathbf{p}_2) + \Psi_1^{(3)}(\mathbf{k}_{12}, \mathbf{p}_3), \tag{2.116}$$

$$\Psi^{(1)}(\mathbf{k}_{23}, \mathbf{p}_1) = \Phi_1(\mathbf{k}_{23}, \mathbf{p}_1)$$
$$+\left(Z - \frac{k_{23}^2}{2\mu_{23}} - \frac{p_1^2}{2\mu_1}\right)^{-1}\int\frac{d\mathbf{k}'_{23}}{(2\pi)^3}\left\langle\mathbf{k}_{23}\left|\,t_{23}\left(Z - \frac{p_1^2}{2\mu_1}\right)\right|\mathbf{k}'_{23}\right\rangle$$

$$\times\left\{\Psi_1^{(2)}\left(\mathbf{k}'_{31},\mathbf{p}'_2\right)+\Psi_1^{(3)}\left(\mathbf{k}'_{12},\mathbf{p}'_3\right)\right\},$$

$$\begin{aligned}\Psi_1^{(2)}(\mathbf{k}_{31},\mathbf{p}_2)&=\left(Z-\frac{k_{31}^2}{2\mu_{31}}-\frac{p_2^2}{2\mu_2}\right)^{-1}\\&\times\int\frac{d\mathbf{k}'_{31}}{(2\pi)^3}\left\langle\mathbf{k}_{31}\left|t_{31}\left(Z-\frac{p_2^2}{2\mu_2}\right)\right|\mathbf{k}'_{31}\right\rangle\\&\times\left\{\Psi_1^{(3)}\left(\mathbf{k}'_{12},\mathbf{p}'_3\right)+\Psi_1^{(1)}\left(\mathbf{k}'_{23},\mathbf{p}'_1\right)\right\},\end{aligned}\tag{2.117}$$

$$\begin{aligned}\Psi^{(3)}(\mathbf{k}_{12},\mathbf{p}_3)&=\left(Z-\frac{k_{12}^2}{2\mu_{12}}-\frac{p_3^2}{2\mu_3}\right)^{-1}\\&\times\int\frac{d\mathbf{k}'_{12}}{(2\pi)^3}\left\langle\mathbf{k}_{12}\left|t_{12}\left(Z-\frac{p_3^2}{2\mu_3}\right)\right|\mathbf{k}'_{31}\right\rangle\\&\times\left\{\Psi_1^{(1)}\left(\mathbf{k}'_{23},\mathbf{p}'_1\right)+\Psi_1^{(2)}\left(\mathbf{k}'_{31},\mathbf{p}'_2\right)\right\},\end{aligned}$$

$$Z\equiv\frac{p_1^{0^2}}{2\mu_1}-\frac{\kappa_{23}^2}{2\mu_{23}}+iO.$$

The quantities $\mathbf{k}'_{31},\mathbf{p}'_2$ and $\mathbf{k}'_{12},\mathbf{p}'_3$ entering the intergrand in the first equation must be expressed in terms of $\mathbf{p}_1$ and $\mathbf{k}'_{23}$. The function $\Phi_1(\mathbf{k}_{23},\mathbf{p}_1)$, by virtue of (2.111), is given by

$$\Phi_1(\mathbf{k}_{23},\mathbf{p}_1)=(2\pi)^3\delta(\mathbf{p}-\mathbf{p}_1^0)\varphi_{\kappa_{23}}(\mathbf{k}_{23}),\tag{2.118}$$

where $\varphi_{\kappa_{23}}(\mathbf{k}_{23})$ is the wave function of the two-particle bound state with the binding energy $\kappa_{23}^2/2\mu_{23}$; $\mathbf{p}_1^0$ is the initial momentum of relative motion of the system.

If all three particles are identical and have zero spin and isotopic spin, then the total wave function of the system Ψ_1 must be symmetric with respect to permutations of any pairs of particles. In this case we have

$$\Psi_1^{(1)}(\mathbf{k},\mathbf{p})=\Psi_1^{(2)}(\mathbf{k},\mathbf{p})=\Psi_1^{(3)}(\mathbf{k},\mathbf{p})\equiv\psi(\mathbf{k},\mathbf{p})$$

and the wave function can be written as

$$\Psi_1=\psi\Big(\mathbf{k}_{23},\mathbf{p}_1\Big)+\psi\Big(\mathbf{k}_{31},\mathbf{p}_2\Big)+\psi\Big(\mathbf{k}_{12},\mathbf{p}_3\Big).\tag{2.119}$$

[The relative momenta $\mathbf{k}_{ij}$ and $\mathbf{p}_k$ (where ijk=123, 231, and 312) are given by (2.90) with $m_1 = m_2 = m_3 = m$.] Then, instead of the system of three equations (2.117), we obtain a single *integral equation for the function* $\psi(\mathbf{k}, \mathbf{p})$:

$$\psi(\mathbf{k},\mathbf{p}) = \varphi(\mathbf{k},\mathbf{p}) + \left(Z - \frac{k^2}{2m} - \frac{3}{4}\frac{p^2}{m}\right)^{-1}$$
$$\times \int \frac{d\mathbf{p}'}{(2\pi)^3}\left\{\left\langle \mathbf{k} \left| t\left(Z - \frac{3}{4}\frac{p^2}{m}\right) \right| \frac{\mathbf{p}}{2} + \mathbf{p}' \right\rangle\right.$$
$$+ \left.\left\langle \mathbf{k} \left| t\left(Z - \frac{3}{4}\frac{p^2}{m}\right) \right| -\frac{\mathbf{p}}{2} - \mathbf{p}' \right\rangle\right\}\psi(\mathbf{p} + \frac{\mathbf{p}'}{2}, \mathbf{p}'), \qquad (2.120)$$

where $\varphi(\mathbf{k}, \mathbf{p})$ is a function whose symmetrization yields the initial wave function of the system

$$\Phi_1 = \varphi(\mathbf{k}_{23}, \mathbf{p}_1) + \varphi(\mathbf{k}_{31}, \mathbf{p}_2) + \varphi(\mathbf{k}_{12}, \mathbf{p}_3). \qquad (2.121)$$

For a three-particle bound state we have $\varphi = 0$, $z = E_0 < 0$. In the case of particle scattering by a two-particle bound state,

$$\varphi(\mathbf{k},\mathbf{p}) = (2\pi)^3\delta(\mathbf{p} - \mathbf{p}^0)\varphi_{N_0}(\mathbf{k}), \qquad Z = \frac{3}{4}\frac{p^{0^2}}{m} - \frac{\kappa^2_{N_0}}{m} + iO. \qquad (2.122)$$

Note that both initial and final states of the matrix elements $< \mathbf{k}|t(Z - \frac{3}{4}(p^2/m))|\mathbf{k}' >$ entering (2.120) lie off the energy shell, i.e., $k^2/m \neq k'^2/m \neq Z - \frac{3}{4}(p^2/m)$.

In the general case the wave function of the system described by the integral equation (2.120) depends on six variables (two relative vectors). For a spherically symmetric interaction potential, the three-dimensional integral equation (2.120) can be reduced to a system of two-dimensional integral equations by expanding the wave function in terms of angular functions and separating the angle variables. Making use of the separable representation of the two-particle scattering amplitude, one reduces the systems of two-dimensional integral equations to systems of one-dimensional integral equations.

Impossibility of separating the interaction region in the three-particle configuration space. If the interaction range is finite, the two-particle interaction is manifested only within a finite configuration region. For three-particle systems, however, one cannot separate in the configuration space an interaction region whose boundaries would be independent of the particle relative energy. To show this, let us consider the scattering details in the two- and three-particle systems. In the simplest case of a collision of two elementary particles, there exists the single elastic scattering channel, the wave function off the interaction region being the sum of the incident plane wave and the diverging spherical wave. If at least one of the colliding particles is composite, there exist, along with the elastic scattering, other channels associated with inelastic processes. For binary processes, the wave functions in the inelastic scattering or reaction exit channels are diverging spherical waves. It is peculiar of binary processes that the number of open channels is restricted (due to the discrete energy spectrum of scattered particles) and the interaction is manifested within a limited configuration region. The binary processes are described by the S-matrix which is completely determined by the behaviour of the system in the interaction region. With the three particles being produced in the final state, the description becomes much more involved.

Let us consider the *scattering in a three-particle system* . Suppose the particle 1 is scattered by a bound state of the other two particles 2 and 3. If the incident particle energy is higher than the scatterer binding energy, there may occur, along with the binary processes of elastic scattering and rescattering, also the disintegration after which all the three particles turn out to be free.The wave function asymptotics may be presented in this case in the form

$$\Psi(\mathbf{x},\mathbf{r}) = \begin{cases} \left\{ e^{i\mathbf{p}_0\mathbf{x}} + f(\mathbf{p}_0,\mathbf{p}_0')\dfrac{e^{ip_0x}}{x} \right\}\varphi_0(r), & \text{at } \begin{array}{l} x\to\infty \\ r=\text{const} \end{array} \\ \tilde{f}(\mathbf{p}_0,\tilde{\mathbf{p}}_0)\dfrac{e^{i\tilde{p}_0\tilde{x}}}{\tilde{x}}\tilde{\varphi}_0(\tilde{r}), & \text{at } \begin{array}{l} \tilde{\mathbf{x}}\to\infty \\ \tilde{\mathbf{r}}=\text{const} \end{array} \\ F(\mathbf{p}_0;\mathbf{p},\mathbf{k})\left(\dfrac{\mu_{23}^2E}{2\pi^2\hbar^2\mu_1}\right)^{3/4}\dfrac{e^{i\sqrt{\frac{2\mu E}{\hbar^2}(x^2+\frac{\mu_{23}}{\mu_1}r^2)}}}{(x^2+\frac{\mu_{23}}{\mu_1}r^2)^{5/4}}, & \text{at } \begin{array}{l} x,r\to\infty \\ x/r=\text{const} \end{array} \end{cases} \tag{2.123}$$

where f, $\tilde{f}$, and F are the elastic scattering, rescattering, and disintegration amplitudes respectively ($E = \frac{\hbar^2p_0^2}{2\mu_1} - \varepsilon_0 = \frac{\hbar^2\tilde{p}_0^2}{2\tilde{\mu}_1} - \tilde{\varepsilon}_0 = \frac{\hbar^2p^2}{2\mu_1} + \frac{\hbar^2k^2}{2\mu_{23}}$). The coordinate dependence of the three-particle disintegration wave function is governed by the *asymptotic Green function for three noninteracting particles* . Since all particle velocities off the interaction region are fixed, the triangle formed by their radius-vector ends must increase with time remaining of the same shape. Therefore, in the free motion region, $r = \beta x$ $(x \to \infty)$, where $\beta = \frac{\mu_1}{\mu_{23}}\frac{k}{p}$. It follows immediately from the asymptotics (2.123) that if the process produces three particles, one cannot separate a configuration region in which the interaction is manifested (in contrast to the binary processes). Indeed, suppose the interaction between the particles 2 and 3 is characterized by a finite range R, and interaction between the particle 1 and the bound state of the particle system 2-3 is characterized by the range X. Then the wave function describing the three-particle interaction takes asymptotic form only provided $x \geqslant X$, R/β or $r \geqslant \beta X, R$. And since the quantity β depends on the relative momenta p and k, one can always find such momenta for which the interaction is manifested in the whole infinite space.

Another distinguishing feature of the processes under consideration is that any fixed total energy E is associated with an infinite number of states corresponding to various probable energy redistributions over the subsystems. This follows from the fact that the *three-particle disintegration wave function* may be presented as an expansion in terms of the scattering system continuum states:

$$\Psi(\mathbf{x}, \mathbf{r}) = \sum_{\mathbf{p},\mathbf{k}} a_{\mathbf{pk}} \psi_{\mathbf{p}}(\mathbf{x}) \varphi_{\mathbf{k}}(\mathbf{r}), \tag{2.124}$$

where the range of allowed $\mathbf{p}$ and $\mathbf{k}$ is determined by the condition $E =$ const (the energy level E is infinitely degenerated). An example suitable for illustrating the peculiarities of the three-particle processes under consideration is the deuteron diffraction dissociation by a black nucleus.

Inasmuch as the interaction region cannot be distinguished, there occurs another peculiar feature of the three-particle systems: their properties essentially depend on the off energy behaviour of the two-particle interaction. Indeed, the conservation laws for the two-particle scattering provide the interaction to be manifested only on the energy shell. The scattering amplitude is directly expressed in terms of scattering phases which describe the two-particle wave function asymptotics. Under the three-particle scattering, the particle interaction is manifested off energy shell too. Intermediate state momenta and energies of two interacting particles do not satisfy usual relations due to the third particle influence, and, hence, the three-particle dynamical equations contain two-particle off energy amplitudes. The off energy behaviour of the two-particle amplitude is governed not only by the two-particle scattering state asymptotics, but also by the wave function coordinate-dependence at short distances. That is why the three-particle problem turns out to be much more sensitive to the particle interaction details than the two-particle problem.

With the two-particle interaction being described by some potential, the two-particle amplitude off energy behaviour may be considered in terms of the Schrödinger equation. So, the data of the two-particle problem are sufficient for obtaining the off energy two-particle amplitude. However, the two-nucleon interaction data are unsufficient and restricted and hence the three-particle problem provides an important additional information source concerning nucleon interactions.

2.3. Three-Nucleon Systems

The main characteristics of three-nucleon nuclei. The study of the *three-nucleon systems* provides additional information on the nucleon-nucleon interaction. The two simplest three-nucleon nuclei are triton, ^3_1H, and nucleus ^3_2He. The triton (the nucleus of superheavy hydrogen atom) contains one proton and two neutrons; the *nucleus* ^3_2He contains two protons and one neutron. One might assume that, in principle, the three-particle forces could be manifested in the three-nucleon nuclei. However, the study of the properties of the mirror nuclei ^3_1H and ^3_2He, nucleon-deuteron interaction, and systems of three unbound nucleons show that the above mentioned systems can be satisfactorily described within the range of experimental errors under the assumption of two-particle nature of the nuclear forces with finite range of action. The latter assumption makes the basis for the further analysis.

Unlike the two-nucleon problem, the problem of three nucleon motion is more sensitive to the shape of the two-nucleon potentials. The other peculiarities of nuclear forces manifest themselves more clearly in the three-nucleon rather than the two-nucleon systems. E.g., in the case of scattering of an incident nucleon by a nucleon that is bound in a deuteron or another complex nucleus, the energies of the colliding nucleons before and after the collision may be different . This is due to the fact that other nucleons of the target nucleus may obtain or transfer some amount of energy in the scattering process so that the energy of the whole system is conserved. Such scattering of an incident nucleon by the bound nucleon is referred to as the process off the energy shell (scattering of two free nucleons occurs on the energy shell). The study of the off-energy-shell scattering in the three-nucleon systems provides additional information on the nucleon-nucleon interaction.

Let us consider the experimental data concerning the main properties of the three-nucleon nuclei ^3_1H and ^3_2He. The nucleus ^3_2He is stable; the mirror nucleus ^3_1H is unstable and decays with the decay period of 12.3 years. The decay results in the formation of the nucleus ^3_2He, i.e.,

$$^3H \to {}^3He + e^- + \bar{\nu}_e.$$

Therefore, one of neutrons in the nucleus ^3H is transformed into a proton, electron and antineutrino. The process of the electron β-decay of ^3H nucleus is associated with the weak interaction and is rather slow. Other physical properties of the mirror nuclei ^3H and ^3He are close since they are associated mainly with the strong nuclear interaction.

The binding energies of the nuclei ^3H and ^3He are equal to 8.432 MeV and 7.718 MeV, respectively. The difference in the binding energies,

$$\Delta E = 0.764\,\text{MeV}, \tag{2.125}$$

can be explained by the Coulomb repulsion between two protons in the nucleus ^3He. We assume that the electric charge within the nucleus ^3He is distributed homogeneously over the volume. Then the energy of electrostatic (Coulomb) interaction between the proton provided the constant charge density within the nucleus ($\rho = e/(\frac{4}{3}\pi R^3)$, where R is the nuclear radius), and another proton, that produces the potential $\varphi(r) = 2\pi\rho(R^2 - \frac{1}{3}r^2)$, is given by

$$E_c = 4\pi\rho \int_0^R dr\, r^2 \varphi(r) = \frac{6e^2}{5R}. \tag{2.126}$$

Substituting the known value $R \approx 2\cdot10^{-13}$ cm, for the radius of ^3He nucleus in (2.126), we find that E_c is close to 0.764 MeV (slightly larger). On the other hand, for

$$E_c = \frac{6e^2}{5R} = 0.764\,\text{MeV},$$

we obtain a reasonable value for the nucleus ^3He radius, $R = 2.3\cdot10^{-13}$ cm, which is in agreement with the value obtained from the data on the high-energy electron scattering by ^3He nuclei (assuming that the charge density is distributed according to the Gauss law). This approximate correspondence can be regarded as the evidence confirming that the nuclear interaction between two protons and two neutrons is basically the same.

The *binding energy per a nucleon pair* is equal to 2.8 MeV in the ^3H nucleus, and to 2.6 MeV in the ^3He. Both values are considerably greater than the deuteron binding energy. Inasmuch as linear dimensions of a deuteron and a three-nucleon nucleus are slightly different, (apparently, a deuteron is even slightly larger), we can deduce that both ^3H and ^3He nuclei are more compact than the deuteron. Therefore, three nucleons in three-nucleon nuclei are more close to each other (in average) than two nucleons in a more loose deuteron. It is worth noting that nuclei ^3H and ^3He, as well as deuteron and ^4He nucleus, have no excited states.

The measured spin J of each nuclei ^3H and ^3He is 1/2. The parities w of the ground states of ^3H and ^3He nuclei and the deuteron are positive ($w = +1$). Let us assume that the ground state of the three-nucleon nucleus is the S-state and that all three nucleons are in the states with zero orbital moments. Then, according to the Pauli principle, two identical nucleons in the nucleus (nn or pp) must have opposite spins. Therefore, the total spin of three nucleons is equal to 1/2, i.e., to the spin of the third nucleon that is not identical to the first two. Since the total orbital moment of three nucleons in the S-state is $L = 0$, the total angular momentum of the three-nucleon nucleus, i.e., its spin J coincides numerically with the total spin S and equals to 1/2.

It is clear that the properties of nuclei ^3H and ^3He are determined mainly by the central nuclear interaction between nucleons. However, inasmuch as three-nucleon nuclei can contain pairs of nonidentical nucleons, which can be in the triplet states, i.e., the states with the total spin 1, we can expect the manifestation of non-central (tensor) forces in the properties of both ^3H and ^3He nuclei and the deuteron. The comparison binding energies and magnetic moments of the nuclei ^3H and ^3He shows that the ground states of these nuclei are not the pure S-states but contain the D-states ($L = 2$) with weights $p_D \approx 0.04$.

The measured *magnetic (dipole) moments of nuclei* ^3H and ^3He (in nuclear magnetons) are equal, respectively, to

$$\mu_{^3H} = 2.9788 \text{ and } \mu_{^3He} = -2.1274. \tag{2.127}$$

If the S-states were pure, the magnetic moments of the nuclei ^3H and ^3He would have numerical values $\mu_p = 2.7928$ and $\mu_n = -1.9130$, respectively, by virtue of the fact that the magnetic moments of two identical neutrons in both nuclei are compensated. The cited numerical values, however, differ from the experimental values of the magnetic moments of the nuclei ^3H and ^3He approximately by 10%. The addition of the D-wave (with the weight $p_D = 0.04$) can only slightly improve the situation. There still remains a descrepancy in magnitudes of the nuclear magnetic moments which can be explained taking into account the meson interchange currents between nucleons which induce additional magnetic moments.

The isotopic spin of the three-nucleon systems. If in a system of three nucleon there are two or three protons, then along with the nuclear forces the Coulomb forces appear. On taking account only the nuclear forces, the total isotopic spin T and its projection T_z are conserved. Three nucleons can be in the states with the total value of the isotopic spin $T = 3/2$ and $T = 1/2$. The states with $T = 3/2$ are related to the three-nucleon systems nnn, nnp, npp, ppp with isotopic spin projections T_z=-3/2, -1/2, +1/2, +3/2, correspondingly. Three identical nucleons nnn or ppp can be only in the states with $T = 3/2$ because $T \geqslant |T_z|$. The states with $T = 1/2$ are related to the systems nnp and npp with the isospin projections $T_z = -1/2$ and $T_z = 1/2$.

As we have seen, the nuclear interaction in the two-particle systems strongly depends on a magnitude of the isotopic spin. Because of this the properties of the three-nucleon systems depend on the isospin magnitude as well. This follows from the general simple considerations. In particular, the spatial structure of the three-nucleon systems strongly depends on the total isospin T. In fact, in the system of three neutrons nnn $(T = 3/2)$, only two from three identical nucleons can be in the states with zero orbital moments, the third nucleon, accordingly the Pauli principle, has nonzero orbital moment. The same spatial structure is typical for the systems nnp and npp in the states with $T = 3/2$, whose energy is larger than the energy of ^3H and ^3He, where all three nucleons have zero orbital moments. Since, there are no the bound systems consisting of identical nucleons nnn and ppp, then the systems nnp and npp cannot be bound in the states with $T = 3/2$. From this follows that the nuclei ^3H and ^3He, which are the systems nnp and npp, must have the isotopic spin $T = \frac{1}{2}$ but not $T = 3/2$. We will see further, that there are different considerations confirming this conclusion. The mirror nuclei ^3H and ^3He along with a neutron and proton belong to the isotopic doublet with the isospin projections $T_z = -\frac{1}{2}$ and $T_z = +\frac{1}{2}$.

Incorporation of the Coulomb interaction between protons in the three-nucleon systems npp and ppp results in violation of the total isotopic spin T conservation law. However, its projection T_z is preserved in this case as well. Because of this, the states of nuclei ^3H and the system neutron-deuteron (accordingly Table 1.2, the deuteron isospin equals to zero) are described by the strictly defined value of the total isotopic spin $T = \frac{1}{2}$. At the same time, the wave functions of the ground state of nuclei ^3He and the system proton-deuteron are the superpositions of the states with isospins $T = \frac{1}{2}$ and $T = 3/2$. However, the contribution from the terms with $T = 3/2$ in the total wave functions of these systems turns out to be very small.

The spin and isospin states of three-nucleon system. Three nucleons can be in the state with the total spin $S = \frac{1}{2}$ and $S = 3/2$. A normalized spin wave function of the three-nucleon system with the definite total spin S and its projection M_S can be presented in the form of superposition of products of the ortho-normalized *one-nucleon spin wave functions*

$$\begin{aligned} \chi_{SM_S}^{(S_{23})}(123) &= \sum_{m_1 M_{23}} \left(\frac{1}{2}m_1 S_{23} M_{23}|SM_S\right) \chi_{\frac{1}{2}m_1}(1)\chi_{S_{23}M_{23}}(23), \\ \chi_{S_{23}M_{23}}(23) &= \sum_{m_2 m_3} \left(\frac{1}{2}m_2 \frac{1}{2}m_3|S_{23}M_{23}\right) \chi_{\frac{1}{2}m_2}(2)\chi_{\frac{1}{2}m_3}(3), \quad (2.128) \end{aligned}$$

where in the spin function instead of arguments we put simply the nucleon numbers. By using the Clebsch-Gordan coefficients, we interrelated the spin states of the second and third nucleons in the two-particle state with the total spin S_{23} and projection M_{23}. After this we connected the obtained state with the first nucleon spin state. It could be done in a different order as well.

We consider firstly the spin states of the three-nucleon system, when their total spin $S = \frac{1}{2}$. The intermediate total spin of two nucleons S_{23} at $S = \frac{1}{2}$ can be 0 or 1. Accordingly, we have introduced the following denotations of two *spin functions of the three-nucleon system* in the case $S = \frac{1}{2}$:

$$\chi' \equiv \chi'_{\frac{1}{2}M_S}(1,23) = \chi^{(S_{23}=0)}_{\frac{1}{2}M_S}(123), \; \chi'' \equiv \chi''_{\frac{1}{2}M_S}(1,23) = \chi^{(S_{23}=1)}_{\frac{1}{2}M_S}(123).$$

Using the denotations $\alpha(j) = \chi_{\frac{1}{2},+\frac{1}{2}}(j)$ and $\beta(j) = \chi_{\frac{1}{2},-\frac{1}{2}}(j)$ for the one-particle spin functions (the functions $\alpha(j)$ and $\beta(j)$ related to positive and negative signs of the spin projections) and substituting numerical values of the Clebsh-Gordan coefficients in (2.128), we obtain the following expressions for functions χ' and χ'' with all possible values M_S:

$$\begin{aligned}
\chi'_{\frac{1}{2},+\frac{1}{2}}(1,23) &= \frac{\alpha(1)}{\sqrt{2}}[\alpha(2)\beta(3) - \beta(2)\alpha(3)], \\
\chi'_{\frac{1}{2},-\frac{1}{2}}(1,23) &= \frac{\beta(1)}{\sqrt{2}}[\alpha(2)\beta(3) - \beta(2)\alpha(3)], \\
\chi''_{\frac{1}{2},+\frac{1}{2}}(1,23) &= \frac{\alpha(1)}{\sqrt{6}}[\alpha(2)\beta(3) + \beta(2)\alpha(3)] \\
&- \sqrt{\frac{2}{3}}\beta(1)\alpha(2)\alpha(3), \\
\chi''_{\frac{1}{2},-\frac{1}{2}}(1,23) &= \frac{\beta(1)}{\sqrt{6}}[\alpha(2)\beta(3) + \beta(2)\alpha(3) \\
&+ \sqrt{\frac{2}{3}}\alpha(1)\beta(2)\beta(3). \qquad (2.129)
\end{aligned}$$

As it must be, the functions χ' are antisymmetric with respect to permutations of the second and third nucleons, since these two nucleons are in the singlet spin state with the total spin $S_{23} = 0$, and the functions χ'' are symmetric with respect to permutation of the second and third nucleons in the triplet spin state ($S_{23} = 1$).

If the total *spin of the system of three nucleons* is $S = \frac{3}{2}$, then the corresponding spin wave function is denoted as follows

$$\chi^s \equiv \chi^s_{\frac{3}{2}M_S}(123) = \chi^{(S_{23}=1)}_{\frac{3}{2}M_S}(123). \tag{2.130}$$

Apparently, in this case the intermediate total spin of two nucleons S_{23} can be equal only to unity, and the spin wave function χ^s is completely symmetric with respect to permutations of three nucleons (a sign s stands for this purpose in its upper part). The symmetry properties of function (2.130) can be seen also from the explicit expressions $\chi^s_{\frac{3}{2}M_S}(123)$ in terms $\alpha(j)$ and $\beta(j)$:

$$\begin{aligned}
\chi^s_{\frac{3}{2},+\frac{3}{2}}(123) &= \alpha(1)\alpha(2)\alpha(3), \\
\chi^s_{\frac{3}{2},+\frac{1}{2}}(123) &= \frac{1}{\sqrt{3}}[\alpha(1)\alpha(2)\beta(3) + \alpha(1)\beta(2)\alpha(3) + \beta(1)\alpha(2)\alpha(3)], \\
\chi^s_{\frac{3}{2},-\frac{1}{2}}(123) &= \frac{1}{\sqrt{3}}[\alpha(1)\beta(2)\beta(3) + \beta(1)\alpha(2)\beta(3) + \beta(1)\beta(2)\alpha(3)], \\
\chi^s_{\frac{3}{2},-\frac{1}{2}}(123) &= \beta(1)\beta(2)\beta(3).
\end{aligned} \tag{2.131}$$

All eight functions (2.129) and (2.131) are ortho-normalized.

The spin functions of different symmetries (2.129) and (2.131) may be constructed by using relevant permutation operators. Let the permutation of two nucleons i and j be denoted by (ij). We introduce the following *three-nucleon permutation operators* :

$$\begin{aligned}
T^s &= (12) + (23) + (31), \\
T' &= \frac{\sqrt{3}}{2}\left[(12) - (23)\right], \\
T'' &= -(31) + \frac{1}{2}\left[(12) + (23)\right].
\end{aligned} \tag{2.132}$$

Applying operators (2.132) to a function symmetric with respect to the coordinates of particles 1and 2, one obtains three functions. The first one is completely symmteric with respect to the coordinates 1,2, and 3, the other two are transformed under the coordinate permutations as a two-dimensional irreducible representation of the permutation group S_3. The *spin wave functions* of the states $S = \frac{1}{2}$ and $S = \frac{3}{2}$ (the total spin projection is taken to be equal to $\frac{1}{2}$) are given by

$$S = \frac{1}{2}, \begin{cases} \chi' = \sqrt{\frac{2}{3}}T'\alpha(1)\alpha(2)\beta(3), \\ \chi'' = \sqrt{\frac{2}{3}}T''\alpha(1)\alpha(2)\beta(3), \end{cases} \tag{2.129'}$$

$$S = \frac{3}{2}, \chi^s = \frac{1}{\sqrt{3}} T^s \alpha(1)\alpha(2)\beta(3). \tag{2.131'}$$

The factors in (2.129′) and (2.131′) are chosen to provide the functions to be normalized to unity.

The functions χ' and χ'' are the basis functions of the *two-dimensional irreducible representation of the permutation group* of three particles, i.e., they have the mixed permutation symmetry. The functions χ^s are the basis functions of the one-dimensional (symmetric) representation of the same permutation group. Simultaneously, the functions χ' and χ'' are the basis functions of the two-dimensional irreducible representation $D^{1/2}$, and the functions χ^s are the basis functions of the four-dimensional representation $D^{3/2}$ of the three-dimensional rotations.

The isospin wave functions of three nucleons ζ', ζ'' and $\zeta^s$for the total isotopic spins $T = \frac{1}{2}$ and $T = \frac{3}{2}$ can be constructed in perfect analogy to the above reported. Introducing denotations of the proton and neutron isospin functions $x(j) = \zeta_{\frac{1}{2},+\frac{1}{2}}(j)$, $y(j) = x_{\frac{1}{2},-\frac{1}{2}}(j)$, we can write, for example, the *isospin wave functions of the three-nucleon system* :

$$\begin{aligned}
\zeta'_{\frac{1}{2},+\frac{1}{2}}(1, 23) &= \frac{x(1)}{\sqrt{2}}[x(2)y(3) - y(2)x(3)], \\
\zeta''_{\frac{1}{2},+\frac{1}{2}}(1, 23) &= \frac{x(1)}{\sqrt{6}}[x(2)y(3) + y(2)x(3)] - \sqrt{\frac{2}{3}}\, y(1)x(2)x(3), \\
\zeta_{\frac{3}{2},+\frac{3}{2}}(123) &= x(1)x(2)x(3).
\end{aligned} \tag{2.133}$$

These isospin functions have the analogous form to the corresponding spin wave functions $\chi'_{\frac{1}{2},+\frac{1}{2}}(1,23)$, $\chi''_{\frac{1}{2},+\frac{1}{2}}(1,23)$, $\chi^s_{\frac{3}{2},+\frac{3}{2}}(123)$, given in (2.129) and (2.131).

We call attention to the fact that it is impossible to construct the antisymmetric (with respect to permutations of any nucleon pair) spin wave function of the system of three nucleons χ^a from the product of one-particle spin wave functions. In fact, the function χ^a must be proportional to a determinant of the third order, which is built from the functions $\chi_{\frac{1}{2}m_j}(j)$. However, the spin projection m_j takes only two values $\pm\frac{1}{2}$ (there are only two different spin functions of a nucleon α and β). Because of this the above-mentioned determinant and the function χ^a are equal to zero. Due to the same reason, the isospin function of the three nucleon system antisymmetric with respect to permutations of any nucleon pair isospin does not exist $\zeta^a = 0$. In this connection it should be noted that in the case of the spatial symmetric wave function of three-nucleon nuclei ^3H and ^3He

with the total spin $S = \frac{1}{2}$ the isospin of these nuclei cannot be equal to $\frac{3}{2}$. At $T = \frac{3}{2}$ the isospin state is completely symmetric, i.e., the spin state must be completely antisymmetric, but, as we have seen above, this state is impossible since $\chi^a = 0$.

We construct now the spin-isospin wave functions of the three-nucleon system with definite values of the total spin S and isospin T and different symmetry with respect to the nucleon permutations. First of all, we consider the case $S = T = \frac{1}{2}$, which is possible both for bound and unbound states of three nucleons. Omitting the indexes of spin, isospin, and their projections, we can write down the *spin-isospin wave functions of the three-nucleon system* with different permutation symmetry at $S = \frac{1}{2}$ and $T = \frac{1}{2}$ in the following form

$$\xi^s = \frac{1}{2}\left(\chi'\zeta' + \chi''\zeta''\right), \; \xi^a = \frac{1}{\sqrt{2}}\left(\chi'\zeta'' - \chi''\zeta'\right), \qquad (2.134)$$

$$\xi' = \frac{1}{\sqrt{2}}\left(\chi'\zeta'' + \chi''\zeta'\right), \; \xi'' = \frac{1}{\sqrt{2}}\left(\chi'\xi' - \chi''\xi''\right). \qquad (2.135)$$

The functions ξ^s and ξ^a, respectively, are completely symmetric and antisymmetric functions with respect to permutations of any pair of nucleons (it can be checked directly). The functions ξ^s and ξ^a are the basis functions of two one-dimensional (symmetric and antisymmetric, respectively) representations of the permutation group of three nucleons. And functions ξ' and ξ'' are the basis functions (with different *Yamanouchi symbols*) of the two-dimensional irreducible representation of the same group and are transformed in terms of each other at permutations of nucleons. They are the functions with the mixed (intermediate) symmetry. All functions (2.134) and (2.135) are normalized by unity and mutually orthogonal.

For a system of three unbound nucleons, we use also the spin-isospin wave functions in the product form $\chi^s\zeta^s$ at $S = T = \frac{3}{2}$, which is a completely symmetric function, and in the forms $\chi^s\zeta''$ and $\chi^s\zeta'$ at $S = \frac{3}{2}, T = \frac{1}{2}$ or $\chi''\zeta^s$ and $\chi'\zeta^a$ at $S = \frac{1}{2}, T = \frac{3}{2}$, which have the mixed permutation symmetry. For the nucleon-deuteron system the total spin S can be equal $\frac{1}{2}$ or $\frac{3}{2}$, and the total isospin T equals to $\frac{1}{2}$ (neglecting the small contribution of the term with $T = \frac{3}{2}$ in the system wave function).

Total wave functions of three-nucleon systems in the case of the central forces. We construct now the total wave function of a three-nucleon system, which supposed to be antisymmetric with respect to simultaneous permutations of the spatial, spin and isospin coordinates of any nucleon pair. The antisymmetricity of the total wave functions of nucleon systems follows from the Pauli principle according to which a neutron and proton become identical fermions after introduction the isotopic spin formalism. We take now into account the Coulomb interaction between protons in system *npp* and *ppp*. This results in an uncertainty of the system total isotopic spin, but the corresponding distorting additions to the wave functions are very small. The central forces give the main contribution in the interactions in three-nucleon systems determining their basic properties, as well as in a deuteron.

Assuming that the interaction between nucleons has a two-particle type, we can present the interaction potential between k-th and j-th nucleons as follows

$$\begin{aligned}\hat{V}_{kj} &= V^{ts}(kj)\pi_t^\sigma(kj)\pi_s^\tau(kj) + V^{tt}(kj)\pi_t^\sigma(kj)\pi_t^\tau(kj) \\ &+ V^{st}(kj)\pi_s^\tau(kj)\pi_t^\tau(kj) + V^{ss}(kj)\pi_s^\sigma(kj)\pi_s^\tau(kj) \\ &+ V_c(kj)\Lambda_p(k)\Lambda_p(j), \end{aligned} \qquad (2.136)$$

where the upper signs t and s are related to the triplet (σ) and singlet spin and isospin (τ) states of two nucleons. Here $V^{ts}(kj) \equiv V^{ts}(|\mathbf{r}_k - \mathbf{r}_j|)$ and so on, $V_c(kj) = \frac{e^2}{|\mathbf{r}_k - \mathbf{r}_j|}$ is the Coulomb potential, $\pi_t^\sigma(kj) \equiv \frac{1}{4}(3 + \boldsymbol{\sigma}_k\boldsymbol{\sigma}_j)$, $\pi_s^\sigma = \frac{1}{4}(1 - \boldsymbol{\sigma}_k\boldsymbol{\sigma}_j)$, $\pi_t^\tau(kj) = \frac{1}{4}(3 + \boldsymbol{\tau}_k\boldsymbol{\tau}_j)$, $\pi_s^\tau(kj) = \frac{1}{4}(1 - \boldsymbol{\tau}_k\boldsymbol{\tau}_j)$ are the projection operators into the triplet spin and singlet isospin states, $\Lambda_p(k) = \frac{1+\tau_{kz}}{2}$ is the project operator into a proton state. Magnitudes of the potentials V^{tt} and V^{ss} at small relative energies of two nucleons are much less than V^{ts} and V^{st}, which enables the terms with V^{tt} and V^{ss} in (2.136) to be neglected.

Since the total Hamiltonian of the three-nucleon system $\hat{H} = \hat{H}_0 + \hat{V}_{12} + \hat{V}_{23} + \hat{V}_{31}$ with the interactions from (2.136) commutes with operators $\mathbf{S}^2$, $\hat{S}_z$ and $\hat{T}_z$, then the total spin S, spin projection S_z and isospin projection T_z have definite values (conserved). The operator $\mathbf{T}^2$ does not commute with the Hamiltonian $\hat{H}$ and, that is why, the total isotopic spin T has a definite value only if there is no the Coulomb interaction. Thus, the total isotopic spin T is an exact integral of motion only for the systems *nnn* and *nnp*.

We neglect for the moment a small influence of the electromagnetic interaction on the three-nucleon system structure (or we consider the systems *nnn* and *nnp*). In this case the total isospin T has an exact value. The total antisymmetric wave function of the relative motion is denoted as Ψ^{ST}. (For brevity sake, we omit arguments of the wave functions and its quantum numbers.) This function, in a general case, can be presented in the form of a finite sum of products of the spatial and spin-isospin wave functions of different symmetry with respect to the nucleon permutation.

We begin with the construction of the total wave function $\Psi^{\frac{1}{2}\frac{1}{2}}$ with $S = T = \frac{1}{2}$. As the possible candidates, that appear in the function $\Psi^{\frac{1}{2}\frac{1}{2}}$, we can immediately chose two already antisymmetric products $\psi^a\xi^s$ and $\psi^s\xi^a$, where ψ^a and ψ^s are the antisymmetric and symmetric with respect to permutations of any nucleon pair spatial wave functions and ξ^s and ξ^a are the symmetric and antisymmetric spin-isospin wave functions. Moreover, there are two more spin-isospin wave functions ξ' and ξ'' of the intermediate symmetry for $S = T = \frac{1}{2}$. We can construct two spatial functions ψ' and ψ'', which are transformed along with the functions ξ' and ξ'' according to the two-dimensional irreducible representation of the three-particle permutation group. From the functions of the intermediate symmetry we can construct for $S = T = \frac{1}{2}$ one more antisymmetric combination, namely, $\psi'\xi'' - \psi''\xi'$, in the same manner as we constructed the antisymmetric spin-isospin wave function. Since we exhausted all spin-isospin wave functions of the three-nucleon system with $S = T = \frac{1}{2}$, the total antisymmetric wave function of the system in this case is a superposition of four terms:

$$\Psi^{\frac{1}{2}\frac{1}{2}} = \psi^a\xi^s + \psi^s\xi^a + \psi'\xi'' - \psi''\xi'. \tag{2.137}$$

This wave function can describe the nucleus ^3H state, as well as unbound systems *nnp*, in particular, the neutron-deuteron system. Wave function (2.137) can be used for the description of ^3He nucleus and the systems *npp* and *ppp*, if the Coulomb interaction between protons is neglected.

We consider now the state of the *three-nucleon system* with $S = \frac{3}{2}$ and $T = \frac{1}{2}$. A total antisymmetric wave function $\Psi^{\frac{3}{2}\frac{1}{2}}$, in this case, can be presented as the product of the completely symmetric spin wave function χ^s (as we have seen it exists for $S = \frac{3}{2}$) and the antisymmetric spatial isospin wave function. The latter cannot be presented in the form of the product of the symmetric spatial function ψ^s and the antisymmetric isospin function ξ^a, since the function ξ^a does not exist in systems of three nucleons ($\xi^a = 0$). It is impossible to construct the spatial isospin wave function as the product of the antisymmetric spatial function ψ^a and the symmetric isospin function ξ^s at $T = \frac{1}{2}$. The point is that the function ξ^s corresponds to $T = \frac{3}{2}$, but not to $T = \frac{1}{2}$. Because of this at $S = \frac{3}{2}$ and $T = \frac{1}{2}$ there is only one way to construct the antisymmetric spatial isospin wave function from the functions of the intermediate symmetry, namely, $\psi'\zeta'' - \psi''\zeta'$. Thus, in the case $S = \frac{3}{2}$ and $T = \frac{1}{2}$, the *total antisymmetric wave function of the three-nucleon system* takes the following form

$$\Psi^{\frac{3}{2}\frac{1}{2}} = \psi'\chi^s\zeta'' - \psi''\chi^s\zeta'. \tag{2.138}$$

The analogous reasoning gives the following general form of the total antisymmetric wave function $\Psi^{\frac{1}{2}\frac{3}{2}}$ of the three-nucleon system for the total spin $S = \frac{1}{2}$ and isospin $T = \frac{3}{2}$:

$$\Psi^{\frac{1}{2}\frac{3}{2}} = \psi'\chi''\zeta^s - \psi''\chi'\zeta^s. \tag{2.139}$$

The structure of the total antisymmetric wave function at $S = T = \frac{3}{2}$ is the simplest:

$$\Psi^{\frac{3}{2}\frac{3}{2}} = \psi^a\chi^s\zeta^s. \tag{2.140}$$

Wave functions (2.138) – (2.140) describe only the continuous spectrum of the three-nucleon systems, in addition, functions (2.139) and (2.140) are related to the systems of identical nucleons *nnn* and *ppp*.

If in the three-nucleon systems *npp* and *ppp* we take into account the Coulomb interaction between protons, then the total isotopic spin does not have an exact value any more. Because of this, for example, the nucleus ^3He is described by the total wave function $\Psi(^3He)$, which is a superposition of the wave functions with $T = \frac{1}{2}$ (as for a nucleus ^3H) and $T = \frac{3}{2}$:

$$\Psi(^3He) = \psi^a\xi^s - \psi^s\xi^a + \psi'\xi'' - \psi''\xi' + \Big(\varphi'\chi'' - \varphi''\chi'\Big)\zeta^s. \tag{2.141}$$

A similar wave function describes the system proton-deuteron, as well as the system *npp* when three nucleons are unbound. The spatial wave functions φ' and φ'', on taking account the Coulomb interaction, are transformed at permutations as the functions ψ' and ψ''. Because of this the spatial spin wave function $\varphi'\chi'' - \varphi''\chi'$ is antisymmetric, and the correction to wave function (2.141), caused by the proton Coulomb interaction $(\varphi'\chi''-\varphi''\chi')\zeta^s$ related to $S = \frac{1}{2}$ and $T = \frac{3}{2}$, has the same structure as function (2.139). Note that the space functions ψ^a, ψ^s, ψ', and ψ'', appearing in (2.141), are deformed by the Coulomb interaction and differ from the corresponding spatial functions in (2.137). It is clear the wave function of the system *ppp* preserves the form (2.140) when switching on the Coulomb interaction but its spatial part is deformed by the Coulomb interaction.

The spatial functions ψ^a, ψ^s, ψ', ψ'', φ', and φ'' every so often are considered in the momentum space with the help of the Fourier transform with respect to all relative space coordinates. At the same time, the symmetry properties of the functions with respect to the nucleon permutations do not change. We do not write arguments of the spatial wave functions, which, in fact, depend on $\mathbf{r}_{ij}$, $\mathbf{x}_k$ or the corresponding relative momenta $\mathbf{k}_{ij}$, $\mathbf{p}_k$ (ijk=123, 231, 312).

Since the explicit form of the spin-isospin wave functions, appearing in the total wave function of the three-nucleon systems, is known, the further task is to obtain the spatial parts of the total wave functions or their equations of motion. The latter can be obtained by substitution of the total wave functions in the Schrödinger equation and eliminating the known spin and isospin wave functions.

The wave functions of bound states (2.137) and (2.141) are normalized by unity. Taking into account the normalization of the spin-isospin wave functions and by using (2.137), we obtain the following normalization condition for the spatial wave functions of ^3H nucleus :

$$\int d\mathbf{r}_{23} \int d\mathbf{x}_1 (|\psi^a|^2 + |\psi^s|^2 + |\psi'|^2 + |\psi''|^2) = 1. \qquad (2.142)$$

In the case of ^3He nucleus, we have

$$\int d\mathbf{r}_{23} \int d\mathbf{x}_1 (|\psi^a|^2 + |\psi^s|^2 + |\psi'|^2 + |\psi''|^2 + |\varphi'|^2 + |\varphi''|^2) = 1. \qquad (2.143)$$

The main contribution (more than 90%) in these six-dimensional normalization integrals gives the symmetric spatial wave function ψ^s. A few per cents contribution is given by the functions of the mixed symmetry ψ' and ψ'', the contribution of the functions ψ^a, φ' and φ'' is very small and every so often can be neglected. Thus, in a good approximation, the total wave function of three-nucleon nuclei in many cases can be written in the form

$$\Psi = \psi^s \xi^a + \psi' \xi'' - \psi'' \xi' \qquad (2.144)$$

with the corresponding normalization

$$\langle \Psi | \Psi \rangle = \int d\mathbf{r}_{23} \int d\mathbf{x}_1 (|\psi^s|^2 + |\psi'|^2 + |\psi''|^2) = 1. \qquad (2.145)$$

The main problem when obtaining the wave functions of three-nucleon systems is associated with the coordinate dependence (in the coordinate representation) or with the momentum dependence (in the momentum representation). There are different approximate, in particular, variational methods of solving the three-body problem when one begins with the given nucleon-nucleon potentials in the nonrelativistic Schrödinger equation and constructs more simple approximate systems of coupled differential (or integral, or integro-differential) equations. These equations can be solved numerically with the help of modern computers.

The *Faddeev equations* can be used to find the wave functions of the three-particle bound and unbound systems. They can be generalized on systems of four and more particles. Note that the Faddeev method was fruitfully used in the problems concerning to the three-particle effects in the nuclear matter.

Integral equations for a three-nucleon system with spin-dependent nucleon-nucleon interaction. The Faddeev equations may be extended to the case of a *three-nucleon system with spin-dependent interaction* (Sitenko and Kharchenko, 1963). One has to employ the equations (2.117) describing an arbitrary three-particle system with regard for the operator nature of the two-particle t-matrix in the nucleon spin-isospin space and spin- and isospin-dependences of the wave function. Since the nucleons are identical, the set of equations (2.117) reduces to a single equation

$$\Psi(\mathbf{p},\mathbf{k}) = \Phi(\mathbf{p},\mathbf{k}) + \left(z_p - \frac{\hbar^2 k^2}{M}\right)^{-1} \int \frac{d\mathbf{p}'}{(2\pi)^3}$$
$$\times\left\{\left\langle \mathbf{k}|t_{23}(z_p)| - \frac{\mathbf{p}}{2} + \mathbf{p}'\right\rangle \Psi^{(2)}\left(\mathbf{p}', -\mathbf{p} - \frac{\mathbf{p}'}{2}\right)\right.$$
$$\left. + \left\langle \mathbf{k}|t_{23}(z_p)| - \frac{\mathbf{p}}{2} - \mathbf{p}'\right\rangle \Psi^{(3)}\left(\mathbf{p}', \mathbf{p} + \frac{\mathbf{p}'}{2}\right)\right\}, \qquad (2.146)$$

where the functions $\Psi^{(2)}$ and $\Psi^{(3)}$ differ from $\Psi \equiv \Psi^{(1)}$ by a cyclic permutation of nucleon spatial, spin, and isospin coordinates ($z_p = Z - 3\hbar^2 p^2/4M$, M is the nucleon mass). The central nuclear interaction between two nucleons (i and j) is described by the potential

$$V_{ij} = \sum_{\nu=1}^{4} V^{(\nu)}(\mathbf{r}_{ij}) P_{ij}^{(\nu)}(\sigma, \tau), \qquad (2.147)$$

where $P_{ij}^{(\nu)}(\sigma,\tau)$ is the projection operator into a ν-spin-isospin state. The values ν=1,2,3, and 4 label respectively triplet-singlet ts, triplet-triplet tt, singlet-triplet st, and singlet-singlet ss spin-isospin two-nucleon states. For the central interaction (2.147), the two-nucleon t-matrix may be presented in the form

$$\langle \mathbf{k}|t_{ij}(z)|\mathbf{k}'\rangle = \sum_{\nu=1}^{4} \langle \mathbf{k}|t^{(\nu)}(z)|\mathbf{k}'\rangle P_{ij}^{(\nu)}(\sigma,\tau), \qquad (2.148)$$

where $t^{(\nu)}(z)$ is the t-matrix eigenvalue in the spin-isospin state ν. Projections of the equation (2.146) at probable three-nucleon spin-isospin states yield a set of integral equations for the spatial functions Ψ^s, Ψ^a, Ψ', and Ψ''.

It is convenient to pass from the functions Ψ^s, Ψ^a, Ψ', and Ψ'', which are transformed according to irreducible representations of the permutation group S_3, to the functions $\psi_\nu(\mathbf{p}_k, \mathbf{k}_{ij})$ associated with definite spin-isospin states ν=1,2,3, and 4 of the nucleon pair ij. The functions $\psi_\nu(\mathbf{p}_k, \mathbf{k}_{ij})$ are even, for ν=1 and 3, and odd, for ν=2 and 4, with respect to the permutation of particles i and j ($\mathbf{k}_{ij} \to \mathbf{k}_{ji} = -\mathbf{k}_{ij}$): $\psi_\nu(\mathbf{p}, -\mathbf{k}) = (-1)^{\nu+1}\psi_\nu(\mathbf{p}, \mathbf{k})$. In the case $S = 1/2$, $T = 1/2$, the spatial functions Ψ^s, Ψ^a, Ψ', and Ψ'' may be expressed in terms of four functions ψ_ν with ν=1,2,3, and 4; in the case $S = 1/2$, $T = 3/2$, the spatial functions Ψ' and Ψ'' are expressed in terms of the two functions ψ_2 and ψ_3; for $S = 3/2$, $T = 1/2$, the spatial functions are expressed in terms of two functions ψ_1 and ψ_2; at last, for $S = 3/2$, $T = 3/2$, the spatial function Ψ^a is expressed in terms of the single function ψ_2:

$$S = \frac{1}{2}, \; T = \frac{1}{2}, \; \begin{cases} \sqrt{2}\Psi^s = T^s(\psi_1 + \psi_3), \\ \sqrt{2}\Psi^a = T^s(\psi_2 + \psi_4), \\ \sqrt{2}\Psi' = T'(\psi_1 - \psi_3) + T''(\psi_2 - \psi_4), \\ \sqrt{2}\Psi'' = T''(\psi_1 - \psi_3) - T'(\psi_2 - \psi_4); \end{cases} \tag{2.149}$$

$$S = \frac{1}{2}, \; T = \frac{3}{2}, \; \begin{cases} \Psi' = -T'\psi_3 + T''\psi_2' \\ \Psi'' = -T''\psi_3 - T'\psi_2; \end{cases} \tag{2.150}$$

$$S = \frac{3}{2}, \; T = \frac{1}{2}, \; \begin{cases} \Psi' = T'\psi_1 + T''\psi_2, \\ \Psi'' = T''\psi_1 - T'\psi_2; \end{cases} \tag{2.151}$$

$$S = \frac{3}{2}, \; T = \frac{3}{2}, \; \Psi^a = T^s\psi_2, \tag{2.152}$$

where the denotation

$$\psi_\nu \equiv \psi_\nu(\mathbf{p}_3, \mathbf{k}_{12}), \; \nu = 1, 2, 3, \text{ and } 4$$

is introduced in order to shorten the formulas. In the general case, different three-nucleon spin-isospin states associated with different S and T are described by different systems of functions ψ_ν.

With the three-nucleon spin and isospin being arbitrary the set of integral equations for the functions $\psi_\nu(\mathbf{p}, \mathbf{k})$ may be presented in the form

$$\psi_\nu(\mathbf{p}, \mathbf{k}) = \varphi_\nu(\mathbf{p}, \mathbf{k}) + \left(z_p - \frac{\hbar^2 k^2}{M}\right)^{-1} \sum_{\nu'=1}^{4} \int \frac{d\mathbf{p}'}{(2\pi)^3}$$
$$\times \left\{ \langle \mathbf{k}|t^{(\nu)}(z_p)| - \frac{\mathbf{p}}{2} + \mathbf{p}'\rangle + (-1)^{\nu+1} \langle \mathbf{k}|t^{(\nu)}(z_p)| - \frac{\mathbf{p}}{2} - \mathbf{p}'\rangle \right\}$$
$$\times C_{\nu\nu'}\psi_{\nu'}\left(\mathbf{p}', \mathbf{p} + \frac{\mathbf{p}'}{2}\right), \tag{2.153}$$

The matrices $C^{(ST)}_{\nu\nu'}$ associated with different spin-isospin states of the three-nucleon system are the following:

$$C^{(\frac{1}{2}\frac{1}{2})}_{\nu\nu'} = \frac{1}{4}\begin{pmatrix} 1 & \sqrt{3} & 3 & -\sqrt{3} \\ \sqrt{3} & -1 & -\sqrt{3} & -3 \\ 3 & -\sqrt{3} & 1 & \sqrt{3} \\ -\sqrt{3} & -3 & \sqrt{3} & -1 \end{pmatrix},$$

$$C^{(\frac{1}{2}\frac{3}{2})}_{\nu\nu'} = \frac{1}{2}\begin{pmatrix} 0 & 0 & 0 & 0 \\ 0 & 1 & -\sqrt{3} & 0 \\ 0 & -\sqrt{3} & -1 & 0 \\ 0 & 0 & 0 & 0 \end{pmatrix}, \tag{2.154}$$

$$C^{(\frac{3}{2}\frac{1}{2})}_{\nu\nu'} = \frac{1}{2}\begin{pmatrix} -1 & -\sqrt{3} & 0 & 0 \\ -\sqrt{3} & 1 & 0 & 0 \\ 0 & 0 & 0 & 0 \\ 0 & 0 & 0 & 0 \end{pmatrix}, \; C^{(\frac{3}{2}\frac{3}{2})}_{\nu\nu'} = -\delta_{\nu 2}\delta_{\nu' 2}. \tag{2.155}$$

In order to consider three-nucleon bound states, one has to put the free term in (2.153) equal to zero. It is to be noted that integral equation (2.120) for a system of three identic spinless particles is reproduced with (2.153) if one puts in the latter

$$C_{\nu\nu'} = \delta_{\nu 1}\delta_{\nu' 1}.$$

The equation (2.153) was derived regardless of the Coulomb interaction. Therefore, it describes only the systems consisting of either three neutrons or two neutrons and one proton. The latter system allows for the formation of a bound state (triton) as well as neutron scattering by a bound state of two other particles. In the latter case the free term in (2.153) is to be written in the form

$$\varphi_\nu(\mathbf{p}, \mathbf{k}) = (2\pi)^3\delta(\mathbf{p} - \mathbf{p}_0)\varphi(\mathbf{k})\delta_{\nu 1}, \tag{2.156}$$

where $\varphi(\mathbf{k})$ is the deuteron ground state wave function.

Calculation of the three-nucleon system characteristics on the Faddeev equation basis. The assumed two-particle character of the nuclear forces between nucleons in the above-reported method of the integral equations allows one to calculate the parameters of three-nucleon systems and describe the processes in these systems. There are many calculations of the binding energy and other characteristics of ^3H and ^3He nuclei within the framework of different models of the nucleon-nucleon interaction.

It is clear that the computation process complexity considerably depends on the relevant model potential. If we take into account the two-particle interaction in the S-states only and describe it by the separable potential, the problem of the binding energy determination reduces to solving of the system of one-dimensional equations. In the more general case when considering the interaction in the states with larger angular momentum, system (2.153) transforms into the system of coupled two-dimensional integral equations.

The qualitatively correct values of the binding energy of three-nucleon systems can be obtained even for the simplest potentials of two-nucleon interactions. However, the central attractive potentials reconciled with the data on the interaction of two nucleons at low energies lead to overestimates of the binding energy of triton compared with the experimental value. The correct behavior of the scattering amplitude of two nucleons on the energy shell at low energies can be ensured by describing the interaction between the nucleons by means of potentials containing two parameters, which are determined by specifying two experimental quantities – the scattering length and effective radius. To find the two-particle scattering amplitude on the energy shell in a large interval of energies, it is necessary to use more complicated potentials, characterized by a larger number of parameters. As is known, data on nucleon-nucleon scattering in the high-energy region point to the existence of a strong repulsion between the nucleons at small distances. A potential with repulsion at small distances and with subsequent short-range attraction should contain at least three parameters. Introduction of the repulsive core radius – an additional parameter characterizing the repulsion – makes it possible to describe correctly the scattering amplitude both at low and high energies. The values of the binding energy of the triton for different potentials correctly describing the phases of the two-nucleon scattering in a large energy interval are quite close to one another.

As an example we cite the results of calculations of some characteristics of ^3H nuclei obtained by using one of the modifications of the *Reyd potential* with a soft repulsive core (Akaishi, 1986). The calculated binding energy of a triton equals to E_T=7.023 MeV, the mean squared radius of the charge distribution in the nucleus $< r^2 >_c^{1/2}$=1.70 fm, the *probability of the D-state* happened to be relatively high P_D=0.093. We also obtained the probabilities of the symmetric S-state

$$P_S \equiv \int d\mathbf{r}_{23} \int d\mathbf{x}_1 |\Psi^s|^2 = 0.889$$

and state with the intermediate symmetry

$$P_{\tilde{S}} \equiv \int d\mathbf{r}_{23} \int d\mathbf{x}_1 \Big(|\Psi'|^2 + |\Psi''|^2 \Big) = 0.017.$$

The use of the Paris nucleon-nucleon potential gives the following numerical parameters of ^3H nucleus (Sasakawa, 1986): the binding energy $E_T = 7.64\,\text{MeV}$, the charge distribution radius $\langle r^2 \rangle_c^{1/2}$=1.79 fm,

$$P_S = 0.901, \; P_{\tilde{S}} = 0.013, \; P_D = 0.085.$$

For ^3He nucleus: the Coulomb energy E_c=0.66 MeV, the charge distribution radius $\langle r^2 \rangle_c^{1/2}$=2.00 fm. It is shown that the addition of the three-nucleon nuclear forces can change the calculated parameters by a few per cents.

As an illustration we give the results of the differential cross-section of the elastic neutron-deuteron scattering at relatively low energies by using the phenomenological separable potentials. The dependencies of the calculated cross-section $\sigma(\vartheta) = \frac{\partial \sigma}{\partial \Omega}$ of the elastic neutron-deuteron scattering on the scattering angle ϑ (in the centre-of-mass coordinate system) at two different energies of incident neutrons E_n=10 MeV and E_n=22.7 MeV and the corresponding experimental data (Adhikari and Tomio, 1995) are given in Fig. 2.4. The upper continuous curves correspond to the *Yamaguchi potential* in both the triplet and singlet states of two nucleon (the calculated binding energy of a triton in this model is E_T=10.65 MeV). The lower continuous curves correspond the Yamaguchi potential in the triplet state and the *Tabakin potential* in the singlet state (the triton binding energy is E_T=7.69 MeV).

The curves corresponding to the different versions are pretty close. The considerable departures appear in the vicinity of a minimum energy E_n=22.7 MeV. The calculated distributions are satisfactorily correspond to the experimental angular distributions. Note that the polarization phenomena are more sensitive to the interaction detail.

Basically, we can construct a system of integral equations analogous to (2.146) even for a larger number of nucleons with the pair interaction. However, if calculations of the scattering amplitude in four-nucleon systems (amplitudes of nucleon-three-nucleon system in a bound state or deuteron-deuteron scattering) can be justified, the use of these equations to calculate the four-nucleon system binding energy is not to the purpose. In fact, for nuclei with mass-numbers $A > 4$ and even for the ^4He nucleus the effect of saturation of the nuclear interaction becomes essential. Due to this effect the total binding energy of a nuclear system is proportional to a number of nucleons (i.e., to the mass-number A) but not to a number of the interacting pairs.

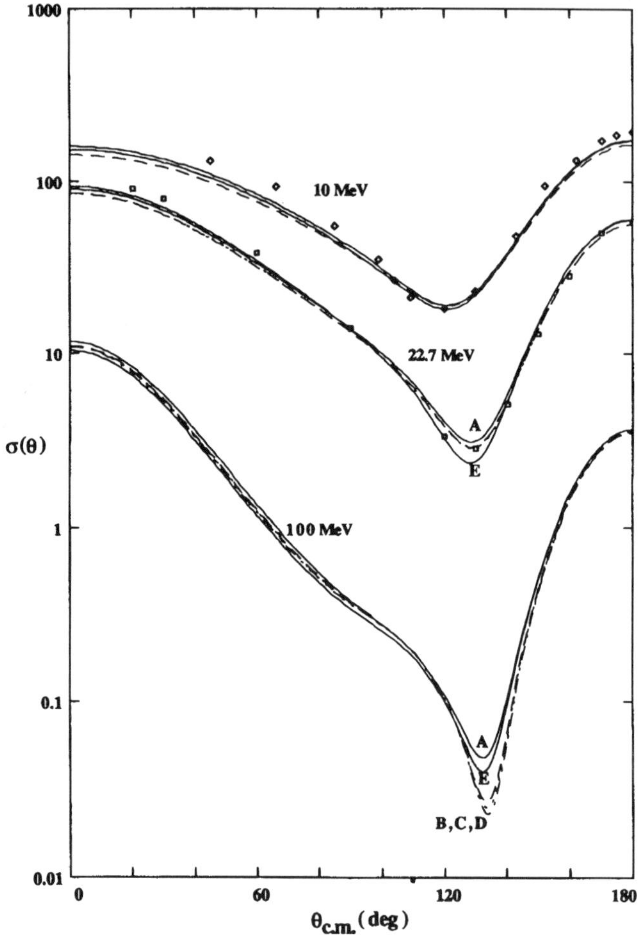

Figure 2.4. The $n - d$ elastic differential scattering cross-section in mb/sr at different incident nucleon energies

2.4. K-Harmonic Method

Hyperspherical functions. To find the spatial wave functions of few-particle systems from the Schrödinger equation, the method of K-harmonic or harmonic polynomials is frequently used. The idea of this method lies in the special representation of the spatial wave functions of few-particle systems in the form of infinite series with respect to some complete sets of the known functions of angle variables which are called *K-harmonics* or *hyperspherical functions*. The method of harmonic polynomials is especially efficient when solving the problems concerning bound states of few-particle systems. We consider a system of A particles (nucleons) with equal masses M, with radius-vectors denoted by $\mathbf{r}_j$ $(j = 1, 2, \ldots, A)$. Let us denote the center-mass radius-vector as $\mathbf{R}$,

$$\mathbf{R} = \frac{1}{A}\sum_j \mathbf{r}_j, \tag{2.157}$$

and introduce $A - 1$ relative radius-vectors $\boldsymbol{\xi}_k$, which are conventionally called *Jacoby's vectors*:

$$\boldsymbol{\xi}_k = \sqrt{\frac{k}{k+1}}\left(\mathbf{r}_{k+1} - \frac{1}{k}\sum_{j=1}^{k}\mathbf{r}_j\right), \qquad k = 1, 2, \ldots, A-1. \tag{2.158}$$

The complete $3A$-dimensional Laplace operator can be written in the form

$$\sum_{j=1}^{A}\frac{\partial^2}{\partial\mathbf{r}_j^2} = \frac{1}{A}\frac{\partial^2}{\partial\mathbf{R}^2} + \sum_{j=1}^{A-1}\frac{\partial^2}{\partial\boldsymbol{\xi}_j^2} \equiv \frac{1}{A}\frac{\partial^2}{\partial\mathbf{R}^2} + \Delta_{3(A-1)}, \tag{2.159}$$

where $\Delta_{3(A-1)}$ - is the Laplace operator in a space $3(A - 1)$ relative coordinates. The elementary volume in $3A$-dimensional space is given by

$$\Pi_{j=1}^{A}d\mathbf{r}_j = A^{3/2}d\mathbf{R}\Pi_{j=1}^{A-1}d\boldsymbol{\xi}_j. \tag{2.160}$$

We now introduce a vector $\boldsymbol{\rho} = (\boldsymbol{\xi}_1, \boldsymbol{\xi}_2, \ldots, \boldsymbol{\xi}_{A-1})$ in $3(A-1)$-dimensional space of relative coordinates, which is called the *global radius*, and its modulus ρ is determined by the expression

$$\rho^2 = \sum_{j=1}^{A-1}|\boldsymbol{\xi}_j|^2 = \frac{1}{2A}\sum_{i,j}^{A}(\mathbf{r}_i - \mathbf{r}_j)^2 = \sum_{j=1}^{A}(\mathbf{r}_j - \mathbf{R})^2. \tag{2.161}$$

A direction of this vector in $3(A - 1)$-dimensional space, apparently, is specified by a set of $3(A - 1) - 1$ angular variables, which are denoted by one letter Ω. They can be chosen by different ways. The global radius ρ is the unique dimensional variable in $3(A-1)$-dimensional space. Introducing the spherical system of coordinates, we can present the elementary volume as follows

$$d\boldsymbol{\rho} \equiv \Pi_{j=1}^{A-1}d\boldsymbol{\xi}_j = \rho^{3A-4}d\rho d\Omega, \tag{2.162}$$

where $d\Omega$ is an element of a solid angle with the vector $\boldsymbol{\rho}$ inside.

Note that at any choice of the angular variables Ω the Laplace operator in $3(A - 1)$-dimensional space can be written in the form

$$\Delta_{3(A-1)} = \frac{1}{\rho^{3A-4}}\frac{\partial}{\partial\rho}\left(\rho^{3A-4}\frac{\partial}{\partial\rho}\right) + \frac{1}{\rho^2}\Delta_\Omega, \tag{2.163}$$

where Δ_Ω is the operator angular part containing derivatives in $3A - 4$ variables in the corresponding spherical coordinate system.

We introduce now the so-called *harmonic polynomials* P_{K_n} choosing them in the form

$$P_{Kn} = \rho^K u_{Kn}(\Omega), \tag{2.164}$$

where functions $u_{Kn}(\Omega)$ (K and n are some quantum numbers) depend on the angular variables Ω. They satisfy the equation

$$\Delta_{3(A-1)} P_{Kn} = 0. \tag{2.165}$$

At the same time, the angular wave functions $u_{Kn}(\Omega)$ are the eigenfunctions of the operator Δ_Ω in the multi-dimensional angular space:

$$\Delta_\Omega u_{Kn}(\Omega) = -K(K + 3A - 5) u_{Kn}(\Omega). \tag{2.166}$$

The angular functions $u_{Kn}(\Omega)$ are called the K-harmonics and normalized by the condition:

$$\int d\Omega u^*_{Kn}(\Omega) u_{K'n'}(\Omega) = \delta_{KK'} \delta_{nn'}. \tag{2.167}$$

The quantum number K, which coincides with the polynomial of order P_{Kn}, describes the value of the total angular momentum in $3(A-1)$-dimensional space and can take any nonnegative integers.

The rest quantum numbers are denoted by a letter n and contain the total angular momentum L of the relative motion of the system in a conventional three-dimensional space, its projection M on a selected axis, the quantum number ν describing the symmetry of the function $u_{Kn}(\Omega)$ with respect to permutations of particle coordinates and include additional quantum numbers, which are necessary in the case of large values of K and L. At $A = 2$, the number K coincides with the quantum numbers of the angular momentum L, and the functions $u_{Kn}(\Omega)$ are the well known spherical functions $Y_{LM}(\vartheta, \varphi)$. Thus, the K-harmonics (hyperspherical functions), i.e., the angular functions $u_{K_n}(\Omega)$, at $A > 2$ can be treated as natural generalization of the spherical functions.

The functions u_{Kn} form a complete set in $3A - 4$-dimensional space of angular variables. Because of this the wave function of the system relative motion can be presented in the expansion form with respect to these functions. The coefficients of the expansions depend on the global radius ρ.

Application of the K-harmonic method in the three-particle problem. We use now the K-harmonic method to the description of a bound state of three particles. Two relative three-dimensional vectors of three particles

$$\begin{aligned}\boldsymbol{\xi}_1 &= \frac{1}{\sqrt{2}}(\mathbf{r}_2 - \mathbf{r}_1) \equiv \frac{1}{\sqrt{2}}\mathbf{r}_{21} \qquad \text{and} \\ \boldsymbol{\xi}_2 &= \sqrt{\frac{2}{3}}\left(\mathbf{r}_3 - \frac{1}{2}(\mathbf{r}_1 + \mathbf{r}_2)\right) \equiv \sqrt{\frac{2}{3}}\mathbf{x}_3 \end{aligned} \tag{2.168}$$

are united in one six-dimensional vector $\boldsymbol{\rho}(\boldsymbol{\xi}_1, \boldsymbol{\xi}_2)$, whose modulus is given by the expression

$$\begin{aligned}\rho &= (\boldsymbol{\xi}_1^2 + \boldsymbol{\xi}_2^2)^{1/2} = \left(\frac{1}{2}\mathbf{r}_{12}^2 + \frac{2}{3}\mathbf{x}_3^2\right)^{1/2} \equiv \left(\frac{1}{2}\mathbf{r}_{23}^2 + \frac{2}{3}\mathbf{x}_1^2\right)^{1/2} \\ &\equiv \left(\frac{1}{2}\mathbf{r}_{13}^2 + \frac{2}{3}\mathbf{x}_2^2\right)^{1/2}. \end{aligned} \tag{2.169}$$

In a spherical coordinate system of the introduced six-dimensional space, the global variable has dimension of length. Rest five variables are the angles determining an orientation of the six-dimensional vector $\boldsymbol{\rho}$.

The kinetic energy operator of the relative motion of three particles in the spherical coordinate system takes the form

$$H_0 = -\hbar^2/2M\left\{\frac{1}{\rho^5}\frac{\partial}{\partial\rho}\left(\rho^5\frac{\partial}{\partial\rho}\right) + \frac{1}{\rho^2}\Delta_\Omega\right\}, \tag{2.170}$$

where Δ_Ω is the angular part of the Laplace operator containing derivatives with respect to five angular variables. The *hyperspherical functions* $u_{Kn}(\Omega)$ satisfy the equation

$$\Delta_\Omega u_{Kn}(\Omega) = -K(K+4)u_{Kn}(\Omega). \tag{2.171}$$

The quantum number K can take nonnegative integers, number n includes values of the total angular momentum of the relative motion of three particles L and its projection on a particular axis M, and quantum number ν characterizes the function permutation symmetry. Out of the quantum number n we can separate an angular momentum of a pair particles l_{23} and angular momentum of one particle l_1 with respect to the inertia center of two other particles. At $L = 0$, the quantum numbers K and ν completely determine a set of the angular orthonormalized functions $u_{Kn}(\Omega)$. They present the complete set of functions of five angular variables Ω. The wave function of relative motion of three nucleons $\Psi \equiv \Psi(\boldsymbol{\xi}_1, \boldsymbol{\xi}_2)$ can be expanded with respect to these functions:

$$\Psi = \sum_{Kn} \varphi_{Kn}(\rho)u_{Kn}(\Omega). \tag{2.172}$$

The explicit form of the angular functions $u_{Kn}(\Omega)$ specified by relations (2.171) and (2.167) can be treated as known and independent on the properties of a nucleon system. The radial functions $\varphi_{Kn}(\rho)$ in expansion (2.172) depend on these properties and, first of all, on the nucleon-nucleon interaction.

Let us elucidate how one can introduce five angular variables Ω and, in particular, write down in the explicit form $d\Omega$ in formula (2.167). For this purpose we start with the expression of the volume element in six-dimensional space

$$d\mathbf{r}_{21}d\mathbf{x}_3 = 3^{3/2}d\boldsymbol{\xi}_1 d\boldsymbol{\xi}_2, \tag{2.173}$$

which is necessary to perform an integration in the six-dimensional normalization integral for the bound state wave function and when calculating different matrix elements of the three-nucleon system. In the spherical coordinate system we have

$$d\boldsymbol{\xi}_1 d\boldsymbol{\xi}_2 = \xi_1^2 d\xi_1 d\Omega \xi_2^2 d\xi_2 d\Omega_2, \qquad \xi_1 \equiv |\boldsymbol{\xi}_1|, \qquad \xi_2 \equiv |\boldsymbol{\xi}_2|, \tag{2.174}$$

where $d\Omega_1 = \sin\vartheta_1 d\vartheta_1 d\varphi_1$ and $d\Omega_2 = \sin\vartheta_2 d\vartheta_2 d\varphi_2$ are elements of the solid angles where the three-dimensional vectors $\boldsymbol{\xi}_1$ and $\boldsymbol{\xi}_2$ lie. Keeping in mind that $\rho^2 = \xi_1^2 + \xi_2^2$, we introduce the polar coordinates ρ and Θ in the plane ξ_1, ξ_2:

$$\xi_1 = \rho\cos\Theta, \qquad \xi_2 = \rho\sin\Theta. \tag{2.175}$$

Then the area element $d\xi_1 d\xi_2$ should be changed by $\rho d\rho d\Theta$, and the volume element in the six-dimensional space can be presented as follows

$$d\boldsymbol{\rho} = 3^{3/2}\rho^5 d\rho d\Omega, \qquad d\Omega = \sin^2\Theta\cos^2\Theta d\Theta d\Omega_1 d\Omega_2. \tag{2.176}$$

Since $\xi_1 \geqslant 0$ and $\xi_2 \geqslant 0$, the angle Θ changes from 0 to $\pi/2$. At this, as usually, $0 \leqslant \vartheta_1$, $\vartheta_1 \leqslant \pi$ and $0 \leqslant \varphi_1$, $\varphi_2 \leqslant 2\pi$. In this way we completely determined the element of the six-dimensional solid angle $d\Omega$ and the integration limits in (2.167) with respect to five angular variables. Thus, accordingly (2.176) in the spherical coordinates in the six-dimensional space of the relative spatial radius-vectors for a system of three nucleons there are one dimensional variable ρ and five angular variables $\Theta, \vartheta_1, \varphi_1, \vartheta_2, \varphi_2$.

Table 2.1 shows the explicit form of the first K-harmonics $u_{Kn}(\Omega)$ in the simplest case $A = 3$ corresponding to K=0,1,2 (Simonov, 1966). In the columns of Table 2.1 there are the total number $n_K = \frac{(K+3)!(K+2)}{12K!}$ of K-harmonics of the K-th order, total number m_K ($\sum m_K = n_K$) of a given

TABLE 2.1.

K	n_K	m_K	L	l_{23}	l_1	$u_{K_n}(\Omega)$
0	1	1	0	0	0	$1/\sqrt{\pi^3}$
1	6	3	1	1	0	$\frac{1}{\pi}\sqrt{8}\cos\Theta Y_{1m}(\vartheta_1,\varphi_1)$
		3	1	0	1	$\frac{1}{\pi}\sqrt{8}\sin\Theta Y_{1m}(\vartheta_2,\varphi_2)$
2	20	5	2	2	0	$\frac{8}{\sqrt{5\pi}}\cos^2\Theta Y_{2m}(\vartheta_1,\varphi_1)$
		5	2	0	2	$\frac{8}{\sqrt{5\pi}}\sin^2\Theta Y_{2m}(\vartheta_2,\varphi_2)$
		5	2	1	1	$\frac{1}{\rho^2}(\xi_{1i}\xi_{2k}+\xi_{2i}\xi_{1k}-\frac{2}{3}\vec{\xi}_1\vec{\xi}_2\delta_{ik})\frac{1}{\sqrt{\pi^3}} \cdot \begin{cases} ß\sqrt{24}, & ßi \neq k \\ ß\sqrt{18}, & ßi = k \end{cases}$
		3	1	1	1	$\frac{1}{\rho^2}\sqrt{\frac{12}{\pi^3}}(\xi_{1i}\xi_{2k}-\xi_{2i}\xi_{1k}),\ i,k=x,y,z$
		1	0	0	0	$\frac{2}{\sqrt{\pi^3}}\cos 2\Theta$
		1	0	1	1	$\frac{4}{\sqrt{\pi^3}}\frac{\vec{\xi}_1\vec{\xi}_2}{\rho^2}$

type K-harmonics, the total angular three-dimensional relative momenta L (L runs odd values from 0 to K at K odd), momentum l_{23} of a pair and momentum l_1 of the first nucleon with respect to the inertia center of a pair.

A value $L = 0$ of the total angular momentum corresponds to K-harmonics with even K only, moreover, at $L = 0$ n in $u_{Kn}(\Omega)$ reduces to one number ν. (Recall that for bound states of three nucleons, i.e., for nuclei ^3H and ^3He, a contribution of terms with $L = 0$ in the total wave function is dominant.) In this important particular case the functions $u_{K\nu}(\Omega)$ depend not on five variables but on two variables only: $\lambda\,(0 \leqslant \lambda \leqslant 2\pi)$ and $\mu\,(0 \leqslant \mu \leqslant 1)$ connected with the former angles Θ and φ between three-dimensional vectors $\boldsymbol{\xi}_1$ and $\boldsymbol{\xi}_2$ by the relations

$$\mu = \sqrt{\cos^2 2\Theta + \sin^2 2\Theta \cos^2 \varphi}, \qquad \cos\lambda = \frac{\cos 2\Theta}{\mu}. \tag{2.177}$$

Normalized accordingly (2.167) at $n = \nu$ functions u_{K_ν} can be expressed in terms of the *Jacobi polynomials* $P_N^{ab}(z)$:

$$u_{K\nu} = \sqrt{\frac{K+2}{2\pi^3}}\mathrm{e}^{-i\nu\lambda}\mu^{|\nu|}P^{|\nu|,0}_{1/2(K/2-|\nu|)}(1-2\mu^2), \tag{2.178}$$

$$\nu = -\frac{K}{2}, -\frac{K}{2}+2, \ldots, +\frac{K}{2}.$$

After integration over three (out of five) angular variables in (2.167) the normalization condition looks much simpler ($\Omega \to \lambda, \mu$)

$$\int d\Omega u^*_{K\nu}(\lambda,\mu)u_{K'\nu'}(\lambda,\mu) = \pi^2\int_0^1 d\mu\mu\int_0^{2\pi} d\lambda u^*_{K\nu}(\lambda,\mu)u_{K'\nu'}(\lambda,\mu) = \delta_{KK'}\delta_{\nu\nu'}. \qquad (2.179)$$

To determine the radial parts $\varphi_{Kn}(\rho)$ of the relative motion of the three-nucleon function, we substitute (2.172) in the Schrödinger equation $\{\hat{H}_0+\hat{V}_{12}+\hat{V}_{23}+\hat{V}_{31}-E\}\Psi=0$. Multiplying the both parts of the obtained equation from the left by $u^*_{Kn}(\Omega)$ and integrating over the angular variables, we obtain the infinite chain of the second order ordinary differential equations

$$\left[\frac{d^2}{d\rho^2}+\frac{5}{\rho}\frac{d}{d\rho} - \frac{K(K+4)}{\rho^2}+\frac{2ME}{\hbar^2}\right]\varphi_{Kn}(\rho) = \frac{2M}{\hbar^2}\sum_{K'n'} V_{Kn,K'n'}(\rho)\varphi_{K'n'}(\rho), \qquad (2.180)$$

$$V_{Kn,K'n'}(\rho)=\int d\Omega u^*_{Kn}(\Omega)(\hat{V}_{12}+\hat{V}_{23}+\hat{V}_{31})u_{K'n'}(\Omega). \qquad (2.181)$$

For the bound system ($E<0$) consisting of a proton and two neutrons (^3H nucleus) we can write the asymptotic expression at $\rho\to\infty$ for an arbitrary radial function $\varphi_{Kn}(\rho)$:

$$\varphi_{Kn}(\rho)\underset{\rho\to\infty}{\longrightarrow}\rho^{\frac{5}{2}}\exp\left(-\rho\sqrt{\frac{2M|E|}{\hbar^2}}\right), \qquad (2.182)$$

following directly from equations (2.180), which become uncoupled at large ρ.

The infinite system of equations (2.180) is the exact one, but for all practical purposes it must be cutted, as well as expansion (2.172), at a finite value K. For bound states of three nucleons this can be justified, since, in this case, the series in K-harmonics (2.172) converges fast and even the first term with $K=0$ gives the dominant contribution in the ground state wave functions of nuclei ^3H and ^3He for the realistic nucleon-nucleon potentials. The terms with $K\neq 0$ give only small corrections.

Note that for the model nucleon-nucleon potentials of an oscillator type

$$\hat{V}_{ij}=\gamma\mathbf{r}_{ij}, \qquad \mathbf{r}_{ij}=\mathbf{r}_i-\mathbf{r}_j \qquad (2.183)$$

matrix elements (2.181) have the diagonal form

$$V_{Kn,K'n'} = 3\gamma\rho^2\delta_{KK'}\delta_{nn'}. \tag{2.184}$$

At the same time, system equations (2.180) decays into independent differential equations for each radial function $\varphi_{Kn}(\rho)$, and K becomes a good (exact) quantum number. However, in a general case, for the different than (2.183) potentials $\hat{V}_{ij}$, the nondiagonal matrix elements (2.181) are not equal to zero and K is not an integral of motion any more.

When obtaining the wave functions of bound three-nucleon systems, i.e., ^3H and ^3He nuclei, one can use a few first terms of expansion (2.172) with $L = 0$ (the contribution of terms with $L > 0$ turns out to be very small). At $L = 0$ the set of the quantum numbers n reduces to one number ν.

Since the nuclei ^3H and ^3He are characterized by the positive parity, the expansions of their spatial wave functions in the K-harmonics consist of only terms with even K. We give here the expressions of spatial wave functions Ψ^s, Ψ' and Ψ'', keeping in the expansions terms with small K and indicating the explicit dependence on angular variables

$$\Psi^s = \frac{1}{\sqrt{\pi^3}}\varphi_{00}(\rho) + \sqrt{\frac{3}{\pi^3}}\left[1 - 2\left(\frac{\xi_1^2 - \xi_2^2}{\rho^2}\right)^2 - 8\frac{(\boldsymbol{\xi}_1\boldsymbol{\xi}_2)^2}{\rho^4}\right]\varphi_{40}(\rho), \tag{2.185}$$

$$\Psi' = \frac{4}{\sqrt{\pi^3}}\frac{\boldsymbol{\xi}_1\boldsymbol{\xi}_2}{\rho^2}\varphi_{21}(\rho), \qquad \Psi'' = \frac{2}{\sqrt{\pi^3}}\frac{\xi_1^2 - \xi_2^2}{\rho^2}\varphi_{21}(\rho). \tag{2.186}$$

The factors standing before the radial functions $\varphi_{K\nu}(\rho)$ are the K-harmonics of even order K=0,2,4 (the small contributions of terms with $K = 4$ in Ψ' and Ψ'' are neglected). The radial functions $\varphi_{K\nu}(\rho)$ can be obtained by the numerical solving of the system of differential equations (2.180) and (2.181). At the same time, we use the potentials obtained on the basis of data on the deuteron and nucleon-nucleon scattering. Note that the radial functions are interrelated by the normalization condition

$$3^{3/2}\int_0^\infty d\rho\rho^5\left[\varphi_{00}^2(\rho) + 2\varphi_{21}^2(\rho) + \varphi_{40}^2(\rho)\right] = 1. \tag{2.187}$$

Usually in the total wave functions of the nuclei ^3H and ^3He, the main term $\sim \Psi^s\xi^a$ is kept. This means that the spatial function Ψ^s is presented by the term with the first K-harmonic $K = 0$ (the next term corresponds to $K = 4$). In this case for rough evaluations, the function Ψ^s can be approximated by the Gauss dependence on ρ:

$$\Psi^s = \frac{2^{3/2}\alpha^3}{3^{3/4}\sqrt{\pi^3}}\exp^{-\alpha^2\rho^2}, \tag{2.188}$$

where the structure parameter α, determined from the data on the elastic electron scattering by three-nucleon nuclei, equals to $0.37\,\mathrm{fm}^{-1}$.

Interpolation model. The K-harmonic method is convenient when obtaining the bound state wave functions of three-nucleon systems. However, it is impossible to apply the K-harmonic method directly for the obtaining of wave functions related to the continuous spectrum of the system. In fact, the direct expansion of the three-particle wave function $\Psi(\boldsymbol{\xi}_1, \boldsymbol{\xi}_2)$, describing an unbound state, with respect to K-harmonics, similar to (2.172), is useless. The corresponding expansion converges very slowly and it requires to take into account basically an entire infinite series.

Usually for the description of bound and two-cluster states of the continuous spectrum of three-particle systems the so-called interpolation model is used. The idea of this method lies in the following. First of all, we have to construct the explicit form of the asymptotics $P(\boldsymbol{\xi}_1, \boldsymbol{\xi}_2)$ of the continuous spectrum wave function $\Psi(\boldsymbol{\xi}_1, \boldsymbol{\xi}_2)$ at large ρ. Next, the difference $\Psi(\boldsymbol{\xi}_1, \boldsymbol{\xi}_2) - \Phi(\rho)P(\boldsymbol{\xi}_1, \boldsymbol{\xi}_2)$ is expanded with respect to K-harmonics $u_{Kn}(\Omega)$. At this $\Phi(\rho)$ is an unknown function of one variable ρ, $\Phi(\rho) \to 1$ at $\rho \to \infty$. As an example we consider the nucleon-deuteron system. The function $P(\boldsymbol{\xi}_1, \boldsymbol{\xi}_2)$ is constructed as a product of the deuteron wave function $\varphi_d(\boldsymbol{\xi}_1)$ and the wave function of the relative motion of the deuteron and unbound nucleon $\Psi(\boldsymbol{\xi}_2)$. In the first (rough) approximation, we can take the plane wave as the function $\Psi(\boldsymbol{\xi}_2)$.

Thus, the wave function of the *nucleon-deuteron system* , from the beginning, is presented in the form

$$\Psi(\boldsymbol{\xi}_1, \boldsymbol{\xi}_2) = \sum_{Kn} C_{Kn}(\rho)u_{Kn}(\Omega) + \Phi(\rho)P(\boldsymbol{\xi}_1, \boldsymbol{\xi}_2), \tag{2.189}$$

where functions $u_{Kn}(\Omega)$ and $P(\boldsymbol{\xi}_1,\ \boldsymbol{\xi}_2)$ are known; functions $C_{Kn}(\rho)$ and $\Phi(\rho)$ do not depend on angular variables and can be found numerically from the equations which obtained when whole function (2.189) is substituted in the Schrödinger equation with given the nucleon-nucleon potentials. The part of wave function (2.189), namely $\sum_{Kn} C_{Kn}(\rho)u_{Kn}(\Omega)$ is important and gives the main contribution in $\Psi(\boldsymbol{\xi}_1, \boldsymbol{\xi}_2)$ in the domain where the nuclear forces are essential, i.e., at sufficiently small ρ. Because of this it is called the inner part of the function $\Psi(\boldsymbol{\xi}_1, \boldsymbol{\xi}_2)$. The series in K-harmonics of this part of the total function $\Psi(\boldsymbol{\xi}_1, \boldsymbol{\xi}_2)$ converges fast. Usually, only the first term with $K = 0$ is kept.

The second part of the wave function $\Psi(\boldsymbol{\xi}_1, \boldsymbol{\xi}_2)$, namely $\Phi(\rho)P(\boldsymbol{\xi}_1, \boldsymbol{\xi}_2)$, known as the outer part of total wave function (2.189), gives a dominant contribution at large ρ, where all functions $C_{Kn}(\rho)$ tend to zero. If, at least, one nucleon of the system moves far away at large distances from the system nucleons, then, according to (2.169), the global variable ρ also becomes large. When obtaining functions $C_{Kn}(\rho)$ and $\Phi(\rho)$, the variational principle is usually used. It is clear that in the external part of function (2.189), unlike its inner part, the dominant contribution is given by a large set of K-harmonics with high K. Sometime in the external part of the wave function the first K-harmonics are omitted (which correspond to K-s that are kept in the expansion of the inner part of function (2.189)). In this case the both parts of function (2.189) are mutually orthogonal that considerably simplifies calculations.

The described approach to the obtaining the continuous spectrum function $\Phi(\boldsymbol{\xi}_1, \boldsymbol{\xi}_2)$ in form (2.189) is called the interpolation model. This model can also be used when studying systems with $A > 3$ nucleons to construct the wave functions of bound and unbound states.

Calculation of three-nucleon system parameters. We report now the calculation results of the main parameters of three-nucleon nuclei on the basis of the K-harmonic method. In particular, with the obtained wave functions we have calculated the binding energies of ^3H and ^3He nuclei, magnetic moments, mean squared radii, form factors of charge distribution and magnetic moment, Coulombic energy of the ^3He nucleus.

Table 2.2 gives the calculated values of the binding energy E of ^3H and ^3He nuclei, calculated Coulomb energy of the ^3He nucleus, calculated magnetic moments μ (in nuclear magnetons), calculated mean squared radii of the electric charge $\sqrt{< r^2 >_c}$ and magnetic moment $\sqrt{< r^2 >_m}$ distributions, and the experimental values of these quantities.

To obtain the theoretical data of Table 2.2, we calculated preliminary the radial functions in (2.186) (see also (2.172)) for the simplest rectangular nucleon-nucleon potential. We also took account of the clasterization of ^3H and ^3He into a nucleon and deuteron. Because of this we found wave function (2.189) within the framework of the interpolation model. (The clusterization is taken into account by the second term of (2.189) with the radial function $\Phi(\rho)$.) The theoretical data of Table 2.2 obtained with these wave functions are given in the main approach, i.e., on keeping both in (2.172) and in the first term of (2.189) only one main harmonic $K = 0$. It is seen that even such rather simplified calculations with the help of the K-harmonic method give the nuclear parameters close to the experimental values. The clusterization effect in many cases improves the correspondence between calculations and experiment.

The K-harmonic method was also used to calculate the binding energies of nuclei with a larger nucleon number. However, as well as in the case of use the Faddeev integral equations, these calculations cannot explain the saturation effect of the nuclear interaction.

As we already noted before, the saturation of the nuclear interaction manifests itself in the fact that the total binding energy of nuclei happens to

TABLE 2.2.

	Nucleus ^3H		
	Theory		Experiment
	without clasterization	with clasterization	
E, MeV	7.875	8.959	8.482
E_C, MeV	–	–	–
μ/M_N	2.793	2.715	2.979
$< r^2 >_c^{1/2}$, fm	1.77	1.69	1.70±0.05
$< r^2 >_m^{1/2}$, fm	1.79	1.76	1.70±0.05
	Nucleus ^3He		
	Theory		Experiment
	without clasterization	with clasterization	
E, MeV	7.088	8.240	7.718
E_C, MeV	0.783	0.726	0.764
μ/M_N	-1.913	-1.829	-2.127
$< r^2 >_c^{1/2}$, fm	1.79	1.87	1.88±0.05
$< r^2 >_m^{1/2}$, fm	1.83	1.86	1.95±0.11

be proportional to a nucleon number A, but not a number of the interacting pairs. At the same time, nuclei have finite dimensions, and their volumes are also proportional A. In other words, the nucleon density inside nuclei is almost the same from the lightest nucleus ^4He up to the heaviest ones.

This property of nuclei allows the nuclear substance to be treated as incompressible liquid whose properties do not depend on the mass-number A. Apparently, the nuclear matter properties must be determined by the character of the nuclear interaction. Therefore, the study of the nuclear substance must give additional information on the character of the nucleon-nucleon interaction. At the same time, we have to note that, in spite of the similarity of the properties of the nuclear substance and incompressible liquid, the different nuclear properties can be described by reducing the two-particle interaction to some self-consistent field and treating a nucleus as a system of quasiparticles in this self-consistent field. This approach lies in the basis of the shell nuclear model.

Problems

2.1. Calculate the *Green function for a system of three noninteracting particles* .

Let us denote the particle coordinates and masses by $\mathbf{r}_1$, $\mathbf{r}_2$, $\mathbf{r}_3$ and m_1, m_2, m_3, respectively. We transform to the c.m.s. and introduce the relative coordinates

$$\mathbf{x} = \mathbf{r}_1 - \frac{m_2\mathbf{r}_2 + m_3\mathbf{r}_3}{m_2 + m_3}, \qquad \mathbf{r} = \mathbf{r}_2 - \mathbf{r}_3. \tag{2.190}$$

The Hamiltonian of the system may be written as

$$H_0 = -\frac{\hbar^2}{2m}\nabla^2_{\mathbf{x}} - \frac{\hbar^2}{2\mu}\nabla_{\mathbf{r}}, \tag{2.191}$$

where m and μ are the reduced masses,

$$m = \frac{m_1(m_2 + m_3)}{m_1 + m_2 + m_3}, \qquad \mu = \frac{m_2 m_3}{m_2 + m_3}. \tag{2.192}$$

According to the general definition $G_0^{(+)}(E) = (E - H_0 + i\mathcal{O})^{-1}$, the Green function in the coordinate representation $G_0(E)$ is determined by

$$\langle \mathbf{x}', \mathbf{r}' | G_0(E) | \mathbf{r}, \mathbf{x} \rangle = \int \frac{d\mathbf{p}}{(2\pi)^3} \int \frac{d\mathbf{q}}{(2\pi)^3} \times \frac{\exp[i\mathbf{p}(\mathbf{x} - \mathbf{x}') + i\mathbf{q}(\mathbf{r} - \mathbf{r}')]}{E - E_p - E_q + i\mathcal{O}}, \tag{2.193}$$

$$E_p = \frac{\hbar^2 p^2}{2m}, \qquad E_q = \frac{\hbar^2 q^2}{2\mu}.$$

After integration in the space of the vector $\mathbf{q}$ and over the angle variables of the vector $\mathbf{p}$ is performed, we obtain

$$\langle \mathbf{x}', \mathbf{r}' | G_0(E) | \mathbf{r}, \mathbf{x} \rangle = \frac{i}{8\pi^3} \frac{\mu}{\hbar^2} \frac{1}{|\mathbf{r} - \mathbf{r}'|} \frac{1}{|\mathbf{x} - \mathbf{x}'|} \int_{-\infty}^{\infty} dp\, p \times \exp\left[i\sqrt{\frac{2}{\hbar^2}(E - E_p + i\mathcal{O})}|\mathbf{r} - \mathbf{r}'| + ip|\mathbf{x} - \mathbf{x}'| \right]. \tag{2.194}$$

We introduce the dimensionless quantities x and y, defined by

$$x \equiv \sqrt{\frac{2m}{\hbar^2}E}|\mathbf{x} - \mathbf{x}'|, \qquad y \equiv \sqrt{\frac{2\mu}{\hbar^2}E}|\mathbf{r} - \mathbf{r}'|,$$

and the dimensionless integration variable ξ,

$$\xi = p\Big/\sqrt{\frac{2m}{\hbar^2}E}.$$

Then

$$\langle \mathbf{x}', \mathbf{r}' | G_0(E) | \mathbf{r}, \mathbf{x} \rangle = \frac{1}{2\pi^3}(m\mu)^{3/2}\frac{E^2}{\hbar^6}\frac{1}{xy} \times \frac{\partial}{\partial x}\int_{-\infty}^{\infty} d\xi \exp\left(-y\sqrt{\xi^2 - 1 - i\mathcal{O}} + ix\xi \right) \tag{2.195}$$

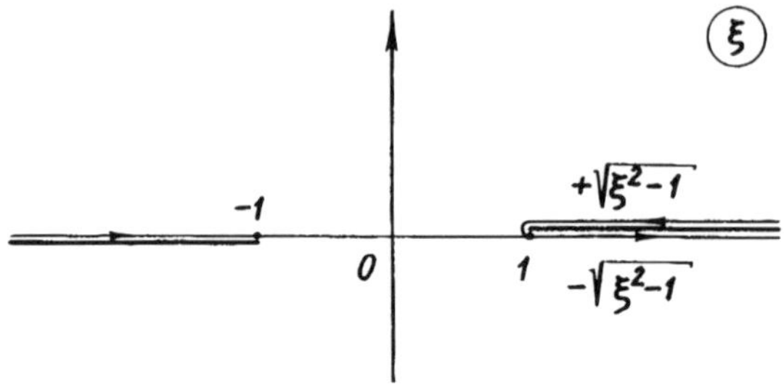

Figure 2.5. Cuts and integration contour in the complex ξ plane for (2.195).

To calculate the integral in (2.195), it is convenient to introduce a complex variable ξ. Since $\xi = \pm(1 + i\mathcal{O})$ are the branch points, it is necessary to cut the complex ξ plane from $-\infty$ to -1 and from 1 to ∞. (The left cut lies somewhat lower than the real axis, the right one lies above it.) We deform the integration contour as shown in Fig. 2.5 and take the value of the root lying on the upper edge of the cut with the opposite sign. Thus we have

$$I = \int_1^\infty d\xi \exp(ix\xi)\left[\exp\left(-y\sqrt{\xi^2-1}\right) - \exp\left(y\sqrt{\xi^2-1}\right)\right].$$

This expression easily reduces to the integral representation of the Hankel function, so that

$$I = i\pi\frac{y}{\sqrt{x^2+y^2}}H_1^{(1)}\left(\sqrt{x^2+y^2}\right).$$

Because

$$\frac{\partial}{\partial x}\frac{y}{\sqrt{x^2+y^2}}H_1^{(1)}\left(\sqrt{x^2+y^2}\right) = -\frac{xy}{x^2+y^2}H_2^{(1)}\left(\sqrt{x^2+y^2}\right),$$

we obtain

$$\langle \mathbf{x}', \mathbf{r}'|G_0(E)\mathbf{r}, \mathbf{x}\rangle = -\frac{i}{2\pi^2}\frac{(m\mu)^{3/2}}{\hbar^2}\frac{E}{x^2+y^2}H_2^{(1)}\left(\sqrt{x^2+y^2}\right), \qquad (2.196)$$

where

$$x^2+y^2 = \frac{2E}{\hbar^2}\{m(\mathbf{x}-\mathbf{x}')^2 + \mu(\mathbf{r}-\mathbf{r}')^2\}.$$

Equation (2.196) determines the Green function for a system of three noninteracting particles.

2.2. Write down the equations of motion of a *system of three bound particles* in the momentum representation.

Let us rewrite equation of motion $(H_0 + \hat{V}_{12} + \hat{V}_{23} + \hat{V}_{31} - E)\Psi = 0$ in the from

$$(E - H_0)\Psi = (\hat{V}_{12} + \hat{V}_{23} + \hat{V}_{31})\Psi. \tag{2.197}$$

For the bound state of the three-nucleon particles the factor $(E - H_0)$ in the left-hand side of equation (2.197) is negative (because $H_0 > 0$), i.e., always nonzero. For the continuous energy spectrum this factor can have both signs or be zero. Thus, in the case of the bound state of all three particles equation (2.197) can be divided by $E - H_0$ and rewritten in the form

$$\Psi_0 = G_0(\hat{V}_{12} + \hat{V}_{23} + \hat{V}_{31})\Psi_0, \qquad G_0 = (E - H_0)^{-1} < 0. \tag{2.198}$$

For an unbound state of three particles such a representation of the motion equation is impossible in a general case.

We introduce now three auxiliary functions

$$\Psi_0^{(ij)} = G_0 \hat{V}_{ij} \Psi_0, \qquad ijl = 123, 231, 312. \tag{2.199}$$

If these three functions are found, then from (2.199) we can obtain the total desired wave function Ψ_0:

$$\Psi_0 = \Psi_0^{(12)} + \Psi_0^{(23)} + \Psi_0^{(31)}. \tag{2.200}$$

From definition (2.199) it follows that to find the functions $\Psi_0^{(ij)}$, we have to solve the following system of coupled integral equations

$$\begin{cases} \Psi_0^{(12)} = G_0 \hat{V}_{12}(\Psi_0^{(12)} + \Psi_0^{(23)} + \Psi_0^{(31)}), \\ \Psi_0^{(23)} = G_0 \hat{V}_{23}(\Psi_0^{(12)} + \Psi_0^{(23)} + \Psi_0^{(31)}), \\ \Psi_0^{(31)} = G_0 \hat{V}_{31}(\Psi_0^{(12)} + \Psi_0^{(23)} + \Psi_0^{(31)}), \end{cases} \tag{2.201}$$

In the momentum representation the results of action of the integral operators $\hat{V}_{ij}$ on each of the functions $\Psi_0^{(kl)}$ (kl=12,23,31)can be presented in the same form as the result of action of the operators $\hat{V}_{ij}$ on the total function Ψ. For an example we write down three functions $\hat{V}_{12}\Psi_0^{(ij)}$, which appear in the right-hand side of the first equation of system (2.201) by using the connections between momenta (we assume that the particle masses are the same and equal to M):

$$\hat{V}_{12}\Psi_0^{(12)} = \int d\mathbf{k}'_{12} V_{12}(\mathbf{k}'_{12} - \mathbf{k}_{12})\Psi_0^{(12)}(\mathbf{k}'_{12}, \mathbf{p}_3)\Big), \tag{2.202}$$

$$\hat{V}_{12}\Psi_0^{(23)} = \int d\mathbf{k}'_{12} V_{12}(\mathbf{k}'_{12} - \mathbf{k}_{12})\Psi_0^{(23)}\left(-\frac{1}{2}\mathbf{k}'_{12} - \frac{3}{4}\mathbf{p}_3, \mathbf{k}'_{12} - \frac{1}{2}\mathbf{p}_3\right), \tag{2.203}$$

$$\hat{V}_{12}\Psi_0^{(31)} = \int d\mathbf{k}'_{12} V_{12}(\mathbf{k}'_{12} - \mathbf{k}_{12})\Psi_0^{(31)}\left(-\frac{1}{2}\mathbf{k}'_{12} + \frac{3}{4}\mathbf{p}_3, -\mathbf{k}'_{12} - \frac{1}{2}\mathbf{p}_3\right). \tag{2.204}$$

As a result the first equation of system (2.201) can be written as follows:

$$\Psi_0^{(12)}(\mathbf{k},\mathbf{p}) = -\frac{1}{|E_0| + (k^2 + \frac{3}{4}p^2)/M} \int d\mathbf{k}' V_{12}(\mathbf{k}' - \mathbf{k})$$
$$\times \left\{ \Psi_0^{(12)}(\mathbf{k}',\mathbf{p}) + \Psi_0^{(23)}\left(-\frac{1}{2}\mathbf{k}' - \frac{3}{4}\mathbf{p}, \mathbf{k}' - \frac{1}{2}\mathbf{p}\right)\right.$$
$$\left. + \Psi_0^{(31)}\left(-\frac{1}{2}\mathbf{k}' + \frac{3}{4}\mathbf{p}, -\mathbf{k}' - \frac{1}{2}\mathbf{p}\right)\right\}. \qquad (2.205)$$

Acting in the similar manner, we can write down two other equations of system (2.201).

We consider now a particular case when the two-particle potentials V_{ij} are the same for all three pairs of particles. Then we can omit indexes ij in the potentials V_{ij} and, according to (2.199), in the functions $\Psi_0^{(ij)}$. Since in this case in equations (2.201) the functions Ψ_0 differs only by arguments, we can consider only one equation of the system. Two other equations can be reduced to it. Thus, the unique function $\Psi_0(\mathbf{k},\mathbf{p})$, which determines the total wave function Ψ_0

$$\Psi_0 = \Psi_0(\mathbf{k}_{12},\mathbf{p}_3) + \Psi_0(\mathbf{k}_{23},\mathbf{p}_1) + \Psi_0(\mathbf{k}_{31},\mathbf{p}_2), \qquad (2.206)$$

we can find from a single integral equation, which follows from (2.205)

$$\Psi(\mathbf{k},\mathbf{p}) = -\frac{1}{|E| + (\mathbf{k}^2 + \frac{3}{4}\mathbf{p}^2)/M} \int d\mathbf{k}' V(\mathbf{k}' - \mathbf{k})$$
$$\times \left\{ \Psi(\mathbf{k}',\mathbf{p}) + \Psi\left(-\frac{1}{2}\mathbf{k}' - \frac{3}{4}\mathbf{p}, \mathbf{k}' - \frac{1}{2}\mathbf{p}\right)\right.$$
$$\left. + \Psi\left(-\frac{1}{2}\mathbf{k}' + \frac{3}{4}\mathbf{p}, -\mathbf{k}' - \frac{1}{2}\mathbf{p}\right)\right\}. \qquad (2.207)$$

Note that if the wave function Ψ is completely symmetric with respect to permutations of particles, then from (2.206) follows $\Psi_0(-\mathbf{k},\mathbf{p}) = \Psi_0(\mathbf{k},\mathbf{p})$.

We consider now a system of three bound nucleons and take into account their spins and isospins. In the case of triton ^3H, the total wave function is expressed in terms of the known spin-isospin functions and the spatial functions Ψ^s, Ψ^a, Ψ' and Ψ'', we use the same denotations for these functions in the momentum space as well. Let us assume for simplicity that the nucleon-nucleon interaction is nonzero only for the S-states. In the center mass coordinate system, the spatial wave functions, on taking account of their symmetry with respect to permutations of nucleus, can be written in the form

$$\Psi^s = u(k_{23},p_1) + u(k_{31},p_2) + u(k_{12},p_3), \ \Psi^a = 0,$$
$$\Psi' = -\frac{\sqrt{3}}{2}\Big[f(k_{31},p_2) - f(k_{12},p_3)\Big],$$
$$\Psi'' = -f(k_{23},p_1) + \frac{1}{2}\Big[f(k_{31},p_2) + f(k_{12},p_3)\Big], \qquad (2.208)$$

in addition $u(-k,p)=u(k,p)$, $f(-k,p)=f(k,p)$. (In the case of the S-interaction and zero total angular moment, the functions depend only on the modulus of vectors $\mathbf{k}$ and $\mathbf{p}$.)

Substituting the total wave function of the ^3H nucleus in the Schrödinger equation, on taking account (2.208), and introducing the denotation $v=u+f$ and $w=u-f$, we obtain the coupled system of two integral equations ($V^{tt}=V^{ss}=0$):

$$\begin{aligned} v(k,p) &= -\frac{1}{|E|+(\mathbf{k}^2+\frac{3}{4}\mathbf{p}^2)/M}\int d\mathbf{p}'\Bigg\{V^{ts}(\mathbf{p}'-\mathbf{k})v(p',p) \\ &+V^{ts}\left(\mathbf{p}'+\frac{\mathbf{p}}{2}-\mathbf{k}\right)\left[\frac{1}{2}v\left(|\mathbf{p}+\frac{\mathbf{p}'}{2}|,p'\right)\right. \\ &\left.\left.+\frac{3}{2}w\left(|\mathbf{p}+\frac{\mathbf{p}'}{2}|,p'\right)\right]\right\}, \end{aligned} \tag{2.209}$$

$$\begin{aligned} w(k,p) &= -\frac{M}{M|E|+\mathbf{k}^2+\frac{3}{4}\mathbf{p}^2}\int d\mathbf{p}'\Bigg\{V^{st}(\mathbf{p}'-\mathbf{k})w(p',p) \\ &+V^{st}\left(\mathbf{p}'+\frac{\mathbf{p}}{2}-\mathbf{k}\right)\left[\frac{1}{2}w\left(|\mathbf{p}+\frac{\mathbf{p}'}{2}|,p'\right)\right. \\ &\left.\left.+\frac{3}{2}v\left(|\mathbf{p}+\frac{\mathbf{p}'}{2}|,p'\right)\right]\right\}. \end{aligned} \tag{2.210}$$

Adding these equations in the approximation $V^{ts}=V^{st}$, we obtain the same equation for the function $u(k,p)$ as equation (2.207) of the function $\Psi(\mathbf{k},\mathbf{p})$ for the S-interaction at zero total orbital moment of the system. The integral equations (2.207), (2.209), (2.210) can be solved numerically by using modern computers.

2.3. Establish the connection between *two-particle Green's operator* $\hat{G}_{ij}$ and the *total Green operator* $\hat{G}$ *for the three-particle system* .

Along with the Green operators $\hat{G}$ and $\hat{G}_0$ for a system of three particles, we consider the two-particle Green operators

$$\hat{G}_{ij}=(z-\hat{H}_{ij})^{-1}, \qquad \hat{H}_{ij}=\hat{H}_0+\hat{V}_{ij}, \qquad z=E\pm i\varepsilon, \tag{2.211}$$

which take into account the interaction only between two particles i and j. At the same time, it is assumed that the third particle does not interact with the i-th and j-th particles (ijl=123, 231, 312). We establish now the connection between the Green operators $\hat{G}$, $\hat{G}_{ij}$ and the operators of two-particle interaction $\hat{V}_{ij}$. For this we employ the operator equality

$$\hat{A}^{-1}-\hat{B}^{-1}=\hat{A}^{-1}(\hat{B}-\hat{A})\hat{B}^{-1}=\hat{B}^{-1}(\hat{B}-\hat{A})\hat{A}^{-1}. \tag{2.212}$$

Setting here $\hat{A}=z-\hat{H}$, $\hat{B}=z-H_{ij}$ and noting that $\hat{H}-\hat{H}_{ij}=\hat{V}-\hat{V}_{ij}$, we obtain the operator relations

$$\hat{G}(z) = \hat{G}_{ij}(z) + \hat{G}_{ij}(z)(\hat{V} - \hat{V}_{ij})\hat{G}(z), \tag{2.213}$$

or

$$\hat{G}(z) = \hat{G}_{ij}(z) + \hat{G}(z)(\hat{V} - \hat{V}_{ij})\hat{G}_{ij}(z), \tag{2.214}$$

which establish the connections between the Green operators of the three particle system $\hat{G}$ and the two-particle operators $\hat{G}_{ij}$ (note that the two-particle operator $\hat{G}_{ij}$ contains the interaction between the i-th and j-th particles and the third particle appears in the operator $\hat{H}_0$). Since in the three-particle system there are three different pairs of interacting particles, we obtain three different relations

$$\hat{G} = \hat{G}_{12} + \hat{G}_{12}(\hat{V}_{23} + \hat{V}_{31})\hat{G}, \tag{2.215}$$

$$\hat{G} = \hat{G}_{23} + \hat{G}_{23}(\hat{V}_{31} + \hat{V}_{12})\hat{G}, \tag{2.216}$$

$$\hat{G} = \hat{G}_{31} + \hat{G}_{31}(\hat{V}_{12} + \hat{V}_{23})\hat{G}, \tag{2.217}$$

We call attention to the fact that all the Green operators in these equations are related to the same total energy of the system E.

First of all, we consider the action of the Green operators multiplied by $\varepsilon \to 0$ ($\varepsilon \geqslant 0$) on different wave functions. Let us act by the total Green operator, which is specified by the general formula (2.63) on an arbitrary function Ψ. We expand the function Ψ in terms of the eigenfunctions Ψ_λ of the system Hamiltonian $\hat{H}$, i.e.,

$$\Psi = \sum_\lambda a_\lambda \psi_\lambda,$$

where $\hat{H}\psi_\lambda = E_\lambda \psi_\lambda$. Thus,

$$\hat{G}\Psi = \sum_\lambda a_\lambda (E + i\varepsilon - \hat{H})^{-1}\psi_\lambda \equiv \sum_\lambda a_\lambda (E + i\varepsilon - E_\lambda)^{-1}\psi_\lambda. \tag{2.218}$$

If the energy E does not equal to any one of the eigenvalues of the Hamiltonian E_λ, then

$$\lim_{\varepsilon \to 0} i\varepsilon \hat{G}(E + i\varepsilon)\Psi = 0, \qquad E \neq E_\lambda. \tag{2.219}$$

But if $E = E_\lambda$, then

$$\lim_{\varepsilon \to 0} i\varepsilon \hat{G}(E_\lambda + i\varepsilon)\Psi = a_\lambda \psi_\lambda. \tag{2.220}$$

We consider an unbound state of the three-particle system with the wave function $\Psi_k = \Psi_k^{(+)}$, which satisfies the Schrödinger equation $(\hat{H} - E_k)\Psi_k^{(+)} = 0$ with the boundary condition for scattered particles or the *Lippmann–Schwinger equation* :

$$\Psi_k^{(+)} = \varphi_k + \hat{G}_0(E_k + i\varepsilon)\hat{V}\Psi_k^{(+)}, \tag{2.221}$$

where φ_k is a solution of the equation $(\hat{H}_0 - E_k)\varphi_k = 0$. It is easy to check that

$$\lim_{\varepsilon\to 0} i\varepsilon\hat{G}(E_k + i\varepsilon)\varphi_k = \Psi_k^{(+)}, \tag{2.222}$$

i.e., with help of the total Green operator $\hat{G}(E_k+i\mathcal{O})$ the solution of the Lippmann–Schwinger equation $\Psi_k^{(+)}$ with the energy E_k or the solution of the corresponding Schrödinger equation with a given boundary condition is expressed directly in terms of the incident wave φ_k for the same energy E_k.

The analogous relations can be obtained for the two-particle Green operators $\hat{G}_{ij}$ and the corresponding wave functions $\psi_k^{(ij)}$ satisfying the Schrödinger equation

$$\hat{H}_{ij}\psi_k^{(ij)} = E_k\psi_k^{(ij)}, \qquad \hat{H}_{ij} = \hat{H}_0 + \hat{V}_{ij}, \qquad ij = 12, 23, 31, \tag{2.223}$$

when only two particles out of three the i-th and j-th interact. At infinity the functions $\psi_k^{(ij)}$ contain divergent waves. The wave function $\psi_k^{(ij)}$ can describe the state of three particles when two of them (i-th and j-th) are bound, i.e., interact, and the third particle (l-th) is free (i.e., does not interact with the i-th and j-th), or the state when the mutual scattering of the i-th and j-th particles occurs, i.e., they are unbound, but interact, and the third l-th particle is free. Similarly to (2.219) and (2.220) the following relations hold true:

$$\lim_{\varepsilon\to 0} i\varepsilon\hat{G}_{ij}(E_k + i\varepsilon)\psi_{k'}^{(i'j')} = \delta_{kk'}\delta_{ij,i'j'}\psi_k^{(ij)}, \tag{2.224}$$

in addition, one of two indexes ij of the Green operator always coincides with one of the indexes $i'j'$ of the three-particle function $\psi_{k'}^{(i'j')}$.

If we "switch off" the interaction of the l-th particle with i-th and j-th ones (the interaction between two latter particles is unchanged), then from (2.222) we immediately obtain the following relation

$$\lim_{\varepsilon\to 0} i\varepsilon\hat{G}_{ij}(E_k + i\varepsilon)\varphi_k = \psi_k^{(ij)}. \tag{2.225}$$

When writing the operator connections (2.215) – (2.217) we used relation (2.213). Now we employ relation (2.214). Acting by the left and right parts of this operator equality on the incident wave φ_k and multiplying the obtained equation by $i\varepsilon$, we pass to the limit $\varepsilon \to 0$. Utilizing the formulas (2.222) and (2.225), we obtain the following connection between the function $\Psi_k^{(+)}$ and the functions $\psi_k^{(ij)}$

$$\Psi_k^{(+)} = \psi_k^{(ij)} + \hat{G}(E_k + i\varepsilon)(\hat{H} - \hat{H}_{ij})\psi_k^{(ij)}. \tag{2.226}$$

Using (2.223) and the explicit form of the operator $\hat{G}^{-1} \equiv \hat{G}^{-1}(E_k + i\varepsilon) = E_k + i\varepsilon - \hat{H}$, we obtain from (2.226) the useful general formula

$$\lim_{\varepsilon\to 0} i\varepsilon\hat{G}(E_k + i\varepsilon)\psi_k^{(ij)} = \Psi_k^{(+)}. \tag{2.227}$$

Here the wave function $\Psi_k^{(+)}$ of a three interacting particles system with a given boundary condition at infinity is connected with the three-particle wave function

$\psi_k^{(ij)}$ by the total Green operator $\hat{G}$. The asymptotics of $\psi_k^{(ij)}$ (divergent waves at infinity) is given, but only two particles interact (the i-th and j-th).

2.4. Write down the *integral equation for the three-particle system* with the help of the two-particle Green operators $\hat{G}_{ij}$.

First of all we consider the scattering problem of a particle by two other particles in a bound state. Prescribe a number 1 to the scattered particle, and let the 2-nd and 3-d particles be bound. The corresponding asymptotic Green function is Φ_1.

Substituting in turn the right parts of three equalities (2.215) – (2.217) in the left-hand side of general formula (2.227) and by using (2.224) and (2.227), we obtain three equations

$$\Psi_{\text{kb}}^{(+)} = \Phi_1 + \hat{G}_{23}(E_k + i0)(\hat{V}_{31} + \hat{V}_{12})\Psi_{\text{kb}}^{(+)}, \tag{2.228}$$

$$\Psi_{\text{kb}}^{(+)} = \hat{G}_{12}(E_k + i0)(\hat{V}_{23} + \hat{V}_{31})\Psi_{\text{kb}}^{(+)}, \tag{2.229}$$

$$\Psi_{\text{kb}}^{(+)} = \hat{G}_{31}(E_k + i0)(\hat{V}_{12} + \hat{V}_{23})\Psi_{\text{kb}}^{(+)}, \tag{2.230}$$

Each two-particle operator $\hat{G}_{ij}$, which appears in these equations, when acting a function that stands behind it, gives the divergent waves with respect to the coordinates of the particle whose interaction is not included into the operator $\hat{G}_{ij}$. The waves diverge relatively coordinates of the particles whose interaction is taken into account in the operator $\hat{G}_{ij}$. Note that individual Green's operator $\hat{G}_{ij}$ and the corresponding equation for this operator do not contain all the boundary conditions of the three-particle problem. In fact we have to simultaneously solve all three equations. The equations should be treated as the coupled system of consistent equations. At the same time, each of three equations, which are solved simultaneously, is responsible for the absence of the convergent waves and appearance of divergent waves at infinity with respect to corresponding relative space coordinate.

In the case of a system of three unbound particles the analogous equations for the wave function Ψ_{123} can be obtained substituting the right parts of equations (2.215) – (2.217) in (2.227) and by using (2.224) and (2.227). The obtained integral equations, unlike to equations (2.228) – (2.230), are completely symmetric with respect to coordinates of three particles

$$\Psi_{\text{ks}}^{(+)} = \Psi_{\text{ks}}^{(12)} + \hat{G}_{12}(E_k + i0)(\hat{V}_{23} + \hat{V}_{31})\Psi_{\text{ks}}^{(+)}, \tag{2.231}$$

$$\Psi_{\text{ks}}^{(+)} = \Psi_{\text{ks}}^{(23)} + \hat{G}_{23}(E_k + i0)(\hat{V}_{31} + \hat{V}_{12})\Psi_{\text{ks}}^{(+)}, \tag{2.232}$$

$$\Psi_{\text{ks}}^{(+)} = \Psi_{\text{ks}}^{(31)} + \hat{G}_{31}(E_k + i0)(\hat{V}_{12} + \hat{V}_{23})\Psi_{\text{ks}}^{(+)}. \tag{2.233}$$

To obtain the single-valued solution for the wave function $\Psi_{\text{ks}}^{(+)}$ with the correct asymptotics at infinity, equations (2.231) – (2.233), as well as equations (2.228) – (2.230), must be solved simultaneously. Then all boundary conditions on three relative vector coordinates $\mathbf{r}_{12}$, $\mathbf{r}_{23}$, $\mathbf{r}_{31}$ must be taken into account.

In conclusion we write down the integral equation of the wave function ψ_0 of a bound state of the three-particle system which correspond to the energy $E_0 < 0$.

Unlike the above considered three-particle systems, where at least one particle is not bound with others, in the case of three bound particles the problem of the solution ambiguity is automatically removed. Since, according to (2.220), for the wave function ψ_0 we have

$$\lim_{\varepsilon \to +0} i\varepsilon \hat{G}(E_0 + i\varepsilon)\psi_0 = \psi_0, \tag{2.234}$$

because $\hat{H}\psi_0 = E_0\psi_0$ and

$$\lim_{\varepsilon \to +0} i\varepsilon G_0(E_0 + i\varepsilon)\psi_0 = \lim_{\varepsilon \to +0} i\varepsilon \hat{G}(E_0 + i\varepsilon)\psi_0 = 0, \tag{2.235}$$

due to $\hat{H}_0\psi_0 \neq E_0\psi_0$ and $\hat{H}_{ij}\psi_0 \neq E_0\psi_0$, then after action by both parts of operator equalities (2.215) – (2.217) on the wave function of the bound state ψ_0 multiplied by $i\varepsilon$ (with $\varepsilon \to +0$), we obtain the homogeneous with respect to the function ψ_0, three equations

$$\begin{aligned} \psi_0 &= \hat{G}_{12}(\hat{V}_{23} + \hat{V}_{31})\psi_0, \\ \psi_0 &= \hat{G}_{23}(\hat{V}_{31} + \hat{V}_{12})\psi_0, \\ \psi_0 &= \hat{G}_{31}(\hat{V}_{12} + \hat{V}_{23})\psi_0. \end{aligned} \tag{2.236}$$

To find the function ψ_0, we can now solve only one out of three equations (in addition, the boundary condition at infinity where $\psi_0 \to 0$ is satisfied).

References

Adhikari, S.K. and Tomio, L. (1955) *Phys.Rev. C*, **51** , 70.

Akaishi, Y. (1986) *Theoretical and Experimental Investigations of Hadronic Few-Body Systems*, Rome. P.120.

Amado, R.D. (1979) *Phys.Rev. C*, **19**, 1473.

Faddeev, L.D. (1960) *Zh.Eksp.Theor.Fiz.*, **39**, 1459. [Sov.Phys. – *JETP*, **12**, 1014 (1961)].

Sasakawa, T. (1986) *Theoretical and Experimental Investigations of Hadronic Few-Body Systems*, Rome. P.104.

Simonov, Yu.A. (1966) *Sov.J.Nucl.Phys.*, **3**, 461.

Sitenko, A.G. and Kharchenko, V.F. (1963) *Nucl.Phys.*, **49**, 15.

Sitenko, A.G., and Kharchenko, V.F. (1971) *Sov.Phys. – Uspekhi*, **14**, 125.

CHAPTER 3

NUCLEAR MATTER

3.1. Saturation of the Nuclear Interaction

Dimensions of nuclei and the density distribution inside nuclei. The *spatial structure of atomic nuclei* , which are composite particles consisting of simpler particles, namely, neutrons and protons, is determined by the spatial structure of the neutrons and protons and by the nature of the interaction between them. As we have already mentioned, a distinguishing feature of nuclei is the fact that they have sharply defined boundaries. Since the nuclear interaction between nucleons is much stronger than the Coulomb interaction, the dimensions of nuclei and the distribution of the nucleons inside them are mainly determined by the nuclear forces. Because of the charge independence of the nuclear forces, the spatial distributions of neutrons and protons inside nuclei are almost identical. A certain increase of the volume occupied by the protons, due to the Coulomb repulsion, is matched by approximately the same increase of the volume occupied by the neutrons, arising from the increase in the number of excess neutrons with increasing nuclear charge. Therefore, it is usually assumed that the dimensions of the nucleus are determined with very good accuracy by the distribution of the charge density.

Nuclear radii and the density distribution inside nuclei are determined most accurately from experiments on the scattering of high-energy electrons by nuclei. Relations characterizing the effect on the scattering of the spatial structure of the nuclei can be obtained in a simple form if the Born approximation is applicable. For the Coulomb scattering of ultrarelativistic electrons, the Born approximation is valid when the condition

$$Ze^2/\hbar c \ll 1, \tag{3.1}$$

is fulfilled, where Z is the charge of the nucleus and c is the speed of light. This condition is well fulfilled for not very heavy nuclei.

The cross-section for elastic scattering of high-energy electrons in the Coulomb field of a point charge Z is determined, when the condition (3.1) is fulfilled, by the *Mott formula*

$$\sigma_0(\vartheta) = (Ze^2/2E)^2(\cos^2 \tfrac{1}{2}\theta/\sin^4 \tfrac{1}{2}\theta), \tag{3.2}$$

where ϑ is the scattering angle and E is the energy of the electrons. In the case of scattering of electrons by a nucleus with charge Z characterized by the spatial distribution $\varrho^{(Z)}(r)$, the cross-section is given by the expression

$$\sigma(\vartheta) = \sigma_0(\vartheta)[F(\mathbf{q})]^2, \tag{3.3}$$

where $F(q)$ is the *form factor of the charge distribution* :

$$F(\mathbf{q}) = (1/Z) \int d^3 r e^{i\mathbf{q}\cdot\mathbf{r}} \varrho^{(Z)}(r). \tag{3.4}$$

Here $\mathbf{q}$ is the momentum transferred to the nucleus by the electron; for ultra-relativistic electrons ($E \gg mc^2$; m is the electron mass),

$$q = (2E/\hbar c)\sin(\vartheta/2). \tag{3.5}$$

The formulae (3.2) and (3.3) are written in the centre-of-mass frame of the electron and the nucleus.

Comparing (3.3) with (3.2), we see that the deviation of the law for scattering of electrons by a nucleus from the law for electron scattering by a point charge is directly connected with the spatial distribution of charge in the nucleus. According to (3.4), the form factor $F(\mathbf{q})$ is simply the Fourier transform of the charge distribution in the nucleus; therefore, knowing $F(\mathbf{q})$, it is not difficult to find $\varrho^{(Z)}(\mathbf{r})$ by means of the inverse Fourier transformation:

$$\varrho^{(Z)}(\mathbf{r}) = Z \int (d^3 q/(2\pi)^3) e^{-i\mathbf{q}\cdot\mathbf{r}} F(\mathbf{q}). \tag{3.6}$$

Since the charge distribution coincides with the distribution of nucleons in the nucleus, the nucleon density distribution $\varrho(\mathbf{r})$ in the nucleus,

$$\varrho(\mathbf{r}) = (A/Z)\varrho^{(Z)}(\mathbf{r}), \tag{3.7}$$

will also be determined by a function of the form (3.6).

If the electron wavelength is large compared with the nuclear dimensions, the exponential in formula (3.4) can be expanded in a series in powers of $\mathbf{q}\cdot\mathbf{r}$; we then find

$$F(q) = 1 - (1/6)q^2 R^2 + \ldots \quad (qR < 1), \tag{3.8}$$

where R is the *root mean square of the nuclear radius* :

$$R^2 = \int d^3 r r^2 \varrho(r) / \int d^3 r \varrho(r). \tag{3.9}$$

Clearly, for it to be possible to detect effects associated with the finite dimensions of the nucleus, it is necessary that the electron wavelength be less than or at least comparable with the nuclear dimensions. With decreasing wavelength (increasing energy) of the electrons, it is possible to study the spatial structure of the nucleus in more detail.

In the case of heavy nuclei, the Born approximation is inapplicable. To obtain information on the spatial structure of such nuclei from electron scattering data, it is necessary to use numerical methods.

Analysis of the numerous experimental data shows (Hofstadter, 1956; Elton, 1961) that the spatial distribution of the nucleon density in nuclei with mass numbers $A > 20$ is best described by the *Fermi function* [1]

$$\varrho(r) = \frac{\varrho_0}{1 + \mathrm{e}^{(r-R')/a}}, \tag{3.10}$$

where the quantity R' is proportional to $A^{1/3}$:

$$R' = R'_0 A^{1/3}. \tag{3.11}$$

If $R' \gg a$, the distribution described by the function (3.10) is characterized by almost uniform density $\varrho \approx \varrho_0$ in the whole nuclear volume apart from a surface layer of thickness of order a, in which the density falls rapidly to zero. The quantity ϱ_0 is the nucleon density at the centre of the nucleus and R' corresponds to the distance from the centre at which the density has decreased to half its maximum value. The quantity a characterizing the thickness of the surface layer in which the rapid decrease of density occurs is usually called the *diffuseness of the nuclear surface.* If $a \to 0$, the distribution described by the Fermi function (3.10) goes over into a uniform distribution with radius R'. Since

$$\int d^3r\,\varrho(r) = A,$$

the quantities ϱ_0, a and R'_0 occurring in (3.10) are connected by the relation

$$\frac{4\pi}{3} R_0'^3 \left(1 + \pi^2 \frac{a^2}{R_0'^2} A^{-2/3}\right) \varrho_0 = 1. \tag{3.12}$$

The numerical values of the quantities ϱ_0, a and R'_0 are the same for all nuclei (with $A > 20$) within the limits of the experimental accuracy and, according to the data of Hofstadter, are equal to

$$\varrho_0 = 1.68 \times 10^{38}\,\mathrm{cm}^{-3}; \quad a = 0.57 \times 10^{-13}\,\mathrm{cm}; \quad R'_0 = 1.10 \times 10^{-13}\,\mathrm{cm}. \tag{3.13}$$

[1] In the case of very light nuclei, it is important to take account of the internal structure of the nucleons in the determination of the nucleon density distribution from the electron scattering data.

By the *nuclear radius* R, we usually mean the root mean square radius defined by the expression (3.9). As we did for R', we represent the nuclear radius R in the form

$$R = R_0 A^{1/3}, \tag{3.14}$$

where

$$R_0 = 1.2 \times 10^{-13}\,\text{cm}. \tag{3.15}$$

The uniform distribution that is characterized by the same value of the root mean square radius R as the true distribution is usually called the *equivalent uniform distribution* . The radius R'_{equ} of such a distribution is obviously connected with the radius R by the relation

$$R'_{\text{equ}} = \sqrt{5/3}R.$$

The equivalent uniform distribution leads to the same scattering at low energies as the real distribution.

If the nucleus is represented in the form of a sphere with a uniform density distribution, the quantity R_0 will be proportional to the cube root of the volume per nucleon. The constancy of this quantity for all nuclei indicates the constancy of the density. The small decrease of R_0 (by approximately 12%) means that the equivalent uniform density increases slowly with increasing A.

The quantity R'_0 defining the Fermi distribution (3.10) is practically constant for all nuclei (to within 2%). According to (3.12), this means that the nucleon density in the central region of the nucleus (that is, except in a thin surface layer) is practically the same for all nuclei with $A \gg 1$. Therefore, if we neglect surface effects, the volumes of nuclei are indeed found to be proportional to the number A of nucleons. This property, the incompressibility of the nuclear substance, lies at the basis of the so-called liquid-drop model of the nucleus.

Binding energy of nuclei. The mass of a stable nucleus in its ground state is always less than the sum of the masses of the protons and neutrons composing the nucleus. We write the energy of the nucleus in the ground state:

$$E(A, Z) = \{M(A, Z) - ZM_p - (A - Z)M_n\}c^2, \tag{3.16}$$

where $M(A, Z)$ is the mass of the nucleus and M_p and M_n are the masses of the proton and neutron. The quantity $E(A, Z)$ taken with the opposite sign is called the *binding energy*. Clearly, the binding energy is the work that must be expended to break up the nucleus into its constituent particles. The binding energy per nucleon is characteristic of the stability of the nucleus. Figure 3.1 shows the dependence of the *mean binding energy per nucleon* on the mass number A for different nuclei. The binding energy

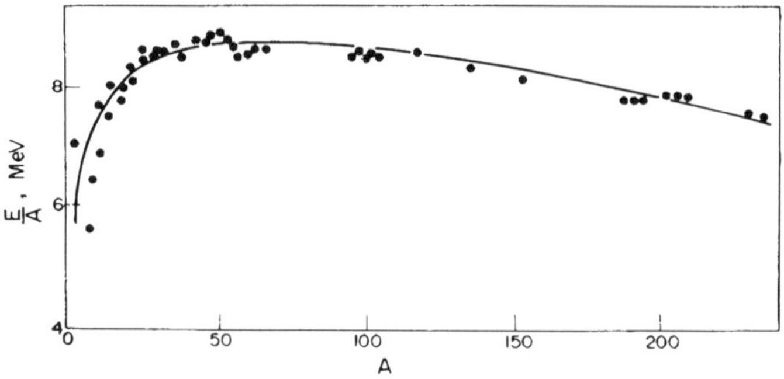

Figure 3.1. Dependence of the mean binding energy per nucleon on the mass number A.

per nucleon first increases rapidly with increasing A, attaining a value of 8 MeV by $A = 16$, then increases very slowly to a value of 8.8 MeV at $A \approx 60$, and then decreases very slowly, taking the value 7.4 MeV for the heaviest nuclei. The most characteristic feature of this dependence is the fact that the mean binding energy per nucleon is approximately constant (≈ 8 MeV) over almost the whole range of variation of A, with the exception of the region of very light nuclei. Consequently, the total nuclear binding energy, like the nuclear volume, is proportional to the number of nucleons in the nucleus. Small deviations from this proportionality are connected with surface effects, the Coulomb repulsion between the protons, the difference between the numbers of protons and neutrons, and the effect of the parity of the numbers of protons and neutrons.

The fact that the mean binding energy per nucleon, like the volume per nucleon, is independent of the number of nucleons in the nucleus can be explained by assuming that the nuclear forces between the nucleons possess the property of saturation, that is, each nucleon in the nucleus can interact only with a small number of nearest nucleons. The *saturation of the nuclear forces* points to an analogy between the properties of the nuclear substance and those of an ordinary liquid. On the basis of this analogy, a liquid-drop model of the nucleus is introduced, according to which the nucleus can be regarded as a drop of nuclear liquid. The *liquid-drop model of the nucleus* makes it possible to explain the dependence of the nuclear binding energy on the mass number A and the charge Z of the nucleus.

The dependence of the nuclear binding energy on A and Z is described by the *Weizsäcker semi-empirical formula*

$$\begin{aligned} E(A,Z) &= -\varepsilon_1 A + \varepsilon_2 A^{2/3} + \varepsilon_3 (Z^2/A^{1/3}) \\ &\quad + \varepsilon_4 [(A-2Z)^2/A] + \varepsilon_5 A^{-3/4}\delta, \end{aligned} \tag{3.17}$$

where (Green, 1955) the numerical values of the coefficients are $\varepsilon_1 = 15.75$ Mev, $\varepsilon_2 = 17.8$ Mev, $\varepsilon_3 = 0.71$ Mev, $\varepsilon_4 = 23.7$ Mev and $\varepsilon_5 = 34$ Mev, and the quantity δ depends on the parity of the numbers of neutrons and protons and is equal to

$$\delta = \begin{cases} -1 & A \text{ even, } Z \text{ even;} \\ 0 & A \text{ odd;} \\ +1 & A \text{ even, } Z \text{ odd.} \end{cases}$$

The first three terms in formula (3.17), which represent the volume, surface and Coulomb energies, are predicted by the liquid-drop model of the nucleus. The *volume energy*, which takes into account the proportionality of the nuclear binding energy to the total number of nucleons, corresponds to complete saturation of the nuclear interaction between the nucleons in the nucleus. The presence of the *surface* leads to a decrease of the binding *energy*, since the nucleons at the surface interact with a smaller number of neighbours than do the nucleons situated inside the nucleus. The *Coulomb energy* is the energy of the electrostatic repulsion in the nucleus, which is regarded as a drop of a uniformly charged liquid. The fourth and fifth terms (3.17) characterize the energy associated with the symmetry of the forces between the neutrons and protons and with the effect of pairing between the neutrons and protons.

The liquid-drop model makes it possible to explain not only the experimental dependence of the binding energy of nuclei on A and Z, but also certain other properties of nuclei. The representation of a nucleus as a liquid drop accords with the presence of a comparatively sharp boundary in nuclei. The liquid-drop model of the nucleus makes it possible to explain the phenomenon of nuclear fission and also the existence in nuclei of excited states associated with deformation of the surface of the nuclei.

3.2. Fermi Gas Model of the Nucleus

Saturation of the nuclear interaction and independence of the motion of individual nucleons in the nucleus. We shall consider the properties of the nuclear substance in rather more detail. As we have seen, it follows from analysis of the numerous experimental data on the properties of nuclei that the density of the nuclear substance and the mean binding energy per nucleon are approximately the same for all except the lightest nuclei. The constancy of the density and of the binding energy per nucleon is connected with the *saturation of the nuclear forces*, which can be explained qualitatively by treating the nuclear substance as an incompressible liquid. In addition, nuclei possess certain properties which can be explained by assuming that the nucleons in nuclei move independently, that is, by regarding nuclei as systems of independent particles. Indeed, knowing the density of the nuclear substance, it is not difficult to determine the mean distance between the nucleons in the nucleus: this amounts to about 1.8×10^{-13} cm. But the spatial dimensions of nucleons are equal to 0.45×10^{-13} cm. Consequently, the nucleons occupy only about one fiftieth of the whole volume of the nucleus. Therefore, despite the saturation of the nuclear interaction, nucleons in a nucleus retain their individual properties, which, in turn, are manifested in the properties of nuclei. The existence of especially stable nuclei, containing well-defined so-called *magic numbers of nucleons* , indicates the existence of nucleon shells in nuclei. The values of the spins, magnetic moments and certain other characteristics of nuclei can be explained by regarding the nucleus as a system of independent nucleons moving in a certain *self-consistent field* and distributed over well-defined shells in accordance with the Pauli principle.

Thus, the nuclear substance can be regarded, on the one hand, as a system of strongly interacting particles in a condensed liquid-like state, and, on the other hand, as a system of almost non-interacting particles with properties characteristic of the gaseous state. This duality is a characteristic property of the nuclear substance and is unconnected with the distinguishing features of individual nuclei.

In order to eliminate effects associated with the distinguishing features of nuclei when studying the properties of the nuclear substance, we introduce the concept of *nuclear matter* . By nuclear matter we shall mean the substance of a hypothetical nucleus consisting of equal numbers of neutrons and protons (but without the Coulomb repulsion between the protons) and so large that surface effects can be neglected.

In reality, Coulomb effects increase in proportion to Z^2 and become important before the mass number A reaches a value at which it would be possible to neglect surface effects. In particular, the Coulomb repulsion is responsible for the fact that nuclei with mass numbers A appreciably greater than 200 do not exist.

The most characteristic property of nuclear matter is connected with the saturation of the nuclear interaction and is manifested in the fact that the *nucleon density* ϱ and the energy ε per nucleon do not depend on A. In accordance with the relations (3.10), (3.12) and (3.17) and the value of R'_0 (3.13), we choose the values of ϱ and ε for nuclear matter ($A \gg 1$) to be

$$\varrho \equiv \frac{1}{(4\pi/3)R_0'^3} = 1.8 \times 10^{38}\,\mathrm{cm}^{-3}, \quad \varepsilon = -15.75\,\mathrm{MeV}, \tag{3.18}$$

that is, we shall assume that ϱ coincides with the value of the nuclear density at the centre of the nuclei ($A \gg 1$) and that the energy ε is associated only with the volume term in the *Weizsäcker formula.*

Another, no less important, property of nuclear matter is the fact than, dynamically, nuclear matter can be regarded as a system of independent particles. Since the particle density is fixed, for a given number of particles the system has a completely defined volume. For definiteness, in the following we shall understand by *nuclear matter* a system consisting of A nucleons ($A/2$ neutrons and $A/2$ protons) occupying a volume Ω, A being assumed to be so large that surface effects can be neglected.

Assuming that only two-particle forces act between the nucleons, we can write the Hamiltonian of the system in the form

$$H = \sum_i [-(\hbar^2/2M)\nabla_i^2] + \sum_{i<j} V_{ij}, \tag{3.19}$$

where $-(\hbar^2/2M)\nabla_i^2$ is the kinetic energy of an individual nucleon and V_{ij} is the potential of the two-particle interaction between the nucleons i and j. Obviously, the total energy of the nuclear matter is equal to

$$E = \left\langle \Phi \left| \sum_i [-(\hbar^2/2M)\nabla_i^2] + \sum_{i<j} V_{ij} \right| \Phi \right\rangle, \tag{3.20}$$

where Φ is the wave function of the ground state of the system.

The fundamental problem of the theory is to explain the properties of nuclear matter on the basis of the forces acting between the nucleons. The existence of stable nuclei means that the forces acting between the nucleons in nuclei are mainly attractive. If, however, only attractive forces acted between the nucleons, these would lead to collapse of the nuclei, that is, the linear dimensions of nuclei would coincide with the range of the nuclear forces. In this case the density would be proportional to the total number of particles in the nucleus. In reality, the density ϱ does not depend on A (saturation of the nuclear forces), and this means that the forces acting between the nucleons must be partly repulsive. As we know, the nuclear forces are indeed repulsive at very short distances. In addition, nuclear forces have a partly exchange character and, consequently, will be forces of attraction or repulsion, depending on the state of the interacting particles.

Nuclear forces are characterized by extremely short range and very great strength. At first glance, this feature of nuclear forces is incompatible with the idea of nuclear matter as a system of independent particles. Actually, however, by virtue of the Pauli principle, the interaction between the nucleons is substantially weakened, and this explains why the independent-particle model is applicable for the description of nuclear matter.

Properties of the ideal Fermi gas. Certain properties of nuclear matter can be described satisfactorily by means of the *Fermi-gas model*, in

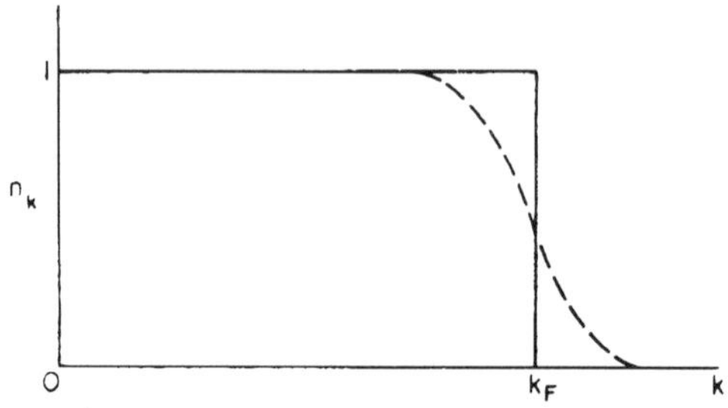

Figure 3.2. Momentum distribution of the nucleons in an ideal Fermi gas (the dashed curve corresponds to the distribution of nucleons at a non-zero temperature).

which the interaction between the particles is completely neglected (Bethe and Bacher, 1936). According to this very simple model, nuclear matter is represented in the form of an ideal nucleon gas, occupying a definite volume Ω. Clearly, the total energy in the case of an ideal Fermi gas is described by the general expression (3.20), in which, however, the interaction between the nucleons must be assumed equal to zero. The wave function Φ_0 for the ideal Fermi gas is represented in the form of an anti-symmetrized product of one-particle wave functions

$$\Phi_0 = P_a \prod_{i=1}^{A} \varphi_{\alpha_i}(\mathbf{r}_i). \tag{3.21}$$

Here, $\varphi_{\alpha_i}(\mathbf{r}_i)$ is the wave function of an individual nucleon in a given state α_i. If the volume Ω is assumed to be sufficiently large, the one-particle wave functions $\varphi_\alpha(\mathbf{r})$ can be chosen in the form of plane waves

$$\varphi_{\alpha_i}(\mathbf{r}) = (1/\sqrt{\Omega})\mathrm{e}^{i\mathbf{k}_\alpha \cdot \mathbf{r}}, \tag{3.22}$$

where $\mathbf{k}_\alpha$ is the momentum of the nucleon (we have not written out explicitly the charge and spin parts of the function).

The kinetic energy of a particle with well defined momentum is equal to $\hbar^2 k_\alpha^2/2M$. By virtue of the *Pauli principle*, no more than four nucleons, differing in the values of their components of spin and isotopic spin, can be found in each of the states with a given value of the energy. The minimum energy corresponds to the ground state of the system. This means that in the ground state all the one-particle levels corresponding to values of the momentum less than a certain *limiting Fermi momentum* k_F are occupied, and the higher levels are vacant. The momentum distribution of the nucleons in this case has the form of a rightangled step and corresponds to zero temperature of the system (Fig. 3.2).

The limiting Fermi momentum k_F is determined by the nuclear density ϱ. In fact, the number of possible nucleon states with momentum values within the range d^3k is equal to

$$4\Omega(d^3k/(2\pi)^3). \tag{3.23}$$

The numerical factor in (3.23) takes account of the degeneracy associated with the existence of the spin and isospin of the nucleons. Since in the ground state all levels with $k \leqslant k_F$ are occupied and the number of nucleons is equal to A, we have

$$4\Omega \int_{k \leqslant k_F} (d^3k/(2\pi)^3) = A,$$

whence it follows that

$$k_F = (3\pi^2\varrho/2)^{1/3}, \quad \varrho \equiv (A/\Omega). \tag{3.24}$$

Furthermore, since $\varrho = \left(\frac{4}{3}\pi R_0'^3\right)^{-1}$, we have

$$k_F = \sqrt[3]{(9\pi/8)}(1/R_0'). \tag{3.25}$$

Noting that $R_0' = 1.10 \times 10^{-13}$ cm, we have $k_F = 1.38 \times 10^{13}\,\text{cm}^{-1}$ whence for the *limiting Fermi energy* E_F we find

$$E_F \equiv \hbar^2 k_F^2/2M \approx 39\,\text{MeV}. \tag{3.26}$$

The total energy of the whole system is equal to

$$E = \sum_\alpha (\hbar^2 k_\alpha^2/2M), \tag{3.27}$$

where the summation is extended over all the filled one-particle levels. If the system is in the ground state, then

$$E_0 = \sum_{k_\alpha \leqslant k_F} \frac{\hbar^2 k_\alpha^2}{2M} = 4\Omega \int_{k \leqslant k_F} \frac{d^3k}{(2\pi)^3} \frac{\hbar^2 k^2}{2M} = \frac{3}{5} A \frac{\hbar^2 k_F^2}{2M}. \tag{3.28}$$

Thus, in the case of an ideal Fermi gas, the mean energy per nucleon is equal to

$$\frac{E_0}{A} = \frac{3}{5}\frac{\hbar^2 k_F^2}{2M}. \tag{3.29}$$

Using the numerical value (3.18) for ϱ, we find

$$E_0/A \approx 24\,\text{MeV}. \tag{3.30}$$

The positive sign of the quantity E_0/A indicates the unsatisfactory nature of the *ideal Fermi gas model*, inasmuch as this model, while explaining the uniformity of nuclear matter and the proportionality of the total energy to the volume of the system, does not explain the existence of the bound state of nuclear matter.

The ideal Fermi gas model can be somewhat improved, and the bound state of nuclear matter explained, by considering nuclear matter as a gas of non-interacting nucleons moving in a certain potential well occupying a volume Ω. The depth of the well should be chosen such that the energy required to detach an individual nucleon coincides with the experimental value. According to the experimental data, the separation energy S of an individual nucleon is, on average, the same for all nuclei and is approximately equal to 8 MeV. Since the least strongly bound nucleons are those with energies close to the limiting energy E_F, the depth of the well must be assumed equal to

$$V_0 = E_F + S. \tag{3.31}$$

According to (3.26), the magnitude of V_0 amounts to about 47 MeV.

This model for nuclear matter (the *Fermi gas in a potential well*) is similar to the free-electron model in the theory of metals. Like the latter, it describes well those properties of nuclei that are associated with the momentum distribution of the nucleons, but not the values of the energies or the spatial distribution of the nucleons in nuclei. The chief defect of this model is the necessity of introducing an external field.

Collapse of nuclei under the influence of the attractive forces between the nucleons. We shall now take into account the interaction between the nucleons, confining ourselves to the first approximation of perturbation theory. In this approximation, the wave function Φ retains the form (3.21), and the total energy of the system can be represented in the form

$$E = E_0 + \sum_{\alpha>\beta} U_{\alpha\beta}, \tag{3.32}$$

where $U_{\alpha\beta}$ is the interaction energy of a pair of nucleons in the states α and β:

$$U_{\alpha\beta} = \langle \psi^0_{\alpha\beta} | V | \psi^0_{\alpha\beta} \rangle \tag{3.33}$$

($\psi^0_{\alpha\beta}$ is the anti-symmetrized wave function of the two nucleons), and the summation in (3.32) extends over all the occupied states.

Introducing the centre-of-mass coordinates and relative coordinates for the two nucleons, $\mathbf{R} = \frac{1}{2}(\mathbf{r}_1 + \mathbf{r}_2)$ and $\mathbf{r} = \mathbf{r}_1 - \mathbf{r}_2$ we can represent the space part $\varphi^0_{\alpha\beta}$ of the wave function in the form of a product of the wave function of the centre-of-mass motion of the two nucleons with the wave function of the relative motion:

$$\varphi^0_{\alpha\beta}(\mathbf{r}_1,\mathbf{r}_2) = (1/\sqrt{\Omega})\mathrm{e}^{i\mathbf{K}\cdot\mathbf{R}}\varphi^0_k(\mathbf{r}), \quad \mathbf{K}=\mathbf{k}_\alpha+\mathbf{k}_\beta, \quad \mathbf{k}=\frac{1}{2}(\mathbf{k}_\alpha-\mathbf{k}_\beta). \tag{3.34}$$

Although the complete wave function (including the charge and spin parts) must, by virtue of the Pauli principle, be anti-symmetric with respect to the interchange of two nucleons, the wave function $\varphi^0_k(\mathbf{r})$ of the relative motion can be either symmetric or antisymmetric:

$$\varphi^{0(\pm)}_k(\mathbf{r}) = (1/\sqrt{2\Omega})(\mathrm{e}^{i\mathbf{k}\cdot\mathbf{r}} \pm \mathrm{e}^{-i\mathbf{k}\cdot\mathbf{r}}). \tag{3.35}$$

If there are equal numbers of neutrons and protons in the system, and the numbers of neutrons (and protons) with opposite spins are equal, then the statistical weights of the symmetric and anti-symmetric states $\varphi^{(+)}$ and $\varphi^{(-)}$ are equal to $\frac{3}{8}$ and $\frac{5}{8}$, respectively. In fact, taking into account the weights of the singlet and triplet isospin and ordinary-spin states, we have

$$\tfrac{1}{4}\{\tfrac{3}{4}\mid\varphi^{(+)}\mid^2_{S=1}+\tfrac{1}{4}\mid\varphi^{(-)}\mid^2_{S=0}\}_{T=0}+\tfrac{3}{4}\{\tfrac{1}{4}\mid\varphi^{(+)}\mid^2_{S=0}+\tfrac{3}{4}\mid\varphi^{(-)}\mid^2_{S=1}\}_{T=1}$$
$$=\frac{3}{8}|\varphi^{(+)}|^2+\frac{5}{8}|\varphi^{(-)}|^2.$$

Therefore, the two-nucleon interaction energy (3.33), averaged over the charge and spin states of the nucleons, is equal to

$$\langle U_{\alpha\beta}\rangle = (1/\Omega)\int d^3r\{1-\frac{1}{4}\cos 2\mathbf{k}\cdot\mathbf{r}\}V(r). \tag{3.36}$$

We shall find the total potential energy of the system by summing (3.36) over all pairs of nucleons. If the nucleon density is sufficiently large ($k_F r_{\text{eff}} \gg 1$), the term with $\cos 2\mathbf{k}\cdot\mathbf{r}$ in the relation (3.36) can be neglected. Therefore, the two-nucleon interaction energy will be the same for all pairs of nucleons. In all, there are $\frac{1}{2}A(A-1)$ pairs. Using (3.29) and (3.32), we obtain the following expression for the energy per nucleon:

$$\frac{E}{A} \approx \frac{3}{10}\frac{\hbar^2k_F^2}{M}+\frac{1}{3\pi^2}k_F^3\int d^3rV(r) \quad (A\gg 1). \tag{3.37}$$

If the interaction between the nucleons corresponds to forces of attraction at all distances ($V(\mathbf{r})<0$), then, according to (3.37), the energy will decrease with increasing density $\varrho=(2/3\pi^2)k_F^3$, and, consequently, collapse of the nucleus should occur (Blatt and Weisskopf, 1952). This conclusion is not connected with the use of first-order perturbation theory. In fact, first-order perturbation theory gives for the ground-state energy of any system a value greater than or equal to the true value. Therefore, the true energy of the nucleus will be even lower than the value obtained from (3.37) and, consequently, there cannot be saturation.

Clearly, in the presence of attractive forces only, collapse of the nucleus will proceed until all the nucleons occupy a region of linear dimensions equal to the range of the nuclear forces. In this case, all the nucleons interact with each other, and the total energy of the nucleus is proportional to the number of interacting pairs, equal to $\frac{1}{2}A(A-1)$. All this is true for very light nuclei. For example, the binding energies of the nuclei ^{2}H, ^{3}He and ^{4}He are equal to 2.2, 7.6 and 28.1 MeV, respectively. That the increase in binding energy is somewhat more rapid than would be indicated by the numbers of interacting pairs (which, for these nuclei, are respectively 1, 3 and 6) is due to the fact that in ^{4}He the nucleons spend a larger part of their time within the range of the nuclear forces than they do in ^{2}H and ^{3}He. However, in the nuclei following ^{4}He we do not observe any significant increase in the energy per nucleon. This means that, in nuclei heavier than ^{4}He, each nucleon interacts with only a limited number of neighbours, that is, saturation of the nuclear interaction occurs.

The saturation condition in the case of exchange nuclear forces. One can attempt to explain the saturation of nuclear forces by making an assumption about the exchange character of the nuclear interaction between nucleons (Heisenberg, 1932).

We shall assume that, in addition to the ordinary forces characterized by a potential $V(\mathbf{r})$, there are also exchange forces acting between the nucleons, which can be characterized by a potential $\widetilde{V}(\mathbf{r})\mathbf{P}_r$, where $\widetilde{V}(\mathbf{r}) < 0$, and $\mathbf{P}_r$ is the operator interchanging the spatial coordinates. The eigenvalues of the operator $\mathbf{P}_r$ for the states $\varphi^{(+)}(\mathbf{r})$ and $\varphi^{(-)}(\mathbf{r})$ are equal to ± 1 respectively. Thus, in the case of the exchange interaction, the forces will be attractive in spatially symmetric states and repulsive (but of the same magnitude) in spatially antisymmetric states.

The two-nucleon exchange-interaction energy averaged over the different charge and spin states of the nucleons is equal to

$$\langle \widetilde{U}_{\alpha\beta} \rangle = (1/\Omega) \int d^3r \left\{ -\frac{1}{4} + \cos 2\mathbf{k}\cdot\mathbf{r} \right\} \widetilde{V}(r). \tag{3.38}$$

By summing (3.38) over all pairs of nucleons and combining the result with (3.37), we obtain in the high-density case the following expression for the energy per nucleon:

$$\frac{E}{A} \approx \frac{3}{10}\frac{\hbar^2 k_F^2}{M} + \frac{1}{3\pi^2} k_F^3 \int d^3r \left\{ V(r) - \frac{1}{4}\widetilde{V}(r) \right\} \quad (A \gg 1). \tag{3.39}$$

If $V(r)$ and $\widetilde{V}(r)$ do not change sign as functions of distance, then saturation is possible only when the exchange interaction is more than four times greater in magnitude than the non-exchange interaction and has the same sign.

It follows from an analysis of the experimental data on nucleon-nucleon scattering that the strengths of the ordinary and exchange forces are almost equal. In particular, the Serber potential (1.140) leads to good agreement with the experimental data on nucleon-nucleon scattering. Since the condition $\widetilde{V} > 4V$ is not fulfilled for the real potential (1.140), allowance for only the exchange nature of the nuclear forces does not lead to an explanation of the saturation.

3.3. Independent-pair Nuclear Model

Repulsion between nucleons at short distances. To explain the saturation of the nuclear forces, we shall assume that the nuclear interaction is repulsive at very short distances between the nucleons. Evidence for the existence of repulsive forces between nucleons at short distances is furnished by experiments on nucleon-nucleon scattering at high energies. The high-energy nucleon-nucleon scattering data agree with the assumption that nucleons have a strongly repulsive core that does not permit them to come together more closely than a distance equal to the core radius $\mathbf{r}_c \approx 0.4 \times 10^{-13}$ cm.

In the case of infinite forces of repulsion at short distances between the nucleons, the independent-particle model is completely inapplicable and we cannot use the anti-symmetrized product (3.21) of one-particle wave functions for the wave function of the ground state of nuclear matter. Indeed, the repulsion at short distances does not permit two nucleons to be at a distance apart of less than the core radius $\mathbf{r}_c$, whereas the wave function (3.21) takes into account only the correlation between nucleons due to the effect of the Pauli principle.

The existence of strong correlation between the nucleons, associated with the repulsion at short distances, is taken into account in the so-called *independent-pair model* (Brueckner, 1959; Gomes, Walecka and Weisskopf, 1958). In this model, the interaction between any pair of nucleons is treated as accurately as possible, but the interaction of all the other nucleons, with each other and with the given pair, is neglected.

Choosing the one-particle states in the same way as for the free Fermi gas, we can write the energy of the system in the form

$$E = E_0 + \sum_{\alpha > \beta} U_{\alpha\beta}, \tag{3.40}$$

where $U_{\alpha\beta}$ is the contribution to the energy from the interaction of a pair of nucleons in the states α and β, and the summation extends over all the occupied states. The quantity $U_{\alpha\beta}$ is calculated under the assumption that the interaction occurs only between the pair of nucleons in the states α and β. Because of the interaction between the nucleons, the wave function $\varphi_{\alpha\beta}$ of the pair of nucleons is not the product of the one-particle nucleon wave functions. All the other particles in states $\gamma \neq \alpha$ or β are assumed to be completely non-interacting and their wave functions are chosen in the form of plane waves. Although, by assumption, all the other nucleons do not interact with the selected pair, they nevertheless exert a considerable influence, because of the Pauli principle, on the interaction of the pair. To determine the wave function $\varphi_{\alpha\beta}$ and the correction $U_{\alpha\beta}$ we first establish the equation describing the motion of an interacting pair of nucleons in the Fermi distribution.

The Bethe-Goldstone equation. First we shall consider the Schrödinger equation describing the scattering of an isolated pair of particles:

$$\{-(\hbar^2/2M)(\nabla_1^2+\nabla_2^2)+V(\mathbf{r}_1-\mathbf{r}_2)-E^0_{\alpha\beta}\}\varphi^{(i)}_{\alpha\beta}(\mathbf{r}_1,\mathbf{r}_2)=0, \qquad (3.41)$$

where $E^0_{\alpha\beta}=(\hbar^2/2M)(k_\alpha^2+k_\beta^2)$ is the energy and $\mathbf{k}_\alpha$ and $\mathbf{k}_\beta$ are the wave vectors of the colliding particles at infinity. By finding the solution of this equation, we can determine the cross-section for scattering of the particles from the state with wave vectors $\mathbf{k}_\alpha$ and $\mathbf{k}_\beta$ to a state with wave vectors $\mathbf{k}_\gamma$ and $\mathbf{k}_\delta$. One can easily convince oneself that the scattering cross-section is determined by the square of the modulus of the Fourier component of the second term in eqn. (3.41) corresponding to the final values $\mathbf{k}_\gamma$ and $\mathbf{k}_\delta$ of the wave vectors:

$$\sigma(\mathbf{k}_\alpha,\mathbf{k}_\beta;\mathbf{k}_\gamma,\mathbf{k}_\delta)=\frac{M^2}{4\pi^2\hbar^4}\left|\int d^3r_1d^3r_2 e^{-i(\mathbf{k}_\gamma\cdot\mathbf{r}_1+\mathbf{k}_\delta\cdot\mathbf{r}_2)}V(\mathbf{r}_1-\mathbf{r}_2)\varphi^{(i)}_{\alpha\beta}(\mathbf{r}_1,\mathbf{r}_2)\right|^2. \qquad (3.42)$$

If the pair of interacting particles belongs to the Fermi distribution, eqn. (3.41) is not valid. Clearly, scattering processes in which the pair of interacting particles undergo transitions into states that are already occupied must be excluded. To do this, in the second term of eqn. (3.41) it is necessary to exclude Fourier components corresponding to states occupied by other particles.

We introduce the projection operator $Q_{\alpha\beta}$ which excludes all Fourier components in the region inside the Fermi sphere, with the exception of the Fourier component corresponding to the chosen states α and β. If Φ is an arbitrary function, then, clearly,

$$Q_{\alpha\beta}\Phi(\mathbf{r}_1,\mathbf{r}_2)=\sum_{k_\gamma,k_\delta>k_F}\varphi_\gamma(\mathbf{r}_1)\varphi_\delta(\mathbf{r}_2)(\gamma\delta|\Phi)+\varphi_\alpha(\mathbf{r}_1)\varphi_\beta(\mathbf{r}_2)(\alpha\beta|\Phi), \qquad (3.43)$$

where

$$(\gamma\delta|\Phi) = \int d^3r_1 d^3r_2 \varphi_\gamma^*(\mathbf{r}_1)\varphi_\delta^*(\mathbf{r}_2)\Phi(\mathbf{r}_1,\mathbf{r}_2)$$

(the summation in (3.43) extends only over the states outside the Fermi sphere).

Thus, we can obtain the following equation, which is usually called the *Bethe-Goldstone equation* and describes the motion of two interacting particles belonging to the Fermi distribution (Bethe and Goldstone, 1957):

$$\{-(\hbar^2/2M)(\nabla_1^2+\nabla_2^2) + Q_{\alpha\beta}V(\mathbf{r}_1-\mathbf{r}_2) - E_{\alpha\beta}\}\varphi_{\alpha\beta}(\mathbf{r}_1,\mathbf{r}_2) = 0, \quad (3.44)$$

where $E_{\alpha\beta}$ is the energy eigenvalue of the pair of particle with the interaction and the Pauli principle taken into account. Since the second term has no Fourier components corresponding to states that are energetically allowed in the scattering, apart from the initial state, the scattering cross-section vanishes. This makes it possible to find the asymptotic form of the solution of eqn. (3.44) at large distances.

In fact, if there is no scattered wave, the wave function $\varphi_{\alpha\beta}(\mathbf{r}_1,\mathbf{r}_2)$ at large distances ($|\mathbf{r}_1-\mathbf{r}_2|k_F \gg 1$) coincides with the unperturbed wave function, which is the anti-symmetrized product of the one-particle functions:

$$\varphi_{\alpha\beta}(\mathbf{r}_1,\mathbf{r}_2) \xrightarrow[|\mathbf{r}_1-\mathbf{r}_2|\to\infty]{} P_a\varphi_\alpha(\mathbf{r}_1)\varphi_\beta(\mathbf{r}_2). \quad (3.45)$$

At short distances, however, the Pauli principle does not prevent distortion of the wave function arising from the interaction. Indeed, short distances correspond to momenta lying considerably above the limiting Fermi momentum, so that at very short distances ($|\mathbf{r}_1-\mathbf{r}_2|k_F \ll 1$) it is not important to take the Pauli principle into account, and the wave function $\varphi_{\alpha\beta}$ will coincide with the wave function $\varphi_{\alpha\beta}^{(i)}$ for an isolated pair.

Introducing centre-of-mass coordinates and relative coordinates and taking into account the conservation of the total wave vector, $\mathbf{K} = \mathbf{k}_\alpha + \mathbf{k}_\beta$, it is not difficult to obtain from (3.44) the following equation for the wave function $\varphi_\mathbf{k}(\mathbf{r})$ of the relative motion of the two particles:

$$\{-(\hbar^2/M)\nabla^2 + (\hbar^2/4M)K^2 + Q_{\alpha\beta}V(\mathbf{r}) - E_{\alpha\beta}\}\varphi_\mathbf{k}(\mathbf{r}) = 0. \quad (3.46)$$

Using the condition (3.45) at infinity we can reduce this equation, by means of the Fourier transformation, to the integral equation

$$\varphi_\mathbf{k}(\mathbf{r}) = (1/\sqrt{\Omega})e^{i\mathbf{k}\cdot\mathbf{r}} + \int d^3r' G(\mathbf{r}-\mathbf{r}')V(\mathbf{r}')\varphi_\mathbf{k}(\mathbf{r}'), \quad (3.47)$$

where

$$G(\mathbf{r}-\mathbf{r}') = \int\limits_{|\frac{1}{2}\mathbf{K}\pm\mathbf{k}'|>k_F} \frac{d^3\mathbf{k}'}{(2\pi)^3} \frac{e^{i\mathbf{k}\cdot(\mathbf{r}-\mathbf{r}')}}{E_{\alpha\beta} - (\hbar^2K^2/4M) - (\hbar^2k'^2/M)}. \quad (3.48)$$

If we extend the integration over $\mathbf{k}'$ in (3.48) over the whole space (having previously specified the rules for going round the pole of the integrand), we obtain the usual Green function for free particles. Clearly, the function $G(\mathbf{r}-\mathbf{r}')$ defined by the expression (3.48) can be regarded as the Green function for particles belonging to the Fermi distribution. Since the state $\alpha\beta$ corresponds to momenta $\mathbf{k}_\alpha$ and $\mathbf{k}_\beta$ inside the Fermi sphere and the integration in (3.48) is performed outside the Fermi sphere, the integrand in (3.48), unlike that in the case of free particles, has no singularities within the range of integration.

We find the complete function $\varphi_{\alpha\beta}$ by multiplying (3.47) by the wave function of the centre-of-mass motion of the two particles:

$$\varphi_{\alpha\beta}(\mathbf{r}_1,\mathbf{r}_2) = (1/\sqrt{\Omega})e^{i\mathbf{K}\cdot\mathbf{R}}\varphi_{\mathbf{k}}(\mathbf{r}). \tag{3.49}$$

Since the second term in the right-hand side of (3.47) contains only Fourier components outside the Fermi sphere, the function $\varphi_{\alpha\beta}$ satisfies the normalization condition

$$(\varphi^0_{\alpha\beta}, \varphi_{\alpha\beta}) = 1 \tag{3.50}$$

The limits of integration in the expression (3.48) for the Green function depend on the momentum $\mathbf{K}$ of the centre-of-mass motion, and so the wave function $\varphi_{\mathbf{k}}(\mathbf{r})$ of the relative motion of the particles in the Fermi distribution also depends on $\mathbf{K}$ in the general case. We shall confine ourselves below to the simple case $\mathbf{K}=0$. Then the integration over $\mathbf{k}'$ in (3.48) must be performed outside the Fermi sphere ($k' > k_F$), and the Green function takes the form

$$G(\mathbf{r}-\mathbf{r}') = \frac{1}{2\pi^2}\int_{k_F}^{\infty} dk' \frac{k'}{\varepsilon - (\hbar^2 k'^2/M)} \frac{\sin k'|\mathbf{r}-\mathbf{r}'|}{|\mathbf{r}-\mathbf{r}'|}, \tag{3.51}$$

where $\varepsilon = \hbar^2k^2/M$ is the energy of the relative motion.

If $\mathbf{K}=0$, the relative angular momentum is a constant of the motion and the function $\varphi_{\mathbf{k}}(\mathbf{r})$ can be expanded in partial waves corresponding to definite values of the relative angular momentum:

$$\varphi_{\mathbf{k}}(\mathbf{r}) = \sum_l \varphi_l(r) P_l(\cos\vartheta). \tag{3.52}$$

The study of the particle motion in the S-state ($l=0$) is of greatest interest. From (3.47) we obtain for the wave function $\varphi(r)$ of the S-state

$$\varphi(r) = \frac{1}{\sqrt{\Omega}}\frac{\sin kr}{kr} + 4\pi\int_{k_F}^{\infty} dr' r'^2 g(r,r')V(r')\varphi(r'), \tag{3.53}$$

where

$$g(r,r') = \frac{M}{2\pi^2\hbar^2}\frac{1}{rr'}\int_{k_F}^{\infty} dk' \frac{\sin k'r \sin k'r'}{k^2-k'^2}. \tag{3.54}$$

The corresponding differential equation for $\varphi(\mathbf{r})$ has the form

$$\left\{\frac{\hbar^2}{M}\left(\frac{d^2}{dr^2}+\frac{2}{r}\frac{d}{dr}\right)+\varepsilon\right\}\varphi(r) = V(r)\varphi(r) - \int_0^{\infty} dr' r'^2 \chi(r,r')V(r')\varphi(r'),$$

$$\chi(r,r') = \frac{1}{\pi r r'}\left\{\frac{\sin k_F(r-r')}{k_F(r-r')} - \frac{\sin k_F(r+r')}{k_F(r+r')}\right\}. \tag{3.55}$$

We note that the last term in the right-hand side of (3.55) is a consequence of taking the *Pauli principle* into account.

We shall examine the soliton of eqns. (3.53) or (3.55) in the simplest case when the potential $V(r)$ is only a repulsive core of radius r_c:

$$V(r) = \begin{cases} \infty, & r < r_c, \\ 0 & r > r_c. \end{cases} \tag{3.56}$$

Obviously, $\varphi(r) = 0$ for $r < r_c$. Since $V(r)$ goes to infinity inside the core, the product $V(r)\varphi(r)$ is indeterminate for $r < r_c$. We introduce the notation $w(r) \equiv V(r)\varphi(r)$. In the outer region $r > r_c$, $w(r) = 0$. To determine $w(r)$ in the inner region $r < r_c$ we consider eqn.(3.55) for $r < r_c$. The function $\varphi(r)$ is continuous in the whole region, but the derivative $\varphi'(r)$ is discontinuos at the point $r = r_c$; therefore, the second derivative $\varphi''(r)$ has a delta-function singularity at the point $r = r_c$. Let $(d\varphi(r)/dr)_{r=r_c} = \lambda$; then eqn.(3.55) for $r \leqslant r_c$ can be written in the form

$$(\hbar^2/M)\{\lambda\delta(r-r_c) + (2/r)\varphi'(r)\} + \varepsilon\varphi(r) = w(r) - \int_0^{\infty} dr'\, r'^2 \chi(r,r') w(r').$$

Since the integral in the right-hand side is a continuous function, we have

$$w(r) = (\hbar^2/M)\lambda\delta(r-r_c) + w_1(r), \tag{3.57}$$

where the function $w_1(r)$ has no singularities. Substituting the relation (3.57) into (3.53) and neglecting the contribution of w_1, which is found to be of order of $(k_F r_c)^3$, we find

$$\varphi(r) = \frac{1}{\sqrt{\Omega}}\frac{\sin kr}{kr} + \frac{4\pi\hbar^2}{M}\lambda r_c^2 g(r,r_c), \quad r \geqslant r_c.$$

The constant λ determined from the condition $\varphi(r_c) = 0$. Thus, we obtain the solution of the *Bethe-Goldstone equation* for the S-state in the form

$$\varphi(r) = \frac{1}{\sqrt{\Omega}}\left\{\frac{\sin kr}{kr} - \frac{\sin kr_c}{kr_c}\frac{g(r,r_c)}{g(r_c,r_c)}\right\}, \quad r \geqslant r_c, \tag{3.58}$$

where $g(r, r_c)$ is defined by the expression (3.54).

We note that the function $g(r, r_c)$ decreases rapidly with increadsing r. If $r \gg r_c$ then

$$\begin{aligned} g(r,r_c) \approx g(r,0) \quad = \quad & \frac{M}{4\pi^2\hbar^2}\frac{1}{r}\{\cos kr[\mathrm{si}\,(k_F - k)r + \mathrm{si}\,(k_F + k)r] \\ & + \sin kr[\mathrm{ci}\,(k_F - k)r - \mathrm{ci}\,(k_F + k)r]\}, \end{aligned} \tag{3.59}$$

where $\mathrm{si}\,x$ and $\mathrm{ci}\,x$ are the integral sine and integral cosine:

$$\mathrm{si}\,x = -\int_x^\infty dt(\sin t/t), \quad \mathrm{ci}\,x = -\int_x^\infty dt(\cos t/t).$$

Therefore, the solution (3.58) of the Bethe-Goldstone equation goes over at large distances ($k_F r \gg 1$) into the inperturbed solution corresponding to the absence of interaction between the particles:

$$\varphi_0(r) = (1/\sqrt{\Omega})(\sin kr/kr). \tag{3.60}$$

In the case of an isolated pair of particles not belonging to the Fermi distribution, the Green function has the form

$$g^{(i)}(r,r') = -\frac{M}{4\pi\hbar^2}\cdot\frac{1}{k}\begin{cases} \dfrac{\sin kr}{r}\dfrac{\cos kr'}{r'}, & r < r', \\ \dfrac{\cos kr}{r}\dfrac{\sin kr'}{r'}, & r > r', \end{cases} \tag{3.61}$$

while the corresponding solution of the Schrödinger equation for particles in the presence of a repulsive core is given by

$$\varphi^{(i)}(r) = \frac{1}{\sqrt{\Omega}}\cdot\frac{\sin k(r - r_c)}{kr\cos kr_c}, \quad r > r_c. \tag{3.62}$$

Figure 3.3 shows for comparison the dependence on r of the functions $\varphi(r)$, $\varphi^{(i)}(r)$ and $\varphi_0(r)$. The solution $\varphi(r)$ of the Bethe-Goldstone equation, like the solution $\varphi^{(i)}(r)$ of the Schrödinger equation, is non-zero starting from the point $r = r_c$. Then, however, $\varphi(r)$ becomes closer to the unperturbed solution $\varphi_0(r)$, and, after a few small oscilations, practically coincides with the unperturbed solution. Whereas for an isolated pair the

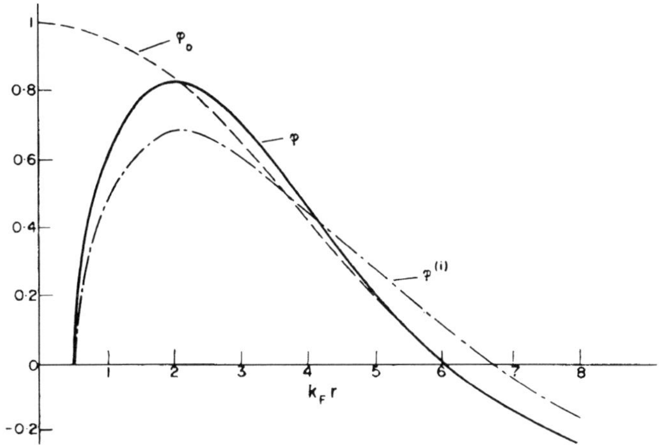

Figure 3.3. Dependence of the functions φ, $\varphi^{(i)}$ and φ_0 on $k_F r$.

change in the wave function caused by the interaction persists, in the form of a phase shift, right out to infinite distances, in the case of a pair of particles belonging to the Fermi distribution these changes are rapidly washed out and the function has already attained the unperturbed form at distances of the order of $1/k_F$.

If, along with the repulsive core, there is also an attractive potential $V_a(r)$ for $r > r_c$, then the integral equation for the function $\varphi_a(r)$, can be represented in the form

$$\varphi_a(r) = \varphi(r) + 4\pi \int_{r_c}^{\infty} dr' r'^2 \left\{ g(r, r') - \frac{g(r, r_c) g(r_c, r')}{g(r_c, r_c)} \right\} V_a(r') \varphi_a(r'), \quad (3.63)$$

where $\varphi(r)$ is given by the expression (3.58). At $r = r_c$ the expression in the curly brackets is equal to zero. Therefore, for values of r close to r_c, the function $\varphi_a(r)$, differs little from $\varphi(r)$. With increasing r, because of the rapid decrease of $g(r, r_c)$, the correction (for finite values of $V_a(r)$) in (3.63) is also small. Thus, in the presence of attractive forces too, the wave function is almost the same as that for the hard-core problem, and almost coincides with the free-particle wave function except in the region bordering directly on the repulsive core. We should make the reservation that this conclusion is based on the relative weakness of the attractive forces. If the attractive forces were considerably stronger, they could modify the two-particle wave function appreciably even in the presence of other particles belonging to the Fermi distribution.

Determination of the nuclear density and binding energy. Knowing the solution of the Bethe-Goldstone equation, we can calculate the nuclear binding energy directly. In fact, using the *Bethe-Goldstone equation* (3.44) and the normalization condition (3.50) for the wave function $\varphi_{\alpha\beta}$, it is not difficult to show that the additional energy $U_{\alpha\beta}$ associated with the interaction of a pair of nucleons is equal to

$$U_{\alpha\beta} = (\varphi^0_{\alpha\beta}|V(\mathbf{r}_1 - \mathbf{r}_2)|\varphi_{\alpha\beta}). \tag{3.64}$$

Assuming that each pair of nucleons interacts independently of the others, we write the total energy of nuclear matter in this case in the form

$$E = E_0 + \frac{1}{2}\sum_{\alpha\neq\beta} U_{\alpha\beta}. \tag{3.65}$$

The total energy E is a function of the limiting Fermi momentum and, consequently, is a function of the density. From the condition that the energy be a minimum, we can find the equilibrium value of the density; this will characterize the density of nuclear matter.

To calculate $U_{\alpha\beta}$ we shall assume that Serber exchange forces (1.140) act between the nucleons. These forces are non-zero only in states with even values of the relative orbital angular momentum. We shall neglect the tensor forces, and also the spin dependence of the nuclear forces, and choose the potential $V(r)$ of the central forces in the form of an attractive square well with a core corresponding to infinite repulsive firces at short distances:

$$V(r) = V_c(r) + V_a(r), \tag{3.66}$$

where

$$V_c(r) = \begin{cases} \infty, & r < r_c; \\ 0, & r > r_c; \end{cases} \qquad V_a(r) = \begin{cases} 0, & r < r_c; \\ -V_0, & r_a < r < r_0; \\ 0, & r_0 < r. \end{cases} \tag{3.67}$$

The core radius r_c and also the range r_0 and magnitude V_0 of the attractive potential, must be chosen so that the properties of a system of two nucleons at low energies are correctly described and, at the same time, so that the high-energy scattering data are reproduced.

Using the exchange character of the Serber potential (1.140), we find for the interaction energy of a pair of nucleons in the states α and β

$$U_{\alpha\beta} = (\varphi^{0(+)}_{\alpha\beta}(\mathbf{r})|V_S(r)|\varphi^{(+)}_{\alpha\beta}(\mathbf{r})), \tag{3.68}$$

where $\varphi^{0(+)}(\mathbf{r})$ and $\varphi^{(+)}(\mathbf{r})$ are symmetrized wave functions:

$$\varphi^{(+)}(\mathbf{r}) \equiv (1/\sqrt{2})\{\varphi(\mathbf{r}) + \varphi(-\mathbf{r})\}. \tag{3.69}$$

Since the potential $V(r)$ depends only on the relative distance, the quantity (3.68) will depend only on the relative momentum k of the nucleons:

$$U_{\alpha\beta} = U(k), \quad \mathbf{k} = \frac{1}{2}(\mathbf{k}_\alpha - \mathbf{k}_\beta). \tag{3.70}$$

Averaging of (3.68) over the charge and spin states of the nucleons reduces to multiplying the expression (3.68) by the statistical weight $\frac{3}{8}$ of the symmetric state:

$$\langle U_{\alpha\beta} \rangle = \frac{3}{8} U(k). \tag{3.71}$$

We find the total interaction energy ΔE of all the pairs by multiplying (3.71) by the number $n(k)$ of pairs of nucleons with relative momentum k (see Problem 3.2)

$$n(k) = 12A(A-1)\left(1 - \frac{3}{2}\frac{k}{k_F} + \frac{1}{2}\frac{k^3}{k_F^3}\right)\frac{k^2}{k_F^3} \tag{3.72}$$

and integrating the expression obtained over the possible values of the relative momentum k:

$$\Delta E = \frac{3}{8}\int_0^{k_F} dk\, n(k)\, U(k). \tag{3.73}$$

In correspondence with (3.66), the total interaction energy is also represented in the form of a sum

$$\Delta E = \Delta E_c + \Delta E_a. \tag{3.74}$$

To calculate ΔE_c, we shall make use of the fact that the wave function φ practically coincides at short distances with the wave function (3.58) corresponding to the presence of only a repulsive core. According to (3.57), we have

$$V_C(r)\varphi(r) \approx (\hbar^2/M)\lambda\delta(r - r_c), \tag{3.75}$$

where

$$\lambda = -\frac{M}{\hbar^2}\frac{1}{4\pi r_c^2}\frac{\varphi_0(r_c)}{g(r_c, r_c)},$$

and, consequently,

$$U_c(k) = -2\frac{\varphi_0^2(r_c)}{g(r_c, r_c)}. \tag{3.76}$$

Since the radius of the repulsive core is small, the energy (3.76) depends very weakly on the relative momentum k, and so we can confine ourselves to calculating the energy for $k = 0$. We write the *Green function* $g(r_c, r_c)$ for this case:

$$g(r_c, r_c) = -\frac{M}{2\pi^2\hbar^2 r_c^2}\int_{k_F}^{\infty} dk' \frac{\sin^2 k' r_c}{k'^2}. \tag{3.77}$$

Replacing the variable $k' r_c$ by x, we obtain

$$g(r_c, r_c) = -\frac{M}{2\pi^2\hbar^2 r_c}\int_{k_F r_c}^{\infty} dx \frac{\sin^2 x}{x^2}.$$

For $k_F r_c \ll 1$, this integral can be represented conveniently in the form of a series in powers of $k_F r_c$. We rewrite the integral as follows:

$$g(r_c, r_c) = -\frac{M}{2\pi^2\hbar^2 r_c}\left\{\int_0^{\infty} dx \frac{\sin^2 x}{x^2} - \int_0^{k_F r_c} dx \frac{\sin^2 x}{x^2}\right\}. \tag{3.78}$$

After integration, the first term in the brackets gives $\pi/2$, and for small values of $k_F r_c$ we can replace $\sin^2 x/x^2$ unity in the second term. Thus, the Green function (3.78) takes the form

$$g(r_c, r_c) \approx -\frac{M}{2\pi^2\hbar^2 r_c}\left(1 - \frac{2}{\pi}k_F r_c + \ldots\right). \tag{3.79}$$

Using this expression and taking into account that $\varphi_0^2(r_c) \approx 1/\Omega$ for $k \to 0$, from (3.76) we obtain

$$U_c(k) \approx U_c(0) = \frac{8\pi}{\Omega}\frac{\hbar^2 r_c}{M}\left(1 + \frac{2}{\pi}k_F r_c + \ldots\right).$$

Then, substituting this expression into a formula of the form (3.73) and noting that, for $A \gg 1$, the total number of interacting pairs is equal to $\frac{1}{2}A^2$, we obtain the following approximate formula for the total energy ΔE_c:

$$\frac{\Delta E_c}{A} = \frac{1}{\pi}\frac{\hbar^2 k_F^2}{M}k_F r_c\left(1 + \frac{2}{\pi}k_F r_c + \ldots\right), \quad k_F r_c \ll 1. \tag{3.80}$$

We observe that the quantity ΔE_c is a part of the kinetic energy (since the potential $V_c(r)$ is zero in the region $r > r_c$ in which the wave function is non-zero) and aries from the deviation of the *Bethe-Goldstone wave function* from the free-particle wave function.

The expression (3.80), obtained on the basis of the independent-pair model, is valid, generally speaking, only for sufficiently low densities. In the general case, the energy ΔE_c has been calculated in the form of an exact expansion in powers of $k_F r_c$ by Lee and Yang (1957):

$$\frac{\Delta E_c}{A} = \frac{1}{\pi}\frac{\hbar^2 k_F^2}{M} k_F r_c \left[1 + \frac{6}{35\pi}(11 - 2\ln 2) k_F r_c + 0.41 k_F^2 r_c^2 + \ldots\right]. \quad (3.81)$$

We now calculate the energy ΔE_a associated with the attractive potential $V_a(r)$. Since the wave function $\varphi(\mathbf{r})$ almost coincides with the free-particle wave function $\varphi_0(\mathbf{r})$ in the region where the potential V_a is non-zero, to calculate $U_a(k)$ we can use the Born approximation. Then

$$U_a(k) = (\varphi_0^{(+)}(\mathbf{r})|V_a(r)|\varphi_0^{(+)}(\mathbf{r})) = \frac{1}{\Omega}\int d^3r(1 + \cos 2\mathbf{k}\cdot\mathbf{r})V_a(r). \quad (3.82)$$

We note that the applicability of the Born approximation in the estimation of $U_a(k)$ is due entirely to the effect of the Pauli principle. Substituting (3.82) into the formula (3.73) and performing the interaction, we find

$$\Delta E_a/A = -(V_0/\pi)[\mathcal{F}(k_F r_0) - \mathcal{F}(k_F r_c)], \quad (3.83)$$

where the function $\mathcal{F}(x)$ is given by the expression

$$\begin{aligned} \mathcal{F}(x) &= x^3/6 + (3/2)\mathrm{Si}\,(2x) + (3/2x^2)\sin 2x \\ &\quad +(3/4x)(1 + 1/x^2)\cos 2x - (9/4x) - (3/4x^3), \quad (3.84) \\ \mathrm{Si}\,(x) &= \mathrm{si}\,(x) + \pi/2. \end{aligned}$$

Thus, the *total energy per nucleon* is equal to

$$\begin{aligned} \frac{E}{A} &= \frac{3}{10}\frac{\hbar^2 k_F^2}{M} + \frac{1}{\pi}\frac{\hbar^2 k_F^2}{M} k_F r_c \left[1 + \frac{6}{35\pi}(11 - 2\ln 2) k_F r_c \right. \\ &\quad \left. + 0.41 k_F^2 r_c^2 + \ldots\right] - \frac{V_0}{\pi}[\mathcal{F}(k_F r_0) - \mathcal{F}(k_F r_c)]. \quad (3.85) \end{aligned}$$

With increase of the density, the absolute value of the energy $\Delta E_a/A$ associated with the attractive forces increases linearly with ϱ and can exceed the kinetic energy $E_0 A \approx \varrho^{2/3}$; it could thus cause *collapse of the nucleus* if the repulsion at short distances did not exist. In reality, the additional kinetic energy $\Delta E_c/A$ due to the decrease of volume on account of the repulsive core increases with increasing density like $\varrho^{4/3}$ and prevents collapse.

The total energy E/A as a function of the density ϱ has a minimum, which determines both the equilibrium density of the nucleus and the binding energy. Because of the relative weakness of the attractive potential, equilibrium is reached at a fairly low density, at which the average distance

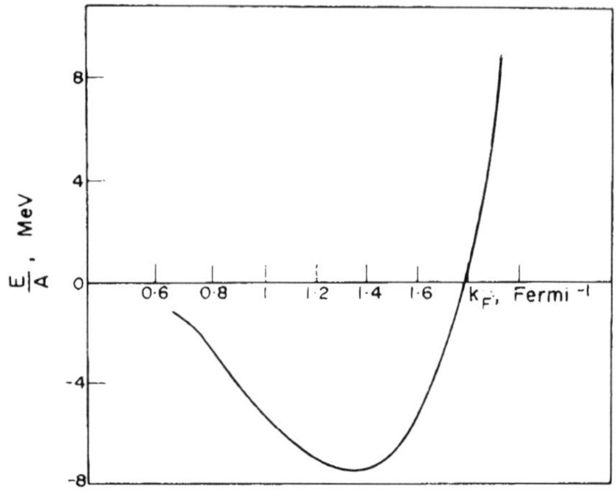

Figure 3.4. Dependence of the energy E/A on the quantity k_F.

$d \approx \varrho^{-1/3}$ between the particles is large compared with the core radius ($d/r_c \approx 4$).

Figure 3.4 shows the dependence, calculated from formula (3.85), of the energy E/A on the magnitude of k_F. In the calculations, the following values of the parameters of the two-nucleon potential were used:

$$r_c = 0.4 \times 10^{-13}\,\mathrm{cm}; \quad r_0 = 2.3 \times 10^{-13}\,\mathrm{cm}; \quad V_0 = (\pi^2/4)\hbar^2/M(r_0 - r_c)^2. \tag{3.86}$$

(The depth of the potential well is chosen such that the binding energy of two nucleons is equal to zero.) In addition, in the expression for ΔE_c, the *effective nucleon mass* $M^* = 0.70\,M$ was used in place of M. For the chosen values of the parameters, the energy E/A has a minimum at $k_F = 1.35 \times 10^{13}\,\mathrm{cm}^{-1}$, corresponding to an equilibrium density $\varrho = 1.65 \times 10^{38}\,\mathrm{cm}^{-3}$ of the nuclear matter.

Calculations using more realistic potentials lead to values of the mean binding energy per nucleon and equilibrium density that agree with the experimental results (3.18).

Applicability of the independent-particle model. Investigation of the behaviour of the *Bethe-Goldstone wave function* at large distances enables us to determine the reasons for the applicability of the independent-particle model in the description of nuclei. We have seen that, as a consequence of the Pauli principle, the interaction between particles belonging to the Fermi distribution is manifested only at distances less than k_F^{-1} and is completely unimportant at distances greater than k_F^{-1}. Therefore, the Bethe-Goldstone wave function differs from the free-particle wave function only at short distances ($< k_F^{-1}$). These distances are found to be considerably shorter than the mean distance $d \approx \varrho^{-1/3}$ between the nucleons in nuclei. At distances of the order of the mean distance between the nucleons in nuclei, the *Bethe-Goldstone wave function* is not essentially different from the free-particle wave function. From this it is clear why the independent-pair model is such a good approximation. If one of the particles of the pair collides with some third particle, the distance between the original pair of particles is on average of order d. But at this distance, because of the effect of the Pauli principle, the wave function of the ralative motion practically coincides with the free-particle wave function. Thus, a collision of one of the particles of the pair with a third particle occurs as if the interaction between the particles of the original pair were completely absent.

The explanation of the applicability of the independent-particle model also follows from this property of the Bethe-Goldstone wave functions. In fact, despite the nuclear interaction, the wave functions almost coincide with the free-particle wave functions, except in those regions of configuration space in which the particles come together to within distances much less than the mean distance between the particles. Therefore, nucleons in the nucleus can be described with sufficient accuracy by free-particle wave functions, except in those cases in which close contact of two nucleons plays an important role.

We have confined ourselves to treating only the symmetric part ($l = 0$) of the Bethe-Goldstone wave function. It is not difficult to convince oneself that higher values of the orbital angular momentum do not play an important role. The wave function of the P-state is not affected by the potential, since we have assumed that interaction is completely absent in odd states. In the D-state, the screening effect of the centrifugal forces is so great that the effect of the nuclear forces on the wave function can be neglected even when the effect of the Pauli principle is not taken into account.

3.4. Hydrodynamical Model of Nucleus

Classical hydrodynamical model of nuclei. The *hydrodynamical model of nucleus* , that is discussed in this section, is well developed at present. There are classical and quantum versions of this model describing various collective properties of middle and heavy nuclei. The classical model is based on the hydrodynamic theory of classical fluid that is treated as a continuous medium. In the classical model it is assumed that the nucleon mean free path in a nucleus is much smaller than its dimensions. The *quantum hydrodynamical model of nuclei* within certain limits is a generalization of the quantum liquid theory. The classical and quantum versions of the model pretend to describe only the averaged (collective) nucleus properties. The main relations of the quantum model are obtained by naturally using some averaging (roughening) of the quantum mechanical equations of motion.

The *quantum hydrodynamical model of nuclei* is associated with microscopic approach, namely, with the time-dependent Hartree-Fock approximation. That is why the derivation of the main equation of motion is also microscopic.

A set of the observed general properties of multinucleon nuclei can be explained and qualitatively described within the framework of comparatively simple dynamical collective model, namely, the phenomenological *classical hydrodynamic model of nuclei* . In this model, the collective motion of nucleons in a nucleus are treated as irrotational motion of nuclear fluid with a constant density $\varrho_0 = \frac{A}{V} = \frac{3}{4\pi R_0^3}$, where A is a number of nucleons in a nucleus, $V = \frac{4\pi}{3} R_0^3 A$ is the nucleus volume, $R_0 = 1.2\,\mathrm{fm}$.

In particular, the classical hydrodynamical model of nuclei makes it possible to describe not only qualitatively but quantitatively the *giant resonance in nuclei*, photonuclear and electronuclear processes accompanied by excitations of the nuclear giant resonance states.

Speaking about the nuclear giant resonance, we usually mean the most intensive photonuclear (and electronuclear as well) giant dipole resonance whose maximum position for light nuclei corresponds to the excitation energy of about 22 MeV, and for heavy nuclei – about 15 MeV with a few MeV in width. This considerably exceeds the resonance widths of the compound nucleus. At these energies of excitation nuclei absorb intensively the incident dipole electric radiation, i.e., E1-transitions occur in nuclei. That is why the corresponding giant resonance is called E1-resonance. On increasing the energy of incident radiation or incident charged particles up to the threshold of the meson formation the giant resonces of higher multipolarity with absorption of E2-, M1-, E3-, M2-, E4-, M3-radiation can be excited. However, these transitions are considerably less intensive than dipole ones, i.e., E1-transitions. The pion formation occurs at a sufficiently high energy of M1-radiation or incident magnetic dipole photons.

The giant dipole resonance phenomena can be explained in a simple way both within the framework of the shell model and the framework of the collective hydrodynamical model of nuclei.

According to the shell model argumentation the absorption of a hard enough photon (or scattering of charged particles with absorption of a virtual photon) by a nucleus can be accompanied by the direct photonuclear process with liberation of one nucleon, as a rule from the peripheral nucleus shell, or formation of the compound highly excited nucleus. The latter can decay in a number of different ways with liberation of nucleons and photons of smaller energy. The existence of many possible channels of the process results in the considerable widening of highly excited levels of nuclei are formed when a large amount of the electromagnetic energy is absorbed. Even before using the shell model, A. Migdal (1945) has shown for the first time that the giant dipole resonance in nuclei can be explained by joint vibrational motion of the nucleus protons above neutrons under action of the time dependent electric field of an incident photon or a scattered charged particle by nuclei. This collective motion of protons about neutrons results in the appearance of the large induced dipole nucleus moment. The intensive electric dipole transitions in nuclei are connected with this dipole moment. Later this approach has been developed by other authors. As a consequence of this specific classical hydrodynamical model of nuclei was developed. According to this model, a nucleus consists of two kinds of liquids – proton and neutron. Their densities $\rho_p \equiv \rho_p(\mathbf{r},t)$ and $\rho_n \equiv \rho_n(\mathbf{r},t)$ are functions of coordinates and time in a general case, but their sum, i.e., the total density of nuclear substance ρ_0 is a constant quantity in a nucleus. It means that the nuclear substance is incompressible:

$$\rho_p(\mathbf{r},t) + \rho_n(\mathbf{r},t) = \rho_0 = \text{Const.} \tag{3.87}$$

Here we consider this liquid or hydrodynamical model of nuclei in greater detail. It is clear that such a model can be a good approximation only for nuclei with a large mass number $A \gg 1$.

In the classical hydrodynamical model of nuclei the local velocities of the small volume elements (containing, however, large numbers of nucleons) of proton and neutron liquids $\mathbf{v}_p(\mathbf{r},t)$ and $\mathbf{v}_n(\mathbf{r},t)$ and the relative velocity $\mathbf{v} - \mathbf{v}_p(\mathbf{r},t) - \mathbf{v}_n(\mathbf{r},t)$ are introduced. The latter quantity vibrates with time along with the densities ρ_p and ρ_n. We assume that these vibrations are small, and in the classic equations of motion of nuclear liquid take into account only linear (in the velocity $\mathbf{v}$) terms. The nonlinear terms responsible for vortexes of the nuclear liquid are neglected. The irrotational motion in hand satisfies the conditions

$$\operatorname{rot}\mathbf{v}_p = \operatorname{rot}\mathbf{v}_n = \operatorname{rot}\mathbf{v} = 0.$$

This enables us to introduce the potential of the relative velocity $\varphi \equiv \varphi(\mathbf{r},t)$:

$$\mathbf{v} = -\nabla\varphi. \tag{3.88}$$

In this simple version of the hydrodynamical model we also neglect the contribution of electrostatic forces in the equations. The influence of the external electromagnetic field on the nucleon motion is included in the boundary conditions only.

Suppose that the dependence of proton and neutron liquid density on time is conditioned by the external electromagnetic perturbation of incident photons or charged particles. If there are no such perturbations, we deal with the equilibrium densities of protons and neutrons $\rho_p^{(0)}$ and $\rho_n^{(0)}$ which do not depend on time t (they do not depend on $\mathbf{r}$ in the considered version of the model). Thus, the external action results in local departures of ρ_p and ρ_n from the constant equilibrium values of the proton and neutron densities in a nucleus $\rho_p^{(0)}$ and $\rho_n^{(0)}$. In the classical hydrodynamical model of nuclei we assume that the above indicated departures of density are repressed by the force associated with the symmetry energy between a number of protons Z and number of neutrons $N = A - Z$ in a nucleus

$$E_s = \varepsilon_4 \frac{(A-2Z)^2}{A}, \quad \varepsilon_4 \approx 20\,\text{MeV}, \tag{3.89}$$

that appears in semiempirical formula (3.17) describing the dependence of the nucleus bound energy on A and Z. It is possible to generalize (3.89) rewriting the symmetry energy E_s in the form

$$E_s = \int d\mathbf{r} U_s(\mathbf{r},t), \quad U_s(\mathbf{r},t) = \varepsilon_4 \frac{(\rho_p(\mathbf{r},t) - \rho_n(\mathbf{r},t))^2}{\rho_0}, \tag{3.90}$$

where $U_s(\mathbf{r},t)$ is the symmetry energy density in a nucleus. Energy (3.90) has a minimum for the equilibrium values of densities $\rho_p = \rho_p^{(0)}$ and $\rho_n = \rho_n^{(0)}$ and transforms into (3.89) for the homogeneous density distributions $\rho_p = Z/V$ and $\rho_n = N/V$.

According to (3.87) inside a nucleus the following equalities are true: $\nabla \rho_p = -\nabla \rho_n$ and $\frac{\partial \rho_p}{\partial t} = -\frac{\partial \rho_n}{\partial t}$. Thus, the equations for $\rho_p(\mathbf{r},t)$ and $\rho_n(\mathbf{r},t)$ are the same and the system of the proton and neutron liquids are described by only two quantities $\rho_p(\mathbf{r},t)$ and $\varphi(\mathbf{r},t)$. In this approximation, in particular, for the small vibrations of the densities and relative velocity, the scalar quantities ρ_p and φ satisfy the convention wave equations (as for sonic waves), if the nonconservative friction forces are neglected:

$$\frac{\partial^2 \rho_p}{\partial t^2} - u^2 \Delta \rho_p = 0, \quad \frac{\partial^2 \varphi}{\partial t^2} - u^2 \Delta \varphi = 0, \tag{3.91}$$

where u is the propagation velocity of the proton density $\rho_p(\mathbf{r},t)$ and potential of the relative velocity $\varphi(\mathbf{r},t)$ vibrations in a nucleus.

Since these vibrations are conditioned by the symmetry energy E_s within the framework of the model under consideration, it is clear that $u^2 \sim \varepsilon_4/M$, where M is the nucleon mass. For vibrations in the nucleus with $N = Z$ some averaged kinetic energy of a nucleon is approximately equal to $\frac{Mu^2}{2} \approx \varepsilon_4$. There are no the above considered vibrations of protons with respect to neutrons in the hypothetic nuclei with $Z \to 0$ or $N \to 0$ and $u^2 \to 0$. That is why at $N \neq Z$, we can assume that u^2 is proportional to a product ZN.

Thus, at $N \neq Z$ the kinetic energy $\frac{Mu^2}{2} \approx \varepsilon_4 \frac{4ZN}{A^2}$, where the factor $\frac{4ZN}{A^2}$ is equal to one at $N = Z$ and zero at $Z = 0$ or $N = 0$. As a result we have approximately

$$u^2 = \frac{8\varepsilon_4}{M}\frac{ZN}{A^2}. \tag{3.92}$$

If we take into account the damping of the proton vibrations around neutrons, that is caused by the nonconservative forces, then in equation (3.91) we have to introduce terms proportional to the first derivatives of quantities ρ_p and φ with respect to time. The classic equations of motion of the system of two nucleon liquids in this case take the form

$$\frac{\partial^2 \rho_p}{\partial t^2} - u^2 \Delta \rho_p + \Gamma \frac{\partial \rho_p}{\partial t} = 0, \tag{3.93}$$

$$\frac{\partial^2 \varphi}{\partial t^2} - u^2 \Delta \varphi + \Gamma \frac{\partial \varphi}{\partial t} = 0. \tag{3.94}$$

Here Γ is a positive parameter whose magnitude describes the friction intensity between proton and neutron liquids activating the nucleus heating (excitation). Note that in the considered model we deal only with the friction forces associated with motion of the proton liquid above neutron liquid.

Now we formulate the boundary conditions for hydrodynamic equations (3.93) and (3.94). As it was noted above, they must consist of the electromagnetic force $\mathbf{F}_{em} = \mathbf{F}_{em}(\mathbf{r}, t)$ of the external action on protons in a nucleus from the incident photons or charged particles. Here, the nucleus shape is treated as nonspherical whose surface equation in the spherical coordinates is given as a distance from the nucleus center to its surface $r = R(\theta, \varphi)$ in the form of a function of angles θ and φ. If $\mathbf{F}_{em} = 0$, then on the nucleus surface $r = R(\theta, \varphi)$ $\mathbf{n}\nabla\rho_p = 0$ and $-\mathbf{n}\dot{\mathbf{v}} = \mathbf{n}\nabla\dot{\varphi} = 0$ ($\mathbf{n}$ is a unit vector perpendicular to the surface of the nucleus, a dot stands for the time derivative). If the external force $\mathbf{F}_{em} \neq 0$, then we obtain nonzero values of $\mathbf{n}\nabla\rho_p$ and $\mathbf{n}\nabla\dot{\varphi}$ on the nucleus surface. To write down the boundary conditions in this more complex case, we note that the external force acting on unit volume of the proton liquid $\rho_p(\mathbf{r}, t)\mathbf{F}_{em}(\mathbf{r}, t)$ must be balanced on the nucleus border by the force $\mathbf{F}_s(\mathbf{r}, t) = -\nabla U_s(\mathbf{r}, t)$ associated with the density of the symmetry energy in a nucleus. The formula for the total force according to (3.90) is given by the expression

$$\int d\mathbf{r}\mathbf{F}_s(\mathbf{r}, t) = -\frac{4\varepsilon_4}{\rho_0}\int d\mathbf{r}(2\rho_p - \rho_0)\nabla\rho_p = \frac{8\varepsilon_4}{\rho_0}\int d\mathbf{r}\rho_p\nabla\rho_p,$$

Here we have carried out the integration by parts. The last expression allows one to present the force acting on protons in unit volume of a nucleus in the form

$$\mathbf{F}_s(\mathbf{r}, t) = \frac{8\varepsilon_4}{\rho_0}\rho_p(\mathbf{r}, t)\nabla\rho_p(\mathbf{r}, t).$$

Thus, the *boundary condition for the proton density* $\rho_p(\mathbf{r},t)$ can be written as follows

$$\frac{8\varepsilon_4}{\rho_0}\mathbf{n}\nabla\rho_p = \mathbf{n}\mathbf{F}_{\text{em}}, \quad r = R(\theta,\varphi). \tag{3.95}$$

The corresponding boundary condition for the velocity potential $\varphi(\mathbf{r},t)$ can be obtained from the second Newton law:

$$-M\mathbf{n}\nabla\dot{\varphi} = \mathbf{n}\mathbf{F}_{\text{em}}, \quad r = R(\theta,\varphi). \tag{3.96}$$

It is worth noting that the main formulas of the classical hydrodynamical model of nuclei may be obtained more strictly with the help of the general Hamiltonian principle. At the same time, the condition of small amplitude of vibrations of the nucleon density and relative velocity becomes unimportant.

Hydrodynamical model for a spherical nucleus. Let us apply the relations of the *classical hydrodynamical model* (3.93) and (3.95) to a spherical nucleus when $R(\theta,\varphi) = R$ (Eisenberg and Greiner, 1970). We present the proton density $\rho_p(\mathbf{r},t)$ in the form

$$\rho_p(\mathbf{r},t) = \rho_p^{(0)}[1+\eta(\mathbf{r},t)], \tag{3.97}$$

where $\rho_p^{(0)}\eta(\mathbf{r},t)$ is the fluctuation density. The equation of motion of the quantity $\eta \equiv \eta(\mathbf{r},t)$ and boundary condition are obtained by substitution (3.97) in (3.93) and (3.95):

$$u^2\Delta\eta - \frac{\partial^2\eta}{\partial t^2} - \Gamma\frac{\partial\eta}{\partial t} = 0, \tag{3.98}$$

$$\mathbf{n}\nabla\eta = \frac{\rho_0}{8\varepsilon_4\rho_p^{(0)}}\mathbf{n}\mathbf{F}_{\text{em}}, \quad r = R. \tag{3.99}$$

Assuming that the wave length of the incident radiation on a nucleus $\lambda \gg R$, we obtain that the electric field $\mathbf{E}(t) = \frac{1}{e}\mathbf{F}_{em}(t)$ within the nucleus region can be treated as practically homogeneous and depending only on time. We assume that it is a periodical one with a frequency ω, i.e., $\mathbf{E}(t) = \mathbf{E}e^{-i\omega t}$. The time dependence of $\eta(\mathbf{r},t)$ is periodic (harmonic) too:

$$\eta(\mathbf{r},t) = \eta(\mathbf{r})\mathrm{e}^{-i\omega t}. \tag{3.100}$$

Substituting (3.100) in (3.98), we obtain the following equation for $\eta(\mathbf{r})$

$$\Delta\eta(\mathbf{r}) + k^2\eta(\mathbf{r}) = 0, \quad k^2 = \frac{\omega^2}{u^2}\left(1+i\frac{\Gamma}{\omega}\right). \tag{3.101}$$

Choosing z-axis along the vector $\mathbf{E}$, we can rewrite boundary condition (3.99) in the form

$$\frac{\partial \eta(\mathbf{r})}{\partial r} = \frac{eE\rho_0}{8\varepsilon_4 \rho_p^{(0)}} \cos\theta, \quad r = R, \tag{3.102}$$

where θ is an angle between $\mathbf{E}$ and $\mathbf{n}$.

We seek a solution of equation (3.101) in the form of a series with respect to the spherical functions

$$\eta(\mathbf{r}) = \sum_{l,m} C_{lm} j_l(kr) Y_{lm}(\theta, \varphi),$$

where $j_l(kr)$ is the Bessel spherical functions. Substituting this expansion in boundary condition (3.102), we obtain

$$\begin{aligned} \eta(\mathbf{r}) &= C(z) j_1(kr) \cos\theta, \\ C(z) &= \frac{eE\rho_0 R}{8\varepsilon_4 \rho_p^{(0)}} \frac{1}{j_1(z) - z j_2(z)}, \quad z = kR. \end{aligned} \tag{3.103}$$

We can see that the constant within the nucleus region electric field initiates only dipole ($l = 1$) vibrations of the proton density in nuclei. As it was noted before and can be seen from (3.103) at $E \to 0$, $\eta(\mathbf{r}) \to 0$. At the same time, $\rho_p(\mathbf{r}, t) \to \rho_p^{(0)}$, i.e., ρ_p does not depend on $\mathbf{r}$ and t. At $E \neq 0$, the neutron density of nuclei vibrates according the same periodic law as the vibration of proton density. But the neutron displacements with respect to the proton displacements go on in antiphase according to (3.87). The amplitude $C(z)$ in (3.103) is the resonance type and tends to infinity in points $z = z_n$ when $j_1(z) = z j_2(z)$ or when

$$\tan z = \frac{2z}{2 - z^2}. \tag{3.104}$$

The roots of this equation z_n ($n = 1, 2, \ldots$) (with exception $z = 0$) determine the resonance frequency according (3.101) and (3.103)

$$\omega_n = \pm \nu_n \sqrt{1 - \left(\frac{\Gamma}{2\nu_n}\right)^2} - \frac{i}{2}\Gamma, \quad \nu_n = \frac{u}{R} z_n. \tag{3.105}$$

The external periodic in time electric field induces a dipole moment in the nuclear system

$$\mathbf{d}(t) = e \int d\mathbf{r}\, \mathbf{r} \rho_p(\mathbf{r}, t) = e\rho_p^{(0)} \int d\mathbf{r}\, \mathbf{r} \eta(\mathbf{r}, t), \tag{3.106}$$

According to (3.100), it depends on time by the periodic (harmonic) law: $\mathbf{d}(t) = \mathbf{d} e^{-i\omega t}$. Apparently, that only the component of the vector $\mathbf{d}(t)$ along the vector of the external electric field does not equal to zero. It is directed along z-axis, and will be denoted simply as $d(t) = de^{-i\omega t}$. Substituting expressions (3.100) and (3.103) in (3.106) and performing the integration, we obtain

$$\begin{aligned} d &= \frac{4\pi}{3} R^3 \frac{e\rho_p^{(0)}}{k} C(z) j_2(z) \equiv \frac{e^2 E A R^2}{8\varepsilon_4} f(z), \\ f(z) &= \frac{j_2(z)}{z j_1(z) - z^2 j_2(z)}, \qquad z = kR. \end{aligned} \tag{3.107}$$

It is convenient to present the function $f(z)$ in the form of the infinite sum

$$f(z) = \sum_n \frac{a_n}{z - z_n},$$

where coefficient a_n can be determined with the help of the Cauchy theorem. As a result we have the expression for the induced nucleus dipole moment

$$d = \frac{e^2 E A u^2}{4\varepsilon_4} \sum_n \frac{1}{2 - z_n^2} \frac{\omega^2 - \nu_n^2 - i\Gamma\omega}{(\omega^2 - \nu_n^2)^2 + \Gamma^2\omega^2}, \qquad z_n > 0. \tag{3.108}$$

Hydrodynamical model and giant nuclear resonance. The resonance portions of the radiation energy $\hbar\omega_n$ absorbed by a nucleus proportional to frequencies (3.105), and in the case $\Gamma \ll \nu_n$ are expressed in terms of the known nuclear parameters ($\nu_n > 0$):

$$\hbar\omega_n = \hbar\nu_n = \hbar u \frac{z_n}{R} = \frac{\hbar z_n}{R} \left(\frac{8\varepsilon_4}{M} \frac{ZN}{A^2} \right)^{1/2}. \tag{3.109}$$

At $N \approx Z$ and $n = 1$, when the first root of equation(3.104) is $z_1 \simeq 2.08$, we get

$$\hbar\omega_1 \approx 70\, A^{-1/3}\, \text{MeV}. \tag{3.110}$$

This is close enough to the energy position of the nucleus giant resonance that is described by the empirical formula $\hbar\omega_{\text{exp}} \approx 80\, A^{-1/3}\,\text{MeV}$. The dipole type vibrations of the proton and neutron densities in nuclei, as it will be shown below, can be identified with the relative giant nuclear resonances because the maximum cross-section of the absorption of the incident radiation corresponds to $n = 1$.

The cross-section $\sigma(\omega)$ of absorption of the photon with a frequency ω equals to ratio between the probability of absorption of the photon with a frequency ω per unit time and the incident photon flow or to ratio between the averaged energy absorbed per unit time $I(\omega)$ and the averaged electromagnetic energy flow $S(\omega)$ (i.e., a magnitude of the Poyting vector).

Here we mean average with respect to time. The quantity $I(\omega)$ can be determined also as the work done by the external electric field per unit time for a displacement of the nucleus charges $\overline{e\mathbf{E}(t)\int d\mathbf{r}\rho_p(\mathbf{r},t)\dot{\mathbf{r}}(t)} = \overline{\mathbf{E}(t)\dot{\mathbf{d}}(t)}$, an over-bar means the average with respect to time. Since we deal with complex variables, the cross-section under consideration, that is expressed quadratically in terms of these quantities, can be written in the following way

$$\begin{aligned} \sigma(\omega) &= I(\omega)/S(\omega), \quad I(\omega) = \frac{1}{2}\mathrm{Re}\,\overline{\{\mathbf{E}^*(t)\dot{\mathbf{d}}(t)\}}, \\ S(\omega) &= \frac{c}{4\pi}\cdot\frac{1}{2}\mathrm{Re}\,\overline{\{\mathbf{E}^*(t)\mathbf{E}(t)\}}. \end{aligned} \tag{3.111}$$

The *photoabsorption cross-section* in the classical hydrodynamical model of nuclei with taking account of (3.106)–(3.108) and (3.111) is presented now in the form

$$\sigma(\omega) = \sum_n \sigma_n(\omega), \quad \sigma_n(\omega) = \frac{4\pi e^2}{cM}\frac{ZN}{A}\frac{2\Gamma\omega^2}{(z_n^2-2)[(\nu_n^2-\omega^2)^2+\Gamma^2\omega^2]}. \tag{3.112}$$

It follows from (3.112) that the photoabsorption cross-section structure is the resonance type. At this time, the shape of the separate partial cross-section $\sigma_n(\omega)$ as a function of ω is described by the so called Lorentz curve (formula). The quantity $\hbar\Gamma$ plays the part of the resonance width. The integral photoabsorption cross-section σ is obtained by integration (3.112) over ω:

$$\sigma = \int_0^\infty d\omega\sigma(\omega) = \sum_n \sigma_n, \tag{3.113}$$

$$\sigma_n = \frac{2\pi^2 e^2}{Mc}\frac{ZN}{A}\frac{2}{z_n^2-2}. \tag{3.114}$$

It happens to be that series (3.113) can be summarized, and we find σ in the explicit form. Multiplying (3.107) and (3.108) by z^2 and taking the limit $z \to 0$, we can use the known asymptotic properties of the Bessel functions and equate the obtained results. Canceling by quantities that do not depend on n, we obtain

$$\sum_n \frac{2}{z_n^2-2} = 1, \tag{3.115}$$

Therefore, for the *integral photoabsorption cross-section* we can get the simple expression (the classical sum rule):

$$\sigma = \frac{2\pi^2 e^2}{Mc}\frac{ZN}{A}. \tag{3.116}$$

A magnitude of the total cross-section of the dipole radiation (3.116) theoretically obtained is in a good agreement with the corresponding experimental data. Formula (3.116) presents the so called dipole sum rule without taking account of the interchange effects and the velocity depending nucleon-nucleon nuclear forces (the particular case of the *Thomas-Reiche-Kuhn sum rule*). The known quantity standing in the right-hand side of (3.116) is the upper bound of the total integral photoabsorption cross-section associated with the electric dipole transitions in nuclei. It is interesting that the contribution of the first term with $n = 1$ in total sum (3.115) is dominant and as much as 84%: $2/(z_1^2 - 2) = 2/(2.08^2 - 2) = 0.84$. This confirms the above formulated statement that the dipole vibrations of the proton density in nuclei with $n = 1$ give the main contribution in the *photonuclear giant dipole resonance* .

Thus, the above discussed rather simple classical hydrodynamical model of nuclei explains and describes good enough the following facts:

1) a position of the *giant dipole resonance in nuclei* ;
2) the classical dipole sum rule is practically completely becomes exhausted within the model framework; at this time, 84% of the total cross-section comes from the particular dipole vibration $n = 1$;
3) the experimental dependence of the giant nuclear resonance energy on their mass number ($\sim A^{-1/3}$).

Note that these facts are explained and described quantitavely by the quantum hydrodynamical model of nuclei. Many conclusions of this theory coincide with the corresponding results of the classical hydrodynamical model considered here. In particular, the photoabsorption cross-section in the giant nuclear resonance region in one of versions of the quantum model has the resonance character as in (3.112). The main contribution in the cross-section gives the first resonance (the analog of the contribution from the term with $n = 1$ in (3.112)). The difference of (3.112) in the quantum theory lies in the only fact that the dependence of partial cross-sections on the radiation energy $\hbar\omega$ is not the Lorentz type anymore, as it takes place in the classical model, but is given by the *Breit-Wigner formula*. At small widths $\hbar\Gamma$, the cross-section dependencies on energy of accident photons in both the quantum and classical models are practically the same.

We considered the collective motion of protons around neutrons in nuclei that sometime are called the isospin vibrations. These vibrations corresponding, to the giant nuclear resonance can emerge, as it was above mentioned, in the inelastic electron scattering when small momenta q are transferred to a nucleus. However, on increasing q, when the magnetic electron scattering associated with the current distribution in nuclei becomes important, the other type of the collective motion may be excited in a nucleus, as the theoretical study shows.

This is the so called spin-isospin vibrations when protons with spin up and neutrons with spin down joinly vibrate about protons with spin down and neutrons with spin up. The corresponding experimental nuclear levels located in a rather narrow band of the excitation energy overlap with the photonuclear giant resonance domain.

Nuclear density matrices. When analysing the current quantum hydrodynamical model of nuclei, to which we pass now, we need some new physical ideas. One of them is the *nuclear density matrix* and the *density matrix of an individual nucleon* (in a general case of the group of nucleons) in a nucleus (Lane, 1964).

Assuming that a nucleus of A nucleons is described by the definite wave function (pure state) $\Psi \equiv \Psi(\mathbf{r}_1, \mathbf{r}_2, \ldots, \mathbf{r}_A, t)$, we determine the total nuclear density matrix ρ_A of a system of A interacting particles in a particular state by using the conventional way

$$\begin{aligned} \rho_A &\equiv \rho_A(\mathbf{r}_1, \mathbf{r}_2, \ldots, \mathbf{r}_A, \mathbf{r}'_1, \mathbf{r}'_2, \ldots, \mathbf{r}'_A, t) \\ &= \Psi(\mathbf{r}_1, \mathbf{r}_2, \ldots, \mathbf{r}_A, t)\Psi^*(\mathbf{r}'_1, \mathbf{r}'_2, \ldots, \mathbf{r}'_A, t). \end{aligned} \tag{3.117}$$

The spur of this matrix for the normalized wave function Ψ is equal to one:

$$\mathrm{Sp}\,\rho_A \equiv \mathrm{Sp}_{\,123\ldots A}\rho_A =$$

$$= \int d\mathbf{r}_1 \int d\mathbf{r}_2 \ldots \int d\mathbf{r}_A \rho_A(\mathbf{r}_1, \mathbf{r}_2 \ldots, \mathbf{r}_A, \mathbf{r}_1, \mathbf{r}_2, \ldots, \mathbf{r}_A, t) =$$

$$= \int d\mathbf{r}_1 \int d\mathbf{r}_2 \ldots \int d\mathbf{r}_A \Psi(\mathbf{r}_1, \mathbf{r}_2 \ldots, \mathbf{r}_A, t)\Psi^*(\mathbf{r}_1, \mathbf{r}_2 \ldots, \mathbf{r}_A, t) = 1. \tag{3.118}$$

Sometimes the density matrix is normalized so that $\mathrm{Sp}\,\rho_A$ equals to a total number of particles in the system, i.e., A. Then the factor A should be introduced in the right-hand side of equality (3.117). For the sake of brevity we indicate explicitly only the nucleon space coordinates $\mathbf{r}_j$ $(j = 1, 2. \ldots, A)$ and time t, but do not indicate spin and isospin variables of the nucleus nucleons and the summation with respect to them. But here and further we mean that they exist.

The nuclear density matrix for pure state (3.117) contains the same information on the nucleus as the nucleus wave function Ψ. The obtaining of the nucleus wave function and appropriate density matrix (3.117), even in some approximations, is an extremely complex many-body problem. However, if we are interested in some averaged integral characteristics of nuclei associated with computations of averaged values or matrix elements of one-particle and two-particle operators, it is possible to pass to quantities which are obtained by integration and summation with respect to a large share of variables of the wave function Ψ. It is obviously that the part of the information on the nucleus is lost in this procedure, but in some cases, it is not important. In particular, the *partly integrated nuclear density matrices* can be related to these quantities:

$$\begin{aligned}
\rho_k &\equiv \rho_k(\mathbf{r}_1, \mathbf{r}_2, \ldots, \mathbf{r}_k, \mathbf{r}'_1, \mathbf{r}'_2, \ldots, \mathbf{r}'_k, t) = \\
&= \int d\mathbf{r}_{k+1} d\mathbf{r}_{k+2} \ldots d\mathbf{r}_A \times \\
&\quad \times \rho_A(\mathbf{r}_1, \mathbf{r}_2, \ldots, \mathbf{r}_k, \mathbf{r}_{k+1}, \ldots, \mathbf{r}_A, \mathbf{r}'_1, \mathbf{r}'_2, \ldots, \mathbf{r}'_k, \mathbf{r}_{k+1}, \ldots, \mathbf{r}_A, t) = \\
&= \mathrm{Sp}_{\,k+1,k+2,\ldots,A}\rho_A. \qquad (3.119)
\end{aligned}$$

These k-particle nuclear density matrices ρ_k at $k < A$ describe, in a general case, mixed states of the nuclear subsystem for the first k nucleons in a nucleus. At $k = A$ density matrix (3.119) transforms to the density matrix of pure state (3.117). Thus, definition (3.119) is more general. (Formally, from (3.119) at $k = 0$, we obtain normalization condition (3.118)).

The most-used are the one-particle ρ_1 and two-particle ρ_2 nuclear density matrices

$$\rho_1 \equiv \rho_1(\mathbf{r}_1, \mathbf{r}'_1, t) = \mathrm{Sp}_{\,2,3,\ldots,A}\rho_A, \qquad (3.120)$$

$$\rho_2 \equiv \rho_2(\mathbf{r}_1, \mathbf{r}_2, \mathbf{r}'_1, \mathbf{r}'_2, t) = \mathrm{Sp}_{\,3,4,\ldots,A}\rho_A. \qquad (3.121)$$

It is clear that $\rho_1 = \mathrm{Sp}_{\,2}\rho_2$. Because of the fact that the nucleus wave function Ψ is antisymmetric, the one-particle density matrices $\rho_1(\mathbf{r}_j, \mathbf{r}'_j, t)$ for all j nucleons in a nucleus $(1 \leqslant j \leqslant A)$ are the same. That is why we omit the nucleon index j and write simply $\rho_1(\mathbf{r}, \mathbf{r}', t)$. At $\mathbf{r}' = \mathbf{r}$, the one-particle density matrix $\rho_1(\mathbf{r}, \mathbf{r}', t)$ gives the spatial nucleon density in a nucleus $\rho_1(\mathbf{r}, t) \equiv \rho_1(\mathbf{r}, \mathbf{r}, t)$. This follows directly from definition (3.119) at $k = 1$. Whereas the exact many-particle wave function of a nuclei $\Psi(\mathbf{r}_1, \mathbf{r}_2, \ldots, \mathbf{r}_A, t)$ along with the relative density matrix ρ_A of pure state (3.117) contains all the necessary information about the nucleus properties, the one-particle density matrix $\rho_1 = \rho(\mathbf{r}, \mathbf{r}', t)$ contains the exhausted information on the one-particle nucleus properties. The two-particle density matrix $\rho_2 = \rho_2(\mathbf{r}_1, \mathbf{r}_2, \mathbf{r}'_1, \mathbf{r}'_2, t)$ provides the information regarding to the two-particle characteristics of a nucleus, and so on.

The many-particle nucleus Hamiltonian $\widehat{H}$ is a sum of the one-particle operators of the kinetic energy $\widehat{T}_j$ and the two-particle operators of the nucleon-nucleon interactions $\widehat{V}_{ij}$:

$$\widehat{H} = \widehat{T} + \widehat{V}, \quad \widehat{T} = \sum_{j=1}^{A} \widehat{T}_j = \sum_{j=1}^{A}\Big(-\frac{\hbar^2}{2M}\Big)\Delta_j, \quad \widehat{V} = \sum_{j<i} \widehat{V}_{ij} = \sum_{j<i} V(\mathbf{r}_i, \mathbf{r}_j). \qquad (3.122)$$

Averaged values of these operators on the basis of the nuclear density matrices can be written by using the antisymmetric nucleus wave function Ψ in the following way

$$\begin{aligned}\langle\widehat{T}\rangle &\equiv \langle\Psi|\widehat{T}|\Psi\rangle = \mathrm{Sp}\,(\widehat{T}\rho_A) = A\mathrm{Sp}\,(\widehat{T}_1\rho_A)\\ &= A\mathrm{Sp}_{\,1}\{(\widehat{T}_1\mathrm{Sp}_{\,2,3,\ldots,A}\rho_A\} = A\mathrm{Sp}_{\,1}\{\widehat{T}_1\rho_1\},\end{aligned} \tag{3.123}$$

$$\begin{aligned}\langle\widehat{V}\rangle &= \mathrm{Sp}\,(\widehat{V}\rho_A) = \frac{A(A-1)}{2}\mathrm{Sp}\,(\widehat{V}_{12}\rho_A)\\ &= \frac{A(A-1)}{2}\mathrm{Sp}_{\,1,2}\{\widehat{V}_{12}\mathrm{Sp}_{\,3,4,\ldots,A}\rho_A\}\\ &= \frac{A(A-1)}{2}\mathrm{Sp}_{\,1,2}\{\widehat{V}_{12}\rho_2\}.\end{aligned} \tag{3.124}$$

(The spurs do not depend on the coordinate representation chosen from the beginning).

Evidently, that similar formulas of the averages hold true for arbitrary the one-particle and two-particle operators. With the help of the density matrices ρ_k defined in (3.119) we calculate the partial average with respect to some part of nucleus variables.

The total density matrix ρ_k for the system of k nucleons in a nucleus (3.119) in the coordinate representation can be treated as the matrix of an operator $\widehat{\rho}^{(k)}(t)$ called the density operator of the system of k nucleons. The matrix ρ_k can be written in the form

$$\rho_k = \langle\mathbf{r}_1, \mathbf{r}_2, \ldots, \mathbf{r}_k|\widehat{\rho}^{(k)}(t)|\mathbf{r}'_1, \mathbf{r}'_2, \ldots, \mathbf{r}'_k\rangle. \tag{3.125}$$

Being time-dependent, but not depending on the representation, the nucleon density operator $\widehat{\rho}^{(k)}(t)$ satisfies the general quantum mechanics equation of motion

$$i\hbar\frac{\partial\widehat{\rho}^{(k)}(t)}{\partial t} = [\widehat{H}^{(k)}, \widehat{\rho}^{(k)}(t)] \equiv \widehat{H}^{(k)}\widehat{\rho}^{(k)}(t) - \widehat{\rho}^{(k)}(t)\widehat{H}^{(k)}, \tag{3.126}$$

where $\widehat{H}^{(k)}$ is the Hamiltonian related to a subsystem of k nucleons which is not characterized by a definite wave function ($k < A$). This subsystem is described by the density matrix ρ_k because the nucleons of the system under consideration interact with external nucleons. Thus, the system of k nucleons is nonisolated. In other words, we deal with the Hamiltonian $\widehat{H}^{(k)}$ of the system in a mixed state. Because of this, the operator $\widehat{H}^{(k)}$ of the system in a mixed state depends on the density matrix ρ_k. Therefore, the Hamiltonian $\widehat{H}^{(k)}$ at $k < A$, as well as ρ_k and $\widehat{\rho}^{(k)}(t)$, depends on t in a general case. Only at $k = A$, i.e., for the whole nucleus characterized by a definite wave function Ψ (pure state), the Hamiltonian $\widehat{H}^A \equiv \widehat{H}$ in equation (3.126) does not depend on the operator of the nucleon density in a nucleus $\widehat{\rho}^{(A)}(t)$ and on time.

Operator equation (3.126) is called the *quantum Liouville equation.* In fact, it is the quantum analog of the classic Liouville theorem. In the particular case $k = 1$, equation (3.126) is known as the *generalized time-dependent Hartree-Fock equation* . We use it further. Equation (3.126) for $\widehat{\rho}^{(A)}(t)$ can be obtained by differentiation of the density matrix (3.117) for a pure state and by using the general wave equation $i\hbar\frac{\partial\Psi}{\partial t} = \widehat{H}\Psi$ for the nucleus wave function Ψ depending on time. After this, exploiting the fact that the nucleus Hamiltonian is Hermitian, we can pass to the nucleon density operator $\widehat{\rho}^{(A)}(t)$. Equation (3.126) for a mixed state ($k < A$) can be obtained in a similar manner if, for example, we introduce a mixed ensemble where the mixture of pure states of the system of k nucleons are presented with different probabilities.

The Hartree and Hartree-Fock approximations. Many approaches in the quantum theories of the systems with finite number of identical fermions, such as atomic nuclei, are based on the *Hartree* and *Hartree-Fock approximations* (theories or methods) explicitly or not explicitly (March, Young, Sampanthar, 1967; Eisenberg, Greiner, 1972). The wave function of a nucleus Ψ in these approximations is presented in the form of a product of one-particle wave functions $\psi_{\alpha_i}(i)$:

$$\Psi = \prod_{i=1}^{A} \psi_{\alpha_i}(i) \tag{3.127}$$

in the Hartree theory or in the *Slater determinant* form

$$\Psi_{\{\alpha\}} = (1/\sqrt{A!}) \begin{vmatrix} \psi_{\alpha_1}(1) & \psi_{\alpha_1}(2) & \ldots & \psi_{\alpha_1}(A) \\ \psi_{\alpha_2}(1) & \psi_{\alpha_2}(2) & \ldots & \psi_{\alpha_2}(A) \\ \ldots & \ldots & \ldots & \ldots \\ \ldots & \ldots & \ldots & \ldots \\ \psi_{\alpha_A}(1) & \psi_{\alpha_A}(2) & \ldots & \psi_{\alpha_A}(A) \end{vmatrix} \tag{3.128}$$

in the Hartree-Fock theory. In the both approximations we seek one-particle wave functions $\psi_{\alpha_i}(i)$, substituting the appropriate total wave functions of the nucleus stationary states Ψ in the many-particle Schrödinger equation $\widehat{H}\Psi = E\Psi$ with Hamiltonian (1.195), that contains two-particle interaction potentials $V_{ij} \equiv V_{ij}(\mathbf{r}_i, \mathbf{r}_j)$ between nucleons. They are considered as given. At the same time, in the *Hartree approximation*, the wave function of the system Ψ is not symmetrized. But in the *Hartree-Fock approximation* , the wave function is antisymmetric with respect to permutations of any pair of particles. That is why the Hartree-Fock approximation for a system of identical fermions is more realistic, than the Hartree approximation, and used more frequently.

We, however, consider from the beginning the more simple Hartree approximation. This is useful from the method view point before pass to the study of the more complex Hartree-Fock approximation. As a result of substituting of (3.127) in the Schrödinger equation for one-particle wave functions $\psi_{\alpha_k}(k) \equiv \psi_{\alpha_k}(\mathbf{r}_k)$, we obtain the following self-consistent system of the integro-differential Hartree equations

$$\left\{-\frac{\hbar^2}{2M}\Delta_k + V_k(\mathbf{r}_k)\right\}\psi_{\alpha_k}(\mathbf{r}_k) = \varepsilon_k\psi_{\alpha_k}(\mathbf{r}_k), \quad k = 1, 2, \ldots A. \tag{3.129}$$

$$V_k(\mathbf{r}_k) = \sum_{j \neq k} \int d\mathbf{r}_j \psi^*_{\alpha_k}(\mathbf{r}_j) V_{jk}(\mathbf{r}_j, \mathbf{r}_k) \psi_{\alpha_k}(\mathbf{r}_j). \tag{3.130}$$

Here and further, we explicitly indicate only dependence on space coordinates of nucleons and do not indicate spin and isospin variables, but their existence is meant. In the one-particle Schrödinger equation for k-th nucleon (3.129), there is some average of the one-particle potential $V_k(\mathbf{r}_k)$ that acts on the nucleon with number k from the rest $A-1$ nucleons of the nucleus. At the same time, the potential $V_k(\mathbf{r}_k)$ itself is determined by unknown one-particle wave functions $\psi_{\alpha_j}(\mathbf{r}_j)$ with numbers $j \neq k$. They must be found from the system of equations (3.129) and (3.130). Therefore, we have the self-consistent problem for the wave functions $\psi_{\alpha_k}(\mathbf{r}_k)$ and corresponding energy eigenvalues ε_k (the one-particle Hartree energies) and one-particle potentials $V_k(\mathbf{r}_k)$, as well. That is why the quantity $V_k(\mathbf{r}_k)$ is called the self-consistent Hartree potential for k-th nucleon in the nucleus.

Equations (3.129), (3.130) can be solved by the iteration method: it is necessary to substitute as one-particle wave functions $\psi_{\alpha_k}(\mathbf{r}_k)$ particular well chosen model functions in (3.130) and find the potentials $V_k(\mathbf{r}_k)$ for the given two-particle potentials $V_{jk}(\mathbf{r}_j, \mathbf{r}_k)$, then $V_k(\mathbf{r}_k)$ should be substituted in equations (3.129); when obtained the refined one-particle wave functions from equations (3.129), we can calculate potentials (3.130) with the help of these functions and substitute them again in equations (3.129) to determine a new set of the one-particle wave functions and corresponding energies. This iteration procedure (successive approximations) must be continued until the set of one-particle functions $\psi_{\alpha_k}(\mathbf{r}_k)$ and the potential $V_k(\mathbf{r}_k)$ do not change noticeably at consequent iterations. We can say that in this case we realize the self-consistency. In the same way Hartree calculated for the first time the states of atomic electrons for the known two-particle electromagnetic potentials. Note that, when solving the system equations (3.129), (3.130), we can start iterations with substitution in equation (3.129) some model potentials as the potentials $V_k(\mathbf{r}_k)$. By using this method for nucleon systems in nuclei, we may only hope to obtain the refined sets of one-particle functions $\psi_{\alpha_k}(\mathbf{r}_k)$ with energies ε_k and potentials $V_k(\mathbf{r}_k)$ in any new iteration. That is why in action, the definite model one-particle nuclear potentials and wave functions with parameters are used and fitted by comparing the theoretical results with experimental data.

In the Hartree-Fock theory (method) the identity of all nucleons in a nucleous is taken into account and $V_{ij} = V(\mathbf{r}_i, \mathbf{r}_j)$. The nuclear wave function Ψ is antisymmetrized in this theory, i.e., it takes form (3.128). After substitution Ψ in the Schrödinger equation $\widehat{H}\Psi = E\Psi$ with two-particle interactions $V_{ij} = V(\mathbf{r}_i, \mathbf{r}_j)$ between nucleons, instead of the Hartree equations (3.129), (3.130), we obtain the following self-consistent system of the *Hartree-Fock equations*

$$\left\{-\frac{\hbar^2}{2M}\Delta_k + V(\mathbf{r}_k)\right\}\psi_{\alpha_k}(\mathbf{r}_k) - \int d\mathbf{r}' W(\mathbf{r}', \mathbf{r}_k)\psi_{\alpha_k}(\mathbf{r}') = \varepsilon_k \psi_{\alpha_k}(\mathbf{r}), \tag{3.131}$$

$$\begin{aligned} V(\mathbf{r}_k) &= \sum_{j=1}^{A} \int d\mathbf{r}\psi^*_{\alpha_j}(\mathbf{r}) V(\mathbf{r}, \mathbf{r}_{k},)\psi_{\alpha_j}(\mathbf{r}), \\ W(\mathbf{r}', \mathbf{r}_k) &= W^*(\mathbf{r}_k, \mathbf{r}') = \sum_{j=1}^{A} \psi^*_{\alpha_j}(\mathbf{r}') V(\mathbf{r}', \mathbf{r}_k)\psi_{\alpha_j}(\mathbf{r}_k). \end{aligned} \tag{3.132}$$

With the help of this system of integro-differential equations we can find one-particle wave functions and the Hartree-Fock energies $\psi_{\alpha_k}(\mathbf{r})$ and ε_k, by using, for example, the iteration method. The wave function are chosen so to satisfy the orthonormalization condition

$$\int d\mathbf{r}\psi^*_{\alpha_k}(\mathbf{r})\psi_{\alpha_j}(\mathbf{r}) = \delta_{\alpha_k, \alpha_j}. \tag{3.133}$$

(A quantum indexes α_k and α_j are a set of the quantum-mechanical quantities and must be treated as the complex indexes. The Kronecker symbol in the right side of equality (3.133) is a product of the Kronecker symbols.)

The Hartree-Fock equations (3.131), (3.132) differ from the Hartree equations (3.129), (3.130), firstly, by the presence of the nonlocal interaction in (3.131), which is associated with the appearance of the interchange term in (3.131) with the operator $W(\mathbf{r}', \mathbf{r}_k)$ determined in (3.132) and, secondly, by the one-particle interaction potentials $V(\mathbf{r}_k)$ in (3.132) instead of $V_k(\mathbf{r}_k)$ in (3.130). In sum (3.132) with respect to j, the term with $j = k$ is added absent in (3.130). Because of this potential (3.132) happens to be the same for all nucleons in the nucleus. (Note that terms with $j = k$ in (3.132) when (3.132) substituted in (3.131) cancel out). It is obviously that these differences are brought about by the effect of antisymmetrization of the total nucleon wave function in a nucleus.

In the Hartree and *Hartree-Fock approximations* , every nucleon in a nucleus is described by the definite one-particle wave function $\psi_{\alpha_k}(\mathbf{r}_k)$. The total wave function of the nucleus is a product (as in (3.127)) or a sum of products (as in (3.128)) of functions $\psi_{\alpha_k}(\mathbf{r}_k)$. The approximate model Hamiltonian of the nucleus $\hat{H}_0$ corresponding to these wave functions and, as well as exact Hamiltonian (1.195), is approximated by a sum of the one-particle Hamiltonians $\hat{h}_j^{(1)}$. For example, in the Hartree approximation

$$\hat{H}_0 = \sum_{j=1}^{A} \hat{h}_j^{(1)}, \qquad \hat{h}_j^{(1)} = -\frac{\hbar^2}{2M}\Delta_j + V_j(\mathbf{r}_j), \tag{3.134}$$

Hence, we have the Hamiltonian $\hat{H}_0$ of the independent particle model, which in the case of the spherically symmetric self-consistent interaction potential is the Hamiltonian of the nucleus shell model. In the Hartree-Fock approximation, we have to take into account the interchange term. It is associated, as we have seen, with the additional nonlocal interaction potential. The Hartree-Fock approximation is the best one-particle approximation for the description of a nucleus with the Hamiltonian (1.195), which contains the two-particle interactions between nucleons. It can be strictly proved by using the variational principle and minimizing the nucleus energy as the average value of Hamiltonian (1.195) with the help of trial wave functions (3.128) obtained in the Hartree-Fock approximation. Thus, the Hartree-Fock method allows one to find the best shell wave functions of individual nucleons in a nucleus in spite of that the phenomenological nucleon wave functions are used. The Hartree and Hartree-Fock methods played an important part in the study of atoms, molecules, atomic nuclei, and nuclear matter.

The *residual interaction* $\hat{H} - \hat{H}_0$ is neglected in the Hartree and Hartree-Fock approximations, as well as in the nucleus shell model. In particular, very important short-range correlations between nucleons connected with incomplete consideration of the nucleon-nucleon interaction are also neglected in these approximations. Here we would like to note that there is the Bogolyubov generalization of the Hartree-Fock method, which takes account of the residue interaction between nucleons in nuclei. In this method a minimum of the system energy is calculated by using the more wide class of wave functions in comparison with the Hartree-Fock method, taking into account the pair wave functions along with the one-particle wave functions. This allows one to pick up even more complex correlations between particles. Usually this method is called the *Hartree-Fock-Bogolyubov method* . Currently, this generalized method is the most accurate and wide spreaded method for solving the nuclear many-body problem. Many modern methods used in the nuclear theory can be treated as the limiting cases of the Hartree-Fock-Bogolyubov method (Bogolyubov, 1958; Solov'ev, 1971). However, we will not dwell on this problem here.

Further, when describing the many particle shell nuclear model, we assume that the one-particle nucleon functions in the nucleus $\psi_{\alpha_k}(\mathbf{r}_k)$ are known and construct the many-particle shell wave functions with definite values of different integrals of motion in the nucleus.

The time-dependent Hartree-Fock method. We considered the Hartree and Hartree-Fock approximations known as the conventional time-independent Hartree and Hartree-Fock methods. The many-particle nucleus wave function Ψ was presented in the form of ordinary product (3.127) or in the form of the Slater determinant (3.128) constructed on the basis of the time-independent one-particle wave functions $\psi_{\alpha_k}(\mathbf{r}_k)$. They satisfy systems of integro-differential equations (3.129), (3.130) or (3.131), (3.132) that follow from the stationary many-particle Schrödinger equation $\widehat{H}\Psi = E\Psi$. At this time, the Hamiltonian $\widehat{H}$ and the total wave function does not depend on time.

In the *time-depend Hartree and Hartree-Fock approximations* we start with a more general many-particle time-dependent wave equation

$$i\hbar\frac{\partial\Psi(t)}{\partial t} = \widehat{H}\Psi(t) \tag{3.135}$$

to obtain a more exact many-particle time-dependent wave function. The latter is presented in the form of a product, as well as in the usual Hartree and Hartree-Fock methods:

$$\Psi(t) = \prod_{j=1}^{A} \psi_{\alpha_j}(\mathbf{r}_j, t) \tag{3.136}$$

or in the *Slater determinant form*

$$\Psi(t) = \frac{1}{\sqrt{A!}}\sum_{p}(-1)^p P\{\psi_{\alpha_1}(\mathbf{r}_1,t)\psi_{\alpha_2}(\mathbf{r}_2,t)\ldots\psi_{\alpha_A}(\mathbf{r}_A,t)\}, \tag{3.137}$$

Here P denotes all possible permutations of the nucleon coordinates; $(-1)^P$ is equal to $+1$ for even permutations and -1 for an odd ones. Functions (3.136) and (3.137) differ from the corresponding functions (3.127) and (3.128) because they are constructed from time-dependent one-particle wave functions $\psi_{\alpha_k}(\mathbf{r}_k, t)$ but not from the time-independent functions $\psi_{\alpha_k}(\mathbf{r}_k)$. Substituting the trial functions constructed from the one-particle wave functions $\psi_{\alpha_k}(\mathbf{r}_k, t)$ in equation (3.135), we obtain the system of equations that are similar in many aspects to equations (3.129), (3.130) in the Hartree theory or equations (3.131), (3.132) in the *Hartree-Fock theory* (Brown, 1967).

As a result in the more simple time-dependent Hartree approximation, we have the following system of integro-differential equations

$$i\hbar\frac{\partial\psi_{\alpha_k}(\mathbf{r}_k,t)}{\partial t} = \left\{-\frac{\hbar^2}{2M}\Delta_k + V_k(\mathbf{r}_k,t)\right\}\psi_{\alpha_k}(\mathbf{r}_k,t), \tag{3.138}$$

$$V_k(\mathbf{r}_k,t) = \sum_{j\neq k}\int d\mathbf{r}_j\psi^*_{\alpha_j}(\mathbf{r}_j,t)V_{jk}(\mathbf{r}_{j'}\mathbf{r}_k)\psi_{\alpha_j}(\mathbf{r}_j,t). \tag{3.139}$$

The one-particle wave functions $\psi_{\alpha_j}(\mathbf{r}_j, t)$ that appear in this system of equations do not necessarily depend on time like the stationary state functions, i.e., $\sim \mathrm{e}^{-\frac{i}{\hbar}\varepsilon_j t}$, where ε_j are the one-particle Hartree energies. In fact, among the solutions of system (3.138), (3.139) there is a partial solution $\psi_{\alpha_k}(\mathbf{r}_k, t) = \psi_{\alpha_k}(\mathbf{r}_k)\mathrm{e}^{-\frac{i}{\hbar}\varepsilon_k t}$, $k = 1, 2, \ldots, A$. However, these (stationary) one-particle wave functions do not necessarily provide the lower energy of the system of nucleons in the nucleus in the ground state. Because of this, another, more better set of functions $\psi_{\alpha_k}(\mathbf{r}_k, t)$ that minimizes the total energy of nucleons in the nucleus leads to the time-dependent averaged self-consistent Hartree potential (3.139). In a general case of the many-nucleon nucleus, it is not easy to solve of system equations (3.138), (3.139). However, for the realistic potentials of two-particle interactions $V_{jk}(\mathbf{r}_{j'}\mathbf{r}_k)$, there are a set of one-particle functions $\psi_{\alpha_k}(\mathbf{r}_k, t)$ that are not much different from a set of the stationary wave functions $\psi_{\alpha_k}(\mathbf{r}_k)\mathrm{e}^{-\frac{i}{\hbar}\varepsilon_k t}$. Thus, the average Hartree potential $V_k(\mathbf{r}_k, t)$ specified in (3.139) becomes a slowly varying function of time. This time-dependence is connected with the special choice of the trial wave function of the whole nucleus in form (3.136).

Acting in a similar manner, we can obtain the system of integro-differential equations for the time-dependent wave functions $\psi_{\alpha_k}(\mathbf{r}_k, t)$ in the more complex time-dependent Hartree-Fock approximation on the basis of equation (3.135) and total trial function (3.137). This system of equations is obtained formally from system (3.131), (3.132) for the usual Hartree-Fock approximation by the substitution of functions $\varphi_{\alpha_j}(\mathbf{r}_j)$ for functions $\varphi_{\alpha_j}(\mathbf{r}, t)$, the potential $V(\mathbf{r}_k)$ and the interchange operator $W(\mathbf{r}, \mathbf{r}_k)$ by $V(\mathbf{r}_k, t)$ and $W(\mathbf{r}, \mathbf{r}_k, t)$ respectively, the expression $\varepsilon_k\psi_k(\mathbf{r})$ in the right hand-side of equation (3.131) by $i\hbar\frac{\partial}{\partial t}\psi_{\alpha_k}(\mathbf{r}_k, t)$. As a result instead of equations (3.131), (3.132), we obtain

$$i\hbar\frac{\partial\psi_{\alpha_k}(\mathbf{r}_k,t)}{\partial t} = \left\{-\frac{\hbar^2}{2M}\Delta_k + V(\mathbf{r}_k,t)\right\}\psi_{\alpha_k}(\mathbf{r}_k,t) -$$

$$-\int d\mathbf{r}'W(\mathbf{r}',\mathbf{r}_k,t)\psi_{\alpha_k}(\mathbf{r}',t) \equiv \int d\mathbf{r}'H^{(1)}(\mathbf{r}',\mathbf{r}_k,t)\psi_{\alpha_k}(\mathbf{r}',t), \tag{3.140}$$

$$H^{(1)}(\mathbf{r}',\mathbf{r}_k,t) \equiv \langle \mathbf{r}'|\widehat{H}^{(1)}(t)|\mathbf{r}_k\rangle =$$

$$= \delta(\mathbf{r}'-\mathbf{r}_k)\left\{-\frac{\hbar^2}{2M}\Delta_k + V(\mathbf{r}_k,t)\right\} - W(\mathbf{r}',\mathbf{r}_k,t), \tag{3.141}$$

$$V(\mathbf{r}_k,t) = \sum_{j=1}^{A}\int d\mathbf{r}\psi^*_{\alpha_j}(\mathbf{r},t)V(\mathbf{r},\mathbf{r}_k)\psi_{\alpha_j}(\mathbf{r},t), \tag{3.142}$$

$$W(\mathbf{r}',\mathbf{r}_k,t) = \sum_{j=1}^{A}\psi^*_{\alpha_j}(\mathbf{r}',t)V(\mathbf{r}',\mathbf{r}_k)\psi_{\alpha_j}(\mathbf{r}'_k,t). \tag{3.143}$$

The operator $\widehat{H}^{(1)}(t)$ is called the *Hartree-Fock self-consistent operator.* In the coordinate representation it is given in the form of the matrix $\langle \mathbf{r}|\widehat{H}^{(1)}(t)|\mathbf{r}_k\rangle$ specified in (3.141). The best set of the Hartree-Fock time-dependent one-particle wave functions $\psi_{\alpha_k}(\mathbf{r}_k,t)$ satisfying the system of equations (3.140)–(3.143) and minimizing the ground state energy $\langle\Psi(t)|\widehat{H}|\Psi(t)\rangle$ with functions $\Psi(t)$ in the Slater determinant form (3.137) differs from a set of the stationary wave functions $\psi_{\alpha_k}(\mathbf{r}_k)\mathrm{e}^{-\frac{i}{\hbar}\varepsilon_k t}$ ($\psi_{\alpha_k}(\mathbf{r}_k)$ and ε_k are the wave functions and energies of the usual Hartree-Fock method). There is a set of functions which slightly differs from the corresponding stationary functions $\psi_{\alpha_k}(\mathbf{r}_k)\mathrm{e}^{-\frac{i}{\hbar}\varepsilon_k t}$. In this case, the Hartree-Fock average self-consistent field, as well as the whole one-particle self-consistent Hamiltonian $\widehat{H}^{(1)}(t)$, slowly depends on time.

Now we consider the nuclear density matrices in the usual and time-dependent Hartree-Fock approximations. The k-particle nuclear density matrix ρ_k and equation (3.126) for the k-particle density operator of nucleons in a nucleus $\widehat{\rho}^{(k)}(t)$ can be presented in a rather general form. The matrix ρ_k is expressed in terms of the exact many-particle nucleus wave function Ψ satisfying the general time-dependent wave equation $i\hbar\frac{\partial\Psi}{\partial t} = \widehat{H}\Psi$. Thus, the wave function Ψ, in a general case, does not correspond to the stationary state with definite energy (otherwise the density matrices ρ_k for all $k = 1, 2, \ldots, A$ do not depend on time). It is clear, that to obtain the wave functions and density matrices for a practical implementation we have to use some model ideas about the many-nucleon nuclei. Here we dwell on the use of the usual and time-dependent Hartree-Fock approximation in the most important particular case of the one-particle ($k = 1$) density matrices ρ_1 when one nucleon is considered as a subsystem in a nucleus consisting of A interacting nucleons.

The one-particle time-dependent density matrix in the coordinate representation in the Hartree-Fock approximation can be expressed in terms of the one-particle wave functions $\psi_{\alpha_j}(\mathbf{r}_j, t)$ of the time-dependent Hartree-Fock approximation by substituting the Slater determinant (3.137) instead of the exact many-particle wave function Ψ in the general definition of the density matrix (3.119) at $k = 1$ (the nucleon index j is omitted in α_j and $\mathbf{r}_j$):

$$\rho_1(\mathbf{r}, \mathbf{r}', t) \equiv \langle \mathbf{r}|\rho^{(1)}(t)|\mathbf{r}'\rangle = \sum_\alpha n_\alpha \psi_\alpha(\mathbf{r}, t)\psi^*_\alpha(\mathbf{r}', t), \tag{3.144}$$

Introduced here the population numbers n_α are equal to 1 for the occupied one-particle state α by a nucleon or 0 for the unoccupied state α. It is clear, that for the usual (time-independent) Hartree-Fock method, the one-particle density matrix $\rho(\mathbf{r}, \mathbf{r}')$ does not depend on time, and can be obtained from (3.144) when substituted the stationary one-particle wave functions $\psi_\alpha(\mathbf{r}, t) = \psi_\alpha(\mathbf{r})\mathrm{e}^{-\frac{i}{\hbar}\varepsilon_\alpha t}$:

$$\rho_1(\mathbf{r}, \mathbf{r}') = \sum_\alpha n_\alpha \psi_\alpha(\mathbf{r})\psi^*_\alpha(\mathbf{r}'). \tag{3.145}$$

The corresponding to the one-particle density matrix $\rho_1(\mathbf{r}, \mathbf{r}', t)$ the density operator $\widehat{\rho}^{(1)}(t)$ satisfies operator equation (3.126) at $k = 1$:

$$i\hbar\frac{\partial \widehat{\rho}^{(1)}(t)}{\partial t} = [\widehat{H}^{(1)}, \widehat{\rho}^{(1)}(t)]. \tag{3.146}$$

Further, we assume that the one-particle self-consistent Hamiltonian $\widehat{H}^{(1)}(t)$ in the Hartree-Fock time-dependent approximation is substituted in (3.146) for the operator $\widehat{H}^{(1)}$. At the same time, the Hamiltonian $\widehat{H}^{(1)}(t)$ is given in the coordinate representation by matrix (3.141). In this case, the following equation for the one-particle density matrix $\rho_1(\mathbf{r}, \mathbf{r}', t)$ (see (3.141)) can be associated with operator equation

$$i\hbar\frac{\partial}{\partial t}\rho_1(\mathbf{r}, \mathbf{r}', t) = \langle \mathbf{r}|[\widehat{H}^{(1)}(t), \widehat{\rho}^{(1)}(t)]|\mathbf{r}'\rangle \equiv$$

$$\equiv \int d\mathbf{r}''\{H^{(1)}(\mathbf{r}, \mathbf{r}'', t)\rho_1(\mathbf{r}'', \mathbf{r}', t) - \rho_1(\mathbf{r}, \mathbf{r}'', t)H^{(1)}(\mathbf{r}'', \mathbf{r}', t)\}. \tag{3.147}$$

This equation can also be obtained directly from equation (3.140). To do this, we have to multiply equation (3.140) by $\psi^*_{\alpha_k}(\mathbf{r}_k, t)$ and substract from the result the complex conjugated equation (3.141) multiplied by $\psi_{\alpha_k}(\mathbf{r}_k, t)$ and omit the nucleon index in $\mathbf{r}_k$, $\mathbf{r}'_k$ and α_k. Equations (3.146) and (3.147) are the basic equations of the time-dependent Hartree-Fock approximation. Further, we use these equations to obtain the basic equations of the quantum hydrodynamical model of nuclei.

It is possible to obtain the usual time-independent equation for the Hartree-Fock approximation by using time-dependent equation (3.147). For this we have to pass to the stationary one-particle functions in the same manner as we acted when obtaining (3.145). In this case $\frac{\partial}{\partial t}\rho_1(\mathbf{r},\mathbf{r}') = 0$, the density operator $\widehat{\rho}^{(1)}$ and the one-particle Hartree-Fock Hamiltonian $\widehat{H}^{(1)}$ do not depend on time and, therefore,

$$[\widehat{H}^{(1)}, \widehat{\rho}^{(1)}] = 0. \tag{3.148}$$

This operator equation, as well as the more general equation (3.146), does not depend on a representation.

We can establish the explicit form of the one-particle nuclear density operator in the Hartree-Fock approximation or self-consistent field approximation, expressing this operator in terms of the one-particle self-consistent Hartree-Fock Hamiltonian $H^{(1)}$. For the sake of simplicity, we consider the time-independent approximation. Assuming that the character of a level population in nuclei determined by only the one-particle Hartree-Fock energies ε_α, we can present the population numbers n_α as functions of ε_α for the nucleus ground state in the form of the Heaviside step-function

$$n_\alpha = n(\varepsilon_\alpha) = \theta(\varepsilon_F - \varepsilon_\alpha), \tag{3.149}$$

where $\theta(x) = 1$ at $x \geqslant 0$ and $\theta(x) = 0$ at $x < 0$, ε_F is the Fermi level energy. The density matrix $\rho_1(\mathbf{r},\mathbf{r}')$ with taking into account the density operator $\widehat{\rho}_1$, appearing in (3.149), can be written as follows

$$\rho_1(\mathbf{r},\mathbf{r}') = \langle\mathbf{r}|\widehat{\rho}^{(1)}|\mathbf{r}'\rangle = \int d\mathbf{r}''\delta(\mathbf{r}''-\mathbf{r})\widehat{\rho}^{(1)}_{\mathbf{r}''}\delta(\mathbf{r}''-\mathbf{r}') = \widehat{\rho}^{(1)}_{\mathbf{r}}\delta(\mathbf{r}-\mathbf{r}'), \tag{3.150}$$

where a lower sign $\mathbf{r}$ in the density operator $\widehat{\rho}^{(1)}$ shows that it acts only on $\mathbf{r}$ variable. In (3.150) we use the one-particle wave functions $\psi_{\mathbf{r}}(\mathbf{r}'') \equiv \langle\mathbf{r}|\mathbf{r}''\rangle = \delta(\mathbf{r}''-\mathbf{r})$ which are spatial wave functions in the proper representation. Using the completeness of the Hartree-Fock wave functions $\psi_\alpha(\mathbf{r})$, namely, the relation

$$\sum_\alpha \psi^*_\alpha(\mathbf{r}')\psi_\alpha(\mathbf{r}) = \delta(\mathbf{r}-\mathbf{r}'), \tag{3.151}$$

and expressions (3.145) and (3.150), we get

$$\widehat{\rho}^{(1)}\psi_\alpha(\mathbf{r}) = n_\alpha\psi_\alpha(\mathbf{r}) = n(\varepsilon_\alpha)\psi_\alpha(\mathbf{r}). \tag{3.152}$$

Since $n^2_\alpha = n_\alpha$ for the sharp Fermi border, we have from (3.144), in particular, $[\widehat{\rho}^{(1)}]^2 = \widehat{\rho}^{(1)}$.

Taking into account that (3.130) can be presented formally in the form

$$\widehat{H}^{(1)}\psi_\alpha(\mathbf{r}) = \varepsilon_\alpha\psi_\alpha(\mathbf{r}), \quad \widehat{H}^{(1)} = \widehat{T} + V(r) - \widehat{W}, \quad \widehat{T} = -\frac{\hbar^2}{2M}\Delta, \tag{3.153}$$

where $\widehat{W}$ is the integral operator connected with the nonlocal interaction and determined as

$$\widehat{W}\psi_\alpha(\mathbf{r}) = \int d\mathbf{r}' W(\mathbf{r}', \mathbf{r})\psi_\alpha(\mathbf{r}'), \tag{3.154}$$

we can write

$$n(\widehat{H}^{(1)})\psi_\alpha(\mathbf{r}) = n(\varepsilon_\alpha)\psi_\alpha(\mathbf{r}). \tag{3.155}$$

Here we took into account that a function of an operator has the common (with this operator) system of eigenfunctions and their eigenvalues are connected by the same functional dependence. According to (3.152) and (3.155), we obtain the expression for $\widehat{\rho}^{(1)}$:

$$\widehat{\rho}^{(1)} = n(\widehat{H}^{(1)}). \tag{3.156}$$

By using (3.149) we get the following simple formula for the nuclear density operator in the nuclear self-consistent field approximation

$$\widehat{\rho}^{(1)} = \theta(\varepsilon_F - \widehat{H}^{(1)}). \tag{3.157}$$

The same reasoning leads to the analogous formula in the *time-depending Hartree-Fock representation*

$$\widehat{\rho}^{(1)}(t) = \theta(\varepsilon_F - \widehat{H}^{(1)}(t)). \tag{3.158}$$

This expression for the operator $\widehat{\rho}^{(1)}(t)$ is often used in calculations.

Along with this, the so called point effective interaction between two nucleons j and k in a nucleus, for example, the *Skyrme interactions* proportional to delta function $\delta(\mathbf{r}_k - \mathbf{r}_j)$ is frequently used. In this case, the nonlocal interaction reduces to the local one, so that integrals over $\mathbf{r}'$ in (3.130), (3.140) and (3.154) are simply performed. Since

$$W(\mathbf{r}', \mathbf{r}) = W(\mathbf{r})\delta(\mathbf{r} - \mathbf{r}'), \tag{3.159}$$

expression (3.154) reduces to a simple product of $W(\mathbf{r})$ and $\psi_\alpha(\mathbf{r})$:

$$\widehat{W}\psi_\alpha(\mathbf{r}) = \int d\mathbf{r}'\delta(\mathbf{r} - \mathbf{r}')W(\mathbf{r})\psi_\alpha(\mathbf{r}') = W(\mathbf{r})\psi_\alpha(\mathbf{r}). \tag{3.160}$$

Thus, in this case, we obtain some redefined expression of the nuclear self-consistent potential in the usual Hartree-Fock approximation:

$$U(\mathbf{r}) = V(\mathbf{r}) - W(\mathbf{r}). \tag{3.161}$$

Analogously in the case of the Skyrme force in the time-dependent Hartree-Fock approximation, we have

$$W(\mathbf{r}, \mathbf{r}', t) = W(\mathbf{r}, t)\delta(\mathbf{r} - \mathbf{r}'), \tag{3.162}$$

The redefined self-consistent Hamiltonian is equal to

$$\widehat{H}^{(1)}(t) = \widehat{T} + U(\mathbf{r}, t), \quad U(\mathbf{r}, t) = V(\mathbf{r}, t) - W(\mathbf{r}, t), \tag{3.163}$$

where $U(\mathbf{r}, t)$ is the redefined time-dependent self-consistent local nuclear potential. The corresponding nuclear density operator (3.158) takes the form

$$\widehat{\rho}^{(1)}(t) = \theta(\varepsilon_F - \widehat{T} - U(\mathbf{r}, t)). \tag{3.164}$$

We will use this expression for the density operator further.

Quantum distribution function. The equations of motion of the quantum hydrodynamical model of nuclei for the local quantities, that characterize nuclei, are similar to the equations of the usual classic hydrodynamics and can be obtained from the *quantum kinetic equation* (Balbutsev, Dymarz, Mikhailov and Vaishvila, 1981; Kolomietz and Tang Henry, 1981), that determines the Fourier-component of the one particle nuclear density matrix $\rho_1(\mathbf{r}_1, \mathbf{r}_2, t)$ with respect to difference $\mathbf{s} = \mathbf{r}_1 - \mathbf{r}_2$, but preserves the coordinate dependence on $\mathbf{r} = \frac{1}{2}(\mathbf{r}_1 + \mathbf{r}_2)$:

$$f(\mathbf{r}, \mathbf{p}, t) \equiv \rho(\mathbf{r}, \mathbf{p}, t) = \int d\mathbf{s}\, \mathrm{e}^{-\frac{i}{\hbar}\mathbf{p}\mathbf{s}} \rho_1\left(\mathbf{r} + \frac{\mathbf{s}}{2}, \mathbf{r} - \frac{\mathbf{s}}{2}, t\right) =$$

$$= \int d\mathbf{s}\, \mathrm{e}^{-\frac{i}{\hbar}\mathbf{p}\mathbf{s}} \left\langle \mathbf{r} + \frac{\mathbf{s}}{2} \,\middle|\, \widehat{\rho}^{(1)})(t) \,\middle|\, \mathbf{r} - \frac{\mathbf{s}}{2} \right\rangle \equiv \left(\widehat{\rho}^{(1)}(t)\right)_W. \tag{3.165}$$

This transformation of the one-particle operator $\widehat{\rho}^{(1)}(t)$ is called the *Wigner transformation* and denoted as $(\widehat{\rho}^{(1)}(t))_W$. The transformation inverse to (3.165) can be written as

$$\rho_1(\mathbf{r}_1, \mathbf{r}_2, t) = \int \frac{d\mathbf{p}}{(2\pi\hbar)^3} \mathrm{e}^{\frac{i}{\hbar}\mathbf{p}(\mathbf{r}_1 - \mathbf{r}_2)} f\left(\frac{\mathbf{r}_1 + \mathbf{r}_2}{2}, \mathbf{p}, t\right). \tag{3.166}$$

The quantity $f(\mathbf{r}, \mathbf{p}, t)$ is called the quantum distribution function or the *Wigner distribution function* . It contains the same information on a nucleus as the one-particle density matrix.

Integrating $f(\mathbf{r}, \mathbf{p}, t)$ over $\mathbf{r}$, we obtain the nuclear distribution function over momenta in a nucleus. If it is integrated over momenta $\mathbf{p}$, we obtain the coordinate distribution or the space density of the nucleons in a nucleus $\rho(\mathbf{r}, t)$. This can be seen directly from (3.166) at $\mathbf{r}_1 = \mathbf{r}_2 = \mathbf{r}$:

$$\rho(\mathbf{r}, t) = \rho_1(\mathbf{r}, \mathbf{r}, t) = \int \frac{d\mathbf{p}}{(2\pi\hbar)^3} f(\mathbf{r}, \mathbf{p}, t). \tag{3.167}$$

In the general quantum case the quantity $f(\mathbf{r}, \mathbf{p}, t)$ cannot be treated as the distribution function over coordinates and momenta simultaneously, because this contradicts to the basic principles of quantum mechanics. In addition, the quantum distribution function $f(\mathbf{r}, \mathbf{p}, t)$ according to definition (3.165) may be not only positive but negative, as well. (The function $f(\mathbf{r}, \mathbf{p}, t)$ is real because $\rho_1(\mathbf{r}, \mathbf{r}', t) = \rho_1^*(\mathbf{r}', \mathbf{r}, t)$, that immediately follows from definition (3.119)). Only the classic and quasiclassic limits of the function $f(\mathbf{r}, \mathbf{p}, t)$ are always positive and have the meaning of the distribution function over coordinates and momenta simultaneously.

Let us expand the quantum distribution function $f(\mathbf{r}, \mathbf{p}, t)$ into a series with respect to powers of the Planck constant $\hbar$:

$$f(\mathbf{r}, \mathbf{p}, t) = \sum_{m=0}^{\infty} \hbar^m f_m(\mathbf{r}, \mathbf{p}, t). \tag{3.168}$$

It is obviously, that first terms in the expansion, at $m = 0$, corresponds to the classic distribution function

$$f_0(\mathbf{r}, \mathbf{p}, t) = \lim_{\hbar \to 0} f(\mathbf{r}, \mathbf{p}, t). \tag{3.169}$$

For the self-consistent field model, the quantum distribution function $f(\mathbf{r},\ p, t)$ is obtained by substitution (3.164) in (3.165). In the limit $\hbar \to 0$, we get

$$f_0(\mathbf{r}, \mathbf{p}, t) = \theta\Big(\varepsilon_F - \frac{p^2}{2M} - U(\mathbf{r}, t)\Big) = \theta(p_F^2 - p^2), \tag{3.170}$$

$$p_F = p_F(\mathbf{r}, t) = \sqrt{2M(\varepsilon_F - U(\mathbf{r}, t))}. \tag{3.171}$$

The explicit expressions for $f_m(\mathbf{r}, \mathbf{p}, t)$ at $m \neq 0$, when (3.164) is used, contain derivatives of delta-functions and are rather cumbersome. As it was above mentioned, the quantity $f_0(\mathbf{r}, \mathbf{p}, t)$ in contrast to $f(\mathbf{r}, \mathbf{p}, t)$ is always positive (or equal to zero in some domain). The employment of a few (sometimes even two or three) the first terms of expansion (3.168) is enough to describe satisfactory the real nuclei, i.e., the quasiclassic approximation is correct in this case. After substitution of (3.164) in (3.165) and use expansion (3.168) for the distribution function $f(\mathbf{r}, \mathbf{p}, t)$, where only the second order terms in the powers of $\hbar$ are taken into account, we can obtain the following expression for the nuclear density in a nucleus in the *quasiclassic approximation*

$$\rho(\mathbf{r}, t) = \frac{2p_F^3}{3\pi^2\hbar^3}\Big\{1 - \frac{\hbar^2 M}{4p_F^4}\Big[\Delta U(\mathbf{r}, t) + \frac{M}{2p_F^2}(\nabla U(\mathbf{r}, t))^2\Big]\Big\}\theta(\varepsilon_F - U(\mathbf{r}, t)). \tag{3.172}$$

Here we took into account the degeneration in spin and isospin, assuming that protons and neutrons are identical particles.

The *Wigner transformation* analogous to (3.165) can be introduced for an arbitrary operator $\widehat{A} = \widehat{A}(t)$ which, in a general case, may also depend on time:

$$(\widehat{A})_W \equiv A(\mathbf{r}, \mathbf{p}, t) = \int d\mathbf{s}\, \mathrm{e}^{-\frac{i}{\hbar}\mathbf{ps}} \left\langle \mathbf{r} + \frac{\mathbf{s}}{2} \,\middle|\, \widehat{A}(t) \,\middle|\, \mathbf{r} - \frac{\mathbf{s}}{2} \right\rangle. \qquad (3.173)$$

From this definition we can obtain a useful relation for the Wigner transformation for both a product and commutator of the two one-particle operators $\widehat{A}$ and $\widehat{B}$:

$$\begin{aligned} (\widehat{A}\widehat{B})_W &= (\widehat{A})_W \exp\left\{ \frac{i\hbar}{2}(\overleftarrow{\nabla}_{\mathbf{r}}\overrightarrow{\nabla}_{\mathbf{p}} - \overleftarrow{\nabla}_{\mathbf{p}}\overrightarrow{\nabla}_{\mathbf{r}}) \right\} (\widehat{B})_W, \\ ([\widehat{A}, \widehat{B}])_W &= 2i(\widehat{A})_W \sin\left\{ \frac{\hbar}{2}(\overleftarrow{\nabla}_{\mathbf{r}}\overrightarrow{\nabla}_{\mathbf{p}} - \overleftarrow{\nabla}_{\mathbf{p}}\overrightarrow{\nabla}_{\mathbf{r}}) \right\} (\widehat{B})_W, \end{aligned} \qquad (3.174)$$

where arrows in ∇ operators indicate the direction of differentiation; the lower indexes are the differentiation variables. (Here we used the relation $f(\mathbf{r}+\mathbf{a}) = \exp(\mathbf{a}\nabla_{\mathbf{r}})f(\mathbf{r})$ for an arbitrary function $f(\mathbf{r})$, assuming that $\mathbf{a}$ is a constant vector). The general relations obtained for one-particle operators (3.174) allows one to write down the kinetic equation for the quantum distribution function $f(\mathbf{r}, \mathbf{p}, t)$ in the one-particle approximation neglecting the residial or correlation interactions between nucleons in a nucleus (neglecting the nucleon-nucleon collisions):

$$\frac{\partial f(\mathbf{r}, \mathbf{p}, t)}{\partial t} + \frac{1}{M}\mathbf{p}\nabla_{\mathbf{r}} f(\mathbf{r}, \mathbf{p}, t) - \frac{2}{\hbar}\sin\left(\frac{\hbar}{2}\nabla_{\mathbf{p}}^{f}\nabla_{\mathbf{r}}^{U}\right) U(\mathbf{r}, t) f(\mathbf{r}, \mathbf{p}, t) = 0. \qquad (3.175)$$

Supplementary upper indexes in ∇ operators indicate the function on which these operators act. Quantum kinetic equation (3.175) is equivalent to the initial equation for the one-particle density operator (3.146) in the time-dependent Hartree-Fock approximation. If $\hbar \to 0$, equation (3.175) reduces to the kinetic equation of the classic (quasiclassic) distribution function or to the Vlasov equation and forms a basis for a set of the nuclear models.

Equation (3.175) can be generalized in the case, when the nucleon collisions (correlation) are also taken into account. In this case, in the right-hand side of kinetic equation (3.175) a definite term (instead of zero) depending on the distribution function $f(\mathbf{r}, \mathbf{p}, t)$ appears. It is called the collision term. Usually it is assumed that the collision term gives small corrections to the nuclear distribution function.

The equations of motion for different nuclear quantities that obtained with the help of (3.175) are the basis of the quantum hydrodynamical model of nuclei.

Quantum hydrodynamical model of nuclei. The next step when describing the quantum averaged (collective) nuclear characteristics is introduction the so called moments of the distribution function $f(\mathbf{r}, \mathbf{p}, t)$ and obtaining the equations of these moments, which are the basic equations in the *quantum hydrodynamical model of nuclei* . A moment of the function $f(\mathbf{r}, \mathbf{p}, t)$ represents the integral in the momentum space of a product $f(\mathbf{r}, \mathbf{p}, t)$ and a polynomial (its general form is specified below) depending on the momentum $\mathbf{p}$ projections. The polynomial degree determines the moment order. The zero moment (zero order moment) we already determined in (3.167). It is the nucleon density number in a nucleus $\rho(\mathbf{r}, t)$. The equation of the zero moment of the distribution function $\rho(\mathbf{r}, t)$ can be obtained by a simple integration of kinetic equation (3.175) over momenta $\mathbf{p}$:

$$\frac{\partial}{\partial t}\rho(\mathbf{r}, t) + \nabla[\rho(\mathbf{r}, t)\mathbf{u}(\mathbf{r}, t)] = 0, \tag{3.176}$$

where $\mathbf{u}(\mathbf{r}, t)$ is the collective velocity of the nuclear substance in a point $\mathbf{r}$ at moment time t (i.e., the velocity field). It is the first moment of the distribution function, which is the first rank tensor (or a vector)

$$\mathbf{u}(\mathbf{r}, t) = \frac{1}{\rho(\mathbf{r}, t)} \int \frac{d\mathbf{p}}{(2\pi\hbar)^3} \frac{\mathbf{p}}{M} f(\mathbf{r}, \mathbf{p}, t). \tag{3.177}$$

Equation (3.176) is the known continuity equation. It is clear that for the obtaining of the zero order moment of the distribution functions $\rho(\mathbf{r}, t)$ we need to know its first moment $u_\nu(\mathbf{r}, t)$, where $\nu = 1, 2, 3$ (or x, y, z). Note that in the ground state of nuclei $\mathbf{u}(\mathbf{r}, t) = 0$.

Higher moments of the distribution function $f(\mathbf{r}, \mathbf{p}, t)$, beginning with the second one, are determined by one general formula. The n-th order moment is a tensor of the n-th rank$(n \geqslant 2)$:

$$P_{\nu_1 \nu_2 \ldots \nu_n}(\mathbf{r}, t) = \frac{1}{M^{n-1}} \int \frac{d\mathbf{p}}{(2\pi\hbar)^3} (p_{\nu_1} - M u_{\nu_1}) \times$$
$$\times (p_{\nu_2} - M u_{\nu_2}) \ldots (p_{\nu_n} - M u_{\nu_n}) f(\mathbf{r}, \mathbf{p}, t). \tag{3.178}$$

Here in this definition, the vector $\mathbf{p} - M\mathbf{u}(\mathbf{r}, t)$ is a momentum of the inner motion, which is equal to zero in the nucleus ground state; $\mathbf{p}$ is the total momentum and $M\mathbf{u}(\mathbf{r}, t)$ is the momentum of the collective motion. The second moment $P_{\nu_1 \nu_2}(\mathbf{r}, t)$ presents a set of components of the second rank pressure tensor in a nucleus. It appears in the velocity field equation $\mathbf{u}(\mathbf{r}, t)$ that is obtained by the integration of kinetic equation (3.175) over momenta $\mathbf{p}$ after multiplying it by $p_\nu / (2\pi\hbar)^3$:

$$\frac{\partial}{\partial t}\Big[M\rho(\mathbf{r}, t)u_{\nu_1}(\mathbf{r}, t)\Big] + \nabla_{\nu_2}\Big[M\rho(\mathbf{r}, t)u_{\nu_1}(\mathbf{r}, t)u_{\nu_2}(\mathbf{r}, t)\Big] =$$

$$= -\left[\nabla_{\nu_2} P_{\nu_1\nu_2}(\mathbf{r},t) + \rho(\mathbf{r},t)\nabla_{\nu_1} U(\mathbf{r},t)\right]. \tag{3.179}$$

(There is summation taken over repeating indexes). This equation is analogous to the Euler equation for elastic media and reflects the momentum conservation law. The third moment $P_{\nu_1\nu_2\nu_3}(\mathbf{r},t)$ is the transport energy and momenta tensor (third rank). It appears in the equation for the pressure tensor $P_{\nu_1\nu_2}(\mathbf{r},t)$, that is obtained by the integration of kinetic equation (3.175) over momenta with a weight $p_{\nu_1}p_{\nu_2}$.

Partly convoluted, with respect to a pair of indexes, tensors of the lower ranks than initial ones, have the definite physical meaning: $\bar{P} = P_{\nu_1\nu_1} \equiv \sum_{\nu_1=1}^{3} P_{\nu_1\nu_1}$, $\bar{P}_{\nu_1} = P_{\nu_1\nu_2\nu_2} \equiv \sum_{\nu_2=1}^{3} P_{\nu_1\nu_2\nu_2}$, and so on. For example, the inner energy density in a nucleus is equal to

$$\varepsilon_{\text{in}}(\mathbf{r},t) = \int \frac{d\mathbf{p}}{(2\pi\hbar)^3} \frac{(\mathbf{p}-M\mathbf{u})^2}{2M} f(\mathbf{r},\mathbf{p},t), \tag{3.180}$$

and the heat flow density is given by the expression

$$Q_\nu(\mathbf{r},t) = \int \frac{d\mathbf{p}}{(2\pi\hbar)^3} \frac{p-Mu_\nu}{M} \frac{(\mathbf{p}-M\mathbf{u})^2}{2M} f(\mathbf{r},\mathbf{p},t). \tag{3.181}$$

The equation for $\varepsilon_{in}(\mathbf{r},t)$ associated with the energy conservation law can be obtained by the integration of kinetic equation (3.175) over $\mathbf{p}$ with a weight $\mathbf{p}^2/(2\pi\hbar)^3$:

$$\frac{\partial}{\partial t}\varepsilon_{\text{in}}(\mathbf{r},t) + \nabla_\nu[\varepsilon_{\text{in}}(\mathbf{r},t)u_\nu(\mathbf{r},t)] =$$

$$= -\frac{1}{2}P_{\nu_1\nu_2}(\mathbf{r},t)[\nabla_{\nu_1}u_{\nu_2}(\mathbf{r},t) + \nabla_{\nu_2}u_{\nu_1}(\mathbf{r},t)] - \nabla_\nu Q_\nu(\mathbf{r},t). \tag{3.182}$$

To obtain the equations for higher moments of the distribution function, we must continue the multiplication of kinetic equation (3.175) by more high powers of the vector $\mathbf{p}$ projections and integrate the results over $\mathbf{p}$.

Apparently, that acting in such a way, we can obtain an infinite system of coupling equations for the distribution function moments. If we cut the chain of these equations at definite stage (that is usually done), then the residual finite system of equations becomes incomplete, because it contains more unknown functions (moments) than the equations. To determine the maximum order of a moment (that can be partly convoluted tensors) in the residual equations, we need to know the distribution function $f(\mathbf{r},\mathbf{p},t)$. Basically, the cutted system of equations can be closed, if one notes that the distribution function moment of the maximum order, as well as other moments order of $n \geqslant 1$, appears in the equations under signs of the spatial derivatives (divergency). Because of this, the corresponding terms of the last equation in the system with the maximum order moment (or with a combination of its components) after integration over the volume can be transformed with the help of the Gauss theorem into the integral over the surface. This surface can be chosen infinitely far from a nucleus, where the distribution function and all its moments vanish. Acting in the same manner, we can get rid of the maximum order moment and close the system of equations. Every so often to close the above discussed system of equations, we can act even simpler but less strictly, making some assumptions regarding to the form of the quantum distribution function. By using this function we can obtain its maximum order moment that is left in the equation of motion chain for the local nuclear quantity.

In many cases in the quantum hydrodynamical model of nuclei, one uses only the chain of equations for the moments of the distribution functions with the first three equations of motion: for the nuclear density in a nuclei $\rho(\mathbf{r}, t)$, collective velocity $\mathbf{u}(\mathbf{r}, t)$, pressure density $P_{\nu_1 \nu_2}(\mathbf{r}, t)$ and heat flow density $Q_\nu(\mathbf{r}, t)$. In other words, we deal with the system of the above obtained equations (3.176), (3.179) and (3.182). It is interesting that the Planck constant $\hbar$ does not appear in these equations. This means that they do not change their form in the quasiclassic approximation, when $\hbar \to 0$. The Planck constant appears in the explicit form in the equations for the higher moments beginning with an order $n = 3$, when we take into account one more equation in the system. In (3.179), basically, we have three equations, but consider them as one equation for the vector quantity $\mathbf{u}(\mathbf{r}, t)$). At definite realistic assumptions regarding the quantum distribution function $f(\mathbf{r}, \mathbf{p}, t)$, the properties of the ground state of nuclei, as well as the excited states, follows from the same equations of motion (3.176), (3.179), (3.182) that depend only on the interaction between nucleons in a nucleus.

Sometimes the system of equations (3.176), (3.179) is used. The simplest version of the quantum hydrodynamical model of nuclei is the model based on equations (3.176) and (3.179) for the given *Fermi distribution function*

$$f(\mathbf{r}, \mathbf{p}, t) = \frac{g}{1 + \mathrm{e}^{\frac{p^2 - p_F^2}{2MT}}}, \quad g = (2s+1)(2t+1) = 4, \tag{3.183}$$

with the help of which we determine the pressure tensor $P_{\nu_1 \nu_2}(\mathbf{r}, t)$. Here the Fermi momentum p_F and temperature T are functions of the space coordinates and time. Below we give more detail of this version of the model. The use of (3.176) is justified for the sufficiently large nucleus excitation energy, when the Pauli principle influence is considerably weaken. The distribution function (3.183) corresponds to the Thomas-Fermi approximation at almost constant temperature, i.e., when the quasiclassic approach with a rather slow variations of the self-consistent field within the distances of order of the characteristic length of the nucleon wave in the many-nucleon nucleus are correct.

The pressure tensor (particular case of (3.178) at $n = 2$)

$$P_{\nu_1\nu_2}(\mathbf{r},t) = \frac{1}{M}\int \frac{d\mathbf{p}}{(2\pi\hbar)^3}(P_{\nu_1} - Mu_{\nu_1})(P_{\nu_2} - Mu_{\nu_2})f(\mathbf{r},\mathbf{p},t), \quad (3.184)$$

calculated with the help of distribution function (3.183) is diagonal with respect to two indexes:

$$P_{\nu_1\nu_2}(\mathbf{r},t) = [P_q(\mathbf{r},t) + P_T(\mathbf{r},t)]\delta_{\nu_1\nu_2}, \quad (3.185)$$

where $P_q(\mathbf{r},t)$ and $P_T(\mathbf{r},t)$ are the quantum and conventional heat pressures, respectively, that expressed in terms of the nuclear density in a nucleus

$$P_q(\mathbf{r},t) = \frac{\hbar^2}{5M}\left(\frac{6\pi^2}{g}\right)^{2/3}\rho^{5/3}(\mathbf{r},t), \quad P_T(\mathbf{r},t) = \frac{\hbar^2}{3M}\left(\frac{6}{\pi g}\right)^{2/3}\sigma^2\rho^{5/3}(\mathbf{r},t). \quad (3.186)$$

Here, the quantity σ has the meaning of the entropy, related to one nucleon, and it is equal to

$$\sigma = \frac{\pi^2}{2}\frac{T}{\varepsilon_F}. \quad (3.187)$$

Since, the pressure tensor in this case is expressed in terms of $\rho(\mathbf{r},t)$, then instead of the system of three equations of motion (3.176), (3.179), (3.182), we have the closed system of only two equations of motion (one of them is vectorial) which can be written in the form

$$\frac{\partial}{\partial t}\rho(\mathbf{r},t) + \nabla_\nu[\rho(\mathbf{r},t)u_\nu(\mathbf{r},t)] = 0,$$

$$\frac{\partial}{\partial t}[M\rho(\mathbf{r},t)u_\nu(\mathbf{r},t)] + \nabla_{\nu_1}[M\rho(\mathbf{r},t)u_\nu(\mathbf{r},t)u_{\nu_1}(\mathbf{r},t)] = \quad (3.188)$$

$$= -\nabla_\nu[P_q(\mathbf{r},t) + P_T(\mathbf{r},t)] - \rho(\mathbf{r},t)\nabla_\nu U(\mathbf{r},t).$$

These equations can be used to calculate any quantity of the collective motion (vibration) in nuclei.

Note that for the Fermi distribution function (3.183), the Fermi surface in the momentum space that determine the border of the quantum states, that are occupied by the nucleons in a nucleus, preserves the spherical symmetry. In a general case, the Fermi surface may be considerably deformed. Taking account of this deformation enables a qualitatively new nuclear phenomena to be described. The nucleon motion in a nucleus forms the vortex collective motion ($\mathrm{rot}\,\mathbf{u}(\mathbf{r},t) \neq 0$). The deformation dynamics of the Fermi surface provides also the additional contribution into the nuclear rigidness. This in its own turn must manifest itself in the potential energy of deformation, when the nucleus changes its shape with time. In the case of the strong collective nuclear excitations, we can describe some nuclear phenomena better, taking into account the effect of the nucleus surface deformations. In particular, this effect must be taken into account in the heavy nuclei fission processes.

When studied the ground states of nuclei within the framework of the quantum hydrodynamical model, we can use the above reported system of general equations of motion (3.176), (3.179), (3.182) or (3.188), imposing the condition $\mathbf{u}(\mathbf{r}, t) = 0$. However, sometimes in the quantum and classical hydrodynamical models, the variational principle (the minimum condition of the potential energy in the nucleus ground state) is frequently used, introducing separately, the neutron $\rho_n(\mathbf{r}, t)$ and proton $\rho_p(\mathbf{r}, t)$ densities, treating them as independently varied quantities. In addition the energy of Coulomb repulsion between protons in a nucleus must be also taken into account. In this case, as well as in the Hartree-Fock approximation in the hydrodynamical model, one frequently uses the phenomenological Skyrme nucleon-nucleon interactions or their modifications depending on the neutron and proton densities in every point of a nucleus. In action this model of nuclear forces is quite convenient because it has a zero radius of action that removes some computation problems.

There is no need to focuse our attention on the computation detail (we carried them out on a PC). Here, we cite some of these results obtained within the framework of the above described the quantum hydrodynamical model for spherical nuclei. The computed on the base of the Skyrme forces the equilibrium (time-independent) neutron $\rho_n(r)$ and proton $\rho_p(r)$ densities in a nuclei ^{208}Pb as functions of a distance from the nucleus center are shown in Fig. 3.5. The continuous lines correspond to the densities obtained in the quantum hydrodynamical model in the quasiclassic approximation, the dashed line – the corresponding dependencies obtained in the Hartree-Fock approximation. The nucleon density oscillations in the latter approximation are connected with the shell and boundary effects which disappear in the quasiclassic expansion of the distribution function in powers of $\hbar$ up to terms proportional to $\hbar^4$. Thus, the used version of the hydrodynamical model gives only, in definite sense, the averaged densities of nucleons. The dependencies of the nucleon densities $\rho_n(r)$ and $\rho_p(r)$ for a much lighter nucleus ^{40}Ca with equal numbers of neutrons and protons which, in particular, characterize the Coulombic repulsion influence on the behaviour of $\rho_p(r)$, are depicted in Fig. 3.6. The obtained numerical dependencies of the nucleon densities in nuclei on r can be approximated by the

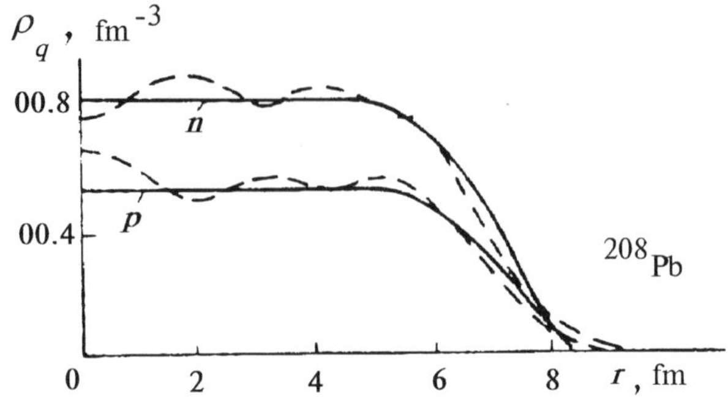

Figure 3.5. Dependencies of the neutron (n) and proton (p) densities of the nucleus ^{208}Pb in the quantum hydrodynamic model (solid lines) and the Hartre-Fock approximation (dotted lines)

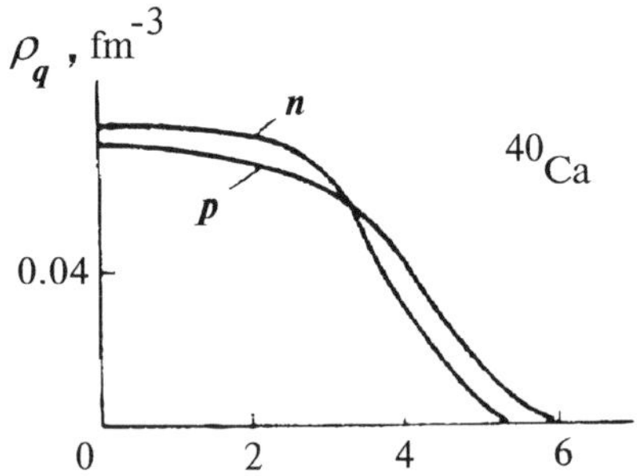

Figure 3.6. Neutron $\rho_n(r)$ and proton $\rho_p(r)$ densities in the nucleus ^{40}Ca ground state calculated in the hydrodynamical model

following simple smooth function (different from (3.10))

$$\rho_N(r) = \rho_{N0} \Big/ \left(1 + e^{\frac{r-R}{c}}\right)^b, \quad N = n, p, \tag{3.189}$$

where four parameters ρ_{N0}, R, c, b in the case $\rho_n(r)$ for nuclei ^{208}Pb and ^{40}Ca are equal respectively: 0.086 fm^{-3}, 7.47 fm, 0.58 fm; 1.84 and 0.074 fm^{-3}, 4.45 fm, 0.45 fm, 2.23. We can see from these graphs that the quantum hydrodynamical model enables one to introduce naturally diffuseness of the nucleus border.

We can obtain the nucleus bound energy within the framework of the quantum hydrodynamical model, neglecting the shell effects by substituting the obtained formulas for the nucleon densities $\rho_n(r)$ and $\rho_p(r)$ in the total nucleus energy. It is interesting that in this way we obtain the semiempirical Weizecker formula (3.17) from first the terms with some dependencies on A and Z and number factors which are close to the empirical ones.

Varying the total potential energy of a nucleus obtained in the hydrodynamical model with respect to the neutron and proton densities independently, we can find the corresponding averaged potential energies separately for neutrons and protons. It happens to be that these averaged nuclear fields are smooth functions of r and they are very close to the *Woods-Saxone phenomenological shell potentials*

$$V_N(r) = -\frac{V_{N0}}{1+e^{\frac{r-R}{a}}}, \quad N = n, p, \tag{3.190}$$

$$V_{n0} = V_0\left(1 - 0.63\frac{N-Z}{A}\right), \quad V_{p0} = V_0\left(1 + 0.63\frac{N-Z}{A}\right),$$

$$V_0 = 53\,\text{MeV}, \quad R = R_0 A^{1/3}, \quad R_0 = 1.2\,\text{fm}, \quad a = 0.6\,\text{fm}.$$

By using the obtained nucleon density, we can compute the collective velocity field in particular excited states of nuclei. With the help of the continuity equation (3.176), considering the small vibrations of the nucleon density around the equilibrium nucleon distribution in a nucleus, we can find the field of the collective velocity absolute value $u(r)$ in the case of the monopole excitations of spherical nuclei. The dependence $u(r)$ for nucleus with the mass number $A = 208$ (Fig. 3.7) is computed both in the quantum hydrodynamic model of nuclei (curve 1) and in the random phase approximation (curve 2) by using the Skyrme nuclear forces (the random phase approximation is described in Chapter 6). We can see that in the both cases the results obtained are close. The quantum hydrodynamical model of nuclei efficiently describes the energy dependence $\hbar\omega_{0^+}$ of the lowest nuclear monopole resonance on a mass number A. The computed in this model product $\hbar\omega_{0^+} \cdot A^{1/3}$ quite slowly increases with a growth in A from 70 MeV at $A = 70$ to 78 MeV at $A = 120$. This is in a good agreement with the experiment.

Thus, the quantum hydrodynamical model of nuclei, and in the limit the classic hydrodynamical model of nuclei, confirms and justifies a set of aspects of other nuclear models. It can independently describe and predict some averaged (collective) characteristics of nuclei and the phenomena, taking place in them.

Problems

3.1. Show that a number of particles $n(k_\alpha k)$ with a given momentum k related to a particle with given momentum k_α, in the *degenerate Fermi gas* of A identical particles is equal to

$$n(k_\alpha k) = 24A\frac{k^2}{k_F^3}\zeta(k_\alpha, k) \tag{3.191}$$

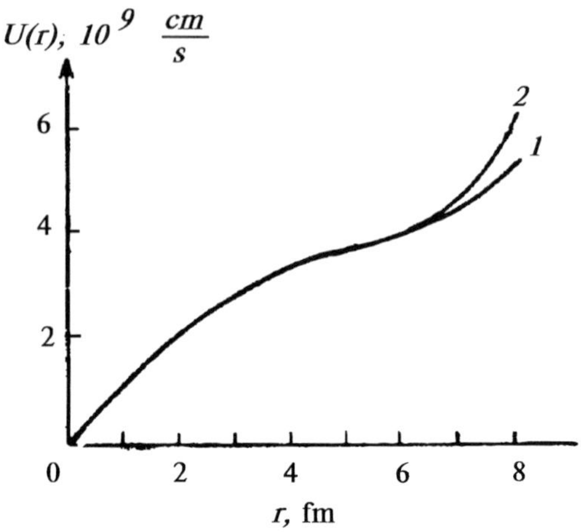

Figure 3.7. Collective velocity field $u(r)$ for the monopole excitations in a nucleus with the mass number $A = 208$ in the hydrodynamical model (curve 1) and in the random phase approximation (curve2)

where

$$\zeta(k_\alpha, k) = \begin{cases} 1, & k < \frac{1}{2}(k_F - k_\alpha); \\ \frac{1}{2}\left(1 + \frac{k_F^2 - k_\alpha^2 - 4k^2}{4k_\alpha k}\right), & \frac{1}{2}(k_F - k_\alpha) < k < \frac{1}{2}(k_F + k_\alpha); \\ 0, & \frac{1}{2}(k_p + k_\alpha) < k. \end{cases}$$

(The Fermi momentum k_F is connected with the total number A of particles and the gas volume Ω by the relation

$$\frac{\Omega}{(2\pi)^3}\frac{4\pi}{3}k_F^3 = A.\Big)$$

The positive sign of the quantity E_0/A indicates the unsatisfactory nature of the ideal Fermi gas model, while explaining the uniformity of nuclear matter and the proportionality of the total energy to the volume of the system, does not explain the existence of the bound state of nuclear matter.

We shall consider two particles with momentum k_α and k_β belonging to the Fermi distribution ($k_\alpha \leqslant k_F$, $k_\beta \leqslant k_F$). We introduce the relative momentum $\mathbf{k}$:

$$\mathbf{k} = \frac{1}{2}(\mathbf{k}_\beta - \mathbf{k}_\alpha). \tag{3.192}$$

Let the momentum $\mathbf{k}_\alpha$ of the first particle be fixed. We shall express the momentum $\mathbf{k}_\beta$ of the second particle in terms of $\mathbf{k}_\alpha$ and the relative momentum $\mathbf{k}$:

$$\mathbf{k}_\beta = \mathbf{k}_\alpha + 2\mathbf{k}. \tag{3.193}$$

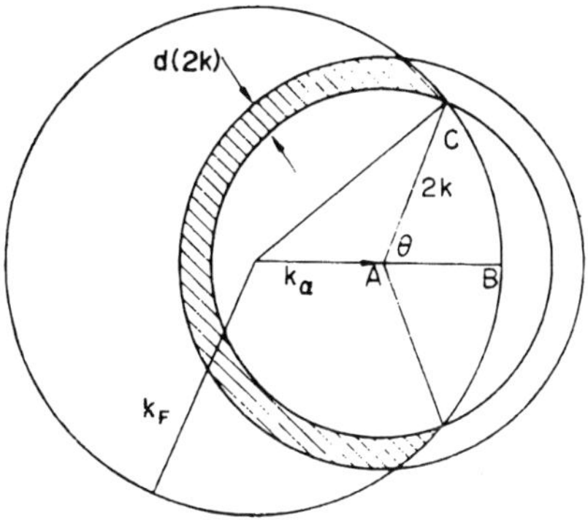

Figure 3.8. Section of the Fermi sphere delimiting the possible values of the momentum $\mathbf{k}_\beta = \mathbf{k}_\alpha + 2\mathbf{k}$.

Since $k_\beta \leqslant k_F$, we have

$$|\mathbf{k}_\alpha + 2\mathbf{k}| \leqslant k_F.$$

Figure 3.8 shows a section of the Fermi sphere delimiting the momentum values k_β. Obviously, for a fixed value of k_α the maximum possible value of k is equal to $(k_F + k_\alpha)/2$. If $k < (k_F + k_\alpha)/2$, the vector $\mathbf{k}$ can be oriented in any direction (in Fig. 3.8) the segment AB is equal to $k_F - k_\alpha$). In this case, the quantity $n(k_\alpha, k)dk$ is equal to the number of particles in a spherical shell of radius $2k$ and thickness $d(2k)$:

$$n(k_a, k)dk = \frac{\Omega}{(2\pi)^3} 4\pi(2k)^2 d(2k) = 24A\frac{k^2 dk}{k_F^3}, \quad k < \frac{1}{2}(k_F - k_\alpha). \tag{3.194}$$

If $k > \frac{1}{2}(k_F - k_\alpha)$, the directions of the vector $\mathbf{k}$ that the lie within the cone formed by rotating the radius AC about the axis AB are forbidden. We denote the angle between the directions of AB and AC by θ; then the solid angle cut out by the cone on the surface of the sphere of radius $2\mathbf{k}$ is equal to $2\pi(1 - \cos\theta)$. As can be seen from Fig. 3.8, the following relation is valid:

$$\cos\theta = \frac{k_F^2 - k_\alpha^2 - 4k^2}{4kk_\alpha}, \quad \frac{1}{2}(k_F - k_\alpha) < k < \frac{1}{2}(k_F + k_\alpha). \tag{3.195}$$

Therefore, in this case we have

$$n(k_\alpha, k)dk = \frac{\Omega}{(2\pi)^3}[4\pi - 2\pi(1 - \cos\theta)](2k)^2 d(2k) = 12A(1 + \cos\theta)\frac{k^2 dk}{k_F^3}, \quad \tfrac{1}{2}(k_F - k_\alpha) < k < \tfrac{1}{2}(k_F + k_\alpha). \tag{3.196}$$

If $k > \frac{1}{2}(k_F - k_\alpha)$, $n(k_\alpha, k) = 0$. Thus, we have proved the relation (3.191).

We note that formula (3.191) is also valid for a system of nucleons (particles possessing spin and isospin); for the limiting momentum in this case, however, we must take the quantity

$$k_F = \left(\frac{3\pi^2}{2}\frac{A}{\Omega}\right)\Bigg)^{1/3}.$$

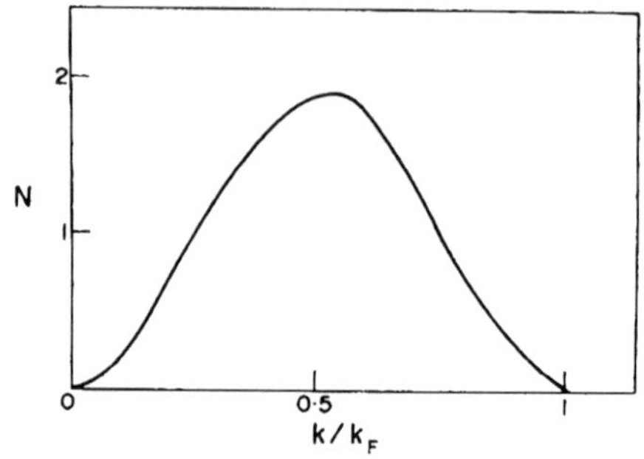

Figure 3.9. Distribution of the probability density for the relative momentum of a pair of particles.

3.2. Show that the number of pairs of particles with a given relative momentum k in the degenerate Fermi gas of A identical particles is equal to

$$n(k) = 12A(A-1)\left(1 - \frac{3}{2}\frac{k}{k_F} + \frac{1}{2}\frac{k^3}{k_F^3}\right)\frac{k^2}{k_F^3} \quad (0 < k < k_F). \tag{3.197}$$

We find the number $n(k)$ of pairs of particle with a given relative momentum k by summing $n(k_\alpha, k)$ over all the occupied states k_α:

$$n(k) = \int\limits_{k_\alpha < k_F} n(k_\alpha, k)\frac{\Omega}{(2\pi)^3} d^3k_\alpha. \tag{3.198}$$

Using (3.191), we obtain (3.197).

By integrating (3.197) over k we find the total number of pairs in the system

$$\int\limits_0^{k_F} n(k)\, dk = \frac{1}{2}A(A-1). \tag{3.199}$$

The ratio (3.197) and (3.199) gives the normalized distribution of the probability density for a given value of the relative momentum of a pair of particles:

$$N(k) = 24\left(1 - \frac{3}{2}\frac{k}{k_F} + \frac{1}{2}\frac{k^3}{k_F^3}\right)\frac{k^2}{k_F^3}. \tag{3.200}$$

The distribution $N(k)$ is shown in Fig. (3.9). The mean magnitude of the relative momentum is found to be of the order of $0.5\, k_F$ which, for the nuclear density (3.13), corresponds to a relative energy of 25 MeV. In the centre-of-mass system of the nucleus, this energy corresponds to the scattering of a nucleon with energy 50 MeV by a nucleon at rest, while the maximum magnitude of the relative momentum corresponds to the scattering of a nucleon with energy of about 160 MeV.

3.3. Find the momentum dependence of the energy per nucleon in nuclear matter, assuming that two-particle forces described by the potential

$$\hat{V} = \frac{1 + P_r}{2} V(r), \quad V(r) = \begin{cases} -V_0, & r < r_0; \\ 0, & r > r_0, \end{cases} \tag{3.201}$$

act between the nucleons.

If the system is uniform, then, even in the presence of strong interaction between the particles, the one-particle wave functions can be chosen in the form of plane waves; however, the one-particle energy in this case does not reduce to the kinetic energy of the particle, but is equal to

$$E_\alpha = \frac{\hbar^2 k_\alpha^2}{2M} + V(k_\alpha), \tag{3.202}$$

where $V(k_\alpha)$ is the additional energy due to the interaction of the particle with all the other particles. This energy is usually called the *one-particle potential*, and in the general case is found to depend on the momentum of the particle. Since the potential (3.201) depends only on the relative distance between the particles, the interaction energy of two nucleons in states with momenta $\mathbf{k}_\alpha$ and $\mathbf{k}_\beta$ depends only on the relative momentum. We choose the wave function of a pair of nucleons in the form (3.34). Then the mean value of the interaction energy of the pair of nucleons will be given by the expression

$$\langle U_{\alpha\beta}\rangle \equiv \frac{3}{8}U(k) = \frac{3}{8}(1/\Omega)\int d^3r(1+\cos 2\mathbf{k}\cdot\mathbf{r})V(r), \tag{3.203}$$

where $\mathbf{k} = \frac{1}{2}(\mathbf{k}_\alpha - \mathbf{k}_\beta)$ is the relative momentum and 3/8 is the statistical weight of the spatially symmetric states in which the interaction (3.201) occurs. By multiplying the energy (3.203) by the number $n(k_\alpha, k)$ of particles with momentum k relative to a particle with a given momentum k_α and integrating the expression obtained over all the relative momenta, we find the total energy of interaction of the chosen particle with all the other particles:

$$V(k_\alpha) = \frac{3}{8}\int_0^\infty dk\, n(k_\alpha, k)\, U(k). \tag{3.204}$$

Using (3.191) and (3.203) and integrating over k, we obtain the following expression for the one-particle potential:

$$V(k_\alpha, r_0) = -\frac{V_0}{12\pi}k_F^3 r_0^3 - \frac{3}{8\pi}V_0\left\{\mathrm{Si}\,(k_F+k_\alpha)r_0 + \mathrm{Si}\,(k_F-k_\alpha)r_0 - 2\sin k_F r_0 \frac{\sin k_\alpha r_0}{k_\alpha r_0}\right\}. \tag{3.205}$$

Figure 3.10 shows the dependence of the one-particle potential $V(k_\alpha)$ on the momentum k_α/k_F for $k_F = 1.38\times 10^{13}\,\mathrm{cm}^{-1}$. (We have used the following values for the potential parameters: $V_0 = (\pi^2/4)\hbar^2/Mr_0^2$, $r_c = 0$ and $r_0 = 2.3\times 10^{-13}\,\mathrm{cm}$.) The potential $V(k_\alpha)$ varies approximately quadratically with k_α and so a good representation for the potential is given by an approximation in the form

$$V(k_\alpha) = V_0 + V_1\left(\frac{k_\alpha^2}{k_F^2}\right), \tag{3.206}$$

where $V_0 \approx -25.2$ MeV and $V_1 = 10.2$ MeV. The expression (3.206) can be regarded as an expansion of the potential $V(k_\alpha)$ in powers of k_α^2. In Fig. 3.10 the dashed line corresponds to the expression (3.206).

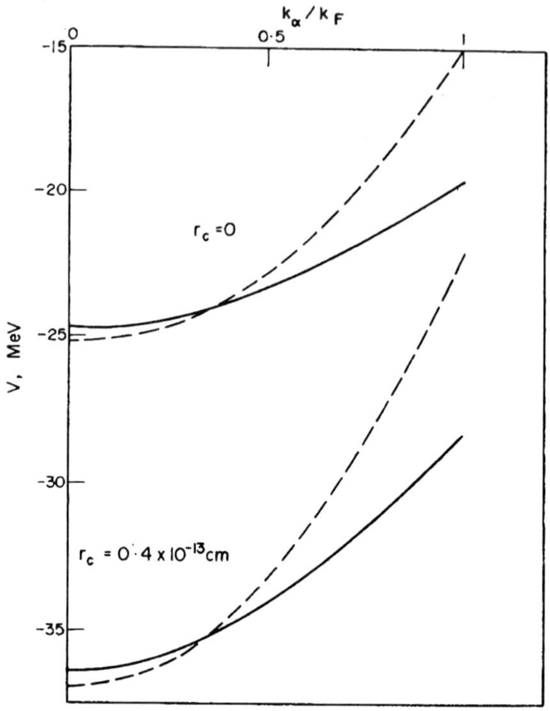

Figure 3.10. Dependence of the one-particle potential $V(k_\alpha)$ on the momentum k_α for $r_c = 0$ and $r_c = 0.4 \times 10^{-13}$ cm.

The above approximation for the potential $V(k_\alpha)$ makes it possible to write the one-particle energy in the form

$$E_\alpha = \frac{\hbar^2 k_\alpha^2}{2M^*} + V_0, \tag{3.207}$$

where M^* is the so-called *effective mass*

$$1/M^* = 1/m + (2/\hbar^2)(V_1/k_F^2). \tag{3.208}$$

By making use of the numerical value of V_1, we find $M^* \approx 0.8\,M$ for the effective mass of a nucleon in nuclear matter. Thus, the interaction between the nucleons leads to a change of the momentum dependence of the energy and can be taken into account formally by introducing an effective nucleon mass. We note that the effective mass introduced in (3.208) takes into account only the attractive forces between the nucleons.

If we assume that the attractive forces appear only at distances greater than the radius $\mathbf{r}_c$ of the repulsive core, then the one-particle potential $V(k_\alpha, r_0, r_c)$ due to the attractive forces will be determined by the difference of two expression of the type (3.205):

$$V(k_\alpha, r_0, r_c) = V(k_\alpha, r_0) - V(k_\alpha, r_c). \tag{3.209}$$

Figure 3.10 also shows the dependence of the potential (3.209) on k_α for $r_c = 0.4 \times 10^{-13}$ cm, $r_0 = 2.3 \times 10^{-13}$ cm and $V_0 = (\pi^2/4) \times \hbar/M(r_0 - r_c)^2)$. Approximating the potential (3.209) by a parabola, we find $V_0 = -37.0$ MeV and $V_1 = 15.2$ MeV. The corresponding value of the effective mass in this case is $M^* = 0.7\,M$.

3.4. Consider the correlation properties of the perfect Fermi gas. The wave function of a system of A identical particles satisfying the Pauli principle can be represented in the form of the *Slater determinant*

$$\Phi(\mathbf{r}_1, \mathbf{r}_2, \ldots, \mathbf{r}_A) = \frac{1}{\sqrt{A}} \begin{vmatrix} \varphi_{\alpha_1}(\mathbf{r}_1) & \varphi_{\alpha_1}(\mathbf{r}_2) & \ldots & \varphi_{\alpha_1}(\mathbf{r}_A) \\ \varphi_{\alpha_2}(\mathbf{r}_1) & \varphi_{\alpha_2}(\mathbf{r}_2) & \ldots & \varphi_{\alpha_2}(\mathbf{r}_A) \\ \ldots & \ldots & \ldots & \ldots \\ \varphi_{\alpha_A}(\mathbf{r}_1) & \varphi_{\alpha_A}(\mathbf{r}_2) & \ldots & \varphi_{\alpha_A}(\mathbf{r}_A) \end{vmatrix}, \tag{3.210}$$

where $\varphi_\alpha(\mathbf{r})$ is a one-particle wave function.

The probability of finding any particle at a given point in space is determined by the square of the modulus of the wave function (3.210), integrated over the coordinates of all the other particles:

$$w(\mathbf{r}_1) = \int d^3r_2\, d^3r_3 \ldots d^3r_A |\Phi(\mathbf{r}_1, \mathbf{r}_2, \ldots, \mathbf{r}_A)|^2. \tag{3.211}$$

Substituting the wave function $\Phi(\mathbf{r}_1, \mathbf{r}_2, \ldots, \mathbf{r}_A)$ in the form (3.210) into (3.211), we find

$$w(\mathbf{r}) = (1/A) \sum_\alpha |\varphi_\alpha(\mathbf{r})|^2, \tag{3.212}$$

where the summation is taken over all the occupied one-particle states. By multiplying (3.212) by the total number A of particles we find the particle distribution function (particle density):

$$\varrho(\mathbf{r}) = \sum_\alpha |\varphi_\alpha(\mathbf{r})|^2. \tag{3.213}$$

The probability of finding any particle at the point $\mathbf{r}_1$ and another particle at the point $\mathbf{r}_2$ is determined by the square of the modulus of the wave function $\Phi(\mathbf{r}_1, \mathbf{r}_2, \ldots, \mathbf{r}_A)$ integrated over the coordinates of the other $A - 2$ particles:

$$w(\mathbf{r}_1, \mathbf{r}_2) = \int d^3r_3 \ldots d^3r_A |\Phi(\mathbf{r}_1, \mathbf{r}_2, \ldots, \mathbf{r}_A)|^2. \tag{3.214}$$

Using (3.210), it is not difficult to obtain

$$w(\mathbf{r}, \mathbf{r}') = \frac{1}{A(A-1)} \sum_{\alpha,\beta} |\varphi_\alpha(\mathbf{r})\varphi_\beta(\mathbf{r}') - \varphi_\beta(\mathbf{r})\varphi_\alpha(\mathbf{r}')|^2. \tag{3.215}$$

By multiplying (3.215) by the total number $A(A-1)/2$ of independent pairs of particles, we find the particle pair distribution function:

$$\varrho(\mathbf{r}, \mathbf{r}') = \frac{1}{2} \sum_{\alpha,\beta} |\varphi_\alpha(\mathbf{r})\beta(\mathbf{r}') - \varphi_\beta(\mathbf{r})\varphi_\alpha(\mathbf{r}')|^2. \tag{3.216}$$

The quantity $\varrho(\mathbf{r}, \mathbf{r}')$ is not equal to the product $\varrho(\mathbf{r})\varrho(\mathbf{r}')$. In fact, using (3.216), it is not difficult to show that

$$\varrho(\mathbf{r}, \mathbf{r}') = \varrho(\mathbf{r})\varrho(\mathbf{r}') - \varrho'(\mathbf{r}, \mathbf{r}'). \tag{3.217}$$

where

$$\varrho'(\mathbf{r}, \mathbf{r}') = \sum_{\alpha} |\varphi_\alpha(\mathbf{r})\varphi_\alpha^*(\mathbf{r}')|^2. \tag{3.218}$$

The difference between $\varrho(\mathbf{r}, \mathbf{r}')$ and $\varrho(\mathbf{r})\varrho(\mathbf{r}')$ is connected with the correlation between identical particles that arises from the Pauli principle.

If the system is uniform, the one-particle wave functions $\varphi_\alpha(\mathbf{r})$ can be chosen in the form of plane waves

$$\varphi_\alpha(\mathbf{r}) = (1/\sqrt{\Omega})e^{i\mathbf{k}_\alpha \cdot \mathbf{r}}. \tag{3.219}$$

As we should expect, the particle density in this case is constant:

$$\varrho(\mathbf{r}) = \varrho_0 = A/\Omega, \tag{3.220}$$

and the pair distribution function can be represented in the form

$$\varrho(\mathbf{r}, \mathbf{r}') = \varrho_0^2\{1 + g(\mathbf{r}' - \mathbf{r})\}, \tag{3.221}$$

where the function $g(\mathbf{r}' - \mathbf{r})$ which depends on the coordinate difference $\mathbf{r}' - \mathbf{r}$, is called the *correlation function* . According to the formulae (3.221) and (3.217), we have

$$g(r) = -\left|\frac{\Omega}{A}\int_{k<k_F} \frac{d^3k}{(2\pi)^3} e^{i\mathbf{k}\cdot\mathbf{r}}\right|^2, \tag{3.222}$$

whence (Wigner and Seitz, 1933)

$$g(r) = -\frac{9}{k_F^4 r^4}\left(\frac{\sin k_F r}{k_F r} - \cos k_F r\right)^2. \tag{3.223}$$

For small values of $k_F r$ the expansion of the correlation function $g(r)$ has the form

$$g(r) = -1 + \frac{k_F^2 r^2}{5} - \frac{3}{175}k_F^4 r^4 + \ldots \tag{3.224}$$

Thus, if $\mathbf{r}'$ coincides with $\mathbf{r}$ the pair distribution function $\varrho(\mathbf{r}, \mathbf{r}')$ goes to zero. This is a natural consequence of the Pauli principle, which forbids identical particles to approach each other. Figure 3.11 shows the dependence the correlation function $g(r)$ on $k_F r$. As can seen from Fig. 3.11, the characteristic correlation length r_F is determined by the Fermi momentum, $r_F \approx k_F^{-1}$.

If two-particle forces act between the particles and can be characterized by a potential $V(\mathbf{r})$ depending on the relative distance between the particles, then the total interaction energy ΔE of the system can be expressed directly, in first-order perturbation theory, in terms of the pair distribution function $\varrho(\mathbf{r}, \mathbf{r}')$:

$$\Delta E = \int d^3r\, d^3r'\, \varrho(\mathbf{r}, \mathbf{r}')V(\mathbf{r} - \mathbf{r}'). \tag{3.225}$$

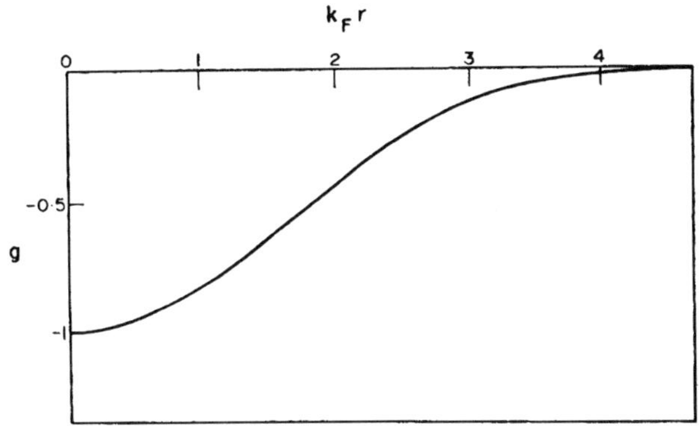

Figure 3.11. Dependence of the correlation function $g(r)$ on $k_F r$.

If we choose the potential $V(r)$ in the form of the square well (3.201), in the case of a uniform system it is not difficult to obtain from (3.225) an expression for the total interaction energy ΔE that coincides with the expression (3.83) obtained previously for $r_c = 0$.

3.5. Establish the connection between the pair correlation function and the particle-density fluctuations. The pair correlation function $g(\mathbf{r} - \mathbf{r}')$ describes the static of the system and is related to the random fluctuations of the particle density. We define the particle density operator by the following relation:

$$\widehat{\varrho}(\mathbf{r}) = \sum_i \delta(\mathbf{r} - \mathbf{r}_i), \tag{3.226}$$

where $\mathbf{r}_i$ is the position vector of an individual particle; the summation is taken over all the particles in unit volume. We denote the mean value of the particle density $\varrho(\mathbf{r})$:

$$\varrho(\mathbf{r}) = \langle \widehat{\varrho}(\mathbf{r}) \rangle, \tag{3.227}$$

where the brackets $\langle \ldots \rangle$ denote averaging over the quantum state of the system:

$$\langle \ldots \rangle = \int d^3r_1\, d^3r_2 \ldots d^3r_A \ldots |\Phi(\mathbf{r}_1, \mathbf{r}_2, \ldots, \mathbf{r}_A)|^2. \tag{3.228}$$

In calculating mean values in the general case, we must also average over the statistical distribution of the different quantum states of the system.

The operator corresponding to the pair distribution function is given by a double sum of a product of delta-functions:

$$\varrho(\mathbf{r}, \mathbf{r}') = \sum_{i \neq j} \delta(\mathbf{r} - \mathbf{r}_i)\delta(\mathbf{r}' - \mathbf{r}_j), \tag{3.229}$$

and the mean value of the pair distribution function is equal to

$$\varrho(\mathbf{r}, \mathbf{r}') = \left\langle \sum_{i \neq j} \delta(\mathbf{r} - \mathbf{r}_i)\delta(\mathbf{r}' - \mathbf{r}_j) \right\rangle. \tag{3.230}$$

In the following, we shall confine ourselves for simplicity to the case of uniform systems. In this case, the mean value of the density is equal to a constant

$$\varrho(\mathbf{r}) = \varrho_0, \tag{3.231}$$

and the mean value of the pair distribution function depends only on the relative distance $\mathbf{r} - \mathbf{r}'$ between the points and is equal to

$$\varrho(\mathbf{r}, \mathbf{r}') = \varrho_0^2\{1 + g(\mathbf{r}' - \mathbf{r})\}, \tag{3.232}$$

where $g(\mathbf{r} - \mathbf{r}')$ is the pair correlation function.

We shall consider random particle-density fluctuations, which we shall characterize by the magnitude of the deviation of the density $\widehat{\varrho}$ from its mean value ϱ_0:

$$\delta\varrho(\mathbf{r}) \equiv \widehat{\varrho}(\mathbf{r}) - \varrho_0. \tag{3.233}$$

Obviously, the mean deviation of the density is equal to zero:

$$\langle \delta\varrho(\mathbf{r}) \rangle = 0.$$

We define the correlation function $\langle \delta\varrho^2 \rangle_{\mathbf{r}'-\mathbf{r}}$ of the particle-density fluctuations as the mean value of the product of the density fluctuations at two different points $\mathbf{r}$ and $\mathbf{r}'$ in space:

$$\langle \delta\varrho^2 \rangle_{\mathbf{r}'-\mathbf{r}} \equiv \langle \delta\varrho(\mathbf{r})\delta\varrho(\mathbf{r}') \rangle. \tag{3.234}$$

Because of the homogeneity of space, the correlation function (3.234) of the particle-density fluctuations depends only on the difference $\mathbf{r}' - \mathbf{r}$ of the coordinates of the points. Substituting the deviation $\delta\varrho$ in the form (3.233) into the relation (3.234) and (3.230), we find

$$\langle \delta\varrho^2 \rangle_{\mathbf{r}'-\mathbf{r}} = \varrho_0\delta(\mathbf{r} - \mathbf{r}') + \varrho(\mathbf{r}, \mathbf{r}') - \varrho_0^2. \tag{3.235}$$

Then, using (3.232), we finally obtain the relation

$$\langle \delta\varrho^2 \rangle_{\mathbf{r}} = \varrho_0\delta(\mathbf{r}) + \varrho_0^2 g(\mathbf{r}), \tag{3.236}$$

connecting the correlation function $\langle \delta\varrho^2 \rangle_{\mathbf{r}}$ of the density fluctuations with the pair correlation functions $g(\mathbf{r})$.

3.6. Determine the space-time correlation function for the particle-density fluctuations.

The spatial correlation function $\langle \delta\varrho^2 \rangle_{\mathbf{r}-\mathbf{r}'} = \langle \delta\varrho(\mathbf{r})\delta\varrho(\mathbf{r}') \rangle$ characterizes the correlation between the particle-density fluctuations at different points in space at the same moment in time, and describes static properties the system. To describe the dynamical properties of the system, we can introduce the space-time correlation function

$$\langle \delta\varrho(\mathbf{r}, t)\delta\varrho(\mathbf{r}', t') \rangle, \tag{3.237}$$

defined as the mean of the product of the particle-density fluctuations at different spatial points $\mathbf{r}$ and $\mathbf{r}'$ and different times t and t'. The averaging in (3.237) is performed both over the quantum state of the system and over the statistical distribution of the different quantum states. If the system is spatially uniform and stationary states of the system are considered, the space-time correlation function

(3.237) will depend only on the relative distance and on the absolute value of the time interval between the points at which the fluctuations are considered:

$$\langle \delta\varrho(\mathbf{r},t)\delta\varrho(\mathbf{r}',t')\rangle = \langle \delta\varrho^2\rangle_{\mathbf{r}'-\mathbf{r},t'-t}. \tag{3.238}$$

The spectral distribution of the fluctuations will be characterized by a spectral correlation function, which we define by the following equality:

$$\langle \delta\varrho^2\rangle_{\mathbf{q}\omega} \equiv \int d^3r\, dt\, \mathrm{e}^{-i\mathbf{q}\cdot\mathbf{r}+i\omega t}\langle \delta\varrho^2\rangle_{\mathbf{r}t}. \tag{3.239}$$

Analogously,

$$\langle \delta\varrho^2\rangle_{\mathbf{q}} = \int d^3r\, \mathrm{e}^{-i\mathbf{q}\cdot\mathbf{r}}\langle \delta\varrho^2\rangle_{\mathbf{r}}. \tag{3.240}$$

It is not difficult to verify that

$$\langle \delta\varrho^2\rangle_{\mathbf{q}} = \int \frac{d\omega}{2\pi}\langle \delta\varrho^2\rangle_{\mathbf{q}\omega}. \tag{3.241}$$

In place of $\langle \delta\varrho^2\rangle_{\mathbf{q}\omega}$ the dynamical form factor $S(\mathbf{q},\omega)$, introduced by van Hove (1954), is often used in the literature:

$$S(\mathbf{q},\omega) \equiv \frac{1}{2\pi\varrho_0}\langle \delta\varrho^2\rangle_{\mathbf{q}\omega}. \tag{3.242}$$

3.7. Find the space-time correlation function for the density fluctuations in the *perfect Fermi gas.*

The particle-density fluctuations in the perfect Fermi gas can be studied most simply using the method of second quantization. We shall define the particle-number density operator by means of the relation

$$\hat{\varrho}(\mathbf{r},t) = \sum_{\sigma=-\frac{1}{2},\frac{1}{2}} \psi_\sigma^+(\mathbf{r},t)\psi_\sigma(\mathbf{r},t), \tag{3.243}$$

where the functions ψ_σ and ψ_σ^+ have the form

$$\left.\begin{aligned} \psi_\sigma(\mathbf{r},t) &= \frac{1}{\sqrt{\Omega}}\sum_{\mathbf{k}} a_{\mathbf{k}\sigma}\mathrm{e}^{i(\mathbf{k}\cdot\mathbf{r}-E_{\mathbf{k}}t)}, \\ \psi_\sigma^+(\mathbf{r},t) &= \frac{1}{\sqrt{\Omega}}\sum_{\mathbf{k}} a_{\mathbf{k}\sigma}^+\mathrm{e}^{-i(\mathbf{k}\cdot\mathbf{r}-E_{\mathbf{k}}t)}. \end{aligned}\right\} \tag{3.244}$$

Here $a_{\mathbf{k}\sigma}$ and $a_{\mathbf{k}\sigma}^+$ are operators which respectively destroy and create a particle in the state with momentum $\mathbf{k}$ and spin component σ; E_k is the energy of a particle with momentum $\mathbf{k}$ (the quadratic dispersion law $E_k = k^2/2M$ is assumed; M is the mass of a particle and $\hbar = 1$). Substituting (2.244) into (3.243), we obtain

$$\hat{\varrho}(\mathbf{r},t) = \frac{1}{\Omega}\sum_{\mathbf{k}\mathbf{k}'\sigma} a_{\mathbf{k}\sigma}^+ a_{\mathbf{k}'\sigma}\mathrm{e}^{-i(\mathbf{k}-\mathbf{k}')\cdot\mathbf{r}+i(E_{\mathbf{k}}-E_{\mathbf{k}'}t)}. \tag{3.245}$$

The diagonal ($\mathbf{k} = \mathbf{k}'$) terms of the operator $\widehat{\varrho}$ determine the mean density ϱ_0 of the particles. The difference between $\widehat{\varrho}$ and its mean value ϱ_0 can be taken as the density-fluctuation operator

$$\delta\widehat{\varrho}(\mathbf{r}, r) \equiv \widehat{\varrho}(\mathbf{r}, t) - \varrho_0. \tag{3.246}$$

The correlation function for the density fluctuations is defined by the expression

$$\langle \delta\varrho^2 \rangle_{\mathbf{r}'-\mathbf{r}, t'-t} = \langle \delta\varrho(\mathbf{r}, t)\delta\varrho(\mathbf{r}', t') \rangle. \tag{3.247}$$

By multiplying the operators $\delta\varrho(\mathbf{r}, t)$ and $\delta\varrho(\mathbf{r}', t')$, we obtain a sum of terms containing different products of the operators $a_{k\sigma}$ and $a'_{k\sigma}$ taken in fours. The mean value of a product of four fermion operators a can be expressed in terms of the mean values of the products of the same operators in pairs:

$$\langle a_1 a_2 a_3 a_4 \rangle = \langle a_1 a_2 \rangle \langle a_3 a_4 \rangle - \langle a_1 a_3 \rangle \langle a_2 a_4 \rangle + \langle a_1 a_4 \rangle \langle a_2 a_3 \rangle. \tag{3.248}$$

The mean value of a product of two operators a_1 and a_2 is non-zero only in the case when these operators are Hermitian conjugates.

Let n_k be the mean value of the product $a^+_{\mathbf{k}\sigma} a_{\mathbf{k}\sigma}$ defining the mean number of particles in a state with given values of $\mathbf{k}$ and σ:

$$n_k \equiv \langle a^+_{\mathbf{k}\sigma} a_{\mathbf{k}\sigma} \rangle. \tag{3.249}$$

Because of the isotropy of space, this number does not depend on the value of the spin component σ. In the case of a completely degenerate Fermi gas at temperature $T = 0$, the distribution function n_k is equal to

$$n_k = \begin{cases} 1, & k \leqslant k_F; \\ 0, & k > k_F. \end{cases} \tag{3.250}$$

there k_F is the Fermi momentum.

By direct inspection, it is not difficult to convince oneself that

$$\langle a^+_{k_1\sigma_1} a_{k'_1\sigma_1} a^+_{k_2\sigma_2} a_{k'_2\sigma_2} \rangle = n_{k_1}(1 - n_{k_2}\delta_{k_1 k'_2}\delta_{k'_1 k_2}\delta_{\sigma_1\sigma_2} \qquad \begin{pmatrix} k'_1 \neq k_1 \\ k'_2 \neq k_2 \end{pmatrix} \tag{3.251}$$

and

$$\langle \delta\varrho^2 \rangle_{\mathbf{r}, t} = \frac{2}{\Omega^2} \sum_{\mathbf{k}_1 \mathbf{k}_2} n_{\mathbf{k}_1}(1 - n_{\mathbf{k}_2}) \mathrm{e}^{i(\mathbf{k}_1 - \mathbf{k}_2)\cdot\mathbf{r} - i(E_{k_1} - E_{k_2})t}. \tag{3.252}$$

In the perfect Fermi gas, the correlation is due to the Pauli principle and so occurs only between particles with the same spin orientations. If we assume the volume Ω to be sufficiently large, we can replace the summation in (3.252) by integration:

$$\sum_k \ldots \to \frac{\Omega}{(2\pi)^3} \int d^3k \ldots \tag{3.253}$$

The positive sign of the quantity E_0/A indicates the unsatisfactory nature of the perfect Fermi gas model, while explaining the uniformity of

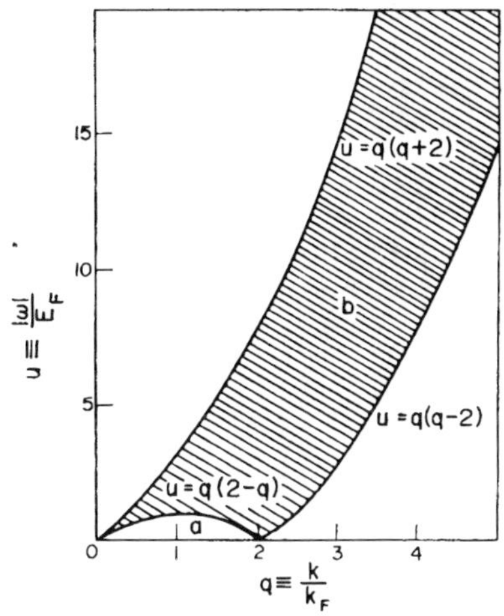

Figure 3.12. Regions of $u \equiv |\omega|/E_F$ and $q \equiv k/k_F$ in which the expressions (3.255) for the spectral distribution of the density fluctuations in an perfect Fermi gas apply.

nuclear matter and the proportionality of the total energy to the volume of the system, does not explain the existence of the bound state of nuclear matter.

Using (3.252), we thus obtain the followind general formula for the spectral distribution of the density fluctuation in the perfect Fermi gas:

$$\langle \delta \varrho^2 \rangle_{\mathbf{q},\omega} = \frac{1}{2\pi^2} \int d^3k \, n_{\mathbf{k}} (1 - n_{\mathbf{k}-\mathbf{q}}) \delta(\omega - E_{\mathbf{k}} + E_{\mathbf{k}-\mathbf{q}}). \tag{3.254}$$

In the case of zero temperature ($T = 0$), the integration in (3.254) is easity performed using (3.250), and as a result we obtain for the spectral distribution of the density fluctuations

$$\langle \delta \varrho^2 \rangle_{\mathbf{k},\omega} = \frac{3\pi}{4} \frac{\varrho_0}{E_F}$$
$$\times \begin{cases} \dfrac{k_F}{k} \dfrac{|\omega|}{E_F}, & -\dfrac{k}{k_F}\left(2 - \dfrac{k}{k_F}\right) < \dfrac{\omega}{E_F} < 0; \\ \dfrac{k_F}{k} \left\{ 1 - \dfrac{1}{4} \left(\dfrac{k}{k_F} + \dfrac{k_F}{k} \dfrac{\omega}{E_F} \right)^2 \right\}, & -\dfrac{k}{k_F}\left(2 + \dfrac{k}{k_F}\right) < \dfrac{\omega}{E_F} < -\dfrac{k}{k_F} \left| \dfrac{k}{k_F} - 2 \right|, \end{cases} \tag{3.255}$$

where E_F is the Fermi energy ($E_F = k_F^2/2M$). We note that, for $T = 0$, only fluctuations with negative frequencies are possible, since the system of Fermi particles is in the ground state.

Figure 3.12 indicates the regions of $u \equiv |\omega|/E_F$ and $q \equiv k/k_F$ in which the different expressions for $\langle \delta \varrho^2 \rangle_{\mathbf{k}\omega}$ apply. Figure 3.13 shows the spectral distribution of the density fluctuation of a degenerate perfect Fermi gas as a function of $u = |\omega|/E_F$ at different values of $q = k/k_F$.

Integrating (3.255) over the frequencies, it is not difficult to obtain the following expression for the correlation function of simultaneous fluctuations in an the

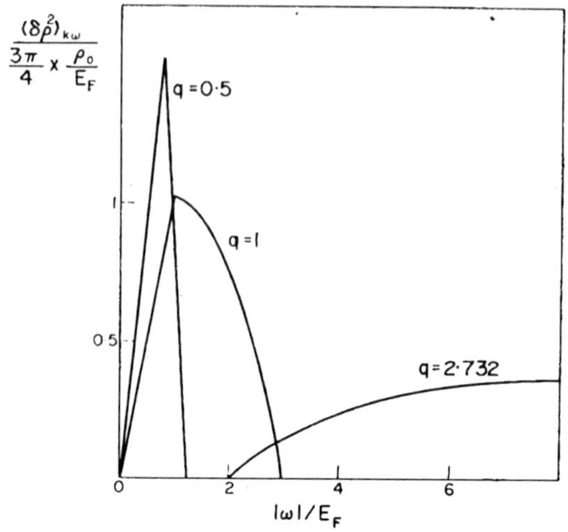

Figure 3.13. Spectral distribution of the density fluctuations in a degenerate perfect Fermi gas.

perfect Fermi gas at $T = 0$:

$$\langle \delta \varrho^2 \rangle_{\mathbf{k}} = \varrho_0 \times \begin{cases} \frac{1}{4}\frac{k}{k_F}\left(3 - \frac{1}{4}\frac{k^2}{k_F^2}\right), & k < 2k_F, \\ 1, & k > 2k_F. \end{cases} \tag{3.256}$$

The fact that $\langle \delta \varrho^2 \rangle_{\mathbf{k}}$ differs ϱ_0 for $k < 2k_F$ is due to the effect of the Pauli principle.

3.8. Determine the optical potential describing the interaction of a nucleon with nuclear substance assuming that the refraction and absorption of the nucleon wave in the nuclear substance is caused by processes of multiple scattering by individual nucleons (Lax, 1951).

We shall consider the motion of a nucleon in a system consisting of randomly distributed scatterers. Denoting the nucleon wave function by Ψ, we writte the Schrödinger equation in the form

$$\left\{ -\frac{\hbar^2}{2M}\nabla^2 + \sum_i V_i + E \right\} \Psi = 0, \tag{3.257}$$

where V_i is the interaction potential of the nucleon with an individual nucleon of the nuclear substance. (We note that the wave function Ψ depends, generally speaking, on the coordinates of both the scattered nucleon and the scattering nucleons.)

We shall assume that, in the absence of scatterers, the nucleon wave function has the form of a plane wave with wave vector $\mathbf{k}$. In the presence of one scattering nucleon, a scattered wave arises, which at large distances from the scatterer has the form of a spherical outgoing wave with amplitude

$$f(\mathbf{k}, \mathbf{k}') = -\frac{M}{4\pi\hbar^2} \int d^3 r e^{-i\mathbf{k}' \cdot \mathbf{r}} V(\mathbf{r}) \Psi_k(\mathbf{r}), \tag{3.258}$$

where $\mathbf{k}'$ is the wave vector of the nucleon after the scattering and $\Psi_k(\mathbf{r})$ is the exact wave function of the nucleon in the presence of one scatterer positioned at the coordinate origin. Fourier-transforming (3.258) with respect to $\mathbf{k}'$ and introducing the new integration variable $\mathbf{q} = \mathbf{k}' - \mathbf{k}$, we obtain

$$V(r)\Psi_k(\mathbf{r}) = -\frac{\hbar^2}{2\pi^2 M}\int d^3q e^{i\mathbf{q}\cdot\mathbf{r}} f(\mathbf{k}, \mathbf{k}+\mathbf{q})\Psi_k^0(\mathbf{r}), \tag{3.259}$$

where $\Psi_k^0(\mathbf{r})$ is the undistorted wave function. If the scattering nucleon is at the point $\mathbf{r}_i$, the scattering potential depends upon the difference $\mathbf{r}-\mathbf{r}_i$ and the equality (3.259) should be rewritten in the form

$$V(\mathbf{r}-\mathbf{r}_j)\Psi_k(\mathbf{r}) = -\frac{\hbar^2}{2\pi^2 M}\int d^3q e^{i\mathbf{q}\cdot(\mathbf{r}-\mathbf{r}_j)} f(\mathbf{k}, \mathbf{k}+\mathbf{q})\Psi_k^0(\mathbf{r}). \tag{3.260}$$

In the presence of a large number of scattering nucleons, the expression (3.260) must summed over all the scattering nucleons:

$$\sum_i V(\mathbf{r}-\mathbf{r}_i)\Psi_k = -\frac{\hbar^2}{2\pi^2 M}\sum_i \int d^3q e^{i\mathbf{q}\cdot(\mathbf{r}-\mathbf{r}_i)} f(\mathbf{k}, \mathbf{k}+\mathbf{q})\Psi_k^{(0)i}. \tag{3.261}$$

In the left-hand side of this equality, we have the exact wave function Ψ_k, which takes account of the effect of all the scatterers and describes the scattering of the incident nucleon in the nuclear substance. The right-hand side of the equality contans the wave function $\Psi_k^{(0)i}$, which does not take into account the effect of the individual scatterer i (it is the effective incident wave with respect to the scatterer i) but takes into account the effect of all the other scatterers. If the number of scattering nucleons is sufficiently great ($A \gg 1$), the influence of the one nucleon on the wave function can be neglected. Therefore, in the right-hand side of the equality (3.261), the wave function $\Psi_k^{(0)i}$, undistorted by the individual nucleon i, can be replaced by the exact wave function Ψ_k:

$$\Psi_k^{(0)i} \to \Psi_k, \quad A \gg 1.$$

Noting furthermore, that the scattering nucleons do not occupy well-defined positions but are randomly distributed in space, we must average the wave function over all possible distributions of the scatterers. The wave function thus averaged will depend only on the coordinates of the scattered nucleon, and will be written as $\Psi_k(\mathbf{r})$.

If we neglect the dependence of the wave function on the coordinates of the scattering nucleons, the sum in the right-hand side of the equality (3.261) reduces to the Fourier transform of the nucleon density in the nuclear substance:

$$\sum_i e^{-i\mathbf{q}\cdot\mathbf{r}_i} = \varrho(\mathbf{q}); \tag{3.262}$$

Thus, we obtain

$$\sum_i V(\mathbf{r}-\mathbf{r}_i)\Psi_k(\mathbf{r}) = \mathcal{V}(\mathbf{r})\Psi_k(\mathbf{r}), \tag{3.263}$$

where

$$\mathcal{V}(\mathbf{r}) = -\frac{\hbar^2}{2\pi^2 M}\int d^3q e^{i\mathbf{q}\cdot\mathbf{r}}\varrho(\mathbf{q})f(\mathbf{k},\mathbf{k}+\mathbf{q}). \tag{3.264}$$

If the relation (3.263) is substituted into (3.257), the Schrödinger equation describing the motion of a nucleon in the nuclear substance reduces to the one-particle equation

$$\left\{-\frac{\hbar^2}{2M}\nabla^2 + \mathcal{V}(\mathbf{r}) - E\right\}\Psi_k(\mathbf{r}) = 0. \tag{3.265}$$

The quantity $\mathcal{V}(\mathbf{r})$ is usually called the *optical potential.* According to (3.264), the optical potential is expressed in terms of the nucleon-nucleon scattering amplitude $f(\mathbf{k},\mathbf{k}+\mathbf{q})$ and the density $\varrho(q)$ of the nucleon distribution. Since, generally speaking, the nucleon-nucleon scattering amplitude is complex, the optical potential is also complex. The real part of the optical potential describes the refraction of the nucleon wave in the nuclear substance and the imaginary part characterizes the absorption of the wave.

In the case of a uniform distribution of nuclear substance, ($\varrho(r) = \varrho_0 = const$) from (3.264) we obtain for the optical potential

$$\mathcal{V}_0 = -\frac{4\pi\hbar^2}{M}\varrho_0 f(0), \tag{3.266}$$

where $f(0)$ is the elastic scattering amplitude at zero angle. The wave function describing the motion of a nucleon with a given momentum in the nuclear substance has in this case the form of a plane wave:

$$\Psi_k = e^{i\mathbf{k}\cdot\mathbf{r}-(i/\hbar)Et}, \quad E = \frac{\hbar^2k^2}{2M} + \mathcal{V}_0. \tag{3.267}$$

Writing the potential (3.266) in the form $\mathcal{V}_0 = -(V_0 + iW_0)$, we have for the refractive potential V_0 and the absorptive potential W_0:

$$V_0 = \frac{4\pi\hbar^2}{M}\varrho_0 \mathrm{Re}\, f(0), \quad W_0 = \frac{4\pi\hbar^2}{M}\varrho_0 \mathrm{Im}\, f(0). \tag{3.268}$$

3.9. Determine the mean free path of a nucleon in nuclear matter.

We shall consider the motion of a nucleon within a nucleus, assuming the nucleus to be sufficiently large ($A \gg 1$) and neglecting surface effects. Assuming also that the distribution of nuclear matter is uniform ($\varrho(r) = \varrho_0$), we can find the mean free path Λ of a nucleon in nuclear matter by considering the scattering of a nucleon wave by random density fluctuations in the nuclear matter (Sitenko,1962). In fact, in an unbounded uniform medium, any wave propagates without scattering; however, any violation of the uniformity of the properties leads to the appearance of scattering. Therefore, the scattering of a nucleon during its motion in nuclear matter can be related to the scattering of the nucleon wave by random density fluctuations in the nuclear matter. This scattering of the nucleon wave can be taken into account by substituting into the expression (3.264) for the optical potential the density ϱ in the form

$$\varrho = \varrho_0 + \delta\varrho, \tag{3.269}$$

where $\delta\varrho$ are the random density fluctuations, which depend both on the coordinates and on the time.

If we assume that $|\delta\varrho| \ll \varrho_0$, the cross-section for scattering of nucleons by the density fluctuations can be calculated without difficulty by using perturbation theory. By performing statistical averaging over the fluctuations, we can finally obtain the following expression for the differential cross-section for scattering of the nucleons (the differential scattering coefficient), per unit volume of nuclear matter.

$$d\Sigma = \frac{2}{\pi M k} |f(\mathbf{k}, \mathbf{k}')|^2 \langle \delta\varrho^2 \rangle_{q\omega} d^3k', \tag{3.270}$$

where $\mathbf{k}'$ and E' are the momentum and energy of the nucleon after scattering; $\mathbf{q} = \mathbf{k}' - \mathbf{k}$, $\omega = E' - E$, and $\langle \delta\varrho^2 \rangle_{q\omega}$ is the spectral distribution of the density fluctuations in nuclear matter. (The nucleon energy is reckoned from the centre of the Fermi sphere.) Integrating (3.270) over $\mathbf{k}'$, we can determine the total nucleon scattering coefficient Σ in nuclear matter. Clearly, it is reciprocal of the coefficient Σ which determines the nucleon mean free path in nuclear matter:

$$\Lambda = 1/\Sigma. \tag{3.271}$$

Assuming for simplicity that the nucleon scattering amplitude is constant ($f(\mathbf{k}, \mathbf{k}') = f_0$), we can represent the total scattering coefficient Σ in the form

$$\Sigma = 4\pi\varrho_0 |f_0|^2 P, \tag{3.272}$$

where P is a dimensionless coefficient characterizing the fluctuation properties of nuclear matter:

$$P = \frac{3}{4Mkk_F^3} \int\limits_{|\mathbf{k}+\mathbf{q}| \geq k_F} d^3q \langle \delta\varrho^2 \rangle_{q\omega}. \tag{3.273}$$

Noting that $\omega = (q^2 + 2\mathbf{k} \cdot \mathbf{q})/2M$, we can reduce the integration over $\mathbf{q}$ to integration over q and ω:

$$P = \frac{3\pi}{2k^2k_F^3} \int\limits_{-\frac{k^2-k_F^2}{2M}}^{0} d\omega \int\limits_{k-\sqrt{k^2+2M\omega}}^{k+\sqrt{k^2+2M\omega}} dq\, q \langle \delta\varrho^2 \rangle_{q\omega}. \tag{3.274}$$

Using the spectral distribution (3.255), we obtain for a degenerate Fermi gas at zero temperature

$$P(\varepsilon) = \begin{cases} 1 - \dfrac{7}{5\varepsilon} + \dfrac{2}{5\varepsilon}(2-\varepsilon)^{5/2}, & \varepsilon < 2, \\ 1 - \dfrac{7}{5\varepsilon} & \varepsilon > 2, \end{cases} \tag{3.275}$$

where ε is the ratio of the energy of the incident nucleon to the Fermi energy ($\varepsilon = E/E_F$). The expression (3.275) characterizes the decrease of the nucleon scattering coefficient Σ in nuclear matter (or the increase of the nucleon mean free path Λ in nuclear matter) due to the effect of the Pauli principle. Figure 3.14 shows the dependence of the quantity $P(\varepsilon)$ on the nucleon energy $\varepsilon = E/E_F$.

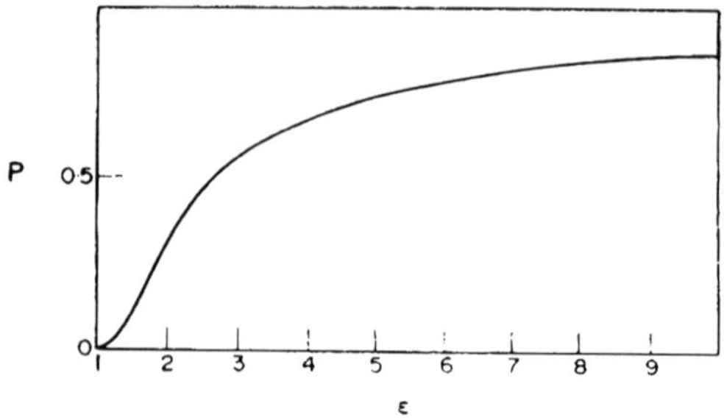

Figure 3.14. Dependence of the quantity $P(\varepsilon)$, which determines the lowering of the nucleon scattering coefficient in nuclear matter due to the Pauli principle, on the nucleon energy.

The differential cross-section for scattering of a nucleon in nuclear matter is determined by the formula

$$\left.\begin{aligned}
&\frac{1}{\Sigma}\frac{d\Sigma}{d\varepsilon' do} = \frac{3}{4\pi P(\varepsilon)}\sqrt{\frac{\varepsilon'}{\varepsilon}}\varphi(\varepsilon',\vartheta);\\
&\varphi(\varepsilon',\vartheta) = \frac{1}{2\sqrt{A(\varepsilon,\varepsilon')}}\times\begin{cases}\varepsilon-\varepsilon', & A(\varepsilon,\varepsilon') > B(\varepsilon,\varepsilon');\\ 1-\dfrac{B(\varepsilon',\varepsilon)}{A(\varepsilon,\varepsilon')}, & B(\varepsilon,\varepsilon') > A(\varepsilon,\varepsilon') > B(\varepsilon',\varepsilon);\end{cases}\\
&A(\varepsilon,\varepsilon') = \varepsilon+\varepsilon'-2\sqrt{\varepsilon\varepsilon'}\cos\vartheta; \quad B(\varepsilon,\varepsilon') = \varepsilon(\sqrt{\varepsilon}-\sqrt{\varepsilon'}\cos\vartheta)^2,
\end{aligned}\right\} \tag{3.276}$$

where ϑ is the nucleon scattering angle and $\varepsilon' = E'/E_F$.

Figure 3.15 shows the energy distribution of the scattered nucleons for different values of the scattering angle, when the incident nucleons have energy $\varepsilon = 2$.

The formulae (3.275) and (3.276) can also be obtained from a microscopic treatment of the successive collisions of the incident nucleon with Fermi-distributed intra-nuclear nucleons (Hayakawa, Kawai and Kikuchi, 1955).

3.10. Find the energy-level distribution of the excited states in nuclear matter.

Nuclear matter is a system with a large number of degrees of freedom and can be excited in many different ways; to describe its properties, therefore, it is convenient to use a statistical approach (Landau, 1937; Bethe, 1947). We shall characterize nuclear matter by a temperature Θ and consider the following thermodynamic functions: the entropy S, the excitation energy U and the free energy F. If the energy spectrum of the system is known, the free energy F is given by the relation

$$e^{-F/\theta} = \sum_\alpha e^{-E_\alpha/\theta}, \tag{3.277}$$

where the summation is taken over all the states of the system. Assuming that the energy spectrum of nuclear matter is continuous and denoting the density of energy levels by $\varrho(E)$, we rewrite the expression (3.277) in the form

$$e^{-F/\Theta} = \int dE\,\varrho(E)e^{-E/\Theta}. \tag{3.278}$$

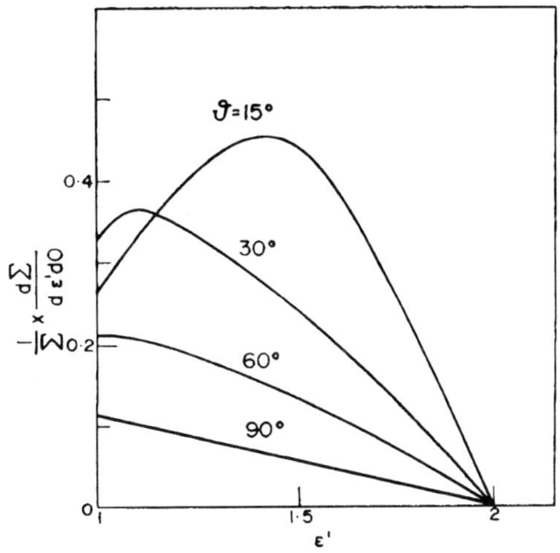

Figure 3.15. Energy distribution of the scattered nucleons in nuclear matter.

The mean excitation energy U of the system is defined by relation

$$U = \frac{\int dE\,\varrho(E)e^{-E/\Theta}E}{\int dE\,\varrho(E)e^{-E/\Theta}}. \tag{3.279}$$

This relation establishes the connection between the excitation energy U of the system and the temperature Θ.

Differentiating the equality (3.278) with respect to the quantity Θ^{-1} and using (3.278), we easily find the usual thermodynamic relation

$$U = F + \Theta S, \tag{3.280}$$

where $S = -dF/d\Theta$ is the entropy of the system. It follows directly from (3.272) that $dU = \Theta\, dS$, or

$$\frac{dS}{dU} = \frac{1}{\Theta}. \tag{3.281}$$

In the integrals (3.278) and (3.279), the main contribution is made by energy values E close to the mean excitation energy U; therefore,

$$e^{-F/\Theta} = \varrho(U)e^{-U/\Theta}\lambda(U), \tag{3.282}$$

where $\lambda(U)$ is some function of U that is slowly varying compared with the functions $e^{-F/\Theta}$, $\varrho(U)$ and $e^{-U/\Theta}$. Using (3.280), we find from (3.282)

$$\varrho(U) = \frac{1}{\lambda(U)}e^{S(U)}, \tag{3.283}$$

where $S(U)$ is the entropy corresponding to the excitation energy U.

In order to find $\lambda(U)$, we substitute $\varrho(E)$ in the form (3.283) into the relation (3.278) and take the slowly varying function $\lambda(E)$ at the point U outside the integral; we then obtain

$$\lambda(U) = \int dE e^{S(E)-(E-F)/\Theta} = \int dE e^{S(E)-S(U)-(E-U)/\Theta}. \tag{3.284}$$

Expanding the entropy $S(E)$ in the integrand of (3.284) in a series in powers of $E-U$ and confining ourselves to the quadratic terms in the expansion, we obtain

$$\lambda(U) = \int dE e^{(1/2)(d^2S/dU^2)(E-U)^2}. \tag{3.285}$$

Since U is a monotonically increasing function of Θ, the quantity $d^2S/d^2U < 0$. According to (3.281), we have

$$\frac{d^2S}{dU^2} = -\frac{1}{\Theta^2}\frac{d\Theta}{dU}. \tag{3.286}$$

Extending the limits of integration in (3.285) to $-\infty$ and $+\infty$, we obtain

$$\lambda(U) = \sqrt{2\pi}\Theta\sqrt{\frac{dU}{d\Theta}}. \tag{3.287}$$

Thus, we finally obtain the following formula for the density of the energy levels of the excited states:

$$\varrho(U) = \frac{1}{\sqrt{2\pi}}\frac{1}{\Theta}\sqrt{\frac{d\Theta}{dU}}e^{S(U)}, \tag{3.288}$$

The mean spacing between the levels of nuclear matter with excitation energy U is given by the expression

$$D(U) = \sqrt{2\pi\Theta}\sqrt{\frac{dU}{d\Theta}}e^{-S(U)}. \tag{3.289}$$

This formula acquires a definite meaning if we are given the relation between the entropy $S(U)$ and the excitation energy, or, which amounts to the same thing, between the energy U and the temperature Θ.

We shall assume that the energy and temperature are equal to zero for the ground state of nuclear matter. Low temperatures corresponds to excited states of nuclear matter with excitation energies small compared with the total binding energy. At low temperatures, the excitation energy of nuclear matter, like that for macroscopic bodies, will be determined by a power dependence on the temperature

$$U = a\Theta^n, \tag{3.290}$$

where a and n are certain constants. Using (3.281), we find

$$S = \frac{n}{n-1}a\Theta^{n-1}. \tag{3.291}$$

According to Nernst's theorem, for $\Theta = 0$ the entropy of the system must equal zero, and so $n > 1$. Using (3.288) and (3.291), we find

$$\varrho(U) = \frac{1}{\sqrt{2\pi n}} \left(\frac{a}{U^{n+1}} \right)^{1/2n} e^{[n/(n-1)]a^{1/n} U^{(n-1)/n}} \tag{3.292}$$

The formula (3.292) explains the characteristic features of the energy spectrum of nuclear matter, and in particular the increase in the density of levels with increasing excitation energy. We note that the formula (3.292) is inapplicable for very low excitation energies, for which we cannot assume that the energy spectrum of the system is continuous.

We find the simplest law for the density of levels by putting $\lambda(U) = \text{const}$. In this cas, $n = 2$ and the distribution of the density of levels is determined by the formula

$$\varrho(U) = C e^{2\sqrt{aU}}. \tag{3.293}$$

The constants a and C can be selected in such a way that the empirical data on the density of levels of nuclei are reproduced (Blatt and Weisskopf, 1952). From the data on the capture of slow neutrons by nuclei with odd A, it follows that, in the range of mass numbers $100 < A < 200$, the constants a and C have the following values: $a \approx A/17\,\text{MeV}^{-1}$, $C \approx 0.01\,\text{MeV}^{-1}$. Thus, for heavy nuclei ($A \approx 200$), for excitation energies $U \approx 10$ MeV, the temperature Θ is found to be of the order of 1 MeV, that is, the nucleus can indeed be assumed to be cold, as we assumed earlier.

We note that the distribution (3.293) corresponds to the case of the degenerate perfect Fermi gas. However, the perfect Fermi-gas model leads to an over-estimate of the constant $a(a_{id} \approx A/5.6\,\text{MeV}^{-1})$, in disagreement with the empirical data.

References

Balbutsev, E.B., Dymarz, R., Mikhailov, I.N. and Vaishvila, Z. (1981) *Phys. Lett*, **105 B**, 84.

Bethe, H.A. (1947) *Elementary Nuclear Theory*, Wiley, N.Y.

Bethe, H.A. and Bacher, R.F. (1936) *Rev. Mod. Phys.* **8**, 82.

Bethe, H.A. and Goldstone, J. (1957) *Proc. Roy. Soc.* **A 238**, 551.

Blatt, J.M. and Weisskopf, V.F. (1952) *Theoretical Nuclear Physics*, Wiley, N.Y.

Bogolyubov, N.N. (1958) *Dokl. Akad. Nauk SSSR* **119**, 224.

Brown, G.E. (1967) Unified Theory of Nuclear Models and Forces, North-Holland Publishing Company, Amsterdam.

Brueckner, K. (1959) "Theory of nuclear matter" (p. 49 in *The Many-body Problem*, ed. C. De Witt, Wiley, N.Y.).

Eisenberg J.M. and Greiner W., (1970) Nuclear Theory. Vol. 1. North-Holland Publishing Compani, Amsterdam, London.

Eisenberg J.M. and Greiner W., (1972) Nuclear Theory. Vol. 3. North-Holland Publishing Compani, Amsterdam, London.

Elton, L. R. B. (1961) *Nuclear Sizes*, Oxford University Press.

Gomes, L., Walecka, J. D. and Weiskopf, V. F. (1953); *Ann. Phys.* . (N.Y.) **3**, 241.

Green, A.E.S. (1955) *Nuclear Physics* , MeGraw-Hill, London.

Hayakawa, S., Kawai, M. and Kikuchi, K. (1955) *Progr. Theor. Phys.*

13, 415.
Heisenberg, W. (1932) *Z. Phys.* **77**, 1.
Van Hove, L. (1954) *Phys. Rev.* **95**, 249.
Hofstadter, R. (1956) *Rev. Mod. Phys.* **28**, 214.
Kolomietz, V.M. and Tang Henry, H.K. (1981) *Physica Scripta* **24**, 915.
Landau, L.D. (1937) *Zh. Eksp. Teor. Fiz.* **7**, 819 (translation in the Collected Papers of L.D.Landau, Peragamon Press, Oxford, 1965).
Lane A.M. (1964) Nuclear Theory, W.A. Benjamin, New York, Amsterdam.
Lax, M., (1951) *Rev. Mod. Phys.* **23**, 287.
Lee, T.D. and Yang, C.N. (1957) *Phys. Rev.* **105**, 1119.
March, N.H., Young W.H. and Sampanthar S. (1976) The Many-Body Problem in Quantum Mechanics. Cambridge, At the University Press.
Sitenko, A.G. (1962) *Zh. Eksp. Theor. Fiz.* **43**, 319 [*Sov. Phys.* – JETP **16**, 228 (1962)].
Solov'ev, V.G. (1971) Theory of Composed Nuclei, Nauka, Moscow.
Wigner, E and Settz, F. (1933) *Phys. Rev.*, **43**, 804.

CHAPTER 4

SHELLS IN NUCLEI

4.1. Motion of a Nucleon in the Self-consistent Field of the Nucleus

Independence of the motion of nucleons in the nucleus and the self-consistent potential. Lying at the basis of the *shell model of the nucleus* is the assumption that the nucleons in the nucleus move almost independently of each other, despite the existence of the strong interaction between free nucleons. The interaction of nucleons with each other reduces in the shell model to the interaction of the individual nucleons with a self-consistent field. Generally speaking, each nucleon moves in the field created by the other nucleons. In the shell model, it is assumed that this field is the same for all the nucleons. The potential of this self-consistent field is assumed to be static and spherically symmetric. Because of the short-range character of the nuclear forces, the potential of the self-consistent field varies in almost the same way as the nuclear density: it is constant inside the nucleus and goes to zero outside the nucleus. As a result, the one-particle wave functions depend on the distance from the centre of the nucleus and go to zero outside the nucleus.

The possibility of introducing the self-consistent field and of treating the motion of the individual nucleons independently in this field is, to a considerable degree, a consequence of the Pauli principle, which is obeyed by the nucleons. The effect of the Pauli principle is that the effective interaction between the nucleons is considerably weakened.

Having chosen the *self-consistent field potential* , we can calculate the spectrum of the energy levels corresponding to the different states of an individual nucleon. Depending on the form of the potential, the one-particle spectrum breaks down into groups of closely spaced levels, separated by large energy intervals. Since the Pauli principle forbids two nucleons to be in the same state, the nuclear ground state, which is characterized by the lowest energy, corresponds to a configuration in which the nucleons occupy the lowest one-particle levels. The most stable configuration (a field shell) is obtained in the case when there is a large energy gap between the highest occupied level and the lowest unfilled level. In the general case, the nucleus consists of a definite number of filled shells and a few outer nucleons in an unfilled shell. [1]

There exist several variants of the shell model of the nucleus. In the simplest of these variants, the *one-particle shell model* , which is used to describe nuclei with odd values of the mass number A, it is assumed that all paired nucleons (including nucleons in the unfilled outer shell) form an inert core with zero spin, and the properties of the nucleus are entirely determined by the state of the last odd nucleon. In the *many-particle shell model* , the properties of the nucleus are determined by all the nucleons, both in the unfilled shell and in the filled shells.

Equation of motion of a nucleon in the self-consistent field of the nucleus. We shall begin our study of the shell model of the nucleus by treating the simplest one-particle model. We shall assume that the potential $V(\mathbf{r})$ of the self-consistent field is characterized by spherical symmetry. We shall consider the motion of an individual nucleon in this field. The wave function $\varphi(\mathbf{r})$ describing the motion of the nucleon satisfies the Schrödinger equation

$$\{-(\hbar^2/2M)\Delta + V(r) - E\}\varphi(\mathbf{r}) = 0, \tag{4.1}$$

where Δ is the Laplacian operator and E is the energy of the nucleon. Because of the spherical symmetry of the potential $V(r)$, the angular momentum of the nucleon is a constant of the motion and the wave function $\varphi(\mathbf{r})$ can be represented in the form of a product of a radial function with an angular function:

$$\varphi_{nlm}(r) = \varphi_{nl}(r)Y_{lm}(\vartheta, \varphi). \tag{4.2}$$

The angular function $Y_{lm}(\vartheta, \varphi)$ is the eigenfunction of a state in which the angular momentum l and its component m have well-defined values. The radial function $\varphi_{nl}(r)$ satisfies the equation

$$\frac{d^2\varphi_{nl}}{dr^2} + \frac{2}{r}\frac{d\varphi_{nl}}{dr} + \frac{2M}{\hbar^2}\left[E_{nl} - V(r) - \frac{\hbar^2}{2M}\frac{l(l+1)}{r^2}\right]\varphi_{nl} = 0 \tag{4.3}$$

and the boundary conditions that the function $\varphi_{nl}(r)$ be finite at the point $r = 0$ and go to zero at infinity ($r \to \infty$). The energy eigenvalues $E_{nl} < 0$ of

[1] The hypothesis of the existence of shell structure in nuclei was first put forward by Elsasser (1934). The modern theory of nuclear shells is based on the work of Goeppert-Mayer (1949, 1950) and Haxel, Jensen and Suess (1949, 1950), who assumed the existence of strong spin-orbit interaction between the nucleons in nuclei. A detailed account of the theory of nuclear shells is given in the monographs by Goeppert-Mayer and Jensen (1955), Heisenberg (1949) and De-Shalit and Talmi (1963), and also in the review by Elliott and Lane (1957), which contains a detailed bibliography.

the bound states of the nucleon are determined from the condition that the radial function $\varphi_{nl}(r)$ go to zero at infinity. Here n is the *principal quantum number*, which characterizes the energy eigenvalue E_{nl} of the nucleon and is equal to the number of nodes of the radial function $\varphi_{nl}(r)$ plus one.

Since the energy eigenvalues E_{nl} do not depend on the quantum number m, each level is characterized by $(2l+1)$-fold degeneracy. In addition, each level is doubly degenerate with respect to the spin, and, if we neglect the Coulomb interaction, is also doubly degenerate with respect to the charge. Thus, the mean spacing between the degenerate levels is found to be considerably greater than the spacing predicted by the statistical theory.

The radial functions $\varphi_{nl}(r)$, like the sequence of one-particle levels E_{nl}, depend on the shape of the potential $V(r)$. Information on the shape of the potential can be obtained from experiments on the scattering of nucleons by nuclei. It follows from these experiments that the depth of the potential well $V(r)$ is almost constant inside the nucleus and decreases rapidly to zero close to the nuclear surface.

The problem of the motion of a nucleon in the self-consistent field $V(r)$ can be solved in explicit form by approximating the true potential $V(r)$ by an oscillator potential well or a square potential well. The results for the true potential can then be obtained by interpolation from the results obtained for these idealized potentials.

Properties of the radial wave functions. We shall examine certain general properties of the radial wave functions of a nucleon moving in a central self-consistent field.

Representing the radial wave function $\varphi_{nl}(r)$ in the form

$$\varphi_{nl}(r) = u_{nl}(r)/r, \tag{4.4}$$

we obtain from (4.3) a differential equation for the function $u_{nl}(r)$

$$\frac{d^2u_{nl}}{dr^2} + \frac{2M}{\hbar^2}\left[E_{nl} - V(r) - \frac{\hbar^2}{2M}\frac{l(l+1)}{r^2}\right]u_{nl} = 0, \tag{4.5}$$

which contains no first derivatives. The function $u_{nl}(r)$ goes to zero at $r = 0$ and, for bound states, must go to zero at infinity.

We note the following interesting circumstance. If we rewrite eqn. (4.5) in the form

$$V(r) = \frac{\hbar^2}{2M}\left[\frac{u''_{nl}(r)}{u_{nl}(r)} - \frac{l(l+1)}{r^2}\right] + E_{nl}, \tag{4.6}$$

it becomes clear that the values of the function $u_{nl}(r)$ for any value of l are sufficient to determine the potential to within an additive constant.

We shall assume that the potential $V(r)$, corresponding to attractive forces ($V(r) \leqslant 0$), varies monotonically with distance

$$| V(r_1) | \geqslant |V(r_2)| \qquad \text{for } r_1 < r_2, \tag{4.7}$$

going to zero at infinity, $V(\infty) = 0$. In addition, we shall assume that

$$\lim_{r\to 0} r^2 V(r) = 0, \tag{4.8}$$

and so the ground state will be characterized by a finite energy value. We introduce the effective potential

$$V_l(r) = V(r) + (\hbar^2/2M)l(l+1)/r^2. \tag{4.9}$$

For $l = 0$ the effective potential (4.9) coincides with $V(r)$, and for $l \neq 0$ the dependence of the effective potential $V_l(r)$ on the distance r is similar to that of the interaction potential of two atoms of a gas for which there is repulsion at short distances (Fig. 4.1). For a given energy $E_{nl} < 0$, for $l = 0$ there is one turning point $r_0(E_{nl} = V_l(r_0))$, and for $l \neq 0$ there are two turning points r_0 and r_0' ($r_0' < r_0$). Obviously,

$$\left.\begin{aligned} &E_{nl} - V_l(r) > 0 \qquad \text{for } r_0' < r < r_0, \\ &E_{nl} - V_l(r) < 0 \qquad \text{for } r > r_0. \end{aligned}\right\} \tag{4.10}$$

In the first case of (4.10), the function $u_{nl}(r)$ is concave to the r-axis, and in the second case is convex to the r-axis. Therefore, in the region $r_0' < r < r_0$ the function $u_{nl}(r)$ has an oscillatory dependence on r, and in the region $r > r_0$ the function $u_{nl}(r)$ decreases in magnitude with increasing r. This means that the nodes of the function $u_{nl}(r)$ can lie only in the region $r < r_0$, that is, within the range of the attractive potential. Outside the range of the potential, the function $u_{nl}(r)$ falls off exponentially with increasing r, with a characteristic decay length equal to $\sqrt{\hbar^2/2M|E_{nl}|}$.

If there are two bound states with the same orbital angular momentum l but different values n_1 and n_2 ($n_2 > n_1$), then $E_{n_2 l} > E_{n_1 l}$; therefore, the state with the larger number of nodes is the least strongly bound. In fact, as the number of nodes (distributed within the range of the potential) of the radial function increases, the wavelength of the particle decreases; this corresponds to an increase of the kinetic energy of the particle, and leads to an increase of the total energy E_{nl} with increasing n.

If there are two states with the same number n of nodes in the radial functions but with different orbital angular momenta l_1 and l_2 ($l_2 > l_1$), then the energy $E_{nl_2} > E_{nl_1}$, that is, the state with the lower value of the orbital angular momentum has the lower energy. (The S-state has the

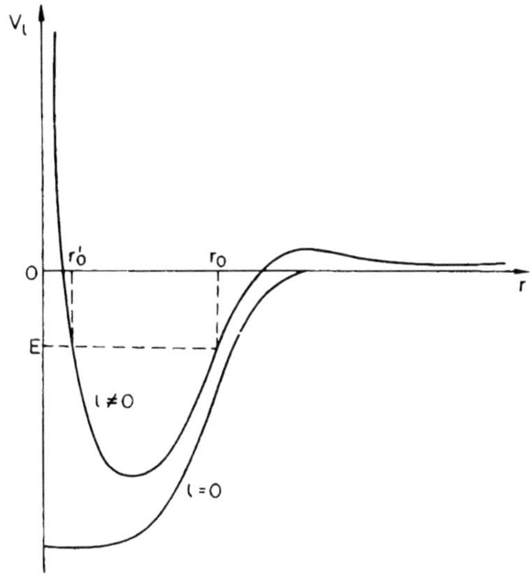

Figure 4.1. Dependence of the effective potential $V_l(r)$, which takes account of the centrifugal energy, on the distance r.

lowest energy.) It is not difficult to convince oneself of the validity of this statement by noting that the centrifugal energy

$$\frac{\hbar^2}{2M}\frac{l(l+1)}{r^2}$$

increases with increasing orbital angular momentum l. Since the potential energy is practically the same in the cases l_1 and l_2, it follows from this that the inequality $E_{nl_2} > E_{nl_1}$ is valid for $l_2 > l_1$.

Spin-orbit interaction. The presence of the spin of the nucleon leads to the appearance of an additional interaction as the nucleon moves in the *self-consistent field of the nucleus* . The energy of this interaction should depend not only on the vectors **s** and **r** but also on the velocity **v** of the nucleon, since it is impossible to construct a scalar expression from the vectors **s** and **r**, which are different in character (**s** is an axial vector and **r** is a true vector). A scalar expression can be constructed from the three vectors **s**, **r** and **v** as follows:

$$f(r)[\mathbf{r} \times \boldsymbol{v}]\cdot\boldsymbol{s} \tag{4.11}$$

where $f(r)$ is some scalar function of the coordinates. Since the vector product $\mathbf{r} \times \mathbf{v}$ is proportional to the nucleon orbital angular momentum $\boldsymbol{l}$, (4.11) can be regarded as a *spin-orbit interaction.* This interaction is relativistic in character (as $v \to 0$ it disappears), and turns out to be considerably weaker than the interaction of the nucleon with the self-consistent field.

To determine the explicit form of the spin-orbit interaction energy, we shall make use of an analogy with electrodynamics (Heisenberg, 1949). We consider the interaction of a particle, moving with momentum $\boldsymbol{p}$ and possessing a magnetic moment $\boldsymbol{\mu} = (e\hbar/Mc)s$, with an electrostatic field $\boldsymbol{E} = -(1/e)\operatorname{grad} V$. In the reference frame associated with the moving particle, the magnetic field intensity is equal to $\boldsymbol{H} = (1/c)[\boldsymbol{p}/M) \times \boldsymbol{E}]$, and, consequently, the magnetic interaction energy is given by the expression

$$U = -\boldsymbol{\mu} \cdot \boldsymbol{H} = -\frac{\hbar}{M^2c^2}[\boldsymbol{p} \times \nabla V]\cdot\boldsymbol{s}. \tag{4.12}$$

Noting that, for a spherically symmetric potential, $\operatorname{grad} V = (\mathbf{r}/r)(\partial V/\partial r)$ and that $\mathbf{r} \times \boldsymbol{p} = \hbar\boldsymbol{l}$, we can represent the energy (4.12) in the form

$$U = (\hbar/Mc)^2(1/r)(\partial V/\partial r)(\boldsymbol{l} \cdot \boldsymbol{s}). \tag{4.13}$$

In the case of a nucleon moving in the self-consistent field $V(r)$ of the nucleus, the spin-orbit interaction can be chosen by analogy with (4.13), provided that we introduce into (4.13) an additional dimensionless factor $-\lambda$, characterizing the magnitude of the spin-orbit coupling:

$$U = -\lambda(\hbar/Mc)^2(1/r)(\partial V/\partial r)(\boldsymbol{l} \cdot \boldsymbol{s}). \tag{4.14}$$

(It follows from an analysis of the experimental data that the spin-orbit coupling constant $\lambda \approx 10$.)

Because of the *spin-orbit interaction*, the *orbital and spin angular momenta* $\boldsymbol{l}$ and $\boldsymbol{s}$ of the *nucleon* are not separately conserved; only the *total angular momentum* $\mathbf{j} = \mathbf{l} + \mathbf{s}$ of the nucleon is conserved. Noting that $2\boldsymbol{j} \cdot \boldsymbol{s} = j(j+1) - l(l+1) - s(s+1)$, we can represent the spin-orbit interaction energy of a nucleon in a state with a well-defined value of the total angular momentum in the form

$$U = -(a/r)(\partial V/\partial r), \tag{4.15}$$

where

$$a = (\lambda/2)(\hbar/Mc)^2 \times \begin{cases} l, & j = l + \frac{1}{2}; \\ -(l+1), & j = l - \frac{1}{2}. \end{cases} \tag{4.16}$$

The Schrödinger equation describing the motion of a nucleon in the self-consistent field $V(r)$ in the presence of spin-orbit interaction has the form

$$\{-(\hbar^2/2M)\nabla^2 + V(r) - (a/r)(\partial V(r)/\partial r) - E\}\psi = 0. \tag{4.17}$$

The wave function, that is, the solution of eqn. (4.17), can be represented in the form of a product of radial and spin-angle functions:

$$\psi_{nljm} = \varphi_{nl}(r)\mathcal{Y}_{l\frac{1}{2}jm}. \tag{4.18}$$

The spin-angle function

$$\mathcal{Y}_{lsjm} = \sum_{m_l+m_s=m} (lm_l sm_s|jm) Y_{lm_l}(\vartheta,\varphi)\chi_{sm}, \tag{4.19}$$

is the eigenfunction of a state in which the total angular momentum j and its component m have well-defined values. In this state, the squares of the orbital and spin angular momenta, characterized by the quantum numbers l and s, also have well-defined values.

The radial function $\varphi_{nl}(r)$ in the presence of spin-orbit interaction satisfies the equation

$$\frac{d^2\varphi_{nl}}{dr^2} + \frac{2}{r}\frac{d\varphi_{nl}}{dr} + \frac{2M}{\hbar^2}\left[E_{nlj} - V(r) + \frac{a_{lj}}{r}\frac{dV(r)}{dr} - \frac{\hbar^2}{2M}\frac{l(l+1)}{r^2}\right]\varphi_{nl} = 0. \tag{4.20}$$

Since the coefficient a_{lj} depends on the value j of the total angular momentum, the degeneracy of the energy levels with respect to j is lifted in the presence of spin-orbit interaction and the nucleon energy levels E_{nlj} depend on three quantum numbers n, l and j.

4.2. One-particle States

The isotropic-oscillator potential. In the case of light nuclei, a good approximation for the nuclear potential in the inner region of the nucleus is the *three-dimensional isotropic oscillator potential*

$$V(r) = -V_0 + \frac{1}{2}M\omega^2 r^2, \tag{4.21}$$

where V_0 is the depth of the potential well at the centre of the nucleus, M is the nucleon mass, and ω is a quantity having the dimensions of a frequency. In the region outside the nucleus the potential $V(r)$ ought to be put equal to zero, but for bound states the wave functions with such a potential differ hardly at all from the wave functions corresponding to the potential (4.21).

At first, we shall neglect the spin-orbit interaction. We substitute the formula (4.21) into (4.3) and replace the variable r by the dimensionless variable $\xi = \frac{M\omega}{\hbar}r^2$; we then rewrite the equation for the radial function in the form

$$\xi\frac{d^2\varphi}{d\xi^2}+\frac{3}{2}\frac{d\varphi}{d\xi}+\frac{1}{2}\left[\varepsilon-\frac{1}{2}\xi-\frac{l(l+1)}{2\xi}\right]\varphi=0, \tag{4.22}$$

where $\varepsilon=(E+V_0)/\hbar\omega$.

To determine the solution of this equation, we investigate the asymptotic behaviour of the function φ at large and small values of ξ. At sufficiently large values of ξ, by retaining only the terms proportional to ξ in eqn. (4.22), we have $d^2\varphi/d\xi^2-\frac{1}{4}\varphi=0$, whence it follows that $\varphi\sim\mathrm{e}^{-\xi/2}$ $(\xi\to\infty)$.

For small values of ξ, retaining only the term with the lowest power of ξ in the square brackets in eqn. (4.22), we have

$$\frac{d^2\varphi}{d\xi^2}+\frac{3}{2\xi}\frac{d\varphi}{d\xi}-\frac{l(l+1)}{4\xi^2}\varphi=0.$$

We seek the solution of the equation obtained in the form $\varphi\sim\xi^2$; in this case, for s we obtain the quadratic equation

$$s(s-1)+\frac{3}{2}s-\frac{l(l+1)}{4}=0.$$

Since φ should be finite at $\xi=0$, we must take the positive root $s=l/2$ of the quadratic equation.

Thus, the solution of eqn. (4.22) can be represented in the form

$$\varphi\sim\mathrm{e}^{-\xi/2}\xi^{l/2}w(\xi), \tag{4.23}$$

where $w(\xi)$ is the polynomial that is finite at $\xi=0$. Substituting (4.23) into (4.22), we obtain for w the equation

$$\xi w''+(l+\frac{3}{2}-\xi)w'+\frac{1}{2}(\varepsilon-l-\frac{3}{2})w=0.$$

The solution of this equation is a confluent hypergeometric function and, consequently,

$$w(\xi)=F\left[-\frac{1}{2}\varepsilon+\frac{1}{2}l+\frac{3}{4},\,l+\frac{3}{2};\,\xi\right]. \tag{4.24}$$

The confluent hypergeometric function $F(\alpha,\gamma;\xi)$ is defined by the series

$$F(\alpha,\gamma;\xi)=1+\frac{\alpha}{\gamma}\frac{\xi}{1!}+\frac{\alpha(\alpha+1)}{\gamma(\gamma+1)}\frac{\xi^2}{2!}+\frac{\alpha(\alpha+1)(\alpha+2)}{\gamma(\gamma+1)(\gamma+2)}\frac{\xi^3}{3!}+\ldots,$$

which is terminated if the argument α is a negative integer or zero. Thus, the solution (4.24) will be a polynomial only in the case when

$$-\frac{1}{2}\varepsilon+\frac{1}{2}l+\frac{3}{4}=-(n-1), \qquad n=1,2,3,\dots, \tag{4.25}$$

where $n-1$ is the integer determining the degree of the polynomial.

The condition (4.25) determines the eigenvalues of the energy of a particle in the oscillator well. Introducing the notation $2(n-1)+l=N$, we have

$$E_N=V_0+(N+\frac{3}{2})\hbar\omega, \qquad N=0,1,2,\dots, \tag{4.26}$$

that is, the energy depends only on the integer N, which is called the *principal quantum number* .

According to (4.26), the energy spectrum of a particle in an oscillator well consists of a series of equally spaced levels, separated from each other by the interval $\hbar\omega$. To each level E_N correspond several states with different values of l; the numbers l and N always have the same parity, and $l\leqslant N$. One can easily convince oneself that the degeneracy of the separate levels is equal to $\sum(2l+1)=\frac{1}{2}(N+1)(N+2)$.

The different states are usually denoted by means of a pair of numbers (nl); the spectroscopic method of notation is used: $1s$; $1p$; $2s, 1d$; $2p, 1f$; $3s, 2d, 1g$;... (the number n implies that the given value of l is appearing for the nth time in the sequence).

The normalized radial wave function corresponding to the state (nl) has the form

$$\varphi_{nl}(r)=C_{nl}\mathrm{e}^{-\alpha^2r^2/2}r^lF(1-n,l+\frac{3}{2};\alpha^2r^2), \tag{4.27}$$

where

$$C_{nl}=\frac{1}{\Gamma(l+\frac{3}{2}}\sqrt{\frac{2\Gamma(1+n+\frac{1}{2})}{\Gamma(n)}}\alpha^{l+3/2}, \qquad \alpha=\left(\frac{M\omega}{\hbar}\right)^{1/2}.$$

Table 4.1 shows the possible states for the first few energy levels characterized by the number N, and also gives the values of the maximum number of particles (either neutrons or protons) permitted by the Pauli principle in each level ($Z_N=(N+1)(N+2)$) and of the total number of particles $\sum_0^N Z_N=\frac{1}{3}(N+1)(N+2)(N+3)$. According to Table 4.1, in the case of an oscillator potential filled shells are obtained when the number of neutrons or protons is equal to 2,8,20,40,70,112, ... The *magic numbers* observed in experiment, however, coincide only with the first four of these values ($N\leqslant 3$) (these are ringed in Table 4.1). To obtain the values of the other magic numbers it is necessary to take the spin-orbit interaction

TABLE 4.1.

N	States (nl)	$Z_N = (N+1)(N+2)$	$\sum Z_N$
0	$1s$	2	2
1	$1p$	6	8
2	$2s, 1d$	12	20
3	$2p, 1f$	20	40
4	$3s, 2d, 1g$	30	70
5	$3p, 2f, 1h$	42	112
6	$4s, 3d, 2g, 1i$	56	168

into account. In the case of the oscillator potential (4.21), allowance for the spin-orbit interaction leads to a constant correction to the energy, equal to $-(a/r)(\partial V/\partial r) = aM\omega^2$. Since the quantity a depends on the orbital and total angular momenta l and j of the particle

$$a_{lj} = (\lambda/2)(\hbar/Mc)^2 \times \begin{cases} l, & j = l + \frac{1}{2}; \\ -(l+1), & j = l - \frac{1}{2}, \end{cases}$$

the degeneracy of the oscillator levels E_N is partially lifted and each level E_N breaks down into a series of sub-levels:

$$E_{Nlj} = -V_0 + (N + \frac{3}{2})\hbar\omega - a_{lj}M\omega^2. \tag{4.28}$$

The magnitude of the *spin-orbit shift* depends on l and increases with increasing l. Because of this, for large values of N the spin-orbit splitting can substantially alter the order in which the shells are filled from that determined by the number N alone. In fact, the sub-level corresponding to the maximum value $J_{\max} = l_{\max} + \frac{1}{2} = N + \frac{1}{2}$ of the total angular momentum of the particle is characterized by the largest shift towards energies typical of smaller values of N. Therefore, if the state with $j_{\max}$ is assigned to the $(N-1)$-shell, by taking into account the degeneracy of these states, equal to $2(N+1)$, we obtain for the number Z'_N of particles in the N-shell; $Z'_N = Z_N - 2(N+1) + 2(N+2)$ (for $N > 0$).

Table 4.2 gives the values of Z'_N and $\sum Z'_N$ for specific values of N (the magic numbers are ringed). Comparing the experimental values of the magic numbers with the sequences of numbers $\sum Z'_N$ and $\sum Z_N$, we see that the order in which the shells are filled is determined for small numbers of particles ($N \lesssim 50$) by the order of the oscillator levels, and for large numbers ($N \gtrsim 28$) by the effect of the *spin-orbit coupling*.

TABLE 4.2.

N	States $(nl)_j$	Z'_N	$\sum Z'_N$
0	$1s_{1/2}$	6	6
1	$1p_{1/2}, 1p_{3/2}$	8	14
2	$2s_{1/2}, 1d_{3/2}, 1d_{5/2}$	14	28
3	$2p_{1/2}, 2p_{3/2}, 1f_{5/2}, 1f_{7/2}$	22	50
4	$3s_{1/2}, 2d_{3/2}, 1g_{7/2}, 1g_{9/2}$	32	82
5	$3p_{1/2}, 3p_{3/2}, 2f_{7/2}, 1h_{9/2}, 1h_{11/2}$	44	126
6	$4s_{1/2}, 3d_{3/2}, 2g_{7/2}, 2g_{9/2}, 1i_{11/2}, 1i_{13//2}$	58	184

Determination of the absolute values of the spacings between the levels. Using the one-particle wave functions found, we can calculate directly the spatial distribution of the nucleon density in a nucleus. Clearly, the density is given by the expression $\varrho(\mathbf{r}) = \sum \varphi^*_{nljm}(\mathbf{r})\varphi_{nljm}(\mathbf{r})$, where the summation is taken over all the occupied states. In the case of filled oscillator shells, the summation over j and m can be performed by making use of the completeness of the spin and angle functions:

$$\varrho_N(r) = \sum_{N'=0}^{N} \sum_{l=0}^{N'} \frac{2l+1}{\pi} \varphi^2_{nl}(r), \qquad n = \frac{N'-l}{2} + 1. \tag{4.29}$$

As an example, we give the *spatial distribution of the nucleon density* for the nuclei ${}^4_2He$, ${}^{16}_8O$ and ${}^{40}_{20}Ca$ with filled neutron and proton shells:

$$\left.\begin{aligned} \varrho_0(r) &= \tfrac{4}{\pi^{3/2}}\alpha^3 e^{-\alpha^2 r^2}; \\ \varrho_1(r) &= \tfrac{4}{\pi^{3/2}}\alpha^3(1+2\alpha^2 r^2)e^{-\alpha^2 r^2}; \\ \varrho_2(r) &= \tfrac{2}{\pi^{3/2}}\alpha^3(5+4\alpha^4 r^4)e^{-\alpha^2 r^2}. \end{aligned}\right\} \tag{4.30}$$

The density is a maximum at the centre of the nucleus and decreases in Gaussian fashion with increasing distance from the centre.

Knowing the nuclear density distribution (4.29) we can find the absolute values of the *spacings between the oscillator levels* by comparing the calculated dimensions of nuclei with the experimental values of the nuclear radii. For example, by calculating the mean-square radii for ${}^4_2He$, ${}^{16}_8O$ and ${}^{40}_{20}Ca$ by means of (4.30) and comparing these with the experimental values of the root mean square radii ($R_{He} = 1{\cdot}61\times10^{-13}$ cm, $R_0 = 2{\cdot}65\times10^{-13}$ cm and $R_{Ca} = 3\cdot 52\times10^{-13}$ cm), we obtain $\hbar\omega_{He}$=24 MeV, $\hbar\omega_0 = 13\cdot 3$ MeV and $\hbar\omega_{Ca} = 10\cdot 1$ MeV.

In the case of nuclei with sufficiently large mass number A, we can establish the dependence of $\hbar\omega$ on A in general form. For this we make use of the fact that, for a nucleon in an oscillator well, the mean kinetic energy and the mean potential energy measured from the bottom of the well are equal. Therefore, the sum of the energies (measured from the bottom of the well) of all the nucleons in the nucleus can be expressed in terms of the value of the mean square radius: $\sum_i (E_i + V_0) = AM\omega^2\langle r^2\rangle$.

If the nucleus is in the ground state, all the lower levels are occupied and, consequently,

$$\sum_i (E_i + V_0) = \sum_0^N 2(N+1)(N+2)(N+\frac{3}{2})\hbar\omega,$$

$$A = \sum_0^N 2(N+1)(n+2).$$

Assuming the density distribution to be uniform (which is justified for large A), we put $\langle r^2\rangle = \frac{3}{5}R^2$ ($R = R_0 A^{1/3}$ is the experimental value of the nuclear radius). As a result, we obtain

$$\hbar\omega = \frac{5}{3}\frac{\hbar^2}{MR^2}\frac{\sum(N+1)(N+2)(N+\frac{3}{2})}{\sum(N+1)(N+2)}$$

and, confining ourselves in the summation to the terms of highest order, we have, approximately,

$$\hbar\omega \approx \frac{5}{4}\left(\frac{3}{2}\right)^{1/3}\frac{\hbar^2}{MR_0^2}A^{-1/3}.$$

Putting $R_0 = 1 \cdot 2 \times 10^{-13}$ cm, we finally obtain the following expression for the spacing between the levels in an oscillator potential:

$$\hbar\omega \approx 41 A^{-1/3}\ \text{MeV}. \tag{4.31}$$

The *spin-orbit interaction constant* λ can be determined by using the experimental data on the shift of the oscillator levels that arises from the spin-orbit interaction. According to (4.28), the energy correction due to the spin-orbit interaction is equal to

$$\Delta E = \beta\hbar\omega \times \begin{cases} -l, & j = l + \frac{1}{2}; \\ l+1, & j = l - \frac{1}{2}, \end{cases} \tag{4.32}$$

where the dimensionless constant $\beta = \frac{1}{2}\lambda\hbar\omega/Mc^2$. The best agreement with the experimental data is obtained with $\beta \approx 1/20$. Since $\hbar\omega \approx 10$ MeV, we find the spin-orbit coupling constant $\lambda \approx 10$.

Square potential well. We shall consider the one-particle model, taking for the nuclear potential a square well of finite depth:

$$V(r) = \begin{cases} -V_0, & r < R; \\ 0, & r > R, \end{cases} \tag{4.33}$$

where R is the nuclear radius. A square potential well characterized by a constant depth inside the nucleus with a sharp change at the nuclear surface turns out to be a good approximation for the self-consistent potential of heavy nuclei.

We write down the equation for the radial function

$$\frac{d^2\varphi}{dr^2} + \frac{2}{r}\frac{d\varphi}{dr} + \left\{ \frac{2M}{\hbar^2}\left[E - V(r) + \frac{a}{r}\frac{dV(r)}{dr} \right] - \frac{l(l+1)}{r^2} \right\}\varphi = 0. \tag{4.34}$$

We shall consider the bound states of the nucleon, which correspond to negative energies $E < 0$ smaller in magnitude than the depth of the potential, $|E| < V_0$.

Because of the discontinuous change of the potential (4.33) at $r = R$, the spin-orbit forces act on a nucleon only at the surface of the nucleus:

$$(a/r)(dV/dr) = (a/R)V_0\delta(r - R). \tag{4.35}$$

Equation (4.34) can be simplified considerably if we consider separately the region inside the nucleus

$$\frac{d^2\varphi}{dr^2} + \frac{2}{r}\frac{d\varphi}{dr} + \left\{ k^2 - \frac{l(l+1)}{r^2} \right\}\varphi = 0, \quad k^2 = \frac{2M}{\hbar^2}(V_0 + E), \quad r < R,$$

and the region outside the nucleus

$$\frac{d^2\varphi}{dr^2} + \frac{2}{r}\frac{d\varphi}{dr} - \left\{ \kappa^2 + \frac{l(l+1)}{r^2} \right\}\varphi = 0, \quad \kappa^2 = \frac{2M}{\hbar^2}|E|, \quad r > R.$$

In this case, the solutions in the inner and outer regions should go over into each other at the point $r = R$, since the wave function is continuous:

$$\varphi(R - 0) = \varphi(R + 0). \tag{4.36}$$

However, the derivative of the wave function at $\mathbf{r} = R$ will change discontinuously because of the delta-function term (4.35) in the original equation (4.34). Retaining the principal terms near the point $\mathbf{r} = R$ in eqn. (4.34),

$$\varphi'' + (2M/\hbar^2)(a/r)V_0\delta(r - R)\varphi = 0$$

and integrating over a small interval containing the point $r = R$, we find

$$\varphi'(R - 0) = \varphi'(R + 0) + (a/R)(2M/\hbar^2)V_0\varphi(R). \qquad (4.37)$$

The relation obtained is the boundary condition determining the discontinuity, due to the effect of the spin-orbit forces, of the derivative of the wave function at the surface of the nucleus. The solution in the inner region, satisfying the condition that it be finite at the point $r = 0$, has the form

$$\varphi(r) = Cj_l(kr), \qquad r < R, \qquad (4.38)$$

where $j_i(x)$ is the spherical Bessel function. Correspondingly, the solution in the outer region that vanishes at infinity is expressed in terms of the spherical MacDonald function (a modified Bessel function of the third kind): $\varphi(r) \propto k_l(\kappa r), r > R$. [2] We find the proportionality coefficient by using the continuity condition (4.36) for the wave function at the surface of the nucleus:

[2]We shall define the spherical functions $j_l(x)$ and $k_l(x)$:

$$j_l(x) = \sqrt{\frac{\pi}{2x}}J_{l+1/2}(x), \quad k_l(x) = \sqrt{\frac{\pi}{2x}}K_{l+1/2}(x).$$

The functions $j_l(x)$ and $k_l(x)$ can be expressed in terms of trigonometric and exponential functions:

$$j_l(x) = (-1)^l x^l \frac{d^l}{(xdx)^l}\left(\frac{\sin x}{x}\right); \; k_l(x) = \frac{\pi}{2x}e^{-x}\sum_{n=0}^{l}\frac{(l+n)!}{n!(l-n)!(2x)^n}.$$

In the particular case $l = 0$, $j_0(x) = \sin x/x$ and $k_0(x) = (\pi/2)(e^{-x}/x)$. The functions $j_l(x)$ and $k_l(x)$ satisfy the following recursion relations:

$$j_l' = j_{l-1} - \frac{l+1}{x}j_l; \; k_l' = -k_{l-1} - \frac{l+1}{x}k_l; \; j_l' = -j_{l+1} + \frac{l}{x}j_l; \; k_l' = -k_{l+1} + \frac{1}{x}k_l.$$

We also give the values of integrals of quadratic functions of j_l and k_l:

$$\int drr^2 j_l^2(\alpha r) = (r^3/2)\{j_l^2(\alpha r) - j_{l-1}(\alpha r)j_{l+1}(\alpha r)\};$$

$$\int drr^2 k_l^2(\beta r) = (r^3/2)\{k_l^2(\beta r) - k_{l-1}(\beta r)k_{l+1}(\beta r)\}.$$

$$\varphi(r) = C\frac{j_l(kR)}{k_l(\kappa R)}k_l(\kappa r),\ r > R. \tag{4.39}$$

The constant C is determined from the normalization condi tion $\int_0^\infty dr \times r^2\varphi^2(r) = 1$:

$$C = \frac{\sqrt{2}}{R^{3/2}}\left\{\frac{j_l^2(kR)}{k_l^2(\kappa R)}k_{l-1}(\kappa R)k_{l+1}(\kappa R) - j_{l-1}(kR)j_{l+1}(kR)\right\}^{1/2}. \tag{4.40}$$

Substituting the solutions (4.38) and (4.39) into the boundary condition (4.37), we obtain an equation for the energy eigenvalues of a nucleon in a square potential well:

$$k\frac{j_l'(kR)}{j_l(kR)} = \kappa\frac{k_l'(\kappa R)}{k_l(\kappa R)} + \frac{a}{R}\frac{2M}{\hbar^2}V_0. \tag{4.41}$$

We shall confine ourselves to finding the solution of eqn. (4.41) in the simplest case of the S-state ($l = 0$). If $l = 0$, the spin-orbit interaction is absent ($a = 0$) and Eqn. (4.41) is simplified:

$$k \cot kR = -\kappa. \tag{4.42}$$

Noting next that $k^2 + \kappa^2 = (2M/\hbar^2)V_0$, we can represent eqn. (4.42) in the form

$$|\sin x| = \frac{x}{\sqrt{\frac{2MR^2}{\hbar^2}V_0}}, \qquad x = kR. \tag{4.43}$$

It is not difficult to solve this equation graphically, as shown in Fig. 4.2. Since k and κ are positive, by virtue of (4.42) we must have $\cot x < 0$, and, consequently, the roots x must lie in the second, fourth, sixth, etc., quadrants. Putting $V_0 = 43\,\text{MeV}$, $R_0 = 1\cdot 2 \times 10^{-13}\,\text{cm}$ and $A = 125$, we obtain the following values of the roots: $x_1 = 2\cdot 9$, $x_2 = 5\cdot 6$ and $x_3 = 8\cdot 2$.

The corresponding energy levels, given by the formula $E_i = -V_0 + (\hbar^2/2MR^2)x_i^2, i = 1,2,\ldots$, are equal to $E_1 = -38\cdot 2\,\text{MeV}$, $E_2 = -25\cdot 1\,\text{MeV}$ and $E_3 = -4\cdot 4\,\text{MeV}$.

For $l \neq 0$, the solution of eqn. (4.41), needed to find the possible energy levels, is considerably more complicated. However, qualitative results on the distribution of the levels can be obtained by considering an infinite square well ($V_0 \to \infty$). This is connected with the fact that, as we go over from a well of finite depth to an infinite well, the order of the levels remains unchanged and only the relative spacing between the levels change.

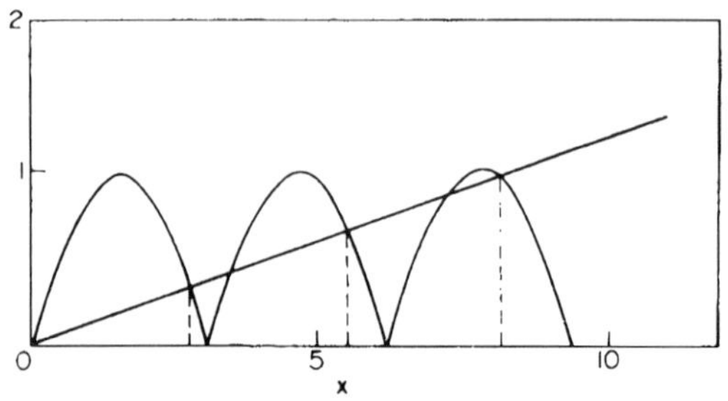

Figure 4.2. Graphical solution of the equation for the energy eigenvalues of a nucleon in a square well, for $l = 0$. The ordinate axis represents the values of the functions on the left- and right-hand sides of eqn. (4.43).

In the case of an infinite well, the boundary condition (4.41) takes the form

$$j_l(kR) = 0, \tag{4.44}$$

and, consequently, the energy levels are uniquely determined by the zeros of the function $j_l(x)$.

Table 4.3 shows the sequence of levels for an infinite square well, the Bessel-function zeros corresponding to the states (nl), the number $2(2l+1)$ of particles in each state, and the total number $\sum_{nl} 2(2l+1)$ of particles.

TABLE 4.3.

States (nl)	Zero of $j_l(x)$	$2(2l+1)$	$\sum_{nl} 2(2l+1)$	States (nl)	Zero of $j_l(x)$	$2(2l+1)$	$\sum_{nl} 2(2l+1)$
$1s$	3 · 142	2	2	$2d$	9 · 095	10	68
$1p$	4 · 493	6	8	$3s$	9 · 425	2	70
$1d$	5 · 763	10	18	$1h$	9 · 356	22	92
$2s$	6 · 283	2	20	$2f$	10 · 417	14	106
$1f$	6 · 988	14	34	$3p$	10 · 904	6	112
$2p$	7 · 725	6	40	$1i$	10 · 513	26	138
$1g$	8 · 183	18	58	$2g$	11 · 705	18	156

The principal difference between the one-particle spectrum in the case of the square well and that in the case of the oscillator potential lies in the lifting of the degeneracy in the latter of levels with different values of the orbital angular momentum l, the level corresponding to the largest value of l being lowered below all the others. This is connected with the fact that with increasing l the centrifugal forces pull the particle out towards the potential-well boundary, where the potential is much lower for the square well than for the oscillator potential. Inclusion of the spin-orbit interaction in the square-well case, as in the case of the oscillator potential, leads to an additional splitting of the levels in accordance with the values of the total angular momentum j of the particles.

The individual levels in a square well can be characterized conveniently not only by the values of l and j but also by the values of N corresponding to the given level when we go adiabatically from the square-well to the oscillator potential. Thus, each level will be uniquely defined by the quantum numbers N, l and j.

Magic numbers. Figure 4.3 shows the level scheme for an isotropic harmonic oscillator (on the left) and a square well (on the right). Levels corresponding to each other are linked by straight lines, which give an interpolation between the two limiting cases. In an intermediate case the levels break down into separate groups, separated by large energy intervals. The spin-orbit interaction leads to the appearance in a given group of levels of a level of the opposite parity and higher total angular momentum from the next oscillator band up.

The *magic numbers* (which are ringed in Fig. 4.3) are determined by the maximum number of states corresponding to a given group of levels. As can be seen from the Figure, these magic numbers are 2, 8, 20, 28, 40, 50, 82 and 126, of which the first three and the fifth arise from the oscillator nature of the potential and the others from the spin–orbit interaction. For heavy nuclei, because of the Coulomb interaction, the sequence of one-particle levels for the protons will differ slightly from the corresponding sequence for the neutrons. Figure 4.4 shows an empirical scheme of the arrangement of the one-particle levels for neutrons and protons (Klinkenberg, 1952).

Nuclei containing a magic number of neutrons or protons differ markedly in their properties from other nuclei. Indeed, the magic numbers are associated with the formation of filled shells, and so nuclei containing a magic number of nucleons are characterized by a large binding energy. The energies of the excited states relative to the ground state are very great for such nuclei. The neighboring nuclei, in which the next shells are beginning to be filled, are characterized by much smaller binding energies. As an example, we can take the doubly magic nucleus ${}^{208}_{82}Pb$, which consists of a magic number (82) of protons and a magic number (126) of neutrons. Whereas the binding energies of a neutron and proton in ${}^{208}Pb$ are equal to 7·36 and 8·97 MeV respectively, the binding energies of a neutron and proton in the neighboring nuclei are found to be equal to 3·86 MeV (${}^{209}Pb$) and 3·76 MeV (${}^{209}Bi$). Whereas for ordinary heavy nuclei the energies of the lowest

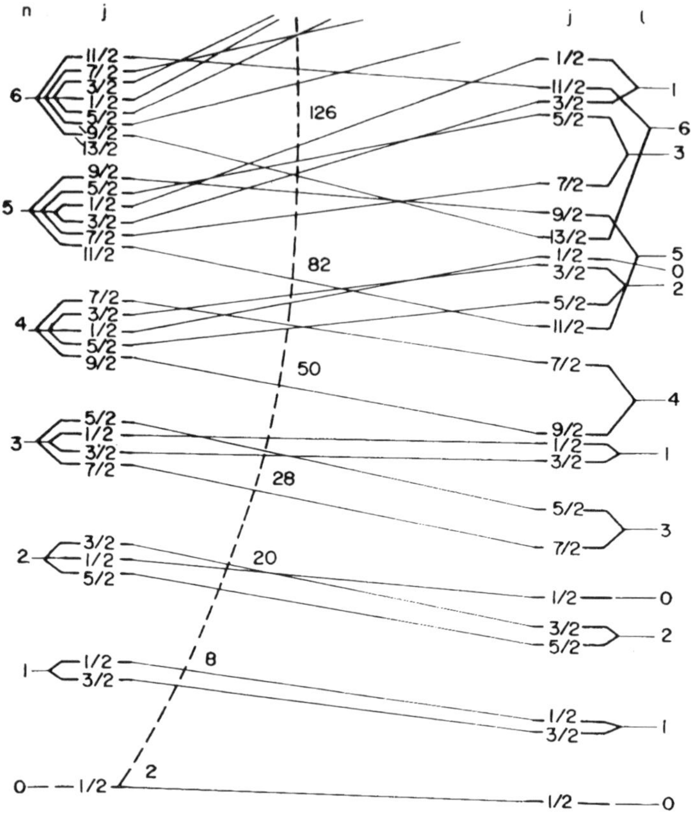

Figure 4.3. Level scheme for the isotropic harmonic oscillator (left) and square (right). The dashed line corresponds to the empirical sequence of one-particle levels.

excited states are of the order of 0·1 MeV, for the doubly magic nucleus ^{208}Pb the energy of the lowest excited state is equal to 2·6 MeV.

The magnitude of the binding energy also explains the distinct status of magic nuclei, manifested in anomalies in the total abundance of the elements and also in the relative abundance of individual isotopes and isotones. Thus, the elements with Z=20, 28, 50 and 82 are found to have the largest number of stable isotopes. (The largest number of stable isotopes (ten) is possessed by the element ^{50}Zn.) Analogously, the largest numbers of stable isotones correspond to N=20, 28, 50, 82 and 126. The large binding energies of the magic nuclei are directly related to the small density of the energy levels, which leads to the very small experimentally observed values of the cross-sections for capture of slow neutrons by the magic nuclei.

Since heavy unstable elements with high numbers Z up to Z=110, placed in the seventh period of the periodic system, have been synthe-

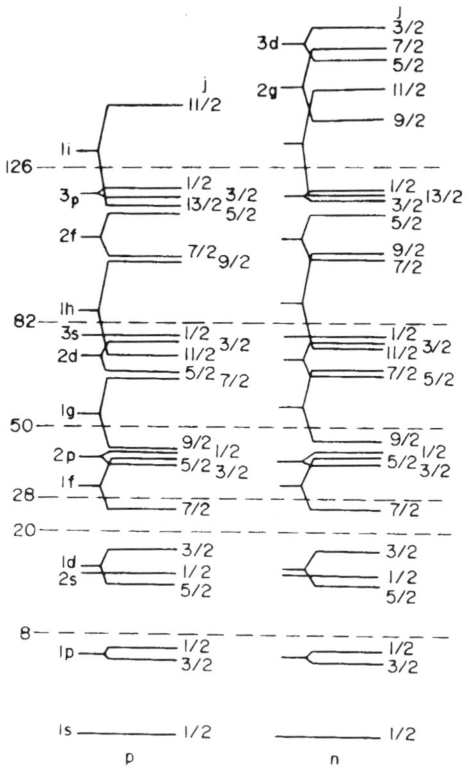

Figure 4.4. Empirical schemes of the one-particle levels for protons and neutrons.

sized under laboratory conditions, and new experiments aimed to discover new superheavy elements are being carried out, it is of interest to extrapolate the shell-model regularities of proton and neutron magic numbers to the poorly known range of nuclei with large numbers Z and N. In this range, the values of nucleon magic numbers may be sensitive not only to the spin-orbit interaction, but also to the shape of the shell-model potential well and the Coulomb repulsion of intranuclear protons. This means that proton and neutron magic numbers may be different. Complicated calculations with realistic potentials, which have been carried out with the use of modern computers, predict the next proton and neutron magic numbers after 82 and 126 to be equal to Z=114 and N=184 respectively. Thus, the nuclei with the proton number Z=114 (the relevant chemical element corresponds to the seventh period of the periodic system and must be analogous to lead) and the neutron number N=184 must be magic, and the nucleus with Z=114 and N=184 must be double magic. If the stability of nuclei with 114 protons or 184 neutrons will be confirmed experimentally, this will not only show that theoretical ideas of the shell model are correct, but also confirm the predictive power of the quantum theory of nuclear shells. The increased stability of nuclei with $Z \approx 114$ and $N \approx 184$ must be manifested in their relatively large lifetimes and increased resistance to spontaneous fission. However, when approaching this stability island for superheavy nuclei, the lifetimes and other properties may change in irregular manner. We note that the magic number 184 has already appeared in Table 4.2 for the oscillatory potential with spin-orbit interaction. As for the other "theoretical" magic number 114, the close number 112 has twice appeared in Tables 4.1 and 4.3 for different versions of the shell potential. The next proton magic number may turn out to be Z=126, or a close (even) number. The relevant chemical element will correspond to the new eigths period of the periodic table. The element with Z=126 must be chemically analogous to some element of the actinic group.

4.3. One-particle Shell Model of Nuclei

Spins and parities of ground states of nuclei. The *one-particle shell model* makes it possible not only to explain the special stability of magic nuclei, but also to predict correctly the spins and parities of the ground states of most nuclei with odd mass numbers A. According to the one-particle shell model, all the paired nucleons form a core with zero spin (this is in agreement with the experimental fact that thc spin of even-even nuclei is always equal to zero) ; therefore, the spin and parity of the nucleus are determined by the state of the last unpaired nucleon. Obviously, this model can only describe the properties of odd nuclei. Evidence for the applicability of the one-particle model is provided by the fact that, in almost all cases, the experimentally determined spins and parities of the ground states of odd nuclei with the values by the model for the last unpaired nucleon.

In a more sophisticated model, the nucleus is represented as consisting of a certain number of filled shells and outer neutrons and protons in unfilled shells. In this case, the properties of the nucleus are determined by these outer nucleons. Since, however, for a system of identical particles there are several different independent states differing in symmetry and, consequently, in total angular momentum, the spin of the nucleus is not uniquely determined. This indeterminacy is removed by taking into account the residual interaction between the nucleons, which causes the degeneracy of the states to be lifted. If the range of the interaction is small compared with the dimensions of the nucleus, the spin J of the ground state of a group of identical particles (neutrons or protons) in the j-shell turns out to be equal to 0 if the number of particles is even and j if it is odd. This is in accordance with the results of the simpler one-particle model.

Thus, knowing the one-particle level scheme determining the order in which the shells are filled, we can determine immediately the spins and parities of the ground states of the odd nuclei. Table 4.4 contains the values, obtained from the level scheme of Fig. 4.3. of the spins and parities for the ground states of a number of nuclei, and also the corresponding experimental values.

TABLE 4.4.

Nucleus (Z odd)	J^π Theory	J^π Experiment	Nucleus (N odd)	J^π Theory	J^π Experiment
${}^{7}_{3}$Li	$\frac{3}{2}^-$	$\frac{3}{2}^-$	${}^{9}_{4}$Be	$\frac{3}{2}^-$	$\frac{3}{2}^-$
${}^{11}_{5}$B	$\frac{3}{2}^-$	$\frac{3}{2}^-$	${}^{13}_{6}$C	$\frac{1}{2}^-$	$\frac{1}{2}^-$
${}^{15}_{7}$N	$\frac{1}{2}^-$	$\frac{1}{2}^-$	${}^{17}_{8}$O	$\frac{5}{2}^+$	$\frac{5}{2}^+$
${}^{17}_{9}$F	$\frac{5}{2}^+$	$\frac{5}{2}^+$	${}^{25}_{12}$Mg	$\frac{5}{2}^+$	$\frac{5}{2}$
${}^{27}_{13}$Al	$\frac{5}{2}^+$	$\frac{5}{2}^+$	${}^{43}_{20}$Ca	$\frac{7}{2}^-$	$\frac{7}{2}^-$
${}^{35}_{17}$Cl	$\frac{3}{2}^+$	$\frac{3}{2}^+$	${}^{61}_{28}$Ni	$\frac{5}{2}^-$	$\frac{3}{2}$
${}^{53}_{25}$Mn	$\frac{7}{2}^-$	$\frac{7}{2}^-$	${}^{69}_{30}$Zn	$\frac{1}{2}^-$	$\frac{1}{2}$
${}^{75}_{32}$As	$\frac{5}{2}^-$	$\frac{3}{2}$	${}^{97}_{42}$Mo	$\frac{5}{2}^+$	$\frac{5}{2}^+$
${}^{79}_{35}$Br	$\frac{5}{2}^-$	$\frac{3}{2}$	${}^{105}_{46}$Pd	$\frac{5}{2}^+$	$\frac{5}{2}^+$
${}^{209}_{83}$Bi	$\frac{9}{2}^-$	$\frac{9}{2}^-$	${}^{131}_{54}$Xe	$\frac{3}{2}^+$	$\frac{3}{2}^+$

Deviations from this rule occur for nuclei in which the next shell up has a higher spin and begins to be filled. Since the pairing energy of two nucleons with high spin is greater than the pairing energy of nucleons with low spin, the order of occupation of the shells is inverted and occupation of the higher shell by one or more pairs of nucleons occurs. (For example, as a result of occupation of the shell $1f_{5/2}$, which succeeds the shell $2p_{3/2}$, the spins of many nuclei in the region $28 < (N \text{ or } Z) < 38$ are found to be equal to 3/2.) When inversion occurs, the state corresponding to the normal order of occupation of the shells is found to he a low-lying metastable state, since its spin differs greatly from that of the ground state. Deviations from the general rule are also observed for heavy nuclei, this being connected with violation of the spherical symmetry of the nuclei.

Magnetic dipole moments of nuclei. According to the one-particle model, the magnetic dipole moment of an odd nucleus is determined by the magnetic dipole moment of the last unpaired nucleon.[3] We shall calculate the magnetic moment of an individual nucleon moving in the self-consistent field of the nucleus. We represent the operator of the total magnetic moment of the nucleon, expressed in nuclear magnetons $e\hbar/2Mc$, in the form of a sum of two terms:

$$\boldsymbol{\mu} = g_i \mathbf{l} + g_s \mathbf{s}, \tag{4.45}$$

where $\mathbf{l}$ and $\mathbf{s}$ are the operators of the orbital and spin angular momenta of the nucleon, and g_l, and g_s are the orbital and spin gyromagnetic ratios. Because of the spin-orbit interaction, the orbital and spin angular momenta l and s are not separately conserved in the motion of the nucleon; only the total angular momentum of the nucleon is conserved:

$$\mathbf{j} = \mathbf{l} + \mathbf{s}. \tag{4.46}$$

The *magnetic dipole moment* μ is defined as the mean value of the projection of the vector $\boldsymbol{\mu}$ on the direction of the vector $\mathbf{j}$ of the total angular momentum of the nucleon in the state with the maximum value of the projection of the vector $\mathbf{j}$:

$$\mu \equiv \left\langle \frac{\boldsymbol{\mu}\cdot\mathbf{j}}{\mathbf{j}^2} j_z \right\rangle_{m=j}, \tag{4.47}$$

or, noting that $\langle j_z \rangle = j$ and $\langle j^2 \rangle = j(j+1)$, we obtain

$$\mu = \frac{1}{j+1} \langle \boldsymbol{\mu} \cdot \mathbf{j} \rangle_{m=j} . \tag{4.48}$$

Substituting the expression (4.45) for $\boldsymbol{\mu}$ into (4.48) and using the identities

$$2\mathbf{l}\cdot\mathbf{j} = \mathbf{j}^2 + \mathbf{l}^2 - \mathbf{s}^2, \quad 2\mathbf{s}\cdot\mathbf{j} = \mathbf{j}^2 - \mathbf{l}^2 + \mathbf{s}^2,$$

we find

$$\mu = \begin{cases} (j - \frac{1}{2})g_l + \frac{1}{2}g_s, & j = l + \frac{1}{2}, \\ \left[1 + \frac{1}{2(j+1)}\right] j g_l - \frac{j}{2(j+1)} g_s, & j = l - \frac{1}{2}. \end{cases} \tag{4.49}$$

The orbital gyromagnetic ratio g_l, is equal to zero for the neutron and one for the proton. The spin gyromagnetic ratio g_s can be put equal to twice the

[3]The theory of nuclear magnetic moments is described in detail in the review by Blin-Stoyle (1956).

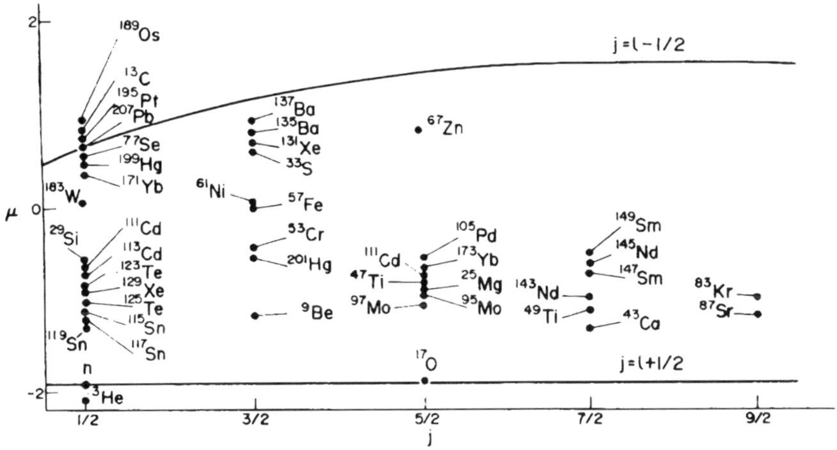

Figure 4.5. Schmidt curves and experimental values of the magnetic moments for nuclei with an odd number of neutrons.

value of the intrinsic magnetic moment of the neutron or proton: $g_s = 2\mu_n$ or $2\mu_p$. Thus, the one-particle magnetic moments of nuclei are given by the formulae

$$\mu = \begin{cases} \mu_n & j = l + \frac{1}{2}, \\ -\dfrac{j}{j+1}\mu_n, & j = l - \frac{1}{2}, \end{cases} \tag{4.50}$$

if the unpaired particle is a neutron, and by

$$\mu = \begin{cases} j - \frac{1}{2} + \mu_p, & j = l + \frac{1}{2}, \\ \left[1 + \dfrac{1}{2(j+1)}\right] j - \dfrac{j}{j+1}\mu_p, & j = l - \frac{1}{2}, \end{cases} \tag{4.51}$$

if the unpaired particle is a proton. Each of the formulae (4.50) and (4.51) leads to two curves (*Schmidt curves*) for the dependence of the magnetic moment μ of the nucleus on the spin j. Figures 4.5 and 4.6 show the Schmidt curves and the experimental values of the magnetic moments for odd nuclei. In the main, the experimental values of nuclear magnetic moments do not lie on the Schmidt curves (indicating the approximate character of the one-particle model), but always (with a few exceptions, for example, ^{3}H, ^{3}He, ^{15}N and ^{13}C) lie between the two curves, being in most cases close to one of the curves. This makes it possible to select the correct value of the nuclear spin in accordance with the experimental value of the magnetic moment.

Finally, we note that the existence of the spin-orbit interaction leads to the appearance of a certain additional magnetic moment associated with

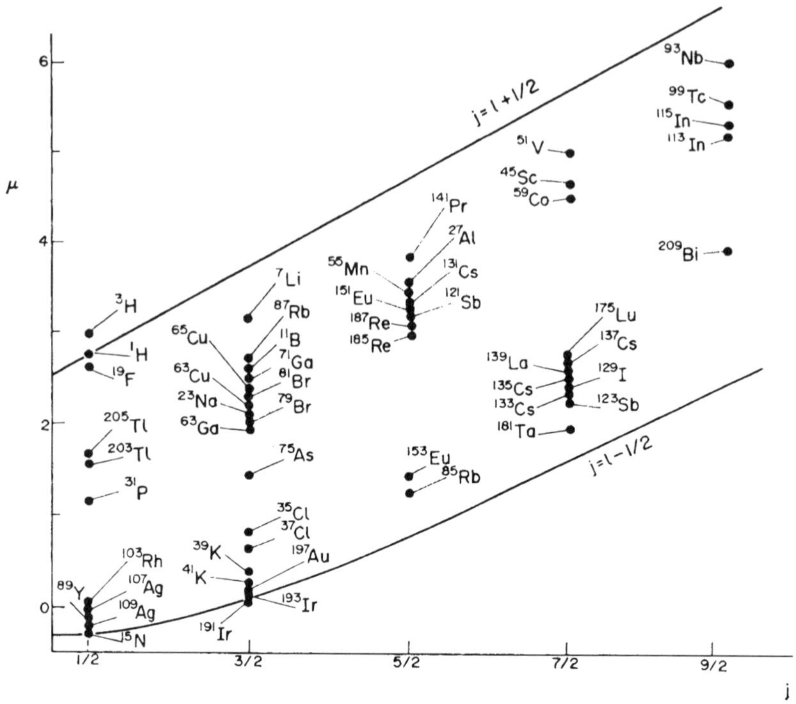

Figure 4.6. Schmidt curves and experimental values of the magnetic moments for nuclei with an odd number of protons.

the protons in the nucleus. In fact, we rewrite the spin-orbit interaction energy (4.11) in the form

$$U = \frac{1}{M} f(r)[\boldsymbol{r} \times \boldsymbol{p}] \cdot \boldsymbol{s} \tag{4.52}$$

which indicates explicitly the dependence on the momentum p. In the presence of an external electromagnetic field characterized by a vector potential $\mathbf{A}$, the particle momentum in (4.52) must be replaced by $\mathbf{p} - (e/c)\mathbf{A}$, where e is the charge of the particle. Consequently, in the Hamiltonian describing the motion of a proton in an electromagnetic field, there is an additional term

$$U' = -\frac{e}{Mc} f(r)[\mathbf{r} \times \mathbf{A}]\cdot\mathbf{s}. \tag{4.53}$$

In the case of a uniform magnetic field, the vector potential $\mathbf{A}$ can be expressed directly in terms of the magnetic field intensity $\mathbf{H} : \mathbf{A} = \frac{1}{2}[\mathbf{H} \times \mathbf{r}]$; therefore, the additional interaction energy (4.53) can be written in the form $U' = -\boldsymbol{\mu}' \cdot \mathbf{H}$, where $\boldsymbol{\mu}'$ is the operator of the additional proton magnetic moment due to the spin-orbit interaction:

$$\boldsymbol{\mu}' = (1/\hbar) f(r)[\mathbf{r} \times [\mathbf{s} \times \mathbf{r}]]. \tag{4.54}$$

The magnetic moment $\boldsymbol{\mu}'$ is expressed in nuclear magnetons. The mean value of the additional proton magnetic moment in a state with a well-defined total angular momentum j is defined analogously to (4.47), and is equal to

$$\mu' = \pm \frac{2j+1}{4(j+1)} \frac{1}{\hbar} \left\langle r^2 f(r) \right\rangle_{nl}, \tag{4.55}$$

where the plus sigh corresponds to the value $j = l + \frac{1}{2}$ and the minus sign corresponds to the value $j = l - \frac{1}{2}$.

Since the function $f(\mathbf{r})$ differs appreciably from zero only on the surface of the nucleus, the factor $\mathbf{r}^2$ inside the averaging symbol can be replaced by the square of the nuclear radius, R^2. The mean value of the function $f(\mathbf{r})$ in this case can be expressed directly in terms of the spin-orbit splitting of the level:

$$\langle f(r) \rangle_{nl} = \frac{M}{\hbar} \frac{2}{2l+1} (E_{nlj=l+1/2} - E_{nlj=l-1/2}).$$

Using (4.32) and noting that $R^2 = \hbar^2 / M\omega$, we find

$$\mu' \approx \frac{2j+1}{2(j+1)} \beta, \quad \beta \approx 1/20.$$

Consequently, the additional magnetic moment (4.55) is found to be small compared with (4.51).

Electric quadrupole moments of nuclei. We next study the *electric quadrupole moments of nuclei* on the basis of the one-particle shell model. The tensor of the electric quadrupole moment of a system of identical charges is defined by the formula

$$Q_{ij} = \sum (3x_i x_j - r^2 \delta_{ij}), \tag{4.56}$$

where $\mathbf{r}$ is the position vector of an individual charge and the summation is performed over all the charges.

For a state of the system characterized by a well-defined *total angular momentum* $\mathbf{J}$, the quadrupole-moment tensor Q_{ij} of the system can be related directly to the vector $\mathbf{J}$. The tensor Q_{ij} is symmetric and the sum of its diagonal elements is equal to zero; therefore, Q_{ij} can be expressed uniquely in terms of the J_i: $Q_{ij} = A(J_i J_j + J_j J_i - \frac{2}{3}\mathbf{J}^2 \delta_{ij})$, where A is some constant.

Strictly, the quadrupole moment Q of a system is defined as the mean value of the component Q_{zz} of the tensor in the state in which the projection of the angular momentum $\mathbf{J}$ of the system on to the z-axis is a maximum. Putting, $J_z = J$ and $\mathbf{J}^2 = J(J+1)$, we have

$$Q = \frac{2}{3} AJ(2J-1). \qquad (4.57)$$

It follows from this formula that the quadrupole moment of the system is zero when $J = 0$ or $J = \frac{1}{2}$. By expressing the constant A in terms of Q by means of(4.57), we can represent the quadrupole-moment tensor of the system in the form

$$Q_{ij} = \frac{3Q}{2J(2J-1)} \{J_i J_j + J_j J_i - \frac{2}{3} J(J+1)\delta_{ij}\}. \qquad (4.58)$$

In the one-particle approximation, the nuclear quadrupole moment is determined by the quadrupole moment of the last unpaired proton. We write down the component O_{zz} of the quadrupole-moment tensor in the spherical coordinate system:

$$Q_{zz} = \sqrt{(16\pi/5)} r^2 Y_{20}(\theta). \qquad (4.59)$$

Choosing the one-particle wave function in the form

$$\psi_{nljm} = \varphi_{nl}(r) \sum_{m'\mu} (lm' \frac{1}{2} \mu | jm) Y_{lm'} \chi_{1/2,\mu},$$

we obtain the following expression for the quadrupole moment in the state with $m = j$:

$$Q = \sqrt{16\pi/5} \left\langle r^2 \right\rangle_{nl} \sum_{m+\mu=j} (lm \frac{1}{2} \mu | jj)^2 \langle lm | Y_{20} | lm \rangle, \qquad (4.60)$$

where $\langle r^2 \rangle_{nl}$ is the mean square radius:

$$\left\langle r^2 \right\rangle_{nl} = \int_0^\infty dr r^4 \varphi_{nl}^2(r).$$

Noting next that the matrix element of the spherical harmonic is equal to

$$\langle lm | Y_{20} | lm \rangle = \sqrt{\frac{5}{4\pi}} \frac{l(l+1) - 3m^2}{(2l+3)(2l-1)}. \qquad (4.61)$$

it is not difficult to perform the summation in (4.60) by making use of the values of the Clebsch-Gordan coefficients:

$$(lm \tfrac{1}{2} j - m | jj)^2 = \begin{cases} \begin{array}{ccc} m = j - \frac{1}{2}, & m = j + \frac{1}{2}, & \\ 1, & 0, & j = l + \frac{1}{2}, \\ \dfrac{1}{2(j+1)}, & \dfrac{2j+1}{2(j+1)}, & j = l - \frac{1}{2} \end{array} \end{cases} \qquad (4.62)$$

Thus, for the quadrupole moment of the unpaired proton we obtain the formula

$$Q = -\frac{2j-1}{2(j+1)} \left\langle r^2 \right\rangle_{nl}. \tag{4.63}$$

It follows from this formula that one-particle quadrupole moments are always negative.

Because of the recoil motion of the nuclear core, an unpaired neutron also possesses a certain effective quadrupole moment. The magnitude of the effective quadrupole moment of the neutron amounts to Z/A^2 of the magnitude of the proton quadrupole moment.

The electric quadrupole moments calculated using the one-particle model agree with the experimental values of the quadrupole moments only in the case of nuclei containing one nucleon outside a filled shell. The quadrupole moments of almost all other nuclei differ from the one-particle moments both in sign (they take positive values) and in magnitude (the absolute values of the quadrupole moments are found to be many times greater than the values calculated using the one-particle model). The largest quadrupole moments are observed in nuclei with half-filled shells.

Figure 4.7 shows the experimental values of the electric quadrupole moments of odd nuclei as a function of the number of neutrons or protons. Although the periodic dependence of the nuclear quadrupole moments on the number of nucleons is evidence of the shell structure of nuclei, the one-particle model is found to be completely useless for determining the magnitudes of the nuclear quadrupole moments.

Magnetic octupole moments of nuclei. In addition to the magnetic dipole moment, values of the *magnetic octupole moment* have been determined for certain nuclei.

The magnetic multipole moment operator $\hat{M}_{\lambda\nu}$ is a spherical tensor whose rank is given by the order λ of the multipole moment; in the one-particle model of the nucleus, it is given by the expression

$$\hat{M}_{\lambda\nu} = \nabla(r^{\lambda} Y_{\lambda\nu}(\mathbf{n})) \left(\frac{2}{\lambda+1} g_l \mathbf{l} + g_s \mathbf{s} \right). \tag{4.64}$$

If $\lambda = 1$, (4.64) goes over directly into the expression (4.45) for the dipole-moment operator. Like the magnetic dipole moment, the multipole moment (4.64) is expressed in nuclear magnetons.

The multipole moment corresponding to the value $\lambda = 3$ is called the *magnetic octupole moment.* The magnetic octupole moment is a spherical tensor of rank 3.

The nuclear magnetic octupole moment Ω is usually defined as the mean value of the component M_{30} of the tensor in the state with the maximum

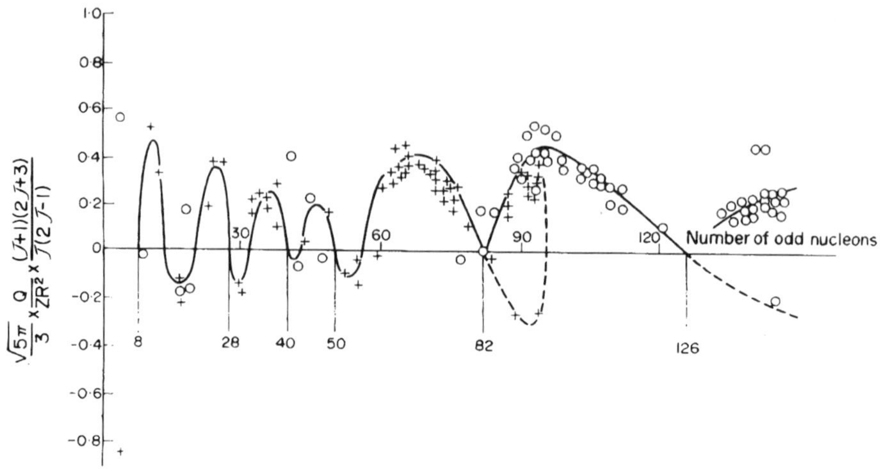

Figure 4.7. Experimental values of the quadrupole moments of odd nuclei as functions the number of neutrons or protons (+ denoted the dependence on Z and o the dependence on N).

possible value of the projection of the total angular momentum of the nucleon ($m = j$):

$$\Omega = -\sqrt{(4\pi/7)}\langle jm = j|\hat{M}_{30}|jm = j\rangle. \tag{4.65}$$

(The factor $-\sqrt{(4\pi/7)}$ has been introduced for convenience.) According to (4.64) and (4.65), the magnetic octupole moment of the nucleus, like the nuclear magnetic dipole moment, is associated with the orbital and spin angular momenta $\mathbf{l}$ and $\mathbf{s}$ of the nucleon.

To determine Ω, we shall make use of the explicit form of the spherical harmonic

$$Y_{30}(\mathbf{n}) = \sqrt{(7/16\pi)}(5\,cos^2\theta - 3)\,cos\,\theta. \tag{4.66}$$

Substituting (4.66) into (4.64) and (4.65) and taking into account that $\mathbf{r}\cdot\mathbf{l} = 0$, we have

$$\begin{aligned}\Omega = -\Big\langle jm = j|r^2\{\tfrac{1}{2}g_l l_z + \sqrt{5\pi}g_l Y_{20}(\mathbf{n})l_z + g_s s_z \\ +2\sqrt{3\pi}g_s(\mathbf{n}\cdot\mathbf{s})Y_{10}(\mathbf{n}) + 2\sqrt{5\pi}g_s Y_{20}(\mathbf{n})s_z\}|jm = j\Big\rangle,\end{aligned} \tag{4.67}$$

where the trigonometric functions arising from the differentiation of $Y_{30}(\mathbf{n})$ have been reduced again to spherical harmonics. We perform the subsequent calculations in (4.67) using the one-particle function (4.18):

$$\psi_{nljm} = \varphi_{nl}(r)y_{l1/2jm}.$$

We note that we can find the result of the action of the operator $\mathbf{n}\cdot\mathbf{s}$ on the function $\mathcal{Y}_{l1/2jm}$ if we take into account that $\mathcal{Y}_{l1/2jm}$ and $(\mathbf{n}\cdot\mathbf{s})\mathcal{Y}_{l1/2jm}$ are eigenfunctions of the operator $\mathbf{s}\cdot\mathbf{l}=\frac{1}{2}(\mathbf{j}^2-\mathbf{l}^2-\mathbf{s}^2)$:

$$\left.\begin{aligned} (\mathbf{s}\cdot\mathbf{l})\mathcal{Y}_{l1/2jm} &= \tfrac{1}{2}[j(j+1)-l(l+1)-\tfrac{3}{4}]\mathcal{Y}_{l1/2jm}; \\ \mathbf{s}\cdot\mathbf{l})(\mathbf{n}\cdot\mathbf{s})\mathcal{Y}_{l1/2jm} &= -\tfrac{1}{2}[j(j+1)-l(l+1)+\tfrac{5}{4}](\mathbf{n}\cdot\mathbf{s})\mathcal{Y}_{l1/2jm}. \end{aligned}\right\} \tag{4.68}$$

One can easily convince oneself that the latter equality is correct by using the following equalities:

$$(\mathbf{n}\cdot\mathbf{s})(\mathbf{s}\cdot\mathbf{l})(\mathbf{n}\cdot\mathbf{s}) = -\tfrac{1}{4}((\mathbf{s}\cdot\mathbf{l})+1), \quad (\mathbf{n}\cdot\mathbf{s})^2=\tfrac{1}{4}.$$

Comparing the equalities (4.68), we see that

$$(\mathbf{n}\cdot\mathbf{s})\mathcal{Y}_{l=j\mp1/2,1/2,jm} = \frac{1}{2}\mathcal{Y}_{l=j\pm1/2,1/2,jm}. \tag{4.69}$$

Thus, the integration over the angles in (4.67) can be performed explicitly and in (4.67) we shall be concerned with integrals of products of not more than three spherical harmonics:

$$\begin{aligned} &\int do Y^*_{lm}(\mathbf{n})Y_{l_1m_1}(\mathbf{n})Y_{l_2m_2}(\mathbf{n}) \\ &= \sqrt{\frac{(2l_1+1)(2l_2+1)}{4\pi(2l+1)}}(l_10l_20|l0)(l_1m_1l_2m_2|lm). \end{aligned} \tag{4.70}$$

Next we use relations for the spherical components of the vector $\mathbf{l}$ (and analogous relations for the spin vector $\mathbf{s}$)

$$\hat{l}_\nu|lm\rangle = (-1)^m\sqrt{l(l+1)}(lm+\nu l-\nu|lm)|lm+\nu\rangle, \quad \nu=0,\ \pm1. \tag{4.71}$$

We can find the resulting sums of products of three *Glebsch–Gordan coefficients* by means of the *Racah formula*:

$$\begin{aligned} &\sum_{m_1}(l_1m_1l_2m_2|l_3m_1+m_2)(l_3m_1+m_2l_4m-m_1-m_2|lm) \\ &\times(l_2m_2l_4m-m_1-m_2|l_5m-m_1) \\ &= \sqrt{(2l_3+1)(2l_5+1)}(l_1m_1l_5m-m_1|lm)W(l_1l_2ll_4;l_3l_5). \end{aligned} \tag{4.72}$$

By taking the values of the Clebsch–Gordan and *Racah coefficients* from tables, we find the following mean values:

$$\left.\begin{aligned}
&\langle jj|Y_{10}(\mathbf{n}\cdot\mathbf{s})|jj\rangle = -\frac{1}{8(j+1)}\sqrt{\frac{3}{\pi}}, && l=j\pm\tfrac{1}{2};\\
&\langle jj|Y_{20}s_z|jj\rangle = \begin{cases} -\dfrac{2j-1}{16(j+1)}\sqrt{\dfrac{5}{\pi}}, & l=j-\tfrac{1}{2};\\ \dfrac{2j^2+j+3}{16(j+1)(j+2)}\sqrt{\dfrac{5}{\pi}}, & l=j+\tfrac{1}{2};\end{cases}\\
&\langle jj|s_z|jj\rangle = \begin{cases} \dfrac{1}{2}, & l=j-\tfrac{1}{2};\\ -\dfrac{j}{2(j+1)}, & l=j+\tfrac{1}{2};\end{cases}\\
&\langle jj|Y_{20}|jj\rangle = -\frac{2j-1}{8(j+1)}\sqrt{\frac{5}{\pi}}, && l=j\mp\tfrac{1}{2}.
\end{aligned}\right\} \tag{4.73}$$

Collecting the calculated mean values, we finally obtain the following expression for the magnetic octupole moment of the nucleus in the one-particle approximation (Schwartz, 1955):

$$\Omega = -\frac{3}{8}\frac{2j-1}{j+1}\times\begin{cases}\left[\left(j-\frac{3}{2}\right)g_l+g_s\right]\langle r^2\rangle_{nl}, & l=j-\frac{1}{2};\\ \dfrac{j-1}{j+2}\left[\left(j+\frac{5}{2}\right)g_l-g_s\right]\langle r^2\rangle_{nl}, & l=j+\frac{1}{2}.\end{cases} \tag{4.74}$$

As in the case of the magnetic dipole moment, allowance for the spin-orbit interaction leads to insignificant corrections to Ω. Figure 4.8 shows the dependence of the magnetic octupole moments of nuclei on the spin, and also the experimental values of the magnetic octupole moments for certain nuclei. As with the magnetic dipole moments, the experimental values of the octupole moments are also confined between lines calculated on the basis of the one-particle model.

4.4. Wave Functions of Many-particle Shells

Slater determinants. In the shell model of the nucleus it is assumed that the interaction between the individual nucleons can be reduced to the interaction of the nucleons with the self-consistent field of the nucleus; in this case, the nucleons can be regarded as independent particles moving in the given field. According to this simplified model, which is sometimes called the *independent-particle model* , the Hamiltonian H_0 of the nucleus is represented in the form of the sum of the Hamiltonians of the individual nucleons:

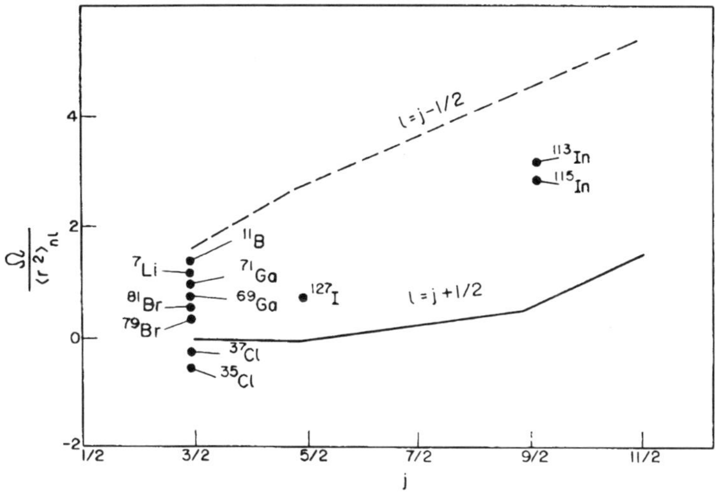

Figure 4.8. Schwartz curves and experimental values of magnetic octupole moments.

$$H_0 = \sum_i \{-(\hbar^2/2M)\nabla_i^2 + V_i\}, \tag{4.75}$$

where $-(\hbar^2/2M)\nabla_i^2$ is the kinetic energy and V_i is the interaction energy of an individual nucleon with the self-consistent field.

The eigenstates of a system of independent particles are usually called the *configurations of the system* . A definite wave function corresponds to each configuration of the system. Clearly, the wave functions of a nucleus described by the Hamiltonian (4.75) can be represented in the form of anti-symmetrized products of one-particle wave functions.

If the spin-orbit interaction is unimportant, the wave function of an individual nucleon in the self-consistent field can be represented in the form of a product of a space function, a spin function and an isospin function:

$$\psi_\alpha(i) \equiv \varphi_{nlm}(\mathbf{r}_i)\chi_\mu(\boldsymbol{\sigma}_i)\zeta_\nu(\boldsymbol{\tau}_i), \tag{4.76}$$

where m is the orbital angular-momentum component, μ is the spin component and ν is the isospin component (α is an abbreviated notation for the set of quantum numbers n, l, m, μ and ν). In the presence of spin–orbit interaction, the one-particle wave functions can be represented in the form of a product of a space-spin function and an isospin function:

$$\psi_\alpha(i) \equiv \varphi_{nljm_j}(\mathbf{r}_i, \boldsymbol{\sigma}_i)\zeta_\nu(\boldsymbol{\tau}_i), \tag{4.77}$$

where m_j is the component of the total angular momentum (α is the set of quantum numbers n, l, j, m_j and ν). The functions (4.77, like (4.76), form a complete set of one-particle functions. We shall assume that each set of functions (4.76) and (4.77) is an orthonormal set, in accordance with the condition

$$\int d\tau \psi_\alpha \psi^*_{\alpha'} = \delta_{\alpha\alpha'}. \tag{4.78}$$

The wave function of a system of A nucleons can be represented in the form of a product of A one-particle wave functions (4.76) or (4.77), which, however, must be anti-symmetrized with respect to interchanges of any pair of nucleons. This anti-symmetrization will be ensured automatically if we write the wave function of the system of A nucleons in the form of a *Slater determinant*:

$$\Psi_{\{\alpha\}} = (1/\sqrt{A!}) \begin{vmatrix} \psi_{\alpha_1}(1) & \psi_{\alpha_1}(2) & \dots & \psi_{\alpha_1}(A) \\ \psi_{\alpha_2}(1) & \psi_{\alpha_2}(2) & \dots & \psi_{\alpha_2}(A) \\ \dots & \dots & \dots & \dots \\ \dots & \dots & \dots & \dots \\ \psi_{\alpha_A}(1) & \psi_{\alpha_A}(2) & \dots & \psi_{\alpha_A}(A) \end{vmatrix}, \tag{4.79}$$

where $\{\alpha\}$ denotes the set of quantum numbers of the occupied one-particle states. Clearly, $\Psi_{\{\alpha\}}$ is an anti-symmetrized eigenfunction of the Hamiltonian H_0.

The Slater determinants $\Psi_{\{\alpha\}}$ for all the different values of $\{\alpha\}$ form a complete set of orthonormal functions. Each eigenfunction Ψ_J of the operator of the total angular momentum of the system can be constructed by means of a linear combination of Slater determinants

$$\Psi_J = \sum_{(\alpha)} C_J^{\{\alpha\}} \Psi_{\{\alpha\}}, \tag{4.80}$$

where $C_J^{\{\alpha\}}$ are the appropriate geometric coefficients.

The Pauli principle for nucleons, according to which there can be no more than one particle in a given one-particle state, follows directly from the determinant (4.79). As a consequence of the Pauli principle, the configuration to the ground state of the nucleus consists of several filled inner shells and one unfilled outer shell. In general, the properties of a nucleus (for example, its dimensions, density distribution, total binding energy, etc.) are determined by the state of all the nucleons, including those in the filled shells. However, certain properties of the nucleus (for example, the symmetry, spin, parity, isotopic spin and other characteristics of the ground state and also of excited states associated with transitions in the unfilled shell) are determined entirely by the state of the nucleons in the unfilled outer shell. Therefore, if we are interested in these properties of nuclei, we can confine ourselves to studying only the nucleons in the unfilled outer shell. Even in this case, however, the use of Slater determinants to construct states corresponding to well-defined values of the constants of the motion is an extremely cumbersome procedure and so the anti-symmetrized many-particle wave functions are usually constructed in some other way.

Schemes for constructing many-particle wave functions. In the independent-particle model, two schemes for constructing many-particle wave functions are used: the LS-coupling and jj-coupling schemes. In the case of light nuclei, when the spin-orbit interaction is unimportant, the *LS-coupling scheme* is applicable. In this case, the orbital angular momenta of the individual nucleons combine to give the total orbital angular momentum $\mathbf{L}$ of the system; the spin angular momenta of the nucleons are combined analogously into the total spin angular momentum $\mathbf{S}$ of the system; the angular momenta $\mathbf{L}$ and $\mathbf{S}$ are then combined to give the total angular momentum $\mathbf{J}$, usually called the *spin of the nucleus* . In this coupling scheme, $\mathbf{L}$ and $\mathbf{S}$ are constants of the motion, that is, the corresponding operators commute with the Hamiltonian H_0 of the system.

If spin-orbit interaction is important, the spin and orbital angular momenta of an individual nucleon combine to give a resultant angular momentum $\mathbf{j} = \mathbf{l} + \mathbf{s}$, which is a constant of the motion. The vectors $\mathbf{j}$ of the individual nucleons are then combined into the total angular momentum $\mathbf{J}$ of the nucleus. This scheme is called the *jj-coupling scheme* . In this case, the square, but not the projection, of the orbital angular momentum of a nucleon will be a constant of the motion (defined by the quantum number l). In the jj-coupling scheme, for each $l > 0$ two values of j are possible ($j = l \pm \frac{1}{2}$), so that, in place of one (nl) shell in the case of LS-coupling, in the case ofjj-coupling we have two shells, characterized by the quantum numbers n, l and j.

Nucleons in a given shell of the nucleus are usually called *equivalent nucleons.* The maximum number of nucleons that can be found in the shell (nl) coincides with the degeneracy of the level and is equal to $4(2l + 1)$. Similarly, the maximum number of nucleons in the shell (nlj) is equal to $2(2j + 1)$. In the following, we shall confine ourselves to studying many-particle wave functions for a system of k nucleons in an unfilled outer shell characterized by the quantum numbers n and l, or n, l and j. Since the particle states and hole states in the shell are equivalent, it makes sense to consider the case when the number of equivalent nucleons in the shell does not exceed half the maximum possible number, that is $k \leqslant 2(2l + 1)$ or $k \leqslant 2j + 1$.

In constructing completely anti-symmetrized many-particle wave functions in the LS-coupling scheme, sets of space and spin-isospin functions possessing all the possible symmetry properties with respect to interchange of particles are considered separately. Analogously, sets of space-spin and isospin functions are considered in the jj-coupling scheme. In this case, the complete anti-symmetric functions of the whole system are represented in the form of linear combinations of products of these functions.

If $k = 2$ the construction is trivial, since in this case the space and spin-isospin (or space-spin and isospin) functions can be either symmetric or anti-symmetric (if we assume that they are characterized by well-defined angular momenta), and so the complete set of anti-symmetric wave functions of two particles will consist only of products of space and spin-isospin (or space-spin and isospin) functions of opposite symmetry. If $k > 2$ the construction of the complete anti-symmetric wave functions is much more complicated; in this case, the sets of space and spin-isospin (or space-spin and isospin) functions contain not only symmetric and anti-symmetric functions but also functions of intermediate symmetry.

We shall consider the set of $(2l+1)^k$ products of one-particle space functions describing k nucleons in the (nl)-shell:

$$\Phi_{m_1 m_2 \ldots m_k}(\mathbf{r}_1, \mathbf{r}_2, \ldots, \mathbf{r}_k) = \varphi_{m_1}(\mathbf{r}_1)\varphi_{m_2}(\mathbf{r}_2)\ldots\varphi_{m_k}(\mathbf{r}_k) \tag{4.81}$$

(for brevity, we have omitted the subscripts nl). The set (4.81) forms a complete set of functions. An arbitrary interchange of the coordinates $\mathbf{r}_1$,$\mathbf{r}_2$,...,$\mathbf{r}_k$ in the general case converts any function $\Phi_{m_1 m_2 \ldots m_k}$ of this set into another function belonging to the same set.

The set of 4^k products of one-particle spin-isospin functions

$$\begin{aligned} &\Omega_{\mu_1\nu_1,\mu_2\nu_2,\ldots,\mu_k\nu_k}(\boldsymbol{\sigma}_1\boldsymbol{\tau}_1, \boldsymbol{\sigma}_2\boldsymbol{\tau}_2, \ldots, \boldsymbol{\sigma}_k\boldsymbol{\tau}_k) \\ = \; &\chi_{\mu_1}(\boldsymbol{\sigma}_1)\xi_{\nu_1}(\boldsymbol{\tau}_1)\chi_{\mu_2}(\boldsymbol{\sigma}_2)\xi_{\nu_2}(\boldsymbol{\tau}_2)\ldots\chi_{\mu_k}(\boldsymbol{\sigma}_k)\xi_{\nu_k}(\boldsymbol{\tau}_k) \end{aligned} \tag{4.82}$$

also forms a complete set of spin-isospin functions. In the jj-coupling case, complete set of the space-spin and isospin functions can be constructed analogously.

Any complete set of functions can be classified according to the irreducible representations of the groups of transformations under which the properties of the system are invariant. Since the Hamiltonian (4.75) is invariant under permutations of nucleons, the sets of functions (4.81) and (4.82) can be classified conveniently according to the irreducible representations of the permutation group.

Irreducible representations of the permutation group. We shall recall certain fundamental properties of the permutation group, for application to our problem (Wigner, 1959; Bayman, 1957). We shall label the k equivalent nucleons in the unfilled outer shell by the integers 1,2,...,k, using these to mean the space, spin and isospin coordinates of the nucleons.

Clearly, from these numbers we can form $k!$ permutations. The transition from the original permutation 1,2,...,k to any other $p_1, p_2, \ldots, p_k$ is effected by replacing 1 by p_1, 2 by p_2, and so on. This operation can be depicted thus:

$$\begin{pmatrix} 1 & 2 & 3 & \ldots & k \\ p_1 & p_2 & p_3 & \ldots & p_k \end{pmatrix} = \mathbf{P} \tag{4.83}$$

The set of all $k!$ permutations $\mathbf{P}$ forms a *group*. The *permutation group* is also called the *symmetric group* and denoted by Π_k or S_k.

The functions (4.81) and (4.82) respectively form bases of $(2l+1)^k$-dimensional and 4^k-dimensional representations of the permutation group S_k. Both these representations of the dimensional group are *reducible.*

Irreducible representations of the permutation group are realized by functions with a well-defined symmetry with respect to permutations of the particles. To find functions with different types of this symmetry, we break down the set of variables 1,2,...,k into $2l+1$ subsets. We denote this partition by

$$[\lambda] \equiv [\lambda_1, \lambda_2, \ldots, \lambda_{2l+1}], \tag{4.84}$$

where $\lambda_1 \geqslant \lambda_2 \geqslant \ldots \geqslant 0$ and $\lambda_1 + \lambda_2 + \ldots + \lambda_{2l+1} = k$. This partition can be depicted clearly by means of a *Young tableau*, in which each component λ_i is represented by a row of λ_i squares, with the rows decreasing in length from top to bottom. We associate one of the variables 1,2,...,k with each square. For example, if k=5, one of the possible Young tableaux has the form

$[3, 2] =$ 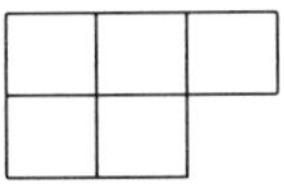 (4.85)

By symmetrizing the function $\Phi(1, 2, \ldots, k)$ (which is a product of k one-particle functions) with respect to the variables occurring in each of the rows, and anti-symmetrizing with respect to the variables of each column of the Young tableau, we obtain a function of a well-defined symmetry type; we shall denote this function by $\Phi(l^k[\lambda])$. Since for a given value of l there are only $2l+1$ different one-particle functions (differing in their values of m), the number of rows in the tableau $[\lambda]$ cannot exceed $2l+1$. In particular, if the Young tableau consists of only one row we obtain a completely symmetric function, while if it consists of only one column we obtain a completely anti-symmetric function.

By assigning the k variables in different ways to the squares of a Young tableau $[\lambda]$ we obtain a series of functions $\Phi(l^k[\lambda])$ which transform into each other under arbitrary permutations of the variables. These functions realize the irreducible representation $D^{[\lambda]}$, corresponding to the given Young tableau $[\lambda]$, of the permutation group S_k. The number n_λ of independent functions, which characterizes the dimensionality of the representations, is given by the number of arrangements in which the variables 1,2,...,k are arranged in the rows (from left to right) and columns (from top to bottom) in the standard order, while the arrangements themselves correspond to the basis functions of the irreducible representation.

In fact, the basis functions of an irreducible representation of some group will be uniquely determined if we know the irreducible representations of the successively smaller subgroups according to which these functions transform. The group S_{k-1} is a subgroup of the group S_k, S_{k-2} is a subgroup of S_{k-1}, and so on. If we have a certain irreducible representation of the group S_k, corresponding to a particular Young tableau of k squares, it is simultaneously also a representation of the subgroup S_{k-1}, but is now, in general, reducible. By means of a similarity transformation this representation can be reduced and we can obtain irreducible representations of the subgroup S_{k-1}. This similarity transformation corresponds to forming, from the original basis functions, certain linear combinations which transform according to irreducible representations of S_{k-1}. In this procedure, the irreducible representation of the group S_k is decomposed into those irreducible representations of the group S_{k-1} that correspond to the Young tableaux of $k-1$ squares obtained by removal of one square from the original tableau.

For example, the irreducible representation of the group S_5 corresponding to the Young tableau (4.85) can be decomposed into two irreducible representations of the group S_4, corresponding to the tableaux:

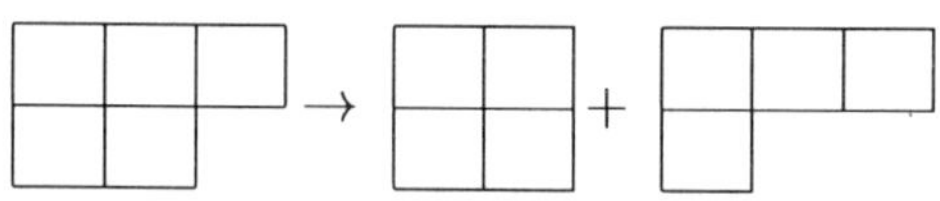

This process can be continued until we obtain the trivial representation of the group S_1, corresponding to a Young tableau of one square:

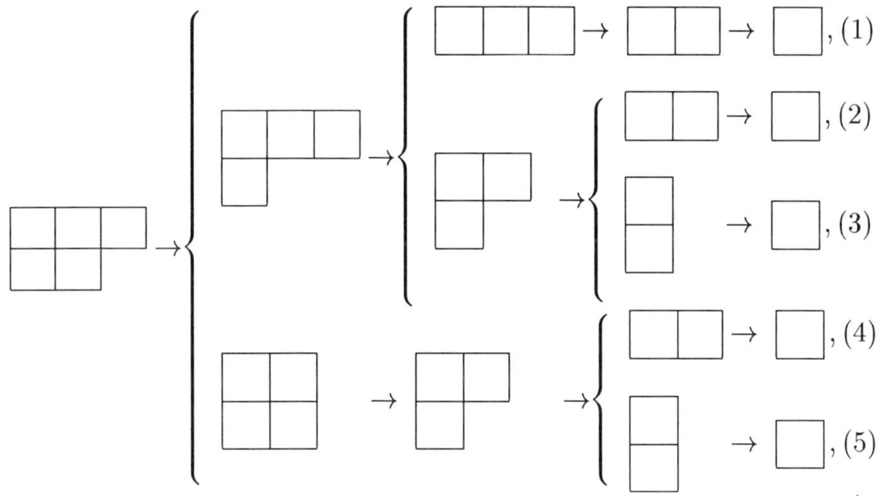

(4.86)

Thus, each basis function of an irreducible representation of the group S_k can be characterized by those irreducible representations of the subgroups S_{k-1}, S_{k-2},..., S_2 and S_1 according to which this function transforms at each stage of the reduction.

To denote these functions we can use a Young tableau with the variables arranged in the standard order in the rows and columns. By successively removing the square with the last variable (the highest number), we obtain chains of Young tableaux, which uniquely determine the required functions. Thus, for example, the basis functions of the irreducible representation of the group S_5 characterized by the Young tableau (4.85) are determined by the chains (4.86); they can be denoted in the following way:

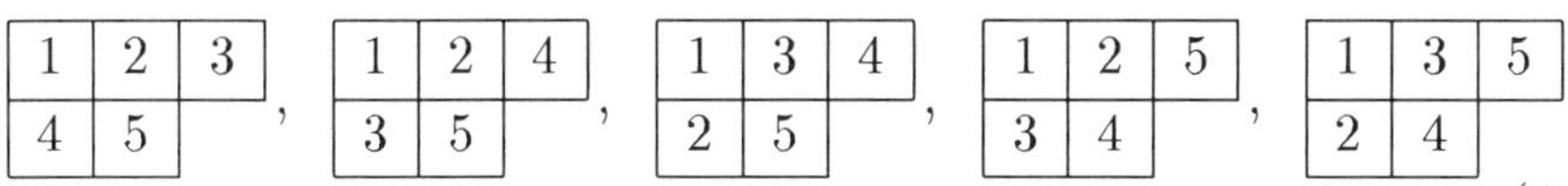

(1) (2) (3) (4) (5) (4.87)

There exists another, more convenient method of denoting the basis functions, namely by means of *Yamanouchi symbols* . The Yamanouchi symbol r is the set of k numbers, $r \equiv (r_k, r_{k-1}, \ldots, r_m, \ldots, r_1)$, indicating the rows r_m to which the variables m belong in the Young tableau $[\lambda]$. For example,

$(3, 2, 1, 2, 1) \equiv$

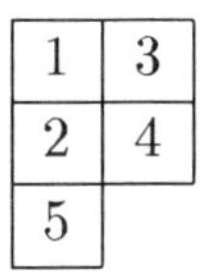

Since for the basis functions the variables are arranged in the standard order in the rows and columns of the Young tableau, the number of ones to the right of any position in the Yamanouchi symbol is not less than the number of twos, and so on. The Yamanouchi symbol r is written in the form of a subscript of the function and indicates that the function $\Phi_r(l^k[\lambda])$ belongs to specific irreducible representations of the permutation groups S_k, S_{k-1},..., and S_1.

Clearly the number of different Yamanouchi symbols corresponding to a given Young tableau $[\lambda]$ is equal to the dimensionality n_λ of the representation. For example, the dimensionality of the irreducible representation of the group S_5 characterized by the Young tableau (4.85) is equal to 5 (it coincides with the number of chains in (4.86)); the basis functions of this representation are characterized by the following Yamanouchi symbols:

$$(22111), (21211), (21121), (12211) \text{ and } (12121). \tag{4.88}$$

There is only one Yamanouchi symbol corresponding to a Young tableau consisting of only one row or only one column, and so the corresponding basis functions are symmetric or anti-symmetric with respect to interchange of any two particles.

In the general case, corresponding to each Young tableau $[\lambda]$ there are n_λ independent functions

$$\Phi_r(l^k[\lambda]), \tag{4.89}$$

of definite symmetry, which form the basis of the irreducible representation $D^{[\lambda]}$ of the permutation group S_k. The number n_λ characterizing the dimensionality of the representation is given by the number of different Yamanouchi symbol r. The functions $\Phi_r(l^k[\lambda])$ corresponding to the different Yamanouchi symbols form an orthogonal set of functions.

By partitioning the set of k variables in all possible ways, that is, by constructing all possible Young tableaux, we find all the irreducible representations of the permutation group S_k. As an example, we give the sets of basis functions of all the irreducible representations of the group S_4:

(1) $D^{[4]}$: $\begin{array}{|c|c|c|c|}\hline 1&2&3&4\\ \hline\end{array} \to (1111);$

(2) $D^{[31]}$: $\begin{array}{|c|c|c|}\hline 1&3&4\\ \hline \multicolumn{1}{|c|}{2}\\ \cline{1-1}\end{array} \to (1121),$ $\begin{array}{|c|c|c|}\hline 1&2&4\\ \hline \multicolumn{1}{|c|}{3}\\ \cline{1-1}\end{array} \to (1211),$

$\begin{array}{|c|c|c|}\hline 1&2&3\\ \hline \multicolumn{1}{|c|}{4}\\ \cline{1-1}\end{array} \to (2111);$

(3) $D^{[22]}$: $\begin{array}{|c|c|}\hline 1&2\\ \hline 3&4\\ \hline\end{array} \to (2211),$ $\begin{array}{|c|c|}\hline 1&3\\ \hline 2&4\\ \hline\end{array} \to (2121);$

(4) $D^{[211]}$: $\begin{array}{|c|c|}\hline 1&2\\ \hline \multicolumn{1}{|c|}{3}\\ \cline{1-1} \multicolumn{1}{|c|}{4}\\ \cline{1-1}\end{array} \to (3211);$ $\begin{array}{|c|c|}\hline 1&3\\ \hline \multicolumn{1}{|c|}{2}\\ \cline{1-1} \multicolumn{1}{|c|}{4}\\ \cline{1-1}\end{array} \to (3121);$

$\begin{array}{|c|c|}\hline 1&4\\ \hline \multicolumn{1}{|c|}{2}\\ \cline{1-1} \multicolumn{1}{|c|}{3}\\ \cline{1-1}\end{array} \to (1321);$

(5) $D^{[1111]}$: $\begin{array}{|c|}\hline 1\\ \hline 2\\ \hline 3\\ \hline 4\\ \hline\end{array} \to (4321).$

The representations 1 and 5 are one-dimensional,the representations 2 and 4 are three-dimensional, and the representation 3 is two-dimensional.

We must distinguish inner and outer products of irreducible representations of the permutation groups. A product of irreducible representations of the same permutation group (that is, for the same system of particles) is called an inner product and is indicated by the symbol $\times$. The sum of irreducible representations into which the inner product can be decomposed is called, as in the case of the rotation group, a *Clebsch-Gordan sum.*

A product of irreducible representations of two different permutation groups (that is, for different systems of particles) is called an *outer product* and is indicated by the symbol $\otimes$. An outer product of irreducible representations of two different permutation groups S_{k_1} and S_{k_2} is a reducible representation of the permutation group $S_{k_1+k_2}$. *Littlewood's rule* enables us to determine which irreducible representations $D^{[\lambda]}$ of the group $S_{k_1+k_2}$ (and with what weight ν_λ) appear in the decomposition of the outer product of the irreducible representations $D^{[\lambda_1]}$ and $D^{[\lambda_2]}$ of the groups S_{k_1} and S_{k_2}:

$$D^{[\lambda_1]} \bigotimes D^{[\lambda_2]} = \sum_{[\lambda]} \nu_\lambda D^{[\lambda]}. \tag{4.90}$$

To find the possible Young tableaux $[\lambda]$ of $k_1 + k_2$ squares given the partial tableaux $[\lambda_1]$ of k_1 squares and $[\lambda_2]$ of k_2 squares, we must:

1. Write in each square of the first row of the Young tableau $[\lambda_2]$ symbol $\alpha_1, \alpha_2,...$ with the subscript labelling the column of the tableau, write in each square of the second row of the tableau $[\lambda_2]$ another subscripted symbol $\beta_1, \beta_2,...$, and so on.

2. Add a square with α_1 to the Young tableau $[\lambda_1]$ in all possible ways such that an allowed Young tableau is obtained; then add, in the same way, the other squares of the first row of the tableau $[\lambda_2]$, but in such a way that no two symbols from the same row of the tableau $[\lambda_2]$ appear in the same column and no symbol with a larger subscript appears in a higher row than a symbol with any smaller subscript.

3. In a similar way, add the squares of the second row of the tableau $[\lambda_2]$, with the symbol β, to the tableaux obtained in (2), in such a way that the subscript of the symbol from the second row of λ_2 appears in a lower row than does the same subscript from the first row, and so on.

For example, we shall multiply $[\lambda_1]$=[11] and $[\lambda_2]$=[21]:

$D^{[11]} \bigotimes D^{[21]} \equiv$ 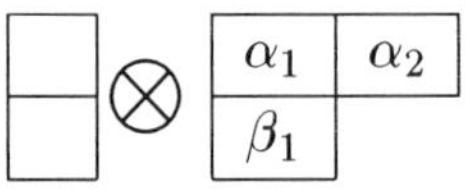

We add the square α_1 to $[\lambda_1]$ in all possible ways:

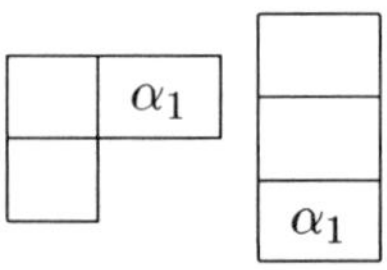

We then add the square α_2:

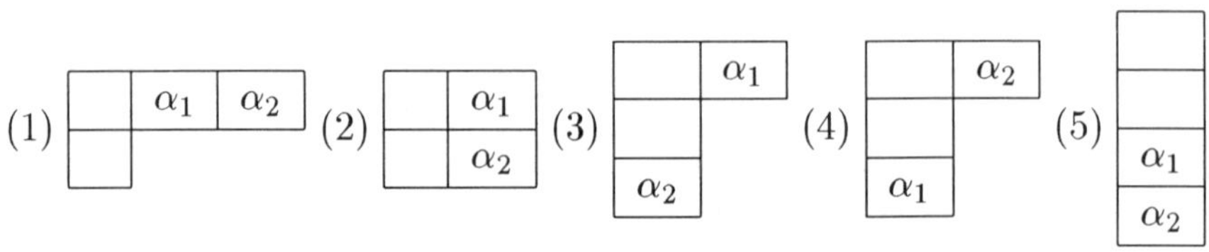

According to Littlewood's rule, the tableaux 2, 4 and 5 must be rejected. We now add the square β_1 to the remaining tableaux 1 and 3:

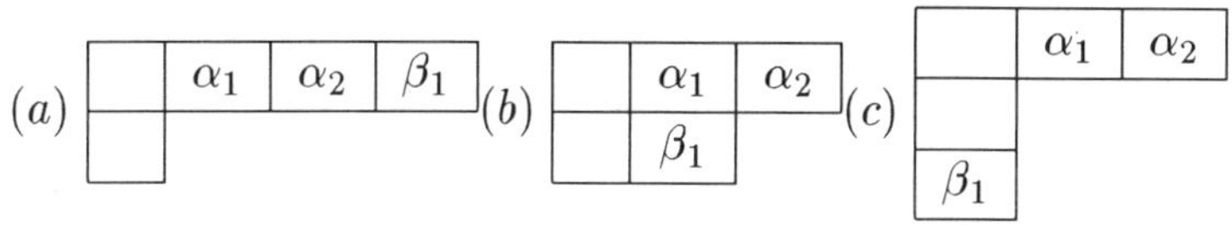

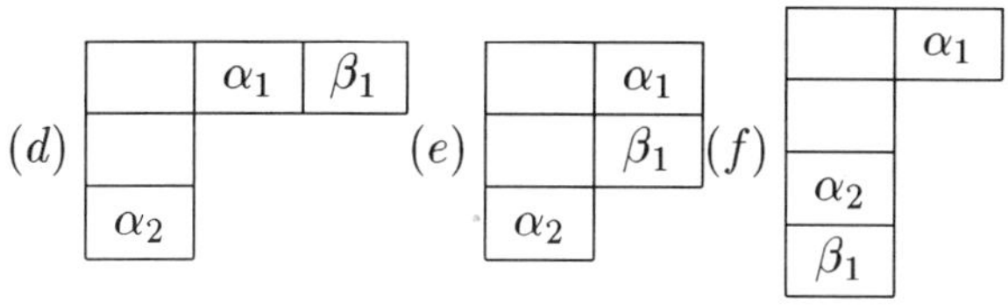

The tableaux a and d must be rejected. Thus, we finally obtain

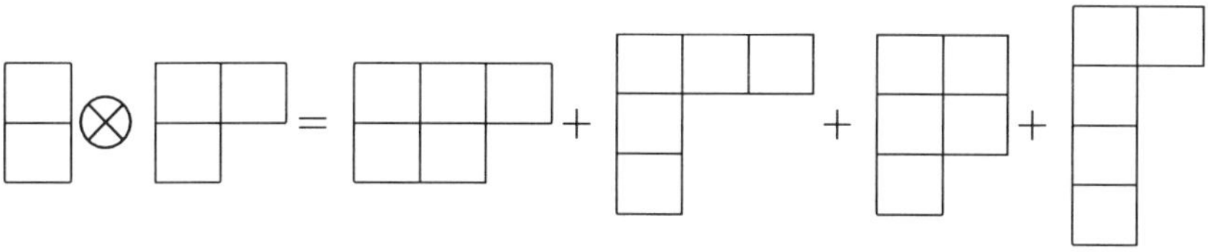

or

$$D^{[11]} \bigotimes D^{[21]} = D^{[32]} + D^{[311]} + D^{[221]} + D^{[2111]}.$$

In a similar way, one can show:

$$\begin{aligned}
D^{[2]} \bigotimes D^{[1]} &= D^{[3]} + D^{[21]}; \\
D^{[2]} \bigotimes D^{[2]} &= D^{[4]} + D^{[31]} + D^{[22]}; \\
D^{[2]} \bigotimes D^{[11]} &= D^{[31]} + D^{[211]}; \\
D^{[11]} \bigotimes D^{[11]} &= D^{[22]} + D^{[211]} + D^{[1111]}; \\
D^{[21]} \bigotimes D^{[21]} &= D^{[42]} + D^{[411]} + D^{[33]} + 2D^{[321]} \\
&+ D^{[3111]} + D^{[222]} + D^{[2211]}.
\end{aligned}$$

It can be seen from the last example that an irreducible representation can appear more than once in a decomposition. We note also that an outer product does not depend on the order in which the representations are multiplied.

The expansions of the above-mentioned *inner (Kronecker) direct products* of presentations of the permutation group S_k in terms of irreducible representations are used less frequently in the nonrelativistic nuclear theory than outer products. We outline this question here because in Chapter 7, where the quark structure of nucleons and the lightest nuclei is discussed, the outer products are used rather frequently. We consider now the simple method of expansion of the inner product of two irreducible representations of the permutation group S_k in the irreducible representations $D^{[f]}$ of the same group

$$D^{[\lambda]} \times D^{[\lambda']} = \sum_{[f]} \nu_f^{\lambda\lambda'} D^{[f]}$$

. It is based on the known characters of the irreducible representations of the symmetrical group. At the same time, the order of multiplied representations $D^{[\lambda]}$ and $D^{[\lambda']}$ makes no difference. Every representations $D^{[\lambda]}$, $D^{[\lambda']}$ and $D^{[f]}$ correspond now to the same k particles.

We recall that the character $\chi(P)$ of an arbitrary representation (reducible or irreducible) is determined for a given element P of the group as a sum of diagonal elements, i.e., as the spur or trace of the matrix $D_{\alpha\beta}(P)$ of this group representation:

$$\chi(P) = \sum_{\alpha} D_{\alpha\alpha}(P).$$

It follows after the definition that characters of all equivalent representations, corresponding to the same group element P, are the same. The characters of all elements P from the same class of the group are also equal to each other.

It can be shown in a general case, that a number ν_f of appearance of a given irreducible representation D^f in the expansion of an arbitrary representation of a finite group in the irreducible representation, is equal to a sum over all elements of the product group of the character representation under consideration and the complex conjugated character of the irreducible representation D^f divided by the group order. Since, the character of the Kronecker product of representations is just equal to a product of the characters of multiplied representations, then in the case of the permutation group S_k (matrices and representation characters can be chosen real), a number of appearance $\nu_f^{\lambda\lambda'}$ of the representation $D^{[f]}(P)$ in the expansion of the inner (Kronecker) product $D^{[\lambda]}(P) \times D^{[\lambda']}(P)$ is given by the expression

$$\nu_f^{\lambda\lambda'} = \frac{1}{k!} \sum_{P} \chi^{[f]}(P) \chi^{[\lambda]}(P) \chi^{[\lambda']}(P).$$

Apparently, $\nu_f^{\lambda\lambda'}$ are nonnegative integers. Since, the characters of all elements P from the same class C of a group are equal (it means that the character is a function of the class and denoted as $\chi(C)$), then the last formula can be also rewritten in the form

$$\nu_f^{\lambda\lambda'} = \frac{1}{k!}\sum_C n_C \chi^{[f]}(C)\chi^{[\lambda]}(C)\chi^{[\lambda']}(C).$$

where n_C is a number of elements in the class C, the sum is taken over all classes of the group S_k.

To find the characters of irreducible representation of the finite groups and the permutation groups, as well, we can use some known methods. There are tables of these characters in the literature.

As an example we consider a few expansions of inner products $D^{[\lambda]} \times D^{[\lambda']}$ or the irreducible representations for different permutation groups S_k obtained by a simple calculation of corresponding numbers $\nu_f^{\lambda\lambda'}$ according to the above-cited formulas, by using the tables of *characters of irreducible representations* :

$$D^{[3]} \times D^{[3]} = D^{[3]}, \qquad D^{[3]} \times D^{[21]} = D^{[21]},$$

$$D^{[3]} \times D^{[111]} = D^{[111]}, \qquad D^{[111]} \times D^{[111]} = D^{[3]},$$

$$D^{[21]} \times D^{[111]} = D^{[21]}, \qquad D^{[21]} \times D^{[21]} = D^{[3]} + D^{[21]} + D^{[111]},$$

$$D^{[31]} \times D^{[22]} = D^{[31]} + D^{[211]},$$

$$D^{[41]} \times D^{[2111]} = D^{[311]} + D^{[221]} + D^{[2111]} + D^{[11111]},$$

$$D^{[32]} \times D^{[311]} = D^{[41]} + D^{[32]} + 2D^{[311]} + D^{[221]} + D^{[2111]},$$

$$\begin{aligned} &D^{[311]} \times D^{[311]} \\ &= D^{[5]} + D^{[41]} + 2D^{[23]} + D^{[311]} + 2D^{[221]} + D^{[2111]} + D^{[11111]}. \end{aligned}$$

In Chapter 7, where we will consider once more the inner products, some of the cited relations become more clear. We only note here that the one-dimensional antisymmetric representation $D^{[f]} = D^{[111...11]} \equiv D^{[1^k]}$ of S_k group with the Young tableaux of one column consisting k squares appears (only one time) in the inner product expansion $D^{[\lambda]} \times D^{[\lambda']}$ only if the Young tableaux can be obtained from each other by change of rows and columns and vice versa. This can be seen, in particular, from the above cited examples. It can be used when constructing the completely antisymmetric wave function of a system of k equivalent nucleons.

Construction of anti-symmetrized wave functions. We shall consider the system of $(2l+1)^k$ space functions (4.81) describing k equivalent nucleons in the unfilled outer shell (nl). In general, these functions transform according to a reducible representation of the permutation group S_k. Symmetrization and anti-symmetrization of these functions with respect to certain space variables in accordance with a particular Young tableau $[\lambda]$ means that we are forming linear combinations of these functions, which transform according to the irreducible representation corresponding to $[\lambda]$. By choosing all possible partitions $[\lambda]$ we obtain a new system of $(2l+1)^k$ linearly independent functions; this system can be broken down into separate sets, each of which transforms according to one of the irreducible representations of the group S_k under permutations.

The wave functions of one such set, characterized by the Young tableau $[\lambda]$, can be written in the form

$$\Phi_r(l^k[\lambda]\xi), \tag{4.91}$$

where r is the Yamanouchi symbol, which takes all possible values for the given tableau $[\lambda]$. The functions of a given set $[\lambda]$ differ in their Yamanouchi symbols. The label $[\lambda]$ of the Young tableau and the Yamanouchi symbol r belong to the set of quantum numbers characterizing the state. If $[\lambda]$ and r do not form a complete set, further quantum number ξ must be introduced to specify the state. It is found to be necessary to introduce the additional quantum numbers ξ in the case when the irreducible representation $[\lambda]$ appears more than once in the reduction of (4.81).

Under permutations, the functions (4.91) transform according to the relation

$$\Phi'_r(l^k[\lambda]\xi) = \sum_{r'} D^{[\lambda]}_{rr'}(P)\Phi_{r'}(l^k[\lambda]\xi), \tag{4.92}$$

where $D^{[\lambda]}_{rr'}$ is the matrix of the representation $[\lambda]$; the dimensionality of this matrix is equal to the number n_λ of basis functions of this representation.

Analogously, by forming linear combinations of the spin-isospin functions (4.82), we can obtain a set of functions

$$\Omega_{r'}((st)^k[\lambda]\eta), \tag{4.93}$$

transforming according to the irreducible representation $D^{[\lambda']}$ of the group S_k, where $[\lambda']$ is a Young tableau of k squares and not more than four rows; r' is the Yamanouchi symbol and η represents the additional quantum numbers. (For nucleons, $s = \frac{1}{2}$ and $t = \frac{1}{2}$). Clearly, the total wave function Ψ of a system of k nucleons should be represented in the form of a linear superposition of products of the space and spin-isospin functions (4.91) and (4.93). $[4(2l+1)]^k$ such products, with all possible Young tableaux, form a complete set of wave functions of the configuration l^k. Because of the Pauli principle, the total wave function Ψ of a system of k nucleons must be anti-symmetric with respect to interchange of any pair of nucleons, and so the spin-isospin functions (4.93) should be described by the Young tableau $[\lambda']$ conjugate to the Young tableau $[\lambda]$ for the space functions (4.91). (Young tableaux are said to be conjugate if they are obtained from each other by converting the rows into columns and vice versa. The Young tableau conjugate to a given tableau $[\lambda]$ is denoted by $[\tilde{\lambda}]$, and the corresponding Yamanouchi symbol by $\tilde{r}$. The dimensionalities of the representations $[\lambda]$ and $[\tilde{\lambda}]$ are clearly the same). Thus, in (4.93) we must put $[\lambda'] = [\tilde{\lambda}]$ and $r' = r$.

The complete anti-symmetric (with respect to a permutation of any pair of nucleons) wave function of a system of k equivalent nucleons Ψ is transformed under action of operations P from the permutation group (symmetrical group) S_k according to the irreducible one-dimensional anti-symmetric representation of S_k group. It is characterized by the Young tableaux consisting of one column with k squares (such Young's tableaux is denoted as $[1^k]$), and the function Ψ gains only a factor $(-1)^P$. The latter equals to +1 for even permutations P or -1 for odd permutations P:

$$P\Psi = (-1)^P \Psi.$$

To express the function Ψ in terms of spatial and spin-isospin wave functions (4.91) and (4.93), we start with the general case when the Young tableaux $[\lambda]$ and $[\lambda']$ in (4.91) and (4.93) do not connected by the complex conjugation condition. The total wave function $\Psi([f]\tau(f)\xi\eta r_f)$ is a basis function of one of the possible irreducible representations $D^{[f]}$ of S_k group that is characterized by definite Young's tableaux $[f]$ with k squares. Here, we introduce the corresponding Yamanouchi symbol $\mathbf{r}_f$ and a new additional quantum number $\tau(f)$ for the total wave function. It is clear that the irreducible representation $D^{[f]}$ must appear in the expansion of the inner (Kronecker) product of representations $D^{[\lambda]}$ and $D^{[\lambda']}$ in terms of the irreducible representations of S_k group. If the representation $D^{[f]}$ appears in this expansion more than one time, we have to introduce the addition quantum number $\tau(f)$, which equals to a number of independent sets of the total wave function. This means that the total wave function $\Psi([f]\tau(f)\xi\eta r_f)$ similarly to the spin-angular wave function (4.19) can be presented, by using the Clebsch-Gordan coefficients of the permutation group $C^{[f]\tau(f)r_f}_{[\lambda]r,[\lambda']r'}$ (the denotations $\langle[\lambda]r,[\lambda']r'|[f]\tau(f)r_f\rangle$ and $S^{[f]\tau(f)[\lambda][\lambda']}_{r_f rr'}$ are also used) in the form of the following superposition

$$\Psi([f]\tau(f)\xi\eta r_f) = \sum_{rr'} C^{[f]\tau(f)r_f}_{[\lambda]r[\lambda']r'} \Phi_r(l^k[\lambda]\xi)\Omega_{r'}((st)^k[\lambda']\eta).$$

If and only if the Young tableaux $[\lambda]$ and $[\lambda']$ are conjugated ($[\lambda'] = [\tilde{\lambda}]$) there is the one-dimensional representation $D^{[f]} = D^{[1^k]}$ with the Young tableaux of one column. It appears only one time ($\tau(f) = 1$). At the same time, the Yamanouchi symbol r_f may have only one value $r_k = (k, k-1, \ldots, 2, 1)$. In this case, the Clebsch-Gordan coefficient of the permutation group S_k takes quite simple form

$$C^{[1^k]\tau(1^k)=1,r_{1^k}}_{[\lambda]r,[\tilde{\lambda}]\tilde{r}'} = \frac{1}{\sqrt{n_\lambda}} \Lambda^{[\lambda]}_r \delta_{rr'},$$

where n_λ is a dimension of the representation $D^{[\lambda]}$; the sign factor is $\Lambda^{[\lambda]}_r = +1$, if the symbol r is obtained from the Yamanouchi symbol with all numbers arranged in descending (nonincreasing) order with the help of an even number permutations; $\Lambda^{[\lambda]}_r = -1$, if a transposition number is odd and the transpositions of equal and the transpositions of equal numbers should not be taken into account (we will consider the inner products of representations coefficients of the Clebsch-Gordan permutation group in Chapter 7).

Thus, the *total wave function of the system of k equivalent nucleons* Ψ anti-symmetric with respect to permutations of any pair of nucleons can be expressed in terms of functions Φ_r and $\Omega_{\tilde{r}}$ in the form of single sum of their products:

$$\Psi(1^k[\lambda]\xi\eta) = \frac{1}{\sqrt{n_\lambda}} \sum_r \Lambda^{[\lambda]}_r \Phi_r(l^k[\lambda]\xi)\Omega_{\tilde{r}}\left((st)^k[\tilde{\lambda}]\eta\right), \tag{4.94}$$

where the total function $\Psi([1^k], \tau(1^k) = 1, \xi\eta, r_k)$ is written as $\Psi(l^k[\lambda]\xi\eta)$. The sum in (4.94) is taken over all Yamanouchi symbols associated with the Young tableaux $[\lambda]$. In this case, the wave function Ψ is normalized by one, if Φ_r and $\Omega_{\tilde{r}}$ are normalized correspondingly. The requirement of anti-symmetricity of the function $\Psi(l^k[\lambda]\xi\eta)$ results in the fact that the Young tableaux $[\lambda]$, which determines the symmetry of the spatial function $\Phi_r(l^k[\lambda]\xi)$, describes the symmetry of the spin-isospin function $\Omega_{\tilde{r}}((st)^k[\tilde{\lambda}]\eta)$ and, that is extremely important, characterizes the total antisymmetric wave function of the whole system of k equivalent nucleons $\Psi(l^k[\lambda]\xi\eta)$.

It is worth noting, that when expanded the inner product of permutations $D^{[\lambda]} \times D^{[\lambda']}$ in irreducible representations of the permutation group $D^{[f]}$, we come across the one-dimensional representation $D^{[k]}$ once (and only once) only in the case, if $[\lambda'] = [\lambda]$ (and not $[\lambda'] = [\tilde{\lambda}]!$). In this case, we would obtain the *total symmetric normalized wave function*

$$\Psi^s = \frac{1}{\sqrt{n_\lambda}} \sum_r \Phi_r(l^k[\lambda]\xi)\Omega_r((st)^k[\lambda]\eta)$$

with the Young tableaux $[f] = [k]$ of one row consisting of k squares but without the sign factor $\Lambda_r^{[\lambda]}$. Unfortunately, it has no physical meaning here.

The spatial and spin-orbit wave functions of the system of k identical nucleons separately may be, in particular, symmetric and anti-symmetric. For example, for three nucleons ($k = 3$), according to the above-cited, the possible symmetric Ω^s and anti-symmetric Ω^a spin-isospin functions can be expressed in terms of spin χ and isospin ξ functions in the following way (see Chapter 2)

$$\Omega^s\left((st)^3[3],[3]\right) = \chi^s\left(s^3[3]\right)\xi^s\left(t^3[3]\right),$$

$$S = T = \frac{3}{2},$$

$$\begin{aligned} \xi^s &\equiv \Omega^s\left((st)^3[3],[21]\right) \\ &= \frac{1}{\sqrt{2}}\left[\chi'\left(s^3[21]\right)\xi'\left(t^3[21]\right) + \chi''\left(s^3[21]\right)\xi''\left(t^3[21]\right)\right], \end{aligned}$$

$$S = T = \frac{1}{2},$$

$$\begin{aligned} \xi^a &\equiv \Omega^a\left((st)^3[1^3],[21]\right) \\ &= \frac{1}{\sqrt{2}}\left[\chi'\left(s^3[21]\right)\xi''\left(t^3[21]\right) - \chi''\left(s^3[21]\right)\xi'\left(t^3[21]\right)\right], \end{aligned}$$

$$S = T = \frac{1}{2},$$

where S and T are the total spin and isospin of the three-nucleon system. The first Young tableaux in the functions Ω^s and Ω^a characterizes the permutation symmetry of the whole spin-isospin function. The second Young tableaux characterizes only the spin function symmetry. The functions with index s are completely symmetric, i.e., correspond to the Young tableaux [3]. The functions with index a are anti-symmetric and correspond to the Young tableaux $[1^3]$. The spin functions of the intermediate symmetry $\chi'(s^3[21]) \equiv \chi'(s^3[21]121)$ and $\chi''(s^3[21] \equiv \chi''(s^3[21]211)$ and analogous isospin functions ξ' and ξ'' for the Young tableaux [21] correspond to two the possible Yamanouchi symbols $r = 121$ and $\tilde{r} = 211$. That is why the related functions $\Omega^s((st)^3[3],[21])$ and $\Omega^a((st)^3[1^3][21])$ consist of two terms. These two terms are the parts the functions $\Omega^a((st)^3[1^3][21])$ with opposite signs, since the factors $\Lambda^{[21]}_{121}$ and $\Lambda^{[21]}_{211}$ have opposite signs. It is clear (see Chapter 2) that functions χ^s and ξ^s correspond to the total values of spin S and isospin T, which equal to 3/2, and functions χ', ξ' and χ'', ξ'' correspond to $S = T = 1/2$. For the wave functions with one prime, the first two nucleons are in the singlet (anti-symmetric) spin and isospin states, and for the functions with two primes, they are in the triplet (symmetric) states.

Note that, if we take formally two terms in the functions $\Omega^s((st)^3[3],[21])$ and $\Omega^a((st)^3[1^3][21])$ with different relative signs, we obtain two *spin-isospin wave functions with the intermediate permutation symmetry* .

$$\begin{aligned}
\xi'' &\equiv \Omega''\left((st)^3[21],[21]\right) \\
&= \frac{1}{\sqrt{2}}\left[\chi'\left(s^3[21]\right)\xi'\left(t^3[21]\right) - \chi''\left(s^3[21]\right)\xi''\left(t^3[21]\right)\right], \\
S &= T = \frac{1}{2},
\end{aligned}$$

$$\begin{aligned}
\xi' &\equiv \Omega'\left((st)^3[21],[21]\right) \\
&= \frac{1}{\sqrt{2}}\left[\chi'\left(s^3[21]\right)\xi''\left(t^3[21]\right) + \chi''\left(s^3[21]\right)\xi'\left(t^3[21]\right)\right], \\
S &= T = \frac{1}{2}.
\end{aligned}$$

They correspond to the Yamanouchi symbols 211 and 121. If Φ' and Φ'' are the spatial basis functions of the two-dimensional irreducible representation $D^{[21]}$ of S_3 group with the Yamanouchi symbols 121 and 211, then, according to (4.94), the combination $\Phi'\Omega'' - \Phi''\Omega'$ is the anti-symmetric function with respect to permutations of any two nucleons.

Since in the total anti-symmetric function (4.94) the Young tableaux for the space and spin-isospin functions are mutually conjugate, the number of columns in the Young tableau for the space function cannot be greater than four, and the number of columns in the Young tableau for the spin-isospin function cannot exceed $2l + 1$. Hence it follows, in particular, that the maximum number of nucleons in a shell with orbital angular momentum l is equal to $4(2l+1)$.

Irreducible representations of groups of unitary transformations. The structure of the functions (4.91) and (4.93) is such that they not only form a basis of an irreducible representation of the permutation group but are simultaneously basis functions of an irreducible representation of a group of unitary transformations.

We shall consider the *group of linear unitary transformations in the* $(2l+1)$*-dimensional space of the functions* φ_{nlm}; we shall denote this group by U_{2l+1}. The one-particle wave functions φ_{nlm} transform into each other under these transformations:

$$\varphi'_{nlm}(\mathbf{r}) = \sum_{m'} U_{m'm}(U_{2l+1})\varphi_{nlm'}(\mathbf{r}). \tag{4.95}$$

(These transformations conserve the orthogonality of the set of functions φ_{nlm}.) The group U_{2l+1} contains as a subgroup the unimodular group SU_{2l+1}, which incorporates all the unitary transformations in the $(2l+1)$-dimensional space for which the determinants of the matrices are equal to $+1$. Clearly, the *rotation group* R_3 in three-dimensional space is contained as a subgroup in the *unitary unimodular group* SU_{2l+1} .

As was shown by Weyl (1946), the basis functions $\Phi_r(l^k[\lambda]\xi)$ of an irreducible representation of the permutation group are simultaneously basis functions of an irreducible representation of the group SU_{2l+1}, characterized, like the irreducible representation of the permutation group S_k, by the Young tableau $[\lambda]$. The dimensionality of the irreducible representation of the group SU_{2l+1} will be determined by the number of possible values of the additional quantum numbers ξ. In the following we shall denote this number by N_λ.. Thus, under the action of an operation of the group SU_{2l+1}, the space functions (4.91) transform according on the rule

$$\Phi'_r(l^k[\lambda]\xi) = \sum_{\xi'} U^{[\lambda]}_{\xi'\xi}(SU_{2l+1})\Phi_r(l^k[\lambda]\xi'). \tag{4.96}$$

It can be seen from (4.92) and (4.96) that the number of times that the representation $D^{[\lambda]}$ appears in the reduction of (4.81) is equal to the dimensionality N_λ of the representation $U^{[\lambda]}$, and conversely. The set of functions $\Phi_r(l^k[\lambda]\xi)$ belonging to the same Young tableau $[\lambda]$, but differing in their values of the quantum numbers r and ξ, can be represented conveniently in the form of a rectangular table, with the rows labelled by the Yamanouchi symbols r and the columns by the quantum numbers ξ:

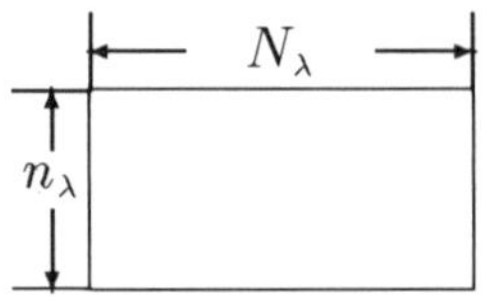

Under permutations functions appearing in the same column transform according to the irreducible representation $D^{[\lambda]}$ of the group S_k, while under an operation from the group SU_{2l+1} functions appearing in the same row transform according to the representation $U^{[\lambda]}$ of the group SU_{2l+1}.

Under an operation of the group SU_4, the spin-isospin functions $\Omega_{\tilde{r}}((st)^k \times [\tilde{\lambda}]\eta)$ transform according to the irreducible representation $U^{[\tilde{\lambda}]}$ of this group. We note that, in general, the dimensionality $N_{[\tilde{\lambda}]}$ of this representation differs from the dimensionality N_λ. The set of functions $\Omega_{\tilde{r}}((st)^k[\tilde{\lambda}]\eta)$ can also be represented conveniently in the form of a rectangular table, with the rows and columns labelled by means of the quantum numbers r and η:

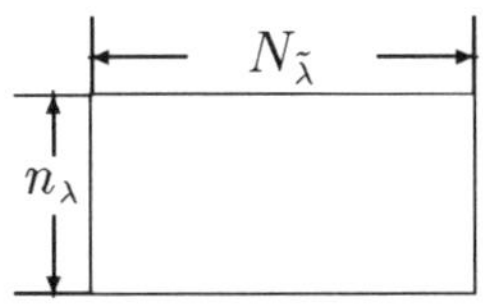

Under permutations functions in the same column transform according to the irreducible representation $D^{[\tilde{\lambda}]}$ of the group S_k, and under an operation of the group SU_4 functions from the same row transform according to the irreducible representation $U^{[\tilde{\lambda}]}$ of the group SU_4.

According to (4.94), the complete wave functions of a system of k nucleons, anti-symmetrized with respect to interchanges of any pair of nucleons, are characterized by particular Young tableaux; functions $\Psi(l^k[\lambda]\xi\eta)$ corresponding to a given Young tableau and differing in their values of the additional quantum numbers ξ and η transform according to irreducible representations of the groups SU_{2l+1} and SU_4. We can further classify the states (4.94) of a system of k nucleons by considering the irreducible representations of the subgroups of the group SU_{2l+1} and of the group SU_4, for example, the irreducible representations of the rotation group R_3, and so on.

4.5. Classification of Many-particle States

Classification of states in the LS-coupling scheme. By forming linear combinations of the functions (4.91) with fixed values of $[\lambda]$ and r and different values of ξ, we can obtain sets of functions which are basis functions of the irreducible representations D^L of the rotation group R_3, that is, which are eigenfunctions of the total orbital angular momentum L and of its projection M. The functions which form a basis of the irreducible representation D^L will be denoted as follows:

$$\Phi_r(l^k[\lambda]\alpha LM), \tag{4.97}$$

where α is the extra quantum number necessary when the representation D^L appears more than once in the reduction of $U^{[\lambda]}$. The possible values of the orbital angular momentum L are determined by direct decomposition of the irreducible representation $[\lambda]$ of the group SU_{2l+1} into irreducible representations of the rotation group R_3.

As an example, we shall consider the simplest case of a system of two equivalent nucleons. The space function in this case can be represented in the form

$$\Phi(l^2[\lambda]LM) = \sum_{m_1 m_2} (lm_1 lm_2|LM)\varphi_{lm_1}(\mathbf{r}_1)\varphi_{lm_2}(\mathbf{r}_2). \tag{4.98}$$

The function (4.98) is symmetric for the Young tableau [2] and anti-symmetric for the tableau [11]. (The Yamanouchi symbol in the case of two particles takes only one value, and the extra quantum number α is absent.) We act on (4.98) with the permutation operator P_r. Using the symmetry property of the Clebsch-Gordan coefficients:

$$(lm_1 lm_2|LM) = (-1)^{2l-L}(lm_2 lm_1|LM)$$

and noting that l is an integer, we obtain

$$P_r\Phi(l^2[\lambda]LM) = (-1)^L\Phi(l^2[\lambda]LM), \tag{4.99}$$

that is, even values of L should correspond to symmetric wave functions and odd values of L to anti-symmetric wave functions. Thus, in the reduction of the representation $U^{[2]}$ representations D^L with even L can appear, while in the reduction of $U^{[11]}$ representations D^L with odd L appear ($0 \leqslant L \leqslant 2l$). For example, in the case of a p-shell ($l = 1$), we have

$$\left.\begin{array}{lll} [2] & \rightarrow & L = 0, 2; \\ [11] & \rightarrow & L = 1. \end{array}\right\} \tag{4.100}$$

For shells containing a larger number of nucleons, the analogous reductions are considerably more complicated. However, the possible values of L for a given tableau $[\lambda]$ can always be found by vector addition of the angular momenta of the individual nucleons. Thus, we find the possible Young tableaux for three nucleons by adding one square to each of the tableaux (4.100):

$$[2]\bigotimes[1] = [3]+[21],$$
$$[11]\bigotimes[1] = [21]]+[111].$$

In the case of a p-shell, $[2]\to D^0+D^2$, $[11]\to D^1$ and $[1]\to D^1$, and so on

$$[3]+[21]\to(D^0+D^2)\bigotimes D^1 = 2D^1+D^2+D^3;$$

$$[21]+[111]\to D^1\bigotimes D^1 = D^0+D^1+D^2,$$

whence we find:

$$\left.\begin{array}{rcl}[3] & \to & L=1,3;\\ [21] & \to & L=1,2;\\ [111] & \to & L=0.\end{array}\right\} \qquad (4.101)$$

We also give the possible Young tableaux and values of L for four nucleons in a p-shell ($k=4$, $l=1$):

$$\left.\begin{array}{rcl}[4] & \to & L=0,2,4;\\ [31] & \to & L=1,2,3;\\ [22] & \to & L=0,2;\\ [211] & \to & L=1.\end{array}\right\} \qquad (4.102)$$

For $l=1$, Young tableaux with four rows are ruled out, since the maximum number of rows in the Young tableau characterizing the anti-symmetric function (4.94) is equal to $2l+1$.

In an analogous way, we can construct a set of spin-isospin functions

$$\Omega_{\tilde{r}}((st)^k[\tilde{\lambda}]\beta STM_SM_T), \qquad (4.103)$$

transforming not only according to an irreducible representation of the group SU_4 but also according to irreducible representations of its subgroups. These representations are characterized by the quantum numbers S and T of the total spin and total isotopic spin (M_S and M_T are their projections). The additional quantum number β is introduced when a pair of numbers (S,T) appears more than once for a given Young tableau $[\lambda]$.

The *complete anti-symmetric wave functions* describing the states of a system *of k nucleons* characterized by a definite space symmetry $[\lambda]$ and well-defined values of the orbital angular momentum L, the spin S, the isotopic spin T, and their projections M, M_S and M_T can be written in the form

$$\Psi(l^k[\lambda]\alpha\beta LSTMM_SM_T) = \frac{1}{\sqrt{n_\lambda}}\sum_r \Lambda_r^{[\lambda]}\Phi_r(l^k[\lambda]\alpha LM)\Omega_{\tilde{r}}((st)^k[\tilde{\lambda}]\beta STM_SM_T). \quad (4.104)$$

TABLE 4.5.

k	$[\lambda]$	L	$(2T+1,\ 2S+1)$
0	[0]	0	(11)
1	[1]	1	(22)
2	[2]	0,2	(13), (31)
	[11]	1	(11),(33)
3	[3]	1,3	(22)
	[21]	1,2	(22),(24),(42)
	[111]	0	(22),(44)
4	[4]	0, 2, 4	(11)
	[31]	1, 2, 3	(13),(31),(33)
	[22]	0, 2	(11),(15),(51),(33)
	[211]	1	(13),(31),(33),(35),(53)
5	[41]	1, 2, 3, 4	(22)
	[32]	1, 2, 3	(22),(24),(42)
	[311]	0, 2	(22),(24),(42),(44)
	[221]	1	(22),(24),(42),(26),(62),(44)
6	[42]	0, 2^2,3,4	(13),(31)
	[411]	1, 3	(11),(33)
	[33]	1, 3	(11),(33)
	[321]	1, 2	(13),(31),$(33)^2$,(15),(51),(35),(53)
	[222]	0	(13),(31),(35),(53),(17),(71)

Table 4.5 shows the complete set of states in the LS-coupling scheme for a system of k nucleons in a p-shell ($l = 1$) (Jahn, 1950). A pair of values of S and T corresponds to each value of L in the rows of Table 4.5. For example, for $k = 2$ for the Young tableau [2] the following states are possible: ${}^{13}S[2]$, ${}^{31}S[2]$, ${}^{13}D[2]$, and ${}^{31}D[2]$. It can be seen from Table 4.5 that, in the general case, for a given state symmetry described by a Young tableau $[\lambda]$ there exist several values of the pairs of numbers (T, S). The aggregate of these degenerate values of (T, S) for a given tableau $[\lambda]$ is

called a *Wigner supermultiplet* . On the other hand, the same values of T and S can correspond to different Young tableaux, that is, to states with different symmetry.

The quantum numbers $[\lambda], L, S$ and T uniquely describe the states of the system, except in the case of six nucleons, for which pairs of states correspond to the sets of quantum numbers $^{13}D[42]$, $^{31}D[42]$, $^{33}P[321]$, and $^{33}D[321]$ (see Table 4.5). For a complete description of the states in this case, the additional quantum numbers α and β are necessary.

If $l > 1$, the space states in the majority of cases are not uniquely determined by specifying $[\lambda]$ and L. To classify such states, we can introduce a certain group which contains R_3, as its subgroup and is simultaneously a subgroup of the group SU_{2l+1}. For this group, it is convenient to take the group R_{2l+1} of rotations in a space of $2l+1$ dimensions, which leaves invariant the symmetric bilinear form constructed from the space functions of two particles:

$$\sum_m (l-mlm|00)\varphi_{nl-m}(\mathbf{r}_1)\varphi_{nlm}(\mathbf{r}_2) \propto \sum_m (-1)^m \varphi_{nl-m}(\mathbf{r}_1)\varphi_{nlm}(\mathbf{r}_2). \tag{4.105}$$

The group R_{2l+1} is called the $(2l+1)$-dimensional orthogonal group.

Like the irreducible representation D^L of the group R_3, which is characterized uniquely by the positive integer L (since l is an integer), an irreducible representation of the group R_{2l+1} is characterized by a set of non-negative integers $(\omega_1, \omega_2, \ldots, \omega_l)$, chosen such that $\omega_1 \geqslant \omega_2 \geqslant \ldots \geqslant \omega_l$. This set of numbers corresponds to the additional number α introduced previously.

With increase of l, it is found necessary to introduce new groups, contained in R_{2l+1} and having R_3 as a subgroup. By this means, it becomes possible to perform a further classification of the states, according to the irreducible representations of these groups. Thus, the space functions of systems of k nucleons in a given shell (nl) can be characterized completely and uniquely by a set of quantum numbers indicating the irreducible representations of the chain of groups

$$\begin{array}{ccccccc} SU_{2l+1} \supset & R_{2l+1} \supset & \ldots \supset & R_3 \supset & R_2, \\ \downarrow & \downarrow & & \downarrow & \downarrow \\ [\lambda] & \alpha & & L & M \end{array} \tag{4.106}$$

according to which these functions transform. We have already considered a problem of this type, that is, the classification of states by means of the irreducible representations of a chain of groups, when we constructed basis functions of the irreducible representations of the permutation groups $S_k, S_{k-1}, \ldots, S_2, S_1$.

We can treat the spin-isospin functions Ω in an analogous way and elucidate the meaning of the quantum numbers β. If we know Φ and Ω, it is not difficult to construct, in accordance with (4.104), a complete orthogonal set of anti-symmetric wave functions Ψ for the configuration l^k.

Classification of states in the jj-coupling scheme. We now consider the classification of the states of a system of k equivalent nucleons in an unfilled outer shell in the presence of spin-orbit interaction (Flowers, 1952). In this case, the shell will be characterized by the quantum numbers n, l and j, and the one-particle wave functions will be represented in the form of products (4.77) of space-spin and isospin functions.

In place of the functions (4.81) and (4.82), we now consider products of k one-particle space-spin and isospin functions:

$$\begin{aligned} &\Phi_{m_1,m_2\ldots m_k}(\mathbf{r}_1\boldsymbol{\sigma}_1;\mathbf{r}_2\boldsymbol{\sigma}_2;\ldots;\mathbf{r}_k\boldsymbol{\sigma}_k) \\ =\ &\varphi_{m_1}(\mathbf{r}_1\boldsymbol{\sigma}_1)\varphi_{m_2}(\mathbf{r}_2\boldsymbol{\sigma}_2)\ldots\varphi_{m_k}(\mathbf{r}_k\boldsymbol{\sigma}_k), \end{aligned} \tag{4.107}$$

$$Z_{r_1r_2\ldots r_k}(\boldsymbol{\tau}_{1_k}\boldsymbol{\tau}_2,\ldots,\boldsymbol{\tau}_k)=\xi_{\nu_1}(\boldsymbol{\tau}_1)\xi_{\nu_2}(\boldsymbol{\tau}_2)\ldots\xi_{\nu_k}(\boldsymbol{\tau}_k). \tag{4.108}$$

From (4.107) and (4.108) we next construct, by means of Young tableaux, linear combinations $\Phi_r(j^k[\lambda]\xi)$ and $Z_{\tilde r}(l^k[\tilde\lambda]\eta)$ (like (4.91) and (4.93)) transforming according to the irreducible representations $D^{[\lambda]}$ and $D^{[\tilde\lambda]}$ of the permutation group S_k. Here ξ and η are additional quantum numbers. The space-spin functions $\Phi_r(l^k[\lambda]\xi)$ will simultaneously be basis functions of the irreducible representations $U^{[\lambda]}$ of the group SU_{2j+1} of all unitary unimodular transformations in the $(2j+1)$-dimensional space of the functions $\varphi_{jm}(\mathbf{r}\boldsymbol{\sigma})$. As in the case of LS-coupling, the representations $U^{[\lambda]}$. regarded as representations of the rotation group R_3, which is a subgroup of SU_{2j+1}, can be reduced and decomposed into irreducible representations D^L of the rotation group R_3. The only difference between this case and the case of LS-coupling lies in the half-integer character of the number j.

Thus, the space-spin and isospin functions of a system of k equivalent nucleons can be chosen in the form

$$\Phi_r(j^k[\lambda]\alpha JM_J), \tag{4.109}$$

$$Z_{\tilde r}(t^k[\tilde\lambda]TM_T), \tag{4.110}$$

where J is the total angular momentum of the shell, M is the projection of the total angular momentum, α represents the additional quantum numbers characterizing the space-spin state, T is the total isotopic spin and M_T is the projection of the isotopic spin (to specify the isospin state no additional quantum numbers are necessary).

We shall consider the simplest case of two nucleons. In this case, the space-spin function

$$\Phi(j^2[\lambda]JM_J) = \sum_{m_1 m_2} (jm_1 jm_2|JM_J)\varphi_{jm_1}(\mathbf{r}_1, \boldsymbol{\sigma}_1)\varphi_{jm_2}(\mathbf{r}_2, \boldsymbol{\sigma}_2) \quad (4.111)$$

can be either symmetric (for the Young tableau [2]) or anti-symmetric (for the Young tableau [11]). By making use of the symmetry of the Clebsch-Gordan coefficients

$$(jm_1 jm_2|JM_J) = (-1)^{2j-J}(jm_2 jm_1|JM_J),$$

we can write

$$P\Phi(j^2[\lambda]JM_J) = (-1)^{J+1}\Phi(j^2[\lambda]JM_J) \quad (4.112)$$

(where we have taken into account that $2j$ is an odd number). Thus, anti-symmetric functions correspond to even values of J and symmetric functions to odd values of J. Therefore, $U^{[11]}$ decomposes into the representations $D^0, D^2, \ldots, D^{2j-1}$, and $U^{[2]}$ into the representations $D^1, D^3, \ldots, D^{2j}$. Similar decompositions can also be performed for systems in which the number of particles $k > 2$. For example, in the case of two nucleons ($k = 2$) in the shell $j = \frac{3}{2}$, we have

$$\left.\begin{array}{lcl} [2] & \rightarrow & J = 1, 3; \\ [11] & \rightarrow & J = 0, 2. \end{array}\right\} \quad (4.113)$$

Analogously, for three nucleons ($k = 3$) in the shell $j = \frac{3}{2}$, we can obtain

$$\left.\begin{array}{lcl} [21] & \rightarrow & J = \frac{1}{2}, \frac{3}{2}, \frac{5}{2}, \frac{7}{2}; \\ [111] & \rightarrow & J = \frac{3}{2}, \end{array}\right\} \quad (4.114)$$

and so on.

The *complete anti-symmetric wave functions of a system of k equivalent nucleons* have the form

$$\Psi(j^k[\lambda]\alpha JTM_J M_T) = \frac{1}{\sqrt{n_\lambda}} \sum_r \Phi_r(j^k[\lambda]\alpha JM_J) Z_{\tilde{r}}(t^k[\tilde{\lambda}]TM_T). \quad (4.115)$$

Since the number of possible values of the projection of the total angular momentum of one nucleon is equal to $2j + 1$, and the projection of the isotopic spin of the nucleon takes only two values, the Young tableau $[\lambda]$ can have no more than $2j+1$ rows and two columns. Since the conjugate Young tableau $[\tilde{\lambda}]$ characterizing the isospin function of the system can have only two rows (and $2j+1$ columns), there exists a unique correspondence between the Young tableau $[\lambda]$ and the value T of the isotopic spin of the system. Indeed, since the isotopic functions is symmetric in the particle coordinates in the rows and anti-symmetric in the particle coordinates in the columns of the tableau $[\tilde{\lambda}]$, the total isotopic spin T of the system will be equal to half the difference between the numbers of squares in the two rows of the tableau $[\tilde{\lambda}]$. (The isotopic spins of the particles whose coordinates are positioned in the protruding squares of the first row, under which there are no squares in the second row, can be assumed to be oriented in the same way, that is, they give the maximum projection T of the total isotopic spin). Clearly, the numbers of squares in the two rows of the tableau $[\tilde{\lambda}]$ (and in the two columns of the tableau $[\lambda]$ are equal to $\frac{1}{2}k \pm T$:

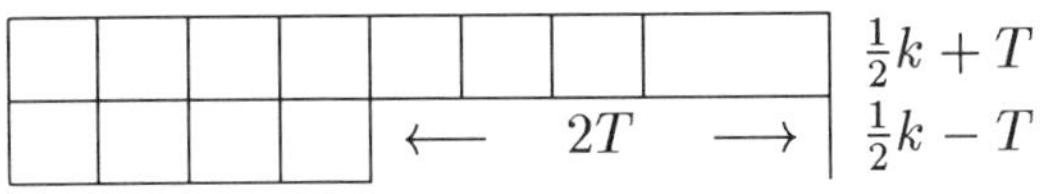

(4.116)

We can also make the converse statement, namely, that for a given number k of particles the isotopic spin T uniquely determines the tableau $[\tilde{\lambda}]$, that is, determines the isotopic permutation symmetry and, consequently, the space-spin permutation symmetry characterized by the tableau $[\lambda]$.

The Young tableau $[\lambda]$ also characterizes the complete anti-symmetric function (4.115) of the system. Clearly, one of the quantum numbers $[\lambda]$ and T in (4.115) is redundant, although both these numbers are usually given in the classification of states in the jj-coupling scheme. The situation here is similar to that encountered in the classification of the electron states of atoms, where the total spin S of the shell uniquely determines the permutation symmetry of the wave function.

Seniority and reduced isotopic spin. If $j \geqslant 3/2$ and the number k of equivalent nucleons is greater than 3, then several different states are possible for given values of the total spin and isospin characterized by the quantum numbers J, M_J, T and M_T. To distinguish these states we introduce two additional quantum numbers, the *seniority* s and the *reduced isotopic spin* t, which characterize the behavior of the wave function under symplectic transformations.

Symplectic transformations, which leave invariant the anti-symmetric bilinear form

$$\sum_m (j-mjm|00)\varphi_{j-m}(\mathbf{r}_1,\boldsymbol{\sigma}_1)\varphi_{jm}(\mathbf{r}_1,\boldsymbol{\sigma}_1)$$

$$\propto \sum_m (-1)^{j+m} \varphi_{j-m}(\mathbf{r}_1, \boldsymbol{\sigma}_1) \varphi_{jm}(\mathbf{r}_1, \boldsymbol{\sigma}_1), \tag{4.117}$$

form the *symplectic group* Sp_{2j+1} in a space of $2j + 1$ dimensions. The symplectic group Sp_{2j+1} is a subgroup of the unitary group U_{2j+1}, and, in its turn, Sp_{2j+1} contains the rotation group R_3 as a subgroup. The symplectic group Sp_{2j+1} plays the same role in the case of jj-coupling as that played by the group R_{2l+1} in the case of LS-coupling.

The irreducible representations of the symplectic group are characterized by Young tableaux obtained from $[\lambda]$ by removing squares corresponding to pairs of nucleons with opposite angular-momentum projections and with isospin equal to unity. We shall denote these tableaux by (σ):

$$(\sigma) = (\sigma_1, \sigma_2, \ldots, \sigma_\nu). \tag{4.118}$$

Clearly, $\sigma_1 \geqslant \sigma_2 \geqslant \ldots \geqslant \sigma_\nu \geqslant 0$ and $\sigma_1 \leqslant 2$, where the integer ν is defined by the condition $2\nu = 2j + 1$.

Since (σ) contains only two columns, to define (σ) it is sufficient to give the numbers of squares in these columns; by analogy with (4.116), we denote these numbers by $\frac{1}{2}s \pm t$. The number s gives the total number of elements in (σ) and is called the *seniority*, and the number t is half the difference between the lengths of the columns of (σ) and is called the *reduced isotopic spin*. Clearly, the relations $s \leqslant k$ and $t \leqslant T$ hold.

The numbers s and t uniquely define the irreducible representations (σ) of the symplectic group Sp_{2j+1}. We denote the basis functions of the irreducible representations thus:

$$\Phi_r(j^k[\lambda](\sigma)JM). \tag{4.119}$$

The different basis functions of a given irreducible representation (σ) differ in their values of J and M. We denote the number of basis functions, which defines the dimensionality of the representation (σ), by n_σ. Clearly, $n_\sigma = \sum(2J + 1)$, where the summation is performed over all the values of J appearing in (σ).

Instead of specifying (σ) in (4.119), we can specify the values of s and t. The seniority s of a state Φ defines the smallest number of particles for which a state possessing the same transformation properties as Φ is still possible. Analogously, the reduced isotopic spin t of a state Φ can be interpreted as the value of the isotopic spin T of the state with the same transformation properties as Φ and with the minimum number s of particles.

Thus, in the jj-coupling scheme the space-spin wave functions of k equivalent nucleons are classified according to the irreducible representations of the chain of groups

$$SU_{2j+1} \supset Sp_{2j+1} \supset R_3 \supset R_2. \qquad (4.120)$$

If the quantum numbers appearing in (4.119) are insufficient for a complete description of the state, it is necessary to introduce the additional quantum numbers α'.

The complete anti-symmetric wave function of a system of k nucleons can be written in the form

$$\begin{aligned}&\Psi(j^k[\lambda]\alpha' stJTM_JM_T)\\&= (1/\sqrt{n_\lambda})\sum_r \Lambda_r^{[\lambda]}\Phi_r(j^k[\lambda]\alpha' stJM_J)Z_{\tilde{r}}(t^k[\tilde{\lambda}]TM_T).\end{aligned} \qquad (4.121)$$

We recall that specifying $[\lambda]$ uniquely determines the isotopic spin T of the system, and so we need not write the quantum number T.

Table 4.6 gives a complete classification of the states of a system of k nucleons for the shells $j = \frac{1}{2}, \frac{3}{2}, \frac{5}{2}$ and $\frac{7}{2}$. The Table shows the possible values of J for each symmetry type defined by the tableaux $[\lambda]$ and (σ). A power attached to the number J indicates the number of times that the given value of J appears in the corresponding state. For a given configuration j^k, the states are arranged in order of increasing T. For a given T, the states are arranged in order of increasing s, and for given s, in order of increasing t. This arrangement of the states corresponds to the sequence of energy levels of the system, if we assume that residual short-range Wigner or Heisenberg forces act between the nucleons. These forces lead to the following values of the spins for the ground states of nuclei:

A and N even,	s=0,	t=0,	J=0;
A odd,	s=1,	$t = \frac{1}{2}$,	$J = j$;
A even, N odd,	s=2,	t=0,	J=1,3,...,2j.

TABLE 4.6.

j	k	$[\lambda]$	T	(σ)	(st)	J
1/2	0	[0]	(0)	(0)	(0, 0)	0
	1	[1]	$\frac{1}{2}$	(1)	$(1, \frac{1}{2})$	$\frac{1}{2}$
	2	[2]	0	(2)	(2, 0)	1
		[11]	1	(0)	(0, 0)	0
3/2	0	[0]	0	(00)	(0, 0)	0
	1	[1]	$\frac{1}{2}$	(10)	$(1, \frac{1}{2})$	$\frac{3}{2}$
	2	[2]	0	(20)	(2, 0)	1, 3
		[11]	1	(00)	(0, 0)	0
				(11)	(2, 1)	2
	3	[21]	$\frac{1}{2}$	(10)	$(1, \frac{1}{2})$	$\frac{3}{2}$
				(21)	$(3, \frac{1}{2})$	$\frac{1}{2}, \frac{5}{2}, \frac{7}{2}$
		[111]	$\frac{3}{2}$	(10)	$(1, \frac{1}{2})$	$\frac{3}{2}$
	4	[22]	0	(00)	(0, 0)	0
				(11)	(2, 1)	2
				(22)	(4, 0)	2,4
		[211]	1	(20)	(2, 0)	1, 3
				(11)	(2, 1)	2
		[1111]	2	(00)	(0, 0)	0
5/2	0	[0]	0	(000)	(0, 0)	0
	1	[1]	$\frac{1}{2}$	(100)	$(1, \frac{1}{2})$	$\frac{5}{2}$

Additional quantum numbers. The introduction of the intermediate groups, namely, the orthogonal group R_{2l+1} in the case of LS-coupling and the symplectic group Sp_{2j+1} in the case of jj-coupling, is a convenient method for performing, by means of the irreducible representations for these groups, a further classification of the states of a system of equivalent nucleons. In addition to the above groups, it is sometimes also possible to introduce other intermediate groups; however, a situation often arises (both in the case of LS-coupling and in the case of jj-coupling) in which

TABLE 4.6. (*cont*)

j	k	$[\lambda]$	T	(σ)	(st)	J
	2	[2]	0	(200)	(2, 0)	1, 3, 5
		[11]	1	(000)	(0, 0)	0
				(110)	(2, 1)	2, 4
	3	[21]	$\frac{1}{2}$	(100)	$(1, \frac{1}{2})$	$\frac{5}{2}$
				(210)	$(3, \frac{1}{2})$	$\frac{1}{2}, \frac{3}{2}, \frac{5}{2}, (\frac{7}{2})^2$
						$\frac{9}{2}, \frac{11}{2}, \frac{13}{2}$
		[111]	$\frac{3}{2}$	(100)	$(1, \frac{1}{2})$	$\frac{5}{2}$
				(111)	$(3, \frac{3}{2})$	$\frac{3}{2}, \frac{9}{2}$
	4	[22]	0	(000)	(0, 0)	0
				(110)	(2, 1)	2, 4
				(220)	(4, 0)	$0, 2^2, 3, 4^2, 5, 6^2, 8$
		[211]	1	(200)	(2, 0)	1,3,5
				(110)	(2, 1)	2,4
				(211)	(4, 1)	$1, 2, 3^2, 4, 5, 6, 7$
		[1111]	2	(000)	(0, 0)	0
				(110)	(2, 1)	2, 4
	5	[221]	$\frac{1}{2}$	(100)	$(1, \frac{1}{2})$	$\frac{5}{2}$
				(210)	$(3, \frac{1}{2})$	$\frac{1}{2}, \frac{3}{2}, \frac{5}{2}, (\frac{7}{2})^2, \frac{9}{2}, \frac{11}{2}, \frac{13}{2},$
				(111)	$(3, \frac{3}{2})$	$\frac{3}{2}, \frac{9}{2}$
				(221)	$(5, \frac{1}{2})$	$\frac{1}{2}, \frac{3}{2}, (\frac{5}{2})^2, (\frac{7}{2})^2$
						$(\frac{9}{2})^2, (\frac{11}{2})^2, \frac{13}{2}$
						$\frac{15}{2}, \frac{17}{2}$

there are no additional intermediate groups of transformations, even though the classification may still be incomplete. In the LS-coupling scheme, this

TABLE 4.6. *(cont)*

j	k	$[\lambda]$	T	(σ)	(st)	J
5/2		[2111]	$\frac{3}{2}$	(100)	$(1, \frac{1}{2})$	$\frac{5}{2}$
				(210)	$(3, \frac{1}{2})$	$\frac{1}{2}, \frac{3}{2}, \frac{5}{2}, (\frac{7}{2})^2,$ $\frac{9}{2}, \frac{11}{2}, \frac{13}{2}$
				(111)	$(3, \frac{3}{2})$	$\frac{3}{2}, \frac{9}{2}$
		[11111]	$\frac{5}{2}$	(100)	$(1, \frac{1}{2})$	$\frac{5}{2}$
	6	[222]	0	(200)	(2, 0)	1, 3, 5
				(211)	(4, 1)	1, 2, 3^2, 4, 5, 6, 7
				(222)	(6, 0)	1, 3^2, 4, 5, 6, 7, 9
		[2211]	1	(000)	(0, 0)	0
				$(110)^2$	$(2, 1)^2$	$2^2, 4^2$
				(220)	(4, 0)	0, 2^2, 3,4^2, 5, 6^2, 8
				(211)	(4, 1)	1, 2, 3^2, 4, 5, 6, 7
		[21111]	2	(200)	(2, 0)	1, 3, 5
				(110)	(2, 1)	2, 4
		[111111]	3	(000)	(0, 0)	0

situation first arises in the classification of states in a p-shell ($l = 1$). Here the orthogonal group R_{2l+1} for $l = 1$ effectively coincides with the group R_3 of ordinary three-dimensional rotations, since the irreducible representations of the orthogonal group for $l = 1$ are characterized by only one parameter ω_1, as are the irreducible representations of the rotation group. At the same time, as we know (see Table 4.5), for the configuration p^6 there are two D-states with the Young tableau [42], and so it is necessary to introduce *additional quantum numbers* . To determine the complete set of quantum mechanical quantities in such cases, it is necessary to find those operators Ω characterizing the system of particles that commute with the operators corresponding to the physical quantities already introduced and also with each other. At the present time, no general method has been developed for finding such complete sets of operators and the physical quantities corresponding to them, and so they are sought separately for each specific system. Naturally, the additional quantum characteristics found in this way rarely have an intuitive physical meaning.

In the classification of the states of a d-shell ($l = 2$) in the LS-coupling scheme, only the orthogonal group R_5, whose irreducible representations are characterized by two numbers ω_1 and ω_2, can be inserted between the unitary unimodular group SU_5 and the rotation group R_3. But for the configuration d^4 the states [31](3,1)H and [31](3,1)F occur in pairs. In the case of an f-shell ($l = 3$), in addition to the orthogonal group R_7, whose irreducible representations are characterized by three numbers ω_1, ω_2 and ω_3, there also exists a group which contains the group R_3 as a subgroup and is itself a subgroup of R_7. This group is denoted by G_2. In certain cases the classification according to the irreducible representations of these groups has a definite physical meaning (for example, in the case of short-range forces between the nucleons).

In the jj-coupling scheme, a set of quantum numbers that includes the total angular momentum J, the isotopic spin T (or the Young tableau $[\lambda]$), the seniority s and the reduced isotopic spin t is sufficient to describe the individual states only in shells with $j = \frac{1}{2}$ or $j = \frac{3}{2}$. But in a shell with $j = \frac{5}{2}$ for three equivalent nucleons, it is necessary to introduce *additional quantum numbers* . For example (see Table 4.6), the states $(\frac{5}{2})^3[21]$ $(s = 3, t = \frac{1}{2})(J = \frac{7}{2})$, $(\frac{5}{2})^4[211](4,1)3$ and $(\frac{5}{2})^6[222](6,0)3$ are encountered twice. With increasing j and increasing numbers of equivalent nucleons it becomes necessary to introduce more and more additional quantum numbers.

Supplementary classification of many-particle states in the case of the oscillator shell potential. We have considered a general classification of many-particle states that does not depend on the explicit form of the shell potential. In the case of the oscillator shell potential there is a further possibility of classification, based on the symmetry properties of the oscillator Hamiltonian with respect to the transformations of the group SU_3. Such a classification was proposed by Elliott (1958 a,b).

We shall consider a particle in a three-dimensional isotopic oscillator well. The Hamiltonian in the second-quantization representation can be written in the form

$$H_0 = \left(\mathbf{a}^+ \cdot \mathbf{a} + \frac{3}{2}\right)\hbar\omega, \tag{4.122}$$

where $\mathbf{a}^+$ and $\mathbf{a}$ are creation and destruction operators, related to the coordinate and momentum of the particle by the expressions

$$\mathbf{a}^+ = \frac{1}{\sqrt{2}}\left(\alpha\mathbf{r} - \mathbf{i}\frac{\mathbf{p}}{\hbar\alpha}\right); \quad \mathbf{a} = \frac{1}{\sqrt{2}}\left(\alpha\mathbf{r} - \mathbf{i}\frac{\mathbf{p}}{\hbar\alpha}\right) \quad \left(\alpha = \sqrt{\frac{M\omega}{\hbar}}\right). \tag{4.123}$$

The scalar product of the vectors $\mathbf{a}$ and $\mathbf{a}^+$ appearing in (4.122) is invariant under unitary transformations in three-dimensional space; in other words, the Hamiltonian (4.122) is invariant under unitary transformations of the group U_3 in the three-dimensional space of the vectors $\mathbf{a}$. The Hamiltonian is also invariant under transformations of the subgroups of the group U_3, namely, under transformations of the unitary unimodular group SU_3 and of the three-dimensional rotation group R_3.

The energy eigenvalues of the isotopic harmonic oscillator are given by a single quantum number N: $E_N = (N + \frac{3}{2})\hbar\omega$, N=0,1,2,... . Each energy level is degenerate, and the degeneracy is equal to $k_N = (N+1)(N+2)/2$, that is, is determined by the same number N. Therefore, we can associate with each energy level of the oscillator a set of k_N linearly independent functions, which transform into each other under transformations that leave the Hamiltonian (4.122) unchanged. This set of k_N functions forms a basis of an irreducible representation of the unitary group U_3. The representation is itself uniquely defined by the quantum number N, and its dimensionality is equal to k_N. An irreducible representation $D^{(N)}$ of the group U_3 is also an irreducible of the group SU_3. However, with respect to the rotation group R_3 this representation will be reducible and decomposes into irreducible representations D^l of the group R_3, where $l = N, N-2, N-4$,... down to $l = 1$ or 0, depending on the parity of N, each representation D^l with a given value of l occurring once. Thus, the states of a particle in an oscillator well (the oscillator states) can be characterized by functions φ_{Nlm} (m is the projection of the orbital angular momentum l) whose quantum numbers N, l and m define the irreducible representations of the chain of groups $U_3(SU_3) \supset R_3 \supset R_2$.

The irreducible representations of the groups U_3 and SU_3 can be characterized also by Young tableaux. As we know, a Young tableau also defines an irreducible representation of the permutation group. In the case of motion of a particle in an oscillator well, we must also keep in mind the permutations of the oscillator quanta. In the state φ_{Nlm} there are N such quanta. A possible Young tableau of N squares will characterize the symmetry properties of the wave function φ_{Nlm} with respect to permutations of the oscillator quanta, that is, will characterize the corresponding irreducible representation of the permutation group. The Young tableau will also characterize the properties of the function φ_{Nlm} under transformations of the group $SU_3(U_3)$, that is, will also define the irreducible representation of the group $SU_3(U_3)$.

If we have several nucleons in the oscillator well, the symmetry properties of the complete wave function under permutations of the nucleons are characterized by a Young tableau $[\lambda]$. Generally speaking, the total Hamiltonian of the system will be invariant under a broader class of transformations than those of the groups U_3 and SU_3, although it also remains invariant under transformations of the groups U_3 and SU_3. A classification of the states of several nucleons moving in an oscillator potential can be performed by using precisely this symmetry of the total Hamiltonian with respect to the group SU_3.

The irreducible representations of the group SU_3 (as distinct from those of the group U_3) can be characterized by the so-called truncated Young tableaux that are obtained from the Young tableaux by removing the filled columns. In the case of the group SU_3, a filled column is three squares in length. If a representation of the group U_3 is characterized by the tableau $[\lambda] = [\lambda_1, \lambda_2, \lambda_3]$, then the representation of the group SU_3 can be characterized by the tableau $[\lambda'] = [\lambda'_1, \lambda'_2]$,that is, by only two numbers $\lambda'_1 \equiv \lambda_1 - \lambda_3$ and $\lambda'_2 \equiv \lambda_2 - \lambda_3$. The representations $D^{[\lambda]}$ and $D^{[\lambda']}$ of the group SU_3 are equivalent. In place of the pair of numbers λ'_1 and λ'_2, Elliott introduced another equivalent pair of numbers to denote an irreducible representation of the group SU_3:

$$(\bar{\lambda}, \bar{\mu}) \equiv (\lambda_1 - \lambda_2, \lambda_2 - \lambda_3). \tag{4.124}$$

We shall use Elliott's notation.

In the case of one nucleon, the function φ_{Nlm} in the occupation-number representation can be obtained from the ground-state function φ_{000} by acting on the latter with the quantum-creation operator to the power N. Since these operators commute with each other, the function φ_{Nlm} is symmetric under permutations of the quanta. Therefore, the Young tableau corresponding to the function φ_{Nlm} has one row, in which the number of squares is equal to the number N of quanta, and the function itself is a basis function of an irreducible representation of the permutation group. The truncated Young tableau has the same form. Therefore, an irreducible representation of the group SU_3 will be described by the two Elliott numbers $\bar{\lambda} = N$ and $\bar{\mu} = 0$, or by the abbreviated form $(N, 0)$. As we should expect, this representation, like the energy and degeneracy of the level, depends only on the oscillator number N.

We shall call a collection of nucleons whose wave functions transform according to one and the same irreducible representation $(\bar{\lambda}, \bar{\mu}) = (N, 0)$ of the group SU_3 an *oscillator shell.* When there are k such nucleons, we shall say that they form a configuration, or oscillator shell, $(N, 0)^k$. Clearly, an oscillator shell contains, in the general case, several of the usual l-shells with which we were concerned earlier, in each of which all the nucleons had the same orbital quantum number. Only the oscillator shells with $N = 0$ and $N = 1$ coincide, respectively, with the usual s- and p-shells with $l = 0$ and 1.

The states of the oscillator shell $(N, 0)^k$ can be classified in the following way. We specify the Young tableau $[\lambda]$, that is, that irreducible representation $D^{[\lambda]}$ of the group of permutations of k nucleons according to which the space part of the wave function transforms; we specify the symbol $(\bar{\lambda}, \bar{\mu})$ for all k nucleons, which characterizes that irreducible representation of the group SU_3 according to which the complete wave function transforms; we specify the total angular momenta L, S and T and their projections M_L, M_S and M_T. Obviously, these quantum characteristics must be compatible with each other. Thus, an oscillator configuration is described by the wave function

$$\Psi((N0)^k[\lambda](\bar{\lambda}, \bar{\mu})LSTM_LM_SM_T). \tag{4.125}$$

For a filled shell, the symbol $(\bar{\lambda}, \bar{\mu})$ will be equal to (0,0).

We now indicate how one can find the possible values of the complete symbols $(\bar{\lambda}, \bar{\mu})$ when the symbols $(\bar{\lambda}_i, \bar{\mu}_i)$ of the states of the individual nucleons are known. For this, it is necessary to transform from the symbols $(\bar{\lambda}_i, \bar{\mu}_i)$ to the corresponding truncated Young tableaux $[\lambda_i'] = [\lambda_1 - \lambda_3, \lambda_2 - \lambda_3] = [\bar{\lambda}_i + \bar{\mu}_i, \bar{\mu}_i]$, multiply them in accordance with Littlewood's rule and from each of the Young tableaux obtained transform back to the symbols $(\bar{\lambda}, \bar{\mu}) = (\lambda_1 - \lambda_2, \lambda_2 - \lambda_3)$. For example, we shall find the possible numbers $(\bar{\lambda}, \bar{\mu})$ for the configuration $(\bar{\lambda}_i, \bar{\mu}_i)^k \equiv (1, 0)^3$. In this case, we must decompose the product $D^{(1,0)} \bigotimes D^{(1,0)} \bigotimes D^{(1,0)}$ into irreducible representations $D^{(\bar{\lambda}, \bar{\mu})}$ of the group SU_3. Since $\bar{\mu}_i = 0$, the truncated Young tableau corresponding to the symbol (1,0) has the simple form [1]. Multiplying in accordance with Littlewood's rule, we have

$$D^{[1]} \bigotimes D^{[1]} \bigotimes D^{[1]} = (D^{[2]} + D^{[11]}) \bigotimes D^{[1]} = D^{[3]} + 2D^{[21]} + D^{[111]}.$$

We next go back to the symbols $(\bar{\lambda}, \bar{\mu})$: $[3] \to (3, 0), [21] \to (1, 1)$ and $[111] \to (0, 0)$. Thus, we have finally

$$D^{(1,0)} \bigotimes D^{(1,0)} \bigotimes D^{(1,0)} = D^{(3,0)} + 2D^{(1,1)} + D^{(0,0)}.$$

Elliott's method enables us to decompose rapidly any irreducible representations $D^{(\lambda,\mu)}$ of the unimodular group SU_3 (characterized by a Young tableau with no more than three rows and no more than four columns, or by the corresponding two-number symbol $(\bar{\lambda}, \bar{\mu})$) into irreducible representations D^L of the three-dimensional rotation group R_3. The essence of the method lies in the fact that, for a given symbol $(\bar{\lambda}, \bar{\mu})$, the orbital angular momentum L can take the following values:

$$L = \Lambda,\ \Lambda + 1,\ \Lambda + 2, \ldots, \Lambda + \Omega, \tag{4.126}$$

where $\Lambda = \min\{\bar{\lambda}, \bar{\mu}\}, \min\{\bar{\lambda}, \bar{\mu}\}-2, \min\{\bar{\lambda}, \bar{\mu}\}-4, \ldots, 1$ or $0, \Omega = \max\{\bar{\lambda}, \bar{\mu}\}$, with the exception of the case $\min\{\bar{\lambda}, \bar{\mu}\} = 0$, when

$$L = \Omega, \quad \Omega - 2, \quad \Omega - 4, \ldots, 1 \text{ or } \quad 0. \tag{4.127}$$

Using this rule, it is easy to generate all the values of L (see Table 4.5). In the case of the Young tableau [42] (with the corresponding symbol $(\bar{\lambda}, \bar{\mu}) = (2, 2)$), we obtain, as we should, two states with the same value $L = 2$.

Elliott's rule is found to be especially useful in the classification of states in the oscillator shell model (classification in the *Elliott* SU_3 *scheme*), and makes it possible to find quickly all possible values of the total orbital angular momentum L consistent with the quantum symbol $(\bar{\lambda}, \bar{\mu})$ characterizing an irreducible representation of the group SU_3.

4.6. Motion of the Centre of Mass of the Nucleus in the Shell Model

Hamiltonian of the nucleus in the oscillator shell model and the Jacobi coordinates. In the study of many-particle shells in nuclei, the most widely applied model has been the shell model with an oscillator potential. This is connected both with the relative simplicity of performing calculations in this model and with the possibility of separating the motion of the centre of mass of the system from the internal motion in the model. We shall consider certain features of the oscillator shell model in more detail.

We choose the one-particle shell potential in the form of a three-dimensional oscillator well

$$V(r) = -V_0 + \frac{1}{2} M\omega^2 r^2. \tag{4.128}$$

Regarding the nucleus as a system of A independent nucleons, all in the field of the same oscillator potential (4.128), we can write the Hamiltonian in the form

$$H_0 = \sum_{i=1}^{A} H_i = \sum_{i=1}^{A} \{-(\hbar^2/2M)\nabla_i^2 - V_0 + \frac{1}{2} M\omega^2 \mathbf{r}_i^2\}. \tag{4.129}$$

An important feature of the Hamiltonian (4.129) is the fact that in it we can separate the part H_R associated with the motion of the nuclear centre of mass from the part H_0' associated with the internal motion in the nucleus:

$$\left.\begin{aligned}
&H_0 = H_R + H_0';\\
&H_R = -\frac{\hbar^2}{2AM}\nabla_R^2 - V_0 + \tfrac{1}{2}AM\omega^2R^2; \quad \mathbf{R} = \tfrac{1}{A}\sum_{i=1}^{A}\mathbf{r}_i;\\
&H_0' = \tfrac{1}{A}\sum_{i<j}^{A}\left\{-\frac{\hbar^2}{2M}(\nabla_i - \nabla_j)^2 - 2V_0 + \tfrac{1}{2}M\omega^2(\mathbf{r}_i - \mathbf{r}_j)^2\right\}.
\end{aligned}\right\} \qquad (4.130)$$

Since all the nucleons move in the same potential field, we can similarly separate out the part of H_0' associated with the motion of the centre of mass of a group of nucleons with any number of particles $k \leqslant A$:

$$\left.\begin{aligned}
&H_0 = H_{R_k} + H_{0k}' + \sum_{i=k+1}^{A} H_i;\\
&H_{R_k} = -\frac{\hbar^2}{2kM}\nabla_{R_k}^2 - V_0 + \tfrac{1}{2}kM\omega^2R_k^2; \quad \boldsymbol{R}_k = \tfrac{1}{k}\sum_{i=1}^{k}\mathbf{r};\\
&H_{0k}' = \tfrac{1}{k}\sum_{i<j}^{k}\left\{-\frac{\hbar^2}{2M}(\nabla_i - \nabla_j)^2 - 2V_0 + \tfrac{1}{2}M\omega^2(\mathbf{r}_i - \mathbf{r}_j)^2\right\}.
\end{aligned}\right\} \qquad (4.131)$$

It can be seen from (4.131) that the centre of mass of any number $k \leqslant A$ of nucleons in the oscillator shell model moves in the same oscillator well as do the individual nucleons.

A system of A nucleons is characterized by a set of $3A$ independent space variables, of which three variables characterize the motion of the nuclear centre of mass and $3A - 3$ variables characterize the internal motion in the nucleus. Because of the separability of the variables, we can write the wave function of the nucleus in the oscillator model in the form of a product of a function $\varphi(\mathbf{R})$ describing the motion of the nuclear centre of mass and a function $\Phi'(3A-3)$ describing the internal motion in the nucleus:

$$\Phi(\mathbf{r}_1, \mathbf{r}_2, \ldots, \mathbf{r}_A) = \varphi(\boldsymbol{R})\Phi'(3A-3). \qquad (4.132)$$

For the $3A - 3$ variables characterizing in the internal motion in the nucleus it is convenient to use the so-called *Jacobi coordinates* . We define the Jacobi coordinates as the set of vectors

$$\boldsymbol{\xi}_k = \sqrt{\frac{k}{k+1}}\left[(1/k)\sum_{i=1}^{k}\mathbf{r}_i - \mathbf{r}_{k+1}\right], \qquad k = 1, 2, 3, \ldots, A-1. \qquad (4.133)$$

The vector $\boldsymbol{\xi}_k$ defines the relative distance between the $(k+1)$th particle and the centre of mass of a system of k particles. In the case of two particles

$(A=2)$, there is only one such vector, $\boldsymbol{\xi}_k = \frac{1}{\sqrt{2}}(\mathbf{r}_1 - \mathbf{r}_2)$, coinciding with the relative distance between the particles.

The Hamiltonian H_0' of the internal motion, in the Jacobi coordinates, has the form

$$H_0' = \sum_{k=1}^{A-1} \left\{ -(\hbar^2/2M)\nabla_k^2 - V_0 + \frac{1}{2}\mu_k\omega^2\boldsymbol{\xi}_k^2 \right\}, \qquad \nabla_k^2 = \frac{\partial^2}{\partial\boldsymbol{\xi}_k^2}, \tag{4.134}$$

that is, breaks down into a such of $A-1$ one-particle three-dimensional oscillator Hamiltonians, each of which is associated with only one of the vectors $\boldsymbol{\xi}_k$. Thus, in Jacobi coordinates, the internal function Φ' can be represented in the form of a linear combination of products of $A-1$ one-particle oscillator functions $\varphi_{n_k l_k m_k}(\boldsymbol{\xi}_k)$, depending on the parameter $\alpha_k = \sqrt{M\omega/\hbar}\,(k = 1, 2, \ldots, A-1)$. We note that the Jacobi coordinates $\boldsymbol{\xi}_1, \boldsymbol{\xi}_2, \ldots, \boldsymbol{\xi}_{A-1}$ characterizing the internal motion in the nucleus can also be defined in another way. For example, an equivalent set of Jacobi coordinates is obtained if we make some permutation of the particle labels in (4.133). In place of the vectors $\boldsymbol{\xi}_k$, another set of vectors can be introduced as follows: divide all the nucleons into two groups, in any way, and introduce the relative position vector joining the centres of mass of these groups; then divide each of the resulting two groups of nucleons into two parts and introduce the relative position vectors of the centres of mass of the corresponding parts, and so on until each part consists of one nucleon. The $A-1$ vectors thus obtained are also called *Jacobi coordinates*, and the Hamiltonian of the internal motion can also be separated in these coordinates.

The Talmi transformation. We shall examine in more detail the case of two nucleons in an oscillator potential. The wave function of the two nucleons can be represented either in the form of a product of one particle functions

$$\varphi_{n_1 l_1 m_1}(\alpha\mathbf{r}_1)\varphi_{n_2 l_2 m_2}(\alpha\mathbf{r}_2), \qquad \alpha = \sqrt{M\omega/\hbar}, \tag{4.135}$$

or in the form of a product of a function of the centre-of-mass motion, depending on the position vector $\mathbf{R} = \frac{1}{2}(\mathbf{r}_1 + \mathbf{r}_2)$ of the centre of mass, and a function of the relative motion, depending on the relative position vector $\mathbf{r} = \mathbf{r}_1 - \mathbf{r}_2$:

$$\varphi_{NLM}(\sqrt{2}\alpha\boldsymbol{R})\varphi_{nlm}((1/\sqrt{2})\alpha\mathbf{r}). \tag{4.136}$$

The latter two functions are respectively eigenfunctions of the Hamiltonians H_{R_2} and H_{02}' appearing in (4.131), for $k=2$.

The functions (4.135) and (4.136) can be expressed in terms of each other, since they are two complete sets of functions. For example, we shall expand the product (4.135) in the functions (4.136).

$$\begin{aligned}
&\varphi_{n_1 l_1 m_1}(\alpha \mathbf{r}_1)\varphi_{n_2 l_2 m_2}(\alpha \mathbf{r}_2) \\
&= \sum_{NLMnlm} \left\langle \begin{matrix} n_1 l_1 m_1 \\ n_2 l_2 m_2 \end{matrix} \middle| \begin{matrix} NLM \\ nlm \end{matrix} \right\rangle \varphi_{NLM}(\sqrt{2}\alpha \boldsymbol{R})\varphi_{nlm} \\
&\times \left(\frac{1}{\sqrt{2}}\alpha \mathbf{r}\right). \qquad (4.137)
\end{aligned}$$

This transformation is called the *Talmi transformation* and the quantities

$$\left\langle \begin{matrix} n_1 l_1 m_1 \\ n_2 l_2 m_2 \end{matrix} \middle| \begin{matrix} NLM \\ nlm \end{matrix} \right\rangle$$

are called *Talmi coefficients*. For fairly small numerical values of the oscillator quantum numbers the Talmi coefficients can be found directly from (4.137), since the explicit form of the oscillator functions appearing in (4.137) is known. The transformation (4.137) may be found useful in calculations of the matrix elements of operators depending only on the relative position vector $\mathbf{r}$, for example, the two-particle interaction operator $V(r)$:

$$\begin{aligned}
&\langle n_1 l_1 m_1; n_2 l_2 m_2 | V(r) | n_1' l_1' m_1'; n_2' l_2' m_2' \rangle \\
&= \sum_{NLMnlm\bar{n}\bar{l}\bar{m}} \left\langle \begin{matrix} n_1 l_1 m_1 \\ n_2 l_2 m_2 \end{matrix} \middle| \begin{matrix} NLM \\ nlm \end{matrix} \right\rangle \left\langle \begin{matrix} n_1' l_1' m_1' \\ n_2' l_2' m_2' \end{matrix} \middle| \begin{matrix} NLM \\ \bar{n}\bar{l}\bar{m} \end{matrix} \right\rangle \\
&\times \langle nlm | V(r) | \bar{n}\bar{l}\bar{m} \rangle. \qquad (4.138)
\end{aligned}$$

In (4.138), we have made use of the fact that the wave functions of the centre-of-mass motion are orthonormal (the operator $V(r)$ does not depend on the centre-of-mass coordinates).

States described by the wave functions (4.135) and (4.136) are not characterized by well-defined angular momenta. States with well-defined angular momenta are easily obtained from (4.135) and (4.136) by means of Clebsch-Gordan coefficients. As a result, in place of (4.137) we obtain

$$\begin{aligned}
&\sum_{m_1 m_2} (l_1 m_1 l_2 m_2 | \lambda\mu) \varphi_{n_1 l_1 m_1}(\alpha \mathbf{r}_1)\varphi_{n_2 l_2 m_2}(\alpha \mathbf{r}_2) \\
&= \sum_{NLnl} \langle [n_1 l_1, n_2 l_2]^\lambda | [NL, nl]^\lambda \rangle \\
&\times \sum_{Mm} (LMlm|\lambda\mu) \varphi_{NLM}(\sqrt{2}\alpha \mathbf{R}) \varphi_{nlm}\left(\frac{1}{\sqrt{2}}\alpha \mathbf{r}\right), \qquad (4.139)
\end{aligned}$$

where $\boldsymbol{\lambda} = \mathbf{l}_1 + \mathbf{l}_2 = \mathbf{L} + \mathbf{l}$ is the total angular momentum, and μ is its projection. The transformation coefficients appearing in (4.139) are also called Talmi coefficients. They do not depend on the angular-momentum components and are convenient to use in calculations. These coefficients are connected with the Talmi coefficients introduced previously, by the relation

$$\begin{aligned} &\langle [n_1l_1, n_2l_2]^\lambda | [NL, nl]^\lambda \rangle \\ &= \sum_{m_1 m_2 M m} (l_1 m_1 l_2 m_2 | \lambda\mu)(LMlm|\lambda\mu) \\ &\times \left\langle \begin{matrix} n_1 m_1 l_1 \\ n_2 l_2 m_2 \end{matrix} \,\middle|\, \begin{matrix} NLM \\ nlm \end{matrix} \right\rangle. \end{aligned} \tag{4.140}$$

The transformation coefficients appearing in (4.139) are partially tabulated in the book by de-Shalit and Talmi (1963).

In studying nuclear processes in which certain aggregates (*clusters*) of nucleons (deuterons, α-particle, etc.) take part as a whole, it is convenient to use the so-called *generalized Talmi coefficients.* These are the coefficients of transformations, similar to (4.137) or (4.139), between functions describing the motion of two particles (or clusters) with different masses in an oscillator potential. The generalized Talmi coefficients depend on the ratio of the masses of these two particles. For these coefficients it is not difficult to obtain general formulae, which, however, are extremely cumbersome.

Separation of the centre-of-mass motion of the nucleus in the oscillator shell model. The translationally invariant shell model. The shell model has the important defect that in it the wave function of the nucleus is represented in the form of a product (or appropriately symmetrized sum of products) of the wave functions of the individual particles, and such a function does not ensure that the centre of mass of the nucleus is stationary. This is illustrated particularly clearly in the oscillator shell model, in which the centre of mass of the nucleus is not at rest or in a state of uniform rectilinear motion (as it should be in the absence of external forces), but performs oscillatory motions. The wave function describing this motion is an eigenfunction of the Hamiltonian H_R (4.130), that is, it is an *oscillator function.*

This circumstance is a direct consequence of the fact that, in the shell model, the motions of all $3A$ degrees of freedom of the nucleus are treated independently, whereas for a fixed centre of mass of the nucleus only $3A-3$ coordinates, describing the internal motion, are independent. Therefore, the shell-model wave functions, in partially reflecting nuclear centre-of-mass motion (introduced by the model itself) that does not in fact exist, can lead, especially for small A, to results that are physically meaningless.

We shall write the wave function of the nucleus in the oscillator shell model in the form of a product of a function $\varphi_{\bar{\mathcal{N}}\bar{L}\bar{M}}(\sqrt{A}\alpha R)$ of the centre-of-mass motion and a function $\Phi'(\boldsymbol{\xi}_1, \boldsymbol{\xi}_2, \ldots, \boldsymbol{\xi}_{A-1})$ describing the internal motion in the nucleus:

$$\Phi(\mathbf{r}_1, \mathbf{r}_2, \ldots, \mathbf{r}_A) = \varphi_{\bar{\mathcal{N}}\bar{L}\bar{M}}(\sqrt{A}\alpha \boldsymbol{R})\Phi'(\boldsymbol{\xi}_1, \boldsymbol{\xi}_2, \ldots, \boldsymbol{\xi}_{A-1}). \tag{4.141}$$

According to (4.141), different states of the centre-of-mass motion, with quantum numbers $\bar{\mathcal{N}}$, $\bar{L}$ and $\bar{M}$, can correspond to each internal state of the nucleus. The number of these centre-of-mass states is determined by the total energy and by other nuclear characteristics depending on these quantum numbers. Clearly, the properties of the nucleus are completely determined by the wave functions Φ' of the internal motion and should not depend on the motion of the centre of mass. In the simplest case, when the wave function of the nucleus is represented in the form (4.141), the calculation of matrix elements of operators acting on the internal coordinates leads to the correct result, which does not depend on the state of the centre-of-mass motion, only in the case when the quantum numbers $\bar{\mathcal{N}}$, $\bar{L}$ and $\bar{M}$ are the same in the initial and final states. Since the complete shell functions of the nucleus (which, when the spins and isospins of the nucleons are taken into account, are anti-symmetrized products of one-particle functions) can be represented in the general case by linear combinations of the functions (4.141) with different quantum numbers $\bar{\mathcal{N}}$, $\bar{L}$ and $\bar{M}$, the results of calculations of matrix elements are found to depend on the statistical weights of the different states of the centre-of-mass motion, and, consequently, have no physical meaning.

To avoid obtaining unphysical results, we can proceed in two ways. If we remain within the framework of the ordinary shell model, in the expansion of the shell functions in terms of the functions (4.141) we must remain only one term, with definite quantum numbers $\bar{\mathcal{N}}$, $\bar{L}$ and $\bar{M}$ (for example, the term with $\varphi_{\bar{\mathcal{N}}=0,\bar{L}=0,\bar{M}=0}$), and omit the remaining terms. The term retained is often called the *spurious-state wave functions.* For the physical state one usually chooses the state corresponding to the zero-point vibrations of the nuclear centre of mass:

$$\Phi(\mathbf{r}_1, \mathbf{r}_2, \ldots, \mathbf{r}_A) = \varphi_{\bar{\mathcal{N}}=0,\bar{L}=0,\bar{M}=0}(\sqrt{A}\alpha \boldsymbol{R})\Phi'(\boldsymbol{\xi}_1, \boldsymbol{\xi}_2, \ldots, \boldsymbol{\xi}_{A-1}). \tag{4.142}$$

The other way is to treat the motion of the nucleons in the nucleus relative to the fixed centre of mass rather than relative to the centre of the field as in the shell model. This model is usually called the *translationally invariant shell model* (although it is no longer a shell model in the ordinary sense), since the Hamiltonian and the wave functions in this model possess the property of invariance under translations in space. The Hamiltonian in the translationally invariant shell model is obtained from the shell Hamiltonian by replacing the part H_R associated with the motion of the centre of mass by the operator of the kinetic energy of the free nucleus:

$$\tilde{H} = -(\hbar^2/2AM)\nabla_R^2 + H_0'. \tag{4.143}$$

The wave function in the translationally invariant shell model has the form

$$\tilde{\Phi}(\mathbf{r}_1, \mathbf{r}_2, \ldots, \mathbf{r}_A) = \mathrm{e}^{\frac{i}{\hbar}\mathbf{P}\cdot\mathbf{R}}\Phi'(\boldsymbol{\xi}_1, \boldsymbol{\xi}_2, \ldots, \boldsymbol{\xi}_{A-1}), \tag{4.144}$$

rather than (4.142). In the formula (4.144) $\mathbf{P}$ is the momentum of the whole nucleus.

Separation of the physical and spurious states in the oscillator shell model. We shall discuss in more detail the question of the elimination of the spurious states from the shell wave functions. Clearly, the problem reduces to representing each shell wave function in the form of a superposition of functions of the form (4.141), which for $\bar{\mathcal{N}} = 0$ correspond to physical states and for $\bar{\mathcal{N}} \neq 0$ correspond to spurious states:

$$\begin{aligned} &\Psi\Big(\mathcal{N}, [\lambda](\bar{\lambda}, \bar{\mu})LSTM_LM_SM_T\Big) \\ =\ &\sum_{\substack{ß\bar{\mathcal{N}}\bar{L}\bar{L}' \\ (\bar{\lambda}', \bar{\mu}')}} a_{\bar{\mathcal{N}}\bar{L}\bar{L}'(\bar{\lambda}\bar{\mu})} \sum_{\bar{M}_L M'_{L'}} (\bar{L}\bar{M}_{\bar{L}}L'M'_{L'}|LM_L)\varphi_{\bar{\mathcal{N}}\bar{L}\bar{M}_{\bar{L}}}(\sqrt{A}\alpha\mathbf{R}) \\ &\times\Psi'\Big(\mathcal{N} - \bar{\mathcal{N}}, [\lambda], (\bar{\lambda}', \bar{\mu}')L'STM'_{L'}M_SM_T\Big), \end{aligned} \tag{4.145}$$

where N is the total oscillator quantum number defining the total energy of the nucleus, $E_{\mathcal{N}} = (\mathcal{N} + \frac{3}{2}A)\hbar\omega$. The total energy $E_{\mathcal{N}}$ is equal to the sum of the internal energy $E'_{\mathcal{N}-\bar{\mathcal{N}}} = [\mathcal{N} - \bar{\mathcal{N}} + \frac{3}{2}(A-1)]\hbar\omega$ and the energy $E_{\bar{\mathcal{N}}} = (\bar{\mathcal{N}} + \frac{3}{2})\hbar\omega$ of the centre-of-mass motion. Clearly, the ground state of the nucleus ($\mathcal{N} = \mathcal{N}_{\min}$) and some of the low-lying excited states, corresponding to $\bar{\mathcal{N}} = 0$, are physical states (the theorem of Bethe and Rose (1937) and Elliott and Skyrme (1955)).

The wave function $\varphi_{\bar{\mathcal{N}}\bar{L}\bar{M}_{\bar{L}}}(\sqrt{A}\alpha\mathbf{R})$ of the centre-of-mass motion, appearing in (4.145), does not depend on the spin and isospin variable and is symmetric in the space variables of all the nucleons. Therefore, the complete function $\Psi(\mathcal{N}, [\lambda](\bar{\lambda}, \bar{\mu})LSTM_LM_SM_T)$ will be characterized by the same Young tableau $[\lambda]$ and by the same numbers STM_SM_T as the internal function Ψ'. Hence it follows that a spurious state with a given $\mathcal{N} \neq 0$ will have the same quantum numbers $[\lambda]STM_SM_T$ as the physical state $\mathcal{N}=0$, if the internal state is the same in both cases. The total orbital angular momentum L can take values from $|\bar{L}-L'|$ to $\bar{L}+L'$ ($\bar{L}=\mathcal{N}, \mathcal{N}-2,..., 1$ or 0), $M_L=\bar{M}_{\bar{L}}+M'_{L'}$, and the possible values of the symbol $(\bar{\lambda}, \bar{\mu})$ are found by multiplying the representations $D^{(\mathcal{N},0)}$ and $D^{(\bar{\lambda}',\bar{\mu}')}$ using *Littlewood's rule*

As an example, we shall consider the classification of the states of a system with a given number of nucleons (not necessarily equivalent) moving in an oscillator potential. We shall consider all the possible one-quantum excited states of a system of five nucleons and determine which of these states are physical, which are spurious, and which are mixtures of pure and spurious states.

The ground state, according to the oscillator shell model (in the Elliott SU_3 scheme), will have the configuration $(0,0)^4(1,0)^1$, in which one of the two oscillator shells, with $N=0$, is completely filled, and the second, with $N=1$, contains one nucleon. We recall that an oscillator shell is defined by one oscillator (principal) quantum number N, and the maximum number of nucleons in this shell is equal to $\frac{1}{2}(N+1)(N+2)(2s+1)(2t+1)$, where $s=\frac{1}{2}$ and $t=\frac{1}{2}$ are the values of the spin and isospin of the nucleon. The oscillator shell with quantum number N is sometimes denoted by indicating all the possible values of the orbital angular momentum: $(N,0)^k \equiv (N, l = N, l = N-2, l = N-4, \ldots, l = 1$ or $0)$. In the case under consideration, $(0,0)^4(1,0)^1 \equiv (0s)^4(1p)^1$. The total energy is determined by the number $\mathcal{N}=\sum_N Nk$. In the ground state for our case, $\mathcal{N}_{\min}=0\times 4+1\times 1=1$. According to the *Bethe-Rose-Elliott-Skyrme theorem* , the ground state is a physical state.

First of all, we shall write the quantum characteristics $[\lambda]$, L and $(\bar{\lambda},\bar{\mu})$ for each of the two shells. For the configuration $(0s)^4$ we have: $[\lambda]=4$, since the length of the columns of the Young tableau must not exceed the number $2l+1$; $L=0$, since all the nucleons in this shell have $l=0$; the symbol $(\bar{\lambda},\bar{\mu})=(0,0)$, since $D^{(0,0)}\otimes D^{(0,0)}=D^{(0,0)}$. For the configuration $(1p)^1$ we have: $[\lambda]=1$, $L=1$ and $(\bar{\lambda},\bar{\mu})=(1,0)$, since there is only one nucleon in the shell.

The quantum numbers of the whole aggregate of five nucleons, with the minimum energy ($\mathcal{N}=1$), will be:

$$[\lambda]=[41], \qquad (\bar{\lambda},\bar{\mu})=(1,0), \qquad L=1, \qquad S=\frac{1}{2}, \qquad T=\frac{1}{2}. \quad (4.146)$$

The state $[\lambda]=[5]$ is excluded, since the Young tableau cannot contain more than four columns. We have not indicated the components of the angular momenta. The state (4.146) corresponds to $\bar{\mathcal{N}} = 0$, that is, it is a physical state.

The one-quantum excited states are defined by the total oscillator quantum number $\mathcal{N} = \mathcal{N}_{\min} + 1 = 2$, and correspond to the energy level

$$E = \hbar\omega\left(\mathcal{N} + \frac{3}{2}\sum_N k\right) = \hbar\omega(2 + \frac{3}{2} \times 5) = 19\hbar\omega/2.$$

With this oscillator energy, two configurations are possible for a system of five nucleons: $(0s)^3(1p)^2$ and $(0s)^4(2d,2s)^1$. In the first of these one nucleon from the oscillator shell with $N = 0$ has been transferred into the shell with $N = 1$, and in the second the only nucleon in the shell with $N = 1$ has been transferred into the shell with $N = 2$.

We shall find all the possible quantum numbers of these configurations. For the configuration $(0s)^3 \equiv (0,0)^3$, we have

$$[\lambda] = [3], \qquad L = 0, \qquad (\bar{\lambda}, \bar{\mu}) = (0,0);$$

analogously, for $(1p)^2 \equiv (1,0)^2$, we have

$$[\lambda] = [2], \quad L = 0 \quad \text{or}\, 2, \quad (\bar{\lambda}, \bar{\mu}) = (2,0),$$

$$[\lambda] = [11], \quad L = 1, \quad (\bar{\lambda}\bar{\mu}) = (0,1).$$

The orbital angular momentum values consistent with a given symbol $(\bar{\lambda}\bar{\mu})$ are determined by Elliott's method. According to Littlewood's rule,

$$D^{[3]} \bigotimes D^{[2]} = D^{[41]} + D^{[32]}, \quad D^{[3]} \bigotimes D^{[11]} = D^{[41]} + D^{[311]};$$

$$D^{(0,0)} \bigotimes D^{(2,0)} = D^{(2,0)}, \quad D^{(0,0)} \bigotimes D^{(0,0)} = D^{(0,1)},$$

and so the possible quantum states of the configuration $(0s)^3(1p)^2$ will be:

$$[\lambda] = [41], \quad (\bar{\lambda}, \bar{\mu}) = (2,0), \quad L = 0, 2, \quad ST = \frac{1}{2}\frac{1}{2}; \tag{4.147}$$

$$[\lambda] = [32], \quad (\bar{\lambda}, \bar{\mu}) = (2,0), \quad L = 0, 2, \quad ST = \frac{1}{2}\frac{1}{2}, \frac{1}{2}\frac{3}{2}, \frac{3}{2}\frac{1}{2}; \tag{4.148}$$

$$[\lambda] = [41], \quad (\bar{\lambda}, \bar{\mu}) = (0,1), \quad L = 1, \quad ST = \frac{1}{2}\frac{1}{2}; \tag{4.149}$$

$$[\lambda] = [311], \quad (\bar{\lambda}, \bar{\mu}) = (0, 1), \quad L = 1, \quad ST = \frac{1}{2}\frac{1}{2}, \frac{1}{2}\frac{3}{2}, \frac{3}{2}\frac{1}{2}, \frac{3}{2}\frac{3}{2}. \quad (4.150)$$

Analogously, it can be shown that the quantum numbers for the configuration $(0s)^4 \equiv (0,0)^4$ are $[\lambda] = [4]$, $L = 0$, and $(\bar{\lambda}, \bar{\mu}) = (0, 0)$, and for the configuration $(2d, 2s)^1 \equiv (2,0)^1$ are $[\lambda] = [1]$, $L = 0, 2$, and $(\bar{\lambda}, \bar{\mu}) = (2, 0)$. Hence we obtain that the state of the complete configuration $(0s)^4(2d, 2s)^1$ is defined by the following quantum numbers:

$$[\lambda] = [41], \quad (\bar{\lambda}, \bar{\mu}) = (2, 0), \quad L = 0, 2, \quad ST = \frac{1}{2}\frac{1}{2}. \quad (4.151)$$

The values of the total spin S and the total isotopic spin T are taken from Table 4.5. We recall that these values depend only on the Young tableau $[\lambda]$.

Since excitation of the centre of mass cannot lead to change of the Young tableau (or of the values of S and T) and since we have confined ourselves to considering only one-quantum excitations, the states (4.148) and (4.150) characterized by the Young tableaux [32] and [311] are physical states, since their Young tableaux differ from the Young tableau [41] of the ground state (4.146). Consequently, in the states (4.148) and (4.150), only internal degrees of freedom of the nucleus are excited; the centre of mass performs only zero-point oscillations, as in the ground state ($\bar{\mathcal{N}} = 0$).

We shall determine the quantum numbers of the spurious states for one-quantum excitations of a system of five nucleons. In this case ($\bar{\mathcal{N}} = 1$), the symbol $(\bar{\lambda}', \bar{\mu}')$ of the state of internal motion coincides with the full symbol $(\bar{\lambda}, \bar{\mu}) = (1, 0)$ of the ground state (4.146), since for the state of the centre-of-mass motion this symbol will be (0, 0) in the ground state ($\bar{\mathcal{N}} = 0$), and $D^{(0,0)} \otimes D^{(1,0)} = D^{(1,0)}$. In view of the fact that $D^{(\bar{\lambda}', \bar{\mu}')} \otimes D^{(\bar{\mathcal{N}}, 0)} = D^{(1,0)} \otimes D^{(1,0)} = D^{(2,0)} + D^{(0,1)}$, the spurious states ($\bar{\mathcal{N}} = 1$) will be characterized by the following quantum numbers:

$$[\lambda] = [41], \quad (\bar{\lambda}, \bar{\mu}) = (2, 0), \quad L = 0, 2, \quad ST = \frac{1}{2}\frac{1}{2}. \quad (4.152)$$

$$[\lambda] = [41], \quad (\bar{\lambda}, \bar{\mu}) = (0, 1), \quad L = 1, \quad ST = \frac{1}{2}\frac{1}{2}. \quad (4.153)$$

Eliminating the resulting spurious states (4.152) and (4.153) from the states (4.147) – (4.151), we obtain the quantum characteristics of all the physical ($\bar{\mathcal{N}} = 0$) one-quantum ($\mathcal{N} = 2$) excited states:

$$[\lambda] = [32], \quad (\bar{\lambda}, \bar{\mu}) = (2, 0), \quad L = 0, 2, \quad ST = \frac{1}{2}\frac{1}{2}, \frac{1}{2}\frac{3}{2}, \frac{3}{2}\frac{1}{2}; \quad (4.154)$$

$$[\lambda] = [311], \quad (\bar{\lambda}, \bar{\mu}) = (0, 1), \quad L = 1, \quad ST = \frac{1}{2}\frac{1}{2}, \frac{1}{2}\frac{3}{2}, \frac{3}{2}\frac{1}{2}, \frac{3}{2}\frac{3}{2}; \tag{4.155}$$

$$[\lambda] = [41], \quad (\bar{\lambda}, \bar{\mu}) = (2, 0), \quad L = 0, 2, \quad ST = \frac{1}{2}\frac{1}{2}. \tag{4.156}$$

It is these states which occur in the translationally invariant shell model. The fact that the states (4.154) and (4.155) are physical states has already been demonstrated. These states coincide with the states (4.148) and (4.150). From the example of the states (4.152) and (4.156), we see that physical and spurious states can be characterized by the same quantum numbers (apart from $\bar{\mathcal{N}}$). These equivalent states, which are encountered twice in the case under consideration, are associated with the different configurations $(0s)^3(1p)^2$ and $(0s)^4(2d,2s)^1$; half of them are spurious states (4.152) and half are the physical states (4.156). Therefore, the states (4.147) and (4.151) are superpositions of the spurious states (4.152) with $\bar{\mathcal{N}} = 1$ and the physical states (4.156) with $\bar{\mathcal{N}} = 0$. (The problem then consists in determining the coefficients in these superpositions.) We see that, amongst all the possible physical states (4.154) – (4.156), there is no set of quantum numbers similar to the set in (4.153). Therefore, the states (4.149) are wholly spurious states with $\bar{\mathcal{N}} = 1$.

Finally, using a simple example, we shall show how in principle one can determine the contributions of the physical and spurious states in the states of the oscillator shell model, or, which is the same thing, how to find the coefficients of the inverse transformations, that is, the coefficients of the expansions of the physical and spurious states in terms of the shell configurations. For example, we shall elucidate how to find these coefficients in the expansions of the function for the physical state with $L = 0$ from (4.156) and of the function for the spurious state with $L = 0$ from (4.152) in terms of the function (4.147) for the configuration $(0s)^3(1p)^2$, with $L = 0$ and the function (4.151) for the configuration $(0s)^4(2d,2s)^1$ with $L = 0$:

$$\begin{aligned} &\Psi_{\text{phys.}}\Big([41](2,0)L = 0,\ S = T = \frac{1}{2}\Big) \\ &= a_1\Psi_1((0s)^3(1p)^2[41]\Big(2,0)\,L = 0,\ S = T = \frac{1}{2}\Big) \\ &+ a_2\Psi_2((0s)^4(2d,2s)^1[41]\Big(2,0)L = 0,\ S = T = \frac{1}{2}\Big), \end{aligned} \tag{4.157}$$

$$\Psi_{\text{spurious}}([41]\Big(2,0)L = 0,\ S = T = \frac{1}{2}\Big)$$

$$= b_1\Psi_1((0s)^3(1p)^2[41](2,0)\,L=0,\ S=T=\frac{1}{2})$$

$$+b_2\Psi_2((0s)^4(2d,2s)^1[41](2,0)L=0,\ S=T=\frac{1}{2}), \qquad (4.158)$$

The functions $\Psi_{\text{phys.}}$ and Ψ_{spurious} are eigenfunctions, corresponding to the energy eigenvalues $E_0 = \frac{3}{2}\hbar\omega$ and $E_1 = \frac{5}{2}\hbar\omega$, respectively, of the Hamiltonian H_R of the centre-of-mass:

$$H_R\Psi_{\text{phys.}} = \frac{3}{2}\hbar\omega\Psi_{\text{phys.}}, \qquad H_R\Psi_{\text{spurious}} = \frac{5}{2}\hbar\omega\Psi_{\text{spurious}}. \qquad (4.159)$$

These conditions, together with the normalization conditions for the functions ($a_1^*a_1 + a_2^*a_2 = 1$, $b_1^*b_1 = b_2^*b_2 = 1$), determine the coefficients a_1, a_2, b_1 and b_2. Actual calculations are performed conveniently in the second-quantization representation. Then, by means of the inverse transformation, we can express the functions Ψ_1 and Ψ_2 in (4.157) and (4.158) in terms of $\Psi_{\text{phys.}}$ and Ψ_{spurious}, that is, we can determine the contributions of the physical and spurious states in states with a given configuration. We note that a physical-state wave function cannot in general be described by a finite shell-model configuration. It is superposition of functions for different configurations but with the same oscillator energy.

There are also other ways of constructing the wave functions of the physical states from the functions of the different shell configurations. For example, we can first construct all the spurious states with any given oscillator energy and then construct functions orthogonal to them; these functions will clearly be physical-state functions.

4.7. Fractional Parentage Coefficients

Calculation of matrix elements by means of many-particle functions. Fractional parentage coefficients. In the preceding section, we studied the classification of states describable by completely anti-symmetrized wave functions of k nucleons in an outer shell, based on the transformation properties of these functions with respect to certain groups of transformations. The expressions obtained for the wave functions can be used in the calculation of matrix elements of the different operators characterizing the system. In the general case of many-particle systems, it is necessary to calculate matrix elements of many-particle operators using many-particle functions. In practice, however, for systems with two-particle interaction we need only concern ourselves with matrix elements of one-particle or two-particle operators (that is, operators acting on the coordinates of one or two particles). Thus, according to (1.195), the Hamiltonian of the nucleus can be represented in the form of a sum

$$H = \sum_i T_i + \sum_{i<j} V_{ij}, \tag{4.160}$$

where $T_i = -(\hbar^2/2M)\nabla_i^2$ is the kinetic energy of an individual nucleon and V_{ij} is the two-nucleon interaction energy. The calculation of matrix elements of the operators in (4.160) between many-particle functions can be simplified considerably by introducing the so-called *fractional parentage coefficients* , which enable us to reduce the wave function of a system of k particles to wave functions of a system of $k-1$ particles, and so on. By means of the fractional parentage coefficients, the calculation of matrix elements of the above operators reduces in practice to the calculation of one- and two-particle matrix elements. Therefore, the explicit form of the many-particle wave functions is very rarely used in practice and to find the matrix elements of physical quantities it is sufficient to know the fractional parentage coefficients.

In the calculation of matrix elements of one-particle operators (for example, of the kinetic energy operator T_i), it is convenient to use the so-called *one-particle fractional parentage coefficients* . The introduction of these is based on the fact that the complete anti-symmetric wave function Ψ of a system of k particles is anti-symmetric with respect to permutations in any subsystem of k' particles ($k' < k$) and, in particular, with respect to permutations of $k-1$ particles. Therefore, the wave function Ψ can be represented in the form of a linear combination of products of anti-symmetric wave functions of $k-1$ particles and wave functions of one particle, the coefficients of this expansion being, by definition, the one-particle fractional parentage coefficients. In calculating matrix elements of two-particle operators, it is necessary to separate out states of $k-2$ particles and two particles in the complete wave function Ψ, and two-particle fractional parentage coefficients can then be defined analogously.

As an example, we shall consider the calculation of a matrix element of the Hamiltonian (4.160). By making use of the form of the operator H and the fact that the complete function Ψ is anti-symmetric, we find

$$(\Psi'|H|\Psi) = k(\Psi'|T_k|\Psi) + [k(k-1)/2](\Psi'|V_{k-1,k}|\Psi). \tag{4.161}$$

The next step consists in reducing $(\Psi'|T_k|\Psi)$ and $(\Psi'|V_{k-1,k}|\Psi)$ respectively to sums of one-particle and two-particle matrix elements by means of the corresponding expansions of the functions Ψ and Ψ' using the fractional parentage coefficients. Using the LS-coupling scheme as an illustration, we shall describe briefly one of the methods by which such expansions of the functions Ψ can be set up and the fractional parentage coefficients determined.

One-particle fractional parentage coefficients. A definition of the complete one-particle fractional parentage coefficients has been given above. The simplest way of finding their explicit form is to separate these coefficients into factors corresponding to the space and spin-isospin parts of the wave functions.

First we consider the space functions Φ and formulate the following problem. Suppose that we know the complete set of functions $\Phi_{r'}(l^{k-1}[\lambda'] \times \alpha' L' M')$ for the configuration l^{k-1}. We shall construct a similar complete set of functions for the configuration l^k containing one more particle. We note that the products $\Phi_{r'}(l^{k-1}[\lambda']\alpha' L' M')\varphi_{lm}(k)$, where $\varphi_{lm}(k)$ is the wave function of the kth particle with orbital angular momentum l and orbital angular-momentum component m, form a complete set of functions describing a system of k equivalent nucleons. Therefore, the k-particle space functions $\Phi_r(l^k[\lambda]\alpha LM)$ with a given symmetry can be expressed in terms of these products. Since the function $\Phi_r(l^k[\lambda]\alpha LM)$ is characterized by well-defined values L and M of the orbital angular momentum and its projection, it must be represented in the form of a certain linear combination of the expressions

$$\sum_m (L'M - mlm|LM)\Phi_{r'}(l^{k-1}[\lambda']\alpha' L' M - m)\varphi_{lm}(k), \tag{4.162}$$

which transform according to the irreducible representation D^L of the rotation group. Clearly, the coefficients of this linear combination should be independent of M. (If the coefficients depend on M, the function introduced in this way would not transform as required under rotations and, consequently, would not correspond to a well defined value of M).

Under permutations from the group S_{k-1}, the function $\Phi_r(l^k[\lambda]\alpha LM)$ transforms according to the irreducible representation $D^{[\lambda']}$ corresponding to the Young tableau $[\lambda']$ obtained from $[\lambda]$ by removing the square with the highest label. Therefore, the function $\Phi_r(l^k[\lambda]\alpha LM)$ should be constructed from those expressions (4.162) that contain $(k-1)$-particle functions transforming according to the same irreducible representation of the group S_{k-1}; the coefficients of the linear combination should be independent of the Yamanouchi symbol r' (in the same way that they must be independent of M). Thus, finally, the expansion of the space function for the configuration l^k can be represented in the following form:

$$\begin{aligned}\Phi_r(l^k[\lambda]\alpha LM) &= \sum_{\alpha' L'} (l^{k-1}[\lambda']\alpha' L'; l|\}l^k[\lambda]\alpha L) \\ &\times \sum_m (L'M - mlm|LM)\Phi_{r'}(l^{k-1}[\lambda']\alpha' L' M - m)\varphi_{lm}(k),\end{aligned} \tag{4.163}$$

where the expansion coefficients $(l^{k-1}[\lambda']\alpha' L'; l|\}l^k[\lambda]\alpha L)$ are called *one-particle orbital fractional parentage coefficients* . These coefficients are determined from the condition that the function (4.163) transform according to the irreducible representation $D^{[\lambda]}$ of the group S_k. The form of the tableau $[\lambda']$ and the symbol r' in the right-hand side of (4.163) are determined by the Yamanouchi symbol r.

The fractional parentage coefficients that we have introduced are real and satisfy the following orthonormalization condition:

$$\sum_{\alpha' L'} (l^{k-1}[\lambda']\alpha' L'; l|\}l^k[\lambda]\alpha L)(l^{k-1}[\lambda']\alpha' L'; l|\}l^k[\bar{\lambda}]\bar{\alpha} L) = \delta_{[\lambda][\bar{\lambda}]}\delta_{\alpha\bar{\alpha}}. \quad (4.164)$$

An analogous expansion with a one-particle state separated out can also be performed for the spin-isospin function:

$$\begin{aligned} \Omega_{\tilde{r}}((st)^k[\tilde{\lambda}]\beta STM_SM_T) = &\sum_{\beta' S' T'} ((st)^{k-1}[\tilde{\lambda}']\beta' S' T'; st|\}(st)^k[\tilde{\lambda}]\beta ST) \\ &\times \sum_{\mu\nu} (S'M_S - \mu s\mu|SM_S)(T'M_T - \nu t\nu|TM_T) \\ &\times \Omega_{\tilde{r}'}((st)^{k-1}[\tilde{\lambda}']\beta' S' T' M_S - \mu M_T - \nu)\chi_\mu(k)\zeta_\nu(k), \end{aligned} \quad (4.165)$$

where, by definition, $((st)^{k-1}[\tilde{\lambda}']\beta' S' T'; st|\}(st)^k[\tilde{\lambda}]\beta ST)$ is a *one-particle spin-isospin fractional parentage coefficient* ($s = \frac{1}{2}$, $t = \frac{1}{2}$). Substituting of (4.163) and (4.165) into (4.104) leads to the following expression for the expansion of the complete anti-symmetric function (since the sum over r reduced to a sum over $[\lambda']$ and r' for a given Young tableau $[\lambda]$):

$$\begin{aligned} &\Psi(l^k[\lambda]\alpha\beta LSTMM_SM_T \\ = &\sum_{[\lambda']\alpha'\beta' L' S' T'} (l^{k-1}[\lambda']\alpha'\beta' L' S' T'; lst|\}l^k[\lambda]\alpha\beta LST) \\ &\times \sum_{m\mu\nu} (L'M - mlm|LM)(S'M_S - \mu\frac{1}{2}\mu|SM_S)(T'M_T - \nu\frac{1}{2}\nu|TM_T) \\ &\times \Psi(l^{k-1}[\lambda']\alpha'\beta' L' S' T' M - mM_S - \mu M_T - \nu)\varphi_{lm\mu\nu}(k), \end{aligned} \quad (4.166)$$

where the summation is taken over all the possible Young tableaux $[\lambda']$ obtained from the tableau $[\lambda]$ by removing one square. The expansion coefficients in (4.166) are the complete one-particle fractional parentage coefficients, which are expressed in terms of the orbital and spin-isospin fractional parentage coefficients in the following way:

$$(l^{k-1}[\lambda']\alpha'\beta' L'S'T'; lst|\}l^k[\lambda]\alpha\beta LST) = \sqrt{\frac{n_{\lambda'}}{n_\lambda}}(l^{k-1}[\lambda']\alpha' L'; l|\}l^k[\lambda]\alpha L)$$
$$\times((st)^{k-1}[\tilde{\lambda}']\beta' S'T'; st|\}(st)^k[\tilde{\lambda}]\beta ST). \quad (4.167)$$

Functions of the form (4.166) are used in the calculation of one-particle matrix elements in concrete problems. The complete fractional parentage coefficients are also real and satisfy the orthonormalization condition

$$\sum_{[\lambda']\alpha'\beta' L'S'T'} (l^{k-1}[\lambda']\alpha'\beta' L'S'T'; lst|\}l^k[\lambda]\alpha\beta LST)$$
$$\times(l^{k-1}[\lambda']\alpha'\beta' L'S'T'; lst|\}l^k[\bar{\lambda}]\bar{\alpha}\bar{\beta} LST) = \delta_{[\lambda][\bar{\lambda}]}\delta_{\alpha\bar{\alpha}}\delta_{\beta\bar{\beta}}. \quad (4.168)$$

Tables of the one-particle fractional parentage coefficients have been compiled. [4]

If necessary, *two-particle fractional parentage coefficients* can be defined in a similar way; in this case, two-particle states are separated out in the wave functions. It is convenient to use these coefficients in calculations of matrix elements of two-particle operators. The two-particle fractional parentage coefficients can be expressed in terms of one-particle fractional parentage coefficients.

Recursion relation for the fractional parentage coefficients. The one-particle fractional parentage coefficients can be calculated by means of recursion relations connecting the fractional parentage coefficients for configurations with different numbers of particles. As an illustration, we shall consider the recursion relation connecting the orbital fractional parentage coefficients $(l^{k-1}[\lambda']L'; l|\}l^k[\lambda]L)$ with the fractional parentage coefficients $(l^{k-2}[\lambda'']L''; l|\}l^{k-1}[\lambda']L')$ corresponding to a system with one less particle.

We first introduce the Young projection operator for a system of k equivalent particles:

$$C_{rs}^{[\lambda]} = \frac{n_\lambda}{k!}\sum_P \langle r|P|s\rangle^{[\lambda]} P, \quad (4.169)$$

where $\langle r|P|s\rangle^{[\lambda]} \equiv D_{rs}^{[\lambda]}(P)$ is a matrix of the irreducible representation $[\lambda]$ of the permutation group S_k, n_λ is the dimensionality of the representation $[\lambda]$ and the summation is taken over all the permutations P of the

[4]Fractional parentage coefficients for p- and d-shells have been calculated by Jahn and van Wieringen (1951). The fractional parentage coefficients for shells with j=3/2,5/2 and 7/2 were calculated by Edmonds and Flowers (1952). Detailed tables of fractional parentage coefficients are given in the monograph by Neudachin and Smirnov (1969).

group S_k. Using the orthogonality condition for the matrices of irreducible representations

$$\sum_P \langle r|P|s\rangle^{[\lambda]*} \langle t|P|u\rangle^{[\mu]} = \frac{k!}{n_\lambda} \delta_{[\lambda][\mu]} \delta_{rl} \delta_{su}, \tag{4.170}$$

it is not difficult to show that

$$C_{rs}^{[\lambda]} \Phi_t(l^k[\mu]) = \delta_{[\lambda][\mu]} \delta_{st} \Phi_r(l^k[\lambda]), \tag{4.171}$$

that is, the Young operator $C_{rs}^{[\lambda]}$ acting on an arbitrary function $\Phi_t(l^k[\mu])$ projects out from in the component $\Phi_s(l^k[\lambda])$, which it then transforms into the function $\Phi_r(l^k[\lambda])$. The Yamanouchi symbol r characterizing a system of k particles determines simultaneously the Young tableau $[\lambda']$ and the Yamanouchi symbol r' of the system of $k-1$ particles, and also the label ϱ_k of the column in which the last particle (the kth) is situated ($r \equiv [\lambda']$, r' and ϱ_k). Thus,

$$C_{r\bar{r}}^{[\lambda]} = C_{[\lambda']r'\varrho_k, [\bar{\lambda}']\bar{r}'\bar{\varrho}_k}^{[\lambda]}. \tag{4.172}$$

We shall divide the $k!$ permutations P of the group S_k into two sets P' and $P_{ik}P'$, where P' are the permutations of the group S_{k-1} and P_{ik} is the operator interchanging two particles, of which particle i belong to the system of $k-1$ particles. Noting that $P_{ik} = P_{i,k-1}P_{k-1,k}P_{i,k-1}$, we can represent the permutations of the group S_k that do not appear in the group S_{k-1} in the form $P_{i,k-1}P_{k-1,k}P'$. Then, by means of this partition, the Young operator can be represented in the form

$$\begin{aligned} C_{r\bar{r}}^{[\lambda]} &= \frac{n_\lambda}{k!} \Bigg\{ \sum_{P'} \langle r'|P'|\bar{r}'\rangle^{[\lambda']} \delta_{[\lambda'][\bar{\lambda}']} \delta_{\varrho_k \bar{\varrho}_k} P' \\ &+ \sum_{P_{ik}P'} \sum_{s't'} \langle r'|P_{i,k-1}|s'\rangle^{[\lambda']} \langle [\lambda']s'\varrho_k|P_{k-1,k}|[\bar{\lambda}']t'\bar{\varrho}_k\rangle^{[\lambda]} \\ &\times \langle t'|P'\bar{\mathbf{r}}'\rangle^{[\bar{\lambda}']} P_{i,k-1}P_{k-1,k}P' \Bigg\}. \end{aligned} \tag{4.173}$$

(We recall that $P_{k-1,k}$ does not appear in the group S_{k-1} and $[\lambda']$ and $[\bar{\lambda}']$ characterize representations of this group, so that a matrix element of the permutation $P_{k-1,k}$ is not diagonal in a Young tableau of $k-1$ squares.) The summation over the elements of the set $P_{ik}P'$ is equivalent to a summation over the elements of the set P' with a fixed value of i, with a subsequent summation over all possible values of i:

$$\sum_{P_{ik}P'} \cdots \to \sum_{i,P'} \cdots .$$

Using the definition (4.169), we can reduce the sums over the elements P' in the right-hand side of (4.173) to Young operators for a system of $k-1$ particles:

$$\begin{aligned} C_{r\bar{r}}^{[\lambda]} &= \frac{1}{k}\frac{n_\lambda}{n_{\bar{\lambda}'}}\Big\{\delta_{[\lambda'][\bar{\lambda}']}\delta_{\varrho_k\bar{\varrho}_k}C_{r'\bar{r}'}^{[\lambda']} \\ &+ \sum_i \sum_{s'=[\lambda'']r''\varrho_{k-1}} \langle r'|P_{i,k-1}|s'\rangle^{[\lambda']}\langle[\lambda'][\lambda'']|P_{k-1,k}|[\bar{\lambda}'][\lambda'']\rangle^{[\lambda]} \\ &\times P_{i,k-1}P_{k-1,k}C_{[\lambda'']r''\bar{\varrho}_{k-1},\bar{r}'}^{[\lambda']}\Big\}. \end{aligned} \tag{4.174}$$

In writing this and the following formulae, we are using the fact that a matrix element of the permutation $P_{k-1,k}$ of the last two particles does not depend on the Yamanouchi symbol for the first $k-2$ particles; moreover, the numbers ϱ_k, ϱ_{k-1}, $\bar{\varrho}_k$ and $\bar{\varrho}_{k-1}$ are uniquely determined by specifying the corresponding pairs of Young tableau $[\lambda][\lambda']$, $[\lambda'][\lambda'']$, $[\lambda][\bar{\lambda}']$ and $[\bar{\lambda}'][\lambda'']$, and so are omitted in the matrix elements of the permutations.

The one-particle orbital fractional parentage coefficients are defined by the relation (4.163). The inverse relation has the form

$$\begin{aligned} &\sum_m (L'M-mlm|LM)\Phi_{r'}(l^{k-1}[\lambda']L'M-m)\varphi_{lm}(k) \\ &= \sum_{[\lambda]}(l^{k-1}[\lambda']L';l|\}l^k[\lambda]L)\Phi_r(l^k[\lambda]LM) \end{aligned} \tag{4.175}$$

and follows directly from (4.163) if we make use of the orthonormalization condition for the fractional parentage coefficients:

$$\sum_{[\lambda]}(l^{k-1}[\lambda']L';l|\}l^k[\lambda]L)(l^{k-1}[\lambda']\bar{L}';l|\}l^k[\lambda]L) = \delta_{L'\bar{L}'}. \tag{4.176}$$

(Like the Clebsch-Gordan coefficients, the fractional parentage coefficients define a unitary transformation between the functions.)

Using the definition (4.171) of the projection operator and (4.175), it is not difficult to obtain

$$\begin{aligned} &\sum_{m\bar{m}}(L'M-mlm|LM)(\bar{L}'M-\bar{m}l\bar{m}|LM) \\ &\times\langle\Phi_{r'}(l^{k-1}[\lambda']L'M-m)\varphi_{lm}(k)|C_{r\bar{r}}^{[\lambda]}|\Phi_{\bar{r}'}(l^{k-1}[\bar{\lambda}']\bar{L}'M-\bar{m})\varphi_{l\bar{m}}(k)\rangle \\ &= (l^{k-1}[\lambda']L';l|\}l^k[\lambda]L)(l^{k-1}[\bar{\lambda}']\bar{L}';l|\}l^k[\lambda]L). \end{aligned} \tag{4.177}$$

We substitute in the left-hand side of this equality the representation (4.174) of the operator $C^{[\lambda]}_{r\bar{r}}$. We then sum the left- and right-hand sides of the equality (4.177) over the Yamanouchi symbols r' and divide both sides of the resulting equality by $n_{\lambda'}$ (the dimensionality of the representation $[\lambda']$). Since the right-hand side of the equality (4.177) does not depend at all on the Yamanouchi symbols, it is unchanged, while the left-hand side of the equality is simplified considerably if we make use of the orthonormalization property of the matrices of an irreducible representation of the permutation group S_{k-1}:

$$\sum_{r'} \langle r'|P_{i,k-1}|s'\rangle^{[\lambda']} \langle r'|P_{i,k-1}|\bar{s}'\rangle^{[\lambda']} = \delta_{s'\bar{s}'}. \tag{4.178}$$

The dependence on i is thus eliminated and the sum over i can be taken into account by introducing a factor $k-1$. Since only matrix elements of the permutation $P_{k-1,k}$ remain, the sum over the Yamanouchi symbol r'' for $k-2$ particles can be replaced by the factor $n_{\lambda''}$, equal to the dimensionality of the representation $[\lambda'']$. Thus, we obtain the following equality:

$$\begin{aligned}
&\frac{1}{n_\lambda} \sum_{r'} \langle \Phi_{r'}(l^{k-1}[\lambda']L'M-m)\varphi_{lm}(k)|C^{[\lambda]}_{r\bar{r}}|\Phi_{\bar{r}'}(l^{k-1}[\bar{\lambda}']\bar{L}'M-\bar{m})\varphi_{l\bar{m}}(k)\rangle \\
&= \frac{1}{k}\frac{n_\lambda}{n_{\bar{\lambda}'}} \Big\{ \delta_{[\lambda'][\bar{\lambda}']}\delta_{\varrho_k\bar{\varrho}_k}\delta_{L'\bar{L}'}\delta_{m\bar{m}} + (k-1) \\
&\times \sum_{[\lambda'']} \frac{n_{\lambda''}}{n_{\lambda'}} \langle [\lambda'][\lambda'']|P_{k-1,k}|[\bar{\lambda}'][\lambda'']\rangle^{[\lambda]} \\
&\times \langle \Phi_{[\lambda'']r''\varrho_{k-1}}(l^{k-1}[\lambda']L'M-m)\varphi_{lm}(k)|P_{k-1,k}|\Phi_{[\lambda'']r''\bar{\varrho}_{k-1}} \\
&(l^{k-1}[\bar{\lambda}']\bar{L}'M-\bar{m})\varphi_{l\bar{m}}(k)\rangle \Big\}.
\end{aligned} \tag{4.179}$$

Here the Yamanouchi symbols have been written out in detail in the $(k-1)$-particle functions, which can differ only in the column label of the $(k-1)$th particle. The symbol r'', on which the matrix element does not depend, is arbitrary.

In the calculation of the matrix element of the operator $P_{k-1,k}$, we shall express the wave functions of the $(k-1)$-particle system in terms of the wave functions of a $(k-2)$-particle system by means of the fractional parentage coefficients:

$$\Phi_{r'}(l^{k-1}[\lambda']L'M') = \sum_{L''} (l^{k-2}[\lambda'']L''; l|\}l^{k-1}[\lambda']L')$$

$$\times \sum_m (L''M'-mlm|L'M')\Phi_{r''}(l^{k-2}[\lambda'']L''M'-m)\varphi_{lm}(k-1). \quad (4.180)$$

Thus, we find

$$\begin{aligned} &\langle \Phi_{r'}(l^{k-1}[\lambda']L'M')\varphi_{lm}(k)|P_{k-1,k}|\Phi_{\bar{r}'}(l^{k-1}[\bar{\lambda}']\bar{L}'\bar{M}')\varphi_{l\bar{m}}(k)\rangle \\ = \; & \delta_{r''\bar{r}''}\sum_{L''}(l^{k-2}[\lambda'']L'';l|\}l^{k-1}[\lambda']L')(l^{k-2}[\lambda'']L'';l|\}l^{k-1}[\bar{\lambda}']\bar{L}') \\ & \times(L''M'-\bar{m}l\bar{m}|L'M')(L'M'-\bar{m}lm|\bar{L}'\bar{M}'). \end{aligned} \quad (4.181)$$

(In the case under consideration, $r'' = \bar{r}''$.) By substituting (4.179) and (4.181) into (4.177) and performing the summation over the magnetic quantum numbers, we finally obtain the recursion relation in the following form:

$$\begin{aligned} &(l^{k-1}[\lambda']L';l|\}l^k[\lambda]L)(l^{k-1}[\bar{\lambda}']\bar{L}';l|\}l^k[\lambda]L) = \frac{1}{k}\frac{n_\lambda}{n_{\lambda'}}\Big\{\delta_{[\lambda'][\bar{\lambda}']}\delta_{L'\bar{L}'} \\ &+(k-1)\sum_{[\lambda'']}\frac{n_{\lambda''}}{n_{\bar{\lambda}'}}\langle[\lambda'][\lambda'']|P_{k-1,k}|[\bar{\lambda}'][\lambda'']\rangle^{[\lambda]}\sum_{L''}(-1)^{L+L''-L'-\bar{L}'} \\ &\times\sqrt{(2L'+1)(2\bar{L}'+1)}W(lL''Ll;L'\bar{L}')(l^{k-2}[\lambda'']L'';l|\}l^{k-1}[\lambda']L') \\ &\times(l^{k-2}[\lambda'']L'';l|\}l^{k-1}[\bar{\lambda}']\bar{L}')\Big\}. \end{aligned} \quad (4.182)$$

where $W(lL''Ll;L'\bar{L}')$ is a Racah coefficient. This formula enables us to find the one-particle fractional parentage coefficients for a system of k particles if the fractional parentage coefficients for a system of $k-1$ particles are known.

Putting $[\lambda'] = [\bar{\lambda}']$ and $L' = \bar{L}'$ in (4.182), we obtain a simpler formula, which, however, gives only the magnitude and not the sign of the fractional parentage coefficients:

$$\begin{aligned} &(l^{k-1}[\lambda']L';l|\}l^k[\lambda]L)^2 = \frac{1}{k}\frac{n_\lambda}{n_{\lambda'}}\Bigg\{1+(k-1)\sum_{[\lambda'']}\sum_{L''}\frac{n_{\lambda''}}{n_{\lambda'}} \\ &\times\langle[\lambda'][\lambda'']|P_{k-1,k}|[\lambda'][\lambda'']\rangle^{[\lambda]}(-1)^{L+L''}(2L'+1)W(lL''Ll;L'L') \\ &\times(l^{k-2}[\lambda'']L'';l|\}l^{k-1}[\lambda']L')^2\Bigg\}. \end{aligned} \quad (4.183)$$

Unlike (4.183), the formula (4.182) enables us to find the relative sign of the fractional parentage coefficients. Noting that the coefficient $(l;l|\}l^2[\lambda]L) = 1$, by successively increasing the number k of particles by unity we can construct by means of the recursion relation (4.182) a complete table of the one-particle fractional parentage coefficients for a given l.

As an illustration, we shall consider the case $k = 3$ and $[\lambda] = [21]$. Let $[\lambda'] = [\bar{\lambda}'] = [2]$, $L' = 0$ and $\bar{L}' = 2$. Then from (4.182) we find for the fractional parentage coefficients $(p^2[2]0; 1|\}p^3[21]1)$ and $(p^2[2]2; 1|\}p^3[21]1)$:

$$(p^2[2]0; 1|\}p^3[21]1)^2 = \frac{2}{3}\{1 + 2\langle P_{23}\rangle W(1111; 00)\};$$

$$(p^2[2]2; 1|\}p^3[21]1)^2 = \frac{2}{3}\{1 + 2\langle P_{23}\rangle 5W(1111; 22)\};$$

$$(p^2[2]0; 1|\}p^3[21]1)(p^2[2]2; 1|\}p^3[21]1) = \frac{4}{3}\langle P_{23}\rangle \sqrt{5}W(1111; 02),$$

where

$$\langle P_{23}\rangle \equiv \langle [2][1]|P_{23}|[2][1]\rangle^{[21]} = \langle (211)|P_{23}|(211)\rangle.$$

Noting that $\langle (211)|P_{23}|(211)\rangle = -\frac{1}{2}$, $W(1111; 00) = W(1111; 02) = \frac{1}{3}$ and $W(1111; 22) = \frac{1}{30}$, we finally obtain

$$(p^2[2]0; 1|\}p^3[21]1) = \frac{2}{3}, \qquad (p^2[2]2; 1|\}p^3[21]1) = -\sqrt{\frac{5}{9}}.$$

The sum of the squares of the fractional parentage coefficients found is equal to unity, in agreement with the normalization condition (4.164).

Two-particle parentage coefficients. In the shell model, along with the one-particle fractional parentage coefficients, two-particle coefficients are also used. In order to define the two-particle fractional parentage coefficients, we use the one-particle fractional parentage coefficients to separate out from the k-nucleon wave function first the nucleon labelled by k and then the nucleon labelled by $k - 1$:

$$\begin{aligned} &\Psi(l^k[\lambda]\alpha\beta LSTM_LM_SM_T) \\ = &\sum_{[\lambda']\alpha'\beta'L'S'T'} (l^{k-1}[\lambda']\alpha'\beta'L'S'T'; l|\}l^k[\lambda]\alpha\beta LST) \\ &\times \sum_{[\lambda'']\alpha''\beta''L''S''T''} (l^{k-2}[\lambda'']\alpha''\beta''L''S''T''; l|\}l^{k-1}[\lambda']\alpha'\beta'L'S'T') \\ &\times \left[[\Psi(l^{k-2}[\lambda'']\alpha''\beta''L''S''T'')\psi_l(k-1)]^{L'S'T'}\psi_l(k)\right]^{LST}_{M_LM_SM_T} \end{aligned} \quad (4.184)$$

For brevity, here we have introduced the following notation for the vector bundle of functions:

$$\left[\Psi(l^{k-1}[\lambda']\alpha'\beta'L'S'T')\psi_l(k)\right]^{LST}_{M_LM_SM_T}$$
$$= \sum_{\substack{M'_LM'_SM'_T \\ m\mu\nu}} (L'M'_Llm|LM_L)(S'M'_S\frac{1}{2}\mu|SM_S)(T'M'_T\frac{1}{2}\nu|TM_T)$$
$$\times\Psi(l^{k-1}[\lambda']\alpha'\beta'L'S'T'M'_LM'_SM'_T)\psi_{lm\mu\nu}(k). \tag{4.185}$$

Changing the order of the vector addition in (4.184) by means of the Racah coefficients and coupling together the wave functions of the nucleons with the labels k and $k-1$ (see Problem 4.1), we obtain

$$\Psi(l^k[\lambda]\alpha\beta LSTM_LM_SM_T)$$
$$= \sum_{\substack{[\lambda'']\alpha''\beta'' \\ \bar{L}\bar{S}\bar{T}}} (l^{k-2}[\lambda'']\alpha''\beta''L''S''T''; l^2\bar{L}\bar{S}\bar{T}|\}l^k[\lambda]\alpha\beta LST)$$
$$\times\left[\Psi(l^{k-2}[\lambda'']\alpha''\beta''L''S''T'')\Psi(l^2\bar{L}\bar{S}\bar{T})\right]^{LST}_{M_LM_SM_T}, \tag{4.186}$$

where

$$(l^{k-2}[\lambda'']\alpha''\beta''L''S''T''; l^2\bar{L}\bar{S}\bar{T}|\}l^k[\lambda]\alpha\beta LST)$$
$$= \sum_{[\lambda']\alpha'\beta'L'S'T'} \sqrt{(2L'+1)(2\bar{L}+1)(2S'+1)(2\bar{S}+1)(2T'+1)(\bar{T}+1)}$$
$$\times W(L''lLl; L'\bar{L})W(S''\frac{1}{2}S\frac{1}{2}; S'\bar{S})W(T''\frac{1}{2}T\frac{1}{2}; T'\bar{T})$$
$$\times(l^{k-2}[\lambda'']\alpha''\beta''L''S''T''; l|\}l^{k-1}[\lambda']\alpha'\beta'L'S'T')$$
$$\times(l^{k-1}[\lambda']\alpha'\beta'L'S'T'; l|\}l^k[\lambda]\alpha\beta LST). \tag{4.187}$$

The coefficients occurring in the expansion (4.186) are called *two-particle fractional parentage coefficients* . The formula (4.187) expresses the two--particle coefficients in terms of the one-particle fractional parentage coefficients. To conclude, we give the definition of the many-particle fractional parentage coefficients. By analogy with the case when one or two particles are separated out, the *many-particle fractional parentage coefficients* are defined as the coefficients of the expansion of the wave function $\Psi(l^k[\lambda]\alpha\beta LSTM_LM_SM_T)$ characterizing a state of k equivalent nucleons in terms of vectorially coupled wave functions of h ($h < k$) and $k-h$ nucleons:

$$\begin{aligned} &\Psi(l^k[\lambda]\alpha\beta LSTM_LM_SM_T) \\ &= \sum_{\substack{[\lambda_1]\alpha_1\beta_1L_1S_1T_1 \\ [\lambda_2]\alpha_2\beta_2L_2S_2T_2}} (l^{k-h}[\lambda_1]\alpha_1\beta_1L_1S_1T_1; l^h[\lambda_2]\alpha_2\beta_2L_2S_2T_2|\}l^k[\lambda]\alpha\beta LST) \\ &\quad \times \left[\Psi(l^{k-h}[\lambda_1]\alpha_1\beta_1L_1S_1T_1)\Psi(l^h[\lambda_2]\alpha_2\beta_2L_2S_2T_2\right]^{LST}_{M_LM_SM_T}. \end{aligned} \tag{4.188}$$

The fractional parentage coefficients will be equal to zero if the irreducible representation $D^{[\lambda]}$ does not appear in the decomposition of the outer product $D^{[\lambda_1]} \otimes D^{[\lambda_2]}$. This is a Young-tableau selection rule, based on Littlewood's condition.

A relation similar to (4.187) can be established between the fractional parentage coefficients when three, four, etc., particles are separated out and the fractional parentage coefficients when any smaller number of particles is separated out.

4.8. Energy Spectra of Light Nuclei

Intermediate-coupling approximation. The classification of many-nucleon states described in the preceding section is applicable both to the description of the ground states of nuclei and to the description of weakly excited states corresponding to the low-energy part of the nuclear spectrum. We shall consider a configuration consisting of a completely filled nucleon shell and k nucleons outside the filled shell. If the nucleus is described by the shell-model Hamiltonian

$$H_0 = \sum_i (T_i + V_i), \tag{4.189}$$

that is, if we assume that the nucleons move in the field of the self-consistent potential V without interacting with each other, then all the states of a given configuration $(nl)^k$ will be degenerate. In reality, the nuclear Hamiltonian

$$H = \sum_i T_i + \sum_{i<j} V_{ij} \tag{4.190}$$

differs from the model Hamiltonian (4.189). The difference between them is due to the presence of residual interaction between the nucleons, which is not taken into account in the shell model. Clearly, the *residual interaction* energy is defined by the difference

$$H - H_0 = \sum_{i<j} V_{ij} - \sum_i V_i. \tag{4.191}$$

Allowance for the residual interaction between the nucleons removes the degeneracy of the levels that is a feature of the shell model. Thus, to make possible the calculation of the spectra of the weakly excited states of nuclei, we must know the energy (4.191) of the residual interaction.

Usually, the energy levels E of the nucleus and the corresponding wave function Ψ are found directly by solving the Schrödinger equation

$$H\Psi = E\Psi. \tag{4.192}$$

For this, the wave function Ψ is represented in the form of an expansion in terms of the complete set, defined by the shell model, of states $\Psi_{\{\alpha\}}$ of a given configuration:

$$\Psi = \sum_{\{\alpha\}} C^{\{\alpha\}} \Psi_{\{\alpha\}}. \tag{4.193}$$

Taking into account that the Hamiltonian H is represented in the form of a sum of one-particle and two-particle operators, and separating out the one-particle and two-particle states in $\Psi_{\{\alpha\}}$ by means of the fractional parentage coefficients, we can reduce a matrix element of the Hamiltonian H to one-particle and two-particle matrix elements. Diagonalizing the calculated matrix of H, we can determine the energy spectrum $E^{(i)}$ and the coefficients $C_i^{\{\alpha\}}$ in the expansion (4.193), that is, the wave functions $\Psi^{(i)}$ for the corresponding states.

If it is known that some shell quantum number is a good quantum number, then states characterized by different values of this quantum number are weakly coupled. In this case, eqn. (4.192) can be solved by means of perturbation theory; for this, however, the off-diagonal matrix elements must be small compared with the difference between the corresponding diagonal elements.

In calculations, one usually uses a Hamiltonian H that includes a two-particle central interaction and a one-particle spin-orbit interaction:

$$H = \sum_i \{-(\hbar^2/2M)\nabla_i^2\} + \frac{1}{2}\sum_{i \neq j} V_{ij} + \xi \sum_i \mathbf{l}_i \cdot \mathbf{s}_i. \tag{4.194}$$

The appearance of the one-particle spin-orbit term in (4.194) arises from the averaged two-particle interaction of the outer nucleons with the nucleons in the filled shell. (The summation in (4.194) is taken only over the nucleons in the unfilled outer shell.) The interaction energy in (4.194) is chosen in a form convenient for numerical diagonalization of the Hamiltonian.

The use of a Hamiltonian with a two-particle central interaction and a one-particle spin-orbit interaction corresponds to thee so-called *intermediate-coupling approximation* . The name is connected with the fact that the LS-coupling scheme is applicable when only the two-particle central interaction is present, whereas the jj-coupling scheme applies in the presence of strong spin-orbit interaction. The states of the nucleus in the intermediate-coupling approximation are found by diagonalizing (4.194).

The two-particle potentials used in calculations in the intermediate-coupling approximation (Inglis, 1953; Kurath, 1956) differ from the potential describing the interaction of free nucleons. Thus, Kurath used for the two-nucleon potential

$$V_{ij} = V(\mathbf{r}_i - \mathbf{r}_j)(P_r + \frac{1}{4}P_\sigma), \qquad V(\mathbf{r}) = -V_0 \mathrm{e}^{-\gamma^2 \mathbf{r}^2}, \tag{4.195}$$

where P_r and P_σ are exchange operators. The parameter γ, which determines the range of the forces, was chosen to be $\gamma \approx (0 \cdot 9 - 1 \cdot 0)\alpha$. The relative weights of the Majorana and Bartlett forces were chosen from the conditions for complete saturation (Rosenfeld, 1948).

Kurath used the intermediate-coupling model with interaction as in (4.194) and (4.195) to explain the energy spectra of light nuclei with a filled p-shell. Despite the qualitative nature of the calculations, the results were found to be in good agreement with the experimental data, especially for levels with excitation energy not exceeding 6 MeV.

It is found that, for the $1p$-shell in the case of the exchange potential (4.195), the results depend only on three parameters ξ, L and K, where ξ is the spin-orbit interaction constant, and L and K are expressed in terms of the so-called *Slater integrals* F^l:

$$L = F^0 + \frac{4}{25}F^2, \qquad K = \frac{3}{25}F^2. \tag{4.196}$$

The Slater integrals are determined by the coefficients of the expansion of the function $V(\mathbf{r}_i - \mathbf{r}_j)$ characterizing the radial dependence of the two-particle potential in Legendre polynomials:

$$V(\mathbf{r}_i - \mathbf{r}_j) = \sum_{l=0}^{\infty} V_l(r_i, r_j) P_l(\mathbf{n}_i \cdot \mathbf{n}_j), \qquad \mathbf{n} = \mathbf{r}/r. \tag{4.197}$$

To be precise, they are equal to

$$F^l \equiv F^l(n_1'l_1', n_2'l_2', n_1l_1, n_2l_2) = \int_0^\infty dr \int_0^\infty dr' r^2 r'^2$$
$$\times \varphi_{n_1'l_1'}(r')\varphi_{n_2'l_2'}(r')V_l(r,r')\varphi_{n_1l_1}(r)\varphi_{n_2l_2}(r'), \qquad (4.198)$$

where $\varphi_{nl}(r)$ are the radial one-particle wave functions normalized to unity. The dependence of the energy levels on the parameters ξ, L and K was investigated, and the explicit form of the radial function $\varphi_{nl}(r)$ was found to be important. If the parameter K is chosen as the unit of energy, the position of the levels depends on only two quantities, namely, the ratios L/K and ξ/K. The calculations showed that the position of the levels depends weakly on the magnitude of L/K, which is determined by the central potential, but has a strong dependence on the ratio ξ/K. For a reasonable choice of the dimensions of the nuclei and of the range of the forces, the ratio L/K is found to be in the region of 6-7. For all the light nuclei, the same value of L/K was used in the calculations. By comparing the calculated level spectrum with the experimental spectrum, one can find the magnitude of the parameter ξ, which increases with the increasing mass number A.

Energy spectra of nuclei with $A = 7$. As an illustration, we shall consider the results of the calculations of the energy spectrum of the nuclei with mass number $A = 7$ (Kurath, 1956) and compare these with the experimental data. Figure 4.9 shows the calculated dependence of the energy spectrum on the magnitude of ξ/K for $L/K = 6 \cdot 8$, and also shows the levels for $\xi/K = 2 \cdot 0$ and $L/K = 8 \cdot 5$.
The energy E is expressed in units of K. It can be seen from Fig. 4.9 that the spacings between the levels depend strongly on the quantity ξ/K and weakly on L/K. Figure 4.10 shows the experimental levels of the mirror nuclei ^7Li and ^7Be and the calculated levels ($L/K = 6 \cdot 8$, $\xi/K = 1 \cdot 0$ and $K = -1 \cdot 2$ MeV). For definiteness, we shall discuss the spectrum of the levels of the nucleus ^6Li.

The ground state of ^7Li in the LS-coupling scheme is described by the Young tableau [3] corresponding to the maximum orbital symmetry. According to Table 4.5, the Young tableau [3] corresponds to a supermultiplet consisting of the doublets $^{22}P[3]$ and $^{22}F[3]$, the energy of the P-state being lower than the energy of the F-state. If we include the spin-orbit interaction as a perturbation, both doublets are split, the state with the larger total angular momentum J being characterized by the lower energy in each doublet. Since the ground state belongs to supermultiplet [3], the states

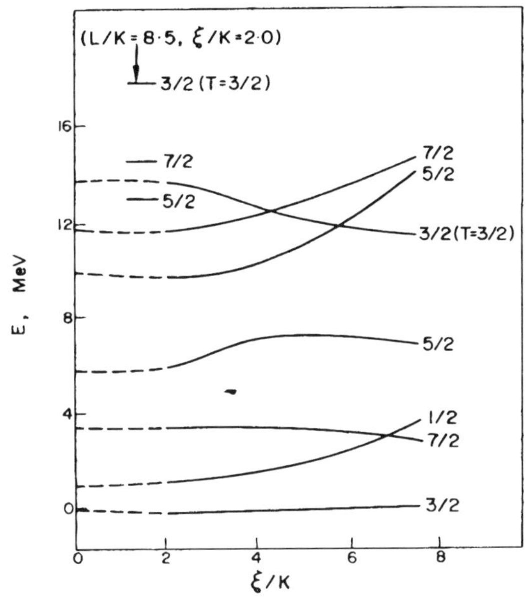

Figure 4.9. Dependence of the energy spectrum on ξ/K in the intermediate-coupling approximation, for the case $A = 7$.

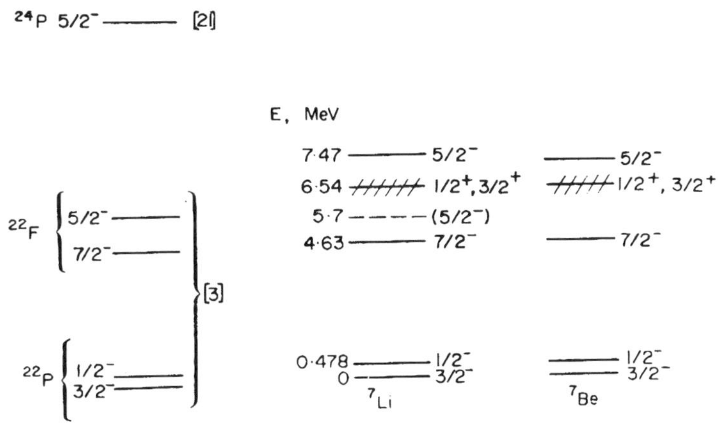

Figure 4.10. Level scheme in the intermediate-coupling approximation for $A = 7$ (left) and experimental levels for the mirror nuclei ^7Li and ^7Be (right).

$^{22}P_J[3]$ and $^{22}F_J[3]$ are the lowest energy states of the ^7Li nucleus. Clearly, the ground state can be taken to be the state $^{22}P_{3/2^-}[3]$; then the first excited state of the ^7Li nucleus will be the state $^{22}P_{1/2^-}[3]$, that is, the second state of the doublet $^{22}P[3]$. Above these are the levels $^{22}F_{7/2^-}[3]$ and $^{22}F_{5/2^-}[3]$, respectively. The parity of all these states, including the ground state, is negative, since it is equal to $\pi = (-1)^{\sum_i l_i}$ and the three nucleons are in a shell with $l = 1$. States of the same parity as the ground state are called *normal-parity states.*

Thus, the above four states of the ^7Li nucleus are described by the configuration $(1s)^4(1p)^3$ and are normal-parity states. The experimental values of the spins and parities of the ground state and first excited state of the ^7Li nucleus are in agreement with the hypothesis that these two states correspond to the doublet $^{22}P[3]$. The next excited level may be associated with the state $^{22}F_{7/2^-}$. If we take into account the spin-orbit interaction in first order of perturbation theory, the ratio of the spacing of the energy levels in the doublet $^{22}F[3]$ to the corresponding spacing in the doublet $^{22}P[3]$ should be 7 : 3. From this, we find that the level $^{22}F_{5/2^-}$ should be characterized by an energy value between 5 and 6 MeV. In Fig. 4.10 the corresponding level is depicted by a dashed line. The spacings between the levels in the two doublets ^{22}P and ^{22}F are considerably smaller than the energy gap between the two doublets, which amounts to about 4 MeV. This confirms the correctness of the assumption that the spin-orbit interaction is small and justifies the use of perturbation theory to take it into account.

Above the levels we have considered, there are states of positive parity with $J = \frac{1}{2}$ and $\frac{3}{2}$. States with parity opposite to the parity of the ground state are called *anomalous-parity* or *opposite-parity* states. These states correspond to well-defined one-particle excitations, with the transfer of a nucleon into another shell. Thus, the opposite-parity states $(\frac{1}{2})^+$ and $(\frac{3}{2})^+$ of ^7Li with energy $6 \cdot 54$ MeV can be described by the configurations $(1s)^4(1p)^2(1d)^1$ or $(1s)^4(1p)^2(2s)^1$, which correspond to a change in the value of the orbital angular momentum l of one of the nucleons by unity.

The theory then predicts a normal-parity state $(\frac{5}{2})^-$ at $12 \cdot 5$ MeV. A level $(\frac{5}{2})^-$ is observed experimentally, but at energy 7·47 MeV. According to the LS-coupling scheme, this state must belong to another supermultiplet, described by the Young tableau [21], and is a P-state. According to the rule for addition of angular momenta, the total spin S of this state cannot be less than $\frac{3}{2}$ $(S = J - L)$; therefore, in accordance with Table 4.5, this state is $^{24}P_{5/2^-}[21]$.

The large difference between the experimental and theoretical values for the energy of this state can be explained if we take into account that, for large excitation energies ($\geqslant 5$ MeV), so-called *configuration mixing* begins to play an important role in light nuclei. We have considered pure states of nucleons, in which the one-particle states were described by well-defined quantum numbers nl (or nlj). Configuration mixing corresponds to the case when it is necessary to consider a superposition of one-particle states nl (or nlj) with closely spaced energies. (It must be borne in mind that states of opposite parity do not mix with each other.) For example, in the case of light nuclei, the states $1d$ and $2s$ are very close in energy, so that they can be considered together by representing the corresponding wave function in the form of a linear combination of the two states. The admixture of other configurations usually leads to a lowering of the energy level compared with the case of the pure configuration. Therefore, the excited state of ^7Li with energy $7 \cdot 47$ MeV can be represented as a mixture of the principal configuration $(\frac{5}{2})^-$ at $12 \cdot 5$ MeV with other configurations.

Problems

4.1. Find the relation between the wave functions of three particles in the different *angular-momentum vector-coupling schemes* .

We shall consider a system of three particles characterized by the angular momenta j_1, j_2 and j_3. This system can be in a state in which the total angular momentum J of the three particles, its projection M, and the total angular momentum J_{12} of the first two particles have well-defined values. We shall denote the wave function of this state by $\left[(j_1 j_2)^{J_{12}} j_3\right]^J_M$. By means of the Clebsch-Gordan coefficients, we can represent this function, which is characterized by the set of quantum numbers j_1, j_2, j_3, J_{12}, J and M, in the form

$$\left[(j_1 j_2)^{J_{12}} j_3\right]^J_M \sum_{M_{12} m_3} (J_{12} M_{12} j_3 m_3 | JM)$$
$$\times \left[\sum_{m_1 m_2} (j_1 m_1 j_2 m_2 | J_{12} M_{12}) \psi_{j_1 m_1} \psi_{j_2 m_2}\right] \psi_{j_3 m_3}. \tag{4.199}$$

If the resultant angular momentum J_{23} of the second and third particles is given, for the same values J and M of the total angular momentum and its projection for the three particles, we have another state of the three-particle system, with another coupling scheme for the angular momenta. This state is characterized by the set of quantum numbers j_1, j_2, j_3, J_{23}, J and M, and its wave function can be represented in the form

$$\left[j_1 (j_2 j_3)^{J_{23}}\right]^J_M = \sum_{m_1 M_{23}} (j_1 m_1 J_{23} M_{23} | JM) \psi_{j_1 m_1}$$
$$\times \left[\sum_{m_2 m_3} (j_2 m_2 j_3 m m_3 | J_{23} M_{23}) \psi_{j_2 m_2} \psi_{j_3 m_3}\right]. \tag{4.200}$$

The functions in both coupling schemes are complete sets of functions, and those of one scheme can be expressed in terms of those of the other. For example,

$$\left[j_1 (j_2 j_3)^{J_{23}}\right]^J_M = \sum_{J_{12}} \sqrt{(2J_{12}+1)(2J_{23}+1)}\, W(j_1 j_2 J j_3; J_{12} J_{23})$$
$$\times \left[(j_1 j_2)^{J_{12}} j_3\right]^J_M \tag{4.201}$$

The transformation coefficients $W(j_1j_2Jj_3;J_{12}J_{23})$ appearing in (4.201) are called Racah coefficients. They can be expressed in terms of the Clebsch-Gordan coefficients. For this, it is sufficient to perform the inverse transformation and express the product of the three one-particle functions $\psi_{j_1m_1}$, $\psi_{j_2m_2}$ and $\psi_{j_3m_3}$, for example, in terms of the functions $\left[(j_1j_2)^{J_{12}}j_3\right]^J_M$, and then substitute the resulting expression into the right-hand side of the equality (4.200). As a result, we obtain

$$\begin{aligned}&\sqrt{(2J_{12}+1)(2J_{23}+1)}\,W(j_1j_2Jj_3;J_{12}J_{23})\\ =\ &\sum_{m_1m_2}(j_1m_1j_2m_2|J_{12}m_1+m_2)(J_{12}m_1+m_2j_3M-m_1-m_2|JM)\\ &\times(j_2m_2j_3M-m_1-m_2|J_{23}M-m_1)(j_1m_1J_{23}M-m_1|JM). \quad (4.202)\end{aligned}$$

In place of the Racah coefficients, one often introduces the so-called Wigner $6j$-symbols, which are characterized by higher symmetry under interchanges of the angular momenta:

$$\begin{Bmatrix} j_1 & j_2 & J_{12} \\ j_3 & J & J_{23} \end{Bmatrix} = (-1)^{j_1+j_2+j_3+J}\,W(j_1j_2Jj_3;J_{12}J_{23}). \quad (4.203)$$

The Wigner $6j$-symbols are invariant with respect to any permutation of columns, and also with respect to interchanges of any two angular momenta of the first row with the corresponding angular momenta of the second row.

4.2. Find the coefficients of the transformation between the wave functions in the LS- and jj-coupling schemes.

Since $\Psi(l^k[\lambda]\alpha\beta LSTM_LM_SM_T)$ and $\varPsi(l^k[\lambda]\alpha' stJTMM_T)$ are complete sets of functions in the LS- and jj-coupling schemes respectively, each of these functions in one coupling scheme can be represented uniquely in the form of a linear combination of wave functions in the other coupling scheme. It is particularly simple to find the coefficients of this transformation in the case of two particles. The treatment can be generalized to the case of inequivalent nucleons with different orbital angular momenta l_i, and also to the case of particles with arbitrary spins s_i.

The coefficients of the transformation between the wave functions of two particles in the jj- and LS-coupling schemes (we denote these wave functions by

$$\left[(l_1s_1)^{j_1}(l_2s_2)^{j_2}\right]^J_M \text{ and } \left[(l_1l_2)^L(s_1s_2)^S\right]^J_M)$$

are easily found by successively permuting the order of the vector couplings of the four angular momenta on which the wave functions depend. Thus the problem reduces to finding the relation between the four particle wave functions in the different vector-coupling schemes for their angular momenta.

Using the transformation formula in which a wave function with the three angular momenta a, b and c vectorially coupled in one order is expressed in terms of the wave functions with the same angular momenta coupled in a different order:

$$\left[\psi^a(\varphi^b\chi^c)^d\right]^f_m = \sum_e \sqrt{(2e+1)(2d+1)}W(abfc;ed)\left[(\psi^a\varphi^b)^e\chi^c\right]^f_m, \quad (4.204)$$

we can write, successively:

$$\left[(l_1s_1)^{j_1}(l_2s_2)^{j_2}\right]^J_M = \sum_a \sqrt{(2a+1)(2j_2+1)}\, W(j_1l_2Js_2;aj_2)$$
$$\times \left[((l_1s_1)^{j_1}l_2)^a s_2\right]^J_M$$
$$= \sum_{a,L} (-1)^{2l_2+j_1-a+l_1-L} \sqrt{(2a+1)(2L+1)(2j_1+1)(2j_2+1)}$$
$$\times W(j_1l_2Js_2;aj_2)W(l_2l_1as_1;Lj_1)\left[((l_1l_2)^L s_1)^a s_2\right]^J_M$$
$$= \sum_{LS} \sqrt{(2j_1+1)(2j_2+1)(2L+1)(2S+1)} \left\{ \begin{matrix} l_1s_1j_1 \\ l_2s_2j_2 \\ LSJ \end{matrix} \right\}$$
$$\times \left[(l_1l_2)^L(s_1s_2)^S\right]^J_M \tag{4.205}$$

and thus obtain the required answer. In writing the latter equality, we have made use of the so-called *Wigner 9j-symbols* , which are related to the Racah coefficients as follows:

$$\left\{ \begin{matrix} l_1s_1j_1 \\ l_2s_2j_2 \\ LSJ \end{matrix} \right\} = \sum_a (-1)^{l_1+2l_2+j_1-L-a}(2a+1)$$
$$\times W(j_1l_2Js_2;aj_2)W(l_2l_1as_1;Lj_1)W(Ls_1Js_2;aS). \tag{4.206}$$

The Wigner $9j$-symbols $\left\{ \begin{matrix} l_1s_1j_1 \\ l_2s_2j_2 \\ LSJ \end{matrix} \right\}$ are invariant under reflections across their two diagonals, and under any odd permutation of the rows or columns they acquire only the factor $(-1)^\sigma$, where σ is the sum of all the nine angular momenta occurring in the $9j$-symbols. If we multiply the transformation formula obtained by

$$\sqrt{(2j_1+1)(2j_2+1)} \left\{ \begin{matrix} l_1s_1j_1 \\ l_2s_2j_2 \\ L'S'J \end{matrix} \right\},$$

sum over j_1 and j_2, and make use of the orthonormalization of the $9j$-symbols:

$$\sum_{j_1j_2} (2j_1+1)(2j_2+1) \left\{ \begin{matrix} l_1s_1j_1 \\ l_2s_2j_2 \\ LSJ \end{matrix} \right\} \left\{ \begin{matrix} l_1s_1j_1 \\ l_2s_2j_2 \\ L'S'J \end{matrix} \right\} = \frac{\delta_{LL'}\delta_{SS'}}{(2L+1)(2S+1)}, \tag{4.207}$$

we obtain the inverse transformation

$$\left[(l_1l_2)^L(s_1s_2)^S\right]^J_M = \sum_{j_1j_2} \sqrt{(2L+1)(2S+1)(2j_1+1)(2j_2+1)}$$
$$\times \left\{ \begin{matrix} l_1l_2L \\ s_1s_2S \\ j_1j_2J \end{matrix} \right\} \left[(l_1s_1)^{j_1}(l_2s_2)^{j_2}\right]^J_M . \tag{4.208}$$

The transformation formulae for two particles are those most often used. In the case of a larger number of particles, the relations are considerably more complicated, since in classifying the states we must use additional quantum numbers, and the coefficients of the transformation from one coupling scheme to another will also depend on these quantum numbers.

4.3. Express the many-particle matrix element of a sum of one-particle spherical tensor operators in terms of reduced one-particle matrix elements.

By a *spherical tensor* of rank I, or an *irreducible tensor operator* of rank I, we mean an aggregate of $2I+1$ operators V_{IM} $(M=-I,-I+1,\dots,I-1,I)$ which transform under rotations according to the irreducible representation D^I of the rotation group R_3:

$$V_I=\sum_{M'} D^I_{M'M}(R)V'_{IM'}, \tag{4.209}$$

that is, transform in the same way as wave functions with angular momentum I. Any operator can be represented in the form of a linear superposition of irreducible tensor operators. A fundamental property of spherical tensors is described by the *Wigner-Eckart theorem*, according to which a matrix element of a spherical tensor operator between wave functions with well-defined angular momenta can always be represented in the form of a product of two factors, one of which contains in explicit form the whole dependence of the matrix element on the magnetic quantum numbers:

$$\langle\alpha'j'm'|V_{IM}|\alpha jm\rangle=\frac{(-1)^{2I}}{\sqrt{2j'+1}}\langle\alpha'j'||V_I||\alpha j\rangle\,(jmIM|j'm')\,, \tag{4.210}$$

where α and α' are the additional quantum numbers, completing the sets in the initial and final states. Here, $\langle\alpha'j'||V_I||\alpha j\rangle$ is a so-called *reduced matrix element*, which can be found from (4.210) by calculating the left-hand side of the equality (4.210) for any possible set of the magnetic quantum numbers m, m' and M.

We shall consider a spherical tensor operator $V_{\tau\nu}(i)$ of rank τ, acting on the space coordinates of one (the ith) nucleon. We shall calculate the matrix element of the sum $\sum_{i=1}^k V_{\tau\nu}(i)$ of one-particle tensor operators between the shell wave functions of k equivalent nucleons. By separating out the state of one nucleon by means of the complete one-particle fractional parentage coefficients, we can represent the matrix element in the form

$$\begin{aligned}
&\langle\Psi(l^k[\lambda]\alpha\beta\, LSTM_LM_SM_T)\,\Big|\sum_{i=1}^k V_{\tau\nu}(i)\,\Big|\,\Psi(l^k[\bar\lambda]\bar\alpha\bar\beta\bar L\bar S\bar T\bar M_L\bar M_S\bar M_T)\rangle\\
&=k\sum_{\substack{[\lambda']\alpha'\beta'\\ L'S'T'}}(l^{k-1}[\lambda']\alpha'\beta'\,L'S'T';lst|\}l^k[\lambda]\alpha\beta LST)\\
&\times(l^{k-1}[\lambda']\alpha'\beta'\,L'S'T';lst|\}l^k[\bar\lambda\bar\alpha\bar\beta\bar L\bar S\bar T)\\
&\times\sum_{M'_Lm\bar m}(L'M'_Llm\,\Big|\,LM_L)(L'M'_Ll\bar m\,\Big|\,\bar L\bar M_L)\\
&\times(nlm|V_{\tau\nu}|nl\bar m)\delta_{S\bar S}\delta_{T\bar T}\delta_{M_S\bar M_S}\delta_{M_T\bar M_T}\,.
\end{aligned} \tag{4.211}$$

By using the properties of a spherical tensor operator, in the one-particle matrix element we can separate out the dependence on the magnetic quantum numbers (the *Wigner-Eckart theorem*):

$$(nlm|V_{\tau\nu}|nl\bar{m}) = \frac{(-1)^{\tau}}{\sqrt{2l+1}}(\tau\nu l\bar{m}|lm)(nl||V_{\tau}||nl), \tag{4.212}$$

where $(nl||V_{\tau}||nl)$ is a reduced matrix element. By expressing the sum over the magnetic quantum numbers of the product of three Clebsch-Gordan coefficients in terms of Racah coefficients:

$$\begin{aligned}
&\sum_{M_L' m\bar{m}} (L'M_L'lm|LM_L)(L'M_L'l\bar{m}|\bar{L}\bar{M}_L)(\tau\nu l\bar{m}|ml) \\
&= \quad (-1)^{\tau}\sqrt{(2l+1)(2\bar{L}+1)}(\bar{L}M_L - \nu\tau\nu|LM_L)W(\tau lLL';l\bar{L}),
\end{aligned}$$

we finally obtain for the matrix element the expression

$$\begin{aligned}
&\langle\Psi(l^k[\lambda]\alpha\beta LSTM_LM_SM_T|\sum_{i=1}^{k}V_{\tau\nu}(i)|\Psi(l^k[\bar{\lambda}]\bar{\alpha}\bar{\beta}\bar{L}\bar{S}\bar{T}\bar{M}_L\bar{M}_S\bar{M}_T)\rangle \\
&= \quad k\sum_{\substack{[\lambda']\alpha'\beta' \\ L'S'T'}} (l^{k-1}[\lambda']\alpha'\beta'L'S'T';lst|\}l^k[\lambda]\alpha\beta\lambda\sigma\tau) \\
&\quad \times(l^{k-1}[\lambda']\alpha'\beta'L'S'T';lst|\}l^k[\bar{\lambda}]\bar{\alpha}\bar{\beta}\bar{L}\bar{S}\bar{T}) \\
&\quad \times\sqrt{2\bar{L}+1}(\bar{L}M_L - \nu\tau\nu|LM_L) \\
&\quad \times W(\tau lLL';l\bar{L})(nl||V_{\tau}||nl)\delta_{S\bar{S}}\delta_{T\bar{T}}\delta_{M_S\bar{M}_S}\delta_{M_T\bar{M}_T}.
\end{aligned} \tag{4.213}$$

4.4. Express the matrix elements of a two-particle central potential in terms of the slater integrals F^l.

We shall calculate a matrix element of the two-nucleon potential $V(\mathbf{r}_1 - \mathbf{r}_2)$ by means of two-nucleon wave functions with well-defined total orbital angular momenta:

$$\begin{aligned}
&\langle n_1'l_1'n_2'l_2'L'M'|V(\mathbf{r}_1 - \mathbf{r}_2)|n_1l_1n_2l_2LM\rangle \\
&\equiv \int d^3r_1\int d^3r_2\left[\sum_{m_1'm_2'}(l_1'm_1'l_2'm_2'|LM)Y^*_{l_1'm_1'}(\mathbf{n}_1)Y^*_{l_2'm_2'}(\mathbf{n}_2)\right] \\
&\times\varphi_{n_1'l_1'}(r_1)\varphi_{n_2'l_2'}(r_2)V(\mathbf{r}_1 - \mathbf{r}_2)\left[\sum_{m_1m_2}(l_1m_1l_2m_2|LM)Y_{l_1m_1}(\mathbf{n}_1)Y_{l_2m_2}(\mathbf{n}_2)\right] \\
&\times\varphi_{n_1l_1}(r_1)\varphi_{n_2l_2}(r_2)\delta_{LL'}\delta_{MM'}.
\end{aligned} \tag{4.214}$$

Such matrix elements are encountered in the diagonalization of the Hamiltonian matrix in the intermediate-coupling approximation. Using the expansion (4.197) of the two-nucleon potential in Legendre polynomials and the addition theorem for spherical harmonics

$$P_l(\mathbf{n}_1 \cdot \mathbf{n}_2) = \frac{4\pi(-1)^l}{\sqrt{2l+1}} \sum_m (lml-m|00) Y_{lm}(\mathbf{n}_1) Y_{l-m}(\mathbf{n}-2),$$

it is not difficult to perform the integration over the angles of the vectors $\mathbf{n}_1$ and $\mathbf{n}_2$. The sums over the magnetic quantum numbers of the product of Clebsch-Gordan coefficients can be reduced to Racah coefficients. Introducing the Slater integrals F^l in accordance with the definition (4.198), we obtain the following formula for the matrix element:

$$\begin{aligned} &\langle n_1' l_1' n_2' l_2' L' M' | V(\mathbf{r}_1 - \mathbf{r}_2) | n_1 l_1 n_2 l_2 LM \rangle \\ &= (-1)^{l_1' + l_2 - L} \sqrt{(2l_1+1)(2l_2+1)} \delta_{LL'} \delta_{MM'} \\ &\times \sum_{l=0}^{\infty} F^l W(l_1' l_2' l_1 l_2; Ll)(l_1 0 l 0 | l_1' 0)(l_2 0 l 0 | l_2' 0). \end{aligned} \tag{4.215}$$

In particular, if both nucleons are in the $1p$-shell ($n_1 = n_2 = n_1' = n_2' = 1$, $l_1 = l_2 = l_1' = l_2' = 1$), the sum over l contains only two terms, $l = 0$ and $l = 2$, so that the matrix element will be expressed in terms of the two Slater integrals F^0 and F^2. (We recall that the coefficient $(10l0|10)$ is non-zero only for $l = 0$ and $l = 2$.)

4.5. Determine the one-particle fractional parentage coefficients in the case of mixed configurations.

We shall consider k_1 nucleons in a shell with orbital angular momentum l_1 and k_2 nucleons in another shell with angular momentum l_2. Using the LS-coupling scheme, we can describe the state of the system of $k_1 + k_2$ nucleons by the function

$$\left[\Psi(l_1^{k_1}[\lambda_1]\alpha_1\beta_1 L_1 S_1 T_1)\Psi(l_2^{k_2}\alpha_2\beta_2 L_2 S_2 T_2)\right]^{LST}_{M_L M_S M_T}, \tag{4.216}$$

which is a bundle of functions for the individual shells corresponding to the pure configurations. The quantum numbers indicated in (4.216) characterize a state of the whole group of $k_1 + k_2$ nucleons. However, a state of the mixed configuration under consideration can also be specified in another way, namely, by specifying the Young tableau $[\lambda]$ of all $k_1 + k_2$ nucleons, the corresponding wave function being represented in the form of a sum of the states (4.216). The complete Young tableau $[\lambda]$ can take only those values which appear in the decomposition of

$$D^{[\lambda_1]} \bigotimes D^{[\lambda_2]} = \sum_{[\lambda]} \nu_\lambda D^{[\lambda]}$$

in accordance with Littlewood's rule.

The wave functions introduced must be anti-symmetrized with respect to permutations of all the nucleons. Within each shell the functions are anti-symmetric, and so anti-symmetrization must be performed with respect to permutations of the nucleons from different shells. We introduce the anti-symmetrization operator

$$A = \sum_P (-1)^P P, \tag{4.217}$$

where the summation is taken over all permutations P of nucleons between different shells; $(-1)^P = 1$ if P is an even permutation and $(-1)^P = -1$ if P is an odd permutation (the labels of the nucleons in each shell should be arranged in increasing order after the permutation). The total number of such permutations is $(k_1 + k_2)!/k_1!k_2!$.

A completely anti-symmetric normalized wave function of the first type has the form

$$\begin{aligned} &\tilde{\Psi}^{LST}_{M_L M_S M_T}(l_1^k[\lambda_1]\alpha_1\beta_1 L_1 S_1 T_1, l_2^{k_2}[\lambda_2]\alpha_2\beta_2 L_2 S_2 T_2) \\ =\ &\sqrt{\frac{k_1!k_2!}{(k_1+k_2)!}}\, A\Big[\Psi(l_1^{k_1}[\lambda_1]\alpha_1\beta_1 L_1 S_1 T_1) \\ &\times \Psi(l_2^{k_2}[\lambda_2]\alpha_2\beta_2 L_2 S_2 T_2)\Big]^{LST}_{M_L M_S M_T}. \end{aligned} \tag{4.218}$$

This formula establishes the connection between the wave functions $\tilde{\Psi}$ and Ψ of the mixed and pure configurations respectively.

We shall now introduce the fractional parentage coefficients for mixed configurations. In separating out the state of one particle in a function of the first type, we need to introduce two one-particle fractional parentage coefficients, to separate a nucleon first from one shell and then from the other:

$$\begin{aligned} &\tilde{\Psi}^{LST}_{M_L M_S M_T}(l_1^{k_1}[\lambda_1]\alpha_1\beta_1 L_1 S_1 T_1, l_2^{k_2}[\lambda_2]\alpha_2\beta_2 L_2 S_2 T_2) \\ =\ &\sum_{\substack{[\lambda_1]\alpha_1'\beta_1'L_1'S_1'T_1' \\ L'S'T'}} ([l_1^{k_1-1}[\lambda_1']\alpha_1'\beta_1' L_1' S_1' T_1', l_2^{k_2}[\lambda_2]\alpha_2\beta_2 L_2 S_2 T_2]^{L'S'T'}; \\ &\times l_1|\}[l_1^{k_1}[\lambda_1]\alpha_1\beta_1 L_1 S_1 T_1, l_2^{k_2}[\lambda_2]\alpha_2\beta_2 L_2 S_2 T_2]^{LST}) \\ &\times \Big[\tilde{\Psi}^{L'S'T'}(l_1^{k_1-1}[\lambda_1']\alpha_1'\beta_1' L_1' S_1' T_1', l_2^{k_2}[\lambda_2]\alpha_2\beta_2 L_2 S_2 T_2)\varphi^{l_1}\Big]^{LST}_{M_L M_S M_T} \\ +\ &\sum_{\substack{[\lambda_2']\alpha_2'\beta_2'L_2'S_2'T_2' \\ L'S'T'}} ([l_1^{k_1}[\lambda_1]\alpha_1\beta_1 L_1 S_1 T_1, l_2^{k_2-1}[\lambda_2']\alpha_2'\beta_2' L_2' S_2' T_2']^{L'S'T'}; l_2|\} \\ &\times [l_1^{k_1}[\lambda_1]\alpha_1\beta_1 L_1 S_1 T_1, l_2^{k_2}[\lambda_2]\alpha_2\beta_2 L_2 S_2 T_2]^{LST}\Big[\tilde{\Psi}^{L'S'T'}(l_1^{k_1}[\lambda_1]\alpha_1\beta_1 L_1 S_1 T_1, \\ &\times l_2^{k_2-1}[\lambda_2']\alpha_2'\beta_2' L_2' S_2' T_2')\varphi^{l_2}\Big]^{LST}_{M_L M_S M_T}. \end{aligned} \tag{4.219}$$

Bundles of the function $\tilde{\Psi}^{L'S'T'}$ for $k_1 + k_2 - 1$ particles and φ^{l_1} or φ^{l_2} for one particle appear in this formula, the structure of the function $\tilde{\Psi}^{L'S'T'}$ being similar to that of the complete wave function $\tilde{\Psi}^{LST}$ of the mixed configuration for k_1+k_2 particles.

4.6. Establish the relation between the fractional parentage coefficients of mixed and pure configurations.

This connection can be derived by using the relation between the wave functions of the mixed and pure configurations. We consider a mixed configuration of two shells, and, for definiteness, calculate the one-particle fractional parentage coefficient for the separation of one nucleon from the second shell, that is, the coefficient in the second sum in the expression (4.219). Taking the scalar product of the bundle in the second sum in(4.219) with the whole expression (4.219) for the wave function, we obtain

$$\begin{aligned}
&([l_1^{k_1}[\lambda_1]\alpha_1\beta_1 L_1S_1T_1, l_2^{k_2-1}[\lambda_2']\alpha_2'\beta_2' L_2'S_2'T_2']^{L'S'T'} : l_2|\}\\
&\times[l_1^{k_1}[\lambda_1]\alpha_1\beta_1 L_1S_1T_1, l_2^{k_2}[\lambda_2]\alpha_2\beta_2 L_2S_2T_2]^{LST}\\
=&\left\langle\left[\tilde{\Psi}^{L'S'T'}(l_1^{k_1}[\lambda_1]\alpha_1\beta_1 L_1S_1T_1, l_2^{k_2-1}[\lambda_2']\alpha_2'\beta_2' L_2'S_2'T_2')\varphi^{l_2}\right]^{LST}|\tilde{\Psi}^{LST}\right.\\
&\left.\times(l_1^{k_1}[\lambda_1]\alpha_1\beta_1 L_1S_1T_1, l_2^{k_2}[\lambda_2]\alpha_2\beta_2 L_2S_2T_2)\right\rangle, \qquad (4.220)
\end{aligned}$$

where in the right-hand side we have dropped the magnetic quantum numbers from the bundle and the function $\tilde{\Psi}^{LST}$ (the full expression no longer depends on them).

Using (4.218) and (4.217) and separating out the state of one nucleon in the second shell of the function $\tilde{\Psi}^{LST}$ by means of the one-particle fractional parentage coefficients, we can represent the function $\tilde{\Psi}^{LST}$ for k_1+k_2 particle in the following form:

$$\begin{aligned}
\tilde{\Psi}^{LST} =& \sqrt{\frac{k_1!k_2!}{(k_1+k_2)!}}\sum_P(-1)^P P \sum_{[\bar{\lambda}_2]\bar{\alpha}_2\bar{\beta}_2\bar{L}_2\bar{S}_2\bar{T}_2}\\
&\times(l_2^{k_2-1}[\bar{\lambda}_2]\bar{\alpha}_2\bar{\beta}_2\bar{L}_2\bar{S}_2\bar{T}_2; l_2|\}l_2^{k_2}[\lambda_2]\alpha_2\beta_2 L_2S_2T_2)\\
&\times\Big[[\Psi(l_1^{k_1}[\lambda_1]\alpha_1\beta_1 L_1S_1T_1)\\
&\times[\Psi(l_2^{k_2-1}[\bar{\lambda}_2]\bar{\alpha}_2\bar{\beta}_2\bar{L}_2\bar{S}_2\bar{T}_2)]^{L'S'T'}\varphi^{l_2}]^{L_2S_2T_2}\Big]^{LST} \qquad (4.221)
\end{aligned}$$

By coupling the angular momenta in a different order by means of the Racah coefficients, we can write the function $\tilde{\Psi}^{LST}$ conveniently as follows:

$$\begin{aligned}
\tilde{\Psi}^{LST} =& \sqrt{\frac{k_1!k_2!}{(k_1+k_2)!}}\sum_P(-1)^P P \sum_{[\bar{\lambda}_2]\bar{\alpha}_2\bar{\beta}_2\bar{L}_2\bar{S}_2\bar{T}_2}\\
&\times(l_2^{k_2-1}[\bar{\lambda}_2]\bar{\alpha}_2\bar{\beta}_2\bar{L}_2\bar{S}_2\bar{T}_2; l_2|\}l_2^{k_2}[\lambda_2]\alpha_2\beta_2 L_2S_2T_2)\\
&\times\sum_{\bar{L}'\bar{S}'\bar{T}'}\sqrt{(2\bar{L}'+1)(2L_2+1)(2\bar{S}'+1)(2S_2+1)(2\bar{T}'+1)(2T_2+1)}
\end{aligned}$$

$$\times W(L_1 \bar{L}_2 L l_2; \bar{L}' L_2) W(S_1 \bar{S}_2 S \frac{1}{2}; \bar{S}' S_2) W(T_1 \bar{T}_2' T \frac{1}{2}; \bar{T}' T_2)$$
$$\times \left[[\Psi(l_1^{k_1}[\lambda_1]\alpha_1\beta_1 L_1 S_1 T_1)\Psi(l_2^{k_2-1}[\bar{\lambda}_2]\bar{\alpha}_2\bar{\beta}_2 \bar{L}_2 \bar{S}_2 \bar{T}_2)]^{L'S'T'} \varphi^{l_2} \right]^{LST} \quad (4.222)$$

We shall also express the $(k_1 + k_2 - 1)$-particle function $\tilde{\Psi}^{L'S'T'}$ occurring in (4.220) in terms of the functions of the pure configurations:

$$\tilde{\Psi}^{L'S'T'} = \sqrt{\frac{k_1!(k_2-1)!}{(k_1+k_2-1)!}} \times \sum_{P'} (-1)^{P'} P' \left[\Psi(l_1^{k_1}[\lambda_1]\alpha_1\beta_1 L_1 S_1 T_1) \times \Psi(l_2^{k_2-1}[\lambda_2']\alpha_2'\beta_2' L_2' S_2' T_2') \right]^{L'S'T'} \quad (4.223)$$

Here the permutations P' affect only the $k_2 - 1$ particles in the second shell.

After substitution of (4.222) and (4.223) into the right-hand side of the equality (4.220), we shall have in the summand scalar products of bundles of the same type, namely, bundles of three functions, of k_1 particles, $k_2 - 1$ particles and one particle respectively. Using the orthonormalization of these bundles, after transformations and summations which are now simple by virtue of the Kronecker symbols, we obtain the following expression for the one-particle fractional parentage coefficient of a mixed configuration in terms of the one-particle fractional parentage coefficient of a pure configuration:

$$\left([l_1^{k_1}[\lambda_1]\alpha_1\beta_1 L_1 S_1 T_1, l_2^{k_2-1}[\lambda_2']\alpha_2'\beta_2' L_2' S_2' T_2']^{L'S'T'}; l_2 | \} \times [l_1^{k_1}[\lambda_1]\alpha_1\beta_1 L_1 S_1 T_1, l_2^{k_2}[\lambda_2]\alpha_2\beta_2 L_2 S_2 T_2]^{LST} \right)$$
$$= \sqrt{\frac{k_2}{k_1+k_2}(2L'+1)(2L_2+1)(2S'+1)(2S_2+1)(2T'+1)(2T_2+1)}$$
$$\times W(L_1 L_2' L l_2; L' l_2) W(S_1 S_2' S \frac{1}{2}; S' S_2) W(T_1 T_2' T \frac{1}{2}; T' T_2)$$
$$\times (l_2^{k_2-1}[\lambda_2']\alpha_2'\beta_2' L_2' S_2' T_2'; l_2^2 | \} l_2^k[\lambda_2]\alpha_2\beta_2 L_2 S_2 T_2). \quad (4.224)$$

In a similar way, we can express the many-particle fractional parentage coefficient of a mixed configuration in the general case of simultaneous separation of nucleons from several shells (pure configurations) in terms of the fractional parentage coefficients of the pure configurations. If we have mixing of two configurations, it will be possible, in the general case, to couple together four angular momenta in deriving the relation analogous to (4.224), and the final formula will contain Wigner $9j$-symbols instead of the Racah coefficients (*Wigner $6j$-symbols*); moreover there will be not one (as in our case), but two fractional parentage coefficients of pure configurations, for the separation of nucleons from each of the (two) shells.

4.7. Determine the magnetic moment of a nucleus containing k nucleons in an unfilled outer shell.

The magnetic moment μ of a nucleus in the ground state is defined as the mean value of the operator of the z-component of the total magnetic moment of the k equivalent nucleons in the unfilled outer shell:

$$\hat{\mu}_z = \sum_{i=1}^{k} (g_s^{(i)} s_z^{(i)} + g_l^{(i)} l_z^{(i)}), \tag{4.225}$$

where

$$g_s^{(i)} = (1 - \tau_z^{(i)})\mu_n + (1 + \tau_z^{(i)})\mu_p, \qquad g_l^{(i)} = \frac{1}{2}(1 + \tau_z^{(i)}).$$

We shall choose the wave function of the ground state in the form

$$\Psi_0(l^k[\lambda]\alpha\beta LSJTMM_T) = \sum_{M_L M_S} (LM_L SM_S|JM)\Psi(l^k[\lambda]\alpha\beta LSTM_L M_S M_T)$$

and determine the mean value of (4.225) in the state in which the z-component M of the total moment is equal to its maximum possible value J.

Since $\hat{\mu}_z$ is a one-particle operator, symmetric with respect to permutations of the nucleons, it is convenient to separate out the wave function of one nucleon from the k-particle wave function by means of the one-particle fractional parentage coefficients. The one-particle function is an eigenfunction of the operators τ_z, s_z and l_z, corresponding to the eigenvalues ν, μ and m. Consequently,

$$\begin{aligned}
\mu &\equiv \langle \Psi_0 | \hat{\mu}_z | \Psi_0 \rangle_{M=J} \\
&= k \sum_{[\lambda']\alpha'\beta' L' S' T'} (l^{k-1}[\lambda']\alpha'\beta' L' S' T'; lst|\} l^k[\lambda]\alpha\beta LST)^2 \\
&\times \sum_{M_L M_S} (LM_L SM_S|JJ)^2 \sum_{M'_T \nu} (T'M'_T \tfrac{1}{2}\nu|TM_T)^2 \\
&\times \sum_{M'_S \mu} (S'M'_S \tfrac{1}{2}\mu|SM_S)^2 \sum_{M'_L m} (L'M'_L lm|LM_L)^2 \\
&\times \{[(1-2\nu)\mu_n + (1+2\nu)\mu_p]\mu + \frac{1}{2}(1+2\nu)m\}.
\end{aligned} \tag{4.226}$$

We shall use the relation

$$\begin{aligned}
\sum_{mm'} m'(jmj'm'|JM)^2 &= M\frac{J(J+1) + j'(j'+1) - j(j+1)}{2J(J+1)}\Delta(jj'J) \\
&\quad \times\Delta(JJ1)\Delta(j'j'1),
\end{aligned}$$

which is easily proved by replacing the factor m' in the left-hand side of the equality by $-(10j'm'|j'm') \times \sqrt{j'(j'+1)}$. By definition, the factor $\Delta(jj'J)$ is equal to unity if the three angular momenta j, j' and J satisfy the triangle relation $|j - j'| \leqslant J \leqslant j + j'$, and equal to zero otherwise.

Thus, for the magnetic moment of a nucleus treated in the shell model with LS-coupling we obtain the following formula:

$$\begin{aligned}
\mu \;=\; & \frac{k}{2J+1}\Delta(JJ1)\Delta(JLS)\sum_{\lambda']\alpha'\beta'L'S'T'} \\
& \times(l^{k-1}[\lambda']\alpha'\beta'L'S'T';lst|\}l^k[\lambda]\alpha\beta LST)^2 \\
& \times\left\{\left[\mu_n+\mu_p-(\mu_n-\mu_p)M_T\frac{T(T+1)-T'(T'+1)+\frac{3}{4}}{T(T+1)}\Delta(TT1)\right]\right. \\
& \times\frac{S(S+1)-S'(S'+1)+\frac{3}{4}}{2S(S+1)}[J(J+1)+S(S+1)-L(L+1)]\Delta(SS1) \\
+ \; & \left[\frac{1}{2}+M_T\frac{T(T+1)-T'(T'+1)+\frac{3}{4}}{2T(T+1)}\Delta(TT1)\right] \\
& \times\frac{L(L+1)+l(l+1)-L'(L'+1)}{2L(L+1)} \\
& \left.\times[J(J+1)+L(L+1)-S(S+1)\Delta(LL1)\Delta(ll1))\right\}. \qquad (4.227)
\end{aligned}$$

According to (4.227), to find the magnetic moment we need to know certain quantum characteristics of the nucleus.. In Table 4.7, numerical values are given for the magnetic moments of a series of light nuclei with a filled $1p$-shell ($l=1$), calculated from this formula. Also indicated in the Table are the states $(p^k[\lambda]^{2T+1,2S+1}L_{J\pi})$ of the nucleons in the $1p$-shell of the nucleus.

TABLE 4.7.

Nucleus	State in the $1p$-shell	Magnetic moment μ
${}^5_2\mathrm{He}$	$p^1[1]^{22}P_{3/2-}$	$\mu_n=-1\cdot 91$
${}^5_3\mathrm{Li}$	$p^1[1]^{22}P_{3/2-}$	$\mu_p+1=3\cdot 79$
${}^6_3\mathrm{Li}$	$p^2[2]^{13}S_{1+}$	$\mu_n+\mu_p=0\cdot 88$
${}^7_3\mathrm{Li}$	$p^3[3]^{22}P_{3/2-}$	$\mu_p+\frac{1}{3}=3\cdot 12$
${}^7_4\mathrm{Be}$	$p^3[3]^{22}P_{3/2-}$	$\mu_n+\frac{2}{3}=-1\cdot 24$
${}^9_4\mathrm{Be}$	$p^5[41]^{22}P_{3/2-}$	$\mu_n+\frac{1}{3}=-1\cdot 58$
${}^{10}_5\mathrm{B}$	$p^6[42]^{13}D_{3+}$	$\mu_n+\mu_p+1=1\cdot 88$
${}^{11}_5\mathrm{B}$	$p^7[43]^{22}P_{3/2-}$	$\mu_p+\frac{2}{3}=3\cdot 46$
${}^{13}_6\mathrm{C}$	$p^9[441]^{22}P_{1/2-}$	$-\frac{1}{3}\mu_n+\frac{4}{9}=1\cdot 08$
${}^{13}_7\mathrm{N}$	$p^9[441]^{22}P_{1/2-}$	$-\frac{1}{3}\mu_p+\frac{2}{9}=-0\cdot 71$
${}^{14}_7\mathrm{N}$	$p^{10}[442]^{13}D_{1+}$	$-\frac{1}{2}(\mu_n+\mu_p)+\frac{3}{4}=0\cdot 31$
${}^{15}_7\mathrm{N}$	$p^{11}[443]^{22}P_{1/2-}$	$-\frac{1}{3}\mu_p+\frac{2}{3}=-0\cdot 26$

4.8. Determine the energy eigenvalues and eigenfunctions for a nucleon moving in the field of a three-dimensional anisotropic oscillator potential. Make the transitions to oscillator potentials with axial and spherical symmetry.

The Hamiltonian of a particle in the field of a three-dimensional anisotropic oscillator potential with different frequencies ω_1, ω_2 and ω_3,

$$H = -\frac{\hbar^2}{2M}\nabla^2 - V_0 + \frac{M}{2}(\omega_1^2 x^2 + \omega_2^2 y^2 + \omega_3^2 z^2), \tag{4.228}$$

is not invariant with respect to rotations, since the potential does not possess central symmetry. The state of a particle in such a field is not characterized by a definite orbital angular momentum. Since the Hamiltonian (4.228) is separable in Cartesian coordinates, the solution of the Schrödinger equation

$$H\psi(\mathbf{r}) = E\psi(\mathbf{r}) \tag{4.229}$$

must be found in the form of a product

$$\psi(\mathbf{r}) = \psi^{(1)}(x)\psi^{(2)}(y)\psi^{(3)}(z). \tag{4.230}$$

On substituting (4.230) into (4.229) we obtain an equation for each of the functions $\psi^{(i)}$; for example,

$$\left\{-\frac{\hbar^2}{2M}\frac{d^2}{dx^2} + \frac{M}{2}\omega_1^2 x^2 - E_1\right\}\psi^{(i)}(x) = 0, \tag{4.231}$$

with analogous equations for $\psi^{(2)}(y)$ and $\psi^{(3)}(z)$. We note that the only quantity with physical meaning is the total energy $E = -V_0 + \sum_{i=1}^{3} E_i$. We denote

$$\frac{E_i}{\hbar\omega_i} = \varepsilon_i; \qquad \sqrt{\frac{M\omega_i}{\hbar}} = \alpha_i; \qquad \sqrt{2}\alpha_1 x = \xi \tag{4.232}$$

and introduce the function

$$H(\xi) = \mathrm{e}^{(1/4)\xi^2}\psi^{(1)}, \tag{4.233}$$

for which we obtain from (4.231) the following equation:

$$\frac{d^2H(\xi)}{d\xi^2} - \xi\frac{dH(\xi)}{d\xi} + \left(\varepsilon_1 - \frac{1}{2}\right)H(\xi) = 0. \tag{4.234}$$

The solutions of the equation obtained are the Hermite polynomials:

$$H_{n_1}(\xi) = (-1)^{n_1}\mathrm{e}^{\xi^2/2}\frac{d^{n_1}}{d\xi^{n_1}}\mathrm{e}^{-\xi^2/2}, \tag{4.235}$$

where $\varepsilon_1 - \frac{1}{2} = n_1$ and n_1 is a non-negative integer. The Hermite polynomials possess the orthogonality property

$$\int_{-\infty}^{\infty} d\xi \mathrm{e}^{-\xi^2/2} H_k(\xi) H_l(\xi) = \sqrt{2\pi}k!\delta_{kl}.$$

Thus, the orthonormal eigenfunctions of the Hamiltonian (4.228) can be represented in the form

$$\begin{aligned}\psi_{n_1 n_2 n_3}(\mathbf{r}) &\equiv \psi^{(1)}_{n_1}(x)\psi^{(2)}_{n_2}(y)\psi^{(3)}_{n_3}(z) = \left(\frac{\alpha_1^2\alpha_2^2\alpha_3^2}{\pi^3}\right)^{1/4} \\ &\times (n_1!n_2!n_3!)^{-(1/2)}\, \mathrm{e}^{-1/2(\alpha_1^2x^2+\alpha_2^2y^2+\alpha_3^2z^2)} \\ &\times H_{n_1}(\sqrt{2}\alpha_1 x)H_{n_2}(\sqrt{2}\alpha_2 y)H_{n_3}(\sqrt{2}\alpha_3 z),\end{aligned} \tag{4.236}$$

and the energy eigenvalues are equal to

$$E_{n_1 n_2 n_3} = -V_0 + \hbar\omega_1(n_1 + \frac{1}{2}) + \hbar\omega_2(n_2 + \frac{1}{2}) + \hbar\omega_3(n_3 + \frac{1}{2}), \tag{4.237}$$

where n_1, n_2 and n_3 are non-negative integers. It can be seen from (4.236) and (4.237) that, in the case of an anisotropic oscillator potential, different sets of the quantum numbers n_1, n_2 and n_3 defining the state correspond, generally speaking, to different energies. Degeneracy can occur only when the frequencies satisfy certain relations, for example, when one of the frequencies is a multiple of another, and, in particular, when two of the frequencies, or all three, are equal.

If $\omega_1 = \omega_2 = \omega'$, the Hamiltonian (4.228) is invariant under rotations about the z-axis (axial symmetry): therefore, the states of a nucleon in such a potential can be characterized by the z-component of the orbital angular momentum, and the solution $\psi(\mathbf{r})$ of eqn. (4.229) must be found in the form of a product

$$\psi(\mathbf{r}) = \Phi(\varphi)\chi(\varrho)\psi^{(3)}(z), \tag{4.238}$$

where $\Phi(\varphi) = (1/\sqrt{2\pi})\mathrm{e}^{im\varphi}$. In this case, eqn. (4.229) is separable in the variables $\varphi, \varrho = \sqrt{x^2+y^2}, z$, that is, in cylindrical coordinates. Functions (4.238) with a given principal quantum number $N = n_1 + n_2 + n_3$ are linear combinations of the function (4.236) with $\alpha_1 = \alpha_2 = \sqrt{M\omega'/\hbar}$. For the characteristics of a state described by such a function, we can choose the three quantum numbers: N, n_3 and m, or $n_\perp (\equiv N - n_3)$, n_3 and m. Here m is the eigenvalue of the projection of the orbital angular momentum of the nucleon along the symmetry axis of the field and takes the values:$\pm(N-n_3), \pm(N-n_3-2), \ldots, \pm 1$ or 0. For these values of m and a given value of N, the functions (4.238) and (4.236) transform in the same way under inversion $\mathbf{r} \to -\mathbf{r}(\varphi \to \pi + \varphi)$. Thus, in the case of an oscillator field with axial symmetry, the eigenfunctions have the form

$$\begin{aligned}\psi_{n_\perp n_3 m}(\mathbf{r}) &= \frac{1}{\sqrt{2\pi}}\,\mathrm{e}^{im\varphi}\chi_{n_\perp m}(\varrho)\frac{1}{\sqrt{n_3!}} \\ &\times \sqrt[4]{\frac{\alpha_3^2}{\pi}}\,\mathrm{e}^{-(1/2)\alpha_3^2z^2}H_{n_3}\left(\sqrt{2}\alpha_3 z\right),\end{aligned} \tag{4.239}$$

where the wave function $\chi_{n_\perp m}(\varrho)$ satisfies the equation

$$\begin{aligned}&\frac{1}{\varrho}\frac{d}{d\varrho}\left(\varrho\frac{d\chi_{n_\perp m}(\varrho)}{d\varrho}\right) + \left(\Lambda - \alpha_\perp^4\varrho^4 - \frac{m^2}{\varrho^2}\right) \\ &\times \chi_{n_\perp m}(\varrho) = 0,\end{aligned} \tag{4.240}$$

$$\Lambda = \frac{2M}{\hbar^2}(E - E_3 + V_0), \quad \alpha_\perp = \sqrt{\frac{M\omega'}{\hbar}}.$$

At small ϱ the function $\chi_{n_\perp m}(\varrho) \sim \varrho^{|m|}$, and at large ϱ $\chi_{n_\perp m}(\varrho) \sim \exp(-\frac{1}{2}\alpha_\perp^2\varrho^2)$. Thus, for all ϱ, a solution of equation (4.240) can be proposed taking the form $\chi_{n_\perp m}(\varrho) = \varrho^{|m|}F(\varrho)\mathrm{e}^{-\frac{1}{2}\alpha_\perp^2\varrho^2}$, where $F(\varrho)$ is a polynomial, which is proportional to the degenerated hypergeometric function. It is the polynomial of a finite degree, if $\frac{1}{2}\left(\frac{\Lambda}{2\alpha_\perp^2} - |m| - 1\right) = \frac{1}{2}(n_\perp - |m|)$ is a nonnegative integer, which is denoted as n. In this case, the degenerate hypergeometric function reduces to the Laguerre polynomial $L_n^{|m|}(\alpha_\perp^2\varrho^2)$, within a factor. The normalized solution of equation (4.240) takes the form

$$\chi_{n_\perp m}(\varrho) = \left(\frac{2\alpha_\perp^2 n!}{(n+|m|)!}\right)^{1/2}\left(\alpha_\perp\varrho\right)^{|m|} L_n^{|m|}\left(\alpha_\perp^2\varrho^2\right)\mathrm{e}^{-\frac{1}{2}\alpha_\perp^2\varrho^2},$$

$$n = \frac{n_\perp - |m|}{2}. \tag{4.241}$$

In the case of axial symmetry, the energy eigenvalues $E = E_{n_\perp, n_3}$ do not depend on m:

$$E = E_{n_\perp, n_3} = -V_0 + \hbar\omega'(n_\perp + 1) + \hbar\omega_3(n_3 + \frac{1}{2}). \tag{4.242}$$

The degeneration multiplicity in the axial symmetric case is determined by $n_\perp + 1 = n_1 + n_2 + 1 = N - n_3 + 1$.

In the case of an isotopic oscillator potential, when $\omega_1 = \omega_2 = \omega_3 = \omega$ (spherical symmetry), the energy eigenvalues (4.237) depend only on the sum of the quantum numbers, $n_1 + n_2 + n_3 = N$, that is, we arrive at the expression (4.26). Corresponding to the energy level E_N are $(N+1)(N+2)/2$ states, each of which is described by a wave function (4.236) with one of the possible sets of three quantum numbers n_1, n_2, n_3 satisfying the condition $n_1 + n_2 + n_3 = N$:

$$\psi_{n_1n_2n_3}(\mathbf{r}) = \left(\frac{\alpha^2}{\pi}\right)^{3/4}(n_1!n_2!n_3!)^{-1/2}\,\mathrm{e}^{-(1/2)\alpha^2r^2}$$

$$\times H_{n_1}\left(\sqrt{2}\alpha x\right)H_{n_2}\left(\sqrt{2}\alpha y\right)H_{n_3}\left(\sqrt{2}\alpha z\right),$$

$$\alpha = \sqrt{\frac{M\omega}{\hbar}}. \tag{4.243}$$

The functions obtained do not correspond to definite values of the orbital angular momentum. However, by selecting certain linear combinations of the functions obtained, we can construct a set of $(N+1)(N+2)/2$ new functions, corresponding to definite values of the orbital angular momentum. According to Table 4.1, several values l of the orbital angular momentum may correspond to a given value of the principal quantum number N in the general case. For small values of N, we can find all the possible values of l by using the properties of the functions $H_{n_i}(\xi)$ and $Y_{lm}(\mathbf{n})$:

$$H_{n_i}(-\xi) = (-1)^{n_i} H_{n_1}(\xi), \qquad Y_{lmm}(-\mathbf{n}) = (-1)^l Y_{lm}(\mathbf{n})$$

and the fact that $l \leqslant N$. For example, for $N = 3$ the functions $\psi_{n_1 n_2 n_3}(\mathbf{r})$ change sign under reflection, and so we can use them to construct linear combinations corresponding to odd orbital angular momenta. In the given case, the possible values of l are $l = 1$ and 3, since the total number of independent functions, $\sum_l (2l+1) = 10$, must be equal to the degeneracy $\frac{1}{2}(N+1(N+2)$, in agreement with the data of Table 4.1. In the general case, for a given N all values of $l \leqslant N$ satisfying the condition $(-1)^{N-l} = 1$, that is, differing in steps of two, are possible.

References

Bayman, B.F. (1957) *Groups and their Applications to Spectroscopy* (Lecture notes, Nordita,Copenhagen; reissued 1960).

Bethe, H.A. and Rose, M.E. (1937) *Phys.Rev.*, **51**, p. 283.

Blin-Stoyle, R.J. (1956) *Rev.Mod.Phys.*, **28**, p.75.

Edmonds, A.R. and Flowers, B.H. (1952) *Proc.Roy..Soc..* **A214**, p.515.

Elliott, J.P. (1958a) *Proc.Roy.Soc.*, **A245**, p.128.

Elliott, J.P. (1958a) *Proc.Roy.Soc.*, **A245**, p.562.

Elliott, J.P. and Lane, A.M. (1957) "The nuclear shell model", *Encyclopaedia of Physics* (ed.S.Flügge), Vol.39, Springer-Verlag, Berlin.

Elliott, J.P. and Skyrme, T.H.R. (1955) *Proc.Roy.Soc.*, **A232**, p.561.

Elsasser, W.M. (1934) *J.de Phys.Radium*, **5**, p.625.

Flowers, B.H. (1952) *Proc.Roy.Soc.* **A212**, p.248.

Goeppert-Mayer, M. (1949) *Phys.Rev.*, **75**, p.1969.

Goeppert-Mayer, M. (1950) *Phys.Rev.*, **78**, p.16.

Goeppert-Mayer, M. and Jensen, J.H.D. (1955) *Elementary Theory of Nuclear Shell Structure*, Wiley, N.Y.

Haxel, O., Jensen, J.H.D. and Suess, H.E. (1949) *Phys.Rev.*, **75**, p.1766.

Haxel, O., Jensen, J.H.D. and Suess, H.E. (1950) *Z.Phys.*, **128**, p.295.

Heisenberg, W. (1949) *Theory of the atomic Nucleus*, Wieweg, Braunschweig (English translation (*Nuclear Physics*) published by Methuen, London, 1953).

Inglis, D.R. (1953) *Rev.Mod.Phys.*, **25**, p.390.

Jahn, H.A. (1950) *Proc.Roy.Soc.*, **A201**, p.516.

Jahn, H. and van Wieringen, H. (1951) *Proc.Roy.Soc.*,**A209**, p.502.

Klinkenberg, P.F.A. (1952) *Rev.Mod.Phys.*, **24**, p.63.

Kurath, D. (1956) *Phys.Rev.*, **101**, p.216.

Neudachin, V.G. and Smirnov, Yu.F. (1969) *Nucleon Clusters in Light Nuclei*"Nauka", Moscow.

Rosenfeld, L. (1948) *Nuclear Forces*, North-Holland, Amsterdam.

Schwartz, C. (1955) *Phys.Rev.*, **97**, p.380.

de-Schalit, A. and Talmi, I. (1963) *Nuclear Shell Theory*, Academic Press, N.Y.

Weyl, H. (1946) *The Classical Groups.* Princeton University Press.

Wigner, E.P. (1959) *Group Theory and its Applications to the Quantum Mechanics of Atomic Spectra*, Academic Press, N.Y.

CHAPTER 5

ROTATION AND VIBRATIONS OF NUCLEI

5.1. Vibrations of Spherical Nuclei

Surface deformations of nuclei. We shall examine the states associated with the excitation of collective degrees of freedom of nuclei. The shell model of the nucleus is based on the simplifying assumption that the individual nucleons move independently, the interaction between them being described by a self-consistent field. The liquid-drop model of the nucleus is based on the assumption of the existence of strong coupling between the nucleons, as a result of which the mean free path of a nucleon in nuclear matter is small compared with the dimensions of the nucleus. Whereas in the shell model of the nucleus one considers one-particle excitations associated with changes of the states of individual nucleons, in the liquid-drop model of the nucleus collective excitations associated with a simultaneous change of the states of many nucleus are taken into account. Lying at the basis of the so-called *generalized model of the nucleus* , which is a synthesis of the shell and liquid-drop models, is the assumption that the individual nucleons move independently in a slowly varying self-consistent field. In this model, as in the shell model, degrees of freedom associated with the motion of one nucleon or several weakly coupled nucleons in the self-consistent field are taken into account. In the generalized model, as in the liquid-drop model, collective degrees of freedom associated with change of the shape of the nucleus and of its orientation in space are taken into account. In the generalized model, it is usually assumed that the internal and collective motions are separable; they are therefore treated independently.[1]

Since nuclear matter is characterized by low compressibility, the collective motion of the nucleons in nuclei reduces principally to deformations of the shapes of the nuclei, with no change in their volume. In the case of spherical nuclei, the collective excitations correspond to vibrations of the surface of the nucleus about the equilibrium shape. In non-spherical nuclei, the collective excitations can be associated with *vibrations of the surface of the nucleus* and with *rotation of the nucleus in space.*

[1]The generalized model of the nucleus was proposed in papers by Bohr (1952) and Bohr and Mottelson (1953). The detailed theory of the collective excited states of nuclei is described in the monograph by Davydov (1967).

We shall consider an arbitrary deformation, described in the spherical system of coordinates by the function $R(\vartheta, \varphi)$, of the surface of a nucleus. The quantity $R(\vartheta, \varphi)$ is the distance from the centre of the nucleus to its surface in the direction of the polar angles ϑ and φ, measured in the laboratory coordinate frame. We shall expand this function as a sum over the spherical harmonics $Y_{\lambda\mu}(\vartheta, \varphi)$:

$$R(\vartheta, \varphi) = \left\{1 + \sum_{\lambda\mu} \alpha_{\lambda\mu} Y_{\lambda\mu}(\vartheta, \varphi)\right\} R, \tag{5.1}$$

where R is the radius of the sphere of the same volume and $\alpha_{\lambda\mu}$ are the deformation parameters characterizing the shape of the deformed nucleus. Since $R(\vartheta, \varphi)$ is real, the *deformation parameters* $\alpha_{\lambda\mu}$ must satisfy the condition

$$\alpha_{\lambda\mu} = (-1)^{\mu} \alpha^*_{\lambda,-\mu}. \tag{5.2}$$

We shall consider small deviations from a spherical nuclear shape. In the case of small deviations from spherical shape, the deformation parameters $\alpha_{\lambda\mu}$ are small quantities. If nuclear matter is incompressible, it follows from the conservation of the volume that

$$\alpha_{00} = -(1/4\pi) \sum_{\lambda\mu} |\alpha_{\lambda\mu}|^2, \tag{5.3}$$

from which it follows that the parameter α_{00} is a second-order small quantity. The parameters $\alpha_{1\mu}$ can be assumed equal to zero, since the deformation (5.1) for $\lambda = 1$ corresponds to a displacement of the nucleus as a whole in space. We note that there is an upper bound on the possible values of λ ($\lambda < A^{1/3}$), inasmuch as the linear dimensions of an element of the nuclear surface should be grater than the mean distance between the nucleons. Thus, surface deformations of the nucleus are characterized by a finite number of deformation parameters $\alpha_{\lambda\mu}$. Ellipsoidal deformations of the nuclear surface, corresponding to the value $\lambda = 2$, are of the greatest interest.

The kinetic energy T and potential energy V of the surface vibrations of the nucleus are given, in the case of small deformations, by the following general formulae:

$$T = \frac{1}{2} \sum_{\lambda\mu} B_\lambda |\dot{\alpha}_{\lambda\mu}|^2; \tag{5.4}$$

$$V = \frac{1}{2} \sum_{\lambda\mu} C_\lambda |\alpha_{\lambda\mu}|^2, \tag{5.5}$$

where the coefficients B_λ and C_λ depend on the specific assumptions about the properties of the nuclear substance.

By regarding the nucleus as a drop of an ideal incompressible charged liquid, we can show that, in the case of irrotational flow of this liquid, the mass coefficient B_λ is equal to

$$B_\lambda = (\varrho/\lambda)R^5, \tag{5.6}$$

where ϱ is the density of the nuclear substance. The deformability coefficient C_λ in the case of a uniform distribution of the nuclear charge over the volume is given by the expression

$$C_\lambda = (\lambda - 1)(\lambda + 2)\sigma R^2 - (3/2\pi)[(\lambda - 1)/(2\lambda + 1)] \cdot (Z^2 e^2/R), \tag{5.7}$$

where σ is the surface-tension coefficient, connected with the coefficient ε_2 in the Weizsäcker formula (3.17) by the relation

$$4\pi\sigma R_0^2 = \varepsilon_2, \;\; R = R_0 A^{1/3}.$$

The potential energy (5.5) of the nucleus is the sum of the surface and Coulomb energies. The surface energy is a minimum for a sphere and increases with deviation from spherical symmetry. Because of the Coulomb repulsive forces, the nucleus, regarded as a drop of charged liquid, tends to take an asymmetric shape. According to (5.5), a nucleus in equilibrium has a spherical shape, stable against deformations, if the deformability coefficients $C_\lambda > 0$. Using (5.7), we obtain the condition for stability of a nucleus against small deformations:

$$(\lambda + 2)\sigma R^2 > 3Z^2 e^2/2\pi(2\lambda + 1)R. \tag{5.8}$$

Putting $\lambda = 2$ and using the values of the parameters of (3.17), we have $Z^2/A < 48$. This condition is fulfilled for all known nuclei. (For example, for elements at the end of the periodic table, the ratio Z^2/A is equal to $35 \cdot 6$ for $^{238}_{92}U$, $36 \cdot 5$ for $^{242}_{94}$Pu, $37 \cdot 8$ for $^{244}_{96}$Cm, $39 \cdot 0$ for $^{246}_{98}$Cf, $40 \cdot 3$ for $^{248}_{100}$Fm and $41 \cdot 0$ for $^{249}_{101}$Mv).

Surface vibrations of spherical nuclei. We shall consider small vibrations of the surface of a spherical nucleus. The dynamical variables in this case are the deformation parameters $\alpha_{\lambda\mu}$, which can be regarded as generalized coordinates. The energy of the surface vibrations

$$E = \frac{1}{2}\sum_{\lambda\mu}\left\{B_\lambda|\dot{\alpha}_{\lambda\mu}|^2 + C_\lambda|\alpha_{\lambda\mu}|^2\right\} \tag{5.9}$$

depends on the generalized coordinates $\alpha_{\lambda\mu}$ and generalized velocities $\dot{\alpha}_{\lambda\mu}$. We can go over from the classical energy E to the quantum Hamiltonian operator $\mathbf{H}$ by introducing the generalized momenta

$$\pi_{\lambda\mu} \equiv \partial E/\partial\dot{\alpha}_{\lambda\mu} = B_\lambda \dot{\alpha}^*_{\lambda\mu} \tag{5.10}$$

canonically conjugate to the coordinates $\alpha_{\lambda\mu}$ and postulating the following commutation relations between the quantities $\alpha_{\lambda\mu}$ and $\pi_{\lambda\mu}$, which must be regarded as operators:

$$\left.\begin{aligned} [\pi_{\lambda\mu}, \alpha_{\lambda'\mu'}] &= (\hbar/i)\delta_{\lambda\lambda'}\delta_{\mu\mu'}; \\ [\alpha_{\lambda\mu}, \alpha_{\lambda'\mu'}] &= [\pi_{\lambda\mu}, \pi_{\lambda'\mu'}] = 0. \end{aligned}\right\} \tag{5.11}$$

The Hamiltonian obtained in this way for the surface vibrations of the nucleus,

$$\mathbf{H} = \frac{1}{2}\sum_{\lambda\mu}\{(1/B_\lambda)|\pi_{\lambda\mu}|^2 + C_\lambda|\alpha_{\lambda\mu}|^2\}, \tag{5.12}$$

has the form of a sum of Hamiltonians for uncoupled harmonic oscillators with eigenfrequencies

$$\omega_\lambda = \sqrt{(C_\lambda/B_\lambda)}. \tag{5.13}$$

The energy eigenvalues of the surface vibrations of the nucleus, according to (5.12), can be represented in the form of a sum of energies of individual oscillators:

$$E = \sum_{\lambda\mu}(n_{\lambda\mu} + \frac{1}{2})\hbar\omega_\lambda, \tag{5.14}$$

where $n_{\lambda\mu}$, are positive integers. When the nucleus undergoes a transition from one state to another, the energy changes by an amount equal to a sum of multiples of the $\hbar\omega_\lambda$. Clearly, the quantity $n_{\lambda\mu}$ determines the number of elementary excitations of a given type $\lambda\mu$, characterized by the energy $\hbar\omega_\lambda$. For a given value of λ, the quantity μ takes $2\lambda + 1$ different values $(-\lambda \leqslant \mu \leqslant \lambda)$; therefore, the excited states with a fixed value of λ are characterized by $(2\lambda + 1)$-fold degeneracy.

The elementary excitations of the surface vibrations of a nucleus can be regarded as quasi-particles – *phonons*. By specifying the numbers of phonons of each type, we are thereby defining the nuclear excited state associated with surface vibrations. In treating nuclear excited states associated with surface vibrations, it is convenient to use the occupation-number representation. For this, in place of the operators $\alpha_{\lambda\mu}$ and $\pi_{\lambda\mu}$, we introduce the new operators $b_{\lambda\mu}$:

$$\left.\begin{aligned}\alpha_{\lambda\mu} &= \left(\frac{\hbar}{2B_\lambda\omega_\lambda}\right)^{1/2}[b_{\lambda\mu} + (-1)^\mu b^+_{\lambda,-\mu}];\\ \pi_{\lambda\mu} &= i\left(\frac{\hbar B_\lambda\omega_\lambda}{2}\right)^{1/2}[b^+_{\lambda\mu} - (-1)^\mu b_{\lambda,-\mu}].\end{aligned}\right\} \tag{5.15}$$

Using (5.11), it is not difficult to obtain the commutation relations

$$\left.\begin{aligned}\left[b_{\lambda\mu}, b^+_{\lambda'\mu'}\right] &= \delta_{\lambda\lambda'}\delta_{\mu\mu'};\\ [b_{\lambda\mu}, b_{\lambda'\mu'}] = \left[b^+_{\lambda\mu}, b^+_{\lambda'\mu'}\right] &= 0,\end{aligned}\right\} \tag{5.16}$$

from which it follows directly that the operators $b_{\lambda\mu}$ and $b^+_{\lambda\mu}$ can be represented in the form of matrices:

$$b_{\lambda\mu} = \begin{bmatrix} 0 & \sqrt{1} & 0\ldots0 & 0 & \ldots\\ 0 & 0 & \sqrt{2}\ldots0 & 0 & \ldots\\ 0 & 0 & 0\ldots0 & 0 & \ldots\\ \ldots & & & & \\ 0 & 0 & 0\ldots0 & \sqrt{n_{\lambda\mu}} & \ldots\\ 0 & 0 & 0\ldots0 & 0 & \ldots\\ \ldots & & & & \end{bmatrix},$$

$$b^+_{\lambda\mu} = \begin{bmatrix} 0 & 0 & 0\ldots & 0 & 0\ldots\\ \sqrt{1} & 0 & 0\ldots & 0 & 0\ldots\\ 0 & \sqrt{2} & 0\ldots & 0 & 0\ldots\\ \ldots & & & & \\ 0 & 0 & 0\ldots & 0 & 0\ldots\\ 0 & 0 & 0\ldots & \sqrt{n_{\lambda\mu}} & 0\ldots\\ \ldots & & & & \end{bmatrix}. \tag{5.17}$$

Substituting the expressions, (5.15) into (5.12), we obtain the Hamiltonian of the system in the form

$$\mathbf{H} = \sum_{\lambda\mu}(b^+_{\lambda\mu}b_{\lambda\mu} + \frac{1}{2})\hbar\omega_\lambda. \tag{5.18}$$

The *wave function in the occupation-number representation* (in the case of phonons) depends on the number $N_{\lambda\mu}$ of phonons in each state $\lambda\mu$, and can be written in the form

$$\begin{aligned}&\Phi_{n_{\lambda_1\mu_1}n_{\lambda_2\mu_2}\ldots n_{\lambda\mu}\ldots}(N_{\lambda_1\mu_1}N_{\lambda_2\mu_2}\ldots N_{\lambda\mu}\ldots)\\ &= \langle N_{\lambda_1\mu_1}N_{\lambda_2\mu_2}\ldots N_{\lambda\mu}\ldots|n_{\lambda_1\mu_1}n_{\lambda_2\mu_2}\ldots n_{\lambda\mu}\ldots\rangle,\end{aligned} \tag{5.19}$$

where $n_{\lambda\mu}$ are the quantum state indices and $N_{\lambda\mu}$ are the representation indices, that is, the arguments of the wave function. The operators $b_{\lambda\mu}$ and $b^+_{\lambda\mu}$ act on the variables $N_{\lambda\mu}$; $b_{\lambda\mu}$ describes the disappearance of one phonon of type $\lambda\mu$, and $b^+_{\lambda\mu}$ describes the creation of one phonon of type $\lambda\mu$:

$$b_{\lambda\mu}|n_{\lambda\mu}\rangle = \sqrt{n_{\lambda\mu}}|n_{\lambda\mu}-1\rangle,\ b^+_{\lambda\mu}|n_{\lambda\mu}\rangle = \sqrt{n_{\lambda\mu}+1}|n_{\lambda\mu}+1\rangle. \tag{5.20}$$

Here we have omitted the arguments of the wave functions, and also the quantum numbers of all states apart from the state $\lambda\mu$. According to (5.20), the wave function of the one-phonon excited state of type $\lambda\mu$ can be represented in the form

$$|1_{\lambda\mu}\rangle = b^+_{\lambda\mu}|0\rangle, \tag{5.21}$$

and the wave function of the state with $n_{\lambda\mu}$ phonons can be represented in the form

$$|n_{\lambda\mu}\rangle = (n_{\lambda\mu}!)^{-1/2}(b^+_{\lambda\mu})^{n_{\lambda\mu}}|0\rangle, \tag{5.22}$$

that is, can be expressed in terms of the wave function $|0\rangle$ of the ground state, in which there are no phonons.

We can extend the analogy between the elementary excitations and particles if we take into account that the operator $b_{\lambda\mu}$ (like the operators $\alpha_{\lambda\mu}$ and $\pi_{\lambda\mu}$) referring to the excitation of the type $\lambda\mu$ is a spherical tensor operator of rank λ. Therefore, each elementary excitation (phonon) of type $\lambda\mu$ has a well defined angular momentum (spin) λ, with component μ along a specified direction, and a well-defined parity $(-1)^\lambda$, in the same way that the spherical function $Y_{lm}(\vartheta,\varphi)$ (a tensor operator of rank l) describes a particle with angular momentum l, angular-momentum component m and parity $(-1)^l$. Since in each state $\lambda\mu$ there can be any numbers of phonons, with integer spins, phonons are bosons. Therefore, the wave function (5.19) should be symmetric with respect to interchanges of phonons of any type.

It follows directly from (5.17) that the operator $b^+_{\lambda\mu}b_{\lambda\mu}$ has diagonal form in the occupation-number representation, with the possible numbers of phonons of type $\lambda\mu$ along the diagonal:

$$b^+_{\lambda\mu}b_{\lambda\mu} = \begin{bmatrix} 0 & 0 & 0 & 0 & \dots & 0 & 0\dots \\ 0 & 1 & 0 & 0 & \dots & 0 & 0\dots \\ 0 & 0 & 2 & 0 & \dots & 0 & 0\dots \\ 0 & 0 & 0 & 3 & \dots & 0 & 0\dots \\ & \dots & & & & & \\ 0 & 0 & 0 & 0 & \dots & n_{\lambda\mu} & 0\dots \\ & \dots & & & & & \end{bmatrix}. \tag{5.23}$$

Therefore, the operator

$$\hat{n}_{\lambda\mu} \equiv b^+_{\lambda\mu} b_{\lambda\mu} \tag{5.24}$$

is the operator of the number of phonons of a given type $\lambda\mu$; this operator is naturally diagonal in its own representation (the occupation-number representation): $\hat{n}_{\lambda\mu}|n_{\lambda\mu}\rangle = n_{\lambda\mu}|n_{\lambda\mu}\rangle$. It follows from (5.16), and also from (5.17), that

$$b_{\lambda\mu} b^+_{\lambda\mu} = \hat{n}_{\lambda\mu} + 1. \tag{5.25}$$

By means of the phonon-number operator (5.24), we can rewrite the Hamiltonian (5.18) of the system in the form

$$\mathbf{H} = \sum_{\lambda\mu} (\hat{n}_{\lambda\mu} + \frac{1}{2}) \hbar\omega_\lambda, \tag{5.26}$$

whence it can be seen directly that excitations of the type $\lambda\mu$ make a contribution to the energy of the collective motions of the nucleus equal to

$$E_{n_{\lambda\mu}} = (n_{\lambda\mu} + \frac{1}{2}) \hbar\omega_\lambda, \tag{5.27}$$

and the total energy E is equal to a sum of expressions of the type (5.27). The operator of the projection of the total angular momentum of the system along the quantization axis z is connected with the phonon-number operator (5.24) by the relation

$$L_z = \hbar \sum_{\lambda\mu} \mu \hat{n}_{\lambda\mu}. \tag{5.28}$$

The eigenvalues of (5.28) in the states $|n_{\lambda\mu}\rangle$ are equal to

$$\langle n_{\lambda\mu}|L_z|n_{\lambda\mu}\rangle = \hbar\mu n_{\lambda\mu}. \tag{5.29}$$

We shall discuss briefly the classification of the *states of a system of several phonons* . The two-phonon wave functions $b^+_{\lambda\mu} b^+_{\lambda'\mu'}|0\rangle$ describe states with well-defined energy $\hbar(\omega_\lambda + \omega_{\lambda'})$, well-defined parity $(-1)^{\lambda+\lambda'}$ and well-defined projection $M = \mu + \mu'$ of the total angular momentum. (In this case, the square of the total angular momentum does not have a well-defined value.) The wave function of a state with well-defined angular momentum Λ lying within the limits $|\lambda - \lambda'| \leqslant \Lambda \leqslant \lambda' + \lambda'$ can be constructed by means of a linear combination of two-particle functions:

$$|\Lambda M\rangle = \sum_{\mu\mu'} (\lambda\mu\lambda'\mu'|\Lambda M) b^+_{\lambda\mu} b^+_{\lambda'\mu'}|0\rangle. \tag{5.30}$$

In the case of phonons of the same type ($\lambda = \lambda'$), it follows from the symmetry of the wave function $|\Lambda M\rangle$ under interchange of phonons and from the symmetry property of the Clebsch-Gordan coefficient $(\lambda\mu\lambda\mu')|\Lambda M) = (-1)^{2\lambda-\Lambda}(\lambda\mu'\lambda\mu|\Lambda M)$ that the only possible values of Λ are the even numbers: $\Lambda = 0, 2, \ldots, 2\lambda$. For example, for $\lambda = 2$ there are three possible values of the total angular momentum of a system of two phonons: $\Lambda = 0, 2$ and 4. If we have a system of three phonons with $\lambda = 2$, the symmetry requirement on the wave function leads to the following values of the total angular momentum: Λ=0,2,3,4 and 6. With increasing numbers of phonons of the same type the states are conveniently classified using group-theoretical methods, in a similar way to that used in the classification of the states of systems of equivalent nucleons. The necessity of invoking group-theoretical methods is connected with the fact that, for a complete description of a system of several collective excitations of the same type, it is no longer sufficient to specify the energy, the angular momentum and its projections, and the parity; additional quantum numbers must be introduced, as, for example, in the study of quadrupolar excitations ($\lambda = 2$) when there are more than three phonons.

Vibrational spectra of nuclei . Even-even nuclei always have spin $J = 0$ in the ground state. If the spin of the system is equal to zero, a spherically symmetric density distribution corresponds to the equilibrium state with lowest energy. However, although a spherically symmetric distribution corresponds to a condition of equilibrium, it is not always the case that the latter is stable.

In fact, if we plot the energy of the ground state of the nucleus as a function of a deformation parameter ε (for example, as a function of the eccentricity of the nucleus in the case when the nucleus has the shape of an ellipsoid of revolution), then the derivative of the energy with respect to the deformation parameter is equal to zero at zero value of the parameter. In the case of stable equilibrium for a spherical shape, the energy curve has a minimum at $\varepsilon = 0$. The doubly magic nuclei and the nearby nuclei have a spherical shape in the ground state (the dependence of the energy on the deformation parameter for these nuclei is depicted schematically by the curves 1 and 2 in Fig. 5.1). However, as the number of nucleons in the unfilled outer shell increases, the polarizing effect of these nucleons leads to violation of the spherical symmetry. In this case, the curve of the energy as a function of the deformation parameter has a minimum at $\varepsilon \neq 0$ (see Fig. 5.1, curve 3), which increases and shifts towards larger deformations ε as the number of nucleons in the unfilled outer shell increases.

In the case of spherical nuclei, the collective excitations correspond to vibrations of the surface of the nucleus about the equilibrium shape. According to (5.13), the square of the frequency ω_λ of the surface vibrations

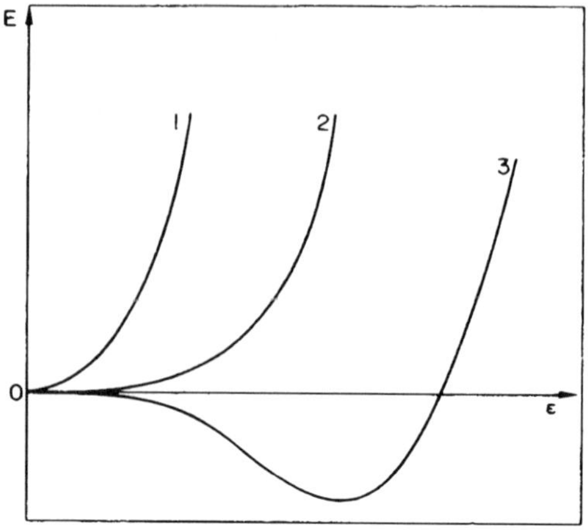

Figure 5.1. Dependence of the energy on the nuclear deformation parameter.

is given by the ratio of the coefficients C_λ and B_λ. The energy $\hbar\omega_\lambda$ of the surface vibrations increases monotonically with increasing λ, in accordance with (5.6) and (5.7). It may be expected, therefore, that the first collective excitations in the low-energy spectra of spherical nuclei will correspond to quadrupolar vibrations with $\lambda = 2$. For the set of quantum numbers defining the excited states, it is convenient to choose the number n of phonons and the angular momentum Λ. For $n = 1$ the spin and parity in the case of quadrupolar vibrations are equal to 2^+; for $n = 2$ there is a degenerate triplet of even states with spin values 0^+, 2^+ and 4^+; for $n = 3$ there are five degenerate even states: 0^+, 2^+, 3^+, 4^+ and 6^+ (Problem 5.16); and so on (Fig. 5.2).

In fact, for the majority of even-even nuclei, the first excited state has spin and parity 2^+. For certain even-even nuclei ($^{110}_{46}Pd$, $^{114}_{48}Cd$, etc.), triplets of the almost degenerate states 0^+, 2^+ and 4^+ predicted by the vibrational model have also been discovered. As an illustration, Fig. 5.2 shows the scheme of the observed levels of the nucleus $^{114}_{48}Cd$. For most nuclei, however, the expected triplets of excited states are not observed, although the second excited state for most nuclei is indeed characterized by values of the spin and parity equal to 0^+, 2^+ or 4^+.

Attempts to perform quantitative calculations of the spectra of the levels of even-even nuclei on the basis of the vibrational model have proved unsuccessful. In fact, using (5.13) with (5.6) and (5.7) and neglecting the Coulomb energy, we have for the quadrupolar vibrations

$$\hbar\omega_2 = a/A^{1/2}, \; a = \hbar\sqrt{(8\sigma/\varrho R_0^3)}.$$

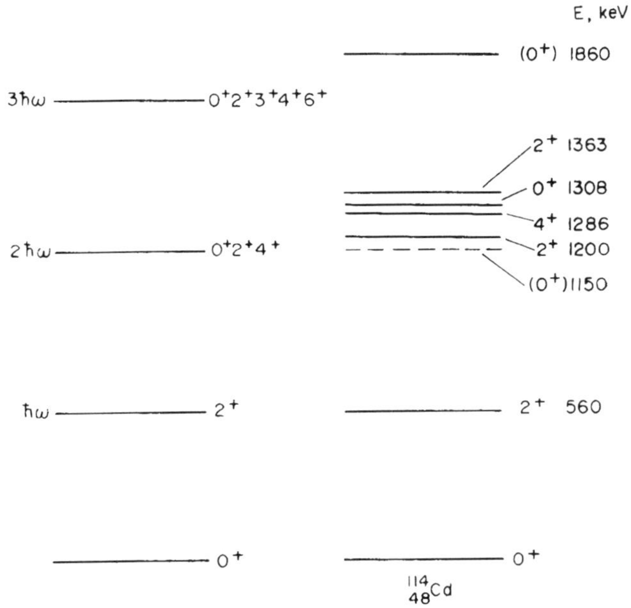

Figure 5.2. Energy spectrum of the quadrupolar vibrations of spherical even-even-nuclei.

Allowance for the Coulomb energy leads to some reduction of this quantity. Consequently, according to the vibrational model of the nucleus, the energy of the first excited level should be a monotonic function of the mass number A. But according to the empirical data, the energy of the first excited state of even-even nuclei depends strongly on the structure of the nuclei and is characterized by very large values for the magic nuclei. This indicates the limited applicability of this model of the nucleus.

Figure 5.3 shows the dependence on the number of neutrons of the ratio E_2/E_1 of the energies of the second and first excited states for even-even nuclei. According to the surface-vibration model, this ratio should be equal to two; in reality, however, there are regions in which this ratio is considerably greater. (Thus, in the ranges of mass numbers $150 < A < 190$ and $A > 220$, this ratio is somewhat greater than three.) It is found that nuclei for which the ratio E_2/E_1 deviates from the characteristic value for vibrational states possess large quadrupole moments; this is connected with the non-spherical equilibrium shape of these nuclei. Therefore, the low-energy levels of these nuclei are associated with the rotation of the nuclei rather than with the surface vibrations.

The quantities $\alpha_{\lambda\mu}$ introduced in (5.1) describe the change of shape of a nucleus relative to a coordinate frame fixed in space. In the case of a non-spherical equilibrium shape, the orientation of the nucleus in space can be

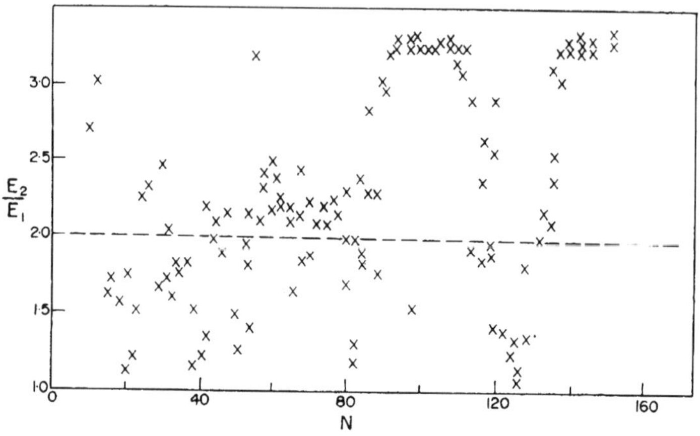

Figure 5.3. Dependence of the ratio of the energies of the second and first excited states on the number of neutrons for even-even nuclei.

described conveniently by the Euler angles, which define the positions of the principal axes of the nucleus relative to a stationary coordinate frame. On rotation of the nucleus, the parameters $\alpha_{\lambda\mu}$ defining the shape of the nucleus in the stationary coordinate frame change with time, although the actual shape of the nucleus remains unchanged. Therefore, to describe surface deformations it is necessary to introduce new parameters defining the shape of the nucleus in the proper coordinate frame associated with the nucleus.

5.2. Rotation of the Coordinate Frame

Rotations of the coordinate frame . We shall consider a certain fixed point P in three-dimensional space. We denote the coordinates of this point in the stationary coordinate frame K by $\mathbf{r} = (x, y, z)$. Correspondingly, we denote the coordinates of this point in the moving coordinate frame K' by $\mathbf{r}' = (x', y', z')$. We shall describe the orientation of the moving coordinate frame K' with respect to the stationary frame K (the origins of the two frames coincide) by the three Euler angles Φ, Θ and ψ, of which Φ and Θ are the polar angles of the z'-axis in the coordinate frame K, and ψ is the angle between the nodal line (the line of intersection of the xy and $x'y'$ planes) and the y'-axis.[2]

The transformation from the coordinate frame K to the coordinate frame K' can be realized by means of three consecutive rotations (Fig.

[2]In Wigner's book (1959), another notation for the Euler angles is used: $(\Phi, \Theta, \psi) \equiv (\gamma, \beta, \alpha)$.

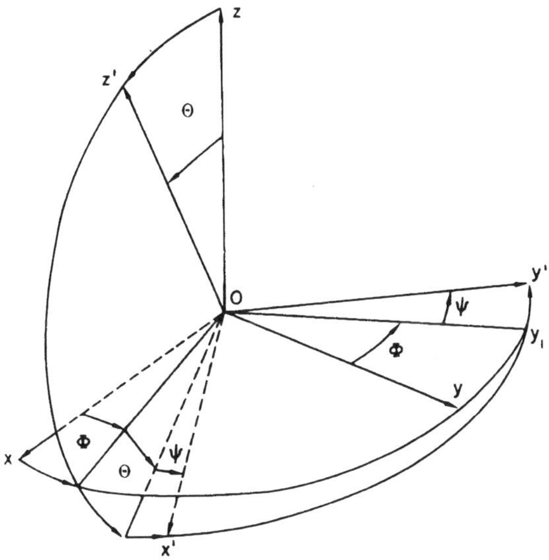

Figure 5.4. Transformation from the coordinate frame K to the coordinate frame K'.

5.4):

1. A rotation through an angle $\Phi (0 \leqslant \Phi < 2\pi)$ about the axis Oz in the direction from the axis Ox to the axis Oy, as a result of which the axis Oy coincides with the nodal line Oy_1. The coordinate frame K thus goes over into the coordinate frame K_1, in which the point P will have the coordinates $\mathbf{r}_1 = (x_1, y_1, z_1)$, which are connected with the coordinates of the same point in the frame K by the relations

$$\mathbf{r}_1 = a^z(\Phi)\mathbf{r},\ a^z(\Phi) = \begin{pmatrix} \cos\Phi & \sin\Phi & 0 \\ -\sin\Phi & \cos\Phi & 0 \\ 0 & 0 & 1 \end{pmatrix}. \tag{5.31}$$

2. A rotation through an angle $\Theta (0 \leqslant \Theta < \pi)$ about the nodal line Oy_1, as a result of which the axis Oz coincides with the axis Oz'. The coordinate frame K_1 thus goes over into the coordinate frame K_2, in which the point P will have the coordinates $\mathbf{r}_2 = (x_2, y_2, z_2)$, connected with the coordinates in the frame K_1 by the relations

$$\mathbf{r}_2 = a^{y_1}(\Theta)\mathbf{r}_1,\ a^{y_1}(\Theta) = \begin{pmatrix} \cos\Theta & 0 & -\sin\Theta \\ 0 & 1 & 0 \\ \sin\Theta & 0 & \cos\Theta \end{pmatrix}. \tag{5.32}$$

3. A rotation through an angle $\psi (0 \leqslant \psi < 2\pi)$ about the axis Oz', as a result of which the coordinate frame K_2 goes over into the coordinate

frame K', in which the point P will have the coordinates $\mathbf{r} = (x', y', z')$, connected with the coordinates in the frame K_2 by the relations

$$\mathbf{r}' = a^{z'}(\psi)\mathbf{r}_2,\ a^{z'}(\psi) = \begin{pmatrix} \cos\psi & \sin\psi & 0 \\ -\sin\psi & \cos\psi & 0 \\ 0 & 0 & 1 \end{pmatrix}. \tag{5.33}$$

Using (5.31), (5.32) and (5.33), the coordinates of the point P in the coordinate frame K' can be expressed directly in terms of the coordinates of the point P in the coordinate frame K:

$$\mathbf{r}' = a(\Phi, \Theta, \psi)\mathbf{r}, \tag{5.34}$$

where the coefficients $a(\Phi, \Theta, \psi)$ of the linear transformation are equal to

$$a(\Phi, \Theta, \psi) \equiv a^{z'}(\psi)a^{y_1}(\Theta)a^{z}(\Phi). \tag{5.35}$$

By inverting the equality (5.34), we can express the coordinates of the point P in the coordinate frame K in terms of the coordinates of the same point in the coordinate frame K':

$$\mathbf{r} = a^{-1}(\Phi, \Theta, \psi)\mathbf{r}', \tag{5.36}$$

where $a^{-1}(\Phi, \Theta, \psi)$ is the inverse of the linear transformation (5.35):

$$a^{-1}(\Phi, \Theta, \psi) = a^{z^{-1}}(\Phi)a^{y_1^{-1}}(\Theta)a^{z'^{-1}}(\psi). \tag{5.37}$$

The finite-rotation operator. We shall consider an arbitrary scalar field, characterized at each point $\mathbf{r}$ in the coordinate frame K by some function $f(\mathbf{r})$. In the coordinate frame K', the same field is characterized, generally speaking, by another function $f'(\mathbf{r}')$ depending on the coordinates of the point in the new coordinate frame. It is obvious that the values of the field calculated at the same spatial point P in the different coordinate frames should coincide. Therefore, the equality

$$f'(\mathbf{r}') = f(\mathbf{r}), \tag{5.38}$$

should be fulfilled if $\mathbf{r}'$ and $\mathbf{r}$ are connected by the relation (5.34).

We introduce the *finite-rotation operator* $D(\Phi, \Theta, \psi)$ defining the scalar-field function $f'(\mathbf{r})$ in the rotated coordinate frame K' when this function $f(\mathbf{r})$ is given in the stationary coordinate frame K:

$$f'(\mathbf{r}) = D(\Phi, \Theta, \psi)f(\mathbf{r}). \tag{5.39}$$

Using the equality (5.38), it is not difficult to obtain a relation enabling us to find an explicit expression for the finite-rotation operator:

$$D(\Phi, \Theta, \psi) f(\mathbf{r}) \equiv f(a^{-1}(\Phi, \Theta, \psi)\mathbf{r}), \tag{5.40}$$

where $a^{-1}(\Phi, \Theta, \psi)$ is the linear transformation defined by the formula (5.37). Inverting the equality (5.40), we obtain

$$D^{-1}(\Phi, \Theta, \psi) f(\mathbf{r}) \equiv f(a(\Phi, \Theta, \psi)\mathbf{r}). \tag{5.41}$$

This relation is convenient for finding the explicit form of the inverse operator $R(\Phi, \Theta, \psi) = D^{-1}(\Phi, \Theta, \psi)$.

We shall find an explicit expression for the operator of a finite rotation of the coordinate frame through an angle ψ about an axis whose direction is specified by the unit vector $\mathbf{n}$. For this, we first consider an infinitesimal rotation of the coordinate frame through an angle $\delta\psi$ about the axis $\mathbf{n}$. In an infinitesimal rotation, $\mathbf{r} \to \mathbf{r}' = a\mathbf{r} = \mathbf{r} + \delta\mathbf{r}$, where $-\delta\mathbf{r} = \delta\psi(\mathbf{n} \times \mathbf{r})$. As a result of this rotation, the scalar-field function $f(\mathbf{r})$ goes over into the function

$$f'(\mathbf{r}) = f(a^{-1}\mathbf{r}) = f(\mathbf{r} - \delta\mathbf{r}) = [1 + \delta\psi\mathbf{n}\cdot(\mathbf{r} \times \nabla)] f(\mathbf{r}).$$

By making use of the definition of the operator of the orbital angular momentum of the system, $\mathbf{l} = -i(\mathbf{r} \times \nabla)$, we can rewrite the equality obtained in the form

$$f'(\mathbf{r}) = (1 + i\delta\psi\mathbf{n} \cdot \mathbf{l}) f(\mathbf{r}). \tag{5.42}$$

The expression in the brackets in the right-hand side of (5.42) can be regarded as the operator of an infinitesimal rotation of the coordinate frame through an angle $\delta\psi$. A rotation through a finite angle ψ about an axis whose direction is specified by the vector $\boldsymbol{n}$ can be represented as an infinite sequence of rotations through infinitesimal angles $\delta\psi \to \psi/k$, with $k \to \infty$:

$$D^{\mathbf{n}}(\psi) f(\mathbf{r}) = \lim_{k\to\infty} \left(1 + i\frac{\psi}{k}\mathbf{n} \cdot \mathbf{l}\right)^k f(\mathbf{r}) = \mathrm{e}^{i\psi\mathbf{n}\cdot\mathbf{l}} f(\mathbf{r}). \tag{5.43}$$

Thus, the operator of a finite rotation of the coordinate frame through an angle ψ about an axis whose direction is specified by the unit vector $\mathbf{n}$ has, in the case of a scalar field, the form

$$D^{\mathbf{n}}(\psi) = \mathrm{e}^{i\psi\mathbf{n}\cdot\mathbf{l}}. \tag{5.44}$$

In the general case, when the physical system possesses intrinsic angular momentum (spin) and the field functions are characterized by well-defined transformation properties, the operator $D^{\mathbf{n}}(\psi)$ of a finite rotation of the coordinate frame is expressed in terms of the operator $\mathbf{j}$ of the total angular momentum of the system:

$$D^{\mathbf{n}}(\psi) = e^{i\psi\mathbf{n}\cdot\mathbf{j}}, \tag{5.45}$$

where $\mathbf{j}$ is the sum of the orbital and spin angular momenta of the system:

$$\mathbf{j} = \mathbf{l} + \mathbf{s}. \tag{5.46}$$

We now find the explicit form of the finite-rotation operator $D(\Phi, \Theta, \psi)$ that effects the transformation from the stationary frame K to the rotated frame K'. Substituting (5.37) into (5.40) and using the definition of the operator of a rotation about a given axis, we obtain

$$\begin{aligned} D(\Phi, \Theta, \psi) f(\mathbf{r}) &= f\left(a^{z^{-1}}(\Phi) a^{y_1^{-1}}(\Theta) a^{z'^{-1}}(\psi)\mathbf{r}\right) \\ &= D^{z}(\Phi) f\left(a^{y_1^{-1}}(\Theta) a^{z'^{-1}}(\psi)\mathbf{r}\right) = D^{z}(\Phi) D^{y_1}(\Theta) f\left(a^{z'^{-1}}(\psi)\mathbf{r}\right) \\ &= D^{z}(\Phi) D^{y_1}(\Theta) D^{z'}(\psi) f(\mathbf{r}). \end{aligned} \tag{5.47}$$

The right-hand side of the equality (5.47) contains operators of rotations about single axes in the different coordinate frames.

According to the identity (5.41), the value of a function of the coordinates of a point in the rotated frame can be expressed in terms of the value of the function of the coordinates of the same point in the stationary frame by means of the operator $D^{-1}(\Phi, \Theta, \psi)$:

$$f(\mathbf{r}') = D^{-1} f(\mathbf{r}), \tag{5.48}$$

where $\mathbf{r}' = a\mathbf{r}$. Clearly, on transforming from $f(\mathbf{r})$ to $f(\mathbf{r}')$, an arbitrary operator O transforms according to the law $O \to O' = D^{-1}OD$. Therefore, in the transformation $K_1 \to K_2$ characterized by the rotation operator $D^{y_1}(\Theta)$, we have

$$D^{z'}(\psi) = D^{y_1^{-1}}(\Theta) D^{z}(\psi) D^{y_1}(\Theta).$$

Analogously, in the transformation $K \to K_1$ characterized by the rotation operator $D^{z}(\Phi)$, we have

$$D^{y_1}(\Theta) = D^{z^{-1}}(\Phi) D^{y}(\Theta) D^{z}(\Phi).$$

Moreover, noting that $D^{z}(\Phi)D^{z}(\psi) = D^{z}(\psi)D^{z}(\Phi)$, we obtain the operator $D(\Phi, \Theta, \psi)$ effecting the transformation from the coordinate frame K to the frame K' in the following form:

$$D(\Phi, \Theta, \psi) = D^z(\psi) D^y(\Theta) D^z(\Phi). \tag{5.49}$$

Using the explicit representation (5.45) for the operators of rotations about the individual axes, we finally obtain the following expression for the finite-rotation operator:

$$D(\Phi, \Theta, \psi) = e^{i\psi j_z} e^{i\Theta j_y} e^{i\Phi j_z}. \tag{5.50}$$

It follows directly from the hermiticity of the operator of the angular momentum of the system that the finite-rotation operator (5.50) is unitary:

$$D^{-1}(\Phi, \Theta, \psi) = D^+(\Phi, \Theta, \psi). \tag{5.51}$$

Using the explicit form (5.50) of the finite-rotation operator, it is not difficult to obtain the following expressions for the derivatives of the operator $D(\Phi, \Theta, \psi)$ with respect to the Euler angles:

$$\begin{aligned} \frac{\partial}{\partial\Phi} D &= iDj_z = i(-j_x \sin\Theta \, \cos\psi + j_y \sin\Theta \, \sin\psi + j_z \cos\Theta) D; \\ \frac{\partial}{\partial\Theta} D &= iD(-j_x \sin\Phi + j_y \cos\Phi) = i(j_x \sin\psi + j_y \cos\psi) D, \\ \frac{\partial}{\partial\psi} D &= iD(j_x \cos\Phi \, \sin\Theta + j_y \sin\Phi \, \sin\Theta + j_z \cos\Theta) = ij_z D. \end{aligned} \tag{5.52}$$

Since $j_{z'} = j_x \cos\Phi \sin\Theta + j_y \sin\Phi \sin\Theta + j_z \cos\Theta$, it follows from the last equality that

$$j_z D = D j_{z'}. \tag{5.53}$$

We introduce the angular-velocity vector

$$\boldsymbol{\omega} = \dot{\boldsymbol{\Phi}} + \dot{\boldsymbol{\Theta}} + \dot{\boldsymbol{\psi}}, \tag{5.54}$$

where the vector $\dot{\boldsymbol{\Phi}}$ is directed along the axis Oz in the coordinate frame K , the vector $\dot{\boldsymbol{\Theta}}$ is directed along the nodal line Oy_1, and the vector $\dot{\boldsymbol{\psi}}$ is directed along the axis Oz' in the coordinate frame K'. Projecting the vector $\boldsymbol{\omega}$ on to the moving and stationary axes, we find

$$\left.\begin{aligned} \omega_x &= -\dot{\Theta} \sin\Phi + \dot{\psi} \cos\Phi \sin\Theta; \\ \omega_y &= \dot{\Theta} \cos\Phi + \dot{\psi} \sin\Phi \sin\Theta; \\ \omega_z &= \dot{\Phi} + \dot{\psi} \cos\Theta \end{aligned}\right\} \tag{5.55}$$

and

$$\left.\begin{aligned}\omega_{x'} &= -\dot{\Phi}\sin\Theta\cos\psi + \dot{\Theta}\sin\psi;\\ \omega_{y'} &= \dot{\Phi}\sin\Theta\sin\psi + \dot{\Theta}\cos\psi;\\ \omega_{z'} &= \dot{\Phi}\cos\Theta + \dot{\psi}.\end{aligned}\right\} \tag{5.56}$$

Using the relations (5.52) and the definition (5.54) of the angular-velocity vector, it is not difficult to find the following expression for the time derivative of the finite-rotation operator:

$$\frac{d}{dt}D = iD(j_x\omega_x + j_y\omega_y + j_z\omega_z) = i(\omega_{x'}j_x + \omega_{y'}j_y + \omega_{z'}j_z)D. \tag{5.57}$$

Generalized spherical harmonics. We shall consider the set of eigenfunctions $\psi_{jm}(\vartheta,\varphi)$ of the operators of the square and projection of the total angular momentum of the system. Since the finite-rotation operator $D(\Phi,\Theta,\psi)$ contains only the operators j_y and j_z, the function $\psi_{jm}(\vartheta',\varphi')$ resulting from the action of the operator $D(\Phi,\Theta,\psi)$ on the function $\psi_{jm}(\vartheta,\varphi)$ can be expanded in the complete set of functions $\psi_{jm'}(\vartheta,\varphi)$. Taking into account the unitarity of the operator $D(\Phi,\Theta,\psi)$ we can write this expansion in the form

$$\begin{aligned}\psi_{jm}(\vartheta'\varphi') &= D^{-1}(\Phi,\Theta,\psi)\psi_{jm}(\vartheta,\varphi) = D^{+}(\Phi,\Theta,\psi)\psi_{jm}(\vartheta,\varphi)\\ &= \sum_{m'} D^{j*}_{mm'}(\Phi,\Theta,\psi)\psi_{jm'}(\vartheta,\varphi),\end{aligned} \tag{5.58}$$

where $D^{j}_{m'm}(\Phi,\Theta,\psi)$ are the matrix elements of the finite-rotation operator, forming the unitary finite-rotation matrix:

$$D^{j}_{m'm}(\Phi,\Theta,\psi) \equiv \langle jm'|D(\Phi,\Theta,\psi)|jm\rangle = \langle jm'|e^{i\psi j_z}e^{i\Theta j_y}e^{i\Phi j_z}|jm\rangle. \tag{5.59}$$

Using the unitarity of the matrix $D^{j}_{m'm}(\Phi,\Theta,\psi)$ it is not difficult to obtain the inverse relation to (5.58):

$$\psi_{jm}(\vartheta,\varphi) = \sum_{m'} D^{j}_{m'm}(\Phi,\Theta,\psi)\psi_{jm'}(\vartheta',\varphi'). \tag{5.60}$$

According to (5.60), the value at a given point in the stationary coordinate frame K of an eigenfunction of the operators of the square and projection of the angular momentum can be expressed in terms of a linear combination of the values of the eigenfunctions of the operators of square and projection of the angular momentum taken at the same point in the rotated coordinate frame K'.

The finite-rotation matrix elements $D^j_{m'm}(\Phi, \Theta, \psi)$ are completely independent of the choice of the coordinate frame in which the functions $\psi_{jm}(\vartheta, \varphi)$ are defined, and for a fixed value of j are well-defined functions of the angles Φ, Θ and ψ. If j is an integer ($j \equiv l$), and either of the indices m and m' equals zero, the functions $D^l_{m'm}(\Phi, \Theta, \psi)$ reduce to ordinary spherical harmonics:

$$\left.\begin{aligned} D^l_{0m}(\Phi, \Theta, \psi) &= \sqrt{\tfrac{4\pi}{2l+1}}\, Y_{lm}(\Theta, \Phi); \\ D^l_{m'0}(\Phi, \Theta, \psi) &= (-1)^{m'}\sqrt{\tfrac{4\pi}{2l+1}}\, Y_{lm'}(\Theta, \psi). \end{aligned}\right\} \tag{5.61}$$

In the particular case $m = m' = 0$, we have

$$D^l_{00}(\Phi, \Theta, \psi) = P_l(\cos\Theta). \tag{5.62}$$

Therefore, the function $D^l_{m'm}(\Phi, \Theta, \psi)$ are usually called *generalized spherical harmonics* , or *Wigner D-functions.* [3]

The generalized spherical harmonics satisfy the following orthonormalization condition:

$$\int_0^{2\pi} d\Phi \int_0^{\pi} d\Theta \sin\Theta \int_0^{2\pi} d\psi D^{j_1 *}_{m_1' m_1}(\Phi, \Theta, \psi) D^{j_2}_{m_2' m_2}(\Phi, \Theta, \psi)$$

$$= \frac{8\pi^2}{2j_1 + 1}\delta_{j_1 j_2}\delta_{m_1 m_2}\delta_{m_1' m_2'}. \tag{5.63}$$

Integrals of products of three generalized spherical harmonics are easily calculated by means of the vector-coupling formula valid for the generalized spherical harmonics:

$$D^{j_1}_{m_1' m_1} D^{j_2}_{m_2' m_2} = \sum_{jm'm} (j_1 m_1' j_2 m_2' | j m')(j_1 m_1 j_2 m_2 | j m) D^j_{m'm}. \tag{5.64}$$

We note that the generalized spherical harmonics $D^j_{m'm}(\Phi, \Theta, \psi)$ form an irreducible unitary representation of the three-dimensional rotation group.

Using the definition (5.59), we can separate out the dependence on the angles Φ and ψ explicitly in the functions $D^j_{m'm}(\Phi, \Theta, \psi)$. Since $|jm>$ and $|jm'>$ are eigenfunctions of the operator of the component j_z of the angular momentum, we have

$$D^j_{m'm}(\Phi, \Theta, \psi) = e^{im'\psi} d^j_{m'm}(\Theta) e^{im\Phi}, \tag{5.65}$$

[3]The properties of the generalized spherical harmonics are considered in detail in the books by Wigner (1959) and Edmonds (1960).

where $d^j_{m'm}(\Theta)$ is a real function defined by the equality

$$d^j_{m'm}(\Theta) \equiv \langle jm'|e^{i\Theta j_y}|jm\rangle. \tag{5.66}$$

It is easy to convince oneself that the function $d^j_{m'm}(\Theta)$ is real by using the expansion

$$e^{i\Theta j_y} = \sum_{n=0}^{\infty} (i\Theta j_y)^n/n!$$

and noting that the matrix elements of the operator j_y are purely imaginary:

$$\begin{aligned}\langle jm'|j_y|jm\rangle &= \frac{i}{2}\{\sqrt{(j+m)(j-m+1)}\delta_{m',m-1} \\ &- \sqrt{(j-m)(j+m+1)}\delta_{m',m+1}\}.\end{aligned}$$

It follows from the fact that the matrices $d^j_{m'm}(\Theta)$ are real and unitary that they are orthogonal:

$$d^{j-1}_{m'm}(\Theta) = d^j_{mm'}(\Theta), \tag{5.67}$$

whence we obtain the following symmetry property:

$$d^j_{m'm}(\Theta) = d^j_{mm'}(-\Theta). \tag{5.68}$$

By making use of the relation

$$\langle jm'|j_y|jm\rangle = -\langle j-m'|j_y|j-m\rangle,$$

we can also easily verify the following symmetry property:

$$d^j_{m'm}(\Theta) = d^j_{-m,-m'}(\Theta). \tag{5.69}$$

Finally, using the operator identity

$$e^{i\Theta j_y} \equiv e^{i\pi j_z}\, e^{-i\Theta j_y}\, e^{-i\pi j_z},$$

it is not difficult to obtain the relation

$$d^j_{m'm}(\Theta) = (-1)^{m'-m} d^j_{mm'}(\Theta). \tag{5.70}$$

The functions $d^j_{m'm}(\Theta)$ for any value of the angle Θ can be expressed in terms of the values of the functions $d^j_{m'm}(\Theta)$ at $\Theta = \pi/2$:

$$d^j_{m'm}(\Theta) = \mathrm{e}^{i(\pi/2)(m'-m)} \sum_{m''} d^j_{m'm''}(\pi/2)\mathrm{e}^{im''\Theta}\, d^j_{m''m}(-\pi/2). \qquad (5.71)$$

In the derivation of this relation, the operator identity

$$\mathrm{e}^{i\Theta j_y} \equiv \mathrm{e}^{i(\pi/2)j_z}\,\mathrm{e}^{i(\pi/2)j_y}\,\mathrm{e}^{i\Theta j_z}\,\mathrm{e}^{-i(\pi/2)j_y}\,\mathrm{e}^{-i(\pi/2)j_z}$$

must be used. We also note the following useful relations for the functions $d^j_{m'm}(\Theta)$:

$$\left.\begin{aligned} d^j_{m'm}(\pi) &= (-1)^{j-m'}\delta_{m',-m};\\ d^j_{m'm}(\pi-\Theta) &= (-1)^{j-m'}d^j_{m',-m}(\Theta);\\ d^j_{m'm}(\pi+\Theta) &= (-1)^{j-m'}d^j_{-m',m}(\Theta). \end{aligned}\right\} \qquad (5.72)$$

To conclude, we also give explicit expressions for the time derivative of the finite-rotation matrix $D^j_{m'm}(\Phi,\Theta,\psi)$:

$$\begin{aligned} \frac{d}{dt}D^j_{m'm} &= i\sum_{k=x,y,z}\omega_k\sum_{m''}D^j_{m'm''}\langle jm''|j_k|jm\rangle\\ &= i\sum_{k'=x',y',z'}\omega_{k'}\sum_{m''}\langle jm'|j_{k'}|jm''\rangle D^j_{m''m}. \end{aligned} \qquad (5.73)$$

These relations follow directly from the operator relations (5.57).

The rotation operator R. We introduce the operator $\mathbf{R}$ an infinitesimal rotation of the moving coordinate frame K'. For this, we choose as the dynamical variables the Euler angles Φ,Θ and ψ and consider an arbitrary scalar function $f(\Phi,\Theta,\psi)$ of these angles. Clearly, in rotations of a moving coordinate frame K' whose orientation is specified by these Euler angles, this function will change in the same way as a scalar-field function changes on rotation of the physical system. (A rotation of the physical system through an infinitesimal angle is equivalent to a rotation of the coordinate frame through the same angle in the opposite direction.) Therefore, in an infinitesimal rotation of the moving coordinate frame K' about an axis whose direction is specified by the unit vector $\mathbf{n}$ (as a result to which the angles Φ,Θ and ψ go over into $\Phi' = \Phi+\delta\Phi$, $\Theta' = \Theta+\delta\Theta$ and $\psi' = \psi+\delta\psi$), the change of the function is equal to

$$\begin{aligned} \delta f(\Phi,\Theta,\psi) &\equiv f(\Phi+\delta\Phi,\Theta+\delta\Theta,\psi+\delta\psi) - f(\Phi,\Theta,\psi)\\ &= i\,\delta\varphi\mathbf{n}\cdot\mathbf{R}f(\Phi,\Theta,\psi), \end{aligned} \qquad (5.74)$$

where $\delta\varphi$ is the infinitesimal angle of rotation and $\mathbf{R}$ is the hermitian rotation operator:

$$\mathbf{n}\cdot\mathbf{R} = -i\frac{\partial}{\partial\varphi}. \tag{5.75}$$

We denote the unite vectors directed along the coordinate axes Oz, Oy_1 and Oz' by $\mathbf{n}_\Phi$, $\mathbf{n}_\Theta$ and $\mathbf{n}_\psi$ respectively. Then the projections of the vector $\mathbf{R}$ along the corresponding directions can be expressed directly in terms of derivatives with respect to the Euler angles:

$$\mathbf{n}_\Phi\cdot\mathbf{R} = -i\frac{\partial}{\partial\Phi}, \quad \mathbf{n}_\Theta\cdot\mathbf{R} = -i\frac{\partial}{\partial\Theta}, \quad \mathbf{n}_\psi\cdot\mathbf{R} = -i\frac{\partial}{\partial\psi}. \tag{5.76}$$

Expressing the scalar products in the left-hand sides of these equalities in terms of the projections of the vectors in the stationary or moving coordinate frame, from (5.76) we can easily find the explicit form of the *rotation operator* $\mathbf{R}$ in the stationary and moving coordinate frames.

The *projections of the rotation operator* $\mathbf{R}$ along the axes of the stationary coordinate frame are given by the expressions

$$\left.\begin{aligned}
\mathbf{R}_x &= -i\left\{-\cos\Phi\cot\Theta\frac{\partial}{\partial\Phi} - \sin\Phi\frac{\partial}{\partial\Theta} + \frac{\cos\Phi}{\sin\Theta}\frac{\partial}{\partial\psi}\right\};\\
\mathbf{R}_y &= -i\left\{-\sin\Phi\cot\Theta\frac{\partial}{\partial\Phi} + \cos\Phi\frac{\partial}{\partial\Theta} + \frac{\sin\Phi}{\cos\Theta}\frac{\partial}{\partial\psi}\right\};\\
\mathbf{R}_z &= -i\frac{\partial}{\partial\Phi}.
\end{aligned}\right\} \tag{5.77}$$

The projections of the rotation operator $\mathbf{R}$ along the axes of the moving coordinate frame are equal to

$$\left.\begin{aligned}
\mathbf{R}_{x'} &= -i\left\{-\frac{\cos\psi}{\sin\Theta}\frac{\partial}{\partial\Phi} + \sin\psi\frac{\partial}{\partial\Theta} + \cot\Theta\cos\psi\frac{\partial}{\partial\psi}\right\};\\
\mathbf{R}_{y'} &= -i\left\{\frac{\sin\psi}{\sin\Theta}\frac{\partial}{\partial\Phi} + \cos\psi\frac{\partial}{\partial\Theta} - \cot\Theta\sin\psi\frac{\partial}{\partial\psi}\right\};\\
\mathbf{R}_{z'} &= -i\frac{\partial}{\partial\psi}.
\end{aligned}\right\} \tag{5.78}$$

Since the square of a vector is an invariant quantity, we find for the operator $\mathbf{R}^2$ in both coordinate frames

$$\mathbf{R}^2 = -\left\{\frac{1}{\sin^2\Theta}\left[\frac{\partial^2}{\partial\Phi^2} - 2\cos\Theta\frac{\partial^2}{\partial\Phi\partial\psi} + \frac{\partial^2}{\partial\psi^2}\right]\right.$$

$$+\frac{1}{\sin\Theta}\frac{\partial}{\partial\Theta}\left(\sin\Theta\frac{\partial}{\partial\Theta}\right)\Bigg\}. \qquad (5.79)$$

Using the explicit form (5.77) of the projections of the rotation operator in the stationary coordinate frame, it is not difficult to show that they satisfy the usual commutation relations for angular-momentum components:

$$\mathbf{R}_x\mathbf{R}_y - \mathbf{R}_y\mathbf{R}_x = i\mathbf{R}_z. \qquad (5.80)$$

Thus, $\mathbf{R}$ can be regarded as the operator of the angular momentum associated with the rotation of a physical system which is rigidly attached to the coordinate frame K'. In the moving coordinate frame K' the projections of the rotation operator satisfy commutation relations differing from (5.80) in the sign in the right-hand side:

$$\mathbf{R}_{x'}\mathbf{R}_{y'} - \mathbf{R}_{y'}\mathbf{R}_{x'} = -i\mathbf{R}_{z'}. \qquad (5.81)$$

The projections of the rotation operator in both the moving and stationary coordinate frames commute with the operator $\mathbf{R}^2$:

$$[\mathbf{R}_i, \mathbf{R}^2] = [\mathbf{R}_{i'}, \mathbf{R}^2] = 0. \qquad (5.82)$$

We now consider a quantum-mechanical system characterized by the total angular-momentum operator J. We denote the wave function of the system in K by $\psi(\vartheta,\varphi)$ and the wave function of the system in K' by $\psi(\vartheta',\varphi')$. According to (5.60), the functions $\psi(\vartheta,\varphi)$ and $\psi(\vartheta',\varphi')$ are connected by the relation

$$\psi_{JM}(\vartheta,\varphi) = \sum_{M'} D^J_{M'M}(\Phi,\Theta,\psi)\psi_{JM'}(\vartheta',\varphi'). \qquad (5.83)$$

It is obvious that the operator $\mathbf{R}$ will act only on the function $\psi(\vartheta',\varphi')$, since a rotation of the coordinate frame K' cannot affect the value the function $\psi(\vartheta,\varphi)$ in the stationary coordinate frame K; that is,

$$\mathbf{R}\psi(\vartheta,\varphi) = 0. \qquad (5.84)$$

Since the functions $\psi(\vartheta',\varphi')$ depend on the Euler angles Φ,Θ and ψ, the change of this function on an infinitesimal rotation of the coordinate frame K' through an angle $\delta\varphi$ about an arbitrary direction $\mathbf{n}$, according to (5.74), is equal to

$$\delta\psi(\vartheta',\varphi') = i\,\delta\varphi\mathbf{n}\cdot\mathbf{R}\psi(\vartheta',\varphi'). \qquad (5.85)$$

Moreover, this change can be expressed in terms of the operator $\mathbf{J}$ of the total angular momentum of the system. According to (5.45) and (5.48), this change is equal to

$$\delta\psi(\vartheta', \varphi') = -i\,\delta\varphi\mathbf{n}\cdot\mathbf{J}\psi(\vartheta', \varphi'). \tag{5.86}$$

Equating the right-hand side of the equalities (5.85) and (5.86), we obtain

$$(\mathbf{R} + \mathbf{J})\psi(\vartheta', \varphi') = 0. \tag{5.87}$$

Using (5.83), we can represent the equality (5.84) in the form

$$\begin{aligned}
&\sum_{M'} \mathbf{R} D^J_{M'M}(\Phi, \Theta, \psi)\psi_{JM'}(\vartheta', \varphi') \\
&\equiv \sum_{M'} \Big\{ \Big(\mathbf{R} D^J_{M'M}(\Phi, \Theta, \psi)\Big) + D^J_{M'M}(\Phi, \Theta, \psi)\mathbf{R} \Big\} \\
&\times \psi_{JM'}(\vartheta', \varphi') = 0,
\end{aligned} \tag{5.88}$$

where $(\mathbf{R}D^J_{M'M}(\Phi, \Theta, \psi))$ denotes the result of the action of the operator $\mathbf{R}$ on $D^J_{M'M}(\Phi, \Theta, \psi)$ and is a certain function of the angles. Taking (5.87) into account, we can rewrite (5.88) in the form

$$\sum_{M'} \Big\{ \Big(\mathbf{R} D^J_{M'M}(\Phi, \Theta, \psi)\Big) - D^J_{M'M}(\Phi, \Theta, \psi)\mathbf{J} \Big\} \psi_{JM'}(\vartheta', \varphi') = 0. \tag{5.89}$$

By projecting the equality (5.89) on to the axis Oz' and noting that $\mathbf{J}_{z'}\psi_{JM'}(\vartheta', \varphi') = M'\psi_{JM'}(\vartheta', \varphi')$, we have

$$\sum_{M'} \Big\{ \Big(\mathbf{R}_{z'} D^J_{M'M}(\Phi, \Theta, \psi)\Big) - M' D^J_{M'M}(\Phi, \Theta, \psi) \Big\} \psi_{JM'}(\vartheta', \varphi') = 0.$$

Since the functions $\psi_{JM'}(\vartheta', \varphi')$ for different values of M' are linearly independent, if this equality is to be fulfilled it is necessary that all the coefficients vanish identically; that is,

$$\mathbf{R}_{z'} D^J_{M'M}(\Phi, \Theta, \psi) = M' D^J_{M'M}(\Phi, \Theta, \psi). \tag{5.90}$$

Analogously, we can obtain

$$\begin{aligned}
&\Big(\mathbf{R}_{x'} \pm i\mathbf{R}_{y'}\Big) D^J_{M'M}(\Phi, \Theta, \psi) \\
&= \sqrt{(J \pm M')(J \mp M' + 1)}\, D^J_{M'\mp 1, M}(\Phi, \Theta, \psi).
\end{aligned} \tag{5.91}$$

Using (5.90) and (5.91), it is not difficult to show that

$$\mathbf{R}^2 D^J_{M'M}(\Phi, \Theta, \psi) = J(J+1) D^J_{M'M}(\Phi, \Theta, \psi). \tag{5.92}$$

To determine the eigenvalue of a projection of $\mathbf{R}$ in the stationary coordinate frame K, we perform the summation over M' in the second term of (5.89)

$$\sum_{M'} \Big(\mathbf{R} D^J_{M'M}(\Phi, \Theta, \psi)\Big) \psi_{JM'}(\vartheta', \varphi') - \mathbf{J}\psi_{JM}(\vartheta, \varphi) = 0. \tag{5.93}$$

By projecting this equality on to the axis Oz and making use of the relation $\mathbf{J}_z \psi_{JM}(\vartheta, \varphi) = M\psi_{JM}(\vartheta, \varphi)$, we find

$$\sum_{M'} \Big\{ \Big(\mathbf{R}_z D^J_{M'M}(\Phi, \Theta, \psi)\Big) - M D^J_{M'M}(\Phi, \Theta, \psi) \Big\} \psi_{JM'}(\vartheta', \varphi') = 0,$$

whence

$$\mathbf{R}_z D^J_{M'M}(\Phi, \Theta, \psi) = M D^J_{M'M}(\Phi, \Theta, \psi). \tag{5.94}$$

Analogously, it can be shown that

$$\Big(\mathbf{R}_x \pm i\mathbf{R}_y\Big) D^J_{M'M}(\Phi, \Theta, \psi) = \sqrt{(J \mp M)(J \pm M + 1)}\, D^J_{M',M\pm 1}(\Phi, \Theta, \psi). \tag{5.95}$$

It is not difficult to verify that the equality (5.92) follows also from (5.94) and (5.95).

Thus, we have shown that the generalized spherical harmonics $D^J_{M'M}(\Phi, \Theta, \psi)$ are simultaneously eigenfunctions of the operators $\mathbf{R}^2$, $\mathbf{R}_z$ and $\mathbf{R}_{z'}$, the corresponding eigenvalues being equal to $J(J+1)$, M and M' respectively.

5.3. Rotation and Surface Vibrations of Deformed Nuclei

Ellipsoidal deformations of nuclei . For small deformations of nuclei, we can confine ourselves to taking into account the second-order deformations only. In this case, the surface of the deformed nucleus is an ellipsoid, arbitrarily oriented in space:

$$R(\vartheta, \varphi) = \Big\{ 1 + \sum_{\mu} \alpha_\mu Y_{2\mu}(\vartheta, \varphi) \Big\} R \tag{5.96}$$

(for simplicity we shall omit the index $\lambda = 2$). In place of the five parameters α_μ describing the deformations, we can introduce three angles defining the orientation of the ellipsoid and two internal parameters defining its shape. (Because the volume of the nucleus is conserved, the product of the three principal axes of the ellipsoid remains constant under deformations).

We introduce the proper coordinate frame K' whose axes coincide with the principal axes of the ellipsoid. The orientation of the frame K' relative to the stationary frame K is defined by the Euler angles Φ, Θ and ψ. We write the equation of the surface in the proper coordinate frame:

$$R(\vartheta', \varphi') = \left\{ 1 + \sum_\nu \alpha'_\nu Y_{2\nu}(\vartheta', \varphi') \right\} R, \tag{5.97}$$

where α'_ν are the deformation parameters in the proper coordinate frame K'. Using the formula (5.60) for the transformation of spherical harmonics on going from K to K', it is not difficult to obtain the following relation between the parameters α_μ and α'_μ:

$$\alpha_\mu = \sum_\nu D^*_{\nu\mu}(\Phi, \Theta, \psi)\alpha'_\nu. \tag{5.98}$$

Because of the symmetry of the ellipsoid about the principal axes in the proper coordinate frame, we have

$$\alpha'_1 = \alpha'_{-1} = 0, \qquad \alpha'_2 = \alpha'_{-2}. \tag{5.99}$$

In fact, on reflection through the plane $z' = 0$ ($\vartheta', \varphi' \to \pi - \vartheta', \varphi'$), the spherical harmonics Y_{20} and $Y_{2\pm2}$ remain unchanged while the functions $Y_{2\pm1}$ change sing. It follows from this that $\alpha'_1 = \alpha'_{-1} = 0$. On reflection through the plane $x' = 0$ ($\vartheta', \varphi' \to \vartheta', \pi - \varphi'$), the spherical harmonics $Y_{2\pm2}$ go over into $Y_{2\mp2}$ whence it follows that $\alpha'_{-2} = \alpha'_2$. Thus, is the proper frame, deformations of the surface are characterized by only two real parameters α'_0 and α'_2. In place of α'_0 and α'_2 it is convenient to introduce two new parameters β and γ by means of the relations

$$\alpha'_0 = \beta \cos\gamma, \qquad \alpha'_2 = (\beta/\sqrt{2}) \sin\gamma. \tag{5.100}$$

By virtue of the unitarity of $D_{\mu\nu}$, we have

$$\sum_\mu |\alpha_\mu|^2 = \alpha_0'^2 + 2\alpha_2'^2 = \beta^2. \tag{5.101}$$

Consequently, the parameter β defines the total deformation of the nucleus. The parameter γ characterizes the deviation of the shape of the nucleus from an axially symmetric shape.

As can be easily verified, the principal semi-axes of the nucleus can be expressed in terms of β and γ in the following way:

$$R_{k'} = \left\{1 + \sqrt{\frac{5}{4\pi}}\beta\cos\left(\gamma - \frac{2\pi}{3}k'\right)\right\}, \qquad k' = 1, 2, 3 \equiv x', y', z'. \quad (5.102)$$

If $\gamma = 0$ and $\beta > 0$, the nucleus is a prolate ellipsoid of revolution:

$$R_{x'} = R_{y'} = \left(1 - \frac{1}{2}\sqrt{\frac{5}{4\pi}}\beta\right)R, \qquad R_{z'} = \left(1 + \sqrt{\frac{5}{4\pi}}\beta\right)R.$$

If $\gamma = \pi$ and $\beta > 0$, we have an oblate ellipsoid of revolution:

$$\mathbf{R}_{x'} = \mathbf{R}_{y'} = \left(1 + \frac{1}{2}\sqrt{\frac{5}{4\pi}}\beta\right)\mathbf{R}, \qquad \mathbf{R}_{z'} = \left(1 - \sqrt{\frac{5}{4\pi}}\beta\right)\mathbf{R}.$$

If $\gamma = \pi/3$ or $2\pi/3$, the symmetry axis will be the axis Oy' or the Ox'.

Thus, in place of the five *deformation parameters* α_μ, we have introduce five new coordinates: the three *Euler angles* Φ, Θ *and* ψ defining the orientation of the principal axes of the ellipsoid in space, and the two *internal parameters* β *and* γ defining the shape of the ellipsoid. Physically, the idea of introducing the new coordinates is to make it possible to study independently the rotation of the nucleus as a whole and the vibrations associated with change of shape of the nucleus.

Vibrational energy and energy of rotation. In the case of *ellipsoidal deformations of the nucleus* , the kinetic energy is given by the general expression

$$T = (B/2)\sum_\mu |\dot{\alpha}_\mu|^2. \quad (5.103)$$

The transformation to the new coordinates ($\alpha_\mu \to \Phi, \Theta, \beta$ and γ) makes it possible to divide the total kinetic energy (5.103) of the surface deformations into rotational and vibrational energies. To find the form of the kinetic energy in the proper coordinate frame, we differentiate the equality (5.98) with respect to the time:

$$\dot{\alpha}_\mu = \sum_\nu (D^*_{\nu\mu}\dot{\alpha}'_\nu + \dot{D}^*_{\nu\mu}\alpha'_\nu).$$

The time derivative of $D_{\mu\nu}$ can be expressed directly, according to (5.73), in terms of the functions $D_{\mu\nu}$ and the time derivatives of the Euler angles. Substituting the expression for $\dot{\alpha}_\mu$ into (5.103), we can represent the kinetic energy as a sum of three terms:

$$T = (B/2) \sum_{\mu\nu\nu'} \left(D^*_{\nu\mu} D_{\nu'\mu} \dot{\alpha}'_\nu \dot{\alpha}'_{\nu'} + 2\,\mathrm{Re}\,[D^*_{\nu\mu} \dot{D}_{\nu'\mu} \dot{\alpha}'_\nu \alpha'_{\nu'}] + \dot{D}^*_{\nu\mu} \dot{D}_{\nu'\mu} \alpha'_\nu \alpha'_{\nu'} \right). \tag{5.104}$$

The first term in (5.104), which is quadratic in $\dot{\alpha}_\nu$, determines the kinetic energy of the vibrations that change the shape of the ellipsoid but conserve its orientation in space:

$$T_v = (B/2) \sum_{\mu\nu\nu'} D^*_{\nu\mu} D_{\nu'\mu} \dot{\alpha}'_\nu \dot{\alpha}'_{\nu'}.$$

Using the unitarity of the functions $D_{\mu\nu}$, it is not difficult to bring the vibrational energy to the form

$$T_v = (B/2)(\dot{\beta}^2 + \beta^2 \dot{\gamma}^2). \tag{5.105}$$

The second term in (5.104), containing the cross-terms, vanishes, as can easily be seen by making use of the explicit form (5.73) of $\dot{D}_{\mu\nu}$:

$$\mathrm{Re} \sum_{\mu\nu\nu'} D^*_{\nu\mu} \dot{D}_{\nu'\mu} \dot{\alpha}'_\nu \alpha'_{\nu'} = \mathrm{Re}\, i \sum_{\nu\nu'} \sum_{k'} \omega_{k'} \langle \nu' | j_{k'} | \nu \rangle \dot{\alpha}'_\nu \alpha'_{\nu'}.$$

Since ν and ν' take only the values -2,0 and 2, only the term $k' = z'$ can give a contribution, and so

$$\mathrm{Re} \sum_{\mu\nu\nu'} D^*_{\nu\mu} \dot{D}_{\nu'\mu} \dot{\alpha}'_\nu \alpha'_{\nu'} = \mathrm{Re}\, i\omega_{z'} \sum_{\nu} \nu \dot{\alpha}'_\nu \alpha'_\nu = 0.$$

The third term in (5.104), which is quadratic in $\dot{D}_{\mu\nu}$, gives the kinetic energy of rotation of the ellipsoid with no change of shape:

$$T_r = (B/2) \sum_{\mu\nu\nu'} \dot{D}^*_{\nu\mu} \dot{D}_{\nu'\mu} \alpha'_\nu \alpha'_{\nu'}.$$

Using the explicit form (5.73) of the derivatives $\dot{D}_{\mu\nu}$ and the unitary of the functions $D_{\mu\nu'}$, we find

$$T_r = (B/2) \sum_{\nu\nu'} \sum_{k'\bar{k}'} \omega_{k'} \omega_{\bar{k}'} \langle \nu' | \mathbf{j}_{\bar{k}'} \mathbf{j}_{k'} | \nu \rangle \alpha'_\nu \alpha'_{\nu'}. \tag{5.106}$$

Since ν and ν' take only even values, only the matrix elements $< \nu' | \mathbf{j}_{\bar{k}'} \mathbf{j}_{k'} | \nu >$ with $k' = \bar{k}'$ make a contribution to (5.106). In fact, if either of k' and $\bar{k}'$ is equal to z' and the other is equal to x' or y', the matrix element $< \nu' | \mathbf{j}_{\bar{k}'} \mathbf{j}_{k'} | \nu >$ vanishes, since $\mathbf{j}_{z'}$ is diagonal, while for $\mathbf{j}_{x'}$ or $\mathbf{j}_{y'}$ the only non-zero matrix elements are for states with quantum numbers ν and ν'

differing by unity. But if either of k' and $\bar{k}'$ is equal to x' while the other is equal to y', the corresponding matrix elements will be purely imaginary and their sum in (5.106) will be equal to zero. Therefore, we can write the kinetic energy of rotation (5.106) in the form

$$T_r = \frac{1}{2}\sum_{k'} \omega_{k'}^2 J_{k'}, \tag{5.107}$$

where $J_{k'}$ is the moment of inertia of the nucleus about the principal axis k':

$$J_{k'} = B\sum_{\nu\nu'} \langle \nu' | \mathbf{j}_{k'}^2 | \nu \rangle \alpha'_\nu \alpha'_{\nu'}.$$

Using the explicit expressions for the matrix elements of the squares of the angular-momentum components, we finally obtain the following formula for the three *moments of inertia of the nucleus* :

$$J_{k'} = 4B\beta^2 \sin^2\left(\gamma - \frac{2\pi}{3}k'\right), \qquad k' = 1, 2, 3. \tag{5.108}$$

The moments of inertia are proportional to the square of the deformation parameter β of the nucleus. In the case of an axially symmetric nucleus, $\gamma = 0$ or $\gamma = \pi$ and the moments of inertia are

$$J_{x'} = J_{y'} = 3B\beta^2, \qquad J_{z'} = 0. \tag{5.109}$$

The angular momentum of the nucleus about the principal axis k' is connected with the angular velocity and moment of inertia of the nucleus by the relation

$$\hbar \mathbf{R}_{k'} = \omega_{k'} J_{k'}. \tag{5.110}$$

Therefore, the energy of rotation of the nucleus can be represented in the form

$$T_r = \sum_{k'} (\hbar^2 / 2J_{k'}) \mathbf{R}_{k'}^2. \tag{5.111}$$

According to (5.5) and (5.101), the potential energy in the case of small deviations of the shape of the nucleus from spherical shape is expressed in terms of the deformation parameter β only:

$$V = \frac{1}{2} C\beta^2. \tag{5.112}$$

Therefore, the collective Hamiltonian H of the nucleus in the case of small deformations can be represented as the sum of two parts H_v and H_r, associated with the vibrations of the surface of the nucleus and with the rotation of the nucleus as a whole:

$$H = H_v + H_r, \quad H_v = \frac{1}{2}B(\dot{\beta}^2 + \beta^2\dot{\gamma}^2) + \frac{1}{2}C\beta^2, \quad H_r = \sum_{k'} \frac{\hbar^2}{2J_{k'}}\mathbf{R}^2_{k'}; \tag{5.113}$$

that is, in the case of small deformations the surface vibrations of the nucleus and the rotation of the nucleus can be completely separated.

Potential energy of vibrations of non-spherical nuclei. We shall consider the surface vibrations of a rotating nucleus whose equilibrium shape is non-spherical. We denote the equilibrium values of the deformation parameters by $\alpha^0_{\lambda\mu}$ and introduce the deviations of the deformation parameters from their equilibrium values:

$$a_{\lambda\mu} = \alpha_{\lambda\mu} - \alpha^0_{\lambda\mu}. \tag{5.114}$$

By $\alpha_{\lambda\mu}$ (which may be large) we shall mean, as before, the coefficients in the expansion in spherical harmonics of the distance from the centre of the nucleus to a point on its surface:

$$R(\vartheta, \varphi) = \left\{1 + \sum_{\lambda\mu}(\alpha^0_{\lambda\mu} + a_{\lambda\mu})Y_{\lambda\mu}(\varphi)\right\}R. \tag{5.115}$$

The equilibrium parameters $\alpha^0_{\lambda\mu}$, defined in the stationary coordinate frame K, also describe the rotation of the nucleus and therefore depend on time. In the rotating coordinate frame K', the parameters $\alpha'^0_{\lambda\mu}$ define only the shape of the nucleus, and are therefore constant.

The kinetic energy associated with the collective degrees of freedom of the nucleus is a quadratic function of the generalized velocities $\dot{\alpha}_{\lambda\mu}$; however, it can also depend on the generalized coordinates $\alpha_{\lambda\mu}$. The kinetic energy is associated with both the rotation of the nucleus and the vibrations of its surface. The potential energy of a non-spherical nucleus will be expressed purely in terms of the deviations of the parameters $\alpha_{\lambda\mu}$ from the equilibrium values $\alpha^0_{\lambda\mu}$, and does not depend on the rotation of the nucleus.

The quantities $a_{\lambda\mu}$, like $\alpha_{\lambda\mu}$ and $\alpha^0_{\lambda\mu}$, transform under rotations of the coordinate frame according to irreducible representations of the rotation group,

$$a_{\lambda\mu} = \sum_{\nu} D^*_{\nu\mu}(\Phi, \Theta, \psi)a'_{\lambda\nu}, \tag{5.116}$$

and are spherical tensor operators of rank λ, satisfying the hermiticity relations $a^*_{\lambda\mu} = (-1)^\mu \times a_{\lambda,-\mu}$ which follow from the reality of (5.115). Since the potential energy of the nucleus is a scalar quantity, it must be expressed in terms of invariant combinations, independent of the rotation of the coordinate frame, of the tensor components $a_{\lambda\mu}$; in other words, the potential energy must be a tensor of rank zero, constructed from the components $a_{\lambda\mu}$. The simplest such invariant is the quadratic invariant (the quadratic invariant expression constructed from the components $a_{\lambda\mu}$):

$$S_2(\lambda) = \sum_{\mu\mu'} (\lambda\mu\lambda\mu'|00) a_{\lambda\mu} a_{\lambda\mu'} = \frac{(-1)^\lambda}{\sqrt{2\lambda+1}} \sum_\mu |a_{\lambda\mu}|^2. \tag{5.117}$$

The invariant containing the components $a_{\lambda\mu}$ to third order has the form

$$S_3(\lambda) = \sum_{\mu_1\mu_2\mu_3} (\lambda\mu_1\lambda\mu_2|\lambda\mu)(\lambda\mu\lambda\mu_3|00)\, a_{\lambda\mu_1} a_{\lambda\mu_2} a_{\lambda\mu_3}. \tag{5.118}$$

For a given λ it is impossible to construct any other second- and third-order invariants from the components $a_{\lambda\mu}$.

The invariants are most easily calculated in the proper coordinate frame K' moving with the nucleus. For $\lambda = 2$, choosing the axes of the coordinate frame to be along the principal axes of the ellipsoid, we introduce, as in the case of a spherical equilibrium shape, the new deformation parameters $(a'_\mu = \alpha'_\mu - \alpha'^0_\mu)$:

$$\left.\begin{array}{lll} \alpha'_0 = \beta\cos\gamma; & \alpha'_2 = \alpha'_{-2} = (\beta/\sqrt{2})\sin\gamma; & \alpha'_{\pm 1} = 0; \\ \alpha'^0_0 = \beta_0\cos\gamma_0; & \alpha'^0_2 = \alpha'^0_{-2} = (\beta_0/\sqrt{2})\sin\gamma_0; & \alpha'^0_{\pm 1} = 0. \end{array}\right\} \tag{5.119}$$

Using these definitions, we obtain the following expressions for the invariants $S_2(2)$ and $S_3(2)$:

$$S_2(2) = \frac{1}{\sqrt{5}}\left\{\left(\beta-\beta_0\right)^2 + 4\beta\beta_0\sin^2\frac{\gamma-\gamma_0}{2}\right\}; \tag{5.120}$$

$$S_3(2) = \sqrt{\frac{2}{35}}\left(\beta\cos\gamma - \beta_0\cos\gamma_0\right)\left\{3\left(\beta\sin\gamma - \beta_0\sin\gamma_0\right)^2 - \left(\beta\cos\gamma - \beta_0\cos\gamma_0\right)^2\right\}. \tag{5.121}$$

We shall consider *small vibrations of the surface of a rotating nucleus* . In this case, the mass coefficients B_λ appearing in the expression for the kinetic energy are constants:

$$T = \frac{1}{2}\sum_{\lambda\mu} B_\lambda |\dot{\alpha}_{\lambda\mu}|^2. \tag{5.122}$$

In the same approximation, the potential energy can be expressed purely in terms of the invariants $S_2(\lambda)$:

$$V = \frac{1}{2}\sum_{\lambda\mu} C_\lambda |a_{\lambda\mu}|^2. \tag{5.123}$$

In the case of *small quadrupolar* ($\lambda = 2$) vibrations, in the kinetic energy (5.122), as in the case of a spherical equilibrium shape ($\beta_0 = 0$), we can separate the energy of rotation of the nucleus from the energy of the vibrations of the nuclear surface. As a result, we obtain for the kinetic energy (for $\lambda = 2$) a sum of the vibrational and rotational energies, $T = T_v + T_r$, given by the expressions (5.105), (5.107) and (5.108), in which, however, in the general case, $\beta_0 \neq 0$ and $\gamma_0 \neq 0$, since only the differences $\beta - \beta_0$ and $\gamma - \gamma_0$ are small while β and γ (like β_0 and γ_0) may be large. For $\beta_0 = 0$ we have a spherical nucleus, and in this case the dependence on γ_0 drops out automatically.

According to (5.120), the potential energy of small quadrupolar vibrations has the form

$$\begin{aligned} V(\beta,\gamma) &= (C/2)\Big\{(\beta-\beta_0)^2 + \beta\beta_0(\gamma-\gamma_0)^2\Big\} \\ &\approx (C/2)\Big\{(\beta-\beta_0)^2 + \beta_0^2(\gamma-\gamma_0)^2\Big\}. \end{aligned} \tag{5.124}$$

Thus, in the case of spheroidal nuclei having non-spherical equilibrium shapes, surface vibrations of two types, usually called *β- and γ-vibrations* , are possible.

In the case of spherical nuclei ($\beta_0 = 0$), the potential energy does not depend on the parameter γ in the quadratic approximation. But if we take into account the anharmonicity of the vibrations, then, according to (5.121), the potential energy will depend on the transverse-deformation parameter γ even in the case of a spherical nucleus:

$$V(\beta,\gamma) = \frac{1}{2}C\beta^2 + C'\beta^3(3 - 4\cos^2\gamma)\cos\gamma, \tag{5.125}$$

where C' is some constant.

5.4. Rotation of Axially Symmetric Nuclei

Rotational model. The simplest version of the generalized model of the nucleus is the so-called *rotational model* , in which the shape of the nucleus is assumed fixed. In the rotational model, in addition to the rotation of the nucleus as a whole, excitations associated with the internal degrees of freedom of the nucleus are also taken into account. [4] Th rotational model gives a good description of the level spectrum and other characteristics of deformed nuclei in the mass-number ranges $150 < A < 190$ and $A > 220$. In the case of deformed even-even nuclei, almost all the weakly excited levels are rotational levels. In the case of deformed nuclei with odd A, low-lying levels associated with the excitation of individual nucleons are also encountered. A separate rotational band is associated with each such level.

We denote the coordinate frame rotating with the deformed nucleus by K'. The Euler angles Φ, Θ and ψ defining the orientation of K' with respect to the stationary coordinate frame K can be regarded as the dynamical variables describing the rotation of the nucleus as a whole. We denote the coordinates characterizing the internal degrees of freedom of the nucleus by $\mathbf{r}'$.

We write the full Hamiltonian of the nucleus in the form

$$H = H_0(\mathbf{r}') + H_r + H', \tag{5.126}$$

where $H_0(\mathbf{r}')$ describes the motion of the internal degrees of freedom, H_r characterizes the rotation of the nucleus and H' takes account of the coupling between the rotation and the internal motion.

Our subsequent treatment will be based on the assumption that the motion is adiabatic, that is, that the wave function of the nucleus can be written in the form of the product

$$\Psi \sim D(\Theta)\chi(\mathbf{r}'), \tag{5.127}$$

where $D(\Theta)$ is the rotational wave function describing the collective motion of the nucleus (by Θ, we shall mean the collection of Euler angles Φ, Θ and ψ); $\chi(\mathbf{r}')$ is the wave function describing the motion of the internal degrees of freedom.

The assumption that the motion is adiabatic means that we can neglect the quantity H' compared with $H_0(\mathbf{r}')$ and H_r. However, the neglect of H' still does not imply complete separation of the collective and internal motions, since these are also coupled in H_r. Therefore, the rotational function $D(\Theta)$ is not, generally speaking, an eigenfunction of the operator H_r.

[4]The theory of the rotation of axially symmetric nuclei was developed in papers by Bohr (1952) and Bohr and Mottelson (1953).

Let **J** be the nuclear angular momentum associated with the internal degrees of freedom and let **R** be the angular momentum of rotation of the nucleus as a whole. The total angular momentum **I** of the nucleus is equal to the sum of the angular momenta **R** and **J**:

$$\mathbf{I} = \mathbf{R} + \mathbf{J}. \tag{5.128}$$

It is clear that the total wave function (5.127) of the nucleus must be an eigenfunction of the square $\mathbf{I}^2$ of the total angular momentum.

Angular-momentum coupling scheme and wave function of the nucleus. We shall determine the wave function (5.127) of the nucleus, taking into account that it is an eigenfunction of the operators of the square $\mathbf{I}^2$ and projection I_z of the total angular momentum. Since the potential of a deformed nucleus does not possess spherical symmetry, the *internal angular momentum* **J** *of the nucleus* is not a constant of the motion. Experiment shows, however, that most deformed nuclei are characterized by axial symmetry. For an *axially symmetric nucleus*, the projection of the internal angular momentum **J** along the symmetry axis of the nucleus will be conserved. Therefore, the internal wave function will correspond to a definite value of $J_{z'}$, which we denote by Ω:

$$< J_{z'} >= \Omega. \tag{5.129}$$

Denoting by τ all the other quantum numbers characterizing the internal state of the nucleus, we can write the internal wave function in the form

$$\chi^{\tau}_{\Omega}(\mathbf{r}'). \tag{5.130}$$

(The wave function $\chi^{\tau}_{\Omega}(\mathbf{r}')$ is assumed to be normalized to unity.)

Since the *total angular-momentum operator* $\mathbf{I} = \mathbf{R} + \mathbf{J}$ is the operator of the simultaneous rotation of the proper coordinate frame K' and of the physical system described by the function $\chi^{\tau}_{\Omega}(\mathbf{r}')$, we have

$$\mathbf{I}\chi^{\tau}_{\Omega}(\mathbf{r}') = 0. \tag{5.131}$$

Consequently, the total angular-momentum operator $\boldsymbol{I}$ acts only on the rotational function $D(\Theta)$, whereas the operator **J** of the internal angular momentum acts on $\chi^{\tau}_{\Omega}(\mathbf{r}')$.

For completeness, we give the commutation relations for the angular momenta **J** and **I**. The operator **J** of the internal angular momentum acts only on the internal coordinates in the proper frame K', and so the components of **J** along the axes of the proper coordinate frame K' satisfy the usual commutation relations

$$\mathbf{J}_{x'}\mathbf{J}_{y'} - \mathbf{J}_{y'}\mathbf{J}_{x'} = i\mathbf{J}_{z'}. \tag{5.132}$$

The operator $\mathbf{I}$ of the total angular momentum acts only on the angles Φ, Θ and ψ defining the orientation of the nucleus in the stationary coordinate frame K. Therefore, the components of $\mathbf{I}$ along the axes of the stationary frame satisfy the same relations:

$$\boldsymbol{I}_x\boldsymbol{I}_y - \boldsymbol{I}_y\boldsymbol{I}_x = i\boldsymbol{I}_z. \tag{5.133}$$

The components of the total angular momentum $\mathbf{I}$ along the axes of the proper coordinate frame K' satisfy the commutation relations

$$\boldsymbol{I}_{x'}\boldsymbol{I}_{y'} - \boldsymbol{I}_{y'}\boldsymbol{I}_{x'} = -i\boldsymbol{I}_{z'}. \tag{5.134}$$

The minus sign in the right-hand side of (5.134) is associated with the fact that, in an arbitrary rotation of the proper coordinate frame K', the function $D(\Theta)$ on which $\mathbf{I}$ acts transforms in the same way as in a rotation of the laboratory coordinate frame K about the same axis and through the same angle, but in the opposite direction. We note that the operators $\mathbf{J}$ and $\mathbf{I}$ commute.

We choose the rotational function $D(\Theta)$ in such a way that it is simultaneously an eigenfunction of the operators $\boldsymbol{I}^2, \boldsymbol{I}_z$ and $\boldsymbol{I}_{z'}$. Inasmuch as the operator $\mathbf{J}$ does not act on $D(\Theta)$, this means that the function $D(\Theta)$ must be simultaneously an eigenfunction of the operators $\mathbf{R}^2$, R_z and $R_{z'}$. According to (5.90), (5.92) and (5.94), these properties are possessed by the *function* $D^I_{KM}(\Phi, \Theta, \psi)$, and so

$$\left.\begin{aligned} \boldsymbol{I}^2 D^I_{KM}(\Phi,\Theta,\psi) &= I(I+1)D^I_{KM}(\Phi,\Theta,\psi); \\ I_z D^I_{KM}(\Phi,\Theta,\psi) &= M D^I_{KM}(\Phi,\Theta,\psi); \\ I_{z'} D^I_{KM}(\Phi,\Theta,\psi) &= K D^I_{KM}(\Phi,\Theta,\psi). \end{aligned}\right\} \tag{5.135}$$

M and K are the projections of the total angular momentum of the nucleus along the axis Oz in the stationary coordinate frame K and along the axis Oz' in the coordinate frame K' moving with the nucleus.

Using (5.63) and assuming that the motion is adiabatic, we can write the normalized wave function of the nucleus in the form

$$\Psi = \sqrt{\frac{2I+1}{8\pi^2}} D^I_{KM}(\Phi,\Theta,\psi)\chi^{\tau}_{\Omega}(\mathbf{r}'). \tag{5.136}$$

Figure 5.5. shows the *angular-momentum coupling scheme in the rotational model* of the nucleus.

Symmetry of the wave function. The above wave function (5.136), however, does not yet satisfy the symmetry requirements stemming from

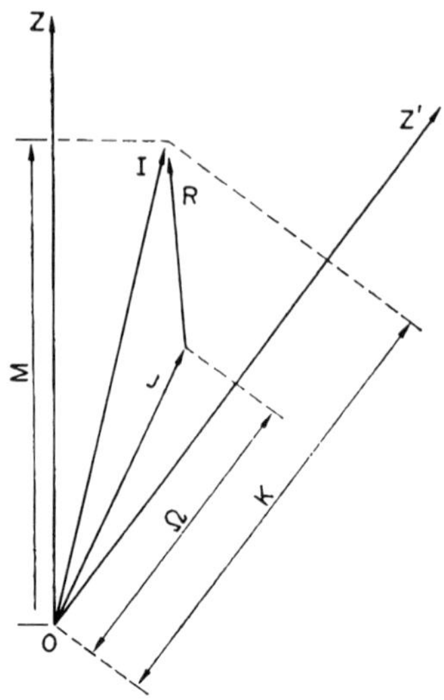

Figure 5.5. Angular-momentum coupling scheme in the rotational model of the nucleus.

the symmetry of the shape of the nucleus. The wave function (5.136) must be modified somewhat in order to satisfy these requirements.

(a) *The axial symmetry of the nucleus* . Because of the axial symmetry of the nucleus, the wave function Ψ must be unchanged on rotation of the coordinate frame associated with the nucleus about the symmetry axis of the nucleus:

$$R_{z'}\Psi = 0. \tag{5.137}$$

Noting that $\mathbf{R} = \mathbf{I} - \mathbf{J}$, we can rewrite this condition

$$\Big(I_{z'}D^I_{KM}(\Phi,\Theta,\psi)\Big)\chi^\tau_\Omega(\mathbf{r}') = D^I_{KM}(\Phi,\Theta,\psi)\Big(J_{z'}\chi^\tau_\Omega(\mathbf{r}')\Big).$$

Consequently, the requirement that the wave function be invariant under rotations about the symmetry axis of the nucleus is equivalent to the condition

$$K = \Omega. \tag{5.138}$$

The condition means that the angular momentum of rotation $\mathbf{R}$ is perpendicular to the symmetry axis for an axially symmetric nucleus.

Thus, for an axially symmetric nucleus the wave function must have the form

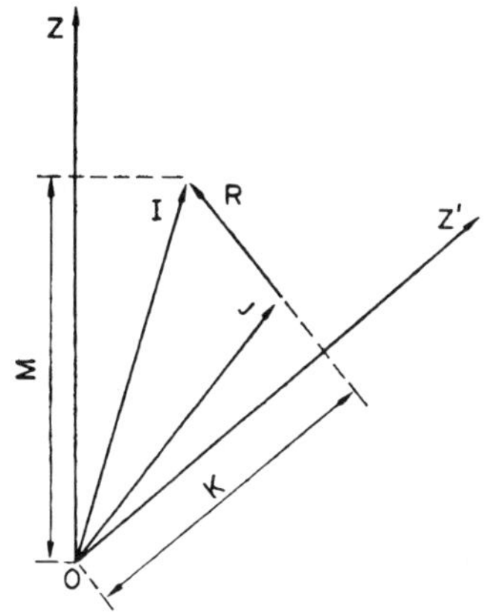

Figure 5.6. Angular-momentum coupling scheme in the case of an axially symmetric nucleus.

$$\Psi_{IKM} = \sqrt{\frac{2I+1}{8\pi^2}} D^I_{KM}(\Phi, \Theta, \psi)\chi^\tau_K(\mathbf{r}'). \tag{5.139}$$

Figure 5.6 shows the angular-momentum coupling scheme for an axially symmetric nucleus.

(b) *The symmetry of the nucleus with respect to the plane $z' = 0$* . If the nucleus is an ellipsoid of revolution, it is transformed into itself on reflection through the principal plane perpendicular to the symmetry axis. The reflection operation p through the plane $z' = 0$ is equivalent for an axially symmetric nucleus to a rotation $R_{y'}(\pi)$ through an angle of 180° about the y'-axis. Under a rotation $R_{y'}(\pi)$, the rotational function $D^I_{KM}(\Phi, \Theta, \psi)$ transforms according to (5.58):

$$R_{y'}(\pi) D^I_{KM} = \sum_{K'} D^{*I}_{K'K}(0, \pi, 0) D^I_{K'M}. \tag{5.140}$$

Noting that

$$D^{*I}_{K'K}(0, \pi, 0) = (-1)^{I-K}\delta_{K',-K},$$

we have

$$R_{y'}(\pi) D^I_{KM} = (-1)^{I-K} D^I_{-KM}. \tag{5.141}$$

Analogously, for the internal wave function χ^τ_K we obtain

$$R_{y'}(\pi)\chi^\tau_K = (-1)^{J-K}\chi^\tau_{-K}, \tag{5.142}$$

where $(-1)^J$ must be regarded as an operator. If

$$\chi_K^\tau = \sum_J c_J \chi_{JK}^\tau,$$

then

$$(-1)^J \chi_K^\tau = \sum_J c_J (-1)^J \chi_{JK}^\tau.$$

Noting that

$$(-1)^{I+J-2K} = (-1)^{I-J+2(J-K)} = (-1)^{I-J},$$

we have

$$R_{y'}(\pi)\Psi_{IKM} = (-1)^{I-J}\Psi_{I-KM}. \tag{5.143}$$

If $K \neq 0$, the requirement of invariance under the rotation $R_{y'}(\pi)$ implies that the wave function should be a linear combination of (5.139) and (5.143). The normalized wave function has the form

$$\Psi(IKM) = \sqrt{\frac{1}{2}}\left\{\Psi_{IKM} + (-1)^{I-J}\Psi_{I-KM}\right\}. \tag{5.144}$$

If $K = 0$, the function (5.139) already satisfies the required invariance property; here it is necessary, however, that I and J have the same parity (otherwise the function vanishes):

$$(-1)^{I-J} = 1. \tag{5.145}$$

(c) *The parity of the wave function* . The total-inversion operator $\boldsymbol{P}$ includes the inversion p' of the internal coordinates with respect to the axes x', y', z' and the inversion p of the collective coordinates (the axes x', y', z'): $\boldsymbol{P} = pp'$. We denote the parity of the internal wave function χ_K^τ by π:

$$p'\chi_K^\tau = \pi\chi_K^\tau. \tag{5.146}$$

Since $p = \mathbf{R}_{y'}(\pi)$ for an axially symmetric nucleus, for states Ψ that are invariant under reflection in the plane $z' = 0$ we have

$$\boldsymbol{P}(\Psi) = \pi(\chi). \tag{5.147}$$

Thus, the parity of the wave function Ψ is determined by the parity of the internal wave function χ.

If $K = 0$, the parity of the internal wave function can be reduced directly to the value $\mathbf{I}$ of the total angular momentum. In fact, for $K = 0$ the function $\chi^{\tau}_{K=0}$ is invariant under rotations about the z'-axis, and so $p' = \mathbf{R}_{y'}(\pi)$. Using (5.142) and (5.145), we find

$$\pi(\chi) = (-1)^J = (-1)^I. \tag{5.148}$$

(d) *The wave functions.* We shall give explicit expressions for the *wave functions of a rotating axially symmetric nucleus* in states with well-defined angular momentum and parity.

If $K \neq 0$, the wave function has the form

$$\begin{aligned}\Psi(IKM) &= \sqrt{\frac{2I+1}{16\pi^2}}\Big\{D^I_{KM}(\Phi,\Theta,\psi)\chi^{\tau}_K(\mathbf{r}')\\ &\quad +(-1)^{I-J}D^I_{-KM}(\Phi,\Theta,\psi)\chi^{\tau}_{-K}(\mathbf{r}')\Big\}.\end{aligned} \tag{5.149}$$

The parity of the function Ψ is the same as the parity of the internal function χ. The set of states (5.149) corresponding to a given internal function χ^{τ}_K but to different values I of the total angular momentum forms a rotational band:

$$I = K, \qquad K+1, \qquad K+2, \ldots. \tag{5.150}$$

If $K = 0$, the wave function has the form

$$\Psi(I0M) = \sqrt{\frac{2I+1}{8\pi^2}}D^I_{0M}(\Phi,\Theta,\psi)\chi^{\tau}_0(\mathbf{r}'), \tag{5.151}$$

or, using (5.61),

$$\Psi(I0M) = \left(1/\sqrt{2\pi}\right)Y_{IM}(\Theta,\Phi)\chi^{\tau}_0(\mathbf{r}'). \tag{5.152}$$

In this case, two rotational bands are possible, for the even and odd states respectively:

$$\left.\begin{aligned} I &= 0^+, 2^+, 4^+, \ldots, \pi = +1,\\ I &= 1^-, 3^-, 5^-, \ldots, \pi = -1.\end{aligned}\right\} \tag{5.153}$$

Rotational spectra of nuclei. Making the assumption of adiabatic motion, we write the Hamiltonian of the rotating nucleus in the form

$$H = H_0(\mathbf{r}') + H_r. \tag{5.154}$$

According to (5.111), the rotational part of the Hamiltonian is given by the expression

$$\mathbf{H}_r = \frac{\hbar^2}{2J_{x'}}\mathbf{R}_{x'}^2 + \frac{\hbar^2}{2J_{y'}}\mathbf{R}_{y'}^2 + \frac{\hbar^2}{2J_{z'}}\mathbf{R}_{z'}^2, \tag{5.155}$$

where $J_{x'}, J_{y'}$ and $J_{z'}$ are the moments of inertia about the principal axes and can be regarded as empirical parameters.

Assuming the nucleus to be axially symmetric and choosing the z'-axis in the direction of the symmetry axis of the nucleus, we put $J_{x'} = J_{y'} = J$. Noting that $\mathbf{R} = \mathbf{I} - \mathbf{J}$, we rewrite (5.155) in the form

$$H_r = \frac{\hbar^2}{2J}\left\{I^2 + \mathbf{J}^2 - 2\mathbf{I}\cdot\mathbf{J}\right\} + \left(\frac{\hbar^2}{2J_{z'}} - \frac{\hbar^2}{2J}\right)(I_{z'} - J_{z'})^2. \tag{5.156}$$

In the case of an axially symmetric nucleus, the projections $I_{z'}$ and $J_{z'}$ along the symmetry axis of the nucleus are equal (cf. (5.138)), and so the Hamiltonian (5.154) of the nucleus can be represented finally in the form

$$H = H_0(\mathbf{r}') + \frac{\hbar^2}{2J}\mathbf{J}^2 + \frac{\hbar^2}{2J}I^2 - \frac{\hbar^2}{J}\mathbf{I}\cdot\mathbf{J}. \tag{5.157}$$

The first two terms (5.157) characterize the motion of the internal degrees of freedom of the nucleus. The internal function χ_K^{τ} can be assumed to be an eigenfunction of this part of the Hamiltonian. The rotational function $D_{KM}^{I}(\Phi, \Theta, \psi)$ is an eigenfunction of the third term in (5.157). If we neglect the fourth term in (5.157), we obtain the following simple formula for the *energy levels of a rotational band* :

$$E_K(I) = E_K^0 + \frac{\hbar^2}{2J}I(I+1). \tag{5.158}$$

where E_K^0 is the internal energy, which does not depend on I.

The term $-\hbar^2\mathbf{I}\cdot\mathbf{J}/J$ in (5.157) takes into account the coupling between the rotational and internal motions in the nucleus. For an estimate of the contribution of this term to the energy, we represent the scalar product of the angular momenta $\mathbf{I}$ and $\mathbf{J}$ in the form

$$\mathbf{I}\cdot\mathbf{J} = I_{z'}J_{z'} + \frac{1}{2}(I_+J_- + I_-J_+), \tag{5.159}$$

where

$$I_\pm = I_{x'} \pm iI_{y'}, \qquad J_\pm = J_{x'} \pm iJ_{y'}. \tag{5.160}$$

Since the function $\Psi(IKM)$ (5.149) is an eigenfunction of the operator $I_{z'}J_{z'}$, corresponding to the eigenvalue K^2, allowance for this term in (5.159) leads only to a displacement of the levels (5.158) of the rotational band, with no change in their relative positions.

The other two terms $\frac{1}{2}(I_+J_- + I_-J_+)$ in (5.159) are not diagonal in the representation (5.149), provided that $K \neq \frac{1}{2}$. Using the commutation relations (5.132) and (5.134), it is not difficult to verify that the only non-zero matrix elements of the operators $I_\pm$ and $J_\pm$ are

$$(K \mp 1|I_\pm|K) = \sqrt{(I \pm K)(I \mp K + 1)};$$
$$(K \mp 1|J_\mp|K) = \sqrt{(J \pm K)(J \mp K + 1)}.$$

Therefore,

$$\begin{aligned} & I_+J_-D^I_{KM}(\Phi, \Theta, \psi)\chi^\tau_{JK}(\mathbf{r}') \\ = & \sqrt{(I+K)(I-K+1)(J+K)(J-K+1)} \\ \times & D^I_{K-1,M}(\Phi, \Theta, \psi)\chi^\tau_{JK-1}(\mathbf{r}'). \end{aligned} \tag{5.161}$$

Thus, the inclusion of I_+J_- and I_-J_+ should cause mixing of states with different values of K. If, however, the difference between the internal energies for the different rotational bands is large compared with the effective value of the coupling energy $-\hbar^2\mathbf{I}\cdot\mathbf{J}/J$, the mixing of states with different values of K can be neglected. In this case, the energy levels in a particular rotational band are given by formula (5.158).

If $K = \frac{1}{2}$, the term $-\hbar^2\mathbf{I}\cdot\mathbf{J}/J$ has a diagonal part, depending on I, which must be taken into account in (5.158) and which in certain cases can change the order of the levels in the rotational band. We shall denote the additional energy in the case $K = \frac{1}{2}$ by ΔE. Taking into account the symmetry of the function (5.149), we have

$$\begin{aligned} \Delta E &= -\frac{\hbar^2}{2J}\left(I\frac{1}{2}M|I_+J_- + I_-J_+|I\frac{1}{2}M\right) \\ &= -\frac{\hbar^2}{J}\left(I\frac{1}{2}M|I_+J_-|I\frac{1}{2}M\right). \end{aligned} \tag{5.162}$$

To calculate ΔE, we note that

$$I_+J_-\Psi(I\frac{1}{2}M) \;=\; \sqrt{\frac{2I+1}{16\pi^2}}\Big\{I_+D^I_{(1/2)M}\sum_J c_J J_-\chi^\tau_{J1/2}$$

$$+ \quad I_+ D^I_{-(1/2)M} \sum_J c_J (-1)^{I-J} J_- \chi^{\tau}_{J-1/2} \Big\}$$

$$= \sqrt{\frac{2I+1}{16\pi^2}} \left\{ \left(I + \frac{1}{2}\right) D^I_{-(1/2)M} \sum_J c_J \left(J + \frac{1}{2}\right) \chi^{\tau}_{J,-1/2} \right.$$

$$+ \left. \left(\text{term with } K = -\frac{3}{2}\right)\right\}.$$

Using the orthonormalization of the functions D^I_{KM} and χ^{τ}_K, we thus obtain

$$\left(I\frac{1}{2}M|I_+J_-|I\frac{1}{2}M\right) = \frac{1}{2}\left(I + \frac{1}{2}\right) \sum_J |c_J|^2 (-1)^{I-J} \left(J + \frac{1}{2}\right). \qquad (5.163)$$

Taking into account that the number $2J+1$ is even for $K = \frac{1}{2}$, we represent the expression for the energy (5.162) in the form

$$\Delta E = \frac{\hbar^2}{2J} a(-1)^{I+1/2} \left(I + \frac{1}{2}\right), \qquad (5.164)$$

where a is the so-called coupling parameter. The magnitude of the *coupling parameter* a depends on the internal state of the odd nucleon and is given by the expression

$$a = \sum_J |c_J|^2 (-1)^{J-1/2} \left(J + \frac{1}{2}\right). \qquad (5.165)$$

Combining (5.158)and (5.165), we finally obtain the following formula for the energy levels of a rotational band:

$$E_K(I) = E^0_K + \frac{\hbar^2}{2J}\left[I(I+1) + a(-1)^{I+1/2}(I + \frac{1}{2})\delta_{K,1/2}\right], \qquad (5.166)$$

which is valid for both the case $K \neq \frac{1}{2}$ and the case $K = \frac{1}{2}$

Comparison with experiment. The rotational model gives a good description of the rotational spectra of strongly deformed nuclei for the rare earths $150 < A < 190$ and the heavy α-active nuclei $A > 220$, (Fig.5.7) and also for the light nuclei $20 < A < 30$.

Even-even nuclei . In the case of even-even nuclei, the lowest rotational band is the band $K = 0^+$, corresponding to the spin sequence I=0,2,4,... We denote the excitation energy, reckoned from the ground state of the nucleus, by E_I:

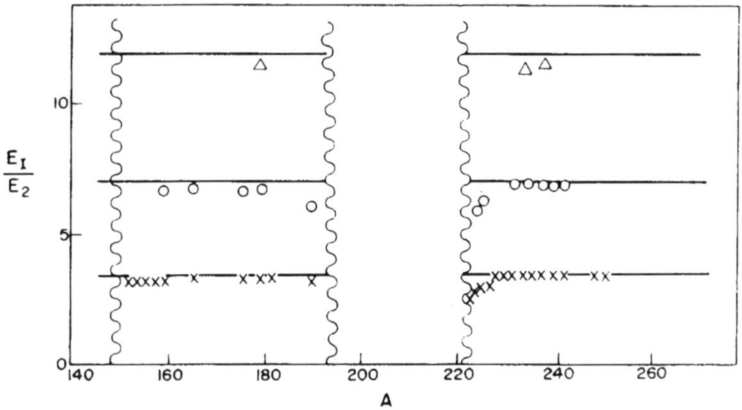

Figure 5.7. Ratios of the rotational energies of strongly deformed nuclei in the regions of the rare-earth nuclei and the heavy α-active nuclei ($\times$ denotes E_4/E_2, $\bigcirc$ denotes E_6/E_2, and Δ denotes E_8/E_2).

$$E_I = \frac{\hbar^2}{2J} I(I+1). \qquad (5.167)$$

According to (5.167), the following interval rule should be fulfilled for the levels of a rotational band:

$$E_2 : E_4 : E_6 : E_8 \ldots = 1 : 10/3 : 7 : 12 \ldots . \qquad (5.168)$$

This rule is well satisfied in the energy spectra of strongly deformed nuclei in the three ranges of A indicated.

As an illustration, the rotational spectrum of the rare-earth element $^{180}_{72}Hf$ is shown in Fig. 5.8. The excitation energy is indicated alongside each level. Using formula (5.167), we can determine the moment of inertia from the position of the first rotational level. For the nucleus $^{180}_{72}Hf$, the moment of inertia found in this way is $J = 2 \cdot 23 \times 10^{-47}$ g.cm^2. The energy values calculated from formula (5.167) are given in brackets in Fig. 5.8. It can be seen that the discrepancy between the values amounts to about 1%.

Figure 5.9 shows the rotational spectrum of the heavy α-active nucleus $^{238}_{94}Pu$. The moment of inertia of the nucleus $^{238}_{94}Pu$ is $J = 4 \cdot 70 \times 10^{-47}$ g.cm^2. This value is roughly twice the moment of inertia of the lighter nucleus $^{180}_{72}Hf$.

Figure 5.10 also shows the rotational spectrum of the light nucleus $^{24}_{12}Mg$. In this case, the moment of inertia is $J = 0 \cdot 152 \times 10^{-47}$ g.cm^2.

Odd nuclei. Rotational spectra are also observed for deformed nuclei with odd A. For the individual levels of a rotational band, the spin values

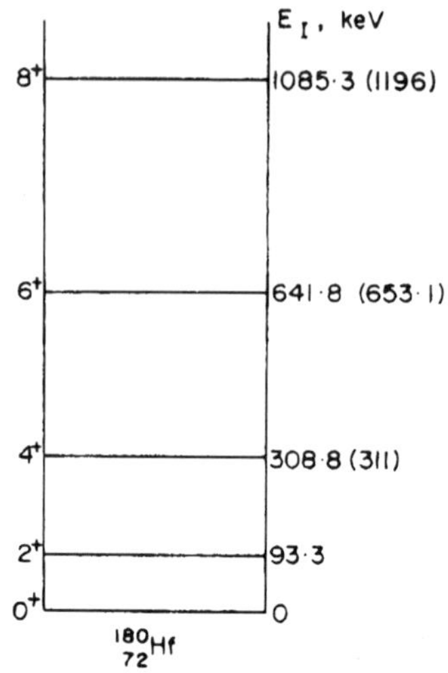

Figure 5.8. Level scheme of the $^{180}_{72}Hf$ nucleus.

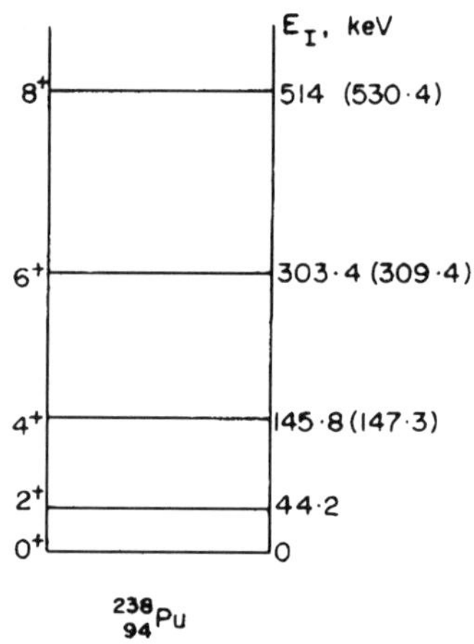

Figure 5.9. Level scheme of the $^{238}_{94}Pu$ nucleus.

are equal to $I_0(=K)$, I_0+1, I_0+2,... .If $K \neq \frac{1}{2}$, we have for the ratio of the excitation energies

$$\frac{E_{I_0+2}}{E_{I_0+1}} = 2 + \frac{1}{I_0+1}. \tag{5.169}$$

An exact measurement of this ratio makes it possible to determine the spin I_0 of the ground state. Figure 5.11 shows the rotational spectrum of $^{181}_{73}Ta$. The observed ratio $E_{I_0+2}/E_{I_o+1} = 2 \cdot 23$; consequently, $I_0 = \frac{7}{2}$.

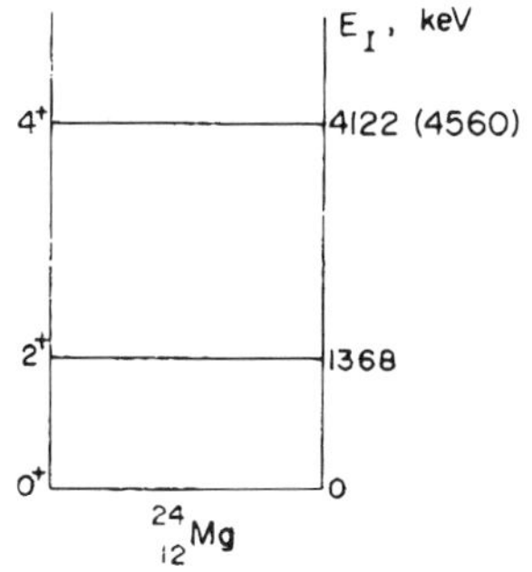

Figure 5.10. Level scheme of the $^{24}_{12}Mg$ nucleus.

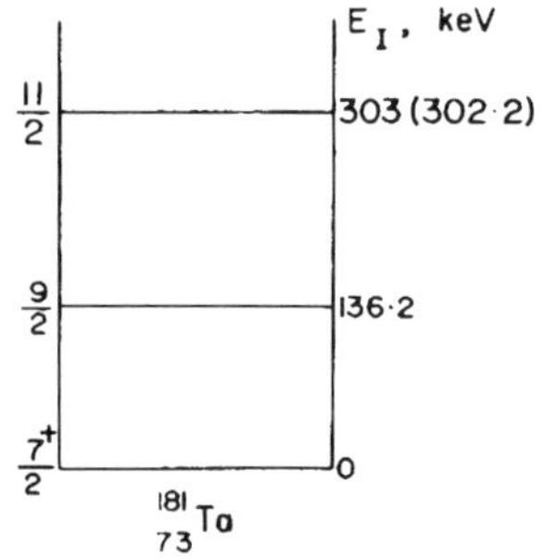

Figure 5.11. Level scheme of the $^{181}_{73}Ta$ nucleus.

From the position of the first rotational level, the moment of inertia $J = 2 \cdot 23 \times 10^{-47}$ g.cm^2, that is, has the same order of magnitude as for $^{180}_{72}Hf$.

The rotational spectrum for $K = \frac{1}{2}$. For $K = \frac{1}{2}$, the spectrum of the levels can differ substantially from a simple rotational spectrum. This is due to the coupling between the internal motion and the rotation, which is taken into account by the term with a in the formula (5.166). The energy levels in this case depend on the two parameters a and J. Figure 5.12 shows the energy-level scheme for $K = \frac{1}{2}$ as a function of the coupling parameter a.

Since $-1 < a < 1$ for the rotational spectra of heavy nuclei with $K = \frac{1}{2}$, allowance for the coupling does not lead to a change in the order of the levels for heavy nuclei. Figure 5.13, for example, shows the rotational spectrum of $^{183}_{74}W$. The empirical values of the parameters a and J, determined from the positions of the first two excited levels, are $a = 0 \cdot 19$ and $J = 2 \cdot 67 \times 10^{-47}$ g.cm^2. Using these values, we can calculate the excitation energies of the higher levels from formula (5.166). For light nuclei, however, the coupling parameter a can take large values (for example, $a = -3 \cdot 2$ for $^{25}_{13}Al$), and this can lead to a change in the order of the levels.

Magnetic moments of deformed nuclei. The operator of the magnetic dipole moment of a rotating nucleus, expressed in nuclear magnetons, is given by the expression

$$\boldsymbol{\mu} = g_R \mathbf{R} + \boldsymbol{\mu}', \tag{5.170}$$

where g_R is the gyromagnetic ratio associated with the rotation of the nucleus as a whole and $\boldsymbol{\mu}'$ is the internal magnetic moment, equal to the sum of the magnetic moments of the individual nucleons:

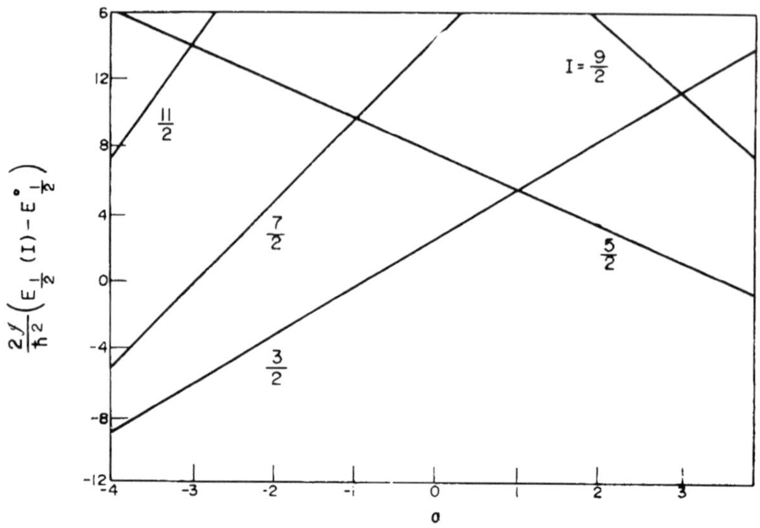

Figure 5.12. Dependence on the coupling parameter a of the rotational energy levels $E_K(I)$ for $K = \frac{1}{2}$.

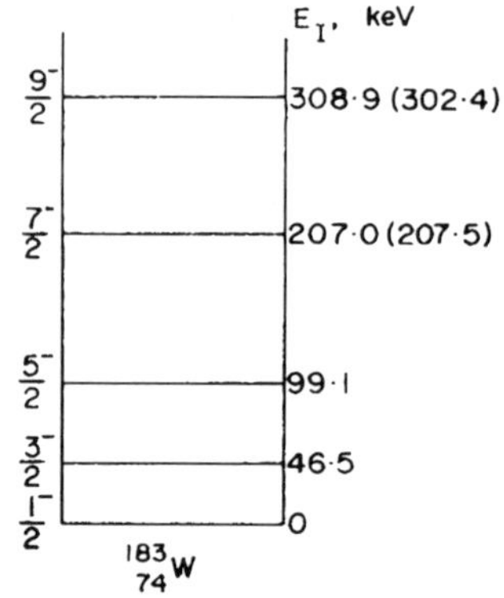

Figure 5.13. Level scheme for the ${}^{183}_{74}W$ nucleus.

$$\mu' = \sum(g_l \mathbf{l}' + g_s \mathbf{s}') \qquad (5.171)$$

(note that $\sum(\mathbf{l}' + \mathbf{s}') = \mathbf{J}$). Expressing the angular momentum of rotation in terms of the total angular momentum of the nucleus, $\mathbf{R} = \mathbf{I} - \mathbf{J}$, we rewrite the relation (5.170) in the form]

$$\mu = g_R \boldsymbol{I} + (\mu' - g_R \mathbf{J}). \qquad (5.172)$$

The magnetic moment μ of the nucleus is defined as the mean value of the projection of the operator $\boldsymbol{\mu}$ along the stationary axis Oz in the state $M = I$. In calculating the mean value, however, the vector $(\boldsymbol{\mu}' - g_R\mathbf{J})$ characterizing the internal motion is more conveniently expressed in the proper coordinate frame K' moving with the rotating nucleus.

Introducing the spherical components of any vector $\mathbf{L}$ by means of the relations

$$L_0 = L_z, \qquad L_{\pm 1} = \mp\sqrt{\frac{1}{2}}L_{\pm}, \tag{5.173}$$

we can write the relation between the components of the vector in the stationary and moving coordinate frames K and K' in the form

$$L_q = \sum_{q'} D^1_{q'q}(\Phi, \Theta, \psi)L_{q'}. \tag{5.174}$$

Thus, for the operator $\boldsymbol{\mu}_z$ of the projection of the magnetic moment we obtain the expression

$$\boldsymbol{\mu}_z = g_R I_z + \sum_{q'=0,\pm 1} D^1_{q'0}(\Phi, \Theta, \psi)(\boldsymbol{\mu}' - g_R\mathbf{J})_{q'}. \tag{5.175}$$

Using the formula (5.149) for the function $\Psi(IKM)$, we shall determine the mean value of μ_z in the state $M = I$. The expression for the mean value of the second term in (5.175) will contain integrals of products of three D-functions. These integrals can be calculated easily by means of the relation (5.64):

$$\int_0^{2\pi} d\Phi \int_0^{\pi} d\Phi \sin\Phi \int_0^{2\pi} d\psi D^{*I}_{\pm KM} D^1_{q'0} D^I_{\pm KM}$$
$$= \frac{8\pi^2}{2I+1}\Big(IM10|IM\Big)\Big(I \pm K1q'|I \pm K\Big). \tag{5.176}$$

Opposite signs of K correspond to the cross-terms arising from the multiplication of the different terms of the function (5.149).

The coefficient $(I_{\pm}K1q'|I_{\pm}K)$ is non-zero only when the condition

$$\pm K = q' \pm K \tag{5.177}$$

is fulfilled. Therefore, if $K \neq \frac{1}{2}$, only the term $q' = 0$ makes a non-zero contribution to (5.175). Thus, for $K \neq \frac{1}{2}$, we have

$$\mu = g_R I + (IM10|IM)(IK10|IK)\Big(\chi_K|\mu'_{z'} - g_R J_{z'}|\chi_K\Big). \qquad (5.178)$$

(The states χ_K and χ_{-K} give the same contribution to (5.178).)Introducing the internal gyromagnetic ratio g_K

$$(\chi_K|\mu'_{z'}|\chi_K) = g_K K \qquad (5.179)$$

and noting that

$$(IK10|IK) = \frac{K}{\sqrt{I(I+1)}},$$

we obtain the following formula:

$$\mu = g_R I + (g_K - g_R)\frac{K^2}{I+1}. \qquad (5.180)$$

If $K = \frac{1}{2}$, the condition (5.177) is also fulfilled when $q' = \pm 1$. In this case, the "cross" matrix elements between the states χ_K and χ_{-K} will be non-zero. Therefore, in (5.178) it is also necessary to take into account the term

$$\frac{1}{2}\Big(II10|II\Big)\Big(I - \frac{1}{2}11|I\frac{1}{2}\Big)\Big(\chi_{1/2}|(\mu'_1 - g_R J_1)(-1)^{I-J}|\chi_{-1/2}\Big)$$

and the complex conjugate term. Introducing the notation

$$\mathrm{Re}\,(\chi_{1/2}|J_1(-1)^{I-J}|\chi_{-1/2}) = \frac{1}{\sqrt{2}}(-1)^{-I+1/2}b,$$

we obtain the correction to (5.178) in the case $K = \frac{1}{2}$ in the form

$$\frac{1}{4}(-1)^{I-1/2}\frac{2I+1}{I+1}(g_K - g_R)b. \qquad (5.181)$$

Here we have used that the fact that

$$\Big(I - \frac{1}{2}11|I\frac{1}{2}\Big) = \frac{1}{\sqrt{8}}\frac{2I+1}{\sqrt{I(I+1)}}.$$

Combining the formulae (5.180) and (5.181), we finally obtain the following expression for the *magnetic moment of a rotating nucleus* :

$$\mu = g_R I + \frac{g_K - g_R}{I+1}\Big\{K^2 + \frac{1}{4}(-1)^{I-1/2}(2I+1)b\delta_{K1/2}\Big\}. \qquad (5.182)$$

For the ground state of the nucleus, we must take $I = K$ in (5.182).

In the general case, the expression (5.182) for the magnetic moment contains three parameters g_K, g_R and b; to calculate these, a more detailed theory is necessary. The experimental determination of the magnetic moment gives one independent relation between these parameters. Simultaneous measurement of the probabilities of transitions between the rotational states makes it possible to find empirical values of the parameters g_K, g_R and b for certain cases.

As an illustration, we shall determine the *magnetic moment of a deformed nucleus with odd A* by means of (5.182). By assuming that the charge distribution is uniform, we put $g_R \approx Z/A$. Assuming also that the internal moment μ' is due to the last odd nucleon, which is in a state with well-defined J, we can put $g_K \approx g_J$, where g_J is the gyromagnetic ratio for the odd nucleon, obtained from the shell model. Thus, for $I > \frac{1}{2}$, we have

$$\mu = g_R I + (g_J - g_R)\frac{I^2}{I+1}, \qquad K = I. \tag{5.183}$$

As an illustration, for a series of nuclei with spin $I = \frac{5}{2}$, we give the experimental values of the magnetic moments and the values calculated from formula (5.183) (Table 5.1).

According to the shell model, the magnetic moment should be equal to $4 \cdot 79$ or $-1 \cdot 91$, depending on whether the odd particle is a proton or a neutron. Consequently, allowance for the additional magnetic moment associated with the deformation of the nuclear core appreciably improves the agreement with experiment.

Quadrupole moments of deformed nuclei . The operator of the electric quadrupole moment of a nucleus is defined by the expression

$$Q_q = \sqrt{\frac{16\pi}{5}} \sum r^2 Y_{2q}(\vartheta, \varphi), \tag{5.184}$$

TABLE 5.1.

Nucleus	μ_{exp}	μ	Nucleus	μ_{exp}	μ
$^{27}_{13}Al$	$3 \cdot 64$	$3 \cdot 77$	$^{25}_{12}Mg$	$-0 \cdot 86$	$-0 \cdot 67$
$^{121}_{51}Sb$	$3 \cdot 36$	$3 \cdot 73$	$^{95}_{42}Mo$	$-0 \cdot 91$	$-0 \cdot 81$
$^{131}_{55}Cs$	$3 \cdot 48$	$3 \cdot 73$	$^{105}_{46}Pd$	$-0 \cdot 6$	$-0 \cdot 81$
$^{187}_{75}Re$	$3 \cdot 20$	$3 \cdot 71$	$^{111}_{48}Cd$	$-0 \cdot 7$	$-0 \cdot 81$

where the summation is extended over all the protons. Going over to the coordinate frame moving with the nucleus, we have, according to (5.60),

$$Q_q = \sum_{q'} D^2_{qq'}(\Phi, \Theta, \psi) Q_{q'}. \tag{5.185}$$

The electric quadrupole moment Q of the nucleus is defined as the mean value of the operator $Q_{q=0}$ in the state $M = I$:

$$Q \equiv < Q_{q=0} >_{M=I} . \tag{5.186}$$

Using the expression (5.149) for the wave function $\Psi(IKM)$, we thus find

$$Q = (II20|II)(IK20|IK)Q_0, \tag{5.187}$$

where Q_0 is the internal quadrupole moment of the nucleus, defined by the expression

$$Q_0 = (\chi_K|Q_{q'=0}|\chi_K). \tag{5.188}$$

(Note that, for $K = \frac{1}{2}$ or 1, in (5.185) it is necessary to include also the terms $q' \neq 0$, whose matrix elements, however, vanish for the states χ_K and χ_{-K}.) Noting that

$$(II20|II)(IK20|IK) = \frac{3K^2 - I(I+1)}{(I+1)(2I+3)},$$

we obtain the following expression for the quadrupole moment:

$$Q = \frac{3K^2 - I(I+1)}{(I+1)(2I+3)} Q_0. \tag{5.189}$$

If I is sufficiently large ($I(I+1) > 3K^2$), the quadrupole moment Q is opposite in sign to the *internal quadrupole moment* .

For the ground state $I = K$ of the rotational band, we have

$$Q = \frac{I(2I-1)}{(I+1)(2I+3)} Q_0. \tag{5.190}$$

The coefficient of Q_0 is usually called the *projecting factor*. The presence of this factor leads to a sharp reduction of the *measured quadrupole moment* Q compared with Q_0. If $I = 0$ or $\frac{1}{2}$, the quadrupole moment Q of the nucleus vanishes, even though the internal quadrupole moment Q_0 may be non-zero. For $I = 1$, the projecting factor is equal to $\frac{1}{10}$, for $I = \frac{3}{2}$ it is equal to $\frac{1}{5}$, and on further increase of I the factor slowly tends to unity.

Assuming the charge distribution to be uniform, we can calculate the internal quadrupole moment of the nucleus:

$$Q_0 = (2Z/V) \int_V d^3r(z^2 - x^2) = (2Z/5)(R_{z'}^2 - R_{x'}^2). \tag{5.191}$$

Using the expression (5.102) for the principal semi-axes of the ellipsoid, we thus obtain

$$Q_0 = \frac{3}{\sqrt{5\pi}} Z R_0^2 \beta \left(1 + \frac{1}{4}\sqrt{\frac{5}{4\pi}}\beta\right), \tag{5.192}$$

where β is the nuclear deformation parameter.

Q_0 can be determined experimentally by measuring the probabilities of γ-transitions of the nucleus from one rotational state to another.

Table 5.2 shows the values of Q_0 and the corresponding values of the deformation parameter β, found from the probabilities of $E2$-transitions, for a series of even-even nuclei.

For nuclei with odd A the quadrupole moment can also be determined from spectroscopic measurements of the hyperfine structure of the spectral lines ($I > \frac{1}{2}$). Table 5.3 shows the values of the internal quadrupole moments Q_0 for a series of odd nuclei, determined from spectroscopic measurements and measurements of $E2$-transition probabilities.

Figure 5.14 shows the values of the quadrupole moments for strongly deformed nuclei in the mass-number range $150 < A < 190$.

One-particle states in deformed nuclei. In studying the shell model, we assumed that the self-consistent nuclear potential in which the nucleons move possesses spherical symmetry. The nuclear shape characterized by this potential was also spherically symmetric. In fact, in the case of filled shells, the equilibrium shape of real nuclei *is* spherically symmetric. However, nuclei in which the last shell is approximately half-filled are characterized by considerable deformations. To explain the properties of deformed nuclei, it is necessary not only to study the collective motions in them but also to

TABLE 5.2.

Nucleus	$Q_0, 10^{-24}\text{cm}^2$	β
$^{150}_{60}Nd$	4·8	0·25
$^{152}_{62}Sm$	5·7	0·28
$^{154}_{62}Sm$	6·7	0·33
$^{160}_{64}Gd$	10	0·46
$^{178}_{72}Hf$	8·1	0·31
$^{184}_{74}W$	6·5	0·24
$^{188}_{76}Os$	5·1	0·18

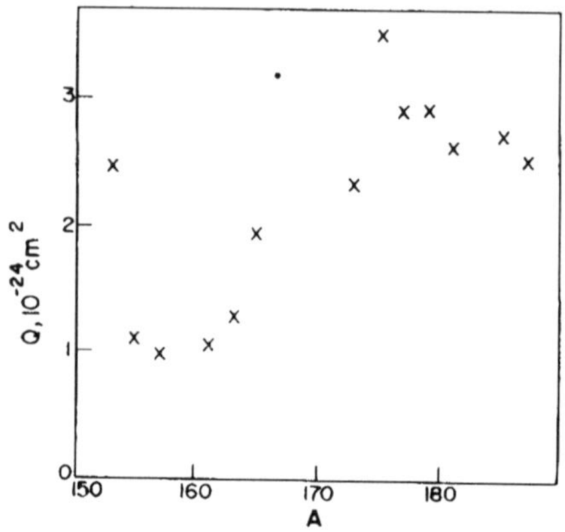

Figure 5.14. Values of the quadrupole moments of strongly deformed nuclei in the region of the rare-earths.

consider the *one-particle states of the nucleons moving in the non-spherical potential* . In the adiabatic approximation, the one-particle and collective motions in the nucleus can be separated, so that the total wave function of the nucleus can be represented in the form of a product of two functions, of which one describes the motion of the nucleus as a whole, that is, the rotation of the nucleus and the vibrations of its surface, and the other describes the internal motion in the non-spherical nucleus.

In treating the internal motion in the nucleus, we shall assume that the individual nucleons move independently in a certain fixed non-spherical field of the nucleus. The Hamiltonian of the internal motion can then be represented, as in the ordinary shell model (without allowance for the residual interaction), in the form of a sum of one-particle Hamiltonians. Thus, the problem reduces to solving the Schrödinger equation for one particle with a one-particle potential which must be chosen in such a way that it reflects the principal features of the motion of nucleons in non-spherical

TABLE 5.3.

Nucleus	I	Q_0 (spectroscopic measurements)	Q_0 (E_2-transition
${}^{153}_{63}Eu$	$\frac{5}{2}$	7·0	7·7
${}^{155}_{64}Gd$		5·5	8·0
${}^{157}_{62}Gd$		5·0	7·7
${}^{175}_{71}Lu$	$\frac{7}{2}$	12	8·2
${}^{185}_{75}Re$	$\frac{5}{2}$	7·8	5·4

nuclei while having at the same time the simplest possible form. This problem has been solved for several potentials by various authors. The so-called *Nilsson potential* (Nilsson, 1955), which applies well to the case of medium and heavy nuclei, has the simplest form.

The Nilsson potential is chosen in the form of a sum

$$V = (M/2)(\omega_1^2 x^2 + \omega_2^2 y^2 + \omega_3^2 z^2) + C\mathbf{l}\cdot\mathbf{s} + D\mathbf{l}^2, \tag{5.193}$$

where the first term is an anisotropic-oscillator potential, which takes into account that the potential well is non-spherical and diffuse at the boundary, the second term describes the spin-orbit interaction, and the third term, proportional to $\mathbf{l}^2$, is introduced to lower the energy of a one-particle state at large values l of the orbital angular momentum. Since the majority of non-spherical nuclei have a shape close to an ellipsoid of revolution, we shall assume that the field of the nucleus is characterized by axial symmetry and $\omega_1 = \omega_2 \neq \omega_3$. As we know, in this case, if the volume of the nucleus remains constant we can describe its shape by introducing only one deformation parameter. In the Nilsson potential this parameter, denoted by δ, is defined as follows:

$$\omega_1^2 = \omega_2^2 = \omega^2\left(1 + \frac{2}{3}\delta\right), \qquad \omega_3^2 = \omega^2\left(1 - \frac{4}{3}\delta\right). \tag{5.194}$$

From the constancy of the nuclear volume, which is inversely proportional to the product of the frequencies ω_1, ω_2 and ω_3, it follows that the quantity ω is not constant and depends on the deformation parameter δ in the following way:

$$\omega \equiv \omega(\delta) = \omega(0)\left(1 - \frac{4}{3}\delta^2 - \frac{16}{27}\delta^3\right)^{-1/6}. \tag{5.195}$$

In the case of small deformations, the quantity δ is connected with the deformation parameter β introduced earlier, by the relation

$$\delta = \frac{3}{2}\sqrt{\frac{5}{4\pi}}\beta. \tag{5.196}$$

We find the total Hamiltonian H of the nucleon by adding the kinetic-energy operator $-\hbar^2\nabla^2/2M$ to (5.193). Separating out in the Hamiltonian the spherically symmetric part

$$H_0 = -\frac{\hbar^2}{2M}\nabla^2 + \frac{M\omega^2(\delta)}{2}r^2, \tag{5.197}$$

we represent the one-particle Hamiltonian in the form

$$H = H_0 + H_\delta + C\mathbf{l}\cdot\mathbf{s} + D\mathbf{l}^2, \tag{5.198}$$

where

$$H_\delta = -\frac{2}{3}\sqrt{\frac{4\pi}{5}}M\omega^2\delta r^2 Y_{20}(\mathbf{n}). \tag{5.199}$$

We note that the operators $\mathbf{j}^2$ and $\mathbf{l}^2$ of the squares of the total angular momentum and orbital angular momentum of the nucleon and the operators l_z and s_z of the components of the orbital and spin angular momenta along the z-axis (the symmetry axis of the field) do not commute with the operator H. We note that the spherically symmetric part H_0 also does not commute with the total Hamiltonian H. However, the operator $j_z = l_z + s_z$ of the component of the total angular momentum along the symmetry axis does commute with H, and so the states of a nucleon in the field (5.193) are characterized by well-defined values Ω of this component of the total angular momentum.

The next step is to diagonalize the one-particle Hamiltonian H. For this, we must choose a convenient representation and write the matrix of the Hamiltonian H in this representation. One usually chooses a representation in which the symmetric Hamiltonian H_0 is diagonal. In other words, for the basis functions in terms of which we shall expand the functions of a nucleon moving in a Nilsson potential and by means of which the matrix of the Hamiltonian H will be defined, we shall use the isotropic oscillator functions $\chi^N_{l\Lambda\Sigma}$ defined by the equation

$$H_0\chi^N_{l\Lambda\Sigma} = \left(N + \frac{3}{2}\right)\hbar\omega\chi^N_{l\Lambda\Sigma}, \tag{5.200}$$

where Λ and Σ are the projections of the orbital and spin angular momenta of the nucleon along the symmetry axis of the nucleus. Although the total one-particle Hamiltonian matrix $\langle N'l'\Lambda'\Sigma'|H|Nl\Lambda\Sigma\rangle$ calculated by means of these functions is not diagonal in any of the quantum numbers N, l, Λ and Σ, it is fairly easily calculated in this representation. In fact, two terms in the total Hamiltonian, H_0 and $D\mathbf{l}^2$, are diagonal in the representation chosen. The non-zero matrix elements of the operator $\mathbf{l}\cdot\mathbf{s}$ can be written as follows:

$$\langle N'l'\Lambda'\Sigma'|\mathbf{l}\cdot\mathbf{s}|Nl\Lambda\Sigma\rangle = \frac{1}{2}\delta_{N'N}\delta_{l'l}$$
$$\times\left\{\sqrt{(l\mp\Lambda)(l\pm\Lambda+1)}\,\delta_{\Lambda',\Lambda\pm1}\delta_{\Sigma',-\Sigma} \pm \Lambda\delta_{\Lambda'\Lambda}\delta_{\Sigma'\Sigma}\right\}. \tag{5.201}$$

The calculation of the matrix elements of H_δ is more cumbersome. The matrix element $\langle l'\Lambda'|Y_{20}|l\Lambda\rangle$ of the angular part of the operator is calculated using (4.70). To calculate the radial matrix elements $\langle N'l'|r^2|Nl\rangle$, we can make use of the explicit form (4.27) of the radial wave function of a particle in the field of an isotropic-oscillator potential. We give the result of the calculation of a more general matrix element using the functions (4.27):

$$\langle N'l'|r^n|Nl\rangle = \left[\frac{\Gamma\left(\frac{N-l+2}{2}\right)\Gamma\left(\frac{N'-l'+2}{2}\right)}{\Gamma\left(\frac{N+l+3}{2}\right)\Gamma\left(\frac{N'+l'+3}{2}\right)}\right]^{1/2} \tag{5.202}$$

$$\times\left(\frac{l-l'+n}{2}\right)!\left(\frac{l'-l+n}{2}\right)!$$

$$\times\sum_\sigma \frac{\Gamma\left(\frac{l+l'+n+3}{2}+\sigma\right)}{\sigma!\left(\frac{N-l}{2}-\sigma\right)!\left(\frac{N'-l'}{2}-\sigma\right)!\left(\frac{l'+n-N}{2}+\sigma\right)!\left(\frac{l+n-N'}{2}+\sigma\right)!}.$$

After matrix $\langle N'l'\Lambda'\Sigma'|H|Nl\Lambda\Sigma\rangle$ has been calculated, the problem reduces to diagonalizing this matrix. The diagonalization has been performed, by means of computers, for several values of the deformation parameter δ and a series of fixed values of the quantum numbers N and Ω. The numerical values of C and D were chosen in such a way as to reproduce the sequence of levels of the shell model for $\delta = 0$. (In the calculations for $\delta = 0$, the quantity ω was chosen in accordance with (4.31), and $C = -0\cdot 1\hbar\omega$.) The ratio $\mu = 2D/C$ was assumed equal to 0 for $N = 0, 1$ and 2, $0\cdot 35$ for $N = 3$, $0\cdot 45$ for $N = 4, 5$ and 6, and $0\cdot 40$ for $N = 7$. We denote the eigenvalues of the part of the Hamiltonian $H - H_0$ by $-(C/2)r_\alpha^{N\Omega}$, where the index α labels these eigenvalues. Then, taking (5.200) into account, we represent the eigenvalues of the one-particle Hamiltonian (5.198) in the form

$$E_\alpha^{N\Omega}(\delta) = \left(N+\frac{3}{2}\right)\hbar\omega(\delta) - (C/2)r_\alpha^{N\Omega}(\delta). \tag{5.203}$$

We denote the eigenfunction of the Hamiltonian H corresponding to the quantum numbers N, Ω and α by $\chi_\Omega^{N\alpha}$. This function is a superposition of the functions $\chi_{l\Lambda\Sigma}^N$:

$$\chi_\Omega^{N\alpha} = \sum_{l\Lambda} a_{l\Lambda}^{N\alpha}(\delta)\chi_{l\Lambda\Sigma}^N, \tag{5.204}$$

where $\Sigma = \Omega - \Lambda$. The superposition coefficients $a_{l\Lambda}^{N\alpha}(\delta)$, which are normalized so that

$$\sum_{l\Lambda} |a_{l\Lambda}^{N\alpha}(\delta)|^2 = 1,$$

are determined simultaneously with the energy eigenvalues in the numerical diagonalization of the Hamiltonian matrix. Tables of the expansion coefficients $a_{l\Lambda}^{N\alpha}(\delta)$ have been compiled, and curves of the single-particle energies $E_\alpha^{N\Omega}$ as functions of the deformation parameter δ have been obtained (Nilsson, 1955).

It can be seen from the relation (5.202) that the Hamiltonian matrix is not diagonal in the quantum number N; N' differs from N by an even number. However, in most cases, the off-diagonal matrix elements in the number N are considerably smaller in magnitude than the nearest matrix elements diagonal in$N$;therefore, the coupling between states with different N can be neglected, except in certain spacial cases. Thus, the one-particle states in the Nilsson potential can be characterized not only by the projection Ω of the total angular momentum but also by the quantum number N. In the case of sufficiently small deformations, when the term H_δ in the Hamiltonian H can be neglected, the operators $\mathbf{l}^2$ and $\mathbf{j}^2$ commute with H, and so the one-particle states can be characterized for $\delta \to 0$ by the set of quantum numbers N, l, j and Ω. For small δ the term H_δ can be taken into account by perturbation theory. With increasing deformation, the quantum numbers l and j lose their meaning. For sufficiently large deformations, when we can neglect the terms with $\mathbf{l} \cdot \mathbf{s}$ and $\mathbf{l}^2$ compared with H_δ in the Hamiltonian, we obtain in the limit the Hamiltonian of an anisotropic oscillator with axial symmetry. In this case, the one-particle states can be described by sets of quantum numbers N, n_3, Λ and Σ (in this case, Σ also has a well-defined value: $\Sigma = \Omega - \Lambda$). For large deformations δ, the terms containing $\mathbf{l} \cdot \mathbf{s}$ and $\mathbf{l}^2$ can be taken into account by perturbation theory. We note that the superposition (5.204) contains terms with values of l differing from each other and from the number N by an even number; therefore, the one-particle states are characterized by well-defined parity π.

As an illustration, we show in Fig. 5.15 the scheme of the levels as a function of the deformation parameter δ for the range $50 < N < 82$, in which many non-spherical nuclei are found. For $\delta > 0$ the nucleus is prolate along the symmetry axis, and for $\delta < 0$ the nucleus is oblate. Quantum numbers $l_j|\Omega|^\pi(Nn_3|\Lambda|)$ are associated with each curve in Fig. 5.15; of these, the projection Ω of the total angular momentum (there is degeneracy in the sign of Ω) and the parity π are good quantum numbers, irrespective of the magnitude of the deformation; the oscillator energy quantum number N is, in the sense indicated above, almost a good quantum number; the quantum numbers l and j are good quantum numbers only when $\delta \to 0$, while n_3 and Λ, on the other hand, are good quantum numbers for

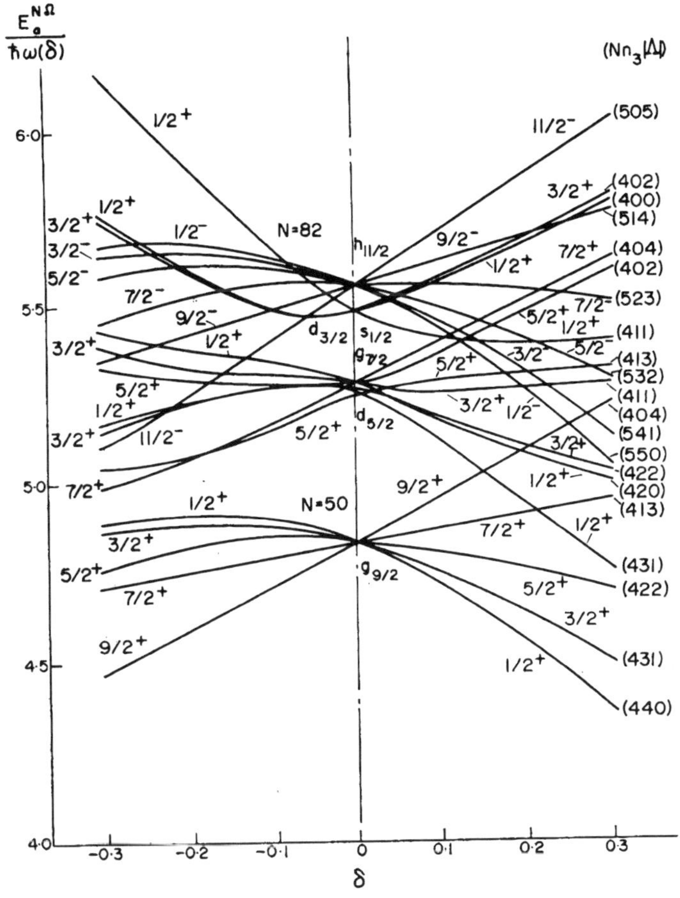

Figure 5.15. Scheme of the energy levels as a function of the deformation parameter δ for nuclei in the range $50 < N < 82$, according to the Nilsson model.

large deformations (asymptotic quantum numbers). For example, one of the curves in Fig. 5.15 is labelled in the following way: $g_{7/2}\frac{5}{2}^+(402)$, which means that $l = 4$ and $j = \frac{7}{2}$ when $\delta = 0$, $|\Omega| = \frac{5}{2}$ and $\pi = +1$, and $N = 4$, $n_3 = 0$ and $|\Lambda| = 2$. (The last two numbers have these values for large deformations δ.) We note that, since the parameters occurring in the one-particle potential H differ somewhat for a proton and a neutron, the level schemes and the sets of coefficients in the superposition (5.204) will also be different.

It is possible to calculate the total energy and the equilibrium deformation of nuclei on the basis of the above model. If we assume that only two-particle forces act between the nucleons, the total Hamiltonian of the nucleus,

$$H = \sum_i T_i + \frac{1}{2} \sum_{i \neq j} V_{ij},$$

can be expressed directly in terms of the Hamiltonian of a single nucleon,

$$H_i = T_i + \sum_{j(\neq i)} V_{ij},$$

in the following way:

$$H = \frac{1}{2} \sum_i (H_i + T_i). \tag{5.205}$$

The total wave function of the nucleus is an anti-symmetrized product of the wave functions (5.204) of the individual nucleons. Therefore, to find the total energy $E(\delta)$ of the nucleus as a function of the deformation parameter δ, we must calculate the mean value of (5.205) using the calculated one-particle functions. Then the equilibrium value δ_0 of the deformation parameter can be determined from the condition that the energy $E(\delta)$ be a minimum:

$$\partial E(\delta)/\partial \delta = 0. \tag{5.206}$$

Figure 5.16 shows the values of the *deformation parameter* δ calculated *in the Nilsson model*, and the experimental values in the region of the rare-earth elements.

The Nilsson model also makes it possible to justify the rule used to determine the spins of nuclei in the ground state. According to (5.203), each one-particle energy level is doubly degenerate ($\pm\Omega$). Neutrons and protons occupy the levels independently in pairs.If the number of identical particles is even, then $\Sigma\Omega_i = 0$; but if this number is odd, the total Ω is equal to the Ω_i of the last, unpaired, particle. If the number of neutrons (protons) is even and the number of protons (neutrons) is odd, then for the ground state the spin is equal to $I_0 = \Omega_p(\Omega_n)$. Thus, the spins of nuclei in the ground state are determined by the properties of the last odd nucleon, and are equal to $I_0 = \Omega = K$, except in the case $\Omega = \frac{1}{2}$. If $\Omega = \frac{1}{2}$, the spin I_0 of the ground state is determined by the magnitude of the coupling parameter a (Table 5.4).

The coupling parameter a, which is defined by the relation (5.165), can be expressed directly in terms of the coefficients $a_{l\Lambda}$ appearing in the formula (5.204):

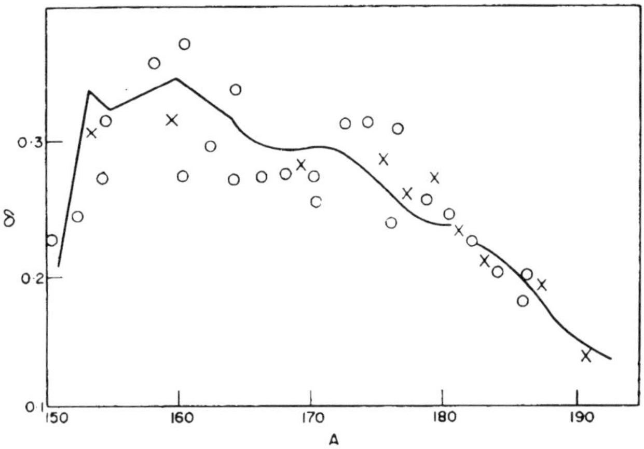

Figure 5.16. Experimental values (the points) of the deformation parameter δ and values calculated from Nilsson model (the solid curve) in the region of the rare-earths ($\times$ denotes odd nuclei and ○ denotes even-even nuclei).

TABLE 5.4.

Range of variation of the coupling parameter a	Ground-state spin I_0	Range of variation of the coupling parameter a	Ground-state spin I_0
$-14 < a < -10$	$\frac{11}{2}$	$4 < a < 8$	$\frac{5}{2}$
$-10 < a < -6$	$\frac{7}{2}$	$8 < a < 12$	$\frac{9}{2}$
$-6 < a < -1$	$\frac{3}{2}$	$12 < a < 16$	$\frac{13}{2}$
$-1 < a < 4$	$\frac{1}{2}$		

$$a = \sum_l (-1)^l (a_{l0}^2 + 2\sqrt{l(l+1)} a_{l0} a_{l1}). \tag{5.207}$$

In an analogous way,we can find the gyromagnetic ratio g_K and the coefficient b occurring in the expression (5.182) for the magnetic moment of the nucleus, in the case $I_0 = K = \frac{1}{2}$:

$$\left.\begin{aligned} g_K &= g_l + (1/2K)(g_s - g_l) \textstyle\sum_l (a_{l,K-1/2}^2 - a_{l,K+1/2}^2); \\ (g_K - g_R) b &= (g_s - g_l) \textstyle\sum_l (-1)^l a_{l0}^2 + (g_l - g_R) a. \end{aligned}\right\} \tag{5.208}$$

The calculated values of the magnetic moments for deformed odd nuclei agree with the experimental data.

Moments of inertia of nuclei. The most important property distinguishing non-spherical from spherical nuclei is the presence in non-spherical nuclei of rotational energy levels. Whereas the general features of the rotational spectra (for example, the ratios of the energies, the values of the spins and parities, the symmetry of the states, etc.) are entirely determined by the rotational nature of the levels, the absolute values of the rotational energies (that is, the magnitudes of the moments of inertia) depend essentially on the internal structure of the nuclei. [5]

We shall assume that an axially symmetric nucleus is rotating slowly with a definite frequency ω about an axis perpendicular to the symmetry axis of the nucleus. The angular momentum associated with the rotation of the nucleus is equal to

$$\hbar R = \omega J,$$

where J is the moment of inertia. To determine the *moment of inertia* J, it is necessary to make use of specific ideas on the internal structure of the nucleus. As limiting cases of the rotational model of the nucleus, we can consider the rotation of a solid and the potential motion of an ideal liquid in a rotating shell. The difference between these limiting cases is very clearly manifested in the example of the rotation of a spherical system. In the case of the solid, on going over to the spherically symmetric case the moment of inertia tends to a finite value, namely, the moment of inertia of a solid sphere. In the case of the hydrodynamic model, in the rotation of the spherical shell the velocity of each point on the surface is directed along a tangent, and the normal component of the velocity is equal to zero. Therefore, only a liquid at rest will satisfy the equations of motion of ideal hydrodynamics in such a vessel (since $v = 0$ satisfies both the equations of motion and the boundary conditions), and, consequently, the moment of inertia of the system will be equal to zero. If the shell has a non-spherical shape,the normal component of the velocity at the surface will be non-zero and the liquid will be dragged along as the shell rotates. The energy of rotation for a given angular velocity will be greater the more the shape of the shell differs from that of a sphere.

We shall represent the nucleus in the form of an ellipsoid of revolution, with semi-axes of length

[5] The relation between the internal structure of the nucleus and the nature of the collective motions is discussed in detail in the review by Moszkowski (1957).

$$R_1 = \left(1 - \frac{1}{2}\sqrt{\frac{5}{4\pi}}\beta\right)R, \qquad R_2 = \left(1 + \frac{1}{2}\sqrt{\frac{5}{4\pi}}\beta\right)R. \tag{5.209}$$

Here β is the deformation parameter. If the nucleus is regarded as a solid, the moment of inertia will be equal to

$$J_0 = \frac{m}{5}\left(R_1^2 + R_2^2\right) = \frac{2}{5}mR^2\left(1 + \frac{1}{2}\sqrt{\frac{5}{4\pi}}\beta + \frac{25}{32\pi}\beta^2\right), \tag{5.210}$$

where m is the mass of the nucleus.

We now determine the moment of inertia for the potential motion of a liquid in an ellipsoidal vessel having the shape of the nucleus. The potential motion of the liquid is described by the potential φ, satisfying the Laplace equation

$$\nabla^2\varphi = 0.$$

The velocity $\mathbf{v}$ of the liquid is given by the gradient of the potential:

$$\mathbf{v} = \operatorname{grad}\varphi.$$

In the case of an ideal liquid, the boundary conditions reduce to the requirement that the normal component of the velocity of the liquid at the surface coincide with the normal component of the velocity of the vessel wall.

The velocity of a point with position vector $\mathbf{r}$ at the surface of the rotating vessel is equal to $\boldsymbol{\omega} \times \mathbf{r}$. Noting that the normal to the surface of the ellipsoid is characterized by the components

$$\mathbf{N} = (x/R_1^2, y/R_1^2, z/R_1^2)$$

(if the symmetry axis is directed along the axis Oz), we write the condition that the normal components of the velocities $\mathbf{v}$ and $\boldsymbol{\omega} \times \mathbf{r}$ be equal at the surface in the form

$$\frac{x}{R_1^2}\left(\frac{\partial\varphi}{\partial x} - \omega z\right) + \frac{y}{R_1^2}\frac{\partial\varphi}{\partial y} + \frac{z}{R_2^2}\left(\frac{\partial\varphi}{\partial z} + \omega x\right) = 0. \tag{5.211}$$

This condition will be satisfied if for the solution of the Laplace equation we choose

$$\varphi = Axz, \tag{5.212}$$

where A is some constant. Substituting (5.212) into (5.211) we find the value of this constant:

$$A = \frac{R_2^2 - R_1^2}{R_2^2 + R_1^2}\omega. \tag{5.213}$$

We shall calculate the kinetic energy of the liquid in the rotating shell. This energy is equal to

$$T_r = (\varrho/2)\int d^3r v^2 = (\varrho/2)A^2\int d^3r(x^2 + z^2),$$

where ϱ is the density of the liquid and the integration is performed over the volume of the ellipsoid. Substituting (5.213) for A and performing the integration, we find

$$T_r = \frac{m}{5}\frac{(R_2^2 - R_1^2)^2}{R_2^2 + R_1^2}\frac{\omega^2}{2}, \tag{5.214}$$

where m is the mass of the liquid. The coefficient at $\omega^2/2$ in (5.214) must be regarded as the *moment of inertia in the hydrodynamic model of the nucleus* :

$$J = \frac{m}{5}\frac{(R_2^2 - R_1^2)^2}{R_2^2 + R_1^2}. \tag{5.215}$$

Using the expressions (5.209) for R_1 and R_2, we can rewrite the moment of inertia in the hydrodynamic model in the form

$$J = \frac{9mR^2}{4\pi}\frac{\beta^2\left(1 + \frac{1}{4}\sqrt{\frac{5}{4\pi}}\beta\right)^2}{2 + \sqrt{\frac{5}{4\pi}}\beta + \frac{25}{16\pi}\beta^2}. \tag{5.216}$$

In the case of small deformations, the ratio of the moments of inertia (5.216) and (4.210) is equal to

$$\frac{J}{J_0} = \frac{45}{16\pi}\beta^2. \tag{5.217}$$

Figure 5.17 (Moszkowski, 1957) depicts the velocity field in the rotation of a solid and in the potential motion of an ideal liquid in a rigid rotating shell. Whereas in the case of a solid the entire system rotates as a whole, in the rotation of a rigid shell filled with an ideal liquid the latter is dragged by the walls only near the surface of the shell.

Figure 5.18 shows the dependence of the moments of inertia on the magnitude of the deformation parameter for some even-even rare-earth nuclei.

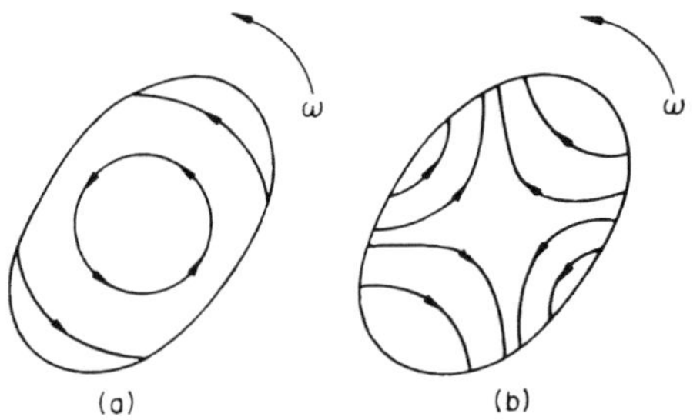

Figure 5.17. Velocity field in the rotation of a solid (a) and in the potential motion of an ideal liquid in a rigid rotating shell (b).

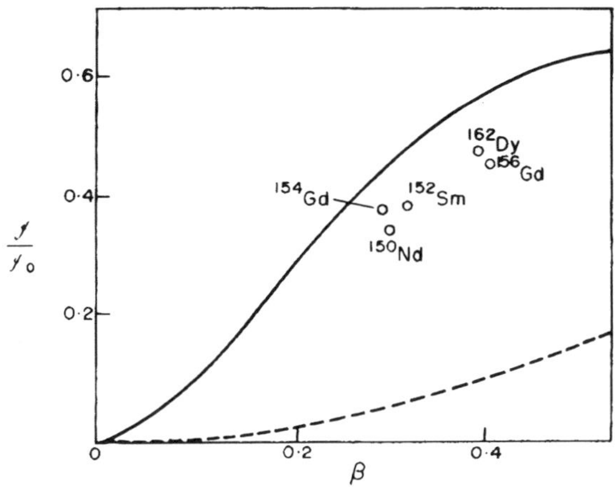

Figure 5.18. Dependence of the moments of inertia J/J_0 on the magnitude of the deformation parameter for even-even rare-earth nuclei, obtained on the basis of the superfluid model (solid curve) and the hydrodynamic model (dashed curve).

As can be seen from Fig. 5.18, for small deformations β the experimentally observed moments of inertia are considerably smaller than the values predicted by the solid model. They are considerably greater, however, than the values predicted by the hydrodynamic model.

The moment of inertia of the nucleus can be calculated more systematically on the basis of the so-called *cranking model* proposed by Inglis (1954, 1956). For this, we consider the motion of a system of particles in a non-spherical potential well that is slowly rotating about an axis that is not coincident with the symmetry axis. The Schrödinger equation describing the motion of the system has the form

$$-\frac{\hbar}{i}\frac{\partial\psi}{\partial t} = H(t)\psi, \tag{5.218}$$

where

$$H(t) = T + V(\mathbf{r}, t)$$

(T is the kinetic energy and $V(\mathbf{r}, t)$ is the potential energy, which varies slowly with time in the stationary coordinate frame). If the rotation proceeds slowly compared with the internal motion, we may speak of the energy of rotation of a system in an unchanged internal state.

We introduce the rotating coordinate frame rigidly attached to the nucleus. We choose the axis Oz along the symmetry axis of the nucleus, and let the rotation occur about the axis Ox. Then, in going over to the rotating coordinate frame, the function $\psi = \psi(\mathbf{r})$ is transformed into the function $\psi' \equiv \psi(\mathbf{r}')$, which, according to (5.48), is connected with ψ by the relation

$$\psi' = \mathrm{e}^{-i\Theta j_x}\psi, \tag{5.219}$$

where $\Theta = \omega t$. The Hamiltonian H of the system is transformed correspondingly into H':

$$H' = \mathrm{e}^{-i\Theta j_x} H \mathrm{e}^{i\Theta j_x}.$$

Since H' is the Hamiltonian of the internal motion in the proper coordinate frame, it does not depend on time. We denote the eigenvalues and eigenfunctions of H' by E_i and χ_i, so that

$$H'\chi_i = E_i\chi_i. \tag{5.220}$$

We find the equation for the function ψ' defined in the rotating coordinate frame. For this, we express the function ψ in terms of ψ' and substitute the result in (5.218). As a result, we obtain

$$-(\hbar/i)(\partial\psi'/\partial t) = \tilde{H}\psi', \tag{5.221}$$

where

$$\tilde{H} = H' + \hbar\omega j_x. \tag{5.222}$$

We introduce the notation

$$\psi' = \tilde{\chi}\mathrm{e}^{-(i/\hbar)\tilde{E}t}. \tag{5.223}$$

Assuming the frequency ω to be sufficiently small, we can calculate the quantity $\tilde{E}$ and the function $\tilde{\chi}$ by means of perturbation theory. Since the diagonal matrix element of j_x is equal to zero, we have

$$\left.\begin{aligned}\tilde{E}_i &= E_i - \hbar^2\omega^2 \textstyle\sum_{j\neq i} \frac{|<j|j_x|i>|^2}{E_j - E_i};\\ \tilde{\chi}_i &= \chi_i - \textstyle\sum_{j\neq i} \frac{<j|j_x|i>}{E_j - E_i}\chi_j.\end{aligned}\right\} \tag{5.224}$$

Knowing the function (5.223), it is not difficult to calculate the energy of the system in the stationary coordinate frame:

$$\begin{aligned}(\psi_i, H\psi_i) &= (\psi_i', H'\psi_i') = \left(\tilde{\chi}_i, (\tilde{H} - \hbar\omega j_x)\tilde{\chi}_i\right)\\ &= E_i + \hbar^2\omega^2 \sum_{j\neq i} \frac{|<j|j_x|i>|^2}{E_j - E_i}.\end{aligned} \tag{5.225}$$

It is clear that the correction to the internal energy E_i must be regarded as the energy of rotation, the moment of inertia being defined as twice the coefficient of ω^2. Thus, we obtain the following formula:

$$J = 2\hbar^2 \sum_{j\neq i} \frac{|<j|j_x|i>|^2}{E_j - E_i}. \tag{5.226}$$

If we use as the internal wave functions the shell functions of independent particles moving in a non-spherical well (for example, in an anisotropic oscillator well), we obtain the same moment of inertia as in the rotation of a solid. This is connected with the fact that the one-particle states of independent (uncorrelated) particles do not change during slow adiabatic rotation of the potential well, and, consequently, a collective rotation of the whole systems occurs, as in the case of a solid. The deviations of the observed values of the moments of inertia of nuclei from the values for the solid model are due to the presence of correlation between the nucleons. The residual interaction between the nucleons (which is neglected in the independent-particle model) leads to correlation between the nucleons, which tends to slow down the collective rotation and reduce the moment of inertia of the system. If the interaction between the nucleons were so strong that the mean free path was small compared with the dimensions of the nucleus, the hydrodynamic model would be applicable and the collective motion would become potential. Correlation between the nucleons due to pairing forces is taken into account in the superfluid model of the nucleus. A systematic method for calculating the moments of inertia of nuclei on the basis of the superfluid model has been developed by Migdal (1959).

5.5. Rotation of Non-axial Nuclei

The non-axial rotator model. A detailed comparison of the observed excitation spectra of deformed nuclei with the predictions of the rotational model in which the nucleus is regarded as an axially symmetric rotator has shown that, although the rotational model does reproduce well the features of the rotational spectra for many nuclei, there is, nevertheless, disagreement between the observed picture and the predictions of the simple rotational model for certain nuclei. Attempts to explain the discrepancy by including corrections for the non-adiabaticity of the rotation have been unsuccessful. Davydov and Filippov (1958) have shown that the difficulties can be removed if the assumption of axial symmetry of the deformed nuclei is discarded. They developed a theory of *non-axial nuclei*, in which the fundamental assumption is that the deformed nucleus is a triaxial ellipsoid whose rotation determines the spectrum of the collective excitations of the nucleus. This theory has made it possible to describe the principal features of the spectra of the excited states of most even-even nuclei.

In the adiabatic approximation, when the rotational motion can be separated from the vibrations of the nuclear surface and from the one-particle motion, the shape of the nucleus can be assumed fixed; then the Hamiltonian describing the rotation of an even-even nucleus can be written in the form

$$H_r = \sum_{k'} \frac{\hbar^2}{2J_{k'}} I_{k'}^2, \tag{5.227}$$

where $I_{k'}$ are the projections of the total angular-momentum operator along the principal axes of the ellipsoid and $J_{k'}$ are the moments of inertia about the principal axes:

$$J_{k'} = 4B\beta^2 \sin^2\left(\gamma - \frac{2\pi}{3}k'\right)$$

(the three moments of inertia of the nucleus are assumed to be all different). The rotational-energy operator (5.227) acts only on the Euler angles and is the Hamiltonian operator of a rigid non-axial rotator (asymmetric top). Before finding the solutions of the Schrödinger equation for the asymmetric top and the corresponding well-defined values I and M of the total angular momentum and its projection along the axis Oz respectively,

$$(H_r - E_{I\tau})\Phi_{IM}^{\tau}(\Phi, \Theta, \psi) = 0, \qquad E_{I\tau} \equiv \frac{\hbar^2}{4B\beta^2}\epsilon_{I\tau} \tag{5.228}$$

TABLE 5.5.

D_2	e	$R_1(\pi)$	$R_2(\pi)$	$R_3(\pi)$	D_2	e	$R_1(\pi)$	$R_2(\pi)$	$R_3(\pi)$
A	1	1	1	1	B_2	1	-1	1	-1
B_1	1	-1	-1	1	B_3	1	1	-1	-1

(τ are the additional quantum numbers), we shall ascertain which conditions, stemming from the symmetry properties of the Hamiltonian H_r, must be imposed on the wave functions $\Phi^{\tau}_{IM}(\Phi, \Theta, \psi)$ of the asymmetric top.

The operator H_r and the commutation relations (5.134) for the projections $I_{k'}$ of the total angular momentum along the principal axes are invariant under the transformations of the group D_2 consisting of the three rotations through an angle of 180° about the axes Ox', Oy' and Oz' ($R'_j(\pi) = e^{i\pi I_{j'}}$) and the identity element e. The invariance is easily verified, if we make use of the relations

$$R_{j'}(\pi) I_{k'} R_{j'}^{-1}(\pi) = (2\delta_{j'k'} - 1) I_{k'} \tag{5.229}$$

(in each rotation $R_{j'}(\pi)$, two of the three operators $I_{j'} (j' = 1, 2, 3)$ change their sign), from which it follows immediately that the Hamiltonian commutes with all the operations of the group D_2. Since each element of the group D_2 forms a class of the group, and the number of classes of a group is equal to the number of irreducible representations, there are in all four one-dimensional irreducible representations of the group D_2, which we denote by the symbols A, B_1, B_2, B_3. In Table 5.5 we give their characters (which coincide in this case with the one-dimensional matrices of the irreducible representations); these can be used to determine the eigenfunctions of the operator H_r.

It is well known that there is a close connection between the eigenfunctions of the stationary states and the irreducible representations of a group of transformations which commute with the Hamiltonian. The energy states of the system can be classified according to the irreducible representations corresponding to these states. It is clear that the degeneracy of an energy level coincides with the dimensionality of the representation, that is, with the number of basis functions of the representation. All these functions belong to the same energy value. Thus, knowing the irreducible representations (of the symmetry group) according to which the stationary-state wave functions transform, we can find certain properties of these functions without solving the Schrödinger equation.

In the case under consideration, all the irreducible representations of the group D_2 are one-dimensional, and so the energy states of the corresponding physical system are non-degenerate. Since there are four irreducible representations in all, the stationary-state wave functions can be classified by their symmetry properties into four types. According to the character Table 5.5, those states which belong to the irreducible representation A are unchanged under the action of any of the operations of the group D_2. These states are called *fully symmetric states* , and the representation A corresponding to them is called the *fully symmetric representation* . Usually, the ground state of the system belongs to the fully symmetric representation.

The wave functions of the collective motions in even-even nuclei depend only on the spatial coordinates of the nucleons. Since the spins of two identical nucleons in the same one-particle (orbital) state have opposite orientations, the coordinate wave function will be symmetric. Inasmuch as the spatial coordinates of the nucleons are interchanged by the operations $R_{j'}(\pi)$, the wave function describing the rotation of the nucleus must belong to the fully symmetric representation A of the group D_2, since only in this case will the wave function of the collective motions be transformed into itself, with the same sign, under any operation of the group D_2.

It is easily verified that the functions (5.149) for a rigid symmetric top (with even K) are invariant under the rotations $R_{j'}(\pi)$, if we take into account that the operators $R_{j'}(\pi)$ act on the generalized spherical harmonics $D^I_{MK}(\Phi, \Theta, \psi)$ in the following way:

$$R_1(\pi) D^I_{MK} = (-1)^K D^I_{MK}; \qquad R_2(\pi) D^I_{MK} = (-1)^{I-K} D^I_{M,-K};$$
$$R_3(\pi) D^I_{MK} = (-1)^I D^I_{M,-K}.$$

(We note that, for any K, the transformed function $\Psi(IKM)$ of the symmetric top will differ from (5.149) only in the replacement of the factor $(-1)^I$ in the second term by the factor $(-1)^{I+K}$, and so the functions $\Psi(IKM)$ with odd K do not possess the above-mentioned invariance.) Taking this fact into account, we can represent the solutions of eqn. (5.228) which belong to the fully symmetric representation A of the group D_2 in the form

$$\Phi^{\tau}_{IM}(\Phi, \Theta, \psi) = \sum_{\text{even } K \geq 0} g^{\tau}_{IK} \Psi(IKM). \tag{5.230}$$

Substituting (5.230) into eqn. (5.228), we obtain a system of linear homogeneous equations for the coefficients g^{τ}_{IK}:

$$\sum_{K' \geq 0} \left\{ \langle IKM | H_r | IK'M \rangle - E_I \delta_{KK'} \right\} g^{\tau}_{IK'} = 0. \tag{5.231}$$

The number of these equations coincides with the number of possible different (but even!) values of K and is equal to $\frac{1}{2}I+1$ for even I and $I-\frac{1}{2}$ for odd I. By equating the determinant of the system of equations to zero, we find the energy eigenvalues $E_{I\tau}$.

As an example, we shall consider the state with angular momentum (spin) $I = 2$. The possible values of K are 0 and 2, and so the system (5.231) will consist of two equations. Using the relations (5.90) and (5.91), it is not difficult to determine the matrix elements $< IKM|H_r|IK'M >$ of the Hamiltonian operator; these can be non-zero only when $K' = K$ and when $K' = K \pm 2$. Thus, in the case $I = 2$, we obtain the system

$$\left.\begin{aligned} \left[\frac{3}{2}(a_1+a_2)-E_2\right]g_{20}+\frac{\sqrt{3}}{2}(a_1-a_2)g_{22}=0; \\ \frac{\sqrt{3}}{2}(a_1-a_2)g_{20}+\left[\frac{1}{2}(a_1+a_2+4a_3)-E_2\right]g_{22}=0, \end{aligned}\right\} \qquad (5.232)$$

where

$$a_{k'} = \frac{\hbar^2}{J_{k'}}.$$

From the condition that the determinant vanish, we find the two energy values:

$$\left.\begin{aligned} E_{2\tau} = {} & (a_1+a_2+a_3) \\ & +(-1)^\tau\sqrt{(a_1+a_2+a_3)^2-3(a_1a_2+a_2a_3+a_1a_3)}. \\ & \qquad\qquad (\tau = 1,2) \\ E_{2\tau} = {} & \frac{\hbar^2}{4B\beta^2}\,\frac{3[3+(-1)^\tau\sqrt{9-8\,\sin^2 3\gamma}]}{\sin^2 3\gamma}. \end{aligned}\right\} \qquad (5.233)$$

According to (5.232) and (5.233), the coefficients g^τ_{IK}, which are orthonormalized by the conditions

$$\sum_{K\geq 0} g^\tau_{IK} g^{\tau'}_{IK} = \delta_{\tau\tau'}; \qquad \sum_\tau g^\tau_{IK} g^\tau_{IK'} = \delta_{KK'}, \qquad (5.234)$$

will be equal to (for $I = 2$)

$$g^{\tau=1}_{2,K} = \sqrt{\frac{1}{2}+(-1)^{K/2}\frac{\sin\gamma\sin 3\gamma+3\cos\gamma\cos 3\gamma}{2\sqrt{9-8\sin^2 3\gamma}}},$$

$$g_{2,K}^{\tau=2} = (-1)^{(2-K)/2} g_{2,2-K}^{\tau=1}, \qquad K = 0, 2. \tag{5.235}$$

For $I = 3$, only the value $K = 2$ is possible; therefore $g_{3,2}^{\tau=1} = 1$ and $\Phi_{3M}^{\tau=1}(\Theta_i) = \Psi(3, 2, M)$. The value of the energy is

$$E_{3,1} = \langle 3, 2, M|H_r|3, 2, M\rangle = 2(a_1 + a_2 + a_3) = \frac{\hbar^2}{4B\beta^2} \frac{18}{\sin^2 3\gamma}. \tag{5.236}$$

(The state with $I = 3$ and $K = 0$ is absent, since $\Psi(3, 0, M) = 0$.)

Principal features of the rotational spectra of non-axial nuclei. The rotational-energy spectrum od axially symmetric nuclei ($\gamma = 0$ or $\pi/3$) is associated with the rotation of the nucleus about an axis perpendicular to the symmetry axis of the nucleus. In non-axial nuclei, however, the rotations are more diverse, and the corresponding energy spectrum is therefore more complicated. For $\gamma \to 0$ or $\gamma \to \pi/3$, some of the levels of non-axial nucleus remain finite and go over into levels of the axially symmetric nucleus. Such levels of a non-axial nucleus form the principal rotational band in the rotational-energy spectrum. The other rotational levels of a non-axial nucleus tend to infinity as $\gamma \to 0$ or $\gamma \to \pi/3$. These levels form anomalous rotational bands.

The ratio of any two *rotational energies of a non-axial nucleus* depends only on the one parameter γ, which can be determined for a given nucleus from the ratio of the observed energies of the two lowest excited states. Knowing γ, we can predict the positions of the other energy levels in the rotational spectrum of non-axial nuclei.

The parameter γ can be found by making use of the ratio of the experimental energy values for the first two spin-2 levels, belonging to the principal and anomalous rotational bands.According to (5.233), this ratio is equal to

$$\frac{E_{2,2}}{E_{2,1}} = \frac{3 + \sqrt{9 - 8\sin^2 3\gamma}}{3 - \sqrt{9 - 8\sin^2 3\gamma}}. \tag{5.237}$$

As the parameter γ changes from 0 to 30°, this ratio decreases monotonically from ∞ to 2. By measuring the ratio (5.237) experimentally, we can calculate γ; knowing γ, it is not difficult to determine the spins and positions of the levels in the principal and anomalous rotational bands.

We shall indicate the following characteristic features of the *spectrum of the excited states of a non-axial rotator* :

1. Deviation of the shape of the nucleus from an axially symmetric shape leads to violation of the interval rule (5.168) in the principal rotational

band. For example, if $\gamma = 30°$, the levels of the principal rotational band will satisfy the following interval rule:

$$E_2 : E_4 : E_6 : E_8 : \ldots = 1 : 2 \cdot 67 : 5 : 8 : \ldots , \qquad (5.238)$$

which is substantially different from (5.168).

2. The presence of non-axiality in nuclei leads not only to a change of the interval rule for the levels of the principal rotational bands but also to the appearance of new (anomalous) rotational bands. The first anomalous band consists of levels with spins 2,3,4,5,..., the second consists of levels with spins 4,5,6,7,..., and so on. If $\gamma > 23°$, the first anomalous-band level with spin 2 lies below the level with spin 4 in the principal rotational band. For the rigid non-axial rotator, the anomalous bands are associated with rotation, although in reality (in a non-adiabatic treatment) they correspond to excitations of a rotational-vibrational character.

As an illustration, we show in Fig. 5.19 the ratio of the excited-state energies to the energy of the first excited state as, a function of the non-axiality parameter γ for a non-axial rotator.

The solid curves correspond to the principal rotational band, and the dashed curves to the first anomalous band. The circles indicate the experimental energy ratios for the nuclei ${}^{166}_{68}Er$, ${}^{186}_{76}Os$, ${}^{190}_{76}Os$ and ${}^{192}_{76}Os$. The experimental spin values are indicated alongside the circles. If, for example, $\gamma = 22°$ for the nucleus ${}^{190}_{76}Os$, the calculated energy ratios in the principal rotational band

$$1 : 3 \cdot 01 : 5 \cdot 76 : 9 \cdot 24$$

will be very close to the experimentally observed ratios

$$1 : 2 \cdot 99 : 5 \cdot 61 : 8 \cdot 90.$$

The value $\gamma = 22°$ also explains the positions of the second spin-2 level and the spin-3 level, both of which belong to the anomalous band.

The non-axial rotator model enables us to determine not only the energies of the excited states but also the wave functions. In the general case, the wave functions (5.230) for the non-axial rotator are represented in the form of superpositions of the states (5.149), which correspond to well-defined values of the projection of the total angular momentum along the axis Oz. The coefficients g^{τ}_{IK} determining the contributions of states with different K depend on the non-axiality parameter γ and are determined by the system of algebraic equations (5.231). If $\gamma < 15°$, the main contribution in the sum (5.230) is made by only one term, and, consequently, K is approximately a constant of the motion. In this approximation, the value $K = 0$ corresponds to the principal rotational band; therefore, the rotation

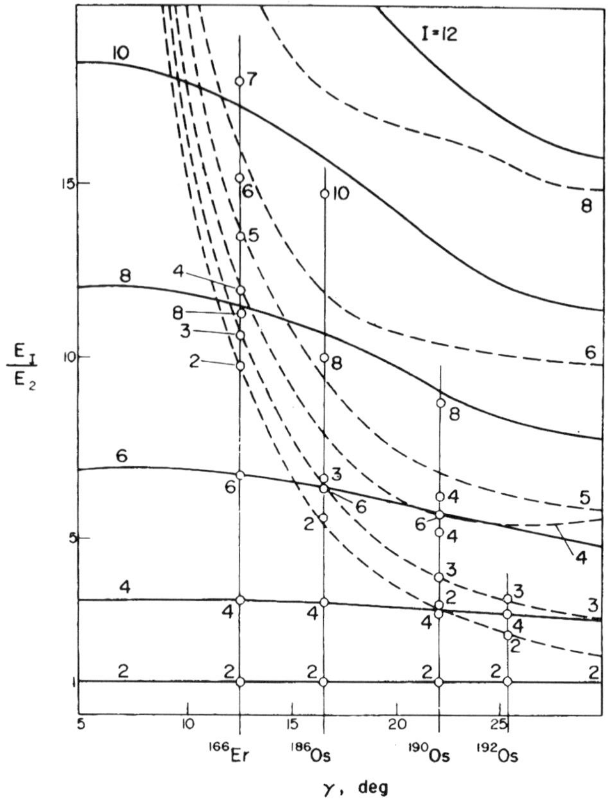

Figure 5.19. Ratios of the energies of the excited states to the energy of the first excited state as a function of the non-axially parameter γ for a non-axial rotator.

occurs about an axis perpendicular to Oz'. For $\gamma > 15°$, one value $K = 2$ corresponds to the first anomalous band. The lowest level $I = 2$ of this band corresponds to rotation of the nucleus about the axis Oz'. For large spin values, a rotation about an axis perpendicular to Oz' is superimposed on the rotation about the axis Oz'. If $\gamma > 15°$, the number K no longer has the meaning of a constant of the motion. The wave function in this case is a superposition of states with different values of K.

5.6. Rotational-vibrational Excitations of Nuclei

General form of the wave equation in the case of quadrupolar deformations. In the adiabatic theory of collective excitations, considered earlier, the coupling between the rotation of the nucleus and the vibrations of its surface was neglected. This neglect is not always justified, since in many cases the energies of the rotational and vibrational excitations are of the same order. We shall consider the *non-adiabatic theory of rotational-vibrational excitations in nuclei* . [6]

When only second-order deformations ($\lambda = 2$) are taken into account, the collective motions in the nucleus are characterized by five degrees of freedom. As the generalized coordinates, it is convenient to choose the three Euler angles Φ, Θ and ψ, specifying the orientation of the deformed nucleus, and two internal parameters β and γ, defining its shape. In order to write the wave function in these variables, it is necessary to transform from the classical energy of the collective motions to the quantum Hamiltonian operator acting in the space of the five variables $\Phi, \Theta, \psi, \beta$ and γ.

The potential-energy operator $V(\beta, \gamma)$, which depends only on the coordinates β and γ, retains its original form in the quantum case. It is elementary to quantize the part of the Hamiltonian associated with the rotation, since the projections of the angular momentum on to the coordinate axes moving with the nucleus and the corresponding angles of rotation are ordinary canonically conjugate operators. Therefore, the quantum Hamiltonian of rotation H_r is obtained from the classical rotational energy (5.111) by simple replacement of the angular-momentum components $R_{k'}$ by the operators of the angular-momentum components along the principal axes of the ellipsoid:

$$H_r = \sum_{k'} \frac{\hbar^2}{2J_{k'}} R_{k'}^2, \qquad J_{k'} = 4B\beta^2 \sin^2\left(\gamma - \frac{2\pi}{3}k'\right). \tag{5.239}$$

In the case of rotation of even-even nuclei, for which the internal angular momentum $J = 0$, the rotational angular momentum $\mathbf{R}$ coincides with the total angular momentum $\mathbf{I}$ of the system:

$$\mathbf{R} \to \mathbf{I}.$$

The transition to the quantum description of the kinetic energy (5.105) of the surface vibrations of nuclei is considerably more complicated.

The total kinetic energy of the system, expressed in terms of the generalized coordinates q_i (the index i labels all the degrees of freedom of the system), is usually written in the form

$$T = (B/2) \sum_{ij} g_{ij} \dot{q}_i \dot{q}_j, \tag{5.240}$$

[6]The non-adiabatic theory of collective excitations in nuclei was developed by Davydov (1959, 1960, 1967).

where B is the mass of the system and g_{ij} are symmetric coefficients which can be arbitrary functions of the coordinates. We introduce the square of the line differential in the space of the generalized coordinates:

$$ds^2 = \sum_{ij} g_{ij} dq_i dq_j; \tag{5.241}$$

then the kinetic energy can be represented in the form

$$T = (B/2)(ds/dt)^2. \tag{5.242}$$

If the square of the line differential is known, the Laplacian operator ∇^2 acting in the space of the generalized coordinates is defined in the following way (Pauli, 1933):

$$\nabla^2 = \frac{1}{\sqrt{g}} \sum_{ij} \frac{\partial}{\partial q_i} \left(\sqrt{g} g_{ij}^{-1} \frac{\partial}{\partial q_j} \right), \tag{5.243}$$

where g is the determinant of the matrix g_{ij}, and g_{ij}^{-1} is the inverse matrix of the matrix g_{ij}. Thus, the kinetic energy operator can be expressed in terms of the Laplacian operator as follows:

$$T = -(\hbar^2/2B)\nabla^2. \tag{5.244}$$

According to (5.105) and (5.107), we can write the total nuclear kinetic energy associated with quadrupolar deformations in the form

$$\begin{aligned} T &\equiv T_r + T_v = (B/2)(ds/dt)^2; \\ ds^2 &= \sum_{k'} (J_{k'}/B) d\varphi_{k'}^2 + d\beta^2 + \beta^2 d\gamma^2; \end{aligned} \tag{5.245}$$

$d\varphi_{k'} = \omega_{k'} dt$ are the angles of rotation about the principal axes Ox', Oy', and Oz' of the ellipsoid in the time interval dt. In the general case, $\beta_0 \neq 0$ and $\gamma_0 \neq 0$. If $\omega_{k'}$ is expressed in terms of derivatives of the Euler angles, in accordance with (5.56), then ds^2 will be the square of the line differential in the five-dimensional space of the variables $\Phi, \Theta, \psi, \beta$ and γ . It is simpler, however, to obtain the explicit form of the Laplacian operator directly in the space of the variables $\varphi_{x'}, \varphi_{y'}, \varphi_{z'}, \beta$ and γ. In fact, in this case, the matrix g_{ij} determining the square (5.245) of the line differential has diagonal form. We denote $\varphi_{x'}, \varphi_{y'}, \varphi_{z'}, \beta$ and g, respectively, by q_i, where i=1,2,3,4 and 5. According to (5.245), we have

$$g_{ij} = g^{(i)} \delta_{ij}, \tag{5.246}$$

where

$$g^{(1)} = J_{x'}/B, \quad g^{(2)} = J_{y'}/B; \quad g^{(3)} = J_{z'}/B, \quad g^{(4)} = 1, \quad g^{(5)} = \beta^2.$$

Therefore, the inverse matrix g_{ij}^{-1} is also diagonal:

$$g_{ij}^{-1} = {g^{(i)}}^{-1}\delta_{ij}. \tag{5.247}$$

It is not difficult to verify that the determinant of the matrix (5.246) is equal to

$$g = 4\beta^8 \sin^2 3\gamma. \tag{5.248}$$

Thus, in accordance with (5.243), the Laplacian operator in the five-dimensional space of the coordinates $\varphi_{x'}, \varphi_{y'}, \varphi_{z'}, \beta$ and γ takes the form

$$\nabla^2 = \sum_{k'} \frac{B}{J_{k'}} \frac{\partial^2}{\partial \varphi_{k'}^2} + \frac{1}{\beta^4} \frac{\partial}{\partial \beta} \beta^4 \frac{\partial}{\partial \beta} + \frac{1}{\beta^2} \frac{1}{\sin 3\gamma} \frac{\partial}{\partial \gamma} \sin 3\gamma \frac{\partial}{\partial \gamma}. \tag{5.249}$$

Since the derivatives with respect to the angles $\varphi_{k'}$ determine directly the components of the angular-momentum operator $\mathbf{R}$ in the proper coordinate frame, we can rewrite (5.249) in the form

$$\nabla^2 = -\sum_{k'} \frac{B}{J_{k'}} \mathbf{R}_{k'}^2 + \frac{1}{\beta^4} \frac{\partial}{\partial \beta} \beta^4 \frac{\partial}{\partial \beta} + \frac{1}{\beta^2} \frac{1}{\sin 3\gamma} \frac{\partial}{\partial \gamma} \sin 3\gamma \frac{\partial}{\partial \gamma}. \tag{5.250}$$

The first three terms in the right-hand side of (5.250) characterize the rotational part of the Laplacian operator, and the last two characterize the vibrational part.

Using (5.244), we obtain the operator of the total kinetic energy in the form

$$T = \sum_{k'} \frac{\hbar^2}{2J_{k'}} \mathbf{R}_{k'}^2 - \frac{\hbar^2}{2B} \left[\frac{1}{\beta^4} \frac{\partial}{\partial \beta} \beta^4 \frac{\partial}{\partial \beta} + \frac{1}{\beta^2} \frac{1}{\sin 3\gamma} \frac{\partial}{\partial \gamma} \sin 3\gamma \frac{\partial}{\partial \gamma} \right]. \tag{5.251}$$

Thus, the quantum Hamiltonian operator for the collective motions of the nucleus with second-order deformation taken into account has the form

$$H = \sum_{k'} \frac{\hbar^2}{2J_{k'}} \boldsymbol{I}_{k'}^2 - \frac{\hbar^2}{2B} \left[\frac{1}{\beta^4} \frac{\partial}{\partial \beta} \beta^4 \frac{\partial}{\partial \beta} + \frac{1}{\beta^2} \frac{1}{\sin 3\gamma} \frac{\partial}{\partial \gamma} \sin 3\gamma \frac{\partial}{\partial \gamma} \right] + V(\beta, \gamma). \tag{5.252}$$

Here, the components of the operator **R** have been replaced by the components of the operator **I**. We write the corresponding wave equation for the stationary states:

$$H\Psi(\Phi, \Theta, \psi, \beta, \gamma) = E\Psi(\Phi, \Theta, \psi, \beta, \gamma). \tag{5.253}$$

The eigenfunctions $\Psi(\Phi, \Theta, \psi, \beta, \gamma)$ of the operator H are studied in the space of the five variables
$\Phi, \Theta, \psi, \beta$ and γ, which vary in the ranges $0 \leqslant \Phi \leqslant 2\pi$, $0 \leqslant \Theta \leqslant \pi$, $0 \leqslant \psi \leqslant 2\pi$, $0 \leqslant \beta < \infty$, and $0 \leqslant \gamma \leqslant 2\pi$. (Sometimes a narrower range is chosen; for example, for a spherical nucleus and small surface vibrations , $0 \leqslant \gamma \leqslant \pi/3$.) The functions $\Psi(\Phi, \Theta, \psi, \beta, \gamma)$ are normalized by the condition

$$\int d\tau \Psi^* \Psi = 1, \tag{5.254}$$

where the volume element $d\tau$, which is proportional to $\sqrt{g}$, is equal to

$$d\tau = \beta^4 |\sin 3\gamma| \sin \Theta \, d\Phi \, d\Theta \, d\psi \, d\beta \, d\gamma. \tag{5.255}$$

The expression (5.252) is valid in the case of small vibrations of the nuclear surface about the equilibrium values of the parameters, $\beta = \beta_0$ and $\gamma = \gamma_0$ (the so-called *β- and γ-vibrations*). The Schrödinger equation (5.253) with the Hamiltonian operator (5.252) has not been solved in the general case. Below, we shall examine the solutions of this equation separately in particular cases, with certain simplifying assumptions.

We have written down the Hamiltonian operator and the corresponding wave equation for the collective motions only, neglecting the one-particle motions of the individual nucleons. However, these motions are coupled and it is not always possible, generally speaking, to separate the collective motions in the nucleus from the one-particle motions. The separation of the collective and one-particle excitations is well fulfilled, for example, for even-even nuclei, in which the minimum one-particle excitation energy is considerably greater than the energy of the collective excitations. In the following, we shall assume that the collective and one-particle motions can be separated, that is, we shall use the adiabatic approximation. In the adiabatic approximation, the total Hamiltonian of the nucleus is represented in the form of a sum of two terms, referring to the collective and one-nucleon motions respectively. The wave function in the adiabatic approximation is written in the form of a product of a wave function depending on the collective variables and a wave function depending on the one-particle variables. Using this function, we can also take into account the coupling between the collective and one-particle motions in the nucleus by perturbation theory, if we can assume this coupling to be a small correction to the Hamiltonian.

Solution of the wave equation for a spherical nucleus. If the equilibrium shape of the nucleus is spherical, then $\beta_0 = 0$ and for small vibrations of the nuclear surface the potential energy is equal to $V = \frac{1}{2}C\beta^2$. In this case, the solution of the Schrödinger equation (5.253) with the operator (5.252) is conveniently chosen in the form of a product of two functions by separating out the dependence of the function on the variable β:

$$\Psi(\Phi, \Theta, \psi, \beta, \gamma) = f(\beta)\Phi(\theta_i, \gamma), \tag{5.256}$$

where θ_i are the three Euler angles. Substituting (5.256) into (5.253), we obtain two equations:

$$\left\{-\frac{1}{\beta^4}\frac{\partial}{\partial\beta}\beta^4\frac{\partial}{\partial\beta} + \frac{BC}{\hbar^2}\beta^2 + \frac{N}{\beta^2} - \frac{2BE}{\hbar^2}\right\} f(\beta) = 0; \tag{5.257}$$

$$\left\{\sum_{j=1}^{3}\frac{I_j^2}{4\sin^2(\gamma - \frac{2\pi}{3}j)} - \frac{1}{\sin 3\gamma}\frac{\partial}{\partial\gamma}\sin 3\gamma\frac{\partial}{\partial\gamma} - N\right\}\Phi(\theta_i, \gamma) = 0, \tag{5.258}$$

where N is the separation parameter, which depends on I.

We first consider the equation for the function $f(\beta)$. We introduce the new function

$$\varphi(\beta) = \beta^2 f(\beta) \tag{5.259}$$

and the new variable

$$x = (\beta/\beta_C)^2, \qquad \beta_C = \sqrt[4]{\hbar^2/BC}.$$

The function $\varphi(x)$ will then satisfy the equation

$$\left\{x\frac{d^2}{dx^2} + \frac{1}{2}\frac{d}{dx} - \frac{x}{4} - \frac{N+2}{4x} + \frac{E}{2\hbar\omega}\right\}\varphi(x) = 0 \tag{5.260}$$

and the boundary conditions

$$\varphi(0) = 0, \qquad \lim_{x\to\infty} x^{-2}\varphi(x) = 0.$$

For small x, this equation has the solution

$$\varphi(x) = x^s, \qquad x \to 0; \qquad s = \frac{1 + \sqrt{9+4N}}{4}, \tag{5.261}$$

while for large x,

$$\varphi(x) = e^{-x/2}, \qquad x \to \infty. \tag{5.262}$$

Therefore, in the whole range of the variable x, the solution of eqn. (5.260) can be represented conveniently in the form

$$\varphi(x) = x^s e^{-x/2} P(x). \tag{5.263}$$

The function $P(x)$ satisfies the equation

$$\left\{ x\frac{d^2}{dx^2} + \left(2s + \frac{1}{2} - x\right)\frac{d}{dx} + n \right\} P(x) = 0, \tag{5.264}$$

where

$$n = (E/2\hbar\omega) - s - \frac{1}{4} = (E/2\hbar\omega) - \frac{1}{2} - \frac{1}{4}\sqrt{9+4N}. \tag{5.265}$$

According to (5.261) and (5.262), the function $P(x)$ should tend to a finite non-zero constant as $x \to 0$, and should behave as a polynomials as $x \to \infty$. Such solutions of eqn. (5.264) are possible for positive integer values of n and can be expressed in terms of the confluent hypergeometric function

$$P(x) = F(-n, 2s + \frac{1}{2}; x). \tag{5.266}$$

Finally, for the function $\varphi(\beta)$, normalized by the condition $\int_0^\infty d\beta \varphi^2(\beta) = 1$, we obtain the expression

$$\begin{aligned} \varphi_{ns}(\beta) \;=\; & \left[\frac{2\Gamma(n + 2s + \frac{1}{2})}{\beta_C n! \Gamma^2(2s + \frac{1}{2})}\right]^{1/2} \left(\frac{\beta}{\beta_C}\right)^{2s} e^{-(1/2)(\beta/\beta_C)^2} \\ & \times F\left(-n, 2s + \frac{1}{2}; \frac{\beta^2}{\beta_C^2}\right). \end{aligned} \tag{5.267}$$

According to (5.265), the corresponding energy values will be

$$E_{ns} = 2\hbar\omega\left(n + s + \frac{1}{4}\right) \tag{5.268}$$

and will depend through s (or N) on the total spin I of the nucleus.

We now consider the solutions of eqn (5.258). Since the total angular momentum I and its z-component M in the laboratory coordinate frame K are conserved, we represent the function $\Phi^{\tau}_{IM}(\theta_i, \gamma)$ in the form

$$\Phi^{\tau}_{IM}(\theta_i, \gamma) = \sqrt{\frac{2I+1}{8\pi^2}} \sum_{K=-1}^{I} g^{\tau}_{IK}(\gamma) D^{I}_{KM}(\theta_i), \tag{5.269}$$

where $D^I_{KM}(\theta_i)$ are the generalized spherical harmonics and the symbol τ denotes the quantum numbers necessary to characterize the state completely. Before proceeding to the determination of the functions $g^\tau_{IK}(\gamma)$, we shall elucidate which symmetry conditions they satisfy.

In transforming from the variables $\alpha_{2\mu}$ to the variables θ_i, β and γ, only the deformation parameter β is uniquely determined by the variables $\alpha_{2\mu}$, whereas the four variables Φ, Θ, ψ and γ depend on the labelling of the axes x', y', and z' and on the choice of the reference directions relative to each of these. The function Φ^τ_{IM}, which depends on these four variables, must be invariant under the above transformations of the coordinate axes. This leads to certain conditions to be imposed on the functions g^τ_{IK}. In particular, using the properties of the functions $D^I_{KM}(\theta_i)$ under transformations of the axes, we can show that the requirement of invariance of the functions Φ^τ_{IM} leads to the following conditions which must be satisfied by the functions $g^\tau_{IK}(\gamma)$:

$$g^\tau_{IK}(\gamma) = \mathrm{e}^{i\pi(I+K)} g^\tau_{I,-K}(\gamma); \tag{5.270}$$

$$g^\tau_{IK}(\gamma) = \mathrm{e}^{i\pi K/2} g^\tau_{IK}(-\gamma); \tag{5.271}$$

$$g^\tau_{IK}(\gamma) = \mathrm{e}^{i\pi K} g^\tau_{IK}(\gamma); \tag{5.272}$$

$$g^\tau_{IK}(\gamma) = \sum_{K'} D^I_{K'K}(0, \pi/2, \pi/2) g^\tau_{IK'}(\gamma - 2\pi/3). \tag{5.273}$$

It follows from (5.272) that

$$g^\tau_{IK}(\gamma) = 0 \qquad \text{for odd } K, \tag{5.274}$$

while from (5.270) we obtain

$$g^\tau_{IK}(\gamma) = (-1)^I g^\tau_{I,-K}(\gamma) \qquad \text{for even } K. \tag{5.275}$$

Using these conditions, we represent the function $\Phi^\tau_{IM}(\theta_i, \gamma)$ in the form

$$\Phi^\tau_{IM}(\theta_i, \gamma) = \sum_{\text{even } K \geq 0} g^\tau_{IK}(\gamma) \Phi(IKM), \tag{5.276}$$

where

$$\Phi(IKM) = \sqrt{\frac{2I+1}{16\pi^2(1+\delta_{0K})}} \left\{ D^I_{KM}(\theta_i) + (-1)^I D^I_{-KM}(\theta_i) \right\} \tag{5.277}$$

are the normalized function of a rigid symmetric top and are invariant under reflection in the plane $z' = 0$ (if K is even).

We consider the states with zero spin ($I = M = K = 0$). In this case, the function $g^{\tau}_{00}(\gamma) \equiv g(\gamma)$ satisfies the equation (there is no rotation of the nucleus)

$$\left(-\frac{1}{\sin 3\gamma}\frac{\partial}{\partial\gamma}\sin 3\gamma\frac{\partial}{\partial\gamma} - N_0\right)g(\gamma) = 0, \tag{5.278}$$

whose solutions, when the conditions (5.271) and (5.273) are taken into account, will be the Legendre polynomials

$$g_m(\gamma) = \sqrt{\frac{3(2m+1)}{2}}P_m(\cos 3\gamma), \tag{5.279}$$

if

$$N_0 = 9m(m+1), \qquad m = 0, 1, 2, \ldots . \tag{5.280}$$

From (5.261), we obtain the connection between s and m:

$$s = 1 + \frac{3}{2}m, \tag{5.281}$$

so that, for the energies of the excited states of zero spin, we have

$$E_{nm} = \hbar\omega\left(2n + 3m + \frac{5}{2}\right), \qquad n, m = 0, 1, 2, \ldots; \tag{5.282}$$

that is, the energy spectrum will be equidistant.

We note that states with spin $I = 1$ are absent, since in this case the only possible value of K is $K = 0$, and from the condition (5.275) we obtain $g^{\tau}_{10}(\gamma) = -g^{\tau}_{10}(\gamma) = 0$.

A study of nuclei with $I > 1$ leads, as in the zero-spin case, to an equidistant energy spectrum for the collective excitations. As we should expect, these results are in complete agreement with the results of the theory of surface vibrations of spherical nuclei (in which internal deformation parameters and Euler angles were not specified; cf. Section 5.1).

The *doubly magic nuclei*, and nuclei close to them, have a spherical equilibrium shape. The experimentally observed energy spectrum of these nuclei differs from the spectrum predicted by the theory of quadrupolar ($\lambda = 2$) vibrations. It is possible that this difference is connected with the neglect of the anharmonicity of the vibrations and also of the interaction between the vibrations of the nuclear surface and the motion of the individual nucleons.

Rotational-vibrational excitations of non-axial nuclei. The rigid non-axial rotator model, in which the interaction between the deformations of the nuclear shape and the rotation of the nucleus is neglected, is a fairly crude approximation to the real picture of the collective excitations in nuclei. In fact, nuclei may experience stretching on rotation, and so, in the general case, it is necessary to take into account the *coupling of the rotation of the nucleus with the deformations of its surface* .

Confining ourselves to second-order deformations ($\lambda = 2$), we shall consider the particular case of the wave equation describing the collective motions in the nucleus when the variable γ of the transverse vibrations is held fixed (we replace it by some effective value Γ) but the variable β of the longitudinal vibrations remains free. This non-adiabatic approach to the study of the *rotational-vibrational excitations in nuclei* was proposed by Davydov and Chaban (1960). In this case, the collective motions in the nucleus are characterized by four degrees of freedom, corresponding to the three Euler angles $\theta_i(\Phi, \Theta, \psi)$ and the variable β. The classical energy of the collective motions in this case has the following form:

$$E = \sum_{k'=1}^{3} 2B\beta^2 \sin^2\left(\Gamma - \frac{2\pi}{3}k'\right)\omega_{k'}^2 + \frac{1}{2}B\dot{\beta} + V(\beta).$$

The associated quantum Hamiltonian operator for the collective motions differs from general expression (5.252). The matrix g_{ij}, which determines the squared differential length in the space of generalized coordinates (5.241), is the fourth order now. Knowing the matrix g_{ij}, we can easily write down the expression of Laplace operator according (5.243) for the case of four variables and the Hamilton operator,as well:

$$H_\Gamma = \sum_{k'=1}^{3} \frac{\hbar^2 I_{k'}^2}{8B\beta^2 \sin^2(\Gamma - \frac{2\pi}{3}k')} - \frac{\hbar^2}{2B}\frac{1}{\beta^3}\frac{\partial}{\partial\beta}\beta^3\frac{\partial}{\partial\beta} + V(\beta). \qquad (5.283)$$

The wave function $\Psi(\theta_i, \beta)$, satisfying the equation

$$(H_\Gamma - E)\Psi(\theta_i, \beta) = 0, \qquad (5.284)$$

normalized by the condition

$$\int d\tau \Psi^*(\theta_i, \beta)\Psi(\theta_i, \beta) = 1, \qquad d\tau = d\Phi \sin\theta d\theta d\psi \beta^3 d\beta.$$

In this approximation we can separate β-variable from Euler angles θ_i. The wave function $\Psi(\theta_i, \beta)$ can be chosen as a product

$$\Psi(\theta_i, \beta) = \beta^{-3/2}\varphi(\beta)\Phi(\theta_i), \qquad (5.285)$$

Substituting in equation (5.284) with Hamiltonian (5.283),we obtain two equations:

$$\left\{\sum_{k'=1}^{3}\frac{I_{k'}^2}{2\sin^2(\Gamma-\frac{2\pi}{3}k')}-\varepsilon_{I\tau}\right\}\Phi_{I\tau}(\theta_i)=0; \tag{5.286}$$

$$\left\{-\frac{\hbar^2}{2B}\frac{d}{d\beta^2}+W_{I\tau}(\beta)-E_{\nu I\tau}\right\}\varphi_{\nu I\tau}(\beta)=0, \tag{5.287}$$

$$W_{I\tau}(\beta)=V(\beta)+\frac{\hbar^2(\varepsilon_{I\tau}+\frac{3}{2})}{4B\beta^2}.$$

Equation (5.286) for the function $\Phi_{I\tau}(\theta_i)$ coincides with the Schrödinger equation for a rigid non-axial rotator considered earlier (see sec. 5.5); $\varepsilon_{I\tau}\equiv\varepsilon_{I\tau}(\Gamma)$ is the rotator dimensionless energy, I is a spin of a nucleus, τ is an additional quantum number determining the rotation band ($\varepsilon_{01}=0$), ν is the energy quantum number associated with the β-vibration of nuclei. It will be specified below. The function $\varphi_{\nu I\tau}(\beta)$ satisfies the condition

$$\varphi_{\nu I\tau}(0)=\varphi_{\nu I\tau}(\infty)=0$$

and is normalized according

$$\int_0^\infty d\beta\varphi_{\nu I\tau}^2(\beta)=1.$$

The quantity $W_{I\tau}(\beta)$ in (5.287) plays a part of the effective potential energy, which depends on the quadrupole nucleus deformations, as well as on its rotations. Note, first of all, that for an arbitrary potential energy of β-vibrations $V(\beta)$, a solution $\varphi_{\nu I\tau}(\beta)$ of equation (5.287) and related energy $E_{\nu I\tau}$ can be written explicitly, near $\beta=\beta_{I\tau}$, which corresponds to a minimum of the effective potential energy $W_{I\tau}(\beta)$. At the same time, the function $\varphi_{\nu I\tau}(\beta)$ takes maximum values ($|\beta-\beta_{I\tau}|\ll 1$):

$$\varphi_{\nu I\tau}(\beta)=N_{\nu I\tau}D_\nu\Big(g_{I\tau}(\beta-\beta_{I\tau})\Big),\qquad g_{I\tau}=\left[\frac{4B}{\hbar^2}\lim_{\beta\to\beta_{I\tau}}\frac{d^2W_{I\tau}(\beta)}{d\beta^2}\right]^{1/4}, \tag{5.288}$$

$$E_{\nu I\tau}\approx\left(\nu+\frac{1}{2}\right)\frac{\hbar^2g_{I\tau}^2}{2B}+W_{I\tau}(\beta_{I\tau}),$$

where $N_{\nu I\tau}$ is the normalization factor, D_ν is the function (we express it in terms of the product of the Gauss function and Hermite function of the first kind) of a parabolic cylinder; ν can be obtained from the boundary condition $D_\nu(-g_{I\nu}\beta_{I\tau})=0$.

Expanding the potential energy (potential) of longitudinal quadrupole deformations $V(\beta)$ in the series with respect to small declinations of the variable β from equilibrium value β_0 and keeping the quadrupole terms only, we obtain

$$V(\beta) = \frac{C}{2}\left(\beta - \beta_0\right)^2, \qquad C > 0, \qquad V(\beta_0) = 0, \tag{5.289}$$

where $C \equiv C_\beta$ is the nucleus elasticity coefficient with respect to β-vibrations. Suppose now that $V(\beta)$ takes form (5.289) not only for small vibrations $\frac{\beta-\beta_0}{\beta_0} \ll 1$. In this case the model of two-parametric potential energy $V(\beta)$ gives the following expression for equation (5.287):

$$\left\{-\frac{\hbar^2}{2B}\frac{d^2}{d\beta^2} + \frac{C}{2}(\beta - \beta_0)^2 + \frac{\hbar^2(\varepsilon_{I\tau} + \frac{3}{2})}{4B\beta^2} - E_{\nu I\tau}\right\}\varphi_{\nu I\tau}(\beta) = 0. \tag{5.290}$$

This equation describes longitudinal vibrations in the general case of a nonspherical ($\beta_0 \neq 0$) non-axial rotating (at $I \neq 0$) even-even nucleus.

On increasing the nucleus spin I, the rotation part of the effective potential energy in (5.290) $W_{I\tau}(\beta) = \frac{C}{2}(\beta - \beta_0)^2 + \frac{\hbar^2\left(\varepsilon_{I\tau}+\frac{3}{2}\right)}{4B\beta^2}$ also increases. It consists of $\varepsilon_{I\tau}$, which leads to a visible displacement of the equilibrium position with respect to β, as well as the minimum of the function $W_{I\tau}(\beta)$ and change in the elasticity coefficient. Expanding $W_{I\tau}(\beta)$ in the series with respect to small departures of β from a new equilibrium value $\beta_{I\tau}$ and keeping only quadratic terms, we get

$$W_{I\tau}(\beta) = W_{I\tau}(\beta_{I\tau}) + \frac{C_{I\tau}}{2}(\beta - \beta_{I\tau})^2,$$

$$\beta_{I\tau} = \beta_0 + \frac{\hbar^2(\varepsilon_{I\tau} + \frac{3}{2})}{2BC\beta_{I\tau}^3},$$

where $C_{I\tau} = C\left[1+\frac{3\hbar^2(\varepsilon_{I\tau}+\frac{3}{2})}{2BC\beta_{I\tau}^4}\right]$ is a new (effective) coefficient of the nucleus elasticity, and $W_{I\tau}(\beta_{I\tau}) = \frac{C}{2}(\beta_{It} - \beta_0)^2 + \frac{\hbar^2(\varepsilon_{I\tau}+\frac{3}{2})}{4BC\beta_{I\tau}^2}$.

Now we introduce a new denotations:

$$p_{I\tau} = \frac{\beta_{I\tau}}{\beta_0} \geqslant 1, \ \mu = \frac{1}{\beta_0}\left(\frac{\hbar^2}{BC}\right)^{1/4}, \ \mu_{I\tau} = \frac{1}{\beta_0}\left(\frac{\hbar^2}{BC_{I\tau}}\right)^{1/4},$$

$$\omega_0 = \sqrt{\frac{C}{B}}, \ \omega_{I\tau} = \sqrt{\frac{C_{I\tau}}{B}}.$$

It is easy to obtain the following interrelations between them

$$2(p_{I\tau}-1)p_{I\tau}^3=\mu^4\Big(\varepsilon_{I\tau}+\frac{3}{2}\Big),\ C_{I\tau}=C\Bigg[1+\frac{3\mu^4\Big(\varepsilon_{I\tau}+\frac{3}{2}\Big)}{2p_{I\tau}^4}\Bigg].$$

Equation (5.290) can be approximately presented in the form

$$\Big\{-\frac{\hbar^2}{2B}\frac{d^2}{d\beta^2}+\frac{C_{I\tau}}{2}\Big(\beta-\beta_{I\tau}\Big)^2-\Big[E_{\nu I\tau}-W_{I\tau}(\beta_{I\tau}\Big]\Big\}\varphi_{\nu I\tau}(\beta)=0. \quad (5.291)$$

This is equation of the parabolic cylinder function (see (5.288)). It is convenient to separate the Gauss type factor, which decreases rapidly with growth in the difference $\beta-\beta_{I\tau}$, in the solution of equation (5.291). Introducing a new variable

$$z=\frac{p_{I\tau}(\beta-\beta_{I\tau})}{\mu_{I\tau}\beta_{I\tau}}=\frac{\beta-\beta_{I\tau}}{\beta_0\mu_{I\tau}},\ -\frac{p_{I\tau}}{\mu_{I\tau}}\leqslant z<\infty$$

and a new function $\psi(z)$:

$$\varphi_{\nu I\tau}(\beta)=\psi(z)\exp(-\frac{1}{2}z^2),$$

for the function $\psi(z)$ from (5.291), we obtain the equation

$$\Big(\frac{d^2}{dz^2}-2z\frac{d}{dz}+2\nu\Big)\psi(z)=0,\ \nu=\frac{E_{I\tau}-W_{I\tau}(\beta_{I\tau})}{\hbar\omega_{I\tau}}-\frac{1}{2},\ \psi\Big(-\frac{p_{I\tau}}{\mu_{I\tau}}\Big)=0.$$

A solution of the last second order differential equation that satisfies the above cited boundary conditions for $\varphi_{\nu I\tau}(\beta)$ and $\psi(z)$ is the first kind Hermite function

$$H_\nu(z)=\frac{1}{2\Gamma(-\nu)}\sum_{k=0}^{\infty}\frac{(-1)^k}{k!}\Gamma\Big(\frac{k-\nu}{2}\Big)(2z)^k.$$

The energy quantum number ν associated above all with longitudinal β-vibrations of the nucleus is a root of the equation $H_\nu(-p_{I\tau}/\mu_{I\tau})=0$.

Thus, the eigenfunctions $\varphi_{\nu I\tau}(\beta)$ and eigenvalues $E_{\nu I\tau}$ for equation (5.291) takes the final form

$$\varphi_{\nu I\tau}(\beta)=\bar{N}_{\nu I\tau}H_\nu\Big(\frac{\beta-\beta_{I\tau}}{\beta_0\mu_{I\tau}}\Big)\exp\Big[-\frac{1}{2}\Big(\frac{\beta-\beta_{I\tau}}{\beta_0\mu_{I\tau}}\Big)^2\Big], \quad (5.292)$$

$$E_{\nu I_\tau}(\beta) = \hbar\omega_0 \left\{ \left(\nu + \frac{1}{2}\right) \sqrt{1 + \frac{3}{2}\left(\frac{\mu}{p_{I_\tau}}\right)^4 \left(\varepsilon_{I_\tau} + \frac{3}{2}\right) + \frac{1}{4}\left(\frac{\mu}{p_{I_\tau}}\right)^2} \right.$$
$$\left. \times \left(\varepsilon_{I_\tau} + \frac{3}{2}\right) + \frac{1}{2}\left(\frac{p_{I_\tau} - 1}{\mu}\right)^2 \right\}, \qquad (5.293)$$

where $\bar{N}_{\nu I_\tau}$ is a normalization factor. The quantum numbers ν, that determined from the boundary conditions, in a general case are not integers. If ν is a nonnegative number n, then the Hermite function $H_\nu(z)$ transforms in the n-th order Hermite polynomial. The quantities ε_{I_τ} and p_{I_τ} depend on the nucleus parameter of non-axiality Γ. Hence, the energy $E_{\nu I_\tau}$ specified by (5.293) depends on the parameter Γ, as well.

The above-specified parameter μ is inversely proportional to the forth power root from the nucleus elasticity coefficient C. For enough rigid (weakly deformable) nuclei when C is large, μ is a small quantity ($\mu \ll 1$). For nuclei that easy change their shape, at small C, the quantity can be order of one. It is naturally to call the parameter μ the nucleus softness parameter. At $\mu \to 0$, when $C \to \infty$, we deal with the limiting case of absolutely rigid nuclei. It corresponds to complete separation of the rotational motion from other nucleus motions, as it is assumed in the adiabatic approximation. That is why μ is called also the parameter of the nucleus unadiabaticity. According to the above-reported, we usually call the nuclei with $\mu < \frac{1}{3}$ rigid nuclei,and nuclei with $\mu > \frac{1}{3}$ soft nuclei. We can approximately divide the collective excited states in rigid nuclei into rotational and vibrational. For soft nuclei such division is quite conditionally. In soft nuclei the collective excite states may be treated as a complex mixture of rotational and vibrational motions. The energy spectrum of soft nuclei, even in the ground rotation band, can considerably differ from the adiabatic theory spectrum. In particular, it happens to be that the energy spectrum of soft nuclei is practically equividistant at some values of parameters in spite of the equilibrium shape can essentially differ from the spherical one.

Knowing the experimental ratio of energies of three levels, we can determine the nucleus parameters of softness and nonaxiality μ and Γ. By using them, we can find ratio between energies of the different nuclear states. Comparison of the theory with experiment shows that nuclei ${}^{56}_{26}Fe$, ${}^{72}_{32}Ge$, ${}^{108}_{46}Pd$, ${}^{198}_{80}Hg$ and others can be related to the soft nuclei. Their softness parameter is $\mu \gtrsim 0.5$. Nuclei ${}^{156}_{64}Gd$, ${}^{166}_{68}Er$, ${}^{186}_{76}Os$, ${}^{188}_{76}Os$, ${}^{230}_{90}Th$, ${}^{238}_{92}U$, ${}^{238}_{94}Pu$ with $\mu < 0.3$ relate to rigid ones. The rotational states of such nuclei are good defined. The available experimental data for some nuclei confirm that their rotational motion and vibrational motion of the surface are closely connected. The corresponding numerical calculations are in a good agreement with these experimental data (Davydov, 1967).

Now we briefly consider solutions of equation (5.287) for other, different from (5.289), model dependencies of the potential energy (potential) of the longitudinal nucleus deformations $V(\beta)$ used in applications. For example, for quite general three-parameter potential energy $V'(\beta)$ with a minimum at $\beta = \beta_0$, we can write:

$$V'(\beta) = \frac{C'}{2}(\beta - \beta')^2 + \frac{\hbar^2 \varepsilon}{4B\beta^2} + D, \qquad C' > 0, \qquad \varepsilon > 0. \qquad (5.294)$$

The wave function $\varphi_{\nu I_\tau}(\beta)$ in the region of $W_{I_\tau}(\beta)$ minimum takes form (5.288) or (5.292) with parameters determined by relations

$$g_{I_\tau} = \frac{\sqrt{2}}{\beta' \mu_{I_\tau}}, \quad \beta' = \beta_0 - \frac{\hbar^2 \varepsilon}{2BC'\beta_0^3}, \quad \beta_{I_\tau} = \beta' + \frac{\hbar^2(\varepsilon + \varepsilon_{I_\tau} + \frac{3}{2})}{2BC'\beta_{I_\tau}^3},$$

$$\mu_{I_\tau} = \frac{1}{\beta'}\left(\frac{\hbar^2}{BC'_{I_\tau}}\right)^{1/4}, \quad C'_{I_\tau} = C' + \frac{3\hbar^2}{2B\beta_{I_\tau}^4}\left(\varepsilon + \varepsilon_{I_\tau} + \frac{3}{2}\right), \quad p'_{I_\tau} = \frac{\beta_{I_\tau}}{\beta'},$$

$$2\left(p'_{I_\tau} - 1\right)p'^3_{I_\tau} = \mu'^4\left(\varepsilon + \varepsilon_{I_\tau} + \frac{3}{2}\right), \quad \mu' = \frac{1}{\beta'}\left(\frac{\hbar^2}{BC'}\right)^{1/4}.$$

In this case, for the energy eigenvalues of the collective motions in a nucleus we obtain the following formula

$$E'_{\nu I_\tau} = D + \hbar\omega'\left\{\left(\nu + \frac{1}{2}\right)\left[1 + \frac{3}{2}\left(\mu'/p'_{I_\tau}\right)^4\left(\varepsilon + \varepsilon_{i\tau} + \frac{3}{2}\right)\right]^{1/2}\right.$$

$$\left. + \ \frac{1}{4}\left(\mu'/p'_{I_\tau}\right)^2\left(\varepsilon + \varepsilon_{I_\tau} + \frac{3}{2}\right) + \frac{1}{2}\left[\left(p'_{I_\tau} - 1\right)/\mu'\right]^2\right\}, \ \omega' = (C'/B)^{1/2},$$

where D is determined from the condition $V(\beta_0) = 0$. Apparently, at $\varepsilon = 0$ $(D = 0)$ and $C' = C$, the potential energy $V'(\beta)$ transforms in (5.289). At the same time, $E'_{\nu I_\tau}$ transforms in (5.293).

For the two-parameter dependencies of the potential energy $V(\beta)$

$$V^{(1)}(\beta) = \frac{C}{8}\beta_0^2\left(\frac{\beta}{\beta_0} - \frac{\beta_0}{\beta}\right)^2, \qquad V^{(2)}(\beta) = \frac{C}{2}\left(\frac{\beta_0}{\beta}\right)^2\left(\beta - \beta_0\right)^2 \quad (5.295)$$

the solutions of equations (5.287) can be written in the explicit form in the whole domain of a change of the parameter $\beta \geqslant 0$ in terms of the Laguere polynomials L_n^{α}:

$$\varphi_{nI\tau}^{(1)}(\beta) \equiv \varphi_{nI\tau}^{(1)}\left(\sqrt{2}\beta_{00}x\right) = \left[\sqrt{2}n!/\beta_{00}\Gamma(n+\sigma)\right]^{1/2} x^{\sigma-\frac{1}{2}} L_n^{\sigma-1}(x^2)\mathrm{e}^{-\frac{x^2}{2}},$$

$$\beta_{00} = \left(\frac{\hbar^2}{BC}\right)^{1/4}, \qquad x = \frac{\beta}{\beta_{00}\sqrt{2}} = \frac{\beta}{\beta_0}\xi,$$
$$\sigma \equiv \sigma_{I\tau} = 1 + \left(1 + \xi^4 + \frac{1}{2}\varepsilon_{I\tau}\right)^{1/2};$$

$$\varphi_{nI\tau}^{(2)}(\beta) \equiv \varphi_{nI\tau}^{(2)}\left(\beta_0 y\right) = \left[\frac{\eta^4 n!}{\beta_0(n+s)^2\Gamma(n+2s)}\right]^{1/2} (2\lambda y)^s L_n^{2s-1}(2\lambda y)\mathrm{e}^{-\lambda y},$$

$$y = \frac{\beta}{\beta_0}, \quad \eta = \beta_0\left(\frac{BC}{\hbar^2}\right)^{1/4}, \quad \lambda = \frac{\eta^4}{n+s},$$
$$s \equiv s_{I\tau} = \frac{1}{2} + \left(1 + \eta^4 + \frac{1}{2}\varepsilon_{I\tau}\right)^{1/2}.$$

The corresponding energy spectra are determined by expressions

$$E_{nI\tau}^{(1)} = \hbar\omega\left(n + \frac{\sigma_{I\tau} - \xi^2}{2}\right), \qquad E_{nI\tau}^{(2)} = \hbar\omega\frac{\eta^2}{2}\left[1 - \frac{\eta^4}{(n+s_{I\tau})^2}\right],$$

$$\omega = (C/B)^{1/2}, \qquad n = 0, 1, 2, \ldots.$$

To study the general wave equation of the collective motion (5.253) with the quantum Hamilton operator (5.252) for five collective dynamical variables $\Phi, \Theta, \psi, \beta, \gamma$, we recall that the potential energy (potential) of the quadrupole deformations $V(\beta, \gamma)$ in (5.252) must be expressed in terms of two invariants (5.120) and (5.121). For small *β-vibrations* around the equilibrium value $\beta = \beta_0$ we can take into account the only one invariant (5.120). That is why in this case,we get

$$V(\beta,\gamma)=\frac{C}{2}\left\{\left(\beta-\beta_0\right)^2+4\beta\beta_0\sin^2\frac{\gamma-\gamma_0}{2}\right\},$$

where $C$is a constant. Further, we deal with an arbitrary $\beta \geqslant 0$, and rewrite $V(\beta,\gamma)$ in the more general form

$$V(\beta,\gamma)=V(\beta)+\beta\beta_0 V(\gamma).$$

Here, we still have a wide range of explicit forms of potentials $V(\beta)$ and $V(\gamma)$. The last expression for $V(\beta,\gamma)$ can also be presented as follows

$$V(\beta,\gamma)=V(\beta)+\frac{\beta_0^4}{\beta^2}V(\gamma)+\delta(\beta,\gamma),$$

$$\delta(\beta,\gamma)=\left(\frac{\beta}{\beta_0}-\frac{\beta_0^2}{\beta^2}\right)\beta_0^2 V(\gamma).$$

Since $\delta(\beta,\gamma)\to 0$ at $\beta\to\beta_0$, we can neglect $\delta(\beta,\gamma)$ and substitute

$$V(\beta,\gamma)\approx V(\beta)+\frac{\beta_0^4}{\beta^2}V(\gamma)$$

in equation (5.252). The contribution of order of $\delta(\beta,\gamma)$ corresponds to the bind between β- and γ-*vibrations* of a nucleus and can be taken into account by the perturbation theory methods. But usually, this bind is weak. Equation (5.253) with five dynamic variables allows one to split out the equation with one variable β.

In fact, let us seek a solution of general wave equation (5.253) in the factorized form

$$\Psi(\Phi,\Theta,\psi,\beta,\gamma)=\beta^{-2}\varphi(\beta)\Phi(\Theta_i,\gamma).$$

Substituting this expression in (5.253) and neglecting by the weak bind between β- and γ-vibrations ($\delta(\beta,\gamma)\approx 0$), we can obtain two independent equations for functions $\Phi(\Theta_i\gamma)$ and $\varphi(\beta)$:

$$\left\{\frac{1}{4}\sum_{k'=1}^{3}\frac{I_{k'}^2}{\sin^2\left(\gamma-\frac{2\pi}{3}k'\right)}-\frac{1}{\sin 3\gamma}\frac{\partial}{\partial\gamma}\left(\sin 3\gamma\frac{\partial}{\partial\gamma}\right)+W(\gamma)-\kappa_{mI\tau}\right\}$$
$$\times\Phi_{mI\tau}(\Theta_i,\gamma)=0, \tag{5.296}$$

$$W(\gamma)=\frac{2B\beta_0^4}{\hbar^2}V(\gamma);$$

$$\left\{-\frac{\hbar^2}{2B}\frac{d^2}{d\beta^2}+V(\beta)+\frac{\hbar^2\left(\kappa_{mI\tau}+2\right)}{2B\beta^2}-E_{\nu mI\tau}\right\}\varphi_{\nu mI\tau}(\beta)=0, \qquad (5.297)$$

where $\kappa_{mI\tau}$ is a separation parameter (eigenvalue of equation (5.296)), m – a quantum number associated with the *nucleus transverse γ-vibrations* , ν is a quantum number of the *nucleus longitudinal β-vibrations* . First of all, note that we can write down straight forward the solution of equation (5.297) $\varphi_{\nu mI\tau}(\beta)$ for the above introduced model potentials $V(\beta)$ of the nucleus longitudinal deformations (5.289), (5.294), (5.295). The point is that equation (5.297) for the function $\varphi_{\nu mI\tau}(\beta)$ is obtained from equation (5.287) after formal substitution the quantity $\varepsilon_{I\tau}+\frac{3}{2}$ by $2(\kappa_{mI\tau}+2)$ in (5.287). Equation (5.296) for the function $\Phi_{mI\tau}(\Theta_i,\gamma)$, in a general case, is quite complex. For the nucleus spin $I\neq 0$ it can be solved numerically. Further, we consider the transverse quadrupole vibrations in a general case of a nonaxial $(\beta_0\neq 0,\gamma_0\neq 0)$ even-even nucleus in the states with zero spin.

At $I=0\,(\tau=1)$ the operator of rotation energy in (5.296) consisting squares of the nucleus spin projections $I^2_{k'}$ give zero contribution. This allows one to simplify equation (5.296), which at $I=0$ takes the form

$$\left\{-\frac{1}{\sin 3\gamma}\frac{d}{d\gamma}\left(\sin 3\gamma\frac{d}{d\gamma}\right)+W(\gamma)-\kappa_m\right\}\Phi_m(\gamma)=0,$$

$$W(\gamma)=\frac{2B}{\hbar^2}\beta_0^4V(\gamma),$$

where $\kappa_m=\kappa_{m01}$ is an eigenvalue, and the desired wave function $\Phi_m(\gamma)\equiv\Phi_{m01}(\gamma)$ does not depend on the Euler angles. The wave function of the transverse nucleus vibration $\Phi_m(\gamma)$ is normalized by the condition

$$\int\limits_0^{\pi/3} d\gamma\sin^3\gamma\Phi_m^2(\gamma)=1.$$

The equation of the function $\Phi_m(\gamma)$ is generalization of equation (5.278) for a nonspherical $(\beta_0\neq 0)$ nonaxial $(\gamma_0\neq 0)$ nucleus (Tartakovsky, 1989).

If we choose the potential energy (potential) of the transverse vibration $V(\gamma)$ as follows

$$W(\gamma)=\frac{9}{4}\left(a\mathrm{ctg}\frac{3\gamma}{2}-b\mathrm{tg}\frac{3\gamma}{2}\right)^2, \qquad ab\geqslant 0, \qquad (5.298)$$

then the solution of the equation for the transverse nucleus vibrations can be presented in the form $(0 \leqslant \gamma \leqslant \frac{\pi}{3})$:

$$\Phi_m(\gamma) = \left[\frac{3m!(2m+a+b+1)\Gamma(m+a+b+1)}{2^{a+b+1}\Gamma(m+a+1)\Gamma(m+b+1)}\right]^{1/2} \times (1-\cos 3\gamma)^{\frac{a}{2}}(1+\cos 3\gamma)^{\frac{b}{2}} P_m^{(a,b)}(\cos 3\gamma), \qquad (5.299)$$

$$\kappa_m = 9[m(m+a+b+1)+\frac{1}{2}(a+b)], \quad m=0,1,2,\ldots, \quad a,b \geqslant -1,$$

where $P_m^{(a,b)}(\cos 3\gamma)$ are the Jacobi polynomials. Note that the potential $V(\gamma)$ has a minimum at $\gamma=\gamma_0$, where $\cos 3\gamma_0 = \frac{b-a}{b+a}$ and $V(\gamma_0)=0$. The nucleus elasticity coefficient with respect to γ-vibrations is

$$C_\gamma \equiv \lim_{\gamma\to\gamma_0} \frac{d^2V(\gamma)}{d\gamma^2} = \frac{81\hbar^2}{4B\beta_0^4}(a+b)^2 > 0.$$

If parameters of (5.298) are equal: $a=b=\xi-\frac{1}{2}$ then we have the one-parameter potential

$$V(\gamma) = \frac{9\hbar^2}{2B\beta_0^4}\left(\xi-\frac{1}{2}\right)^2 \mathrm{ctg}^2 3\gamma$$

with $\gamma_0 = \frac{\pi}{6}$. Its wave functions $\Phi_m(\gamma)$ can be expressed in terms of the ultraspherical polynomials or the Gogenbauer polynomials

$$C_m^\xi(\cos\gamma) = \frac{\Gamma\left(\xi+\frac{1}{2}\right)\Gamma(2\xi+m)}{\Gamma\left(\xi+m+\frac{1}{2}\right)\Gamma(2\xi)} P_m^{(\xi-\frac{1}{2},\xi-\frac{1}{2})}(\cos 3\gamma).$$

In this case, according to (5.299), the wave functions of the nucleus transverse vibrations take the form

$$\Phi_m(\gamma) = 2^{\xi-1}\left[\frac{6m!(m+\xi)}{\pi\Gamma(m+2\xi)}\right]^{1/2} (\sin 3\gamma)^{\xi-\frac{1}{2}}\,\Gamma(\xi)C_m^\xi(\cos 3\gamma),$$

and the corresponding eigenvalues are

$$\kappa_m = 9[m(m+2\xi)+\xi-\frac{1}{2}].$$

At $a = b = -\frac{1}{2}(\xi = 0)$ the functions $\Phi_m(\gamma)$ are expressed in terms of the first kind Tchebyshev polynomials

$$\begin{aligned} T_m(\cos 3\gamma) &= \cos[m \arccos(\cos 3\gamma)] \\ &= \frac{2^{2m}(m!)^2}{(2m)!} P_m^{(-\frac{1}{2},-\frac{1}{2})}(\cos 3\gamma) = \cos 3m\gamma. \end{aligned}$$

Thus, in this case $(\xi = 0)$ the wave functions are equal to:

$$\Phi_m(\gamma) = \left(\frac{6}{\pi \sin 3\gamma}\right)^{1/2} T_m(\cos 3\gamma) = \left(\frac{6}{\pi \sin 3\gamma}\right)^{1/2} \cos m\gamma,$$

$$\kappa_m = 9\left(m^2 - \frac{1}{2}\right).$$

(Note that $\lim_{\xi\to 0} \Gamma(\xi) C_m^{\xi}(\cos 3\gamma) = \frac{2}{m} \cos 3\gamma$.)

At $a = b = 0(\xi = \frac{1}{2})$ the Jacobi and Gogenbauer polynomials reduce to the Legendre polynomials

$$P_m^{(0,0)}(\cos 3\gamma) = C_m^{\frac{1}{2}}(\cos 3\gamma) = P_m(\cos 3\gamma).$$

According to (5.298), this is the case of a spherical nucleus $(\beta_0 = 0)$, when $W(\gamma) = 0$, and we arrive to equation (5.278) with solutions (5.279) and eigenvalues (5.280).

Wave functions(5.299) can be used, in particular, when studying the monopoly electric transitions in nuclei, i.e., EO-transitions, between the nucleus states with zero spin. These O-O-transitions are rather sensitive to the nuclear structure. Thus, we can hope to determine the parameters of the transverse γ-vibrations a and b in (5.298) and (5.299) for particular nuclei by comparison of the calculated nuclear form-factors and other characteristics with the experimental data.

5.7. Dynamical Chaos and Collective Motion in Nuclei

Regular and chaotic motion. Common usage of the term "*regular motion in physical systems*" appeals to those determinate processes where the past and future are uniquely determined by the present state and corresponding equations of motion. One usually considers regular motion in studying the dynamics of various physical systems, including atomic nuclei; however, it has turned out that, under the influence of nonlinear effects, one has to allow for other types of motion, namely, for *chaotic (or stochastic)* *motion* discovered by Henri Poincar é as early as the end of the previous century (Poincar é, 1924). It should be noted that chaotic motion, resulting in complete randomness of behavior of the system under study and the unpredictability of its future development in terms of a deterministic description, arises in rigidly determined systems and follows immediately from the dynamics itself rather than from the complexity of the systems or the limited accuracy in assigning the initial conditions.

Therefore, different nuclear models may be governed by both regular (one-particle nuclear models) and *chaotic patterns of nuclear motion* , with the latter exhibited by some collective models such as that of a liquid droplet. Advances in the statistical theory of multinucleonic nuclei ought to be related not only to the complexity of the nuclear systems which this theory has enabled us to describe, but also to the existence of chaotic types of nuclear motion resulting from properties of nuclear interaction. At the same time, the shell structure arising in nuclei should be related not only to the Pauli principle but also to the regular type of nucleon motion. Under appropriate conditions, regular and chaotic motions in nuclei can co-exist in the same way as in other systems, with the respective contributions depending on the total energy of the system and specific type of the potential energy.

It should be noted that in a quantum system the reasons underlying the observed chaotic motion are closely related to the behaviour of the corresponding classical system with the same number of degrees of freedom and the same potential energy as those in the quantum system. Therefore, it seems advisable to begin our considerations with an examination of a classical physical system, simply to obtain an insight into the chaotic motion therein; at the same time, we will discuss the conditions under which the systems may exhibit regular or *chaotic motion.* It is convenient to proceed from the general canonical Hamilton equations which describe the dynamics of a classical system and are given by a set of ordinary first-order equations with respect to time:

$$\dot{q}_i = \{H, q_i\} = \frac{\partial H}{\partial p_i}, \quad \dot{p}_i = \{H, p_i\} = -\frac{\partial H}{\partial q_i}, \quad i = 1, 2, \ldots, s, \qquad (5.300)$$

where $H \equiv H(q,p)$ is the Hamiltonian of a system with s degrees of freedom. $H(q,p)$ is a function of the independent canonically-conjugate values (variables) $q = (q_1, q_2, \ldots, q_s)$ and $p = (p_1, p_2, \ldots, p_s)$ which are specified by s-dimensional vectors in $2s$-dimensional phase space (q,p). Recall that, in a particular case, the set of q may be spatial coordinates of material particles of the system in question and the set of p may be their momenta. The brackets in (5.300) denote the classical Poisson ones for a pair of the variables f and g, i.e.

$$\{f, g\} = \sum_{i=1}^{s} \left(\frac{\partial f}{\partial p_i} \cdot \frac{\partial g}{\partial q_i} - \frac{\partial f}{\partial q_i} \cdot \frac{\partial g}{\partial p_i} \right). \tag{5.301}$$

The dots above $q_i \equiv q_i(t)$ and $p_i \equiv p_i(t)$ stand, as usual, for derivatives with respect to time t. We confine ourselves to conservative systems where the Hamiltonian $H(q,p)$ does not depend explicitly on time. The motion of such a system is described by a *phase trajectory* in $2s$-dimensional phase space, i.e. by a vector $z(t) = (q(t), p(t))$ with $2s$ components in this space. The value f, which is a function of the *dynamical variables* q *and* p , and does not depend explicitly on time, will remain constant provided that the *Poisson brackets* $\{H, f\}$ for it are zero because, consistent with (5.300),

$$\dot{f} \equiv \sum_{i=1}^{s} (\frac{\partial f}{\partial q_i} \dot{q}_i + \frac{\partial f}{\partial p_i} \dot{p}_i) = \{H, f\}.$$

Such an f is referred to as an integral of motion. In particular, for $f = H$ we get $\{H, H\} = 0$, and this simply means that the total energy E of the system is one of the integrals of its motion, i.e. $H(q,p) = E = \text{const.}$

It is well known that the Hamilton equations (5.300) are integrable if there exist s independent integrals of motion $I_i = I_i(q,p)$ satisfying the following conditions

$$\{I_i, I_k\} = 0, \qquad \{H, I_i\} = 0, \qquad i, k = 1, 2, \ldots, s. \tag{5.302}$$

Needless, to say, one can hardly determine from the general appearance of a system whether it is integrable or not; as first shown by Poincaré, the vast majority of such systems are not integrable. However, if for some physical system there exists a *canonical transformation* $(q,p) \to (\xi, \eta)$ such that the Hamiltonian has the form of a sum of s Hamiltonians of independent one-dimensional oscillators

$$H(q,p) \to H(\xi, \eta) = \sum_{k=1}^{s} \frac{\omega_k}{2} \left(\xi_k^2 + \eta_k^2 \right) \tag{5.303}$$

(here the masses are assumed to be unit), then the *integrability of the system* (5.300) becomes evident because one can readily write down s independent integrals of motion I_k

$$I_k = \frac{1}{2} \left(\xi_k^2 + \eta_k^2 \right), \qquad k = 1, 2, \ldots, s, \tag{5.304}$$

meeting the requirements (5.302). The variables ξ_k and η_k are conveniently substituted by specific canonically-conjugate variables I_k (action) and φ_k (angle):

$$\xi_k = \sqrt{2I_k}\sin\varphi_K, \qquad \eta_k = \sqrt{2I_k}\cos\varphi_k, \qquad k = 1, 2, \ldots, s. \tag{5.305}$$

Hence, it is seen that, in particular, the integrals taken along a closed contour $\oint \eta_k d\xi_k$ are equal to $2\pi I_k$, i.e. the variables I_k are the *adiabatic invariants* , suitable for quantization of the classical system. This being the case, the Hamiltonian of the system will depend only on the *actions* I_k

$$H = \sum_{k=1}^{s} \omega_k I_k, \tag{5.306}$$

and the equations of motion become separable and assume the very simple form

$$\dot{\varphi}_k = \frac{\partial H}{\partial I} = \omega_k, \qquad \dot{I}_k = -\frac{\partial H}{\partial \varphi} = 0. \tag{5.307}$$

Their solutions are given by

$$I_k = C_k, \qquad \varphi_k = \omega_k t + \varphi_{k0}, \tag{5.308}$$

where C_k and φ_{k0} are constants.

With a single degree of freedom ($s = 1$), these solutions are exhibited by a circumference of radius $\sqrt{2I_1}$ in the phase space (or a plane); i.e. $\xi_1^2 + \eta_1^2 = 2I_1 = 2C_1$. For $s = 2$, the phase trajectory $z(t)$ in a three-dimensional space may be conceived as a curve densely wound around the surface of a two-dimensional torus with the radius $\sqrt{2C_1} = \sqrt{2I_1}$ or $\sqrt{2C_2} = \sqrt{2I_2}$. By changing the constants C_1 and C_2 whose interdependence for a given value of energy E is determined, according to (5.306), by the relation $\omega_1 C_1 + \omega_2 C_2 = E$, we can obtain an infinite set of the tori embedded into each other and referred to as the invariant ones. The specific character of motion depends on the ratio of the frequencies ω_1 and ω_2. If these are commensurable, i.e. $\frac{\omega_1}{\omega_2} = \frac{n_1}{n_2}$ where n_1 and n_2 are integers, then the phase trajectory shall close on itself after n_1 rotations by the angle φ_1 and n_2 rotations by the angle φ_2; such tori are termed the rational or resonant ones. If the frequencies are incommensurable, then the phase trajectory will never close on itself in winding a torus (the irrational or nonresonant tori).

In what follows, we restrict our attention to conservative systems with two degrees of freedom ($s = 2$) and show that in this case there can arise chaotic motions for the Hamiltonians of a more complicated form than that given by expression (5.303). This takes place because the system of equations (5.300) is not generally integrable for $s > 2$, whereas at $s = 1$ a conservative system is always integrable. In studying *nonintegrable systems* one usually presents the Hamiltonian as a sum of an integrable part, similar to that given by (5.303), and a nonintegrable remainder which is said to be a perturbation. By and large, the finite motion of an integrable system (without the perturbation) with s degrees of freedom, such as that governed by the Hamiltonian (5.303), is conditionally periodic with the frequencies ω_k, and the phase trajectories lie on the s-dimensional invariant tori; in this case, the system's motion shall be regular. However, with the perturbation switched on, a transition to chaotic motion can take place.

The criteria determining the regular and chaotic motion can be stated in terms of the phase trajectories. In fact, the perturbations mentioned above can give rise to *local instabilities* due to which, in the course of time, initially close trajectories move apart from one another in the phase space. With the regular motion that may take place even given the perturbation, the trajectories drift apart at a low rate, approximately equal to time ($\sim t$). For the *irregular motion* , however, the scattering of initially close trajectories occurs rapidly according to the exponential law ($\sim \mathrm{e}^{\lambda t}$) (Henon, Heiles, 1964).

A rigid *Kolmogorov-Arnold-Moser theorem* (Kolmogorov, 1954; Arnold, 1963; Moser, 1968) has been proven which states that, given a small nonintegrable perturbation, the nonresonant invariant tori with phase trajectories wound around do not generally disintegrate but change their shape, still remaining invariant and embedded into each other; the resonant tori do disintegrate, yet the regular motion still governs the general evolution of the system. With sufficiently strong (nonlinear) perturbations, the nonresonant tori disintegrate as well and this leads to the so-called *global chaos* (a transition from regular to random behaviour) which was first observed in numerical and analytical simulations. Clearly, as the system proceeds from regular motion to chaos, its entropy increases (the so-called *Kolmogorov entropy* (Kolmogorov, 1959).

Procedures for the determination of randomness of motion. There are several general procedures applicable to the tests of arbitrary systems for the chaotic motion; they amount to calculations of the *Lyapunov exponents*, *K-entropy* and so-called *autocorrelation functions*. Yet, for conservative systems with two degrees of freedom ($s = 2$), two simple and illustrative methods, which fall slightly short of the rigorous ideal, are convenient to use for the determination of the domains where motion in the phase space is regular or chaotic, and for the evaluation of the *critical energy of the order-to-chaos transition* . One of these procedures employed for theoretical predictions of the order-to-chaos transition starts from an investigation of the geometric properties of the two-dimensional potential energy surface of the system $V = V(q_1, q_2)$, with the emphasis being usually on the examination of the *Gaussian curvature* and *singular points of* this *surface*. The other procedure, often called that of the *Poincaré cross-sections* , enables one to find out the *domains of order and chaos* in an adequate and straight-forward way. The Poincaré cross-section are known to be given by the set of points of the intersection of a phase trajectory with one of phase planes.

Below we will consider, in more detail, the procedure based on the examination of geometric properties of the potential energy surface. To begin with, we study the stability of an arbitrary phase trajectory $z_0(t) = (q_0(t), p_0(t))$ satisfying eqns. (5.300) with $s = 2$ (Ljapunov, 1954, 1956). Consider the difference $\Delta z(t) = z(t) - z(t + \Delta t)$ where $z(t)$ is an arbitrary phase point, close to $z_0(t)$. For a locally stable trajectory $z_0(t)$ the deviation $\Delta z(t)$ will decrease exponentially with time, and for a locally unstable trajectory it will grow exponentially provided the perturbation is sufficiently great.

We linearize Eqs. (5.300) with respect to $\Delta z(t)$ by having rewritten them in the form

$$\dot{z}_n = \sum_l M_{nl} \frac{\partial H}{\partial z_l}, \qquad M_{nl} = \frac{(-1)^l - (-1)^n}{2},$$
$$n = 1, 2 (l = 1, 2), \quad n = 3, 4 (l = 3, 4), \tag{5.309}$$

where the following designations are used

$$z_1 = p_1, \qquad z_2 = p_2, \qquad z_3 = q_1, \qquad z_4 = q_2.$$

Expanding the Hamiltonian H in the Taylor series with respect to Δz, we get

$$H(z) = H(z_0) + \sum_{k=1}^{4} \frac{\partial H}{\partial z_k} \Delta z_k + \dots \tag{5.310}$$

and having retained the terms of the first power in Δz_k at most, we arrive at the set of four linear equations for the deviations $\Delta z(t)$

$$\Delta \dot{z}_n = \sum_k A_{nk}(t) \Delta z_k, \qquad A_{nk}(t) = \sum_l M_{nl} \frac{\partial H(z_0(t))}{\partial z_l \partial z_k}. \tag{5.311}$$

If the eigenvalues of the matrix $A_{nk}(t)$ are imaginary, $\Delta z(t)$ will not grow in magnitude, therefore, the phase trajectory $z_0(t)$ will be stable. In the opposite case, when the eigenvalues of the matrix A_{nk} are real, the deviation $\Delta z(t)$ exponentially grows in magnitude and the phase trajectory shall be unstable. From here on, we assume that within a small time interval Δt

the value of $z_0(t)$ remains almost constant and, hence, the matrix A_{nk} in (5.311) may be supposed to be independent of time.

For the Hamiltonian with two degrees of freedom, assuming again unit masses and the kinetic energy independent of time

$$H(q,p) = \frac{1}{2}\left(p_1^2 + p_2^2\right) + V(q_1, q_2), \tag{5.312}$$

we get the following equation for the eigenvalues of the matrix A_{nk}

$$\mathrm{Det}\begin{vmatrix} \lambda & 0 & -1 & 0 \\ o & \lambda & 0 & -1 \\ V_{11} & V_{12} & \lambda & 0 \\ V_{21} & V_{22} & 0 & \lambda \end{vmatrix} = 0, \qquad V_{ij} = \frac{\partial^2 V}{\partial q_i \partial q_j}. \tag{5.313}$$

These eigenvalues are equal to

$$\lambda = \pm\frac{1}{\sqrt{2}}\left[-(V_{11}+V_{22}) \pm \sqrt{(V_{11}+V_{22})^2 - 4(V_{11}V_{22} - V_{12}^2)}\right]^{1/2}. \tag{5.314}$$

The roots λ are seen to be real (of both signs possible) with the proviso that

$$\frac{\partial^2 V}{\partial q_1^2} \cdot \frac{\partial^2 V}{\partial q_2^2} - \left(\frac{\partial^2 V}{\partial q_1 \partial q_2}\right)^2 < 0, \tag{5.315}$$

resulting in the exponential instability of the solutions. On the other hand, it follows from geometric definitions that the *Gaussian curvature* $K(q_1, q_2)$ *of the potential energy surface* $V = V(q_1, q_2)$ is determined by the relation

$$\begin{aligned} K(q_1, q_2) = & \left[\frac{\partial^2 V}{\partial q_1^2} \cdot \frac{\partial^2 V}{\partial q_2^2} - \left(\frac{\partial^2 V}{\partial q_1 \partial q_2}\right)^2\right] \\ & \cdot \left[1 + \left(\frac{\partial V}{\partial q_1}\right)^2 + \left(\frac{\partial V}{\partial q_2}\right)^2\right]^{-1}. \end{aligned} \tag{5.316}$$

Comparing the last expression with the instability condition (5.315), we find that the negative sign of the Gaussian curvature of the surface $V = V(q_1, q_2)$ is responsible for the local instability. Therefore, one may relate the conditions, underlying the global chaos in a two-dimensional system, to the domains of the negative Gaussian curvature appearing on the potential energy surface $V = V(q_1, q_2)$. With the energy E of the system growing, generally, the probability of a transition to the chaotic motion increases.

It would be reasonable to determine the critical energy E_C of an order-to-chaos transition as that equal to the minimum value of the potential energy $V(q_1, q_2)$ on the line of zeros of the Gaussian curvature. In most cases, the critical energy thus determined agrees well enough with the value of the critical energy E_C calculated by the Poincaré cross-section technique.

It should be noted, however, that if both the domain of the negative Gaussian curvature on the potential energy surface and the magnitude of this curvature are small, chaos does not necessarily arise. Yet,if chaos has already arisen for the energy $E \gtrsim E_C$, then calculations show that with the energy increasing further, in a number of cases (when $V(q_1, q_2)$ exhibits a localized domain of the negative Gaussian curvature) an opposite transition from chaos to the regular motion may take place; yet, it occurs at sufficiently high energies $E \gtrsim E'_C$, here E'_Cstands for the second critical energy of the chaos-to-order transition; as a rule,$E'_C \gg E_C$ and the former is determined by the upper bound of the negative curvature $K(q_1, q_2)$ domain.

Chaotic oscillations of nuclear surface. According to the Weizsaecker formula (3.17), in a nucleus a considerable fraction of the total binding energy (up to one-third) may be attributed to nuclear surface oscillations and, for this reason, *chaotization of the surface oscillations* becomes of interest, too. Usually one figures the parameters of the collective Hamiltonian out of experimental data but currently some attempts have been made to evaluate these parameters from nucleon-nucleon interactions. At the same time it has turned out that nuclear surface oscillations are described by a collective Hamiltonian resulting in the nonlinear equations of motion which, in their turn, may give rise to the *chaotic motion of nuclear surface* (Bolotin, et al.,1989).

We restrict our consideration to a spherical nucleus of the radius $R = R_0 A^{1/3}$ and investigate the quadrupole oscillations of its surface. Then, in line with (5.97), the oscillating surface should be described by the following equation

$$R(\vartheta', \varphi', t) = R\Big\{1 + \alpha'_0(t) Y_{20}(\vartheta', \varphi') + \alpha'_2(t) \times \Big[Y_{22}(\vartheta', \varphi') + Y_{2-2}(\vartheta', \varphi')\Big]\Big\}, \qquad (5.317)$$

where the time-dependent parameters of the quadrupole deformations $\alpha'_0 = \alpha'_0(t)$ and $\alpha'_2 = \alpha'_2(t)$ act as two collective variables. For the sake of convenience we redesignate these variables as

$$x = \sqrt{2}\alpha'_2, \qquad y = \alpha'_0, \qquad (5.318)$$

introduce the corresponding momenta p_x and p_y, respectively, and arrive at the classical Hamiltonian

$$H \equiv H\left(x, y, p_x, p_y\right) = \frac{1}{2}\left(p_x^2 + p_y^2\right) + V(x, y) \tag{5.319}$$

with two degrees of freedom which is formally similar to (5.312); yet, here we have to employ an explicit form of the potential energy of the quadrupole oscillations $V(x, y)$ and fit its parameters for a specific nucleus to experimental observations. Recall that, according to (5.318) and (5.100), the variables x and y are related to the deformation parameters β and γ by the equalities $x = \beta \sin\gamma$, $y = \beta\cos\gamma$.

As shown in sec. 5.3, even for a nonspherical nucleus the potential energy of nuclear surface oscillations must be expressed in terms of the invariants $S_2(\lambda)$ and $S_3(\lambda)$ derived in the general form, despite the fact that this potential energy is an invariant itself. In the case of the quadrupole surface oscillations ($\lambda = 2$) of a spherical nucleus we dwell upon here, these invariants can be related to the variables α_0' and α_2' or x and y as follows (see (5.117) – (5.125))

$$\begin{aligned} S_2(2) &= \frac{\alpha_0'^2 + 2\alpha_2'^2}{\sqrt{5}} = \frac{x^2+y^2}{\sqrt{5}}, \\ S_3(2) &= \sqrt{\frac{18}{35}}\left(2\alpha_0'\alpha_2'^2 - \frac{1}{3}\alpha_0'^3\right) = \sqrt{\frac{18}{35}}\left(x^2 y - \frac{1}{3}y^3\right). \end{aligned} \tag{5.320}$$

As seen from (5.320), a general polynomial presentation of the potential energy of the quadrupole surface oscillations in a spherical nucleus is given by the expression

$$V(x, y) = \sum_{m+n \geq 1} C_{mn}(x^2+y^2)^n (x^2 y - \frac{1}{3}y^3)^m, \tag{5.321}$$

where m and n are non-negative integers and the constants C_{mn} for real nuclei can be figured out of experimental data.

Now we briefly outline a general approach, irrespective of the given nucleus, to the search of domains of the chaotic and regular patterns of the oscillatory motion of nuclear surface, which stems from the technique suggested above for the studies of the potential energy (5.320) with $n \leqslant 3$ and $m \leqslant 2$:

$$\begin{aligned} V(x, y) &= \frac{a}{2}(x^2+y^2)^2 + b(x^2 y - \frac{1}{3}y^3) + c(x^2+y^2) \\ &+ d(x^2 y - \frac{1}{3}y^3)(x^2+y^2) + e(x^2 y - \frac{1}{3}y^3)^2 + f(x^2+y^2)^3 \end{aligned} \tag{5.322}$$

and describe some of the obtained results. In so doing, we may consider the sum of all terms, except the first term of the oscillatory type $\frac{a}{2}(x^2 + y^2)$, as the perturbation (nonintegrable part of the Hamiltonian) which gives rise to nonlinearities in the equations of motion and, thus, to emerging chaos. In fact, chaos may take place even if we allow for the first three terms ($n \leqslant 2$, $m \leqslant 1$) in (5.322). In Eq. (5.322) the parameters b and d may be assumed positive because the potential energy does not change under the substitution $b \to -b$, $d \to -d$ provided that y is substituted by $-y$ at the same time. In the three-dimensional space $\{x, y, V = V(x, y)\}$, the expression (5.322) is presented by a complicated surface with minima, maxima and saddle points, which is highly sensitive to five parameters (for simplicity, the expression for $V(x, y)$ is conveniently scaled by the parameter a). Hence, using the symmetry of the Hamiltonian, one usually examines the cross-sections of the potential energy (in particular, that at $x = 0$) in the vicinity of characteristic points and investigates the behavior of $V = V(x, y)$ in the most interesting areas in the space of these parameters. If one also takes into account the requirement specifying the lines of zero Gaussian curvature (see (5.315) and (5.316))

$$\frac{\partial^2 V}{\partial x^2} \cdot \frac{\partial^2 V}{\partial y^2} - \left(\frac{\partial^2 V}{\partial x \partial y}\right)^2 = 0, \tag{5.323}$$

one can find the critical energy E_C and single out the domains of positive and negative values of the Gaussian curvature $K(x, y)$.

General numerical and analytical analysis of the potential energy (5.322) with three or six terms has shown that, in the domain of positive Gaussian curvature where the motion should be regular, the Poincaré cross-sections for different energies are presented by an ordered dense set of the points of intersection of the phase trajectory with the phase plane $x = 0$. These points are clustered into isolated, usually dense curves, separated on a phase plane from one another by big voids containing no points at all. Such Poincar é cross-sections are typical for the regular motion of the system. Similar Poincar é cross-sections with ordered pointwise curves are also observed in the domain of negative Gaussian curvature, yet, at much lower energies of the system than the first critical value E_C, i.e. here the motion still remains regular. However, with E approaching E_C, an ordered arrangement of the pointwise curves in the Poincar é cross-sections is gradually destroyed, and at $E \gtrsim E_C$ points in the Poincar é cross-sections will no longer belong to some definite curves but will be chaotically or almost chaotically scattered over the phase plane, thus, being indicative of the arising chaos. In the chaotic case the whole or almost whole phase space is occupied by chaotic trajectories. With the energy E increasing further up to the second critical energy, when $E \gtrsim E'_C$, the regular motion may be restored again, as we have already noted.

The quantitative degree of the chaotic build-up, regarded as a functions of system's energy E, can be determined by that fraction $S(E)$ of the phase space which is occupied by chaotic trajectories, with the Poincaré cross-section technique being pretty illustrative in this respect. The function $S(E)$ is very small at $E < E_C$ and $E > E'_C$; it becomes approximately equal to 1/2 at $E \cong E_C$ and $E \cong E'_C$, whereas in the wide energy range $E_C < E < E'_C$ (remind that $E'_C \gg E_C$!) it is equal to unity almost everywhere. Thus, we may conclude that for a considerable number of the sets of the potential energy parameters (5.322) there exist a pretty narrow transitional energy region, where the regular motion turns into chaos, and the second transitional region, where the regular motion may be restored.

As regards real nuclei, such studies with the potential energy (5.322) were carried out for the isotopes of Kr^{74-80}, with the realistic values of parameters matched to the data reported from measurements. Some experimental results suggested that $Kr^{74,76}$ nuclei should be spherical, whereas other were indicative of a considerable deformation of these nuclei. On the whole, these data may be regarded as an experimental evidence in favor of the theoretically predicted coexistence of nuclear shape isomers (Solovyev, 1971). To put down the potential energy for nonspherical nonaxial nuclei, one ought to make use of the generalized invariants $S_2(2)$ and $S_3(2)$ which can be derived from (5.320) by the substitutions $x \to x - x_0$ and $y \to y - y_0$ where $x_0 = \beta_0 \sin\gamma_0$, $y_0 = \beta_0 \cos\gamma_0$. Here β_0 and γ_0 are the equilibrium nuclear deformation parameters (see (5.119) – (5.121)).

Strongly deformed nuclei exhibit such a geometry of the potential energy surface that nonlinear effects should show up even at low excitation energies; hence, in these nuclei chaotic motion can take place at pretty low energy, i.e. here E_C is not great. It may be also supposed that chaos may also arise at low energies when one type of the motion becomes mixed with another, for example, under the mixing of oscillations and rotations in nonspherical soft nuclei or when there is a mixing of one-particle oscillations with the collective ones.

Yet, there exist some particular special cases. For instance, in Kr^{74} the chaotic regime of surface oscillations was found to occur at $E \cong 2E_C$ rather than at $E \cong E_C$, with the energy $E \cong 2E_C$ corresponding to a saddle-point value of the potential energy. This conclusion is supported by characteristic patterns on the corresponding Poincaré cross-sections. In such cases, of prime importance become the independent evaluations of the critical energy and moment of the onset of chaos, achieved by different techniques.

The chaotic motion of surface oscillations was also reported for

$Kr^{76,78,80}$. In such nuclei this observation was attributed to the existence of a fairly large domain of the negative Gaussian curvature and great magnitudes of the curvature $K(x,y)$ therein. In $Kr^{74,76}$ nuclei, the effect of restoration of the regular motion was observed at the energies greatly exceeding the saddle-point value and later this conclusion was substantiated in studies of the Poincaré cross-sections. Thus, order-chaos-order transitions in the vibrational motion of nuclear surface were noticed to occur in a number of cases. It seems likely that the effect of restoration of the regular motion after a chaotic stage is not uncommon for the dynamical systems with a localized domain of unstable motion. In the isotopes of $Kr^{78,80}$ however, at high energies the oscillatory motion of the nuclear surface does not become regular again, i.e. here it still remains chaotic; this fact may be attributed to a peculiar geometry of the potential energy surface.

Quantum evidence of chaos in nuclei. Before preceding to the particular case of *quantum chaos emerging on a nuclear surface* , we briefly touch upon the distribution function $P(s)$ of the spacings between neighboring energy levels in a nucleus; this function specifies the probability of the neighboring levels with the same spin J and parity Π to be separated by an energy interval s (Dyson, 1962). If a dynamical system is integrable, the distribution function has the Poisson form

$$P_p(s) = \frac{1}{D} \mathrm{e}^{-\frac{s}{D}}, \qquad \int_0^\infty ds P_p(s) = 1, \tag{5.324}$$

where D is an average distance between the levels. The *Poisson distribution* (5.324) can be derived, for example, from the nuclear Fermi-gas model without the configurational mixing when the eigenvalues of energy are distributed stochastically.

Another explicit form of the distribution function $P(s)$ follows from the so-called *model of random matrices* , applicable to nonintegrable systems. In this model, one usually diagonalizes the energy matrix of a great order, supposing that matrix elements of the Hamiltonian are distributed at random. The random matrix model is consistent with the limiting case of a complete mixing of all degrees of freedom, available inside the corresponding energy interval. The energy levels,calculated as the eigenvalues of a random matrix, are governed by the distribution

$$P_w(s) = \frac{\pi s}{2D^2} \exp\left(-\frac{\pi s^2}{4D^2}\right), \qquad \int_0^\infty ds P_w(s) = 1, \tag{5.325}$$

known as the Wigner function and noted for the ultimately strong mixing of configurations.

As $s \to 0$, the eigenvalues of a random matrix behave in a peculiar way: two neighboring levels push apart from each other, i.e. $P_w(0) = 0$ in contrast to $P_p(0) = 1$. In other words, the *Wigner distribution* yields zero probability for two neighboring levels to coincide, whereas for the Poisson distribution the same probability is the greatest. In the simplest case of the Hamilton matrix of the second order H_{ij} $(i, j{=}1{,}2)$ whose eigenvalues $E_{1,2}$ are related to the matrix elements H_{ij} by the well-known formula

$$E_{1,2} = \frac{1}{2}\Big(H_{11} + H_{22}\Big) \pm \frac{1}{2}\sqrt{(H_{11} - H_{22})^2 + 4H_{12}^2}, \tag{5.326}$$

these two levels may coincide ($E_1 = E_2$) if two requirements are simultaneously fulfilled: $H_{11} = H_{22}$ and $H_{12} = 0$, this may happen only by chance as it is suggested by the random matrix model.

Naturally, only an experiment can reliably answer the question which of the two distributions (5.324) or (5.325) applies to reality. Yet, some progress in this respect has been already achieved in theoretical studies. Numerical simulations showed that the energy levels of the system where chaotic motion was supposed to be present drifted away from each other in the classical limit. Some calculations are also indicative of the fact that the following general statement is likely to be true: the distribution function of inter-level energy spacings should be of the Poisson form for systems with the regular classical motion, and it should be given by the Wigner distribution provided the system moves chaotically. Thus, it can be seen that any examination of the quantum properties related, for instance, to the regular or chaotic patterns of motion must start from investigations of the same system within the classical limit, just as it has been attempted in this Section.

In studying *statistical properties of the energy spectrum* of a certain system, it is extremely helpful to account for the symmetry of the Hamiltonian which determines, in particular, the degeneracy of energy levels; note also that a nonintegrable perturbation generally removes the degeneracy, inherent in an integrable part of the Hamiltonian.

To exemplify possible quantum evidence of the classical chaos governed mainly, as shown above, by the geometry of the potential energy surface, consider *quantum surface oscillations in a spherical nucleus* . We restrict the discussion to quadrupole oscillations and truncate the expansion of the potential energy at the terms of the fourth power of the variables x and y; then, the quantum Hamiltonian of surface oscillations of the nucleus can be written as

$$\hat{H} = \frac{1}{2}\Big(\hat{P}_x^2 + \hat{P}_y^2\Big) + V(x, y),$$

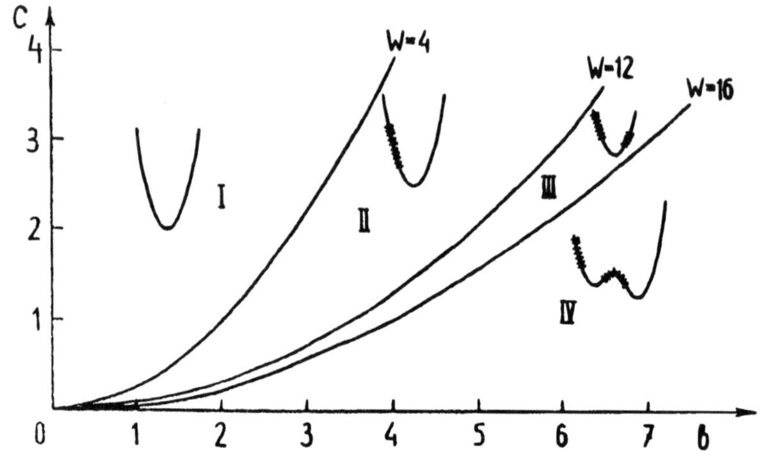

Figure 5.20. Potential energy $V(x=0,9)$ profiles for four domains of the space b and c separated by curves $w \equiv \frac{B^2}{C^2} = const$ with shaded domains of the Gauss negative curvature

$$V(x,y) = \frac{1}{2}(x^2+y^2) + b(x^2y - \frac{1}{3}y^3) + c(x^2+y^2)^2, \qquad (5.327)$$

where $\hat{P}_x = -i\frac{\partial}{\partial x}$, $\hat{P}_y = -i\frac{\partial}{\partial y}$. As above, the parameter a in Eq. (5.322) may be put equal to unity without loss in generality. The parameters b and c, that underlie the occurrence of chaos, specify the degree of nonlinearity of the related equations of motion.

Fig. 5.20 presents, on the b- and c-parameter plane, the shapes of the potential energy $V(x=0,y)$ of Hamiltonian (5.327) for four regions of the parameter space, separated by the curves $w = \frac{b^2}{c}$=const, with the shaded sections of negative Gaussian curvature where chaos may occur. In three regions I, II and III where $0 < w < 16$, the potential energy exhibits a single extremum, i.e. a minimum at the origin; at the same time in region IV there are additional minima arranged at other points. The Gaussian curvature is positive everywhere inside region I and, hence, the corresponding motion should be regular for all energies. The Poincaré cross-sections, typical for region III where $12 < w < 16$, are exemplified in Fig. 5.21 for three values of energy $E = 0.2E_C$, $E = 0.8E_C$ and $E = 1.2E$; they illustrate the order-to-chaos transition for Hamiltonian (5.237). For $E < E_C$, the Poincaré cross-sections are seen to constitute a totality of ordered pointwise curves, peculiar to the regular motion, and at $E = 1.2E_C$ the Poincaré cross-section is presented by the chaotic set of points on the plane $x = 0$, which is innate to chaotic motion (oscillations) of the nuclear shape.

A numerical solution of the Schrödinger equation

Figure 5.21. The Poincaré cross-sections for three values of the system energy E: 1) $0.2E_c$, 2) $0.8E_c$, 3) $1.2E_c$

$$\hat{H}\Psi(x, y) = E\Psi(x, y) \tag{5.328}$$

with Hamiltonian (5.237) was calculated through its diagonalization on the basis of the wave functions of the two-dimensional harmonic oscillator $\varphi_\alpha(x, y)$, corresponding to Hamiltonian (5.237) with $b = c = 0$; the wave functions of nuclear shape oscillations were expanded, respectively, in terms of the basis ones

$$\Psi(x, y) = \sum_\alpha C_\alpha \varphi_\alpha(x, y). \tag{5.329}$$

In that way the eigenvalues of energy E and expansion coefficients C_α were found, and the whole set of the latter gave the wave function in the representation of the eigenfunctions of the $2D$ harmonic oscillator for various values of the parameters b and c of Hamiltonian (5.237) in different regions of both the regular and chaotic classical motion. Then, hundreds of the states of each pattern of motion were analyzed with the aid of computer to an accuracy of $\cong 1\%$.

These preliminary calculations were employed to construct *histograms of the distribution of inter-level spacings* for the Hamiltonian of quadrupole oscillations of the nuclear surface; afterwards, in line with the general assumption, the histograms were fitted with the Poisson distribution for $E < E_C$ (when the motion was regular) and with Wigner distribution for $E > E_C$ (when the motion was chaotic).

Fig. 5.22 presents, in the form of a histogram, the obtained distribution of the inter-level spacings for the chaotic motion region, which was calculated for the potential energy of the nuclear surface oscillations at $w = 13$ (region III in Fig. 5.20) and $E_C < E < E'_C$; for comparison, the Poisson $P_p(s)$ and Wigner $P_w(s)$ distributions are shown in the same figure. It is clearly seen that in the chaotic motion region $P(s)$ is in good agreement with the Wigner distribution. For $E < E_C$ when the motion becomes regular, the calculations conform with the Poisson distribution. The corresponding Poincaré cross-sections were also found and employed to verify the results obtained: the Poisson distributions were consistent with ordered patterns in the Poincaré cross-sections ($E < E_C$) and the Wigner ones correlated with chaotic collections of points in the cross-sections ($E_C < E < E'_C$). With E growing further ($E \gtrsim E'_C$), initially chaotic states of these points were showing some indications of their ordering. Thus, the computer simulations have rendered good grounds to assert that, when the energy of the quantum system increases and the regular motion transforms into the chaotic behaviour, the Poisson distribution changes into its Wigner counterpart and, with the energy increasing further up to the conversion of chaos into a regularity, the Wigner distribution is

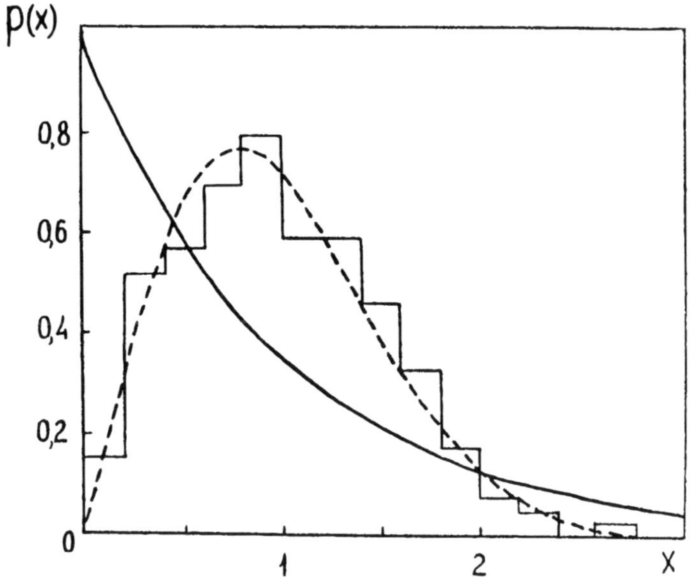

Figure 5.22. A distribution histogram of the distance between neighboring system levels for the chaotic motion (quadrupole nucleus vibrations) and the Poisson (solid line) and Wigner (dotted line) distributions

reconstructed again into the Poisson one. This finding is likely to be general no matter which quantum system namely we study.

At the same time, a peculiar dependence of the wave functions $\Psi(x, y)$ or C_α on the type of the corresponding classical motion was established. It turned out that with the advent of the classical chaos the number of the expansion coefficients C_α in (5.329), which were almost of the same, yet small, value, grew markedly. This fact can be regarded as an indicative criterion of the onset of quantum chaos. If we introduce the function $Q(C_\alpha)$ describing the distribution of the coefficients C_α over their magnitudes, then, this function must show a sharp maximum nearby $C_\alpha = 0$ in the case of the classical regular motion. It is observed for both the energies $E < E_C$ and $E > E'_C$. In the classical chaotic region ($E_C < E < E'_C$) almost all coefficients C_α shall be given by random, almost Gaussian-distributed numbers. In this case, as it was confirmed by the calculations, the distribution $Q(C_\alpha)$ exhibits much broader maximum than that found for the classical regularity. Such a gradually changing distribution $Q(C_\alpha)$ is an inherent feature of quantum chaos. This correlation between statistical properties of the quantum characteristics, such as the energy spectra or wave functions, of stationary states in a quantum system, on the one hand, and nature of the corresponding classical motion was found not only for the Hamiltonian of nuclear surface quadrupole oscillations but also for other Hamiltonians, different from (5.237). As regards Hamiltonian (5.237) itself, it can be readily applied to many other quantum systems.

Investigations of Hamiltonian (5.237) have revealed the effect of multiple intersections of the energy levels under changes of the parameter b when system's energy becomes equal to the classical critical value E_C of the order-chaos transition. On the other hand, for $E < E_C$, when the motion is regular, such intersections are absent even under a considerable nonlinearity. Apparently, this observation is another universal quantum evidence of chaos in classically nonintegrable Hamilton systems.

One more general effect, inherent in quantum systems, has been found, namely that of the *destruction of the shell structure* (or quantum numbers) *in the course of an order-to-chaos transition.* It manifests itself in the variations of quantum numbers (analogues of the classical integrals of motion), which accompany the switch-on of the nonintegrable perturbation resulting itself in destruction of the shell structure of the energy spectrum. Observed under the perturbation switch-on, correlation between the changes of the phase space, governed by the above-mentioned Kolmogorov-Arnold-Moser theorem, and corresponding destruction of the shell structure accompanying the order-to-chaos transition can be regarded as a quantum generalization (analogy) of this theorem. The destruction of the shell structure, occurring when $E \to E_C$,is found to be attended by violations of a quasi-periodic (at $E < E_C$) dependence of system's entropy on its energy E and by a consequent monotonic entropy growth into a plateau of the limiting values, where the energy of levels already becomes much greater than the critical one.

Inasmuch as the main statements above are of a general character, starting from these, we can suggest several assumptions about motion in a multinucleonic atomic nucleus which is quite a complicated dynamical system. For example,one may try to relate the shell structure, arising in the ground and weakly excited states of nuclei, to the assertion that the nuclear system becomes similar to the integrable one due to arising self-consistent field. Then, a residual nucleon-nucleon interaction could be treated as a nonintegrable addition (perturbation) to the self-consistent field and the reason of the shell destruction in a number of nuclei. The existence of shells in many real nuclei at sufficiently strong residual interaction could be interpreted either as the presence of undestroyed nonresonant tori in the phase space, which help the regular motion to survive, or as some other unknown mechanism (apart from the Pauli principle), which restores the regular motion of nucleons in nuclei in the region of well-developed chaos. This new stand point may be helpful in explaining the applicability of drastically different models to the description of nuclei, such as the one-particle models, more closely related to the regular motion (they apply at the excitation energies $E < E_C$), or collective models,mainly characteristic of the chaotic motion (they apply at $E > E_C$ when the internal structure of nuclei changes considerably).

In conclusion it should be noted that the *order-chaos-order transitions* can take place in pretty simple quantum systems, as, for example, the atom of hydrogen. The procedures we have discussed above apply to such systems as well. In the weak-magnetic-field limit, when the Coulomb forces prevail, the problem amounts to the integrable Kepler case and the motion should be regular. For very strong magnetic fields, when one may neglect the Coulomb forces, the problem is also integrable and one deals again with the regular motion. Yet, in the intermediate case, when the intensities of the Coulomb and magnetic forces are of the same order of magnitude, electron begins to move chaotically. Thus, as it follows from the whole foregoing discussion, the chaotic type of motion and order-chaos-order transitions can be due to the dynamics of the process itself rather than to the complexity of the given system, and we are inclined to believe that under appropriate conditions such connection is usual thing rather than an exception.

Problems

5.1. Find the transformation law for a spinor field on rotations of the coordinate frame.

Let $\chi_{s\mu}$ be an eigenfunction of the spin angular momentum and of its projection along a specified direction characterized by the unit vector $\mathbf{N}$

$$\left.\begin{array}{r} s^2\chi_{s\mu} = s(s+1)\chi_{s\mu}; \\ (\mathbf{N}\cdot\mathbf{s})\chi_{s\mu} = \mu\chi_{s\mu}. \end{array}\right\} \tag{5.330}$$

The function $\chi_{s\mu}$ depends on the spatial coordinates $\mathbf{r}$ of the particle and on the spin variable:

$$\chi_{s\mu}(\mathbf{r},\sigma).$$

We shall consider a rotation of the coordinate frame though a certain angle φ about an axis whose direction is specified by the unit vector $\mathbf{n}$. As a result of the rotation, the components of any vector in the rotated coordinate frame K' and in the original coordinate frame K are connected by the linear relation

$$\mathbf{r}' = a\mathbf{r}. \tag{5.331}$$

If a particle is situated at the coordinate origin, its coordinate remain unchanged on rotation of the coordinate frame. The change of the function $\chi_{s\mu}$ in this case will be determined entirely by the dependence on the spin variable σ.

To determine the transformation law for the function $\chi_{s\mu}$ on rotations of the coordinate frame, we shall start from the requirement that the equations of the form (5.330) be invariant in different coordinate frames. In the coordinate frame K', in place of the second equation (5.330) we have

$$(\mathbf{N}'\cdot\mathbf{s})\chi'_{s\mu} = \mu\chi'_{s\mu}. \tag{5.332}$$

(The form of the spin matrices s is the same in all coordinate frames.) The functions $\chi'_{s\mu}$ and $\chi_{s\mu}$ must be related by a certain unitary transformation

$$\chi'_{s\mu} = D_{(s)}\chi_{s\mu}, \tag{5.333}$$

whose form is easily found from the requirement that (5.332) and the second equation (5.330) be equivalent. Substituting (5.333) into (5.332), we find

$$a_{ij}N_j D^{-1}_{(s)} s_i D_{(s)}\chi_{s\mu} = \mu\chi_{s\mu}.$$

This equation is equivalent to the second equation (5.330) if

$$a_{ij} D^{-1}_{(s)} s_i D_{(s)} = s_j$$

or

$$D^{-1}_{(s)} s_i D_{(s)} = a_{ij} s_j, \tag{5.334}$$

since the matrix a_{ij} is orthogonal.

In the case of a rotation through an infinitesimal angle $\delta\varphi$ about an axis specified by the unit vector **n**, the matrix a_{ij} has the form

$$a_{ij} = \delta_{ij} + \delta\varphi\varepsilon_{ijk}n_k, \tag{5.335}$$

where ε_{ijk} is the completely anti-symmetric unit tensor. Since $D_{(s)}(0) = 1$, the unitary operator $D_{(s)}(\delta\varphi)$ can be sought in the form

$$D_{(s)} = 1 + \delta\varphi\mathbf{n}\cdot\mathbf{A},$$

where **A** is an operator to be determined.

We shall consider the case of spin $s = \frac{1}{2}$. The spin operator $\boldsymbol{s}$ is this case is expressed in terms of the Pauli matrices

$$\mathbf{s} = \tfrac{1}{2}\boldsymbol{\sigma},$$

and the function $\chi_{s\mu}$ is a two-component spinor. On an infinitesimal rotation, eqn. (5.334) takes the form

$$(1 - \delta\varphi\mathbf{n}\cdot\mathbf{A})\sigma_i(1 + \delta\varphi\mathbf{n}\cdot\mathbf{A}) = (\delta_{ij} + \delta\varphi\varepsilon_{ijk}n_k)\sigma_j,$$

or

$$\sigma_i A_j - A_j\sigma_i = -\varepsilon_{ijk}\sigma_k. \tag{5.336}$$

Using the commutation relations for the Pauli matrices, it is easily seen that the solution of eqn. (5.336) is $\mathbf{A} = (i/2)\boldsymbol{\sigma}$. Thus, the operator of a rotation through an infinitesimal angle has the form

$$D^{\mathbf{n}}_{(s)}(\delta\varphi) = 1 + \frac{i}{2}\delta\varphi\mathbf{n}\cdot\boldsymbol{\sigma}. \tag{5.337}$$

The operator of a rotation through a finite angle φ about an axis whose direction is specified by the vector **n** can be represented as an infinite sequence of rotations through infinitesimal angles:

$$D^{\mathbf{n}}_{(s)}(\varphi) = \lim_{k\to\infty}\left(1 + \frac{i}{2}\frac{\varphi}{k}\mathbf{n}\cdot\boldsymbol{\sigma}\right)^k = \mathrm{e}^{(i/2)\varphi\mathbf{n}\cdot\boldsymbol{\sigma}}. \tag{5.338}$$

If we now take into account the dependence of the spinor components on the spatial coordinates, then

$$\chi'_{s\mu}(\mathbf{r}', \sigma) = D_{(s)}\chi_{s\mu}(\mathbf{r}, \sigma),$$

or

$$\chi'_{s\mu}(\mathbf{r}', \sigma) = D_{(s)}\chi_{s\mu}(a^{-1}\mathbf{r}', \sigma). \tag{5.339}$$

Then, defining the unitary operator D by means of the equality

$$\chi'_{s\mu}(\mathbf{r}, \sigma) = D\chi_{s\mu}(\mathbf{r}, \sigma)$$

and using (5.339), we have

$$D\chi_{s\mu}(\mathbf{r}, \sigma) = D_{(s)}\chi_{s\mu}(a^{-1}\mathbf{r}, \sigma).$$

Hence, we finally obtain the following expression:

$$D^{\mathbf{n}}(\varphi) = \mathrm{e}^{i\varphi\mathbf{n}\cdot[l+(1/2)\boldsymbol{\sigma}]}. \tag{5.340}$$

The operator $D^{\mathbf{n}}(\varphi)$ defines the transformation law for a two-component spinor on rotations of the coordinate frame.

5.2. Find the transformation law for a vector field on rotations of the coordinate frame.

In a rotation of the coordinate frame, defined by the linear transformation

$$\mathbf{r}' = a\mathbf{r}, \tag{5.341}$$

the vector field $\mathbf{A}(\mathbf{r})$ transforms according to the law

$$\mathbf{A}'(\mathbf{r}') = a\mathbf{A}(\mathbf{r}),$$

or

$$\mathbf{A}'_i(r'_k) = a_{ij}\mathbf{A}_j(a^{-1}_{kl}r'_l). \tag{5.342}$$

We define the finite-rotation operator D by means of the equality

$$\mathbf{A}'(\mathbf{r}) = D\mathbf{A}(\mathbf{r}). \tag{5.343}$$

Using (5.342), we obtain the following equation for D:

$$D\mathbf{A}_i(r_k) = a_{ij}\mathbf{A}_j(a^{-1}_{kl}r_l). \tag{5.344}$$

In the case of a rotation through an infinitesimal angle $\delta\varphi$ about the axis $\mathbf{n}$, the matrix a_{ij} is defined by the expression (5.335). Then, from (5.344) we find

$$D^{\mathbf{n}}(\delta\varphi)\mathbf{A}_i(r_k) = \Big(\delta_{ij} + \delta\varphi\varepsilon_{ijm}n_m\Big)\Big(1 + \delta\varphi\varepsilon_{knl}n_n r_l\frac{\partial}{\partial r_k}\Big)\mathbf{A}_j(r_k). \tag{5.345}$$

We now define three matrices $s_i (i = 1, 2, 3)$ by means of the relation

$$(s_i)_{jk} = -i\varepsilon_{ijk} \tag{5.346}$$

and introduce the orbit angular momentum operator

$$l_i = -i\varepsilon_{ijk} r_j \frac{\partial}{\partial r_k}. \tag{5.347}$$

Keeping the terms linear in $\delta\varphi$ in (5.345), we find

$$D^{\mathbf{n}}(\delta\varphi)\mathbf{A}(\mathbf{r}) = [1 + i\mathbf{n}\cdot(\mathbf{l} + \mathbf{s})]\mathbf{A}(\mathbf{r}). \tag{5.348}$$

In the case of a rotation through a finite angle, we have

$$D^{\mathbf{n}}(\varphi) = e^{i\varphi\mathbf{n}\cdot(\mathbf{l}+\mathbf{s})}. \tag{5.349}$$

It is easily verified that the vector quantity (5.346) satisfies the commutation relations

$$[s_i, s_j] = i\varepsilon_{ijk} s_k, \tag{5.350}$$

and the eigenvalue of the square of the quantity s is equal to

$$s^2 = s(s+1) = 2.$$

Thus, we may associate a spin equal to unity ($s = 1$) with the vector field $\mathbf{A}(\mathbf{r})$. We note that the operator $\mathbf{l}$ occurring in (5.349) characterizes the change of the vector-field function due to the coordinate dependence of the latter, while the operator $\mathbf{s}$ characterizes the change of the function due to the vector character of the field.

If we consider a vector field whose value at each point is proportional to the position vector of the point,

$$\mathbf{A}(\mathbf{r}) = \mathbf{A}\mathbf{r}$$

($\mathbf{A}$ is a constant scalar quantity), then the change of the field function on rotations of the coordinate frame will be equal to zero. This can be seen easily by verifying the relation

$$[\mathbf{n}\cdot(\mathbf{l} + \mathbf{s})]\mathbf{r} = 0, \tag{5.351}$$

where $\mathbf{l}$ and $\mathbf{s}$ are defined by the equalities (5.346) and (5.347).

5.3. Show that the projections of the operator $\mathbf{R}$ along the axis Oz in the stationary coordinate frame and along the axis Oz' in the moving coordinate frame commute:

$$[R_{z'}, R_z] = 0. \tag{5.352}$$

According (5.77) and (5.78),the projections R_z and $R_{z'}$ can be expressed directly in terms of derivatives with respect to the Euler angles Φ and ψ:

$$R_z = -i\frac{\partial}{\partial\Phi}; \qquad R_{z'} = -i\frac{\partial}{\partial\psi}.$$

Using (5.77), we can express the operator $\mathbf{R}_{z'}$ in terms of the projections of the operator $\mathbf{R}$ along the axes of the stationary coordinate frame:

$$R_{z'} = (\cos\Phi\sin\Theta)R_x + (\sin\Phi\sin\Theta)R_y + (\cos\Theta)R_z.$$

Noting that

$$\begin{aligned}
[(\cos\Phi\sin\Theta)R_x, R_z] &= -i\{(\sin\Phi\sin\Theta)R_x + (\cos\Phi\sin\Theta)R_y\};\\
[(\sin\Phi\sin\Theta)R_y, R_z] &= i\{(\sin\Phi\sin\Theta)R_x + (\cos\Phi\sin\Theta)R_y\};\\
[(\cos\Theta)R_z, R_z] &= 0,
\end{aligned}$$

we convince ourselves of the validity of the relation (5.352).

5.4. Prove the operator relation

$$\begin{aligned}
e^{-\tau\hat{b}}\hat{a}e^{\tau\hat{b}} &= \sum_{n=0}^{\infty}\frac{(-t)^n}{n!}\overbrace{\left[\hat{b},[\hat{b},\ldots[\hat{b},\hat{a}]\ldots]\right]}^{n}\\
&\equiv \hat{a} - \tau[\hat{b},\hat{a}] + \frac{\tau^2}{2!}[\hat{b},[\hat{b},\hat{a}]] - \frac{\tau^3}{3!}\left[\hat{b},[\hat{b},[\hat{b},\hat{a}]]\right] + \ldots, \qquad (5.353)
\end{aligned}$$

where $\hat{a}$ and $\hat{b}$ are arbitrary operators.

We represent the operator expression in the left-hand side of (5.353) in the form of a series in powers of the parameter τ:

$$e^{-\tau\hat{b}}\hat{a}e^{\tau\hat{b}} = \sum_{n=0}^{\infty}\tau^n f_n(\hat{b},\hat{a}), \qquad (5.354)$$

where $f_n(\hat{b},\hat{a})$ is some function of the operators $\hat{b}$ and $\hat{a}$. Differentiating the equality (5.354) with respect to τ, we find

$$-e^{-\tau\hat{b}}[\hat{b},\hat{a}]e^{\tau\hat{b}} = \sum_{n=1}^{\infty} n\tau^{n-1}f_n(\hat{b},\hat{a}).$$

Successively differentiating the equality (5.354) k times with respect to τ, we have

$$(-1)^k e^{-\tau\hat{b}}\overbrace{\left[\hat{b},[\hat{b},\ldots[\hat{b},\hat{a}]\ldots]\right]}^{k}e^{\tau\hat{b}} = \sum_{n=k}^{\infty}\frac{n!}{(n-k)!}\tau^{n-k}f_n(\hat{b},\hat{a}).$$

Putting $\tau = 0$ in this equality, we find

$$(-1)^k\overbrace{\left[\hat{b},[\hat{b},\ldots[\hat{b},\hat{a}]\ldots]\right]}^{k} = k!f_k(\hat{b},\hat{a}).$$

Substituting $f_n(\hat{b},\hat{a})$ thus found into (5.354), we obtain the relation (5.353).

5.5. Prove the operator relation

$$e^{-i\Phi j_i} j_k e^{i\Phi j_i} = \cos\Phi j_k + \sin\Phi \varepsilon_{ikl} j_l \qquad (i \neq k), \tag{5.355}$$

where j_i are the components of the angular-momentum operator, satisfying the commutation relations $[j_i, j_j] = i\varepsilon_{ijk} j_k$.

Using the relation (5.353) and separating the even and odd powers of the parameter Φ in the right-hand side, we find

$$\begin{aligned} e^{-i\Phi j_i} j_k e^{i\Phi j_i} = j_k &+ \sum_{n=1}^{\infty} \frac{(-i\Phi)^{2n}}{(2n)!} \overbrace{\Big[j_i, [j_i, \ldots [j_i, j_k] \ldots]\Big]}^{2n} \\ &+ \sum_{n=0}^{\infty} \frac{(-i\Phi)^{2n+1}}{(2n+1)!} \overbrace{\Big[j_i, [j_i, \ldots [j_i, j_k] \ldots]\Big]}^{2n+1}. \end{aligned} \tag{5.356}$$

Since $\Big[j_i, [j_i[j_i, j_k]]\Big] = [j_i, j_k]$, we have

$$\overbrace{\Big[j_i, [j_i, \ldots [j_i, j_k] \ldots]\Big]}^{2n} = j_k, \qquad i \neq k,$$

$$\overbrace{\Big[j_i, [j_i, \ldots [j_i, j_k] \ldots]\Big]}^{2n+1} = [j_i, j_k].$$

Substituting these equalities into (5.356), we obtain the relation (5.355).

5.6. Prove the operator relation

$$e^{-i\Phi j_i} e^{i\Theta j_k} e^{i\Phi j_i} = e^{i\Theta\{(\cos\Phi) j_k + (\sin\Phi)\varepsilon_{ikl} j_l\}}, \tag{5.357}$$

where j_i are the components of the angular-momentum operator.

Using the relation (5.355), we have

$$e^{-i\Phi j_i} (j_k)^n e^{i\Phi j_i} = \{(\cos\Phi) j_k + (\sin\Phi)\varepsilon_{ikl} j_l\}^n.$$

Multiplying this equality by $(i\Theta)^n/n!$ and summing the left-hand and right-hand sides over n from 0 to ∞, we obtain the relation (5.357).

5.7. Show that

$$D^l_{0m}(\Phi, \Theta, \psi) = \sqrt{\frac{4\pi}{2l+1}} Y_{lm}(\Theta, \Phi). \tag{5.358}$$

We shall make use of the relation (5.60), which gives the transformation law for spherical harmonics on rotations of the coordinate frame:

$$Y_{lm}(\vartheta\varphi) = \sum_{m'} D^l_{m'm}(\Phi, \Theta, \psi) Y_{lm'}(\vartheta', \varphi'). \tag{5.359}$$

We choose $\vartheta = 0$, that is, we consider the direction along the axis Oz' in the rotated coordinate frame. Then,

$$Y_{lm'}(0, \varphi') = \sqrt{\frac{2l+1}{4\pi}}\delta_{m'0}.$$

In the stationary coordinate frame, the direction of the axis Oz' is characterized by the angles $\vartheta = 0$ and $\varphi = \Phi$; therefore,

$$Y_l(\Theta, \Phi) = \sqrt{\frac{2l+1}{4\pi}} D^l_{0m}(\Phi, \Theta, \psi),$$

whence the relation (5.358) follows.

5.8. Obtain the addition formula for spherical harmonics by making use of the transformations law for spherical harmonics on rotations of the coordinate frame.

We shall choose an arbitrary point P on a unit sphere, with the polar coordinates ϑ and φ in the original coordinate frame and the coordinates ϑ' and φ' in the rotated coordinate frame. The polar coordinates of the axis Oz' in the original coordinate frame are Θ and Φ. By putting m equal to zero in the inverse formula to (5.359),

$$Y_{lm}(\vartheta', \varphi') = \sum_{m'} D^{l*}_{mm'}(\Phi, \Theta, \psi) Y_{lm'}(\vartheta, \varphi),$$

and substituting $D^{l*}_{0m'}(\Phi, \Theta, \psi)$ in the form (5.358), we obtain the addition formula for spherical harmonics:

$$\sqrt{\frac{2l+1}{4\pi}} Y_{l0}(\vartheta', \varphi') \equiv \frac{2l+1}{4\pi} P_l(\cos\vartheta') = \sum_m Y^*_{lm}(\Theta, \Phi) Y_{lm}(\vartheta, \varphi). \qquad (5.360)$$

5.9. Find the coupling rules for D-functions on addition of angular momenta.

Let $\psi_{j_1m_1}(\mathbf{r}_1)$ and $\psi_{j_2m_2}(\mathbf{r}_2)$ be the wave function for two systems with well-defined values of the square and z-component of the angular momentum. On rotations of the coordinate frame, these functions transform according to the law

$$\left.\begin{aligned} \psi_{j_1m_1}(\mathbf{r}_1) &= \textstyle\sum_{m_1'} D^{j_1}_{m_1'm_1}(\Phi, \Theta, \psi)\psi_{j_1m_1'}(\mathbf{r}_1'); \\ \psi_{j_2m_2}(\mathbf{r}_2) &= \textstyle\sum_{m_2'} D^{j_2}_{m_2'm_2}(\Phi, \Theta, \psi)\psi_{j_2m_2'}(\mathbf{r}_2'). \end{aligned}\right\} \qquad (5.361)$$

According to the rules for addition of angular momenta, we have

$$\psi_{j_1m_1}(\mathbf{r}_1)\psi_{j_2m_2}(\mathbf{r}_2) = \sum_{jm} (j_1m_1j_2m_2|jm)\psi_{jm}(\mathbf{r}_1, \mathbf{r}_2), \qquad (5.362)$$

where $\psi_{jm}(\mathbf{r}_1, \mathbf{r}_2)$ is the wave function of the whole system in a state characterized by well defined values of the square and z-component of the angular momentum. On rotations of the coordinate frame, the functions $\psi_{jm}(\mathbf{r}_1, \mathbf{r}_2)$ transform analogously to (5.361):

$$\psi_{jm}(\mathbf{r}_1, \mathbf{r}_2) = \sum_{m'} D^j_{m'm}(\Phi, \Theta, \psi)\psi_{jm}(\mathbf{r}'_1, \mathbf{r}'_2). \tag{5.363}$$

The addition rule (5.362) for angular momenta is also valid in the rotated coordinate frame; therefore,

$$\psi_{jm'}(\mathbf{r}'_1, \mathbf{r}'_2) = \sum_{m'_1 m'_2} (j_1 m'_1 j_2 m'_2 | jm')\psi_{j_1 m'_1}\mathbf{r}'_1 \psi_{j_2 m'_2}(\mathbf{r}'_2), \tag{5.364}$$

and, consequently,

$$\begin{aligned}&\psi_{j_1 m_1}(\mathbf{r}_1)\psi_{j_2 m_2}(\mathbf{r}_2)\\ &= \sum_{\substack{jmm'\\ m'_1 m'_2}} (j_1 m_1 j_2 m_2 | jm)(j_1 m'_1 j_2 m'_2 | jm') D^j_{m'm}(\Phi, \Theta, \psi)\psi_{j_1 m'_1}(\mathbf{r}'_1)\psi_{j_2 m'_2}(\mathbf{r}'_2).\end{aligned}$$

Substituting (5.361) into the left-hand side of the equality obtained and taking into account that $\psi_{j_1 m'_1}(\mathbf{r}'_1)$ and $\psi_{j_2 m'_2}(\mathbf{r}'_2)$ are linearly independent, we find

$$\begin{aligned}&D^{j_1}_{m'_1 m_1}(\Phi, \Theta, \psi) D^{j_2}_{m'_2 m_2}(\Phi, \Theta, \psi)\\ &= \sum_{\substack{j\\ mm'}} (j_1 m_1 j_2 m_2 | jm)(j_1 m'_1 j_2 m'_2 | jm') D^j_{m'm}(\Phi, \Theta, \psi).\end{aligned} \tag{5.365}$$

Making use of the orthogonality of the Glebsch-Gordan coefficients, we finally obtain the following coupling formula for the D-functions on addition of angular momenta:

$$\begin{aligned}D^j_{m'm}(\Phi, \Theta, \psi) &= \sum_{\substack{m_1 m'_1\\ m_2 m'_2}} (j_1 m_1 j_2 m_2 | jm)(j_1 m'_1 j_2 m'_2 | jm')\\ &\times D^{j_1}_{m'_1 m_1}(\Phi, \Theta, \psi) D^{j_2}_{m'_2 m_2}(\Phi, \Theta, \psi).\end{aligned} \tag{5.366}$$

5.10. Calculate the matrix element

$$d^j_{m'm}(\Theta) = \langle jm' | e^{i\Theta j_y} | jm \rangle, \tag{5.367}$$

by making use of the relation between the angular-momentum operator and the creation and destruction operators for two uncoupled harmonic oscillators.

We shall consider two independent harmonic oscillators, which we characterize by the indices 1 and 2. In the occupation-number representation, with each of the oscillators we can associate operators a_i and a_i^+ (i=1 and 2) destroying and creating excitation quanta and satisfying the following commutation relations:

$$[a_i, a_j^+] = \delta_{ij}, \qquad [a_i, a_j] = [a_i^+, a_j^+] = 0. \tag{5.368}$$

From these operators we can construct four independent quadratic operators:

$$a_1^+ a_1, \quad a_2^+ a_2, \quad a_1^+ a_2 \text{ and } a_2^+ a_1. \tag{5.369}$$

In the occupation-number representation the first two operators are diagonal and determine the numbers of excitation quanta (bosons) of the first or second type. The other two operators, $a_1^+ a_2$ and $a_2^+ a_1$, have non-zero matrix elements for transitions in which the numbers of excitation quanta change by unity. Therefore, we can associate operators of angular-momentum components with the operators (5.369), if the numbers n_1 and n_2 of excitation quanta are associated with the quantum numbers of the square and projections of the angular momentum.

It can be checked directly that the operators j_x, j_y and j_z defined by the equalities

$$\left.\begin{aligned} j_x + ij_y &= a_1^+ a_2; \\ j_x - ij_y &= a_2^+ a_1; \\ j_z &= \tfrac{1}{2}\left(a_1^+ a_1 - a_2^+ a_2\right), \end{aligned}\right\} \tag{5.370}$$

satisfy the usual commutation relations for angular-momentum components. The relations (5.370) can be written compactly if we define formally the two-component spinor operators

$$a^+ = \left(a_1^+, a_2^+\right), \qquad a = \begin{pmatrix} a_1 \\ a_2 \end{pmatrix}$$

and introduce the two-row Pauli matrices

$$\sigma_x = \begin{pmatrix} 0 & 1 \\ 1 & 0 \end{pmatrix}; \quad \sigma_y = \begin{pmatrix} 0 & -i \\ i & 0 \end{pmatrix}; \quad \sigma_z = \begin{pmatrix} 1 & 0 \\ 0 & 1 \end{pmatrix}.$$

Then the operator $\mathbf{j}$ of the angular-momentum vector is written in the form

$$\mathbf{j} = \frac{1}{2} a^+ \boldsymbol{\sigma} a. \tag{5.371}$$

If we substitute the unit matrix in place of $\boldsymbol{\sigma}$ in (5.371), we obtain the scalar operator

$$j = \frac{1}{2} a^+ a. \tag{5.372}$$

It is easily verified that

$$\mathbf{j}^2 = j(j+1). \tag{5.373}$$

The quantities $\mathbf{j}$ and j are operators in the occupation-number representation. The eigenvalues of the operators j and j_z, which we denote by j and m, can be expressed in terms of the numbers n_1 and n_2 of excitation quanta:

$$j = \frac{1}{2}(n_1 + n_2), \qquad m = \frac{1}{2}(n_1 - n_2). \tag{5.374}$$

Consequently, the quantum numbers n_1 and n_2 completely determine the values of the square and projection (along a specified direction) of the angular momentum.

We denote the wave function of the ground state of a system of two oscillators ($n_1 = n_2 = 0$ – the vacuum state) by $|0>$. Obviously, the values

$$j = m = 0$$

correspond to this state.

By means of the creation operators a_1^+ and a_2^+, it is not difficult to construct the wave function for a state with well-defined values of the quantum numbers j and m of the square and projection of the angular momentum. This function, normalized to unity, has the form

$$|jm>= \frac{(a_1^+)^{j+m}(a_2^+)^{j-m}}{\sqrt{(j+m)!(j-m)!}}|0> . \tag{5.375}$$

By means of (5.353), one can prove the following relations for the operators a_k^+ and a_k:

$$\left.\begin{aligned} \mathrm{e}^{-i\Theta j_i} a_k^+ \mathrm{e}^{i\Theta j_i} &= \left(\cos\tfrac{\Theta}{2}\right)a_k^+ - i\left(\sin\tfrac{\Theta}{2}\right)a_j^+(\sigma_i)_{jk}; \\ \mathrm{e}^{-i\Theta j_i} a_k \mathrm{e}^{i\Theta j_i} &= \left(\cos\tfrac{\Theta}{2}\right)a_k + i\left(\sin\tfrac{\Theta}{2}\right)(\sigma_i)_{kj}a_j. \end{aligned}\right\} \tag{5.376}$$

We need two particular cases ($i = y$, k=1,2):

$$\left.\begin{aligned} \mathrm{e}^{i\Theta j_y} a_1^+ \mathrm{e}^{-i\Theta j_y} &= \left(\cos\tfrac{\Theta}{2}\right)a_1^+ - \left(\sin\tfrac{\Theta}{2}\right)a_2^+; \\ \mathrm{e}^{i\Theta j_y} a_2^+ \mathrm{e}^{-i\Theta j_y} &= \left(\cos\tfrac{\Theta}{2}\right)a_2^+ + \left(\sin\tfrac{\Theta}{2}\right)a_1^+. \end{aligned}\right\} \tag{5.377}$$

We calculate the result of the action of the operator $\mathrm{e}^{i\Theta j_y}$ on the function (5.375), using the relation (5.377). Since

$$\mathrm{e}^{i\Theta j_y} = |0>,$$

we have

$$\begin{aligned} \mathrm{e}^{i\Theta j_y}|jm>= &\frac{1}{\sqrt{(j+m)!(j-m)!}} \\ &\times\left[\left(\cos\frac{\Theta}{2}\right)a_1^+ - \left(\sin\frac{\Theta}{2}\right)a_2^+\right]^{j+m}\left[\left(\cos\frac{\Theta}{2}\right)a_2^+ + \left(\sin\frac{\Theta}{2}\right)a_1^+\right]^{j-m}|0> \\ = &\sqrt{(j+m)!(j-m)!}\sum_{n=0}^{j+m}\sum_{k=0}^{j-m}(-1)^{j+m-n}\left(\sin^{2j-k-n}\frac{\Theta}{2}\right) \\ &\times\left(\cos^{k+n}\frac{\Theta}{2}\right)\frac{(a_1)^{j-m-k+n}}{n!(j+m-n)!}\times\frac{(a_2^+)^{j+m+k-n}}{k!(j-m-k)!}|0> . \end{aligned}$$

If for the summation indices we make the replacement

$$k = j - m' - \kappa, \quad n = j + m - \kappa, \quad (-j \le m' \le j),$$

then

$$e^{i\Theta j_y}|jm> \;=\; \sum_{m'=-j}^{j} \sum_{\kappa} (-1)^{\kappa} \times \frac{[(j+m)!(j-m)!(j+m')!(j-m')!]^{1/2}}{\kappa!(j+m-\kappa)!(j-m'-\kappa)!(\kappa+m'-m)!} \times \left(\sin\frac{\Theta}{2}\right)^{2\kappa-m+m'} \left(\cos\frac{\Theta}{2}\right)^{2j+m-m'-2\kappa} |jm'> . \quad (5.378)$$

Finally, we obtain the following representation for the function $d^{j}_{m'm}(\Theta)$ (Wigner, 1959):

$$d^{j}_{m'm}(\Theta) \;=\; \sum_{\kappa} (-1)^{\kappa} \frac{[(j+m)!(j-m)!(j+m')!(j-m')!]^{1/2}}{\kappa!(j+m-\kappa)!(j-m'+\kappa)!(\kappa+m'-m)!} \times \left(\sin\frac{\Theta}{2}\right)^{2\kappa-m+m'} \left(\cos\frac{\Theta}{2}\right)^{2j+m-m'-2\kappa}, \quad (5.379)$$

where the sum is taken over all $\kappa \geq o$.

5.11. Find the explicit form of the functions $d^{j}_{m'm}(\Theta)$ in the case $j = \frac{1}{2}$.

We make use of the definition

$$d^{j}_{m'm}(\Theta) = < jm'|e^{i\Theta j_y}|jm > \quad (5.380)$$

and note that, in the case $j = \frac{1}{2}$, we have

$$j_y = \frac{1}{2}\sigma_y, \quad (5.381)$$

where σ_y is a Pauli matrix. We expand the exponential in (5.380) in a series in powers of Θ and separate the even and odd terms:

$$d^{1/2}_{m'm}(\Theta) = \left\langle \frac{1}{2}m' \mid \sum_{n=0}^{\infty} \left(\frac{i\Theta}{2}\sigma_y\right)^{2n} \frac{1}{(2n)!} + \sum_{n=0}^{\infty} \left(\frac{i\Theta}{2}\sigma_y\right)^{2n+1} \frac{1}{(2n+1)!} \mid \frac{1}{2}m \right\rangle .$$

Since $\sigma_y^2 = 1$, we have

$$d^{1/2}_{m'm}(\Theta) = \left\langle \frac{1}{2}m' \mid \sum_{n=0}^{\infty} (-1)^n \left(\frac{\Theta}{2}\right)^{2n} \frac{1}{(2n)!} + i\sigma_y \sum_{n=0}^{\infty} (-1)^n \left(\frac{\Theta}{2}\right)^{2n+1} \frac{1}{(2n+1)!} \mid \frac{1}{2}m \right\rangle ,$$

or

$$d^{1/2}_{m'm}(\Theta) = \left(\cos\frac{\Theta}{2}\right)\delta_{mm'} + i\left(\sin\frac{\Theta}{2}\right)(\sigma_y)_{m'm} . \quad (5.382)$$

In the representation in which the matrix σ_z is diagonal, the Pauli matrix $(\sigma_y)_{m'm}$ has the form

$$(\sigma_y)_{m'm} = i \begin{pmatrix} 0 & -1 \\ 1 & 0 \end{pmatrix}.$$

Thus, for $d^{1/2}_{m'm}(\Theta)$ we finally obtain

$d^{\frac{1}{2}}_{m'm}(\theta) =$ (5.383)

m' \ m	$\frac{1}{2}$	$-\frac{1}{2}$
$\frac{1}{2}$	$\cos\frac{\theta}{2}$	$\sin\frac{\theta}{2}$
$-\frac{1}{2}$	$-\sin\frac{\theta}{2}$	$\cos\frac{\theta}{2}$

The functions $d^j_{m'm}(\Theta)$ for any j can be obtained from $d^{1/2}_{m'm}(\Theta)$ by successive application of the coupling rules (5.366) for D-functions. Putting $\Phi = \psi = 0$ in (5.366), we obtain

$$d^j_{m'm}(\Theta) = \sum_{\substack{m_1 m_1' \\ m_2 m_2'}} (j_1 m_1 j_2 m_2 | jm)\,(j_1 m_1' j_2 m_2' | jm')\, d^{j_1}_{m_1' m_1}(\Theta) d^{j_2}_{m_2' m_2}(\Theta). \quad (5.384)$$

This relation makes it possible to find $d^j_{m'm}(\Theta)$ for any j from (5.383).

5.12. Find the explicit form of the functions $d^j_{m'm}(\Theta)$ in the case $j = 1$.

We proceed in the same way as in the preceding case:

$$d^1_{m'm}(\Theta) =< 1m'|1 + \sum_{n=0}^{\infty} \frac{(i\Theta j_y)^{2n+1}}{(2n+1)!} + \sum_{n=0}^{\infty} \frac{(i\Theta j_y)^{2n+2}}{(2n+2)!} |1m > .$$

Since , for angular momentum equal to unity,

$$j_y\left(j_y^2 - 1\right) = 0, \quad (5.385)$$

we have

$$j_y^{2n+1} = j_y, \qquad j_y^{2n+2} = j_y^2.$$

Thus, we obtain

$$d^1_{m'm}(\Theta) =< 1m'|1 + i(\sin\Theta) j_y + j_y^2 \sum_{n=1}^{\infty} (-1)^n \frac{\Theta^{2n}}{(2n)!} |1m >,$$

or

$$d^1_{m'm}(\Theta) =< 1m'|1 + i(\sin\Theta) j_y + (\cos\Theta - 1) j_y^2 |1m > . \quad (5.386)$$

In the representation in which the matrix j_z is diagonal, the matrix j_y for $j = 1$ has the form

$$j_y = \frac{1}{\sqrt{2}} \begin{pmatrix} 0 & -i & 0 \\ i & 0 & -i \\ 0 & i & 0 \end{pmatrix}, \quad j_y^2 = -\frac{1}{2} \begin{pmatrix} -1 & 0 & 1 \\ 0 & -2 & 0 \\ 1 & 0 & -1 \end{pmatrix}. \tag{5.387}$$

Thus, for $d^1_{m'm}(\Theta)$ we obtain

$d^1_{m'm}(\theta) =$ (5.388)

$m' \backslash m$	1	0	-1
1	$\frac{1}{2}(1+\cos\theta)$	$\frac{1}{\sqrt{2}}\sin\theta$	$\frac{1}{2}(1-\cos\theta)$
0	$-\frac{1}{\sqrt{2}}\sin\theta$	$\cos\theta$	$\frac{1}{\sqrt{2}}\sin\theta$
-1	$\frac{1}{2}(1-\cos\theta)$	$-\frac{1}{\sqrt{2}}\sin\theta$	$\frac{1}{2}(1+\cos\theta)$

5.13. Verify that, in the case $j = \frac{3}{2}$, the functions $d^{3/2}_{m'm}(\Theta)$ have the form

$d^{\frac{3}{2}}_{m'm}(\theta) =$ (5.389)

$m' \backslash m$	$\frac{3}{2}$	$\frac{1}{2}$	$-\frac{1}{2}$	$-\frac{3}{2}$
$\frac{3}{2}$	$\frac{1}{2}(1+\cos\theta)\cos\frac{\theta}{2}$	$\frac{\sqrt{3}}{2}(1+\cos\theta)\sin\frac{\theta}{2}$	$\frac{\sqrt{3}}{2}(1-\cos\theta)\cos\frac{\theta}{2}$	$\frac{1}{2}(1-\cos\theta)\sin\frac{\theta}{2}$
$\frac{1}{2}$	$-\frac{\sqrt{3}}{2}(1+\cos\theta)\sin\frac{\theta}{2}$	$-\frac{1}{2}(1-3\cos\theta)\cos\frac{\theta}{2}$	$\frac{1}{2}(1+3\cos\theta)\sin\frac{\theta}{2}$	$\frac{\sqrt{3}}{2}(1-\cos\theta)\cos\frac{\theta}{2}$
$-\frac{1}{2}$	$\frac{\sqrt{3}}{2}(1-\cos\theta)\cos\frac{\theta}{2}$	$-\frac{1}{2}(1+3\cos\theta)\sin\frac{\theta}{2}$	$-\frac{1}{2}(1-3\cos\theta)\cos\frac{\theta}{2}$	$\frac{\sqrt{3}}{2}(1+\cos\theta)\sin\frac{\theta}{2}$
$-\frac{3}{2}$	$-\frac{1}{2}(1-\cos\theta)\sin\frac{\theta}{2}$	$\frac{\sqrt{3}}{2}(1-\cos\theta)\cos\frac{\theta}{2}$	$-\frac{\sqrt{3}}{2}(1-\cos\theta)\sin\frac{\theta}{2}$	$\frac{1}{2}(1+\cos\theta)\cos\frac{\theta}{2}$

5.14. Verify that, in the case $j = 2$, the functions $d^2_{m'm}(\Theta)$ have the form

$d^2_{m'm}(\theta) =$

$m' \backslash m$	2	1	0	−1	−2
2	$\frac{1}{4}(1+\cos\theta)^2$	$\frac{1}{2}\sin\theta(1+\cos\theta)$	$\sqrt{\frac{3}{8}}\sin^2\theta$	$\frac{1}{2}\sin\theta(1-\cos\theta)$	$\frac{1}{4}(1-\cos\theta)^2$
1	$-\frac{1}{2}\sin\theta(1+\cos\theta)$	$\frac{1}{2}(1+\cos\theta)(2\cos\theta-1)$	$\sqrt{\frac{3}{2}}\sin\theta\cos\theta$	$\frac{1}{2}(1-\cos\theta)(2\cos\theta+1)$	$\frac{1}{2}\sin\theta(1-\cos\theta)$
0	$\sqrt{\frac{3}{8}}\sin^2\theta$	$-\sqrt{\frac{3}{2}}\sin\theta\cos\theta$	$\frac{1}{2}(3\cos^2\theta-1)$	$\sqrt{\frac{3}{2}}\sin\theta\cos\theta$	$\sqrt{\frac{3}{8}}\sin^2\theta$
−1	$-\frac{1}{2}\sin\theta(1-\cos\theta)$	$\frac{1}{2}(1-\cos\theta)(2\cos\theta+1)$	$-\sqrt{\frac{3}{2}}\sin\theta\cos\theta$	$\frac{1}{2}(1+\cos\theta)(2\cos\theta-1)$	$\frac{1}{2}\sin\theta(1+\cos\theta)$
−2	$\frac{1}{4}(1-\cos\theta)^2$	$-\frac{1}{2}\sin\theta(1-\cos\theta)$	$\sqrt{\frac{3}{8}}\sin^2\theta$	$-\frac{1}{2}\sin\theta(1+\cos\theta)$	$\frac{1}{4}(1+\cos\theta)^2$

(5.390)

5.15. The axes of the stationary coordinate frame can be aligned along the principal axes of inertia of a physical system in twenty-four different ways (if only right-handed coordinate frames are used). In fact, we can choose the orientation of the axis Oz' in six different ways along three mutually perpendicular directions in space, and for each given direction of the axis Oz' we can choose the orientation of the axis Ox' in four different ways (the position of the axis Oy' will then be completely determined in a right-handed coordinate frame).

The only solutions of the Schrödinger equation which have physical meaning are those which are invariant under transformations corresponding to a transition from one orientation of the coordinate frame to another. Transitions from one coordinate frame to another can be described by means of three operators D_1, D_2, D_3 and powers of these operators.

The operator D_1 changes the directions of the axes Oz' and Oy' and corresponds to a rotation of the coordinate frame through an angle π about the axis Ox' ($D_1^2 = 1$). The operator D_2 corresponds to a rotation of the coordinate frame through $\pi/2$ about the axis Oz' ($D_2^4 = 1$). The operator D_3 corresponds to a cyclic permutation of the axes ($D_3^3 = 1$).

The operator D_1 can be represented as the sequence of rotations through the Euler angles $\Phi = 0$, $\Theta = \pi$ and $\psi = \pi$. The operator D_2 can be represented as a rotation through the Euler angle $\psi = \pi/2$. The operator D_3 can be represented as the sequence of rotations through the Euler angles $\Phi = 0$, $\Theta = \pi/2$ and $\psi = \pi/2$.

In any transformation D_k, the function $\psi(\mathbf{r})$ goes over into the transformed function

$$\psi(\mathbf{r}') = R_k \psi(\mathbf{r}),$$

where $R_k = D_k^{-1}$. According to the relations (5.58), on rotations of the coordinate frame the transformed function can be expressed in terms of the functions in the stationary coordinate frame:

$$\psi_{jm}(\mathbf{r}') \equiv R_k \psi_{jm}(\mathbf{r}) = \sum_{m'} D^{j*}_{mm'}\left(\Phi^{(k)}, \Theta^{(k)}, \psi^{(k)}\right)\psi_{jm'}(\mathbf{r}). \tag{5.391}$$

Now calculate the result of the action of the operators R_1, R_2 and R_3 on the D-functions:

$$R_1 D^j_{m'm}(\Phi,\Theta,\psi) = \sum_{m''} D^{j*}_{m'm''}(0,\pi,\pi) D^j_{m''m}(\Phi,\Theta,\psi).$$

Since

$$D^{j*}_{m'm''}(0,\pi,\pi) = \mathrm{e}^{-im'\pi} d^j_{m'm''}(\pi) = (-1)^{-j}\delta_{-m',m''},$$

we have

$$R_1 D^j_{m'm}(\Phi,\Theta,\psi) = (-1)^{-j} D^j_{-m',m}(\Phi,\Theta,\psi). \tag{5.392}$$

By analogy with (5.391), we have

$$R_2 D^j_{m'm}(\Phi,\Theta,\psi) = \mathrm{e}^{-im'(\pi/2)} D^j_{m'm}(\Phi,\Theta,\psi); \tag{5.393}$$

$$R_3 D^j_{m'm}(\Phi,\Theta,\psi) = \sum_{m''} D^j_{m'm''}\left(0,\frac{\pi}{2},\frac{\pi}{2}\right) D^{j*}_{m''m}(\Phi,\Theta,\psi). \tag{5.394}$$

Calculate the result of the action of the operators R_1, R_2 and R_3 on the variables β and γ. Using the transformation formulae (5.58), we find

$$R_1\alpha'_{2\mu} = \sum_{\mu'} D^{2*}_{\mu\mu'}(0,\pi,\pi)\alpha'_{2\mu'} = \alpha'_{2,-\mu}.$$

Since

$$\alpha'_{20} = \beta\cos\gamma, \quad \text{and} \quad \alpha'_{2,2} = \alpha'_{2,-2} = \frac{\beta}{\sqrt{2}}\sin\gamma,$$

we have

$$R_1\{\gamma\} = \{\gamma\}. \tag{5.395}$$

In addition,

$$R_2\alpha'_{2\mu} = \mathrm{e}^{-i\mu(\pi/2)}\alpha'_{2\mu},$$

or

$$R_2\alpha'_{20} = \alpha'_{20}, \qquad R_2\alpha'_{22} = -\alpha'_{22}.$$

Thus,

$$R_2\{\gamma\} = \{-\gamma\}. \tag{5.396}$$

Finally,

$$R_3\alpha'_{2\mu} = \mathrm{e}^{-i\mu(\pi/2)}\left\{ d^2_{\mu 0}(\pi/2)\alpha'_{20} + \left(d^2_{\mu 2}(\pi/2) + d^2_{\mu -2}(\pi/2)\right)\alpha'_{2\mu}\right\}.$$

Using the explicit form of the functions $d^2_{\mu\mu'}(\pi/2)$, we find

$$\begin{aligned} R_3 \cos\gamma &= (\sqrt{3}/2)\sin\gamma - \frac{1}{2}\cos\gamma \equiv \cos(\gamma - 2\pi/3); \\ R_3 \sin\gamma &= -\frac{1}{2}\sin\gamma - (\sqrt{3}/2)\cos\gamma \equiv \sin(\gamma - 2\pi/3), \end{aligned}$$

or

$$R_3\{\gamma\} = \{\gamma - 2\pi/3\}. \tag{5.397}$$

5.16. Show that three-phonon quadrupolar excitations are characterized by the values Λ=0,2,3,4 and 6 of the total angular momentum and have positive parity.

The total angular momentum $\boldsymbol{\Lambda}$ of a three-phonon excitation is equal to the sum of the angular momenta of the individual excitations:

$$\boldsymbol{\Lambda} = \boldsymbol{\lambda}_1 + \boldsymbol{\lambda}_2 + \boldsymbol{\lambda}_3.$$

Correspondingly, the projection Λ_z of the total angular momentum along the axis Oz in the laboratory coordinate frame is equal to

$$\Lambda_z = \lambda_{1z} + \lambda_{2z} + \lambda_{3z}.$$

A one-phonon quadrupolar excitation is characterized by angular momentum equal to 2 with projection μ; the corresponding function $b^+_{2\mu}|0>$, where $b^+_{2\mu}$ is a phonon creation operator, satisfies the equations

$$\boldsymbol{\lambda}^2 b^+_{2\mu}|0> = 2(2+1)b^+_{2\mu}|0>, \qquad \lambda_z b^+_{2\mu}|0> = \mu b^+_{2\mu}|0> .$$

We can choose a complete set of three-phonon functions in the form

$$b^{+(1)}_{2\mu_1} b^{+(2)}_{2\mu_2} b^{+(3)}_{2\mu_3}|0>, \tag{5.398}$$

where μ_1, μ_2 and μ_3 run over all possible values. Clearly, there are 125 different independent functions (5.398). From these we need to construct completely symmetric linear combinations, which must be eigenfunctions of the operators $\boldsymbol{\Lambda}^2$ and Λ_z:

$$\begin{aligned} \boldsymbol{\Lambda}^2 \psi_{\Lambda M} &= \Lambda(\Lambda+1)\psi_{\Lambda M}; \\ \Lambda_z \psi_{\Lambda M} &= M\psi_{\Lambda M}. \end{aligned} \tag{5.399}$$

The rules for vector addition of angular momenta permit, generally speaking, the following angular-momentum values Λ: 0,1,2,3,4,5 and 6. In a system of three phonons, however, not all these values of the total angular momentum are allowed, since the wave function must be completely symmetric with respect to permutations of the phonons. We shall ascertain how many linearly independent functions satisfying the symmetry requirements can be constructed for each value of M.

If M=6 or 5, in each case we can construct only one completely symmetric function:

$$M = 6; \qquad b_{22}^{+(1)} b_{22}^{+(2)} b_{22}^{+(3)} |0>,$$
$$M = 5; \qquad b_{22}^{+(1)} b_{22}^{+(2)} b_{21}^{+(3)} |0> .$$

If $M = 4$, we can construct two functions:

$$M = 4; \quad b_{22}^{+(1)} b_{22}^{+(2)} b_{20}^{+(3)} |0>, \quad b_{22}^{+(1)} b_{21}^{+(2)} b_{21}^{+(3)} |0>$$

If $M = 3$, we can construct three functions:

$$M = 3; \quad b_{22}^{+(1)} b_{22}^{+(2)} b_{2-1}^{+(3)} |0>, \quad b_{22}^{+(1)} b_{21}^{+(2)} b_{20}^{+(3)} |0>,$$
$$b_{21}^{+(1)} b_{21}^{+(2)} b_{21}^{+(3)} |0> .$$

If $M = 2$, or 1, in each case four functions can be constructed:

$$M = 2; \qquad b_{22}^{+(1)} b_{22}^{+(2)} b_{2-2}^{+(3)} |0>, b_{22}^{+(1)} b_{21}^{+(2)} b_{2-1}^{+(3)} |0>;$$
$$b_{22}^{+(1)} b_{20}^{+(2)} b_{20}^{+(3)} |0>, b_{21}^{+(1)} b_{21}^{+(2)} b_{20}^{+(3)} |0>;$$

$$M = 1; \qquad b_{22}^{+(1)} b_{21}^{+(2)} b_{2-2}^{+(3)} |0>, b_{22}^{+(1)} b_{20}^{+(2)} b_{2-1}^{+(3)} |0>,$$
$$b_{21}^{+(1)} b_{21}^{+(2)} b_{2-1}^{+(3)} |0>, b_{21}^{+(1)} b_{20}^{+(2)} b_{20}^{+(3)} |0> .$$

Finally, if $M = 0$, we can construct five functions:

$$M = 0; \qquad b_{22}^{+(1)} b_{2-2}^{+(2)} b_{20}^{+(3)} |0>, b_{22}^{+(1)} b_{2-1}^{+(2)} b_{2-1}^{+(3)} |0>,$$
$$b_{21}^{+(1)} b_{21}^{+(2)} b_{2-2}^{+(3)} |0>, b_{21}^{+(1)} b_{2-1}^{+(2)} b_{20}^{+(3)} |0>,$$
$$b_{20}^{+(1)} b_{20}^{+(2)} b_{20}^{+(3)} |0> .$$

After symmetrization, each of the products gives a linearly independent function with a well-defined value of M. The fact that no additional linearly independent combination appears on going from $M = 6$ to $M = 5$ and from $M = 2$ to $M = 1$ tells us that the values $\Lambda = 5$ and 1 are forbidden.

The positive parity of each of the states follows from the fact that the functions of these states are constructed from quadrupolar one-phonon functions.

An exactly similar elementary treatment shows that two-phonon quadrupolar excitations can have only three values of the total angular momentum: $\Lambda = 0, 2$ and 4.

5.17. Show that four-phonon quadrupolar excitations have positive parity and are characterized by the following values Λ of the total angular momentum: 0,2,4,5,6 and 8; the values $\Lambda = 2$ and 4 are encountered twice and to classify these an additional quantum number – the *seniority* – is introduced.

The solution of this problem is analogous to that of the preceding problem.

5.18. Show that three-phonon octupolar ($\lambda = 3$) excitations have negative parity and the following total angular-momentum values Λ:0,3,4,5,6,7 and 9, the value $\Lambda = 3$ being encountered twice (for three or more phonons, it becomes necessary to introduce an additional quantum number – the seniority – for a complete classification of the states).

The solution of this problem is analogous to the solutions of the preceding problems.

5.19. Quantize the kinetic energy associated with the rotation and surface deformations of a nucleus, selecting the Euler angles Φ, Θ and ψ and the deformation parameters β and γ as the independent variables.

The classical expression for the kinetic energy associated with the rotation and surface deformations of a nucleus has the form

$$T = \frac{1}{2}\sum_{k'} \omega_{k'}^2 J_{k'} + (B/2)(\dot{\beta}^2 + \beta^2\dot{\gamma}^2), \tag{5.400}$$

where the components of the angular velocity along the axes of the moving coordinate frame are expressed in terms of derivatives of the Euler angles:

$$\begin{aligned} \omega_{x'} &= -\dot{\Phi}\sin\Theta\,\cos\psi + \dot{\Theta}\sin\psi; \\ \omega_{y'} &= \dot{\Phi}\sin\Theta\sin\psi + \dot{\Theta}\cos\psi; \\ \omega_{z'} &= \dot{\Phi}\cos\Theta + \dot{\psi}. \end{aligned} \tag{5.401}$$

As the generalized coordinates q_i, we choose the quantities

$$q_1 = \Phi, \qquad q_2 = \Theta, \qquad q_3 = \psi, \qquad q_4 = \beta, \qquad \text{and} \qquad q_5 = \gamma.$$

Then the kinetic energy (5.400) can be represented in the form

$$T = \frac{B}{2}\left(\frac{ds}{dt}\right)^2, \tag{5.402}$$

where ds^2 is the square of the line differential,

$$ds^2 = g_{ij}dq_i dq_j. \tag{5.403}$$

The symmetric matrix g_{ij} in this case has the form

$$g_{ij} = \begin{pmatrix} g_{11} & g_{12} & g_{13} & 0 & 0 \\ g_{21} & g_{22} & 0 & 0 & 0 \\ g_{31} & 0 & g_{33} & 0 & 0 \\ 0 & 0 & 0 & g_{44} & 0 \\ 0 & 0 & 0 & 0 & g_{55} \end{pmatrix}, \tag{5.404}$$

where the non-zero matrix elements are equal to:

$$\left.\begin{aligned}
g_{11} &= \frac{J_1}{B}\sin^2\Theta\cos^2\psi + \frac{J_2}{B}\sin^2\Theta\sin^2\psi + \frac{J_3}{B}\cos^2\Theta;\\
g_{12} &= \frac{1}{B}(J_2 - J_1)\sin\Theta\sin\psi\cos\psi;\\
g_{13} &= \frac{J_3}{B}\cos\Theta;\\
g_{22} &= \frac{J_1}{B}\sin^2\psi + \frac{J_2}{B}\cos^2\psi;\\
g_{33} &= \frac{J_3}{B};\\
g_{44} &= 1;\\
g_{55} &= \beta^2.
\end{aligned}\right\} \tag{5.405}$$

Using (5.405), it is not difficult to find the determinant of the matrix (5.404):

$$g \equiv |g_{ij}| = \frac{J_1 J_2 J_3}{B^3}\beta^2\sin^2\Theta. \tag{5.406}$$

By the rules of linear algebra, we then find the matrix g_{ij}^{-1} inverse to (5.404):

$$g_{ik}^{-1} g_{kj} = \delta_{ij}.$$

This matrix has the form

$$g_{ij}^{-1} = \begin{pmatrix} g_{11}^{-1} & g_{12}^{-1} & g_{13}^{-1} & 0 & 0\\ g_{21}^{-1} & g_{22}^{-1} & g_{23}^{-1} & 0 & 0\\ g_{31}^{-1} & g_{32}^{-1} & g_{33}^{-1} & 0 & 0\\ 0 & 0 & 0 & g_{44}^{-1} & 0\\ 0 & 0 & 0 & 0 & g_{55}^{-1} \end{pmatrix}, \tag{5.407}$$

where

$$\left.\begin{aligned}
g_{11}^{-1} &= B\Big(\frac{\cos^2\psi}{J_1} + \frac{\sin^2\psi}{J_2}\Big);\\
g_{12}^{-1} &= -B\Big(\frac{1}{J_1} - \frac{1}{J_2}\Big)\frac{\sin\psi\cos\psi}{\sin\Theta};\\
g_{13}^{-1} &= -B\Big(\frac{\cos^2\psi}{J_1} + \frac{\sin^2\psi}{J_2}\Big)\cot\Theta;\\
g_{22}^{-1} &= B\Big(\frac{\sin^2\psi}{J_1} + \frac{\cos^2\psi}{J_2}\Big);\\
g_{23}^{-1} &= B\Big(\frac{1}{J_1} - \frac{1}{J_2}\Big)\cot\Theta\sin\psi\cos\psi;\\
g_{33}^{-1} &= B\Big(\frac{\cos^2\psi}{J_1} + \frac{\sin^2\psi}{J_2}\Big)\cot^2\Theta + \frac{B}{J_3};\\
g_{44}^{-1} &= 1; \qquad g_{55}^{-1} = \frac{1}{\beta^2}.
\end{aligned}\right\} \tag{5.408}$$

The Laplacian in the five-dimensional space is defined by the general expression

$$\nabla^2 = \frac{1}{\sqrt{g}} \frac{\partial}{\partial q_i} \left(\sqrt{g}\, g_{ij}^{-1} \frac{\partial}{\partial q_j} \right).$$

Substituting the values of the determinant (5.406) and of the matrix elements of the inverse matrix (5.407) into this formula, we obtain

$$\nabla^2 = -\sum_{k'} \frac{B}{J_{k'}} R_{k'}^2 + \frac{1}{\beta^4} \frac{\partial}{\partial \beta} \beta^4 \frac{\partial}{\partial \beta} + \frac{1}{\beta^2} \frac{1}{\sin\, 3\gamma} \frac{\partial}{\partial \gamma} \sin\, 3\gamma \frac{\partial}{\partial \gamma}, \tag{5.409}$$

where $R_{k'}$ are the components of the angular momentum of rotation along the moving axes:

$$\begin{aligned} R_{x'} &= -i\left(-\frac{\cos\psi}{\sin\Theta} \frac{\partial}{\partial \Phi} + \sin\psi \frac{\partial}{\partial \Theta} + \cot\Theta \cos\psi \frac{\partial}{\partial \psi} \right); \\ R_{y'} &= -i\left(\frac{\sin\psi}{\sin\Theta} \frac{\partial}{\partial \Phi} + \cos\psi \frac{\partial}{\partial \Theta} - \cot\Theta \sin\psi \frac{\partial}{\partial \psi} \right); \\ R_{z'} &= -i \frac{\partial}{\partial \psi}. \end{aligned} \tag{5.410}$$

The operator of the kinetic energy associated with the rotation and surface deformations of the nucleus has the form

$$T \equiv -\frac{\hbar^2}{2B} \nabla^2 = \sum_{k'} \frac{\hbar^2}{2J_{k'}} R_{k'}^2 - \frac{\hbar^2}{2B} \left\{ \frac{1}{\beta^4} \frac{\partial}{\partial \beta} \beta^4 \frac{\partial}{\partial \beta} + \frac{1}{\beta^2} \frac{1}{\sin 3\gamma} \frac{\partial}{\partial \gamma} \sin 3\gamma \frac{\partial}{\partial \gamma} \right\}. \tag{5.411}$$

As we should expect, this formula coincides with the formula (5.251) obtained earlier.

5.20. In a non-adiabatic approximation, consider the nuclear collective motions associated with axially symmetric deformations.

If we confine ourselves to axially symmetric deformations only (that is, put $\gamma = 0$), then

$$J_{x'} = J_{y'} = 3B\beta^2, \quad J_{z'} = 0, \quad \omega_{x'}^2 + \omega_{y'}^2 = \dot{\Phi}^2 \sin^2\Theta + \dot{\Theta}^2,$$

and the kinetic energy of the collective motions takes the form

$$T = \frac{B}{2} \{ 3\beta^2 (\dot{\Phi} \sin^2 \Theta^2 + \dot{\Theta}^2) + \dot{\beta}^2 \}. \tag{5.412}$$

Since in the case of axial symmetry the nucleus can rotate only about directions perpendicular to the symmetry axis, the motion is characterized by only three degrees of freedom: $q_1 = \Phi$, $q_2 = \Theta$ and $q_3 = \beta$. Introducing the square of the line differential, $ds^2 = g_{ij} dq_i dq_j$, and noting that the matrix g_{ij} is diagonal

$$g_{ij} = \begin{pmatrix} 3\beta^2 \sin^2\Theta & 0 & 0 \\ 0 & 3\beta^2 & 0 \\ 0 & 0 & 1 \end{pmatrix},$$

we obtain, using the general rules, the three-dimensional Laplacian

$$\nabla^2 = \frac{1}{\beta^2}\frac{\partial}{\partial\beta}\beta^2\frac{\partial}{\partial\beta} + \frac{1}{3\beta^2}\left(\frac{1}{\sin\Theta}\frac{\partial}{\partial\Theta}\sin\Theta\frac{\partial}{\partial\Theta} + \frac{1}{\sin^2\Theta}\frac{\partial^2}{\partial\Phi^2}\right). \tag{5.413}$$

Taking into account that the potential energy in the case of small deformations is proportional to the square of the deformation parameter β, we write the Hamiltonian of the system in the form

$$H = -\frac{\hbar^2}{2B}\left\{\frac{1}{\beta^2}\frac{\partial}{\partial\beta}\beta^2\frac{\partial}{\partial\beta} + \frac{1}{3\beta^2}\left[\frac{1}{\sin\Theta}\frac{\partial}{\partial\Theta}\sin\Theta\frac{\partial}{\partial\Theta} + \frac{1}{\sin^2\Theta}\frac{\partial^2}{\partial\Phi^2}\right]\right\} + \frac{1}{2}C\beta^2. \tag{5.414}$$

We note that the angular dependence of the Hamiltonian (5.414) coincides with that of the ordinary Laplacian operator, and so the Schrödinger equation

$$H\Psi = E\Psi \tag{5.415}$$

can be solved by the method of separation of variables.

We choose the function $\Psi(\Phi, \Theta, \beta)$ in the form

$$\Psi(\Phi, \Theta, \beta) = f(\beta)Y_{lm}(\Theta, \Phi). \tag{5.416}$$

Then the function $f(\beta)$ will be determined by the equation

$$-\frac{\hbar^2}{2B}\left\{\frac{1}{\beta^2}\cdot\frac{\partial}{\partial\beta}\beta^2\frac{\partial}{\partial\beta} - \frac{l(l+1)}{3\beta^2}\right\}f(\beta) + \frac{1}{2}C\beta^2 f(\beta) = Ef(\beta). \tag{5.417}$$

We introduce the new variables:

$$\left.\begin{aligned} &\beta = \beta_0\xi; \quad \beta_0 = \left(\frac{\hbar^2}{BC}\right)^{1/4}; \\ &E = \frac{\lambda}{2}\hbar\omega; \quad \omega = \sqrt{\frac{C}{B}}. \end{aligned}\right\} \tag{5.418}$$

With these variables,

$$\frac{1}{\xi^2}\frac{\partial}{\partial\xi}\xi^2\frac{\partial}{\partial\xi}f - \frac{1}{3\xi^2}l(l+1)f - \xi^2 f + \lambda f = 0. \tag{5.419}$$

We investigate the behavior of the solution as $\xi \to 0$. Putting $f(\xi) \sim \xi^s$, we obtain for s the characteristic equation

$$s^2 + s - \frac{1}{3}l(l+1) = 0,$$

whence

$$s = -\frac{1}{2} \pm \sqrt{\frac{1}{4} + \frac{1}{3}l(l+1)}.$$

Since $f(\xi)$ must be finite at $\xi = 0$, we choose the first solution. Of the two asymptotic solutions as $\xi \to \infty$, we choose $f(\xi) \sim e^{-(1/2)\xi^2}$ and seek the solution of eqn. (5.419) in the form

$$f(\xi) = \xi^s u e^{-\xi^2/2}, \tag{5.420}$$

where u is a polynomial satisfying the equation

$$u'' + 2\left(\frac{s+1}{\xi} - \xi\right)u' - (2s+3-\lambda)u = 0. \tag{5.421}$$

The solution of this equation is the confluent hypergeometric function

$$u = F\left(\frac{1}{2}(s + \frac{3}{2} - \lambda/2), s + \frac{3}{2}; \xi^2\right),$$

which becomes a polynomial if

$$\frac{1}{2}(s + \frac{3}{2} - \lambda/2) = -n, \tag{5.422}$$

where n is a non-negative integer (n=0,1,2,...). In this case, the confluent hypergeometric function reduces to a Laguerre polynomial

$$F(-n, s + \frac{3}{2}; \xi^2) = \frac{n!}{(s + \frac{3}{2})} L_n^{s+(1/2)}(\xi^2),$$

and the condition (5.422) defines the possible energy values. Finally, the energy levels of the system and the corresponding wave functions can be represented in the form

$$E_{nl} = (\frac{3}{2} + 2n + s)\hbar\omega; \tag{5.423}$$

$$\Psi_{nlm}(\Phi, \Theta, \beta) = N\xi^s L_n^{s+(1/2)}(\xi^2) e^{-\xi^2/2} Y_{lm}(\Theta, \Phi), \tag{5.424}$$

where

$$s = \frac{1}{2}(-1 + \sqrt{1 + \frac{4}{3}l(l+1)}),$$

$\xi = \beta/\beta_0$ and N is a normalization constant:

$$N = \left(\frac{\Gamma(n+1)}{\beta_0^3 \Gamma(n+s+1)}\right)^{1/2}$$

From the requirement that the wave function be symmetric with respect to the plane perpendicular to the symmetry axis of the nucleus and passing though the centre of the nucleus, it follows that only even values of l are possible.

Figure 5.23 depicts the scheme of the energy levels calculated from formula (5.423). The energy levels for the nuclei $^{92}_{40}Zr$, $^{94}_{40}Zr$ and $^{150}_{62}Sm$ are plotted on the right in Fig. 5.23. The magnitude of $\hbar\omega$ for each of the nuclei was determined from the position of the first observable excited level. The model considered enables us to explain the presence of the closely spaced levels 0^+ and 4^+. The ratio, calculated

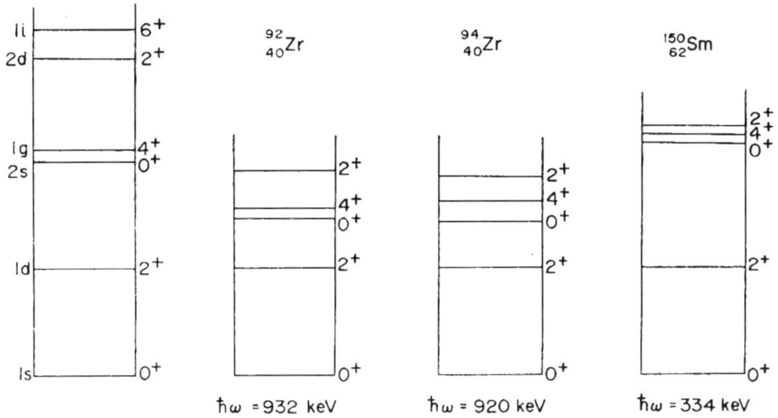

Figure 5.23. Scheme of the rotational-vibrational levels for axially symmetric deformations. Also given are the observed level schemes for the nuclei $^{92}_{40}Zr$, $^{94}_{40}Zr$ and $^{150}_{62}Sm$, for which the parameter $\hbar\omega$ is equal to 932, 920 and 334 keV respectively.

from formula (5.423), of the energies of the closely spaced levels 4^+ and 0^+ is equal to $1 \cdot 06$, whereas the ratios of the experimental values of the levels for the above three nuclei are $1 \cdot 08$ ($^{92}_{40}Zr$),$1 \cdot 12$ ($^{94}_{40}Zr$) and $1 \cdot 05$ ($^{150}_{62}Sm$).

5.21. Assuming that the self-consistent potential of the nucleus differs only slightly from a spherically symmetric potential, use perturbation theory to determine the correction, associated with the deformation of the nucleus, to the one-particle energy.

Unlike the case of large deformations, in the case of small deformations the one-particle energy levels and wave functions of nucleons owing in a weakly deformed nuclear field can be found in explicit form by using perturbation theory. As the perturbation V', we shall take the difference of the non-spherical potential, which depends on the deformation parameters, and the spherical (shell) potential. To determine the first correction ΔE to the energy, we must calculate the diagonal matrix element of the perturbation V' using the wave functions of the shell model

$$\chi_{j\Omega}^{\tau=nl} = R_{nl}(r')\mathcal{Y}_{l(1/2)j\Omega}, \tag{5.425}$$

where $\mathcal{Y}_{l(1/2)j\Omega}$ is the spin-angle function. Choosing the potential in the form of a square well of depth V_0 for simplicity, we represent the correction ΔE to the energy in the form

$$\Delta E \equiv \langle\chi_{j\Omega}^{nl}|V'|\chi_{j\Omega}^{nl}\rangle = \sum_{\text{spins}}\int do_{r'}\left\{-V_0\int_0^{R(\vartheta,\varphi)} dr'r'^2|\chi_{j\Omega}^{nl}|^2 + V_0\int_0^{R} dr'r'^2|\chi_{j\Omega}^{nl}|^2\right\}$$

$$= -V_0\sum_{\text{spins}}\int dO_{r'}\mathcal{Y}^*_{l(1/2)j\Omega}\times\mathcal{Y}_{l(1/2)j\Omega}\int_R^{R(\vartheta,\varphi)} dr'r'^2|R_{nl}(r')|^2. \tag{5.426}$$

Considering only ellipsoidal deformations and taking into account that the deviations of the quantity

$$R(\vartheta,\varphi) = \left(1 + \sum_\nu \alpha'_\nu Y_{2\nu}(\vartheta,\varphi)\right) R$$

from the radius R of a sphere of the same volume are small, after summation over the spin variables we obtain the correction ΔE in the form

$$\Delta E = -V_0 R^3 |R_{nl}(R)|^2 \sum_\nu \alpha'_\nu \langle j\Omega|Y_{2\nu}|j\Omega\rangle. \tag{5.427}$$

If we had not used a square potential V, then in the perturbation V', which for the square potential had the form

$$V' = -V_0 R\delta(r' - R) \sum_\nu \alpha'_\nu Y_{2\nu}(\vartheta,\varphi),$$

it would be necessary to replace $V_0 R\delta(r' - R)$ by the quantity $r'dV/dr'$. Noting that

$$\langle j\Omega|Y_{2\nu}|j\Omega\rangle = -\frac{1}{4}\sqrt{\frac{5}{4\pi}}\,\frac{3\Omega^2 - j(j+1)}{j(j+1)}\delta_{\nu 0},$$

we finally obtain for the correction to the one-particle energy the expression

$$\Delta E = c_j\left[3\Omega^2 - j(j+1)\right]\alpha'_0, \qquad c_j = \frac{1}{4}\sqrt{\frac{5}{4\pi}}\,\frac{V_0 R^3 |R_{nl}(R)|^2}{j(j+1)}, \tag{5.428}$$

or, since $\alpha'_0 = \beta\cos\gamma$, we have ($\gamma < 15°$)

$$\Delta E \equiv \Delta E_{nlj\Omega} = c_j[3\Omega^2 - j(j+1)]\beta\cos\gamma. \tag{5.429}$$

For small deformations ($\beta \ll 1$), the state of a nucleon in the nucleus can be characterized approximately by the quantum number j of the total angular momentum. It follows from (5.429) that for $j = \frac{1}{2}$ the first-order perturbation-theory correction ΔE is equal to zero. There is a degeneracy in the sign of the projection Ω of the total angular momentum.

If we have several nucleons in the shell (nlj), the corresponding correction to the energy will be equal the sum of the separate corrections (5.429) for each of the nucleons. If the shell (nlj) is completely filled, that is, if there are $2j+1$ nucleons with different angular-momentum projections Ω in the shell, the total correction to the energy is equal to zero, since

$$\sum_{\Omega=-j}^{j} [3\Omega^2 - j(j+1)] = 0.$$

5.22. Determine the equilibrium deformation and the order of occupation of the one-particle states in the outer shell of a weakly deformed nucleus.

A nucleus with a completely filled outer shell has spherical shape. If we add one nucleon to such a nucleus, the new nucleus will then differ somewhat from a sphere, the equilibrium deformation being determined by the interaction of the outer nucleon with the nuclear core formed by the filled shells. Because of this interaction, the core is polarized, and deformed, as a result, the field in which the outer nucleon moves will not possess spherical symmetry. The core resists, as it were, deformations; on the other hand, the outer nucleon tends to increase the deformation. The equilibrium deformation will thus depend on the elastic properties of the core and on the state of the outer nucleon in the deformed field of the core.

The dependence of the potential energy of the core on the deformation is determined for small deformations by the quadratic law

$$E'(\beta) = E'(0) + \frac{1}{2}C\beta^2. \tag{5.430}$$

The deformation dependence of the potential energy of the whole nucleus with the one outer nucleon is determined by the expression ($\gamma < 15°$)

$$E(\beta) = E(0) + c_j[3\Omega^2 - j(j+1)]\beta\cos\gamma + \frac{1}{2}C\beta^2. \tag{5.431}$$

We shall consider deformed nuclei with axial symmetry, that is, we put $\gamma = 0$. Then we obtain the equilibrium value β_0 of the parameter β from the condition

$$\left.\frac{\partial E(\beta)}{\partial\beta}\right|_{\beta=\beta_0} = 0. \tag{5.432}$$

Substituting for $E(\beta)$ from (5.431), we obtain for the equilibrium deformation the value

$$\beta_0(\Omega) = -\frac{c_j}{C}[3\Omega^2 - j(j+1)]. \tag{5.433}$$

We obtain the energy of the nucleus in the equilibrium state by substituting (5.433) into (5.431) with $\gamma = 0$:

$$E(\beta_0(\Omega)) = E(0) - \frac{c_j^2}{2C}[3\Omega^2 - j(j+1)]^2. \tag{5.434}$$

It can be seen from this that, for one outer nucleon, the energy in equilibrium will be minimized if we take the largest possible value of $|\Omega|$, that is, if we put $\Omega = \pm j$. Then $\beta_0 < 0$, that is, the nucleus will be oblate along the symmetry axis and will have a negative quadrupole moment.

If there are several nucleons in the shell (nlj), then, since we are assuming that they do not interact with each other, the results can be generalized easily by replacing the quantity $3\Omega^2 - j(j+1)$ in the formulae by a sum over the nucleons:

$$\sum_i [3\Omega_i^2 - j(j+1)].$$

The Pauli principle permits only two identical nucleons to have the same value of Ω_i^2. If the shell is completely filled, the equilibrium deformation is equal to zero, as it should be:

$$\beta_0 = -\frac{c_j}{C} \sum_{\Omega_i=-j}^{j} [3\Omega_i^2 - j(j+1)] = 0.$$

If there are two nucleons in the shell, the energy will be minimized if the projections of their angular momenta are equal in magnitude but opposite in sigh: $\Omega_1 = +j$ and $\Omega_2 = -j$, so that the total projection is equal to zero. If to the two nucleons we add a third, this will have $\Omega_3 = j-1$ or $\Omega_3 = -j+1$, and so on. The total projection $\sum_i \Omega_i$ of the total angular momentum will be equal to zero if there is an even number of nucleons in the unfilled shell; otherwise, it is equal to the projection of the angular momentum of the last odd nucleon.

When the shell is more than half full, it is energetically favorable for nucleons to occupy the states with the minimum projections of the total angular momentum: $\Omega_{1,2} = \pm\frac{1}{2}$, $\Omega_{3,5}4 = \pm\frac{3}{2}$, ... This can be seen clearly from the general expression for the energy of the nucleus in the equilibrium state:

$$E\left(\sum_i \beta_0(\Omega_i)\right) = E(0) - \frac{c_j^2}{2C}\left\{\sum_i [3\Omega^2 - j(j+1)]\right\}^2. \tag{5.435}$$

Suppose, for example, that we have a complete shell less one nucleon. It can be seen from (5.435) that the energy will be minimized if a state with $|\Omega_i| = j$ is not occupied. In fact, in this case we have for the square of the sum in the right-hand side of the equality (5.435):

$$\left\{\sum_i [3\Omega_i^2 - j(j+1)]\right\}^2$$

$$= \left\{\sum_{\Omega_i=-j}^{j} [3\Omega_i^2 - j(j+1)] - [3j^2 - j(j+1)]\right\}^2 = [3j^2 - j(j+1)]^2,$$

which, for $j \geq \frac{3}{2}$, is the largest of all the quantities $[3\Omega_i^2 - j(j+1)]^2$. The equilibrium deformation in this case is equal to

$$\sum_i \beta_0(\Omega_i) = -\frac{c_i}{C}\sum_i [3\Omega_i^2 - j(j+1)] = \frac{c_j}{C}[3j_i^2 - j(j+1)],$$

that is, is a possible quantity, corresponding to a nucleus that is prolate along the symmetry axis, with a positive quadrupole moment.

Thus, we can draw the following general conclusion. If the number of nucleons in the shell does not exceed half the maximum possible number of nucleons in the given shell, the occupation of all states with $|\Omega_i| = j, j-1, j-2$, etc., down to a certain $|\Omega_i| \geq \frac{1}{2}(j+\frac{1}{2})$ is energetically favoured. In this case, the equilibrium deformation is negative, corresponding to an oblate nucleus having a negative quadrupole moment. But if in the shell there are more than half the maximum possible number of nucleons, then states with $|\Omega_i| = \frac{1}{2}, \frac{3}{2}, \ldots$, etc., up to a certain value of $|\Omega_i|$ which is again greater than or equal to $\frac{1}{2}(j+\frac{1}{2})$, are occupied. In this case, the nucleus is prolate (the equilibrium deformation is positive) and its quadrupole moment is positive. It is clear that the maximum equilibrium deformation will be reached when the shell is exactly half-filled.

The simple model described helps us to understand qualitatively the mechanism of the appearance of deformation of nuclei. Quantitatively, however, it does not explain the magnitudes of the quadrupole moments and spins of nuclei. Only for certain nuclei, close to the magic nuclei, does this model correctly predict the spins of the ground states of nuclei. But far from the magic nuclei, where the deformations can no longer be assumed to be small, we cannot assume that the quantum number j of the total angular momentum of a nucleon is, even approximately, a good quantum number. In addition, it is necessary to take into account higher powers in the expansion of the energy of the nucleus in powers of the deformation parameter.

If we plot the energy (5.435) of the nucleus as a function of the number of identical particles in the shell, we obtain a symmetric dependence about the middle of the shell, where the energy has a minimum. As we have seen using the example of the Nilsson potential, the one-particle levels plotted as functions of the deformation parameter intersect more and more frequently as the excitation energy and magnitude of the deformation increase. If we fix the number of nucleons and plot their energy as a function of the magnitude of the deformation, we shall observe discontinuities in this dependence, associated with the fact that individual nucleons will undergo transitions from certain states to others that are energetically more favoured. The change of state will occur exactly on the intersection of the one-particle energy curves. The discontinuous curve describing the energy of several nucleons as a function of the deformation usually has several minima, and it may happen that the magnitude of the energy is almost the same at some of the minima. In this case, a small change of the potential can lead to a sharp change in the order of the levels.

References

Arnold, V.I. (1963) Usp.Math.Nauk **18**, 13.

Bohr, A. (1952) K. Danske Vidensk. Selsk. Mat.-fys. Medd. **26**, No.14.

Bohr, A. and Mottelson, B.R. (1953) K. Danske Vidensk. Selsk. Mat.-fys. Medd.**27**, No.16.

Bolotin Yu.L., Gonchar V.Yu., Inopin E.V. et al. (1989) Phys. Element. Chast. and Atom Yadra **20**, 878.

Davydov, A.S. (1959) Izv. Akad. Nauk SSSR, ser. Fiz. **23**, 792 [Bull. Acad. Sci. USSR Phys. Ser.**23**, 783(1959)].

Davydov, A.S. (1960) Izv. Akad. Nauk SSSR, ser. Fiz. **24**, 820 [Bull. Acad. Sci. USSR Phys. Ser.**24**, 822(1960)].

Davydov, A.S. and Chaban, A.A. (1960) Nucl.Phys. **20**, 499.

Davydov, A.S. and Filippov, G.F. (1958) Zh. Eksp. Teor. Fiz. **35**, 440 [Sov. Phys. - JETP **8**, 303(1959)].

Dyson F.J. (1962) Journ. Math.Phys. **3**, No1. 140, 157, 166.

Edmonds, A.R. (1960)*Angular Momentum in Quantum Mechanics*, Princeton University Press.

Henon M., Heils C. (1964) Astron.J. **69**, 73.

Inglis, D.R. (1954) Phys.Rev. **96**, 1059.

Inglis, D.R. (1956) Phys.Rev. **103**, 1786.

Kolmogorov, A.N. (1954) Dokl.Akad.Nauk SSSR **98**, 527.

Kolmogorov A.N. (1959) Dokl. Akad. Nauk SSSR **124**, 754.

Ljapunov A.M. (1954) Izv. Akad. Nauk SSSR **1**; (1965), **2**.
Migdal, A.B. (1959) Zh.Eksp.Teor.Fiz. **37**, 249
[Sov.Phys. – JETP **10**, 176(1960)].
Moser J.K. (1968) *Lectures on Hamiltonian Systems*, Courant Institute of Math. Science, N.Y.
Moszkowski, S.A. (1957)*Models of Nuclear Structure*, Encyclopaedia of Physics (ed. S. Flügge), **39**, Springer-Verlag-Berlin.
Nilsson, S.G. (1955) K. Danske Vidensk. Selsk. Mat.-fys. Medd. **29**, No.16.
Pauli, W. (1933) *General Principles of Wave Mechanics* (in German),
Poincaré, H (1924) *La Méchanique Nouvelle*, Paris.
Handbuch der Physik, **24**, 1 Chap.2.
Solovyev V.G. (1971) *Theory of Composed Nuclei*, Nauka, Moskow.
Tartakovsky, V.K. (1989) Sov.J.Nucl.Phys. **50**, 335.
Wigner, E.P. (1959) *Group Theory and its Applications to the Quantum Mechanics of Atomic Spectra*, Academic Press, N.Y.

CHAPTER 6

THE PAIRING OF NUCLEONS

6.1. Superfluid States of Nuclei

Short-range pairing forces and long-range multipole forces. The shell model and generalized model considered in the preceding Chapters have made it possible to explain many of the properties of nuclei, and also to establish general trends as the charges and mass numbers of nuclei are varied. There are, however, certain features which cannot be explained in the framework of the shell model or generalized model; this indicates the need to invoke a different approach to the description of the properties of the ground and excited states of nuclei.

Amongst such properties and trends, we list the following as being the most important: the difference in the masses of even and add nuclei; the existence of an energy gap in the spectra of the excited states of even-even nuclei and the absence of such a *gap in the spectra* of odd and odd-odd nuclei; the density of the one-particle levels; the equilibrium deformations of nuclei and their dependence on the mass number; the values of the inertial parameters of nuclei in the case of collective motions.

In a new model-the so-called *superfluid model – of the nucleus* , it is assumed, as in the many-particle shell model, that the nucleons in the nucleus move as predicted by the one-particle shell model, with allowance made for the residual interaction, that is, for the correlation between the nucleons. In the superfluid model, however, the residual interaction is taken into account by completely different, more refined methods. We recall how the residual interaction was taken into account in the many-particle shell model. There, the problem of finding the eigenfunctions and eigenvalues of the Hamiltonian of the nucleus was reduced to a laborious numerical diagonalization of the Hamiltonian with the aid of a computer, and the problem of separating the residual interaction from the two-particle interactions of the nucleons, $\sum_{i<j} V_{ij}$, was not even formulated. As already stated, quite apart from its calculational difficulties, this approach does not enable us to understand the essential physics of the phenomena, since it is almost impossible to distinguish the most important features of the principal physical effects. (Jumping ahead, we point out that the problem of distinguishing the important features is more or less completely solved in the new model by the

introduction of pairing forces.) It is this which prevents us from explaining, within the framework of the shell model, certain features such as those enumerated above. The superfluid model has been able to eliminate these defects of the shell model. In the superfluid model new methods of treating the residual interactions are used, by means of which it is considerably easier to study the properties of many-particle systems.

As will be shown below, it is possible to separate out from the residual interaction a specific short-range pairing interaction which plays an especially important role in the explanation of the properties of the nuclei. Certain experimental data (for example, the binding energy of the last neutron in light nuclei, the zero spins of even-even nuclei in their ground states, etc.) point to the fact that strong correlation exists between two nucleons only when these nucleons are in states with the same energy and the same quantum numbers, except for the components of their total angular momenta.

The nature of the *residual short-range forces* is such as to lead to a much stronger interaction between two nucleons in states in which their resultant angular momentum is equal to zero (the pairing effect).

It is assumed in the superfluid model of the nucleus that the individual nucleons in the nucleus are characterized by the same one-particle states as in the independent-particle model. The *pairing of nucleons*, therefore, can be described by means of the quantum numbers of the independent-particle model. For example, if one nucleon is described by the set of quantum numbers n, l, j and m, the paired nucleon should be described by the set of quantum numbers n, l, j and $-m$. A definite energy is needed to separate the pair of nucleons (the pair interaction is an attractive interaction), and this leads to the appearance of an *energy gap between the ground state and the first excited state* in even-even nuclei, in which all the nucleons are paired (and the total angular momentum of the nucleus is thus equal to zero).

An analogous energy gap also appears in the energy spectra of other fermion systems, for example, electrons in metals. Owing to the presence of the *gap in the spectrum* , the system develops a certain stability against external perturbations, resulting in the phenomenon of superconductivity. This phenomenon is related to the property of superfluidity of systems consisting of identical bosons, since under certain conditions electrons in metals, like nucleons in the nucleus, combine into pairs, the aggregate of which can be regarded as an example of a *Bose gas*, inasmuch as each pair possesses integer spin. At low temperatures this Bose gas can possess the property of superfluidity. The theory of the phenomenon of superconductivity of electrons in metals was constructed using the concept of pairs in this way (Bardeen, Cooper and Schrieffer, 1957; Bogolyubov, 1958a, b; Bogolyubov, Tolmachev and Shirkov, 1958). The analogous property in the case of nuclei is usually called superfluidity rather than superconductivity. The property of the superfluidity of nuclear matter was first pointed out by Bogolyubov (1958c) on the basis of the similarity of the properties of nuclear matter and the properties of electrons in metals.[1]

The property of superfluidity of nuclei arises from the *nucleon-nucleon pair correlations* , which are, in turn, a manifestation of the residual interaction between the nucleons. These correlation, therefore, are often called *superconducting pair correlations* . In this case, the energy of the ground state will be significantly lower than the energy of the state when correlations are not taken into account. This state with lowered energy is called the *superfluid state.*

Lying at the basis of the method of studying the superfluid state is the formalism of *second quantization* . In this formalism, any wave function of a system of fermions will be automatically anti-symmetric. Whereas in the ordinary representation the function is written in a form (a Slater determinant) inconvenient for calculations, in the new representation it has a very simple form and is convenient for calculations. The *creation and destruction operators* (appearing in the Hamiltonian and wave functions) respectively increase and decrease by unity the number of nucleons in a given state. By means of a linear canonical transformation of these creation and destruction operators, we introduce new operators, which, in the general case, are responsible for the creation and destruction not of individual nucleons but of a particular type of quasi-particle, these *quasi-particles being superpositions of nucleon and hole states* . Expressed in terms of the new (quasi-particle) creation and destruction operators, the Hamiltonian operator will depend also on the coefficients of the canonical transformation. These coefficients can be found by minimizing the energy of the system in the ground state using the variational principle. The total Hamiltonian of the system is then represented approximately in the form of a sum of Hamiltonians of noninteracting quasi-particles.

We have seen that it is a very laborious task in the framework of the shell model to take into account the residual interaction between the nucleons, that is, the correlation between the nucleons. Usually, therefore, no account is taken of correlation in the shell model, that is, it is assumed that the nucleons move independently in a self-consistent field. Although the self-consistent field does include a considerate part of the two-particle interaction between the nucleons, it nevertheless tares no account of the *pairing forces*, which have an important effect on certain properties of nuclei. The transformation to quasi-particles in the superfluid model enables

[1]Bohr, Mottelson and Pines (1958) drew attention to the similarity of the excitation spectra of nuclei to the spectra of superconducting metals.

us to take account of a considerable part of the correlation effects even in the case when the interaction between the quasi-particles is neglected.[2]

We proceed now to a general treatment of the *residual interaction*. Assuming, as previously, that the operator U of the potential energy of the nucleus is equal to a sum of two-particle potentials

$$U = \sum_{i<j} V_{ij}, \tag{6.1}$$

we shall elucidate how, in an arbitrary two-particle potential $V_{ij} = V(\mathbf{r}_i - \mathbf{r}_j)$, we can separate the part contributing to the self-consistent potential of the independent-particle model and the part (the pairing forces and multipole forces) characterizing the residual interaction. We shall elucidate the nature and meaning of the appearance of the pairing forces and multipole forces.[3]

Choosing the centre of mass of the nucleus as the coordinate origin, we expand the two-particle potential in a series in Legendre polynomials:

$$V(\mathbf{r}_i - \mathbf{r}_j) = \sum_{\lambda=0}^{\infty} V_\lambda(r_i, r_j) P_\lambda(\cos\Theta_{ij}), \tag{6.2}$$

where Θ_{ij} is the angle between the vectors $\mathbf{r}_i$ and $\mathbf{r}_j$. If the nucleus is spherical, the main contribution to the self-consistent field of the nucleus is made by the first term, which contains the isotropic part $V_0(r_i, r_j)$ of the two-particle potential, while all the other terms, with $\lambda \neq 2$, represent the residual interaction. If the nucleus is deformed, however, the quadrupolar term with $\lambda = 2$ will also give a contribution to the self-consistent potential. In this case, the terms with $\lambda > 2$ make no contribution to the self-consistent potential.

The residual interaction can be divided into two parts: a short-range and a long-range part. We shall examine how this can be done. On variation of the angle Θ_{ij}, the function $P_\lambda(\cos\Theta_{ij})$ falls off sharply from its maximum value $P_\lambda[\cos(\Theta_{ij} = 0)] = 1$ in an angular interval of order $1/\lambda$. Since the distance $|\mathbf{r}_i - \mathbf{r}_j|$ between two nucleons is approximately equal to $R\Theta_{ij}$, where R is the mean value of the quantities r_i and r_j, it is clear that the λ-component of the potential $V(\mathbf{r}_i - \mathbf{r}_j)$ will contribute to the interaction of particles i and j only when $|\mathbf{r}_i - \mathbf{r}_j| \lesssim R/\lambda$. Thus, the range of the forces falls off with increasing λ. The contribution of terms with large λ will grow as the range of the forces decreases. In the *limit of δ-function forces with*

[2]The theory of the superfluid states of nuclei was developed by Belyaev (1959) and Solov'ev (1958, 1959, 1963).

[3]A good account of the question of pairing forces and multipole forces is given in the book by Lane (1964).

zero range , we have

$$V(\mathbf{r}_i - \mathbf{r}_j) \equiv A\delta(\mathbf{r}_i - \mathbf{r}_j) = \sum_{\lambda=0}^{\infty} A\delta(r_i - r_j)\frac{2\lambda+1}{4\pi r_i^2}P_\lambda(\cos\Theta_{ij}). \qquad (6.3)$$

Here, $V_\lambda(r_i r_j) = A\delta(r_i - r_j)(2\lambda+1)/4\pi r_i^2$ increases with increasing λ. But for forces with range greater than the dimensions of the nucleus, the term with $\lambda = 0$ will give the main contribution.

Thus, we may state that terms with small λ are responsible for the long-range part of the two-particle interaction. These terms are associated with the *long-range multipole forces* (*long-range correlations*). They give rise to the self-consistent field ($\lambda = 0$) which keeps the nucleons in the nucleus, and also to the collective motions in the nucleus. Thus, the term with $\lambda = 2$ is principally responsible for the appearance of collective quadrupolar vibrations. On the other hand, the sum of all the terms with large λ is the short-range part of the two-particle potential, and leads to the pairing forces.

Thus, the *superfluid model of the nucleus* is based on the self-consistent mean field of the independent-particle model, but in it account is also taken of the short-range part of the nucleon-nucleon interactions, which leads to pair correlations of the superconducting type. The superfluid model gives a microscopic description of the structure of the nucleus, although phenomenological parameters are used in it. In the superfluid model account is also taken of multipole forces ($\lambda \geqslant 2$) which, like the self-consistent mean field, are associated with the long-range interaction. For simplicity, the multipole forces responsible for collective effects are not taken into account in this Section. We shall take this long-range residual interaction into account in our treatment of the single most general and effective approximate method for studying the structure of the nucleus at the end of this Chapter.

Thus, allowance for pair correlations of the superconducting type has made it possible to explain certain important properties of nuclei and has led to the creation of the superfluid model of the nucleus. We now give a more systematic treatment of the superfluid model of the nucleus.

Hamiltonian of the superfluid model. We shall consider a system of nucleons situated in an axially symmetric self-consistent field. Let the Hamiltonian of an individual nucleon in this field be H_0. The eigenfunctions in this case can be chosen so that they correspond to well-defined values of the projection Ω of the nucleon angular momentum along the symmetry axis of the nucleus. We denote the eigenvalues of H_0 by ε_ν and the corresponding wave functions by $|\nu\rangle$ (ν characterizes the well-defined value of the angular-momentum component and the additional quantum numbers necessary for a complete description of the state of the nucleon). The function $|\nu\rangle$ can be expanded in the eigenfunctions of the angular momentum of the nucleon:

$$|\nu\rangle = \sum_j c_j^\nu \chi_\Omega^j. \tag{6.4}$$

If the system is invariant under time reversal, the levels ε_ν are doubly degenerate. In the case of axial symmetry, the degenerate states differ in the sign off the angular-momentum components (Ω and $-\Omega$). We introduce the time-reversal operator T. Then the state $|-\nu\rangle$ conjugate to the state $|\nu\rangle$ can be defined as $|-\nu\rangle = T|\nu\rangle$.

We go over to the *second-quantization representation* using one-particle functions as the basis functions. The second-quantization formalism makes it possible to take automatic account of the Pauli principle, which the nucleons obey. We introduce the operators creating and destroying a nucleon in the state ν: a_ν^+ and a_ν. These operators satisfy the anticommutation relations

$$a_\nu a_{\nu'}^+ + a_{\nu'}^+ a_\nu = \delta_{\nu,\nu'}. \tag{6.5}$$

We define the *nucleon-vacuum wave function* $|0\rangle$ by the condition

$$a_\nu|0\rangle = 0. \tag{6.6}$$

Then the *one-particle wave functions* $|\nu\rangle$ can be expressed in terms of the vacuum wave function $|0\rangle$ and the creation operators a_ν^+ as follows:

$$|\nu\rangle = a_\nu^+|0\rangle. \tag{6.7}$$

Using the operators a_ν and a_ν^+, we represent the Hamiltonian of a system of interacting nucleons in the form

$$H = \sum_\nu \varepsilon_\nu (a_\nu^+ a_\nu + a_{-\nu}^+ a_{-\nu}) + \frac{1}{2} \sum_{\nu_1\nu_2\nu_1'\nu_2'} \langle \nu_1'\nu_2'|V|\nu_2\nu_1\rangle a_{\nu_1'}^+ a_{\nu_2'}^+ a_{\nu_1} a_{\nu_2}, \tag{6.8}$$

where ε_ν is the energy of an individual nucleon in the self-consistent field and $\langle \nu_1'\nu_2'|V|\nu_2\nu_1\rangle$ is a matrix element of the residual pair interaction between the nucleons. The sum $\sum_{\nu_1\nu_2\nu_1'\nu_2'}$ also includes summation over the conjugate states.

The Hamiltonian (6.8) describes a system with a well defined number n of particles, where n is the eigenvalue of the operator

$$\hat{n} = \sum_\nu (a_\nu^+ a_\nu + a_{-\nu}^+ a_{-\nu}). \tag{6.9}$$

Since the problem is solved more simply if we go over to a system in which the number of particles is not conserved (the canonical transformation used below does not conserve the number of particles), it is convenient to introduce the auxiliary Hamiltonian

$$\widetilde{H} = H - \lambda\hat{n}, \tag{6.10}$$

in which the factor λ can be regarded as the chemical potential (λ characterizes the energy of the last nucleon). The potential λ is determined from the condition

$$\langle\Psi|\hat{n}|\Psi\rangle = n, \tag{6.11}$$

where Ψ is an eigenfunction of (6.10). The results obtained on the basis of the Hamiltonian (6.10) will describe only the averaged properties of nuclei. The averaging should be performed over a small number (the uncertainty in n is small) of the neighbouring even or odd nuclei.

The interaction between the nucleons which remains after the self-consistent field has been separated out is not weak, and so the system characterized by the Hamiltonian (6.8) or (6.10) cannot be studied by means of perturbation theory. However, if we assume (by analogy with the phenomenon of superconductivity in metals) that *strong correlation* exists between nucleons in conjugate states (with angular momenta in opposite directions), then, by means of a canonical transformation (analogous to that introduced by Bogolyubov in the theory of superconductivity), we can go over from a system of interacting nucleons to a system of quasi-particles whose interaction can be neglected (a system of independent elementary excitations).

We shall assume that residual pair interaction appears only between nucleons in states with zero resultant angular momentum. This means that the states ν_1 and ν_2 (ν_1' and ν_2') in (6.8) should be mutually conjugate. Consequently, in the following we shall start from the Hamiltonian

$$\widetilde{H} = \sum_{\nu}(\varepsilon_{\nu} - \lambda)(a_{\nu}^{+}a_{\nu} + a_{-\nu}^{+}a_{-\nu}) + \frac{1}{2}\sum_{\nu\nu'}\langle\nu' - \nu'|V| - \nu\nu\rangle a_{\nu'}^{+}a_{-\nu'}^{+}a_{\nu}a_{-\nu}. \tag{6.12}$$

The matrix element $\langle\nu' - \nu'|V| - \nu\nu\rangle$ is anti-symmetric with respect to the initial and final states of the nucleons. In addition, it follows from the definition of conjugate states that

$$\langle\nu_1'\nu_2'|V|\nu_2\nu_1\rangle = \langle-\nu_1' - \nu_2'|V| - \nu_2 - \nu_1\rangle^{*}.$$

Therefore, the matrix element $\langle\nu' - \nu'|V| - \nu\nu\rangle$ is symmetric in the indices ν and ν'. We note also that, since the self-consistent field has been separated out, the diagonal part of V is equal to zero.

The Bogolyubov canonical transformation. In place of the operators a_{ν} and $a_{-\nu}$, we introduce the new operators

$$\alpha_{\nu} = u_{\nu}a_{\nu} - v_{\nu}a_{-\nu}^{+}; \quad \beta_{\nu} = u_{\nu}a_{-\nu} + v_{\nu}a_{\nu}^{+}, \tag{6.13}$$

where u_ν and v_ν and real numbers connected by the relation

$$u_\nu^2 + v_\nu^2 = 1. \tag{6.14}$$

It is easily verified that the operators $\alpha_\nu(\beta_\nu)$ and $\alpha_\nu^+(\beta_\nu^+)$ satisfy the same commutation relations as the operators a_ν and a_ν^+, and so the transformation (6.13) is canonical.

The transformation (6.13) is a generalization of the usual transformation by means of which one goes over from particles above the *Fermi surface* and holes below the Fermi surface to *elementary excitations.* Indeed, if we put

$$u_\nu = \begin{cases} 1, \\ 0, \end{cases}, \quad v_\nu = \begin{cases} 0, & \varepsilon_\nu > \zeta, \\ 1, & \varepsilon_\nu < \zeta \end{cases}$$

where ζ is the Fermi energy, we obtain

$$\alpha_\nu = \begin{cases} a_\nu, \\ -a_{-\nu}^+, \end{cases} \quad \beta_\nu = \begin{cases} a_{-\nu}, & \varepsilon_\nu > \zeta, \\ a_\nu^+, & \varepsilon_\nu < \zeta. \end{cases}$$

Consequently, α_ν, is an operator destroying an elementary excitation (a particle in the state ν above the Fermi surface or a hole in the state – $-\nu$ below the Fermi surface); correspondingly, α_ν^+, is an operator creating an elementary excitation. The state corresponding to the completely occupied Fermi sea becomes, in this transformation, the vacuum state of the elementary excitations. In the general case, when $(u_\nu, v_\nu) \neq (0, 1)$, the new quasi-particles (elementary excitations) α_ν and β_ν are superpositions of particles and holes, and the vacuum of elementary excitations corresponds to a Fermi sea with a diffuse edge.

The inverse of the transformation (6.13) has the form

$$a_\nu = u_\nu \alpha_\nu + v_\nu \beta_\nu^+; \quad a_{-\nu} = u_\nu \beta_\nu - v_\nu \alpha_\nu^+. \tag{6.15}$$

Substituting (6.15) into (6.12) and bringing the product of operators α and β to normal order (in which creation operators stand to the left of the destruction operators), we represent the Hamiltonian of the system in the following form:

$$\widetilde{H} = U + H_{11} + H_{20} + H'. \tag{6.16}$$

Here, U is a constant quantity which does not depend on the quasi-particle operators α and β:

$$U = \sum_\nu \{(\varepsilon_\nu - \lambda) 2v_\nu^2 - \Delta_\nu u_\nu v_\nu\}, \tag{6.17}$$

where

$$\Delta_\nu = -\frac{1}{2} \sum_{\nu'} \langle \nu' - \nu' | V | - \nu \nu \rangle u_{\nu'} v_{\nu'}. \tag{6.18}$$

The terms H_{11} and H_{20} are quadratic expressions of the operators α and β. H_{11} depends only on the quasi-particle number operators $\alpha_\nu^+\alpha_\nu$ and $\beta_\nu^+\beta_\nu$ and H_{20} depends on the two-quasi-particle creation and destruction operators $\alpha_\nu^+\beta_\nu^+$ and $\beta_\nu\alpha_\nu$:

$$H_{11} = \sum_\nu \{(\varepsilon_\nu - \lambda)(u_\nu^2 - v_\nu^2) + 2\Delta_\nu u_\nu v_\nu\}(\alpha_\nu^+\alpha_\nu + \beta_\nu^+\beta_\nu), \qquad (6.19)$$

$$H_{20} = \sum_\nu \{(\varepsilon_\nu - \lambda)2u_\nu v_\nu - \Delta_\nu(u_\nu^2 - v_\nu^2)\}(\alpha_\nu^+\beta_\nu^+ + \beta_\nu\alpha_\nu). \qquad (6.20)$$

The last term H' in (6.16) contains products of four α or β operators, and describes the interaction between the quasi-particles. This term can be written in the form

$$H' = H_{40} + H_{31} + H_{22}, \qquad (6.21)$$

where

$$\begin{aligned} H_{40} &= -\frac{1}{2}\sum_{\nu\nu'}\langle\nu' - \nu|V| - \nu\nu\rangle u_\nu^2 v_\nu^2(\alpha_\nu^+\beta_\nu^+\alpha_{\nu'}^+\beta_{\nu'}^+ + \alpha_\nu\beta_\nu\alpha_{\nu'}\beta_{\nu'}); \\ H_{31} &= -\frac{1}{2}\sum_{\nu\nu'}\langle\nu' - \nu|V| - \nu\nu\rangle(u_\nu^2 - v_\nu^2)u_{\nu'}v_{\nu'} \\ &\quad \times\{\alpha_\nu^+\beta_\nu^+(\alpha_{\nu'}^+\alpha_{\nu'} + \beta_{\nu'}^+\beta_{\nu'}) - (\alpha_{\nu'}^+\alpha_{\nu'} + \beta_{\nu'}^+\beta_{\nu'})\alpha_\nu\beta_\nu\}; \\ H_{22} &= -\frac{1}{2}\sum_{\nu\nu'}\langle\nu' - \nu'|V| - \nu\nu\rangle\{(u_\nu^2 u_{\nu'}^2 + v_\nu^2 v_{\nu'}^2)\alpha_\nu^+\beta_\nu^+\alpha_{\nu'}\beta_{\nu'} \\ &\quad + u_\nu v_\nu u_{\nu'}v_{\nu'}(\alpha_\nu^+\alpha_{\nu'}^+\alpha_\nu\alpha_{\nu'} + \beta_\nu^+\beta_{\nu'}^+\beta_\nu\beta_{\nu'} + 2\alpha_\nu^+\beta_{\nu'}^+\alpha_\nu\beta_{\nu'})\}. \end{aligned}$$

The subscripts attached to these three terms indicate the numbers of creation and destruction operators appearing in the corresponding products. Thus, H_{40} describes the process of creation of four quasi-particles from the vacuum and the reverse process, and so on. In the following we shall confine ourselves to studying a model of independent quasi-particles; that is, in (6.16) we shall neglect the residual interaction H' between the quasi-particles.

We shall choose the *coefficients* u_ν *and* v_ν *of the canonical transformation* in such a way that the term H_{20} in (6.16) vanishes. Then, in the approximation considered (neglect of H'), the Hamiltonian (6.16) will describe a system of independent quasi-particles:

$$\widetilde{H} = U + H_{11} = U + \sum_\nu\{(\varepsilon_\nu - \lambda)(u_\nu^2 - v_\nu^2) + 2\Delta_\nu u_\nu v_\nu\}(\alpha_\nu^+\alpha_\nu + \beta_\nu^+\beta_\nu). \quad (6.22)$$

Clearly, U has the meaning of the energy of the *quasi-particle vacuum.* From the condition $H_{20} = 0$ we obtain the following equation for u_ν and v_ν:

$$(\varepsilon_\nu - \lambda)2u_\nu v_\nu - \Delta_\nu(u_\nu^2 - v_\nu^2) = 0. \qquad (6.23)$$

It is not difficult to convince oneself that the condition (6.23) for the vanishing of H_{20} is equivalent to the condition for which the vacuum energy U is a minimum. Therefore, in terms of the quasi-particles, the ground state of the system is the vacuum state. The excited states are characterized by well-defined numbers of quasi-particles.

Equation (6.23) can be represented conveniently in another form. For this, we express u_ν and v_ν in terms of Δ_ν, using (6.23) and (6.14):

$$u_\nu^2 = \frac{1}{2}\left\{1 + \frac{\varepsilon_\nu - \lambda}{\sqrt{(\varepsilon_\nu - \lambda)^2 + \Delta_\nu^2}}\right\},$$

$$v_\nu^2 = \frac{1}{2}\left\{1 - \frac{\varepsilon_\nu - \lambda}{\sqrt{(\varepsilon_\nu - \lambda)^2 + \Delta_\nu^2}}\right\}. \tag{6.24}$$

Substituting u_ν and v_ν from (6.24) into (6.18), we obtain in place of (6.23) the following equation for Δ_ν:

$$\Delta_\nu = -\frac{1}{4}\sum_{\nu'} \frac{\langle \nu' - \nu'|V| - \nu\nu\rangle}{\sqrt{(\varepsilon_{\nu'} - \lambda)^2 + \Delta_{\nu'}^2}}\Delta_{\nu'}. \tag{6.25}$$

To elucidate the physical meaning of the quantity Δ_ν we determine the quasi-particle energy E_ν. It follows from (6.22) and (6.24) that

$$E_\nu = \sqrt{(\varepsilon_\nu - \lambda)^2 + \Delta_\nu^2}. \tag{6.26}$$

Thus, in the case of a continuous particle spectrum the quantity Δ_ν characterizes a *gap in the quasi-particle spectrum* .

Equation (6.25) has the trivial solution

$$\Delta_\nu = 0 \quad \text{or} \quad u_\nu v_\nu = 0, \tag{6.27}$$

which corresponds to a sharp Fermi surface. Indeed, if in this case we put

$$u_\nu = \begin{cases} 1, \\ 0, \end{cases} \quad v_\nu = \begin{cases} 0, & \varepsilon_\nu > \lambda, \\ 1, & \varepsilon_\nu < \lambda, \end{cases}$$

then the quasi-particles α and β will corresponds to the original particles above the Fermi surface and holes below the Fermi surface. In the case of sufficiently weak residual interaction between the particles, eqn. (6.25) admits only the trivial solution (6.27). However, if the inequality

$$-\frac{1}{4}\sum_{\nu'} \frac{\langle \nu' - \nu'|V| - \nu\nu\rangle}{|\varepsilon_{\nu'} - \lambda|} > 1 \tag{6.28}$$

is fulfilled, there exists a non-trivial solution of eqn. (6.25), corresponding to a diffuse Fermi surface. According to (6.28), such a solution is possible if the pair interaction is coherent in character, that is, if the matrix element $\langle \nu' - \nu'|V| - \nu\nu\rangle$ has the same sign for most states ν and ν'.

The operator (6.9) of the total number of particles, when expressed in terms of the quasiparticle creation and destruction operators, has the form

$$\hat{n} = \sum_\nu 2v_\nu^2 + \sum_\nu (u_\nu^2 - v_\nu^2)(\alpha_\nu^+ \alpha_\nu + \beta_\nu^+ \beta_\nu) + \sum_\nu 2u_\nu v_\nu (\alpha_\nu^+ \beta_\nu^+ + \beta_\nu \alpha_\nu). \quad (6.29)$$

The *mean number of particles in the ground state* of the system (the quasi-particle vacuum state) is obviously determined by the constant term in (6.29). Using the expression (6.24) for v_ν^2 we have

$$n = \sum_\nu \left\{1 - \frac{\varepsilon_\nu - \lambda}{\sqrt{(\varepsilon_\nu - \lambda)^2 + \Delta_\nu^2}}\right\}. \quad (6.30)$$

This equation enables us to find the chemical potential λ when the number n of particles in the system is specified. Equations (6.25) and (6.30) completely determine Δ_ν and λ for a system in the ground state if the one-particle energy levels ε_ν are known.

Wave functions and energy spectrum. The wave function Ψ_0 of the quasi-particle vacuum is determined from the conditions

$$\alpha_\nu \Psi_0 = \beta_\nu \Psi_0 = 0. \quad (6.31)$$

Using the equalities (6.13), we can rewrite the conditions (6.31) in the form

$$(u_\nu a_\nu - v_\nu a_{-\nu}^+)\Psi_0 = 0; \quad (u_\nu a_{-\nu} - v_\nu a_\nu^+)\Psi_0 = 0.$$

One can easily convince oneself by direct inspection that these conditions are satisfied by the following function:

$$\Psi_0 = \prod_\nu (u_\nu + v_\nu a_\nu^+ a_{-\nu}^+)|0\rangle. \quad (6.32)$$

The wave function Ψ_0 of the quasi-particle vacuum, according to (6.32), is a superposition of states with different numbers of real particles (nucleons in the self-consistent field). Only states with even numbers of nucleons are taken into account, since all the nucleons occur in pairs (ν and $-\nu$) in (6.32).

From (6.32) it is not difficult to establish the physical meaning of the coefficients u_ν and v_ν appearing in the canonical transformation (6.13): v_ν^2 is the probability that there is a pair of nucleons ν and $-\nu$ in the state Ψ_0

corresponding to the quasi-particle vacuum; u_ν^2 is the probability that the pair state $(\nu, -\nu)$ is unoccupied.

The wave functions of the excited states of the system can be found by the action of the *quasi-particle creation operators* α_ν^+ *and* β_ν^+ on the wave function (6.32) of the quasi-particle vacuum. Thus, the wave function of an excited state corresponding to the presence of one quasi-particle has the form

$$\Psi_\nu = \alpha_\nu^+ \Psi_0. \tag{6.33}$$

Using (6.13), it is not difficult to obtain

$$\Psi_\nu = a_\nu^+ \prod_{\nu' \neq \nu} (u_{\nu'} + v_{\nu'} a_{\nu'}^+ a_{-\nu'}^+)|0\rangle. \tag{6.34}$$

In contrast to Ψ_0, the wave function Ψ_ν, according to (6.34), is a superposition of states with an odd number of particles.

The wave function of an excited state corresponding to two quasi-particles is given by the expression

$$\Psi_{\nu,-\nu'} = \alpha_\nu^+ \beta_{\nu'}^+ \Psi_0, \tag{6.35}$$

or

$$\Psi_{\nu,-\nu'} = a_\nu^+ a_{-\nu'}^+ \prod_{\nu'' \neq \nu,\nu'} (u_{\nu''} + v_{\nu''} a_{\nu''}^+ a_{-\nu''}^+)|0\rangle. \tag{6.36}$$

The function $\Psi_{\nu,-\nu'}$, like Ψ_0, is a superposition of states with an even number of particles. The parameters Δ_ν and λ for the ground state of the system are determined from the equations (6.25) and (6.30). In an excited state, the mean number of particles, generally speaking, differs from (6.30). For example, in the case of an excited state with two quasi-particles, we have

$$\langle \Psi_{\nu,-\nu}|\hat{n}|\Psi_{\nu,-\nu}\rangle - \langle \Psi_0|\hat{n}|\Psi_0\rangle = 2\frac{\varepsilon_\nu - \lambda}{\sqrt{(\varepsilon_\nu - \lambda)^2 + \Delta_\nu^2}}.$$

However, this difference can be neglected for small excitations, when $\lambda \approx \varepsilon_\nu$. Therefore, for the low-lying excited states, we can use the same values of Δ_ν and λ as for the ground state.

Wave functions corresponding to states with even and odd numbers of quasi-particles describe different physical systems. The former correspond to systems consisting of even numbers of particles and the latter systems consisting of odd numbers of particles.

In the case of a nucleus with even A, the ground state is described by the quasi-particle vacuum wave function (6.32). The *energy of the ground state* is equal to

$$E_0 = U + \lambda n = \sum_\nu \left[\varepsilon_\nu \left(1 - \frac{\varepsilon_\nu - \lambda}{E_\nu}\right) - \frac{\Delta_\nu^2}{2E_\nu}\right]. \tag{6.37}$$

The excited states of a nucleus with even A are characterized by even numbers of quasi-particles. The lowest excited state (in the case of two quasi-particles) is described by the function (6.35). The *energy of an excited state with two quasi-particles* is equal to

$$E_{(\nu,-\nu')} = E_0 + E_\nu + E_{\nu'}. \tag{6.38}$$

According to (6.38), the energy spectra of nuclei with even A contain a gap separating the first excited state from the ground state; the width of this gap is greater than the quantity $2\Delta_\nu$: $\Delta E \geqslant 2\Delta_\nu$.

In the case of a nucleus with odd A, the ground state is the lowest one-quasi-particle state (6.33). The excited states are states with one, three, etc., quasi-particles. The ground-state energy and the energies of the low-lying excited states (corresponding to one quasi-particle) for a nucleus with odd A are given by the expression

$$E_{(\nu)} = E_0 + E_\nu. \tag{6.39}$$

Since for a nucleus with given A there exist several types of quasi-particle ν, there is no gap in the energy spectrum of nuclei with odd A. Thus, the excitation-energy spectra of nuclei with even and odd A are completely different. At the same time, the ground-state energies differ little, since the energy of the odd particle can be neglected in comparison with the vacuum energy.

6.2. Residual Interaction and Collective Effects in Nuclei

Invariant form of the Hamiltonian with residual interaction. In treating the superfluid states of nuclei in the preceding Section, we confined ourselves to including only the residual pairing forces and neglected the multipole forces, responsible for the collective effects. In transforming to quasi-particles by means of the canonical transformation, we also neglected the interaction between the quasi-particles, that is, we studied a model of independent quasi-particles. This approach enabled us to explain the existence of the energy gap in the spectrum of even-even nuclei. However, to explain certain other features of nuclear spectra it is insufficient to take into account only the pairing forces. We shall now consider the *multipole forces*, but without dividing the residual interaction into two parts corresponding to pairing forces and multipole forces, since such a separation is essentially arbitrary. We shall show that, with this general interaction Hamiltonian, we can write the equations of motion in such a way that they will describe simultaneously both the one-particle excitations, which we considered in the independent-particle model, and the *collective excitations in nuclei.*

This will also enable us to consider *partly collective states*, which cannot be described by other methods.[4]

Since the method can also be applied to heavy nuclei, we shall use jj-coupling and regard the protons and neutrons as different particles. (It is usually also assumed that the interaction between identical nucleons is much stronger than the interaction of neutrons with protons, which is neglected.) The shell-model one-particle states, characterized by the set of quantum numbers n, l, j, m (the component j_z) and q (the charge), will be denoted by the Greek letters, α, β, γ, δ, etc. We shall denote sets of all above quantum numbers except the magnetic quantum number m by the corresponding Latin letters a, b, c, d, etc. The conjugate states α and $-\alpha$ will differ only in the sign of the magnetic quantum numbers: $\alpha \equiv (a, m)$ and $-\alpha \equiv (a, -m)$.

We write the initial Hamiltonian, represented in terms of the nucleon creation and destruction operators a_α^+ and a_α, in such a way that its invariance under rotations and reflections is manifest:

$$H = \sum_\alpha \varepsilon_\alpha a_\alpha^+ a_\alpha + \frac{1}{2} \sum_{\alpha\beta\gamma\delta} \langle\alpha\beta|V|\gamma\delta\rangle a_\alpha^+ a_\beta^+ a_\delta a_\gamma. \tag{6.40}$$

Clearly, only the second sum in (6.40), describing the residual interaction, will undergo transformation. For this, we represent the states α and β of two particles in the form

$$|\alpha\rangle|\beta\rangle \equiv |am_\alpha\rangle|bm_\beta\rangle = \sum_{\mathcal{J}M} (j_\alpha m_\alpha j_\beta m_\beta|\mathcal{J}M)|ab\mathcal{J}M\rangle$$

and proceed analogously for a pair of particles in the states α and δ. In this case, the matrix element of the residual interaction will be equal to

$$\langle\alpha\beta|V|\gamma\delta\rangle =$$

$$= \sum_{\mathcal{J}M} \sum_{\mathcal{J}'M'} (j_\alpha m_\alpha j_\beta m_\beta|\mathcal{J}M)(j_\gamma m_\gamma j_\delta m_\delta|\mathcal{J}'M')\langle ab\mathcal{J}M|V|cd\mathcal{J}'M'\rangle.$$

Since the operator V is invariant under rotations (that is, commutes with the rotation operator), according to the *Wigner-Eckart theorem* we can write the matrix element $\langle ab\mathcal{J}M|V|cd\mathcal{J}'M'\rangle$ in the form

$$\langle ab\mathcal{J}M|V|cd\mathcal{J}'M'\rangle = \delta_{\mathcal{J}\mathcal{J}'}\delta_{MM'}\frac{1}{\sqrt{2\mathcal{J}+1}}\langle ab\mathcal{J}||V||cd\mathcal{J}\rangle,$$

[4]A more detailed account of the theory of nuclear models with inclusion of the residual interaction between the nucleons is given in a lecture by Baranger (1963) and the article by Gillet (1965), and also in the monograph by Brown (1964). A theory of finite systems of Fermi particles with strong interaction, based on the use of the *Green-function method*, is developed in the monograph by Migdal (1965).

whence it follows that only the matrix elements with $\mathcal{J}' = \mathcal{J}$ and $M' = M$ are non-zero. Introducing the notation

$$G(abcd\mathcal{J}) \equiv -\frac{1}{\sqrt{2\mathcal{J}+1}}\langle ab\mathcal{J}||V||cd\mathcal{J}\rangle, \tag{6.41}$$

we finally write

$$\langle\alpha\beta|V|\gamma\delta\rangle = -\sum_{\mathcal{J}M} G(abcd\mathcal{J})(j_\alpha m_\alpha j_\beta m_\beta|\mathcal{J}M)(j_\gamma m_\gamma j_\delta m_\delta|\mathcal{J}M). \tag{6.42}$$

If we couple the α with the state γ, and β with δ, (or α with δ, and β with γ), this leads to another function F, different from G:

$$\langle\alpha\beta|V|\gamma\delta\rangle = -\sum_{\mathcal{J}M} F(acdb\mathcal{J})(-1)^{j_\gamma - m_\gamma}\times$$

$$\times(j_\alpha m_\alpha j_\gamma - m_\gamma|\mathcal{J}M)(-1)^{j_\beta - m_\beta}(j_\delta m_\delta j_\beta - m_\beta|\mathcal{J}M)$$

$$\sum_{\mathcal{J}M} F(adcb\mathcal{J})(-1)^{j_\delta - m_\delta}(j_\alpha m_\alpha j_\delta - m_\delta|\mathcal{J}M)(-1)^{j_\beta - m_\beta}(j_\gamma m_\gamma j_\beta - m_\beta|\mathcal{J}M). \tag{6.43}$$

The functions F and G are related by means of Racah coefficients as follows:

$$F(acdb\mathcal{J}') = -\sum_{\mathcal{J}}(2\mathcal{J}+1)W(j_a j_b j_c j_d;\mathcal{J}\mathcal{J}')G(bacd\mathcal{J}). \tag{6.44}$$

The functions F and G are real and possess certain symmetry properties, which follow from the hermiticity of V and from symmetry properties of the matrix elements $\langle\alpha\beta|V|\gamma\delta\rangle$:

$$\begin{aligned} G(abcd\mathcal{J}) &= G(cdab\mathcal{J}) = -(-1)^{j_a+j_b+\mathcal{J}}G(bacd\mathcal{J}) \\ &= -(-1)^{j_c+j_d+\mathcal{J}}G(abdc\mathcal{J}); \\ F(acdb\mathcal{J}) &= F(dbac\mathcal{J}) = (-1)^{j_a+j_b+j_c+j_d}F(cabd\mathcal{J}). \end{aligned} \tag{6.45}$$

In the case of pure pairing forces, we have

$$G(abcd\mathcal{J}) = \delta_{ab}\delta_{cd}\delta_{\mathcal{J}0}g_{ac}. \tag{6.46}$$

Using the *Bogolyubov transformation*, we now introduce the quasi-particle creation and destruction operators b_α^+ and b_α:

$$\left.\begin{aligned} b_a &= u_a a_a - (-1)^{j_\alpha - m_\alpha}v_a a_{-\alpha}^+; \\ b_\alpha^+ &= u_a a_\alpha^+ - (-1)^{j_\alpha - m_\alpha}v_a a_{-\alpha}, \end{aligned}\right\} \tag{6.47}$$

where u_a and v_a are real coefficients, connected by the relation

$$u_a^2 + v_a^2 = 1. \tag{6.48}$$

The factor $(-1)^{j_\alpha - m_\alpha}$ is introdiced in (6.47) in order that the operators b_α and b_α^+ possess certain transformation properties under coordinate rotations–namely, that they be irreducible tensor operators. In particular, with this definition, the operator b_{jm}^+ is a tensor operator of rank j, and m defines the component of this tensor $(-j \leqslant m \leqslant j)$. When (6.48) is taken into account, the operators b_α and b_α^+ satisfy the same anti-commutation relations as the operators a_α and a_α^+. The *quasi-particle vacuum state* $|0\rangle$ is defined by the condition

$$b_\alpha|0\rangle = 0 \quad \text{or} \quad \langle 0|b_\alpha^+ = 0. \tag{6.49}$$

As we earlier, we introduce the chemical potentials λ_n and λ_p (separately for neutrons and protons) and the number operators $\widehat{n}_n$ and $\widehat{n}_p$ for the neutrons and protons respectively:

$$\widehat{n}_n = \sum_{\alpha(n)} a_\alpha^+ a_\alpha, \quad \widehat{n}_p = \sum_{\alpha(p)} a_\alpha^+ a_\alpha, \tag{6.50}$$

where the summation is performed over the neutron states $\alpha(n)$ or proton states $\alpha(p)$; we consider the operator

$$\widetilde{H} \equiv H - \lambda_n \widehat{n}_n - \lambda_p \widehat{n}_p \tag{6.51}$$

in place of the Hamiltonian operator H. In order to express $\widetilde{H}$ in terms of quasi-particle creation and destruction operators, we use the transformations inverse to (6.47):

$$\left.\begin{aligned} a_\alpha &= u_a b_\alpha + (-1)^{j_\alpha - m_\alpha} v_a b_{-\alpha}^+; \\ a_\alpha^+ &= u_a b_\alpha^+ + (-1)^{j_\alpha - m_\alpha} v_a b_{-\alpha}. \end{aligned}\right\} \tag{6.52}$$

introducing the notation

$$\Delta_a = \frac{1}{\sqrt{2j_a + 1}} \sum_c \sqrt{2j_c + 1} u_c v_c G(aacc0); \tag{6.53}$$

$$\begin{aligned} \mu_a &= \frac{2}{\sqrt{2j_a + 1}} \sum_{b\mathcal{J}} (2\mathcal{J} + 1) v_b^2 G(abab\mathcal{J}) = \\ &= \frac{2}{\sqrt{2j_a + 1}} \sum_b \sqrt{2j_b + 1} v_b^2 F(aabb0); \end{aligned} \tag{6.54}$$

$$\eta_a = \varepsilon_a - \mu_a - \lambda_a, \tag{6.55}$$

we can represent the operator $\widetilde{H}$ in the form

$$\left.\begin{aligned}\widetilde{H} &= U + H_{11} + H_{20} + H';\\ U &= \sum_{\alpha}[v_a^2(\eta_a + \frac{1}{2}\mu_a) - \frac{1}{2}u_a v_a \Delta_a];\\ H_{11} &= \textstyle\sum_{\alpha}[(u_a^2 - v_a^2)\eta_a + 2u_a v_a \Delta_a]b_{\alpha}^{+}b_{\alpha};\\ H_{20} &= \sum_{\alpha}(-1)^{j_{\alpha}-m_{\alpha}}[u_a v_a \eta_a + \frac{1}{2}(v_a^2 - u_a^2)\Delta_a](b_{\alpha}^{+}b_{-\alpha}^{+} + b_{-\alpha}b_{\alpha});\\ H' &= \frac{1}{2}\sum_{\alpha\beta\gamma\delta}\langle\alpha\beta|V|\gamma\delta\rangle N(a_{\alpha}^{+}a_{\beta}^{+}a_{\delta}a_{\gamma}).\end{aligned}\right\} \tag{6.56}$$

where N denotes the *normal product.*

If the ground state of the system coincides with the quasi-particle vacuum state $|0\rangle$, then U is the energy of this ground state, since $U = \langle 0|\widetilde{H}|0\rangle$. The quantities u_a and v_a are chosen in such a way that H_{20} vanishes. Then U takes its minimum value and H_{11} becomes equal to

$$H_{11} = \sum_{\alpha} E_{\alpha} b_{\alpha}^{+} b_{\alpha}, \quad E_{\alpha} = \sqrt{\eta_a^2 + \Delta_a^2}. \tag{6.57}$$

(The same result can be obtained using the variational principle.) We also obtain the relations $2u_a v_a = \Delta_a/E_a$ and $u_a^2 - v_a^2 = \eta_a/E_a$, which, together with the relations (6.53)–(6.55) given earlier, determine the quantities Δ_a, μ_a and λ_a in the case of the most general residual nuclear potential G. The problem is simplified considerably in the pairing forces only are taken into account, when the quantity μ_a can be neglected. In can be seen from the expression (6.57) that E_a can be interpreted as the quasi-particle energy and Δ_a as half the energy gap. The last term H' in (6.56) is the residual interaction between the quasi-particles. It can be expressed in terms of normal products of the quasi-particle operators b_{α}^{+} and b_{α}. If we are considering the collective effects in nuclei, which are associated with multipole forces in the two-particle interaction of the nucleons, it is necessary to take the residual interaction between the quasi-particles into account.

The random-phase approximation. As is well known, to calculate various types of observable physical quantities, we need only know the matrix elements of certain operators; a knowledge of the wave functions by means of which these matrix elements are defined is often unnecessary. In certain cases, therefore, we can formulate the problem of finding the corresponding matrix elements (or the amplitudes in the equations of motion) from the outset, without being concerned with the explicit form of the wave functions of the system. This approach is justified by the fact that it is a very complicated problem to determine the wave functions in the general case.

We shall define operators creating and destroying a pair of quasi-particles in a state with total angular momentum $\mathcal{J}$ with z-component M:

$$\left.\begin{aligned} A^+(ab\mathcal{J}M) &= \sum_{m_\alpha m_\beta} (j_\alpha m_\alpha j_\beta m_\beta|\mathcal{J}M) b_\alpha^+ b_\beta^+; \\ A(ab\mathcal{J}M) &= \sum_{m_\alpha m_\beta} (j_\alpha m_\alpha j_\beta m_\beta|\mathcal{J}M) b_\beta b_\alpha, \end{aligned}\right\} \tag{6.58}$$

Let Ψ_0 be the exact wave function of the ground state of the system of particles; Ψ_0 is not in general the quasi-particle vacuum state $|0\rangle$, but only chose to it in energy. We denote the wave functions of the excited states with angular momentum $\mathcal{J}$ and angular-momentum component M by Ψ_{BM}, where B is the set of all the quantum numbers (apart from M) of the state, including $\mathcal{J}$. We now define two amplitudes:

$$\left.\begin{aligned} \psi_{abB} &\equiv \langle\Psi_0|A(ab\mathcal{J}m)|\Psi_{BM}\rangle; \\ \varphi_{abB} &\equiv (-1)^{\mathcal{J}-M}\langle\Psi_0|A^+(ab\mathcal{J}-M)|\Psi_{BM}\rangle, \end{aligned}\right\} \tag{6.59}$$

which, according to the definitions of the operators A^+, A and b_α^+, b_α, do not depend on M.

Equations for the amplitudes ψ_{abB} and φ_{abB} can be obtained by taking matrix elements, between the wave functions Ψ_0 and Ψ_{BM}, of the quantum equations of motion for the operators A and A^+. The matrix element of the left-hand side of the equation $\dot{A} = (i/\hbar)[\widetilde{H}, A]$ is equal to $i\omega_B\langle\Psi_0|A|\Psi_{BM}\rangle$, where ω_B is the energy of the excited state with wave function Ψ_{BM}, reckoned from the ground-state energy. Thus, we can write the equation

$$\langle\Psi_0|[A, \widetilde{H}]|\Psi_{BM}\rangle = \omega_B\langle\Psi_0|A|\Psi_{BM}\rangle \equiv \omega_B\psi_{abB}. \tag{6.60}$$

We note that Ψ_0 and Ψ_{BM} are eigenfunctions of the Hamiltonian $\widetilde{H}$. An analogous equation can also be obtained with the operator A^+. Using the explicit form of the Hamiltonian (6.56) with $H_{20} = 0$, we calculate the commutators$[A, \widetilde{H}]$ and $[A^+, \widetilde{H}]$:

$$\begin{aligned} &[A(ab\mathcal{J}M), \widetilde{H}] = \sum_{cd} P(abcd\mathcal{J})A(cd\mathcal{J}M) + \\ &+\sum_{cd} R(abcd\mathcal{J})(-1)^{\mathcal{J}-M}A^+(cd\mathcal{J}-M) + O_2^{(-)} + O_4^{(-)}; \end{aligned} \tag{6.61}$$

$$\begin{aligned} &[(-1)^{\mathcal{J}-M}A^+(ab\mathcal{J}-M), \widetilde{H}] = \\ &= -\sum_{cd} P(abcd\mathcal{J})(-1)^{\mathcal{J}-M}A^+(cd\mathcal{J}-M) - \\ &-\sum_{cd} R(abcd\mathcal{J})A(cd\mathcal{J}M) + O^{(+)} + O_4^{(+)}. \end{aligned} \tag{6.62}$$

The functions $P(abcd\mathcal{J})$ and $R(abcd\mathcal{J})$ can be expressed in terms of the functions $G(abcd\mathcal{J})$ and $F(abcd\mathcal{J})$ characterizing the residual nuclear interaction between the nucleons. The terms $O_2^{(\pm)}$ contain products $b_\alpha^+ b_\alpha$ of two quasi-particle operators, while the terms $O_4^{(\pm)}$ contain products of four quasi-particle operators.

The *random-phase approximation* consists in neglecting the contribution of the terms $O_2^{(\pm)}$ and $O_4^{(\pm)}$ in the commutators (6.61) and (6.62). From eqn. (6.61) and the analogous equation for the operator A^+, we then obtain a system of linear homogeneous equations for the amplitudes ψ_{abB} and φ_{abB}:

$$\left.\begin{aligned} \omega_B \psi_{abB} &= \textstyle\sum_{cd}\{P(abcd\mathcal{J})\psi_{cdB} + R(abcd\mathcal{J})\varphi_{cdB}\}, \\ -\omega_B \varphi_{abB} &= \textstyle\sum_{cd}\{P(abcd\mathcal{J})\varphi_{cdB} + R(abcd\mathcal{J})\psi_{cdB}\}, \end{aligned}\right\} \tag{6.63}$$

The random-phase approximation is sometimes called the *method of linearized equations of motion* . A number of arguments can be given in favour of the validity of the random-phase approximation, although it is difficult to justify it in the general case or to establish its range of applicability. We may hope that the discarded terms $O_2^{(\pm)}$ and $O_4^{(\pm)}$ in (6.61) and (6.62) will lead to small corrections for the matrix elements, since in the limit of independent quasi-particles ($H' \to 0$) the commutators (6.60) and (6.62) do not contain any terms $O_2^{(\pm)}$ or $O_4^{(\pm)}$. The name of the random-phase method (approximation) arises from the fact that a matrix element of the operator $O_4^{(\pm)}$ can be represented in the form of a sum of products of two matrix elements of products of two quasi-particle operators, in accordance with the usual rule for finding a matrix element of a product of operators; it is then asserted that this sum is small, inasmuch as the terms cancel each other because of the randomness of the phases. (This explanation is not, of course, fully convincing.)

The basic equations (6.63) of the random-phase method are sometimes represented in another, more convenient form. We define two new amplitudes:

$$f_{abB} = \psi_{abB} + \varphi_{abB}; \quad g_{abB} = \psi_{abB} - \varphi_{abB}. \tag{6.64}$$

Substituting these into (6.63), we obtain two equations for them:

$$\left.\begin{aligned} \omega_B f_{abB} &= (E_a + E_b) g_{abB} - \textstyle\sum_{cd} U_+(abcd\mathcal{J}) g_{cdB}; \\ \omega_B g_{abB} &= (E_a + E_b) f_{abB} - \textstyle\sum_{cd} U_-(abcd\mathcal{J}) f_{cdB}. \end{aligned}\right\} \tag{6.65}$$

Here, the new functions U_+ and U_- can be expressed in terms of the functions G and F as follows:

$$U_\pm(abcd\mathcal{J}) = (u_a u_b \pm v_a v_b)(u_c u_d \pm v_c v_d) G(abcd\mathcal{J})$$

$$+ (u_a v_b \mp v_a u_b)(u_c v_d \mp v_c u_d)[F(abcd\mathcal{J}) \pm (-1)^{j_c+j_d+\mathcal{J}} F(abdc\mathcal{J})], \quad (6.66)$$

and are related to the functions P and R by

$$P(abcd\mathcal{J}) \pm R(abcd\mathcal{J}) = (E_a + E_b)\delta_{ac}\delta_{bd} - U_{\pm}(abcd\mathcal{J}). \quad (6.67)$$

We shall see later that the solutions of eqns. (6.65), and thereby of (6.63), describe not only one-particle properties but also certain collective effects in nuclear systems.

If the interaction between the quasi-particles is small ($H' \to 0$), the function $R(abcd\mathcal{J})$ in (6.61) and (6.62) is much smaller than $P(abcd\mathcal{J})$, and U_+ and U_- differ little from each other in magnitude. In this case, the amplitudes f_{abB} and g_{abB} are approximately equal, since for $U_+ = U_-$ they appear symmetrically in the system of equations (6.65). This means that $|\varphi_{abB}| \ll |\psi_{abB}|$. If we neglect φ_{abB} in eqns. (6.63), we obtain the following equations for the amplitudes ψ_{abB}:

$$\omega_B \psi_{abB} = \sum_{cd} P(abcd\mathcal{J})\psi_{cdB}. \quad (6.68)$$

These equations correspond to the so-called *Tamm-Dancoff method* , which is the particular case of the random-phase method in which the wave function of ground state of the system is put equal to the wave function $|0\rangle$ of the quasi-particle vacuum. Indeed, if $\Psi_0 = |0\rangle$, then, according to the definition (6.59), the amplitude φ_{abB} vanishes, since $\langle 0|A^+ = [A|0\rangle]^+ = 0$, and $\psi_{abB} = \langle 0|A(ab\mathcal{J}M)|\Psi_{BM}\rangle$. In the Tamm-Dancogg approximation, correlations between the particles are taken into account only in the excited states,while the ground state is the same as in the independent-particle shell model.

We have already remarked that in studying the collective motions in nuclei it is necessary to take into account the interactions between the quasi-particles. Allowance for the collective effects leads to an increase of the transition probabilities in nuclei. In the Tamm-Dancoff method, the entire effect of the collective motions is linked with the structure of the excited states. This, however, has been found insufficient for agreement with experimental data; in order to amplify the transitions, it is necessary to take account of particle correlations in the ground state also. This deficiency of the Tamm-Dankoff method is removed in the random-phase method, in which correlations are taken into account in both the ground and excited states of the system.

The ground-state wave function Ψ_0 in the random-phase method can be represented in the form of a superposition

$$\Psi_0 = \left\{\chi_0 + \sum_{\alpha\beta} \chi_2(\alpha\beta) b_\alpha^+ b_\beta^+ + \sum_{\alpha\beta\gamma\delta} \chi_4(\alpha\beta\gamma\delta) b_\alpha^+ b_\beta^+ b_\gamma^+ b_\delta^+ + \cdots\right\}|0\rangle, \quad (6.69)$$

which shows clearly the presence of quasi-particles in the ground state Ψ_0, whereas the wave function $|0\rangle$ corresponds, by definition, to the quasi-particle vacuum. However, the states with wave functions Ψ_0 and $|0\rangle$ are close to each other, so that the coefficient χ_0 in (6.69) is close to unity and all the other coefficients $\chi_{i\geqslant 2}$ are small in magnitude.

The method of approximate second quantization. For a better understanding of how the collective states of a system of nucleons arise, and also of the structure of these states, we shall derive the equations (6.63) of the random-phase approximation in a somewhat different way and give another interpretation of the amplitudes ψ_{abB} and φ_{abB}.

We shall find the operators Q_{BM}, Q_{CM}, etc., having the following properties:

$$\left.\begin{aligned}[Q_{BM}, \widetilde{H}] &= \omega_B Q_{BM};\\ [Q^+_{BM}, \widetilde{H}] &= -\omega_B Q^+_{BM}.\end{aligned}\right\} \tag{6.70}$$

We shall seek these operators in the form of linear combinations of the operators A and A^+ with real coefficients:

$$\left.\begin{aligned}Q_{BM} &= \sum_{ab} X_{abB} A(ab\mathcal{J}M) - \sum_{ab} Y_{abB}(-1)^{\mathcal{J}-M} A^+(ab\mathcal{J}-M);\\ (-1)^{\mathcal{J}-M} Q^+_{B-M} &= \sum_{ab} X_{abB}(-1)^{\mathcal{J}-M} A^+(ab\mathcal{J}-M)-\\ &\quad -\sum_{ab} Y_{abB} A(ab\mathcal{J}M).\end{aligned}\right\} \tag{6.71}$$

If we substitute these expressions into eqns. (6.70) and, as in (6.61) and (6.62), neglect the terms $O_2^{(\pm)}$ and $O_4^{(\pm)}$ in the commutators of Q_{BM} and Q^+_{BM} with the Hamiltonian $\widetilde{H}$, we obtain for the coefficient X_{abB} and Y_{abB} the same equations as eqns. (6.63) for the amplitudes ψ_{abB} and φ_{abB} in the random-phase approximation. Consequently, the coefficients X_{abB} and Y_{abB} are proportional to the amplitudes ψ_{abB} and φ_{abB} respectively.

Thus, the operators Q_{BM} and Q^+_{BM} can be regarded as determined in the random-phase approximation. Accordin to eqns. (6.70), these operators are responsible for the creation or destruction of one vibrational excitation quantum of a harmonic oscillator. The operator Q^+_{BM} creates a vibrational quantum or phonon with energy ω_B, and Q_{BM} destroys this quantum.

The wave function Ψ_0 of the ground state of the system can be determined from the system of equations

$$Q_{BM}\Psi_0 = 0. \tag{6.72}$$

It corresponds to the absence of phonons (but not of quasi-particles). The excited states with energy ω_B are defined as one-phonon states with the wave function

$$\Psi_{BM} = Q^+_{BM}\Psi_0. \tag{6.73}$$

Unlike the one-phonon states, the excited states with wave functions $(Q^+_{BM})^2\Psi_0$, $Q^+_{BM}Q_{CM'}\Psi_0$, etc., with the corresponding energies $2\omega_B$, $\omega_B + \omega_C$, etc., are, in the approximation under consideration, substantially different from the real states and will not be considered here.

From the definition of the operators A and A^+ we obtain the exact commutation relations

$$[A(ab\mathcal{J}M), A(a'b'\mathcal{J}'M')] = [A^+(ab\mathcal{J}M), A^+(a'b'\mathcal{J}'M')] = 0. \tag{6.74}$$

In order that these operators possess all the properties of boson operators, the further commutation condition

$$[A(ab\mathcal{J}M), A^+(a'b'\mathcal{J}'M')] = \delta_{ab\mathcal{J}M,a'b'\mathcal{J}'M'} \tag{6.75}$$

must also be fulfilled.

However, a direct calculation of the commutator in (6.75) gives extra terms, containing the operators $b^+_\alpha b_\beta$ in the right-hand side of the equality (6.75). The mean value of these operators in the quasi-particle vacuum $|0\rangle$ is, however, equal to zero, and the number of quasi-particles appearing as a result of the inclusion of correlations between the nucleons in the ground state Ψ_0 is, as already noted, not great. If this number of quasi-particles is much less than the total number of the possible one-particle states which they could occupy, that is, if the state Ψ_0 is close to the vacuum state $|0\rangle$ (as is usually the case), we can neglect the above-mentioned extra terms in the commutator $[A, A^+]$. Then the relation (6.75) will be filfilled. Thus, we arrive at the quasi-boson approximation, or the method of approximate second quantization. In this approximation the operators Q and Q^+ approximately satusfy the commutation relations

$$\begin{aligned} &[Q_{BM}, Q_{CM'}] = [Q^+_{BM}, Q^+_{CM'}] = 0; \\ &[Q_{BM}, Q^+_{CM'}] = \delta_{BC}\delta_{MM'}, \end{aligned} \tag{6.76}$$

that is, they are operators of the boson type. We remark that the *quasi-boson approximation* is, in essence, another formulation of the random-phase method.

It folliws from (6.75) and (6.76) that there is some violation of the Pauli principle in the quasi-boson approximation. Indeed, the simplicity of the formulation in this approximation is achieved precisely as a result of violations of this principle. It may be said that the condition for the applicability of the quasi-boson approximation, or of the method of approximate second quantization, is that the violation of the Pauli principle in the sense indicated above can be neglected.

Collective states of nuclei in a model with residual interaction. The basic equations of the random-phase method describe both one-particle excitations and collective excitations in nuclei, and this is one of the merits of the method. Using the relations (6.70) and (6.76), we can bring the Hamiltonian operator of a system of nucleons to the form

$$H = \sum_{BM} \omega_B Q^+_{BM} Q_{BM} + \text{const.} \tag{6.77}$$

The Hamiltonian obtained describes a system of Bose particles, which can be situated in different oscillator potentials, depending on the index B. It turns out that of all the possible states B over which the summation is taken in the expression (6.77), only one corresponds to a collective state, this state having an energy $\omega_B = \omega_{\text{coll}}$ that differs considerably from the energies of all the other possible states. The corresponding terms in (6.77) will be the collective Hamiltonian

$$H_{\text{coll}} = \omega_{\text{coll}} Q^+_{\text{coll}} Q_{\text{coll}}. \tag{6.78}$$

If this expression is identifed with the phenomenological collective Hamiltonian, we can establish the connection between the cillective coordinates and the operators Q_{coll} and Q^+_{coll}. In the harmonic-oscillator approximation, this will be a linear dependence. According to (6.58), (6.71) and (6.73), the wave function $\Psi_{\text{coll}} = Q^+_{\text{coll}}\Psi_0$ of the collective excited state is a superposition of functions of two-quasi-particle states.

It is comparatively simple to explain the appearance of the collective state in the Tamm-Dancoff approximation in the case when the residual interaction can be written in separable form (in the form of products of several factors depending on different coordinates). This case is usually called the *Brown-Bolsterli model* (Brown and Bolsterli, 1959). In the familiar notation, the equations in the Tamm-Dancoff approximation (when the interaction between the quasi-particles is taken into account only in the excited states) can be written, according to the formulae (6.65)–(6.68), in the following form ($\psi_{abB} = \text{const} \cdot X_{abB}$, $Y_{abB} = 0$):

$$(\omega_B - E_a - E_b) X_{abB} = -\sum_{cd} U(abcd\mathcal{J}) X_{cdB} \tag{6.79}$$

(in the approximation under consideration, $U_+ = U_- + U$). We now make the principal assumption about the interaction, which is that the function $U(abcd\mathcal{J})$ can be represented in the form of the following product:

$$U(abcd\mathcal{J}) = -\eta A(ab) A^*(cd). \tag{6.80}$$

Here η is a real constant characterizing the strength of the residual interaction. The sign of this constant coincides with that of the interaction, so that $\eta < 0$ will correspond to attractive forces and $\eta > 0$ to repulsive forces. It follows from the symmetry property $U(abcd\mathcal{J}) = U(cdab\mathcal{J})$ that the function $U(abcd\mathcal{J})$ is real.

Substituting (6.80) into (6.79) and noting that the sum $\sum_{cd} A^*(cd)X_{cdB} \equiv \equiv C_B$ is a constant independent of the states a and b, we obtain

$$X_{abB} = C_B \frac{\eta A(ab)}{\omega_B - E_a - E_b}. \tag{6.81}$$

Substituting this expression into (6.79) and using (6.80), we obtain an equation for the energy eigenvalues ω_B:

$$\sum_{cd} \frac{|A(cd)|^2}{E_c + E_d - \omega_B} = -\frac{1}{\eta}. \tag{6.82}$$

The dependence of the left-hand side of the equality (6.82) on the energy ω_B can be represented graphically in the form of tangent-like curves with vertical asymptotes at the values $\varepsilon_i = E_{c1} + E_{d1}, E_{c2} + E_{d2}$, etc. (Fig. 6.1). The eigenvalues are the energies at which these curves intersect the horizontal straight line $-1/\eta$. It can be seen directly from the Figure that, for any change in the strength η of the residual interaction, only the energy of one of the extremal states (the minimum or maximum eigenvalue) can change much. The other energy eigenvalues lie between the two-quasi-particle energies $\varepsilon_i = E_{ci} + E_{di}$, which define the positions of the asymptotes and coincide with the eigenvalues in the absence of residual interaction (when $\eta \to 0$). In the case of attractive forces ($\eta < 0$) it is the minimum eigenvalue which undergoes a strong shift on increase of the strength of the residual interaction, and in the case of repulsive forces ($\eta > 0$) the maximum eigenvalue is strongly shifter as η increases.

It is clear that the energy of the collective state should be more sensitive than the energies of the one-particle states to changes of the residual interaction. This is because correlation effects associated with the residual interaction lead, as already mentioned, to collective motions in nuclei. It is therefore the collective state whose energy undergoes a large change as the residual interaction is "switched on". This state is usually separated from the other states by a considerable energy gap and, in view of the existence of residual interaction in all nuclei, is regularly encountered, with approximately the same energy, in the spectra of many nuclei that are close in mass. (This is one of the indications that observable states of this type are collective in character.)

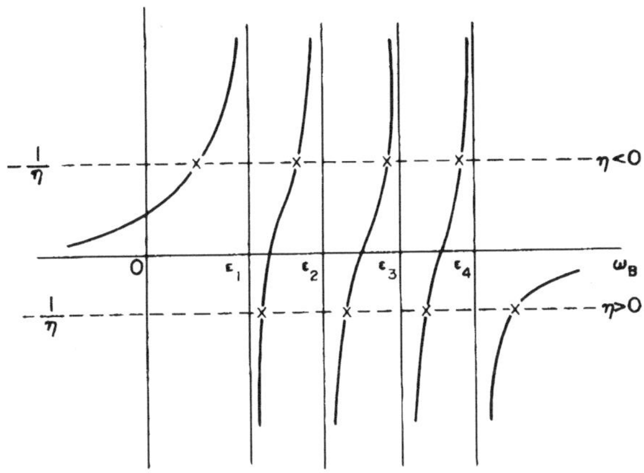

Figure 6.1. Graphical solutation of the equation for the energy eigenvalues in the model with residual interaction.

We now elucidate the reason for the large probability of electromagnetic transitions with excitation of the collective states of nuclei. The transition intensity is proportional to the expression

$$\sum_{\mu M} |\langle\Psi_0|\widehat{T}_{\lambda\mu}|\Psi_{BM}\rangle|^2 \propto |\langle\Psi_0||T_\lambda||\Psi_B\rangle|^2. \tag{6.83}$$

We write the one-particle tensor operator $\widehat{T}_{\lambda\mu}$ in the second-quantized representation

$$\begin{aligned}\widehat{T}_{\lambda\mu} &= \sum_{\alpha\beta}\langle\alpha|\widehat{T}_{\lambda\mu}|\beta\rangle a_\alpha^+ a_\beta = \sum_{ab}\sum_{m_\alpha m_\beta}\langle a j_\alpha m_\alpha|\widehat{T}_{\lambda\mu}|b j_\beta m_\beta\rangle a_\alpha^+ a_\beta \\ &= \sum_{ab}\sum_{m_\alpha m_\beta}\frac{(-1)^{j_\beta - m_\beta}}{\sqrt{2\lambda+1}}(j_\alpha m_\alpha j_\beta - m_\beta|\lambda\mu)\langle a||T_\lambda||b\rangle a_\alpha^+ a_\beta . \end{aligned} \tag{6.84}$$

In the last expression, as in (6.83), we have made use of the Wigner-Eckart theorem. Considering only particle-hole excitations, when the states α and β are respectively above and below the Fermi surface, that is, when u_a and v_a in (6.52) are equal to either unity or zero, and also retaining in (6.84) the term which contributes to the matrix element (6.83) in the *Tamm-Dankoff approximation*, we obtain

$$\widehat{T}_{\lambda\mu} = \sum_{ab}\frac{\langle a||T_\lambda||b\rangle}{\sqrt{2\lambda+1}}A^+(ab\lambda\mu). \tag{6.85}$$

Substituting this relation into (6.83), we see that

$$\left|\sum_{ab}\langle a||T_\lambda||b\rangle X_{abB}\right|^2 \propto \left|\sum_{ab}\frac{\langle a||T_\lambda||b\rangle A_(ab)}{E_a+E_b-\omega_B}\right|^2. \tag{6.86}$$

In the case of an electric transition, the signs of the numerators of all the terms inside the modulus sign in the last expression are the same.

It can be seen directly from (6.86) that if ω_B is equal to one of the extremal eigenvalues, all the terms in the sum (6.86) have the same sign and add coherently for the maximum eigenvalue when $\eta > 0$ and for the minimum eigenvalue when $\eta < 0$. This leads to a sharp increase of the probability of a transition to these extremal states, which are collective in character. In this connection, we note that one of the basic criteria of a collective states is the fact that the probability of an electromagnetic transition involving the state is always several times greater than the value obtained in a one-particle transition. For transitions to all the other (incoherent) states, for which ω_B is confined between values $\varepsilon_i = E_{ai} + E_{bi}$, terms of opposite sign are encountered in the sum in (6.86) and will partially cancel each other (that is, they will add incoherently). In this case the transition probability will be small.

The collective character of the above-mentioned coherent state, the energy of which undergoes a strong shift as the strength of the residual interaction is varied, arises from a coherent superposition of elementary two-quasi-particle states. In the case of attractive forces ($\eta < 0$) we have a low-lying collective state, which, as experiment shows, is vibrational. In the case of repulsive forces ($\eta > 0$) we have a high-energy collective state, usually called a *giant resonance.*

In the limiting case of degeneracy, when all the energies $\varepsilon_i = E_{ai} + E_{bi}$ are the same and equal to ε, the energy gap between the collective state and the incoherent states is easily found, since in this case we have, from formula (6.82),

$$\omega_B = \varepsilon + \eta \sum_{cd} |A(cd)|^2. \tag{6.87}$$

In the random-phase approximation the treatment is considerably more complicated and the differences between the derivations and those in the Tamm-Dankoff approximation reduce, in particular, to the result that, in the case $\varepsilon_i = \varepsilon$, instead of formula (6.87) we have

$$\omega_B = \sqrt{\varepsilon^2 + 2\eta\varepsilon \sum_{cd} |A(cd)|^2}, \tag{6.88}$$

and the transition probability will contain the extra factor ε/ω_B. Thus, in the random-phase approximation, the transition probability increases for transitions to states shifted to lower energies relative to ε ($\eta < 0$), since then $\omega_B < \varepsilon$, and decreases for transitions to states shifted to higher energies ($\eta > 0$).

Problems

6.1. Find the energy gap in the spectrum of a nucleus under the assumption that the interaction between nucleons in different shells can be neglected and that the matrix element of the residual pair interaction is the same for transitions between any levels within a shell (the coherent pair interaction model (Belyaev, 1959)).

The coherent pair interaction model is an example of a very simple pair interaction model in which the condition (6.28) is fulfilled and eqn. (6.25) admits a nontrivial solution. However, to obtain this solution in explicit form it is necessary to make simplifying assumptions not only about the nature of the interaction but also about the nature of the one-particle spectrum ε_ν.

In the case of spherical nuclei, the levels ε_ν are divided into groups corresponding to the individual shells. In the model, the interaction between nucleons in different shells is neglected. (It can be shown that the effect of other shells leads to an increase of the effective interaction between the particles.) We further assume that the matrix elements appearing in (6.25) are the same for transitions between any levels inside a shell:

$$\langle \nu' - \nu' | V | - \nu\nu \rangle = -G. \tag{6.89}$$

The eqn. (6.25) for the gap Δ, which, according to (6.18) and (6.40), does not depend on ν, takes the form

$$1 = \frac{1}{2} G \sum_\nu \frac{1}{\sqrt{(\varepsilon_\nu - \lambda)^2 + \Delta^2}} \tag{6.90}$$

In the most interesting case, when Δ is much greater than the spacing between the levels, the summation in (6.90) can be replaced by integration

$$1 = \frac{G}{2} \int_a^b \varrho(\xi) \frac{d\xi}{\sqrt{\xi^2 + \Delta^2}}, \tag{6.91}$$

where $a = \xi' - \lambda$, $b = \xi'' - \lambda$, and ξ' and ξ'' characterize the edges of the shell. Defining the mean density of states by means of the relation

$$\bar{\varrho} = \int_a^b \frac{\varrho(\xi)\, d\xi}{\sqrt{\xi^2 + \Delta^2}} \bigg/ \int_a^b \frac{d\xi}{\sqrt{\xi^2 + \Delta^2}},$$

we obtain from (6.91)

$$\Delta = \frac{1}{\sinh 2\eta} [b^2 + a^2 - 2ab \cosh 2\eta]^{1/2}, \tag{6.92}$$

where we have introduced the dimensionless quantity

$$\eta = (\bar{\varrho} G)^{-1}. \tag{6.93}$$

The chemical potential can be eliminated from (6.92) by making use of the relation (6.30):

$$n = \int_a^b \left(1 - \frac{\xi}{\sqrt{\xi^2 + \Delta^2}}\right) \varrho(\xi)\, d\xi. \tag{6.94}$$

We consider the case with $b = -a = \frac{1}{2}\varepsilon_0 > 0$. Then we shall have $\Delta = \varepsilon_0(\cosh\eta / \sinh 2\eta)$. Assuming that $\eta > 1$, we obtain

$$\Delta \approx \varepsilon_0 \mathrm{e}^{-\eta} = \varepsilon_0 \mathrm{e}^{-1/\bar{\varrho}G}. \tag{6.95}$$

(at it should, the quantity Δ tends to zero as $G \to +0$.) From (6.95) it can be seen directly that the expression for Δ cannot be found by perturbation theory. In fact, perturbation theory should give corrections to the energy that contain the pair-interaction parameter G to some integer power, whereas $\exp(-1/\bar{\varrho}G)$ cannot be expanded in powers of G, even if it is assumed that the pair interaction between the nucleons is small.

In the case of strongly deformed nuclei (when shell effects can be neglected), the distribution of levels is almost uniform. In this case, the matrix element (6.89) must be assumed to be nonzero for a group of levels that is symmetric about the Fermi level. Therefore, we must put $a = -b$ in (6.91) et seq. (For a spherical nucleus, the interaction region coincides with the shell, and the position of the Fermi level is determined by the number of particles in the shell. It is clear that a deformed nucleus will be equivalent to a spherical nucleus with a half-filled shell when $a = -b$.)

6.2. Find the energy of the ground state in the coherent pair interaction model under the assumption that the density of levels depends linearly on the energy.

We shall assume that the density of state $\varrho(\xi)$ depends linearly on the one-particle energy ξ:

$$\varrho(\xi) = \varrho_0 \left[1 - \varkappa \frac{b+a}{b-a} + \frac{2\varkappa}{b-a}\xi\right], \tag{6.96}$$

where

$$\varrho_0 = \frac{1}{2}[\varrho(b) + \varrho(a)], \quad \varkappa = \frac{\varrho(b) - \varrho(a)}{\varrho(b) + \varrho(a)}.$$

The mean density ϱ_0 is connected with the total number Ω of pair states in the shell by the relation

$$\varrho_0(\xi'' - \xi') = \Omega. \tag{6.97}$$

Performing the integration in (6.94) and using (6.92), we obtain

$$1 - \frac{n}{\Omega} - \frac{b+a}{b-a}\tanh\eta - \frac{\varkappa}{2}\left(1 - \frac{2\eta}{\sinh 2\eta}\right)\coth\eta + $$
$$+ \frac{\varkappa}{2}\left(\frac{b+a}{b-a}\right)^2 \left(1 - \frac{2\eta}{\sinh 2\eta}\right)\tanh\eta = 0. \tag{6.98}$$

We introduce the quantity $\chi_n = b + a/b - a\tanh\eta$, in terms of which the chemical potential can be expressed in the following way:

$$\lambda = \frac{\xi'' + \xi'}{2} - \frac{\xi'' - \xi'}{2}\chi_n \coth\eta. \tag{6.99}$$

From (6.98), we obtain for χ_n

$$\chi_n = 2\frac{1 - \frac{n}{\Omega} - \frac{\varkappa}{2}\gamma(\eta)}{1 + \sqrt{1 - 2\,\varkappa\,\gamma(\eta)\left(1 - \frac{n}{\Omega}\right) + \varkappa^2\,\gamma^2(\eta)}}, \tag{6.100}$$

where

$$\gamma(n) = \coth\eta\left(1 - \frac{2\eta}{\sinh 2\eta}\right).$$

Since $|\,\varkappa\,| \leqslant 1$ and $0 \leqslant \gamma < 1$, according to (5.100) we have $|\chi_n| \leqslant 1$. The limiting values ± 1 are reached at the edges of the shell ($n = 0$ and $n = 2\Omega$):

$$\chi_n = \begin{cases} 1, & n = 0 \\ 0, & n = (1 - \varkappa\,\gamma/2)\Omega \\ -1, & n = 2\Omega. \end{cases}$$

The quantity χ_n characterizes the degree of occupation of the shell. In the case of a uniform distribution of levels ($\varkappa = 0$),

$$\chi_n \to \chi_n^0 = 1 - \frac{n}{\Omega}, \tag{6.101}$$

Eliminating λ from (6.92), we find the energy gap:

$$\Delta = \frac{\xi'' - \xi'}{2\sinh\eta}(1 - \chi_n^2)^{1/2}. \tag{6.102}$$

This result was obtain under the assumption that Δ is greater than the mean spacing between the levels ($\Delta\varrho_0 > 1$). Using (6.97) and (6.102), we can write the condition for the applicability of (6.102) and (6.99) in the form

$$\frac{\Omega}{2\sinh\eta}(1 - \chi_n^2)^{1/2} > 1. \tag{6.103}$$

The ground-state energy is given by the expression

$$E_0 = \int_a^b \left(1 - \frac{\xi}{\sqrt{\xi^2 + \Delta^2}}\right)\varrho(\xi)\xi\,d\xi - \frac{\Delta^2}{G} + \lambda n.$$

Performing the integration, we obtain

$$\begin{aligned} E_0 &= \frac{1}{2}\left[\xi'' + \xi' + \frac{\varkappa}{3}(\xi'' - \xi')\right]n - \frac{\Omega}{4}(\xi'' - \xi')(1 - \chi_n^2) \times \\ &\times \left[\coth\eta + \frac{\varkappa}{3}\chi_n(1 - 3\gamma(\eta)\coth\eta) - \frac{\varkappa^2}{3}\gamma(\eta)\right]. \end{aligned} \tag{6.104}$$

6.3. Calculate the spectroscopic factors of nuclei in superfluid states for reactions with transfer of one nucleon (Belyaev and Zakhar'ev, 1961).

Let the original nucleus, which consists of A nucleons, be in a state of total spin I_0 and be described by the wave function Ψ_{I_0M}. We shall denote the channel function of the system with the same value of the total angular momentum consisting of the daughter nucleus (containing $A-1$ nucleons) with spin I and a nucleon with total angular momentum j by $\Phi_{I_0M}(I,j)$. The spectroscopic factor $S(I_0,I,j)$ is defined by the equality

$$S(I^0,I,j) = |(\Psi_{I_0M},\Phi_{I_0M}(I,j)|^2. \tag{6.105}$$

If the original nucleus is odd, we must take as the wave function Ψ_{I_0M} the one-quasi-particle state (6.33):

$$\Psi_{\nu_0} = a^+_{\nu_0}\prod_{\nu\neq\nu_0}(u_\nu + v_\nu a^+_\nu a^+_{-\nu})|0\rangle, \quad \nu_0 = I_0, M. \tag{6.106}$$

The daughter nucleus is even. If it is in its ground state, we must take as its wave function the wave function (6.32) of the quasi-particle vacuum:

$$\Psi_0 = \prod_\nu (u'_\nu + v'_\nu a^+_\nu a^+_{-\nu})|0\rangle, \quad I = M = 0. \tag{6.107}$$

The primes indicate that the quantities u'_ν and v'_ν refer to the daughter nucleus.

In this case, the channel function of the system with total angular momentum I_0 can be chosen in the form

$$\Phi_{\nu_0} = a^+_{\nu_0}\prod_\nu (u'_\nu + v'_\nu a^+_\nu a^+_{-\nu})|0\rangle. \tag{6.108}$$

It is not difficult to see that

$$\Phi_{\nu_0} = \prod_{\nu\neq\nu_0}(u'_\nu + v'_\nu a^+_\nu a^+_{-\nu})u'_{\nu_0}a^+_{\nu_0}|0\rangle. \tag{6.109}$$

Neglecting the difference between the primed quantities u'_ν, v'_ν and the unprimed quantities u_ν, v_ν we thus find

$$S(I_0,0,I_0) = u^2_{I_0M} = u^2_{I_0}. \tag{6.110}$$

This result has a simple physical meaning: the probability of capture of a nucleon by an even nucleus to give a state I_0, M is determined by the probability that this state is initially unoccupied.

If the original nucleus is even, we must take as the wave function Ψ_{00} the wave function (6.32) of the quasi-particle vacuum. In this case, the channel function can be chosen in the form

$$\Phi_{00} = \sum_M (IMI-M|00)a^+_{IM}a^+_{I-M}\prod_\nu (u'_\nu + v'_\nu a^+_\nu a^+_{-\nu})|0\rangle, \tag{6.111}$$

where $(IMI-M|00)$ is a Clebsh-Gordon coefficient. Again neglecting the difference between the primed and unprimed quantities u and v, we find

$$S(0,I,I) = (2I+1)v^2_I. \tag{6.112}$$

The probability of emission of a nucleon in the state I by an even-even nucleus is determined by the probability that this state is occupied in the quasi-particle vacuum.

Clearly, for the reverse processes, the spectroscopic factor $\widetilde{S}(I_0, I, j)$ will differ from (6.105) by the replacement $I_0 \rightleftarrows I$:

$$\widetilde{S}(I_0, I, j) = S(I, I_0, j). \tag{6.113}$$

6.4. Find an expression for the moment of inertia of a deformed nucleus with allowance for pairing between nucleons (Belyaev, 1959).

According to (5.226), the moment of inertia of a nucleus is determined by the general formula

$$\mathcal{J} = 2\hbar^2 \sum_{n \neq 0} \frac{|\langle n|j_{\varkappa}|0\rangle|^2}{E_n - E_0}, \tag{6.114}$$

where $j_{\varkappa}$ is the operator of the projection of the total angular momentum of the system along the direction of rotation, $|0\rangle$ is the ground state and $|n\rangle$ are the excited states of the system.

We write the operator of the angular-momentum projection in the second-quantization representation:

$$j_{\varkappa} = \sum_{\mu\mu'} \langle \mu'|j_{\varkappa}|\mu\rangle a_{\mu'}^{+} a_{\mu}, \tag{6.115}$$

where a_{μ} and $a_{\mu'}^{+}$ are particle annihilation and creation operators. Using next the canonical transformation (6.15), we express the operator $j_{\varkappa}$ in terms of the quasi-particle annihilation and creation operators α_{μ}, β_{μ}, α_{μ}^{+} and β_{μ}^{+}:

$$j_{\varkappa} = \sum_{\mu\mu'} \langle \mu'|j_{\varkappa}|\mu\rangle (u_{\mu'}\alpha_{\mu'}^{+} + v_{\mu'}\beta_{\mu'})(u_{\mu}\alpha_{\mu} + v_{\mu}\beta_{\mu}^{+}). \tag{6.116}$$

First of all we shall determine the moment of inertia, with allowance for pairing between nucleons, for an even-even deformed nucleus. In this case, we must take as the ground-state wave function $|0\rangle$ the quasi-particle vacuum wave function (6.32):

$$|0\rangle \equiv \Psi_0, \tag{6.117}$$

and for the excited states we must use the two-quasi-particle states described by wave functions of the form (6.35):

$$|n\rangle \equiv \Psi_{\nu,-\nu'} = \alpha_{\nu}^{+}\beta_{\nu'}^{+}\Psi_0 (\nu, \nu' > 0). \tag{6.118}$$

Using the functions (6.117) and (6.118), it is not difficult to calculate the matrix element of the operator (6.116)

$$\langle n|j_{\varkappa}|0\rangle = \langle \nu'|j_{\varkappa}|\nu\rangle (u_{\nu'}v_{\nu} - v_{\nu'}v_{\nu}). \tag{6.119}$$

Here we have used the fact that $u_{-\nu} = u_{\nu}$ and $v_{-\nu} = -v_{\nu}$. According to (6.38), the energy difference between the two-quasi-particle state and the ground state is equal to

$$E_n - E_0 = E_{\nu} + E_{\nu'}, \tag{6.120}$$

where E_ν and $E_{\nu'}$ are the quasi-particle energies. Substituting (6.119) and (6.120) into (6.114), we finally obtain the following expression for the moment of inertia of an even-even deformed nucleus:

$$\mathcal{J} = 2\hbar^2 \sum_{\nu,\nu'} \frac{|\langle \nu' | j_x | \nu \rangle|^2}{E_\nu + E_{\nu'}} (u_{\nu'} v_\nu - v_{\nu'} u_\nu)^2. \tag{6.121}$$

This expression, which takes the pairing between nucleons into account, differs from the corresponding expression for the moment of inertia in the independent-particle model in that it contains the factor $(u_{\nu'} v_\nu - v_{\nu'} u_\nu)^2$ and takes account of the gap in the energy spectrum, which is manifested in the fact that the denominator in (6.121) cannot be less than the quantity $2\Delta_\nu$.

It follows from the form of the operator (6.116) that the matrix element (6.119) characterizes the transition of a particle from the state ν to the state ν'. For such a transition to be possible, it is necessary that the state ν be occupied by a pair (the probability amplitude is v_ν) and that the state ν' be unoccupied (the probability amplitude is $u_{\nu'}$). The reverse transition is also possible. It is the interference of these two effects that leads to the appearance of the factor $(u_{\nu'} v_\nu - v_{\nu'} u_\nu)^2$.

According to (6.121), allowance for the pairing between nucleons leads to a considerable decrease in the value of the moment of inertia compared with the value in the independent-particle model (the rigid-body value). There are two reasons for this decrease. First, for values of ν and ν' corresponding to levels near the Fermi surface, the factor $(u_{\nu'} v_\nu - v_{\nu'} u_\nu)^2$ is less than unity. Secondly, the presence of the gap in the spectrum leads to a considerably greater value of the denominator in (6.121) than in the independent-particle model.

In the case of odd nuclei, the ground state is the lowest state with one quasi-particle. Choosing as the ground-state wave function $|0\rangle$ the one-quasi-particle state (6.33)

$$|0\rangle \equiv \Psi_{\nu_0} = \alpha^+_{\nu_0} \Psi_0, \tag{6.122}$$

and, as the excited states $|n\rangle$, the one- and three-quasi-particle states

$$\Psi_\nu = \alpha^+_\nu \Psi_0 (\nu > \nu_0) \quad \text{and} \quad \Psi_{\nu,\nu'\nu''} = \alpha^+_\nu \alpha^+_{\nu'} \beta^+_{\nu''} \Psi_0 (\nu, \nu'\nu'' > \nu_0), \tag{6.123}$$

we obtain for the moment of inertia an expression differing from (6.121) by an amount

$$\Delta\mathcal{J} = 2\hbar^2 \sum_{\nu \neq \nu_0} \frac{|\langle \nu | j_x | \nu_0 \rangle|^2}{E_\nu - E_{\nu_0}} (u_\nu u_{\nu_0} - v_\nu v_{\nu_0})^2. \tag{6.124}$$

This extra term in the moment of inertia of an odd nucleus is associated with the contribution of the one-quasi-particle excited states. The magnitude of the correction (6.124) depends strongly on the values of the quantum numbers ν_0. Since (6.124) is of the same order of magnitude as (6.124), the moments of inertia of odd deformed nuclei turn out to be much greater (in a number of cases, by almost a factor of two) than the moments of inertia of the neighbouring even-even nuclei. We remark that this increase is many times greater than the contribution of an individual nucleon (as given by the independent-particle-model) to the moment of inertia.

6.5. Find the one-particle Green function for a system of nucleons with allowance for pairing forces. The one-particle causal Green function is defined (Galitskii and Migdal, 1958) by:

$$G^c(\nu, t) = -i\langle 0|T\{a_\nu(t)a_\nu^+(0)\}|0\rangle, \tag{6.125}$$

where T is the chronological operator, which arranges the factors in order of decreasing time arguments from left to right, and the mean value is taken over the ground state of the system.

We denote the total Hamiltonian of the system by H and use the definition

$$a_\nu(t) = e^{iHt/\hbar}a_\nu(0)e^{-iHt/\hbar}.$$

Then the Green function (6.125) can be rewritten in the form

$$G^c(\nu, t) = \begin{cases} -i\langle 0|a_\nu e^{-iHt/\hbar}a_\nu^+|0\rangle e^{iE_0t/\hbar}, & t > 0, \\ i\langle 0|a_\nu^+ e^{iHt/\hbar}a_\nu|0\rangle e^{-iE_0t/\hbar}, & t < 0 \end{cases} \tag{6.126}$$

where E_0 is the energy of the ground state of the system.

We shall calculate the Green function in the approximation of noninteracting quasi-particles. For this, we express the operators a_ν and a_ν^+ in terms of the quasi-particle operators α_ν and β_ν in accordance with (6.15), and by $|0\rangle$ we shall mean the quasi-particle vacuum state. We thus find

$$G^c(\nu, t) = \begin{cases} -iu_\nu^2 e^{-iE_\nu t/\hbar}, & t > 0, \\ iv_\nu^2 e^{iE_\nu t/\hbar}, & t < 0 \end{cases} \tag{6.127}$$

where $E_\nu = \sqrt{\xi_\nu^2 + \Delta_\nu^2}$ is the quasi-particle energy ($\xi_\nu = \varepsilon_\nu - \lambda$ is the particle energy reckoned from the Fermi surface).

By means of a Fourier transformation with respect to the time, it is not difficult to find the energy representation for the causal Green function

$$G^c(\nu, E) = \frac{u_\nu^2}{E - E_\nu + i0} + \frac{v_\nu^2}{E + E_\nu - i0}. \tag{6.128}$$

The poles of the resulting expression determine the quasi-particle energies.

The probability of finding a nucleon in the state ν is determined by the limiting value of the Green function:

$$w_\nu = -iG^c(\nu, t \to -0). \tag{6.129}$$

Using (6.127), we have

$$w_\nu = v_\nu^2, \tag{6.130}$$

that is, in the approximation of noninteracting quasi-particles, the probability of finding a nucleon in the one-particle state ν is equal to v_ν^2.

In an analogous manner, it can be shown that, in the approximation of noninteracting quasi-particles with the pairing forces taken into account, the spectral distribution of the density fluctuations of the nucleons is determined by the following formula (Sitenko and Simenog, 1963):

$$\langle \delta n^2 \rangle_{q\omega} = \frac{1}{8\pi^2} \int d^3k \left\{ 1 - \frac{\xi_k \xi_{k-q} - \Delta^2}{E_k E_{k-q}} \right\} \delta(\omega + E_k + E_{k-q}). \qquad (6.131)$$

We not that, at zero temperature, only negative frequencies are contained in the fluctuation spectrum. Since the quasi-particle energy exceeds the energy gap Δ, for a superfluid state $\langle \delta n^2 \rangle_{q\omega}$ is non-zero only when $\omega < -2\Delta$.

References

Baranger, M. (1963) *Theory of Finite Nuclei*, Cargèse Lectures in Theoretical Physics, 1962, Benjamin, N.Y.

Bardeen, J., Cooper, L.N. and Schrieffer, J.R. (1957) *Phys. Rev.* **108**, 1175.

Belyaev, S. (1959) *K. Danske Vidensk. Selsk. Mat. fys. Medd,* **31**, No. 11.

Belyaev, V.B., and Zakhar'ev, B.N. (1961) *Izv. Akad. Nauk SSSR ser. fiz.* **25**, 1152 [*Bull. Acad. Sci. USSR phys. ser. (USA)* **25**, 1157 (1961)].

Bogolyubov, N.N. (1958a) *Zh. Eksp. Teor. Fiz.* **34**, 58 [*Sov. Phys. – JETP* **7**, 41 (1958)].

Bogolyubov, N.N. (1958b) *Zh. Eksp. Teor. Fiz.* **34**, 73 [*Sov. Phys. – JETP* **7**, 51 (1958)].

Bogolyubov, N.N. (1958c) *Dokl. Akad. Nauk SSSR* **119**, 52 [*Sov. Phys. – Doklady* **3**, 279 (1958)].

Bogolyubov, N.N., Tolmachev, V.V. and Shirkov, D.V. (1958) *A New Method in the Theory of Superconductivity* (translation published by Consultants Bureau, N.Y., 1959).

Bohr, A., Mottelson, B.R. and Pines, D. (1958) *Phys. Rev.* **110**, 936.

Brown, G.E., (1964) *Unified Theory of Nuclear Models*, North-Holland Publishing Co., Amsterdam.

Brown, G.E. and Bolsterli, M. (1959) *Phys. Rev. Lett.* **3**, 472.

Galitskii, V.M., and Migdal, A.B. (1958) *Zh. Eksp. Teor. Fiz.* **34**, 139 [*Sov. Phys. – JETP* **7**, 96 (1958)].

Gillet, V. (1965) "Approximate methods in the theory of nuclear structure" (in French) (Appendix in *Les Modéles en Spectroscopie Nucléaire*, by R.Naraf, Dunod, Paris).

Lane, A.M. (1964) *Nuclear Theory*, Benjamin, N.Y.

Migdal, A.B. (1965) *Theory of Finite Fermi Systems and Properties of Atomic Nuclei*, ("Nauka", Moscow) (translation published by Interscience, N.Y., 1967).

Sitenko, A.G., and Simenog, I.V. (1963) *Ukr. Fiz. Zh.* **8**, 537.

Solov'ev, V.G. (1958) *Zh. Eksp. Teor. Fiz.* **35**, 823 [*Sov. Phys. – JETP* **8**, 572 (1959)].

Solov'ev, V.G. (1959) *Zh. Eksp. Teor. Fiz.* **36**, 1869 [*Sov. Phys. – JETP* **9**, 1331 (1959)].

Solov'ev, V.G. (1963) *Effect of Superconducting Pair Correlations on the Properties of Atomic Niclei*, Gosatomizdat, Moscow.

CHAPTER 7

QUARK MODEL OF NUCLEUS

7.1. Quantum chromodynamics and hadron-hadron interaction

The hadron structure and strong interactions. When studying the nucleon-nucleon scattering and structure of atomic nuclei in the ground and excited states, we everywhere have treated nucleons as point-like structureless particles. In fact, nucleons have finite and comparatively large dimensions, which are of the order of a radius of the nuclear (nucleon-nucleon) forces. It has long been indicated by some facts, for example, the results of the scattering of ultra-relativistic electrons by protons and complex nuclei in the *Hofstadter experiments* in 50's and the consequent experiments. In particular, to describe the observed electron-proton scattering it was necessary to introduce the proton charge distribution form-factor (see (3.4)), (as well as for the complex nuclei), which gave the mean square (electromagnetic) proton radius $R_p \approx 0.8 \times 10^{-13}$ cm[1]. It is worth noting that at low (nonrelativistic) energies of nucleons, particularly in nuclei, the proper nucleon structure manifests itself rather weakly. Thus, the approximation of point-like nucleons used in all the previous sections of this book seems to be justified. In part it is associated with the fact that all complex nuclei are comparatively weakly bound nucleon systems, where the mean distance between nucleons is considerably larger than the nucleon dimensions. Therefore, we can expect that in many cases the proper nucleon structure effect may emerge as small corrections in the nuclear properties and nuclear processes at low energies. However, at high energies of particles colliding with nuclei when the energies and momenta transferred to nucleons may be sufficiently large, the nucleon structure effects can be dominant.

Nucleons (protons and neutrons) belong to the numerous family of the elementary particles called *hadrons* or strongly interacting particles. Hadrons are kept together by the most strong forces in the nature known as the *strong interactions*. This nucleon-nucleon interaction is conventionally called the nuclear interaction because nuclei are built of nucleons. All hadrons are divided into *mesons*, which are bosons with an integer spin (in

[1] A detailed account of the theory of electromagnetic interactions in nuclei is given in the monograph by Akhiezer, Sitenko and Tartakovskii (1994), which contains a detailed bibliography.

$\hbar$ units), and *barions*, which are fermions with a half-integer spin. Therefore, nucleons are barions. All barions possess a nonzero integer positive *barion charge* $B = +1$. Antibarions are characterized by $B = -1$. Mesons do not have the barion charge, i.e., their barion charge $B = 0$. The nuclear barion charge is simply equal to a number of nucleons in the nuclei, i.e., the mass-number A. All barions, with the exception of protons, are unstable particles. Protons, even if they can decay, must have a giant life-time, which is not less than 10^{32} years that exceeds the life-time of our Universe ($\sim 10^{10}$ years) by many orders. At present, more than four hundred hadrons are known.

Along with nucleons, the other hadrons also have a complex structure. They consist of more elementary particles – subparticles or *partons*. The *hadron compound models* appeared even at the end of 40-s after the experimental discovery of π-mesons (*pions*). When analyzing the pion and nucleon properties, Fermi and Yang (Fermi and Yang,1949) suggested the first compound model of hadrons for pions, assuming that each pion consists of two other hadrons – one nucleon and one antinucleon. For example, π^+-meson, accordingly this model, must consist of a proton p and antineutron $\bar{n}$. Since the pion mass $m_\pi \approx 140\,\mathrm{MeV}$, the *Fermi-Yang model* of pions predicts rather large mass excess per nucleon and antinucleon, which considerably exceeds the mass-excess per nucleon in nuclei. Later, after discovering of the so-called *strange mesons* (K-mesons or *kaons*), Markov (1940, 1968) and Sakata (1956) extended the Fermi-Yang model, assuming that a strange meson contains one nucleon (or antinucleon) and one *antilambda-hyperon* (or *lambda-hyperon*). *Hyperons* are the long-lived and more heavy (than nucleons) barions. Their properties are discussed below. In spite of the limited usefulness of the Fermi-Yang and *Markov-Sakata models*, they have played an important role when establishing and formulating the most realistic and, it can be stated now, the commonly accepted hadron *quark model* (not only mesons, but barions as well), which contained a set of the most important features of the first hadron compound models.

The quark model of hadrons was suggested in the middle of 60-s by Gell-Mann (1964) and independently by Zweig (1964). Accordingly this model, hadrons consist of specific subparticles, which are known as quarks. A meson consists of one *quark* and one *antiquark* (i.e., two subparticles, as in the Fermi-Yang and Markov-Sakata models). A barion consists of three quarks, an *antibarion* consists of three antiquarks. These quarks and antiquarks sometime are called the valence ones. Hadrons with another compositions of quarks and antiquarks were not found in the nature so far, although the relevant theories, in fact, admit the existence of different quark structures. It is customarily to denote quarks as q, and antiquarks as $\bar{q}$, therefore, the observed hadrons have one of the possible structures:

$q\bar{q}$ (mesons), qqq (barions), and $\bar{q}\bar{q}\bar{q}$ (antibarions).

We now dwell on the most important quark- and antiquark properties and see how hadrons can be constructed from these particles. It is assumed that at present there are, at least, six different types or *flavors of quarks* q, which are denoted by the latin letters u, d, s, c, b, t. The first five quarks are already experimentally discovered in hadrons. The sixth quark t are still hypothetical, but its existence follows from the current theory. All quarks are fermions with a spin $J = 1/2$. They have a *fractional barion charge* $B = 1/3$ and *fractional electric charge* Q , which equals to 2/3 for u, c, t quarks, and equals to -1/3 for d, s, b quarks (in proton or positron charge units). Quarks u and d, as well as both nucleons, have an isotopic spin (isospin) $T = 1/2$; at this, the isotopic spin projection of u-quark is $M_T = +1/2$, and d-quark - $M_T = -1/2$. Therefore, u-, d-quarks form the *isotopic doublet* , as well as proton and neutron. The rest quarks do not have isotopic spin, but have particular for each quark a new charge (or quantum number): s-quark is called a *strange quark* and characterized by a *strangeness* $s = -1$, c-quark has its own charge $C = +1$, that is known as *charm*, b-quark has a charge $B = -1$, which is called *beauty*, hypothetic t-quark must have a charge $T' = +1$. Recall that a charge is the characteristic, which, for a system of particles (subparticles), equals to the algebraic sum of the corresponding quantities of individual particles (subparticles), preserves, at least, at some particular interactions and can be measured. As for observed hadrons, the *quark charges* are interrelated with the isospin projection M_T by the *Gell-Mann-Nishijima formula* (Gell-Mann,1953; Nishijima, 1955)

$$Q = M_T + \frac{1}{2}Y, \quad Y = B + S + C + B' + T', \tag{7.1}$$

where Y is the so-called *hypercharge*. An antiquark can be obtained (as any antiparticle from a corresponding particle) by changing signs of all the quark (particle) charges. This is the operation of charge conjugation. We can see from (7.1) that a sign of isospin projection of the corresponding antiparticle also changes. As a rule, an antiparticle is denoted by the same letter as the particle but with a bar above the symbol. For example, a proton p and antiproton $\bar{p}$, a quark u and antiquark $\bar{u}$, and so on. If we denote the operator of *charge conjugation* as $\hat{C}$, then $\hat{C}p \to \bar{p}$, $\hat{C}u \to \bar{u}$. In a general case $\hat{C}a \to \bar{a}$, where a denotes a particle, and $\bar{a}$ - the corresponding antiparticle. (Inner parities of the fermion and corresponding antifermion are opposite and inner parities of the boson and corresponding antiboson are the same.) The mass, spin, isospin, and time-life of a particle and the corresponding antiparticle are equal. The same letters, which denote particles (subparticles), often specify the wave functions of these particles (subparticles).

Apparently, the more unusual property, than fractionality of the electric and barion quark charges, is the so-called *quark captivity* in hadrons. This phenomena is also called the *confinement* or the *quark holding* in hadrons. In spite of the fact that the quark existence is firmly established, it was impossible, up to now, to separate a quark from a nucleon or hadron, as an electron from the atom or a nucleon from the nucleus, and record an individual quark experimentally by a counter. Evidently, this hardly could be done in future, if the current theoretical ideas are correct. This results in some unusual properties of the strong quark-quark interaction. It increases with a distance between quarks, and at distances of the order of the hadron sizes ($\sim 10^{-13}\,\text{cm} \equiv 1\,\text{fm}$) becomes so large that quarks cannot move away anymore. The confinement phenomena is explained by namely this property. To the contrary, on decreasing distance between quarks the interaction strongly decreases. In this case, it is customarily to talk about the *asymptotic quark freedom* at small distances. In spite of the unusual behavior of the quark-quark interaction (or qq interaction) with a distance between quarks, the nature of this interaction and nuclear interaction between nucleons (and between hadrons as well) is the same. In either case, we deal with the same strong interaction.

Quarks cannot escape from hadrons, and, because of this, it is possible to speak about the definite quark masses only with some stipulations. There are two magnitudes of the quark masses associated with the so-called constitutive or *block quark* and *current (bare) quark*, whose mass m_q appears in the Lagrangian. The block quark consists of the current quark and a cloud of virtual particles. The *mass of the block quark* is exceeds the current quark mass approximately by of 300 MeV . The block and current quarks have equal charges, spins and other quantum characteristics. The masses of the lightest current quarks u and d are approximately equal to 5 MeV and 7 MeV, respectively. This means that the masses block quark u and d are roughly the same and approximately equal to one third of the mass of nucleon, which consists of the lightest quark u and d. The masses of the rest quarks, in particular, three heavy quarks c, b, and t are considerably larger.

The parameters of six quarks q and masses of current quark m_q (in MeV) are presented in Table 7.1. The q-quark flavors are arranged so that the quark masses increases from up to down. The Table for the corresponding antiquarks $\bar{q}$ can be obtained from Table 7.1 by changing only signs of every charge and isospin projection.

We consider now the quark structure of some hadrons. Nucleons must contain u- and d-quarks because only these quarks are characterized by an isospin. Nucleons cannot contain heavier quarks, for example, s-quarks,

TABLE 7.1.

q	J	T	M_T	Q	B	S	C	B'	T'	m_q, MeV
u	$\frac{1}{2}$	$\frac{1}{2}$	$+\frac{1}{2}$	$+\frac{2}{3}$	$+\frac{1}{3}$	0	0	0	0	5
d	$\frac{1}{2}$	$\frac{1}{2}$	$-\frac{1}{2}$	$-\frac{1}{3}$	$+\frac{1}{3}$	0	0	0	0	7
s	$\frac{1}{2}$	0	0	$-\frac{1}{3}$	$+\frac{1}{3}$	-1	0	0	0	150
c	$\frac{1}{2}$	0	0	$+\frac{2}{3}$	$+\frac{1}{3}$	0	+1	0	0	1400
b	$\frac{1}{2}$	0	0	$-\frac{1}{3}$	$+\frac{1}{3}$	0	0	-1	0	4800
t	$\frac{1}{2}$	0	0	$+\frac{2}{3}$	$+\frac{1}{3}$	0	0	0	+1	>20000

because nucleons are not the strange particles. Since nucleons have integer electric and barion charges (as well as all hadrons) and quarks have fraction charges, it is clear that the proton p must consist of two u-quarks and one d-quark $p = uud$. A neutron consists of one u-quark and two d-quarks: $n = udd$. *Antinucleons* contain the corresponding antiquarks: $\bar{p} = \bar{u}\bar{u}\bar{d}$, $\bar{n} = \bar{u}\bar{d}\bar{d}$. Two quarks of the same flavor in every nucleon are in the triplet spin state, therefore, the spin of the third quark is directed oppositely to spins of the first two quarks so that to obtain the correct spin of the nucleon. Hadrons, which consist of the strange s-quarks, are the called strange hadrons. For example, the above-mentioned the *lambda-hyperon* Λ consisting of u-, d-, s-quarks: $\Lambda = uds$ is a strange particle (with a strangeness $s = -1$), which does not have an electric charge, as well as a neutron.

A positively charged pion π^+ consists of the u-quark and $\bar{d}$-antiquark: $\pi^+ = u\bar{d}$, and its antiparticle the negatively charged pion π^- consists of $\bar{u}$-antiquark and d-quark: $\pi^- = \bar{u}d$. There are barions consisting of three quarks of the same flavor. For example, $\Delta^{++} = uuu$, $\Delta^- = ddd$, $\Omega^- = sss$. These three hadrons have the same spin 3/2. i.e., spins of all three quarks in each of these hadrons are parallel. There are hadrons of the same compound but with different properties. For example, the proton $p = uud$ and the *delta-particle* $\Delta^+ = uud$ with a spin $J = 3/2$ and isospin $T = 3/2$. Hyperons $\Lambda = uds$ and $\Sigma^0 = uds$ have the same spin 1/2, but their isospins are $T = 0$ and $T = 1$, respectively. These *hyperons* have different masses.

All the above-mentioned hadrons have zero relative angular momentum $L = 0$. But the Pauli principle seems to be violated in case hadrons $\Delta^{++}, \Delta^-, \Omega^-$, and p and n, because quarks are described by the Fermi-Dirac statistics. In fact, there is no violation of the Pauli principle because quarks of the same flavor may differ from each other by one more a new quantum characteristic. It was introduced by Bogolyubov, Struminskii and Tavkhelidze (1965), and, independently, by Han and Nambu (1965), and was called the *color*. The quark quantum color number C (don't mix with the charm) can take three different values or three different colors, which are conditionally called red (r), blue (b), and yellow (y) colors analogously with three main colors of the visible spectrum. The mixture of these three real colors results in, as known, the white color or the absence of color at all.

Unusualness of the color quantum number C is connected with the fact that there is no the color degree of freedom (or simply color) in all observable particles-hadrons. In real hadrons both barions and mesons, the quark (antiquark) color components are mutually compensated, so hadrons turn out to be colorless particles ($C = 0$). The quark and antiquark in a meson have complementary colors with respect to each other (color and anticolor), which, being mixed, give a colorless meson.

The captivity of quarks and antiquarks in hadrons is associated with the impossibility of the quark escape from hadron and registration of these particles by relevant devices as colored objects. The impossibility of the quark escape from hadrons must be brought about, as it was above-mentioned, by the quark-quark interaction pecularities. That is why the color is one of the most important quantum parameters, which is connected directly with the strong quark-quark interactions. It turns out to be that the color is a charge characterizing strong interactions and analogous to an electric charge in the electromagnetic interactions. But the color charges are hidden inside hadrons.

Continuing the analogy between the strong and electromagnetic interactions, we have to assume that carriers (or quanta of the strong interactions) exist. They were called *gluons*. That designation, obviously, originates from "glue". Both gluons and photons do not have the rest masses and possess spins, which are equal to unity. The interacting quarks exchange by gluons like electric charges exchange by photons. But photons, the only one type quanta of electromagnetic field, do not have any charge. There are eight different gluons and they have color charges, as well as quarks. A great number of different gluons can be understood from the position that every gluon carries not only one color, as a quark, but simultaneously two of them (color and anticolor). The gluon color charges bring into action the gluon-gluon interaction, which is realized through exchange of new gluons. As a consequence the current relativistic theory of the strong interactions *quantum chromodynamics* (*QCD*), unlike the *quantum electrodynamics* (*QED*) the relativistic theory of the electromagnetic interactions[2], is a nonlinear

[2]The detailed relativistic theory of electromagnetic interactions is described in the monographs by Akhiezer and Berestetskii (1981), and Bogolyubov and Shirkov (1984).

theory. The gluon color charges result in the fact that both gluons and quarks cannot escape from hadrons.

The quark q and antiquark $\bar{q}$ color charges are denoted by upper or lower indexes: q^i and $\bar{q}_i$, where $i = r, b, y$ (red,blue, yellow) or i=1,2,3. Gluons are often specified by a letter g and their color and *anticolor* denoted by two indexes: $g_{i,\bar{j}}$, where $j = \bar{r}, \bar{b}, \bar{y}$ (antired or yellow, antiblue or orange, antiyellow or violet colors) or $\bar{j} = \bar{1}, \bar{2}, \bar{3}$. An index with a bar $\bar{j}$ stands for color, which is supplementary to the color j. The four-vector B^a_μ (μ=0,1,2,3) stands for the gluon field with a color index a=1,2,...,8 (in accordance with the number of different gluons).

It follows from the above-reported that a hadron is a dynamical system of partons – *valence quarks (antiquarks)* , gluons, as well as the "sea" of *virtual quark-antiquark pairs* , which are neutral with respect to all charges. Concerning the quark sizes, we only know that they are considerably smaller than hadrons. In this regard hadrons are as empty as atoms.

Introducing the charge of the strong interaction – the color, we can understand the great difference in the interaction between two quarks and between two hadrons, in particular, between two nucleons. In an analogous way as the electromagnetic interaction between two electrically neutral atoms decreases rapidly with distance between them, the strong interaction between two neutral with respect to the color charge nucleons (and, in general, between any two nucleons) is considerably weaken in comparison with the quark-quark interaction carrying the color charges. By virtue of this, the nucleon-nucleon interaction decreases rapidly with a distance between nucleons, whereas the interaction between quarks can even increase with a distance.

There is the *color charge conservation law* , as well as the conservation law of the electrical charge. Therefore, when interacting quarks change their color charges, interchanging by gluons, which have two colors. For example, the first, assumingly red quark q_r interacting with the second one a blue quark, can eliberate a virtual red-antiblue gluon $g_{r\bar{b}}$, which can be absorbed by the second quark. As a result the first quark becomes blue and the second one – red. This process of the strong interaction can be presented by the simplest *Feynmann diagram* (Fig. 7.1). The straight lines with arrows present a quark and the wavy line corresponds to a gluon. This QCD diagram recalls the corresponding Feynmann diagram in the QED describing the electromagnetic interaction, for example, of two electrons with interchange of one virtual photon. But in the QCD there are diagrams, (which do not exist in the QED) and associated with the gluon-gluon interaction and connected with the interaction between gluons that causes the nonlinear character of the QCD equations (we will see it in the end of this section). For example, the interaction of two gluons with exchange of one

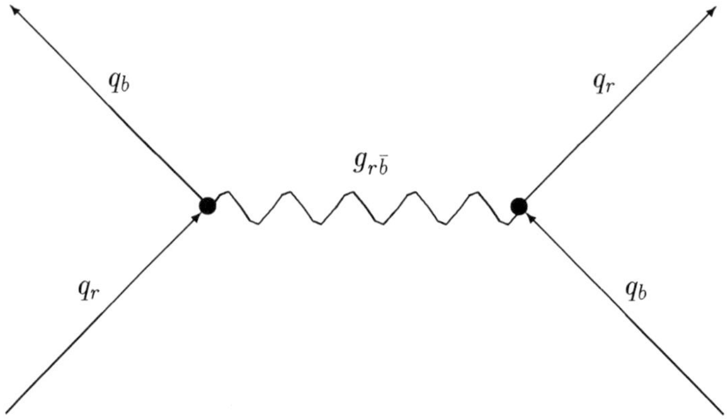

Figure 7.1. The Feynmann diagram of red and blue quarks interaction with one gluon exchange.

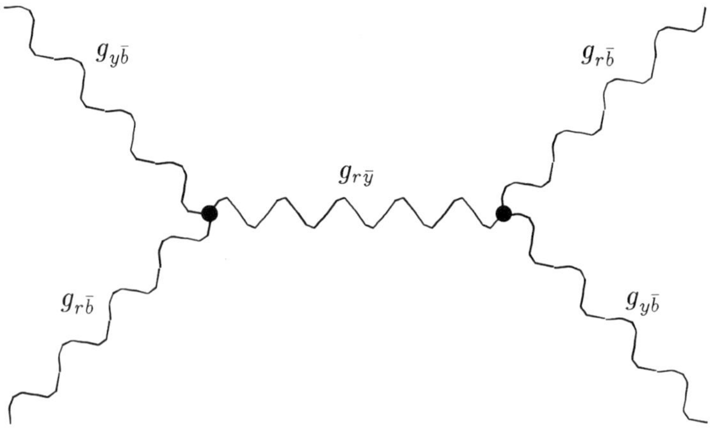

Figure 7.2. The Feynmann diagram of the two-gluon interaction with the third gluon exchange.

(the third) virtual gluon is depicted by the diagram presented in Fig. 7.2.

At small distances between quarks, their interaction is practically negligible (the *quark asymptotic freedom*). In this domain, quarks exchange by one gluon mainly (as it is shown in Fig. 7.1). This process can be described theoretically within the framework of the perturbation theory, as well as the processes in the QED. In this case in the QCD the perturbative approach is permitted. At small distances between quarks or at large relative quark energies and momenta, the bound constant of the strong interaction α_c, which is an analog of the fine structure constant α in the QED, is quite small. It can be explained qualitatively by the above-mentioned nonlinear interaction of gluons, which causes the antisceening of the quark color charge surrounded by a cloud of virtual gluons and quark-antiquark pairs unlike the screening of the electron electric charge by the cloud of virtual photons and electron-positron pairs. That is why the quark color charge decreases, and the electron electric charge increases, when approaching to a quark or electron at very small distances.

When distances between quarks are large, i.e., of the order of the hadron size (when the relative energies and momenta are sufficiently small), quarks interchange by many gluons, and a magnitude of the bound constant of the strong interaction becomes larger. In this case the perturbation theory is unsuitable. In the low energy domain, the strong interaction theory comes across the great difficulties, in particular, when explaining the confinement mechanism, which forbids the escape of quarks and gluons from hadrons. Because of this, the so-called *nonperturbative methods* , which do not assume the interaction smallness and do not use the perturbation theory, have been developed in the QCD.

We give here, in this section, a brief qualitative overview of the hadron structure, the properties of their parts (quarks and gluons) and some properties of the strong interactions. To study these questions in more detail, as well as to study the influence of the quark degrees of freedom on the properties of the lightest nuclei, we have to use the fundamentals of the *continuous transformation theory* . We pass now to a brief description of this theory.

Continuous groups and their representations. In the theoretical physics and, in particular, in the theory of elementary particles, the *continuous groups*, whose elements can be presented by using a finite number of real parameters, are of great importance. Such continuous groups are called the *Lie groups* after the norwegian mathematician Sophus Lie[3]. A minimal number m of independent real continuous, the so-called *group parameters* ξ_k (k=1,2,..., m), which are necessary to specify a current element $g = g$ $(\xi_1, \xi_2, \ldots, \xi_m)$ of the Lie group G is called the order or dimension of the group. The group parameters ξ_k are usually chosen so that the unit group element $g(0, 0, \ldots, 0) = 1$ corresponds to their zero values. In the previous sections, we have already used some of the Lie groups: the three-parameter $(m = 3)$ group of three-dimensional rotations R_3, where the part of the group parameters ξ_1, ξ_2 and ξ_3 is played by three Euler's angles, and its one-parameter $(m = 1)$ rotation subgroup in a plane R_2, as well as some of *unitary* U_n and *unimodular* SU_n *groups.* Here we restrict ourselves mainly by the consideration of the unimodular groups SU_n and their irreducible representation by virtue of their great importance for the hadron physics and the theory of strong interactions.

[3]The theory of the Lie groups is given in the lecture notes by N'guen Van H'eu (1967).

Since an arbitrary quadratic complex matrix of a dimension n, i.e., a matrix with n lines and n columns depends, in a general case, on $2n^2$ real parameters, then n^2 unitary conditions for the unitary group U_n matrices determine uniquely the *dimension (order) of this group* $m = n^2$. For the matrices $U = U(\xi_1, \xi_2, \ldots, \xi_m)$ of the unitary unimodular group SU_n, one more condition must be taken into account, namely, Det$U = 1$. Because of this, a dimension of the SU_3 group is equal to

$$m = n^2 - 1, \qquad n = 2, 3, 4, \ldots . \tag{7.2}$$

(The group SU_1 is trivial and consists of $+1$, and $m = 0$.)

We determine now the most important operators of the continuous Lie group called the *infinitesimal operators* or *group generators*

$$X_k = \frac{1}{i} \lim_{\xi_1, \xi_2, \ldots, \xi_m \to 0} \frac{\partial g(\xi_1, \xi_2, \ldots, \xi_k, \ldots, \xi_m)}{\partial \xi_k}, \quad k = 1, 2, \ldots, m, \tag{7.3}$$

which are, obviously, matrices, if g are matrices, and with the same dimension as g matrices. The group generators X_k can be treated as the expansion coefficients of the group element g in powers of parameters ξ_k corresponding to the first power of these parameters

$$g(\xi_1, \xi_2, \ldots, \xi_m) = I + i \sum_{k=1}^{m} \xi_k X_k + \ldots . \tag{7.4}$$

All generators X_k are linearly independent, and, in a general case, do not commute with each other. Apparently, a number of the Lie group generators equals to the dimension of the group.

In particular, a number of generators X_k of the group SU_n is equal to $n^2 - 1$, according to (7.2) – (7.4). For the group SU_n (the denotation of a current element g of the group is replaced by U) from (7.4) and the condition

$$\text{Det}U = 1 + i \sum_{k=1}^{m} \xi_k SpX_k + \ldots = 1$$

follows that spurs of the group operators (generators) X_k are equal to zero:

$$SpX_k = 0. \tag{7.5}$$

From the unitary condition

$$UU^+ = 1 + i \sum_{k=1}^{m} \xi_k (X_k - X_k^+) + \ldots = 1$$

we deduce that matrices (generators) X_k are Hermitian:

$$X_k^+ = X_k. \tag{7.6}$$

From the expression of a current element $U = 1 + i\sum_{k=1}^{m} \xi_k X_k$ of the group SU_n at small ξ_k, we can obtain U for arbitrary values of the group parameters in the general exponential form

$$U = \exp\{i \sum_{k=1}^{m} \xi_k X_k\}, \quad m = n^2 - 1. \tag{7.7}$$

This expression for $U = U(\xi_1, \xi_2, \ldots, \xi_m)$, which we use further, can be obtained in the same manner as formula (5.44). To do this, we note that the sum $\sum_{k=1}^{m} \xi_k X_k$ can be treated as a scalar product of two vectors in the m-dimensional space with components ξ_k and X_k. Let us make the rotation in this space so that for new projections ξ'_k and X'_k, which are superpositions of the former projections ξ_k and X_k, the scalar product reduces to one term, for example, $\sum_{k=1}^{m} \xi_k X_k = \xi'_1 X'_1$. Now for small ξ_k and ξ'_k, we can present an element U of the group as $U = 1 + i\xi'_1 X'_1$ and an element U for a finite ξ'_1 as the product $U = (1 + i\frac{\xi'_1}{\Lambda} X'_1)^{\Lambda}$, where Λ is a such large number that $\frac{\xi'_1}{\Lambda}$ can be treated as small. Passing to the limit $\Lambda \to \infty$, we get

$$U = \lim_{\Lambda\to\infty} \left(1 + i\frac{\xi'_1}{\Lambda} X'_1\right)^{\Lambda} = \exp(i\xi'_1 X'_1).$$

Coming back to ξ_k and X_k, we obtain (7.7). Expression (7.7) is in accordance with the well known fact that any unitary unimodular matrix U can be always presented in the exponential form $U = \mathrm{e}^{iH}$, where H is the Hermitian matrix with zero spur (see (7.5) and (7.6)). Note that the generators X_k in (7.7) are Hermitian, and, therefore, the group parameters ξ_k are real (to underline this, we isolate a factor i in the exponent (7.7)).

The commutators $[X_j, X_k] \equiv X_j X_k - X_k X_j$ of any two generators X_j and X_k of an arbitrary Lie group can be presented in the form of a linear combination of all m the group generators

$$[X_j, X_k] = \sum_{l=1}^{m} C_{jk}^l X_l, \quad C_{jk}^l = -C_{kj}^l, \tag{7.8}$$

where constants C_{jk}^l are called the *structure group constants* , which are uniquely determined by the structure of a given group. The inverse statement that the structure constants completely specify the particular Lie group is also true. From the *Jacobi equality* (which holds for three arbitrary operators)

$$[X_i,[X_j,X_k]]+[X_j,[X_k,X_i]]+[X_k,[X_i,X_j]]=0 \tag{7.9}$$

one can get the relation for the structure constants

$$\sum_{l=1}^{m}\left(C_{il}^{s}C_{jk}^{l}+C_{jl}^{s}C_{ki}^{l}+C_{kl}^{s}C_{ij}^{l}\right)=0 \tag{7.10}$$

The group, whose all elements g commute with each other, is called the Abelian. All irreducible representations of the *Abelian group* are one--dimensional. It follows from (7.4) that all generators of the continuous Abelian group must commute with each other. In the theory of strong interactions the continuous non-Abelian Lie groups, whose elements do not commute with each other in a general case, are of great interest. The groups SU_n with $n \geqslant 2$ are the *non-Abelian groups.*

Commuting generators $[X_p,X_q]=0$ may exist among the generators of the non-Abelian Lie group. In this case $C_{pq}^{l}=0$. A maximum number of commuting generators is called the *group rank.* If a group does not have, at least, two commuting operators, then we deal with the first rank group, since, in this case, any generator commutes only with itself.

We can construct such operators $\hat{C}_k$, which commute with all the group generators on the basis of the group generators. They are called the *Casimir operators.* A number of the Casimir operators is equal to the group rank. The first rank group has only one Casimir operator $\hat{C}_2$, which is quadratic in the group generators

$$\hat{C}_2=\mathrm{Const}\sum_{i,j=1}^{m}(h^{-1})_{ij}X_iX_j, \tag{7.11}$$

where h is the matrix, which is expressed in terms of the structure group constants:

$$h_{ij}=h_{ji}=\sum_{s,t=1}^{m}C_{it}^{s}C_{js}^{t}. \tag{7.12}$$

The Casimir operators, specified up to a factor, are of great importance in the theoretical quantum physics and, in particular, in the theory of hadrons and strong interactions. It is associated with the fact that the Hamiltonian of the systems under consideration can be expressed in terms of the Lie group generators and, hence, the Casimir operators commute with the Hamiltonian. In other words, the Casimir operators specify the integrals of motion in a physical system.

We consider now the most important for us the particular groups SU_n for $n = 2$ and $n = 3$. The simplest nontrivial SU_2 *group* , whose operators act in a two-dimensional space, presents a set of unitary unimodular matrices of a dimension $n = 2$. Accordingly (7.2), the SU_2 dimension is equal to $m = 3$. A current element of the group SU_3 depends on three real parameters ξ_k (k=1,2,3), as well as a general element of the group of three-dimensional rotations R_3. The group SU_2, thus, has three generators, which, accordingly (7.5) and (7.6), are the Hermite two-dimensional matrices with zero spurs. As generators of the group SU_2 in the case of one particle with a spin 1/2, we can choose three Pauli's spin matrices $\sigma_x, \sigma_y, \sigma_z$, and in the case of one hadron or quark with an isospin 1/2 – three isospin Pauli's matrices τ_j (1.141) or three projections of the isospin operator $T_j = \frac{\tau_j}{2}$, where $j = x, y, z$ or j=1,2,3.

Since the mathematical apparatus of the conventional spin $S = 1/2$ and the isospin $T = 1/2$ is the same, we can introduce two identical in the mathematical sense, but different in the physical meaning, groups – the spin group SU_2^S and isospin group SU_2^T. Further, we will speak mainly about the isotopic group SU_2^T, which is connected with the isotopic symmetry (invariance) of the nuclear (strong) interactions. By using the commutation relations of the Pauli isospin matrices $[\tau_j, \tau_k] = 2i\sum_{l=1}^{3} \varepsilon_{jkl}\tau_l$ (which can be checked directly on the basis of (1.141)), we obtain for the commutators of the group SU_2^T generators $t_j = \frac{1}{2}\tau_j$

$$[t_j, t_k] = i\sum_{l=1}^{3} \varepsilon_{jkl}t_l, \tag{7.13}$$

where ε_{jkl} is the unit completely antisymmetric third rank tensor. Since t_1, t_2, t_3 do not commute with each other, the group SU_2^T is the first rank group. By comparison (7.13) and (7.8), we obtain the explicit form of the structure constants of the group SU_2^T, namely $C_{jk}^l = \varepsilon_{jkl}$. In addition, matrix (7.12) is diagonal $h_{ij} = 2\delta_{ij}$, and $(h^{-1})_{ij} = \frac{1}{2}\delta_{ij}$. As a result, the only one Casimir operator (7.11) of the group SU_2^T reduces to a constant (a unit operator): $\hat{C}_2 = (\frac{\tau_1}{2})^2 + (\frac{\tau_2}{2})^2 + (\frac{\tau_3}{2})^2 = \frac{3}{4}$ for one particle. For a system of N strongly interacting particles with $t = \frac{1}{2}$, the total vector of the isospin $\mathbf{T} = \sum_{\alpha=1}^{N} \frac{1}{2}\boldsymbol{\tau}^{\alpha}$ and the corresponding Casimir operator constructed from the generators $\hat{T}_j$,

$$\begin{aligned} \hat{C}_2 = \hat{\mathbf{T}}^2 &= \hat{T}_1^2 + \hat{T}_2^2 + \hat{T}_3^2, \\ \left[\hat{T}_j, \hat{T}_k\right] &= i\sum_{l=1}^{3} \varepsilon_{jkl}\hat{T}_i, \quad \left[\hat{C}_2, \hat{T}_j\right] = 0 \end{aligned} \tag{7.14}$$

does not reduce to a constant any more.

The three-dimensional rotation group R_3 is not isomorphic but only homomorphic group SU_2. Nevertheless, the structures of these groups are close enough since each element of the group R_3, including the unit element, corresponds only to two elements of the group SU_2. If two group are homomorphic, then a set of the elements of the more rich in a number of elements (senior) group, which corresponds to a unit element of the second (junior) group, is called the *homomorphism kernel* and presents a subgroup of the senior group. The homomorphic continuous groups are called locally isomorphic, if the homomorphism kernel is a discrete group. There is one to one correspondence between the generators of the *locally isomorphic groups* . Further, as it is customary in the elementary particle physics, we identify the rotation group R_3 with the group SU_2 because both of these groups are locally isomorphic, as follows from the above-reported. In the same meaning, the irreducible representations of these groups are identical. In particular, the rotation group R_3 also has three generators, which satisfy the same commutation relations (7.13) or (7.14) as generators of the group SU_2^T. The group R_3, as well as the group SU_2, is the first rank group and has one Casimir operator. Using general formula (7.3) and determining an element of the rotation group R_3 by the axis and corresponding rotation angle, we can obtain the generators of the R_3 group from (5.45) independently. Apparently, they are equal to projections of the total angular momentum $\mathbf{J}$ of the system. Thus, the Casimir operator $\hat{C}_2$ of the three-dimensional rotation group R_3 coincides with the operator of squared total angular momentum:

$$\hat{C}_2 = \hat{\mathbf{J}}^2 = \hat{J}_x^2 + \hat{J}_y^2 + \hat{J}_z^2, \quad \left[\hat{J}_j, \hat{J}_k\right] = i\sum_{l=1}^{3} \varepsilon_{jkl}\hat{J}_l,$$
$$\left[\hat{C}_2, \hat{J}_l\right] = 0. \tag{7.15}$$

The *group* SU_3 presents a set of the unitary unimodular matrices of a dimension $n = 3$ which act in the three-dimension function space. Therefore, a dimension of this group is $m = n^2 - 1 = 8$. Each element of the group SU_3 depends on eight real parameters ξ_k (k=1,2,...,8). The group SU_3 has eight generators, which are Hermitian matrices of dimension 3 with zero spurs according to (7.3) – (7.6). The *Gell-Mann* λ*-matrices* are usually chosen as the generators of the group SU_3

$$\lambda_1 = \begin{pmatrix} 0 & 1 & 0 \\ 1 & 0 & 0 \\ 0 & 0 & 0 \end{pmatrix}, \lambda_2 = \begin{pmatrix} 0 & -i & 0 \\ i & 0 & 0 \\ 0 & 0 & 0 \end{pmatrix}, \lambda_3 = \begin{pmatrix} 1 & 0 & 0 \\ 0 & -1 & 0 \\ 0 & 0 & 0 \end{pmatrix},$$

$$\lambda_4 = \begin{pmatrix} 0 & 0 & 1 \\ 0 & 0 & 0 \\ 1 & 0 & 0 \end{pmatrix}, \lambda_5 = \begin{pmatrix} 0 & 0 & -i \\ 0 & 0 & 0 \\ i & 0 & 0 \end{pmatrix}, \lambda_6 = \begin{pmatrix} 0 & 0 & 0 \\ 0 & 0 & 1 \\ 0 & 1 & 0 \end{pmatrix},$$

$$\lambda_7 = \begin{pmatrix} 0 & 0 & 0 \\ 0 & 0 & -i \\ 0 & i & 0 \end{pmatrix}, \lambda_8 = \frac{1}{\sqrt{3}} \begin{pmatrix} 1 & 0 & 0 \\ 0 & 1 & 0 \\ 0 & 0 & -2 \end{pmatrix}. \tag{7.16}$$

These matrices are analogous to three τ-matrices (or σ-matrices) of Pauli, which are chosen as the group SU_2^T (or group SU_2^S) generators. The Gell-Mann matrices satisfy the following commutation relations

$$[\frac{\lambda_j}{2}, \frac{\lambda_k}{2}] = i \sum_{l=1}^{8} f_{jkl} \frac{\lambda_l}{2}. \tag{7.17}$$

The structure constants $C_{jk}^l = f_{jkl}$ of the group SU_3 are antisymmetric with respect to permutations of three indexes. Nonzero f_{jkl} are equal (it can be shown by using (7.16) and (7.17) directly)

$$f_{123} = 1, \; f_{458} = f_{678} = \frac{\sqrt{3}}{2},$$

$$f_{147} = f_{246} = f_{257} = f_{345} = f_{516} = f_{637} = \frac{1}{2}. \tag{7.18}$$

It follows from (7.16) that $Sp\lambda_j\lambda_k = 2\delta_{jk}$ and an anticommutator $\{\lambda_i, \lambda_j\} \equiv \lambda_i\lambda_j + \lambda_j\lambda_i = 2\sum_{k=1}^{8} d_{ijk}\lambda_k + \frac{4}{3}\delta_{ij}$, where constants d_{ijk} are called sometimes the structure constants. They are completely symmetric with respect to the index permutations and have the following nonzero components:

$$d_{118} = d_{228} = d_{338} = -d_{888} = \frac{1}{\sqrt{3}}, \; d_{448} = d_{558} = d_{668} = d_{778} = -\frac{1}{2\sqrt{3}},$$

$$d_{146} = d_{157} = -d_{247} = d_{256} = d_{344} = d_{355} = -d_{366} = -d_{377} = \frac{1}{2}.$$

The structure constant magnitudes, as in (7.17), do not depend on the particular choice of the group SU_3 generators in form (7.16).

Among all generators of the group SU_3, i.e., λ-matrices (7.16), two matrices only, λ_3 and λ_8, commute with each other (they are diagonal matrices). Because of this, a rank of the group SU_3 is equal to 2 and it has two different Casimir operators. One of them, $\hat{C}_2$, is quadratic in the group generators (see (7.11)), and another one, $\hat{C}_3$, takes the form of a sum of terms proportional to a product of these generators of the group SU_3. The direct checking shows that a sum of squares of the Gell-Mann λ-matrices (7.16) is multiple to a unit matrix $\sum_{j=1}^{8}\lambda_j^2 = \frac{16}{3}$, so this sum commutes with matrix of eight matrices λ_j and can be treated as the Casimir operator $\hat{C}_2$. Thus, the Casimir operator $\hat{C}_2$ of one particle reduces to a constant. For a system of N-particles, the Casimir operator $\hat{C}_2$ of the group SU_3 can be presented in the form (there is an analogy with (7.14) and (7.15)):

$$\hat{C}_2 = \sum_{j=1}^{8}\Big(\sum_{\alpha=1}^{N}\frac{1}{2}\lambda_j^{(\alpha)}\Big)\Big(\sum_{\beta=1}^{N}\frac{1}{2}\lambda_j^{(\beta)}\Big). \tag{7.19}$$

The Casimir operators $\hat{C}_2$ of different groups SU_n will be used further.

The group SU_2 is a subgroup of the group SU_3 (as well as in a general case, the group SU_{n-1} is a subgroup of the group SU_n). It can be constructed by using any two-dimensional subspace of a three-dimensional space, where the group SU_3 is specified, ignoring one basis vector out of three others. By virtue of this the group SU_3 consists of three different subgroups SU_2. The choice of the group SU_3 generators in the form of the Gell-Mann λ-matrix (7.16) is very convenient because it directly shows to which subgroups SU_2 of the group SU_3 the λ-matrices are related. For example, the first three λ-matrices λ_1, λ_2 and λ_3 in (7.16) are simply the group SU_2 generators, which act in a two-dimensional subspace of the three-dimensional space of the group SU_3, when the third basis vector is discarded. It follows from the explicit form of the matrices λ_1, λ_2, λ_3 (as well as from (7.17) and (7.18)) that they satisfy the same commutation relations (7.13) as the Pauli matrices divided by 2. The more general statement can be also proved. If a subset of generators of some Lie's group is closed with respect to commutation operations then this subset is a set of generators of the initial Lie group.

We dwell now briefly on the *irreducible representations of the continuous Lie groups* G . A continuous group have infinite number of elements g and infinite number of finite-dimensional irreducible matrix representations $D(g)$ of different dimension. Since the representation $D(g)$ is the group itself, which is holomorphic to the initial group G, then each its element $D(g)$ depends on the same parameter ξ_k that each (current) element g of the group G. Moreover, for the group-representation $D(g) = D(\xi_1, \xi_2, \ldots, \xi_m)$, we can introduce the infinitesimal operators or generators Y_j, which are matrices of the same dimension as the matrices of the representation $D(g)$:

$$Y_j = \frac{1}{i}\lim_{\xi_1,\xi_2,\ldots,\xi_m\to 0}\frac{\partial D(\xi_1,\xi_2,\ldots,\xi_j,\ldots,\xi_m)}{\partial \xi_j}, \tag{7.20}$$

$$D(\xi_1, \xi_2, \ldots, \xi_m) = I + i \sum_{k=1}^{m} \xi_k Y_k + \ldots . \tag{7.21}$$

It is clear that the dimension (order) m of the group representation $D(g)$ (do not mix with the dimension of matrices of the representation $D(g)$!) and the dimension of the corresponding group G coincide. Thus, a number of the generators Y_j of any representation of $D(g)$ of the group G is equal to a number of the generators X_j of the group G.

Since the dimensions of the matrices of different irreducible representations $D(g)$ of the group, in general, are different, the dimensions of the generators Y_j matrices for different representations are also different. However, the ranks of all irreducible representations of the given group G, which are treated as isomorphic groups, are the same and equal to the rank of the group G. Every irreducible representation of the Lie group is characterized by a set of indexes, whose number coincides with the group rank. For example, the irreducible representations of the group of three dimensional rotations R_3 and groups SU_2^S and SU_3^T, denoting as D^j, D^S, D^T, correspondingly, are characterized by one index j, S and T because they are the first rank groups, and the irreducible representations $D^{(\hat{\lambda},\hat{\mu})}$ of the group SU_3 (see section 4.5) are characterized by two indexes (the Elliott numbers) $\hat{\lambda}$ and $\hat{\mu}$, as well as the second rank group SU_3.

It is obviously that, because of the homomorphism of groups $D(g)$ and G, a set of generators Y_j of any representation $D(g)$ of the group G satisfies the same commutation relations (7.8) as a set of the generators X_j of the group G:

$$[Y_j, Y_k] = \sum_{l=1}^{m} C_{jk}^{l} Y_l \tag{7.22}$$

with the same structure constants C_{jk}^l as in (7.8), which are the same for all representations $D(g)$ of the group G. If two representations of the group G have the same generator, then these representations are equivalent.

According to the above-reported the matrices of the irreducible representation of the group SU_n can be presented by using (7.21) in terms of generators of the representation Y_k in the exponential form similarly to (7.7):

$$D(U) = \exp\left\{ i \sum_{k=1}^{n^2-1} \xi_k Y_k \right\}, \tag{7.23}$$

where, as in (7.5) and (7.6), $SpY_k = 0$, $Y_k^+ = Y_k$, since $D(U)D^+(U) = 1$ and the determinant $\mathrm{Det}D(U) = 1$.

Every Lie's group G has the trivial one-dimension representation when a number $+1$ is a set into correspondence to all elements g of the group G. At the same time, according to (7.21), all generators of this representation $Y_j = 0$. Performing the operation of complex conjugation for all matrices of an arbitrary representation $D(g)$ of the group G, we also obtain the representation $D^*(g)$, which, in a general case, is nonequivalent to the representation $D(g)$ and called the *conjugated representation* with respect to the representation $D(g)$. Apparently, the mutually aconjugated representations $D(g)$ and $D^*(g)$ have the same dimensions and, if $D(g)$ is an irreducible representation, then $D^*(g)$ is irreducible as well. According to (7.21) and due to reality of the group parameters ξ_k, we have for small ξ_k $D^*(g) = I - i\sum_{k=1}^{m} \xi_k Y_k^* \equiv I + i\sum_{k=1}^{m} \xi_k Y_k'$, where Y_k' are the representation $D^*(g)$ generators. Therefore, $Y_k' = -Y_k^*$.

We can also show that it is possible to obtain from a given representation $D(g)$ of the group G one more representation of the same dimension, by taking the matrices $D(g^{-1})$ and then transposing them, i.e., by constructing $\tilde{D}(g^{-1})$ on the basis of $D(g)$. A set of the matrices $\bar{D}(g) = \tilde{D}(g^{-1})$ obtained is the representation of the same group G, which is not equivalent to that of the $D(g)$. The representation $\bar{D}(g)$ is called the *contragradient representation* to $D(g)$, and, if $D(g)$ is an irreducible representation, then the contragradient representation $\bar{D}(g)$ is also irreducible. It follows from (7.21) that the generators $\bar{Y}_k$ of the representation $\bar{D}(g)$ are interrelated to the generators to the generators Y_k of the representation $D(g)$ by $\bar{Y}_k = -\tilde{Y}_k$.

For unitary, and, in particular, unimodular groups, the contragradient to $D(U)$ representation $\bar{D}(U)$ coincides with that conjugated to $D(U)$ the representation of $D^*(U)$:

$$\bar{D}(U) = D^*(U), \quad UU^+ = 1. \tag{7.24}$$

Then the generators of the representations $\bar{D}(U)$ and $D(U)$ are connected by the condition $\bar{Y}_k = -Y_k^*$. We can see further that hadrons with familiar properties are related to one of the irreducible representations of an unimodular group SU_n. In addition, particles and corresponding antiparticles are described by the contragradient irreducible representations, i.e., simply by the conjugated representations of the group SU_n.

Among irreducible representations of unitary unimodular groups SU_n we have to distinguish, first of all, a few following irreducible representations of the smallest dimension (which differ from the trivial one-dimensional) playing a very important part in the physics of hadrons and theory of strong interactions. One of such representations is the *fundamental irreducible representation* , when the group SU_n itself is treated as the representation, i.e., in this case $D_f(U) = U$, and the generators of the representation and group simply coincides: $Y_k = X_k$. The matrices of the irreducible fundamental representation of the group SU_n are the same unimodular matrices of dimension n, which were used for the specification of the group SU_n itself.

The second most important representation of the group SU_n of the same dimension as the fundamental representation is the representation which is contragradient to the fundamental one. For the group SU_n it coincides, according to (7.24), with the conjugated representation: $\bar{D}_f(U) = D_f^*(U) = U^*$. The generators of the representation, which is conjugated to the fundamental irreducible representation $\bar{Y}_k$, are given by the relation $\bar{Y}_k = -\tilde{X}_k = -X_k^*$, where X_k are the generators of the group SU_n.

The third irreducible representation of the group SU_n, which is of great importance, is the so-called *adjoint (regular) representation* $D_c(U)$. The dimension of its matrices coincides with the dimension (order) of the group itself $m = n^2 - 1$. The matrices of the irreducible representation Y_k of the group SU_n have the same dimension $n^2 - 1$ (see (7.21)). There are other irreducible representations of group SU_n associated with the hadron physics, which we will use further.

Some details concerning the group theory will be reported if needed.

Hadron and strong interaction symmetries[4]. The experimental study of the hadron-hadron interactions and properties of hadrons themselves resulted in the discovery of a set of symmetries in the properties of the known and predicted hadrons, which enables one to classify them. All this in turn favored the origin and development of the *current theory of the strong interactions* – quantum chromodynamics. The theory of elementary particles, in general, and the hadron theory, in particular, developed rapidly recently and considerable advances have been made owing to application of the mathematical group methods to study the particle symmetry properties and their interactions.

The charge (isotopic) nucleon symmetry in nuclei and nuclear forces was discovered by Heisenberg(1932) even in 30-s (see section 1.9). After the quark discovery, the *isotopic symmetry of the strong interactions* can be understood and described on the more fundamental (quark) level. A proton and neutron consist of the valence quarks of only two flavors, namely, of the lightest u- and d-quarks. Since the difference of u- and d- quark masses is much less than the nucleon masses (and, in general, smaller than masses of any hadrons), it is possible to assume that the masses of u- and d-quarks are equal. Neglecting, further, the difference in the quark electromagnetic properties, we can treat u- and d-quarks as identical particles. This allows

[4]The theory of the hadron and strong interaction symmetries is descibed in the review by Bogolyubov (1967), the lecture notes by N'guen Van H'eu (1967) and the monograph by Okun' (1990).

us to introduce the isospin wave functions of these quarks ξ_q, where $q = u, d$, in the form of the two-component spinors, as well for a proton and neutron:

$$\xi_u = \begin{pmatrix} 1 \\ 0 \end{pmatrix}, \qquad \xi_d = \begin{pmatrix} 0 \\ 1 \end{pmatrix}. \tag{7.25}$$

It is also possible to introduce the two-component isotopic quark spinor of a general form

$$\xi = \begin{pmatrix} C_u \\ C_d \end{pmatrix} = C_u \xi_u + C_d \xi_d, \qquad |C_u|^2 + |C_d|^2 = 1. \tag{7.26}$$

The coefficients C_u and C_d are often denoted as u and d, so that $\xi = \begin{pmatrix} u \\ d \end{pmatrix}$ similarly to the isotopic spinor $\xi_N = \begin{pmatrix} p \\ n \end{pmatrix}$. Therefore, two quarks u and d, as well as a proton p and neutron n, have the isotopic spin $t = \frac{1}{2}$ and form the isotopic doublet, i.e., here we have the approximate isotopic degeneration with the multiplicity $2t+1$ that is equal to two (a number of quarks in the isotopic doublet).

The isotopic symmetry of the nuclear (strong) interactions lies in the fact that the strong interaction does not distinguish the u- and d-quarks in this approximation. In other words, the corresponding quark Lagrangian $\mathcal{L}$ must not depend on the transformations of the isotopic spinors of quarks, which are realized by the unitary unimodular two-dimensional matrices U (see (7.7))

$$U \equiv U(\xi_1, \xi_2, \xi_3) = \exp\{i \sum_{k=1}^{3} \xi_k \frac{\tau_k}{2}\}, \tag{7.27}$$

which are elements of the group SU_2^T that have been already considered. A set of matrices (7.27) is the fundamental irreducible representation of the group SU_2^T. The isotopic invariance or SU_2^T symmetry of the strong interactions can be formulated as the invariance of the Lagrangian of a system of interacting quarks with respect to transformations of the isotopic group SU_2^T that is the simplest non-Abelian group. The Lagrangians $\mathcal{L}$, as we can see, are bilinear in the system wave functions. At the same time, one of the wave functions appears in $\mathcal{L}$ as complex (Hermitian) conjugated one. The Lagrangian is invariant under a substitution $\xi = \begin{pmatrix} u \\ d \end{pmatrix}$ with $\xi' = \begin{pmatrix} u' \\ d' \end{pmatrix} = U(\xi_1, \xi_2, \xi_3) \begin{pmatrix} u \\ d \end{pmatrix}$, in particular, when changing the spinor $\begin{pmatrix} u \\ d \end{pmatrix}$ with $\begin{pmatrix} \pm d \\ \mp u \end{pmatrix}$, the u- and d-quarks change their places ($u \to \pm d$, $d \to \mp u$), that, according to (7.27), is at $\xi_2 = \pm\pi$ and $\xi_1 = \xi_3 = 0$.

Every irreducible representation D^T of the group SU_2^T is characterized by one number (the first rank group), the isotopic spin T, which can take any nonnegative integer or half-integer number. The dimension $d(T)$ of the matrices of the irreducible representation D^T equals to $d(T) = 2T + 1$ and coincides with a number of different strongly interacting particles (subparticles) with common value of the isospin T and close masses. They form the so-called isotopic multiplet. At $T = \frac{1}{2}$, we have the fundamental irreducible two-dimensional representation $D^{1/2}$ of the group SU_2^T with the isotopic wave functions of u- and d-quarks (7.25) or similar isotopic wave functions of a proton and neutron as the basis functions. At $T = 1$, we have the adjoint (regular) irreducible representation D^1 of the group SU_2^T of a dimension $d(1) = 3$.

The group SU_2^T predicts existence of the hadron isotopic multiplet with an arbitrary hadron number $d(T) \equiv 2T + 1 = 1, 2, 3, \ldots$. However, the nature realizes for hadrons only the multiplet with $T \leq \frac{3}{2}$. This can be explained by the fact that the observed hadrons may consist of not more than three quarks (antiquarks) possessing the isotopic spin, i.e., u- and d-quarks ($\bar{u}$- and $\bar{d}$-antiquarks). Since each of these quarks is associated with the fundamental representation $D^{1/2}$, the isospin wave function of two quarks carrying isospin is transformed accordingly the representation D^0 or D^1, which appear in an expansion of the direct product $D^{1/2} \times D^{1/2}$ of two fundamental representations. In the elementary particle physics these expansions are often written in the form when, instead of the representations D^T, a dimension $d(T) = 2T + 1$ is indicated:

$$2 \bigotimes 2 = 1 \bigoplus 3. \tag{7.28}$$

If a hadron consists of three quarks with the isotopic spin, then instead of the expansion $D^{1/2} \times D^{1/2} \times D^{1/2} = D^{1/2} + D^{1/2} + D^{3/2}$ we can write the following

$$2 \bigotimes 2 \bigotimes 2 = 2 \bigoplus 2 \bigoplus 4, \tag{7.29}$$

thus, the total isospin of three quarks in the hadron is $T = \frac{1}{2}$ or $T = \frac{3}{2}$.

Therefore, we can carry out the partial classification of hadrons in the irreducible representations D^T of the isotopic group SU_2^T. Since this is the first rank group, the classification is associated only with one of generators of the group irreducible representations, namely, with $\hat{T}_3$. Its eigen-value M_T is uniquely connected with hadron charges (see (7.1)). At fixed hypercharge, M_T is determined by only an electric charge Q. Hadrons with different electric charges in a given isotopic multiplet are characterized by the corresponding definite (and different as well) eigen-values of only one generator, since with others it does not commute. The isotopic symmetry is broken in some degree, as it can be seen from examples, since the hadron masses within the same isotopic multiplet differs to some extent from each other. However, a violation of the isotopic SU_2^T symmetry is comparatively small.

We introduce now an important definition associated with the symmetries. If the group parameters ξ_k, appearing in (7.7) and (7.23), are constant numbers, then the symmetry, which corresponds to the group SU_n, is called the *global symmetry* . If the group parameters ξ_k depend on spatial-time coordinates, then the SU_n-symmetry is called the local symmetry. It turns out to be that the isotopic SU_2^T-symmetry, which is caused by the resemblance in the properties of u- and d-quarks, is global one, and the group parameters ξ_1, ξ_2 and ξ_3 in (7.27) are numbers.

So long we have taken into account only isotopic spins of u- and d-quarks. If we take into account the dependence of their wave functions on usual spin, then instead of two-component spinors (7.25) and (7.26), we obtain the four-component spinors that are transformed by the use of the unimodular matrix of the group SU_4^{ST}. This group was firstly introduced by Wigner(1941) when studied the classification of the nuclear states simultaneously with respect to spin and isospin. If the strong interactions would be invariant with respect to the isotopic spin and to the usual spin, then we could treat the group SU_4^{ST} as the more wide symmetry of these interactions which consists of the direct product $SU_2^S \times SU_2^T$ of the spin and isotopic groups as a subgroup. In fact, the strong interactions have the stronger dependence on the spin, thus the strong interaction symmetry associated with the group SU_4^{ST} is quite approximate (strongly broken).

By the early 60's, a number of the discovered strongly interacting particles (there were many strange hadrons among them) along with a number of the isotopic multiplets considerably increased. In this connection, it was necessary to generalize the invariance theory of the strong interactions. In particular, a lot of the so-called *resonances* – strongly interacting particles with a very small life-time of the order of $\sim 10^{-23}$ s., which is typical for the strong interactions, were discovered. This life-time can be obtained by division of a magnitude of the typical linear dimensions of hadrons $\sim 10^{-13}\,cm \equiv 1\,fm$ by the light velocity $c = 3 \times 10^{10}\,cm/s$. There were many attempts to construct the theory of higher (comparing with SU_2^T) symmetries of hadrons and strong interactions. The important achievement of these studies was discovery of the more deep but more hidden (comparing with SU_2^T) symmetry, the so-called unitary *SU_3-symmetry of the strong interactions* made by Gell-Mann(1961) and Ne'eman(1961) (even before the quark discovery) at the beginning of 60's. It was the generalization of the isotopic invariance theory (SU_2^T-symmetry).

We have seen that one of the quantum numbers characterizing hadrons

– an electric charge Q – is associated with one of generators $\hat{T}_3$ of the irreducible representations of the *isotopic symmetry group* SU_2^T. After discovering of the strange particles, one more independent characteristic of hadrons – strangeness S or the connected with it the *hypercharge* Y – appeared. (For hadrons, which do not contain the heavy quarks c, b, t, the hypercharge, accordingly (7.1), is equal to $Y = B + S$.) It would be naturally to generalize the isotopic symmetry so that, at least, two generators of the expanded symmetry group and its irreducible representations would commute with each other. One of these generators must be associated with the hypercharge or strangeness. Then the electric charge and strangeness must have definite values simultaneously.

Thus, a new expanded symmetry group must have a rank not lower then 2. Among the continuous second rank groups (there are only four of them) the group SU_3 best of all satisfies all necessary demands. Many of its predictions were brilliantly confirmed by experiment. It is extremely important that the SU_3 group contains the isotopic group SU_2^T as a subgroup, so that every isotopic multiplet are contained completely in some multiplet (super multiplet) of the SU_3 group. Because of this, the unitary symmetry of the SU_3 group is really a generalization of the isotopic SU_2^T symmetry. The irreducible representations of the SU_3 group are successfully used to classify hadrons. It will be shown that the unitary hadron SU_3-multiplet contains a few isotopic hadron SU_2^T-multiplets. Numbers of hadrons with close properties in the multiplet of SU_3 and SU_2^T groups exactly coincide with dimensions of the corresponding irreducible representations of these groups.

Note that the availability of the SU_3-symmetry after its discovery nearly 1961 was not confirmed solidly up to 1964 when the quarks were introduced. The structure of the hadron SU_3-multiplets looked like the systematization of chemical elements in the Mendeleev periodic Table, and, in analogous manner, the SU_3-multiplets indicate the existence of the hadron structure. The SU_3 symmetry group, which united the isospin and strangeness (hypercharge), has played an important historical part in the quark discovery. But even after the quark introduction, they have been treated by many researchers as mathematical symbols convenient for the hadron classification but not as real subparticles. Above all it was connected with the fact that quarks do not leave hadrons. Only after a set of precision experiments directly indicating the real existence not only quarks but gluons and some other successes in the elementary particle physics, the scientists become less sceptical concerning the quark existence.

The *hadron* SU_3 *-symmetry* , as well as SU_2^T-symmetry, is connected with the quark structure of hadrons. While the SU_2^T-symmetry is conditioned by the closeness of u- and d-quark properties, i.e., it is based on a degeneration only between two quarks. The SU_3-symmetry is based on the approximate degeneration of three the lightest quarks u, d, and s. The symmetry groups based on the degeneration (approximate closeness of some properties) of quarks with different flavor are called the flavor groups. According to the above-reported, the isotopic SU_2^T group under consideration associated with two flavors u, d, and the symmetry group SU_3, associated with three flavors u, d, s, are the flavor groups. Since we will consider the another SU_3-symmetry, which is connected with the color charges, the *flavor symmetry groups* will be supplied with an index f, for example, SU_3^f. The so-called *color group* SU_3^c (with index c) has, certainly, the same mathematical abstract structure as the *flavor group* SU_3^f but reflects the different physics. Speaking about the purely mathematical properties of the groups SU_3^f and SU_3^c, we will omit sometimes the indexes f and c.

The flavor SU_3^f -symmetry in the nature is violated more stronger then the SU_2^T-symmetry, since the strange s-quark is much heavier than the nonstrange u- and d-quarks (see Table 7.1). This results in the strong splitting in mass of hadrons, which appear in the different SU_2^T-multiplets of the same SU_3^f*-multiplet.* On adding the heavy quarks s, b, t, one can analyze the more wide (than SU_3^f) symmetries of the strong interactions and hadrons corresponding to the unitary unimodular flavor groups SU_4^f, SU_5^f, SU_6^f. However, these symmetries are violated much stronger than the SU_3-symmetry, which is caused by the great scattering in the quark masses. The latter, in its turn, appears to be associated with the broken symmetry of weak interactions.

Let us consider first which SU_3^f-multiplets are predicted by an abstract mathematical theory of the SU_3 group and its irreducible representations $D^{(\bar{\lambda},\bar{\mu})}$. A *dimension* $d(\bar{\lambda},\bar{\mu})$ *of the* $D^{(\bar{\lambda},\bar{\mu})}$ *representation* is related to the *Elliott numbers* $\bar{\lambda}$ and $\bar{\mu}$ by the general formula

$$d(\bar{\lambda},\bar{\mu}) = \frac{1}{2}(\bar{\lambda}+1)(\bar{\mu}+1)(\bar{\lambda}+\bar{\mu}+2). \tag{7.30}$$

From this, in particular, it follows that, unlike the dimensions $d(T) = 2T+1$ of matrices of the group SU_2^T irreducible representations, the dimensions $d(\bar{\lambda},\bar{\mu})$ take only the definite nonnegative numbers, for example,

$$\begin{aligned} d(0,0) &= 1,\ d(1,0) = 3,\ d(2,0) = 6,\ d(1,1) = 8 \\ d(3,0) &= 10,\ d(2,1) = d(4,0) = 15,\ d(5,0) = 21,\ d(3,1) = 24, \\ d(2,2) &= 27,\ d(6,0) = 28. \end{aligned} \tag{7.31}$$

The irreducible representations $D^{(\bar{\lambda},\bar{\mu})}$ and $D^{(\bar{\mu},\bar{\lambda})}$ of the SU_3 group are conjugated ($D^{(\bar{\mu},\bar{\lambda})} = \bar{D}^{(\bar{\lambda},\bar{\mu})} = \overset{*}{D}{}^{(\bar{\lambda},\bar{\mu})}$) and have the same dimensions: $d(\bar{\lambda},\bar{\mu}) = d(\bar{\mu},\bar{\lambda})$ according to (7.30). The representation $D^{(\bar{\lambda},\bar{\lambda})}$ with $\bar{\mu} = \bar{\lambda}$ is a self-conjugated: $\overset{*}{D}{}^{(\bar{\lambda},\bar{\lambda})} = D^{(\bar{\lambda},\bar{\lambda})}$.

The fundamental irreducible representation $D^{(1,0)}$ of the SU_3 group has a dimension $d(1,0) = 3$, and an arbitrary matrix of the representation, as well as a current element of the SU_3 group, takes the form

$$U = \exp\{i \sum_{k=1}^{8} \xi_k \frac{\lambda_k}{2}\}, \tag{7.32}$$

where λ_k are the Gell-Mann matrices (7.16). Three linearly independent basis functions of the $D^{(1,0)}$ representation can be written as the following three-component spinors

$$q_u = \begin{pmatrix} 1 \\ 0 \\ 0 \end{pmatrix}, \quad q_d = \begin{pmatrix} 0 \\ 1 \\ 0 \end{pmatrix}, \quad q_s = \begin{pmatrix} 0 \\ 0 \\ 1 \end{pmatrix}, \tag{7.33}$$

then the three-component spinor of a general form is equal to

$$q = \begin{pmatrix} c_u \\ c_d \\ c_s \end{pmatrix} = c_u q_u + c_d q_d + c_s q_s, \quad |c_u|^2 + |c_d|^2 + |c_s|^2 = 1. \tag{7.34}$$

The coefficients c_u, c_d and c_s, whose squared moduli determine the realization probabilities of u-, d-, and s-states, respectively, are often denoted as u, d, and s. Therefore, the general three-component spinor (7.34) can be written in the form $q = \begin{pmatrix} u \\ d \\ s \end{pmatrix}$. Thus, three quarks u, d, s form the fundamental triplet, which is transformed according to the SU_3^f group fundamental representation, like two quarks u, d form the isotopic doublet transforming according to the fundamental representation of the SU_2^T group.

Acting by matrix (7.32) on the spinor q, we obtain another three-component spinor $q' = Uq$. At this, the corresponding Lagrangian does not change, i.e., must be invariant in a transition $q \to q'$. This is the main property of the unitary (flavor) symmetry that does not differ quarks u, d and s. Note the analogy between formulas (7.32) and (7.27), (7.33) and (7.25), (7.34) and (7.26) whose differences are associated with a number of quarks on which the groups SU_3^f and SU_2^T are based. The conjugated to $D^{(1,0)}$ irreducible representation $D^{(1,0)} = \bar{D}^{(1,0)} = D^{*(1,0)}$ is connected with transformations of the antiquark $\bar{u}, \bar{d}, \bar{s}$ triplet. The corresponding three-component spinor wave function $\bar{q}$ is conjugated (Hermitian) with respect to spinor (7.34), as well as three basis functions $\bar{q}_u, \bar{q}_d, \bar{q}_s$ are conjugated with respect to basis spinors (7.33).

Since each barion consists of three quarks, we can consider 27 three-quark functions in the form of a product of one-quark functions (7.33), namely, $q_i(1)q_k(2)q_l(3)$, where $i, k, l = u, d, s$, which are the basis of the irreducible representation $D^{(1,0)} \times D^{(1,0)} \times D^{(1,0)}$ of the SU_3^f group that is the direct product of three fundamental representation $D^{(1,0)}$. Expanding this reducible representation in irreducible representations of the SU_3 group by using the method reported at the end of the section 4.5, we get

$$\begin{aligned} D^{(1,0)} \otimes D^{(1,0)} \bigotimes D^{(1,0)} &= \left(D^{(2,0)} \bigoplus D^{(0,1)}\right) \bigotimes D^{(1,0)} \\ &= D^{(3,0)} \bigoplus D^{(1,1)} \bigoplus D^{(1,1)} \bigoplus D^{(0,0)}. \end{aligned} \tag{7.35}$$

Using the corresponding dimensions of representations $d(\bar{\lambda}, \bar{\mu})$ to denote the representations $D^{(}\bar{\lambda}, \bar{\mu})$ (in the same manner as in (7.28) and (7.29)), we can rewrite (7.35), taking into account (7.31), in the form

$$3 \bigotimes 3 \bigotimes 3 = (6 \bigoplus \bar{3}) \bigotimes 3 = 10 \bigoplus 8 \bigoplus 8 \bigoplus 1. \tag{7.36}$$

This writing clearly shows the dimensions of all representations and their sum: $3 \cdot 3 \cdot 3 = 10 + 8 + 8 + 8 + 1 = 27$. Here and further $\bar{d}(\lambda, \mu)$ stands for the irreducible representation $D^{(\bar{\lambda},\bar{\mu})} = \bar{D}^{(\bar{\mu},\bar{\lambda})} = D^{*(\bar{\mu},\bar{\lambda})}$, if $\lambda < \mu$. Thus, $\bar{3}$ is the representation $D^{(1,0)} = \bar{D}^{(1,0)} = D^{*(1,0)}$.

The irreducible representation $D^{(3,0)}$ of dimension 10 and $D^{(1,1)}$ of dimension 8 obtained in (7.35) or (7.36) are compared with two barion SU_3^f-multiplets: with a decinet containing 10 particles and octet containing 8 particles. To find the isotopic multiplets appearing in these two SU_3^f barion multiplets, it is necessary to expand the representations $D^{(1,1)}$ and $D^{(3,0)}$, which are treated as representations of the subgroup SU_2^T of the group SU_3^f, in the irreducible representations D^T of the subgroup SU_2^T:

$$D^{(1,1)} \to D^1 \bigoplus D^{1/2} \bigoplus D^{1/2} \bigoplus D^0 \, (8 \to 3 + 2 + 2 + 1), \tag{7.37}$$

$$D^{3,0} \to D^{3/2} \bigoplus D^1 \bigoplus D^{1/2} \bigoplus D^0 \, (10 \to 4 + 3 + 2 + 1). \tag{7.38}$$

One can prove (7.37) and (7.38) by use of the following method associated with the construction of complete independent sets of the basis functions of the irreducible representations D^T of the SU_2^T group with $T = 0, \frac{1}{2}, 1, \frac{3}{2}$. Since the group generators, acting on one of the basis functions, transform it either to the rest basis functions of this representation or to their linear combinations, then, to obtain all basis functions of a given irreducible representation, we have to act on the chosen basis function by all generators of the group (in this case the SU_2^T group). Having obtained independent sets of the basis functions for the subgroup SU_2^T of the group SU_3^f from the basis functions of the representations $D^{(1,1)}$ and $D^{(3,0)}$ of the SU_3^f group, we arrived at expansions (7.37) and (7.38).

For a meson that consists of an antiquark and a quark, we can consider 9 products $\bar{q}_i(1)q_k(2)$ $(i, k = u, d, s)$ which are the basis functions of the reducible nine-dimensional representation $D^{(0,1)} \bigotimes D^{(1,0)} \equiv \bar{3} \bigotimes 3$ of the SU_3^f group. This is a so-called nonet of mesons, i.e., a set of nine mesons. Expanding the representation $\bar{3} \times 3$ into the irreducible representations of the SU_3^f group, we obtain (see Section 4.5):

$$D^{(1,0)} \bigotimes D^{(0,1)} = D^{(1,1)} \bigoplus D^{(0,0)} \, (3 \bigotimes \bar{3} = \bar{3} \bigotimes 3 = 8 \bigoplus 1). \qquad (7.39)$$

The representation $D^{(1,1)}$ can be related to the octet, i.e., a set of eight mesons, and the representations $D^{(0,0)}$ – to singlets, i.e., single mesons. It is clear that the one-dimensional representation $D^{(0,0)}$ of the SU_3^f group is irreducible for the subgroup SU_2^T as well, i.e., $D^{(0,0)} \to D^{T=0}$.

Let us correlate, at last, the irreducible representations of the group SU_3^f with the real physical particles – hadrons. There is a surprising correspondence between them, and the sets of hadrons associated with particular irreducible representations of the symmetry group SU_3 are called the SU_3^f-multiplets. The symmetry of hadron properties described by the group SU_3^f is a reflection of the corresponding symmetry of the strong interactions. The flavor SU_3^f-symmetry of hadrons and the strong interactions, caused by the resemblance in u, d, s quarks properties, is the global symmetry, as well as the isotopic SU_2^T-symmetry. As a consequence, all eight parameters of the group ξ_k, appearing in (7.32), do not depend on the spatial coordinates and time, i.e., they are constants.

It is convenient to present the *SU_3^f-multiplet of hadrons* in the $M_T Y$ plane (the isospin projection and hypercharge, as it was noted, take simultaneously the exact values as eigenvalues of two commuting generators of the SU_3 group). Since the hypercharge Y, according to (7.1), is the double average electric charge of the isotopic SU_2^T-multiplet, then the "beginning of coordinates" at point $M_T = 0$ and $Y = 0$ is a center of symmetry of the SU_3^f-multiplet. In Figs. 7.3 – 7.6 we present six SU_3^f-multiplets of the most light hadrons indicating their quark structure: the barion octet and

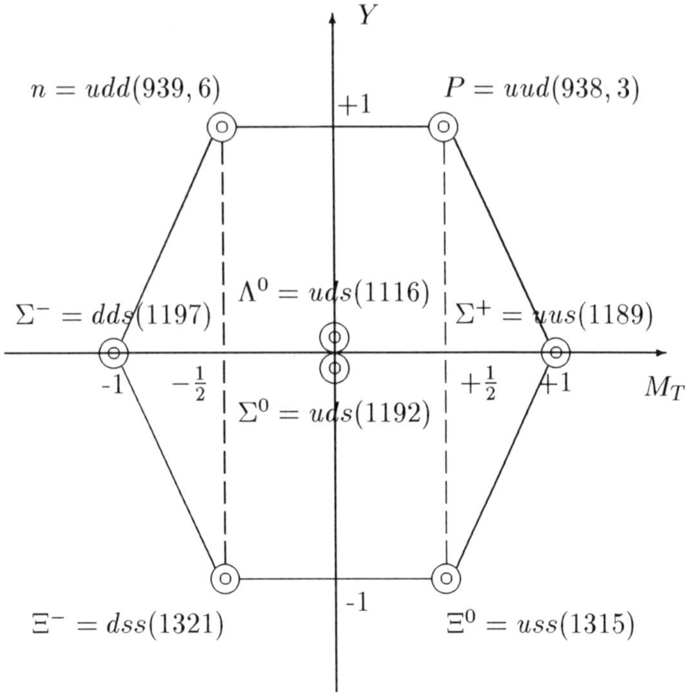

Figure 7.3. An octet of the most light barions.

decinet (Fig. 7.3 and 7.4) and two meson octet and two meson singlet (Fig. 7.5 and 7.6). Values of hadron masses are indicated in MeV in brackets. All particles of the SU_3^f-multiplet have the same spin J, parity Π and angular momentum L of the relative motion of three quarks in a barion, and of two particles – quark and antiquark – in a meson.

All hadrons indicated in Figs. 7.3 – 7.6 have $L = 0$. That is why, for all these hadrons, the spin J of the particle is just equal to the total spin of subparticles. First of all, we note that numbers of hadrons with similar properties in each SU_3^f-multiplet in the figures coincide with the dimensions of the irreducible representations appearing in expansions (7.35) or (7.36) and (7.39). The hadron SU_3^f-multiplets under considerations contain exactly the same isotopic SU_2^T-multiplets, which follow from expansion (7.37) and (7.38). The *barion octet* in Fig. 7.3 consists of one isotopic barion triplet Σ^+, Σ^0, Σ^- with $T = 1$, two isotopic doublets p, n and Ξ^0, Ξ^- with $T = \frac{1}{2}$, and one isotopic singlet Λ^0 with $T = 0$. The spin and parity J^Π of each barion in the octet are equal to $\frac{1^+}{2}$. The spin $J = \frac{1}{2}$ means that spins of two quarks in the barion from the octet are parallel, and the third quark has the opposite spin (since $L = 0$). The positive parity of barions $\Pi = +1$

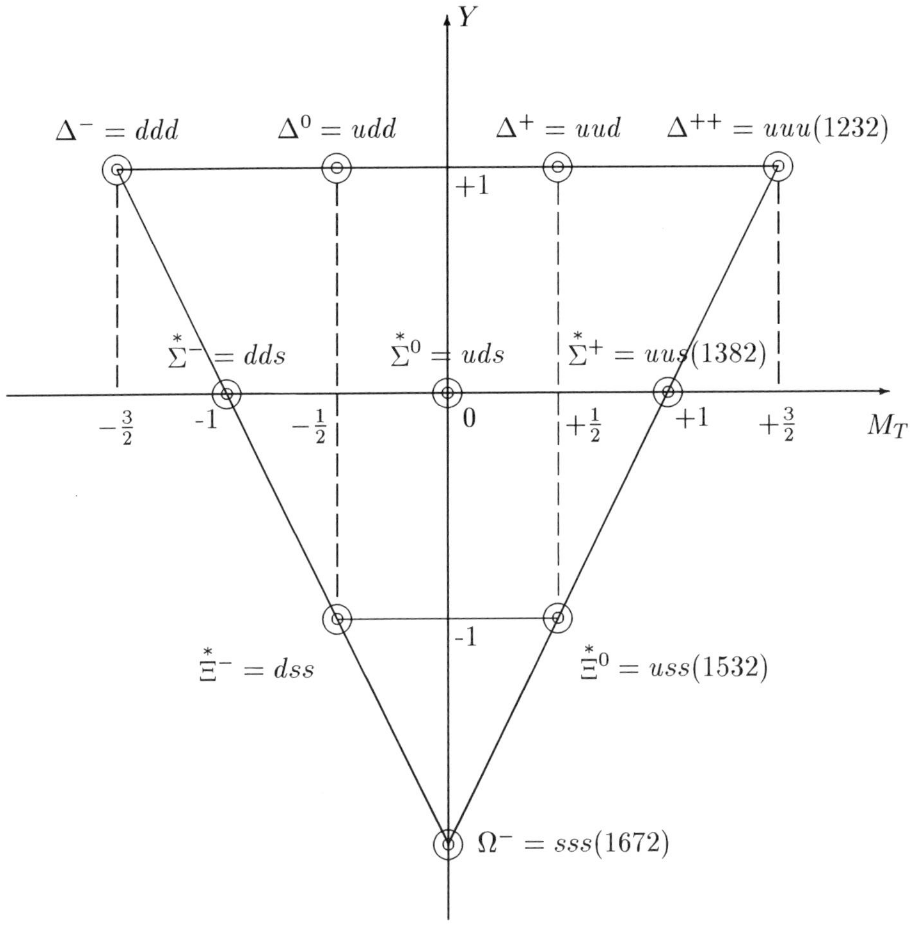

Figure 7.4. A barion decinet.

means that, since $l = 0$, inner parities of all three quarks are the same and positive. The *barion decinet* in Fig. 7.4 contains the following isotopic hadron multiplets: quartet Δ^{++}, Δ^+, Δ^0, Δ^- with $T = \frac{3}{2}$, triplet $\overset{*}{\Sigma}^+$, $\overset{*}{\Sigma}^0$, $\overset{*}{\Sigma}^-$, doublet $\overset{*}{\Xi}^0$, $\overset{*}{\Xi}^-$, and singlet Ω^-. The spin and parity of each barion in the decinet $J^\Pi = \frac{3^+}{2}$, i.e., spins of three quarks in the barion are parallel.

The *meson* SU_3^f*-octets* in Fig. 7.5 and 7.6 consist of the meson triplet π^+, π^0, π^- and ρ^+, ρ^0, ρ^-, two isotopic doublets K^+, K^-; $\bar{K}^0, K^-$ and $\overset{*}{K}^+$, $\overset{*}{K}^0$; $\bar{K}^0, \overset{*}{K}^-$, and isotopic singlets η^0 and φ. The spin and parity of each meson in the octet in Fig. 7.5 are equal to $J^\Pi = 0^-$. This means, since $L = 0$, that spins of a quark and antiquark of each meson in this octet of

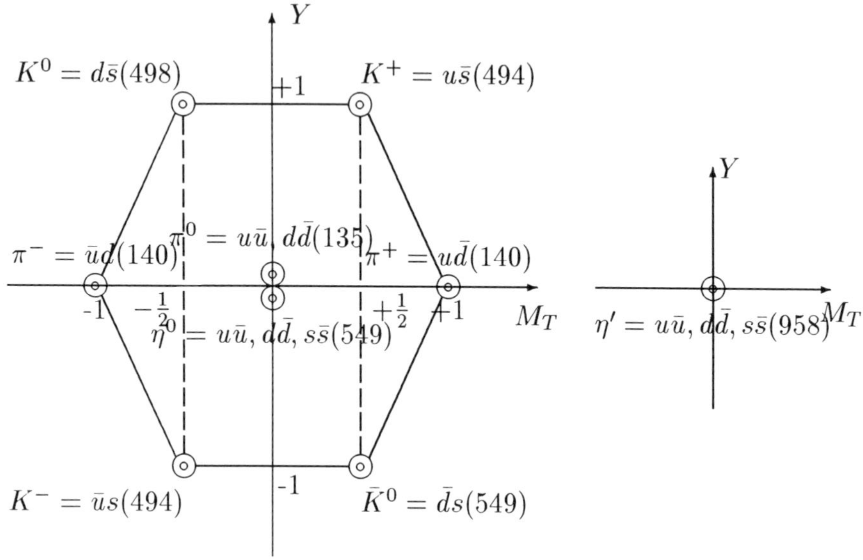

Figure 7.5. An octet and singlet of the pseudoscalar mesons.

pseudoscalar mesons are antiparallel, and intrinsic parities of the quark and antiquark have opposite signs, i.e., the antiquark parity is negative because the parity $\Pi_L = (-1)^{L=0} = +1$ associated with the relative motion of the quark and antiquark are positive. The spins of the quark and antiquark in the second meson octet in Fig. 7.6 are parallel in each meson, i.e., $J^{\Pi} = 1^-$. This is the octet of vector ($J = 1$) mesons. Besides, the SU_3^f-singlets are presented in Fig.7.5 and 7.6: the pseudoscalar ($J^{\Pi} = 0^-$) η'-meson and the vector ($J^{\Pi} = 1^-$) ω-meson, which are the isotopic singlets ($T = 0$). The mesons of the SU_3^f-singlets η' and ω, as well as mesons of the SU_3^f-octets π^0, η^0, and ρ^0, φ with $M_T = Y = 0$, are genuine neutral particles with zero values of all charges. They consist of genuine neutral particles with respect to all charges of the quantum quark-antiquark systems $u\bar{u}$, $d\bar{d}$, and $s\bar{s}$ in different proportions. The particular contribution of each $u\bar{u}$, $d\bar{d}$, $s\bar{s}$ combination into genuine neutral mesons will be determined in the next section where the hadron inner functions are discussed. Here we note only that by virtue of the strong interactions these three quark-antiquark systems can be transformed into each other due to their neutrality. Because of this, the wave functions of genuine neutral mesons are a superposition of such states in a general case.

It is worth noting that while mesons manifest the nonet structure, the barion SU_3^f-octets do not have closely lying singlets, there are no meson decinets, but only barion decinets exist. Moreover, the irreducible representations of the SU_3^f group – $D^{(1,0)}$ and $D^{(2,0)}$ of the minimal dimensions 3

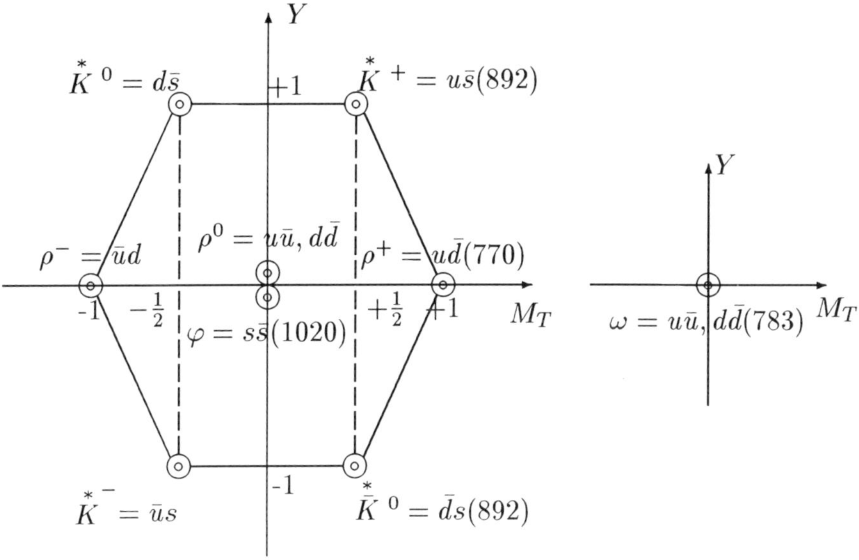

Figure 7.6. An octet and singlet of the vector mesons.

and 6 "are not used" by the nature (they are absent in expansions (7.35) and (7.39) as well!). In fact, as we have seen, the quark triplet u, d, s, is transformed according to the $D^{(1,0)}$ representation, unlike the SU_3^f-triplet of the observed hadrons.

All barions of the octet under consideration and Ω^--hyperon out of the decinet are often called the quasi-stable barions since their life-time is larger by many orders than the strong interactions typical time ($\sim 10^{-23}$ s). These barions would be all stable if we could switch off the weak and electromagnetic interactions (this is the so-called strong interaction approximation). Due to the same reason, we can treat the pseudoscalar mesons as quasistable. The rest hadrons, considered here, with the same quark structure as all barions and mesons in Figs. 7.3 – 7.6, but with $L > 0$, are resonances. The SU_3^f-multiplets of short-lived hadron-resonances with $L \neq 0$, i.e., with $L = 1, 2, \ldots$, are in reality multiplets of the hadron excited states with $L = 0$. The radial excitations of the quark and antiquark motion in hadrons are also possible. This fact explains a great deal of already discovered hadrons. The barion Ω^- quasistability can be explained by the large strangeness of this particle (in modulus) $s = -3$. At the same time, all more light hadrons being decay products of the Ω^- particle have $|s| < 3$. Because of this, the strangeness is not preserved when the Ω^--hyperon decays. The ω^--hyperon decays due to the weak interaction that explains the comparatively large life-time of this particle.

Note that heavy mesons and barions discovered in succeeding years consisting of c- and b-quarks and (or) corresponding $\bar{c}$- and $\bar{b}$-antiquarks decay by virtue of the weak interactions. (The charm is not preserved in this process.) They can be related to the quasistable hadrons since they have comparatively large time-life. But hadrons containing c- and b-quarks have nothing to do with the above-described symmetry SU_3^f.

In the same manner as the spin-isospin symmetry group SU_4^{ST} was introduced, one can introduce the spin unitary symmetry group SU_6^{sf}, when u-, d-, s-quarks are assumed to have the usual spin. It is clear that SU_6^{sf} group, as a subgroup, contains the direct product $SU_2^s \otimes SU_3^f$ of the spin SU_2^s and unitary SU_3^f groups. In spite of the approximate character of the strong interaction symmetry described by the SU_6^{sf} group, it enables one to obtain the ratio between magnetic moments of a neutron and proton $\frac{\mu_n}{\mu_p} = -\frac{2}{3}$, which is close to the experimental value (see (1.1)). This fact was just the one that attracted the attention to the SU_6 group. For an extended discussion see the next section.

Aside from the above-described *global flavor symmetry* of the strong interactions there is, already mentioned, the more fundamental color symmetry of these interactions which forms the basis of quantum chromodynamics. This symmetry is associated with the quark color degeneration: the strong interactions are the same for quarks of all colors. Since there are only three different colors, the corresponding color symmetry is determined by the color group SU_3^c. Its three-row unimodular matrices (the corresponding set is the irreducible fundamental representation $D^{(1,0)}$ of the SU_3^c group) act in the three-dimensional color space. The basis functions of this representation are the components of the color quark triplet of each flavor q^α (α means one of the colors: red, blue, yellow), or α=1,2,3, i.e., q^α represents the color three-component spinor. The antiquark triplet $\bar{q}_\alpha$ (or antitriplet) is transformed, accordingly the conjugated, to the fundamental irreducible representation $D^{(0,1)}$ of the group SU_3^c. The QCD Lagrangian $\mathcal{L}$ is invariant with respect to the SU_3^c-transformations of the quark (antiquark) color triplets. Unlike the approximate global flavor SU_n^f-symmetries, the color symmetry is assumed to be exact and local one. The group parameters ξ_k in (7.32) for the SU_3^c group are now the functions of spatial coordinates and time ones. The *local* SU_3^c*-symmetry* is determined the entire dynamics of the strong interactions. In this respect, the color SU_3^c-symmetry is more deep as the global flavor SU_3^f-symmetry.

The color charge of the observed hadrons $C = 0$, i.e., hadrons are colorless. That is why the color wave function of a hadron, composed of quarks and (or) antiquarks, may be chosen so that it converts into itself in response to the color group SU_3^c operations, i.e. it is transformed according to the one-dimensional (singlet representation) $D^{(0,0)}$ of the SU_3^c group.

Just as the wave function of a system of particles with zero total angular spin momentum converts into itself under the rotation operations (i.e., is transformed according to the irreducible representation D^0 of the group of three-dimensional rotations R_3).

Let us determine how many quarks q or antiquarks $\bar{q}$ (quarks and their wave functions are denoted conventionally by the same letters) are needed to construct the different colorless systems – hadrons. Obviously, we have to find such direct products (first of all their minimal number) of the fundamental representations $D^{(1,0)}$ and the conjugated representations $D^{(0,1)}$, which having been expanded into the irreducible representations of the SU_3^c group, finally give rise to one-dimension al representation $D^{(0,0)}$. This requirement is fulfilled by the expansions (7.35) (or (7.36) and (7.39)) but it is not satisfied, for example, by the expansions $3 \otimes 3 = 6 \oplus \bar{3}, \bar{3} \otimes \bar{3} = \bar{6} \oplus 3, 3 \otimes 3 \otimes \bar{3} = 15 \oplus \bar{6} \oplus 3 \oplus 3, 3 \otimes 3 \otimes 3 \otimes 3 = 15 \oplus 15 \oplus 15 \oplus 15' \oplus \bar{6} \oplus \bar{6} \oplus 3 \oplus 3 \oplus +3$, where $15 = d(2,1)$ and $15' = d(4,0)$, (see (7.31)). As it was noted earlier, the quark colorless systems (hadrons) are the systems of three quarks (barions), three antiquarks (antibarions), one quark and one antiquark (mesons), considering that the expansions $3 \otimes 3 \otimes 3 = 10 \oplus 8 \oplus 8 \oplus 1, \bar{3} \otimes \bar{3} \otimes \bar{3} = \overline{10} \oplus 8 \oplus 8 \oplus 1, 3 \otimes \bar{3} = 8 \oplus 1$ contain the one-dimensional representation $D^{(0,0)}$ of the SU_3^c group. (Note that $\bar{8} = 8$ and $\bar{1} = 1$, since the corresponding irreducible representations are selfconjugated.) In principle, the six-quark and double quark-antiquark colorless systems are also available (as well as the systems of four quarks and one antiquark) since there are the one-dimensional representation $D^{(0,0)}$ in the corresponding expansions $3 \otimes 3 \otimes 3 \otimes 3 \otimes 3 \otimes 3$ and $3 \otimes \bar{3} \otimes 3 \otimes \bar{3}$ (and $3 \otimes 3 \otimes 3 \otimes 3 \otimes \bar{3}$) into the irreducible representations of the SU_3^c group. However, there is no theory that offers a coherent explanation of the naturally occurring systems (except the complex atomic nuclei). Both the flavor and color wave functions of the quark systems will be constructed in the next section.

As it was already noted, the two-color bosons, carrying out the color quark-quark strong interaction, are called gluons. Each gluon must possess a particular color and anticolor. In fact, there are nine color-anticolor combinations ($3 \otimes \bar{3}$) for gluons, i.e., the gluon color nonet exists, in contrast to the quark color triplet. This nonet is the basis of the nine-dimensional reducible representation of the color group SU_3^c, which according to (7.39) can be expanded in two irreducible representations ($3 \otimes \bar{3} = 8 \oplus 1$). One of them is the singlet (one-dimensional) representation. As we have already seen, the singlet states are colorless, therefore, the corresponding "theoretical" colorless gluon must be ruled out. In fact, there are eight physical gluons which belong to the added (regular or octet) irreducible representation $D^{(1,1)}$ of the SU_3^c group of dimension $d(1,1) = 8$. Gluons interact only with the color. Recall that the weak and electromagnetic interactions do not differentiate the color. Due to this reason gluons interact in the same manner with quarks of any flavor.

Quantum chromodynamics (QCD). Quantum chromodynamics is a relativistic theory of quarks and gluons and based, first of all, on the symmetries of the strong interactions. We formulate here the main *postulates of quantum chromodynamics* related to these symmetries which basically sum up the above-reported.

1. It is assumed that the symmetry group of quantum chromodynamics is the color group SU_3^c with generators T_a (a=1,2,...,8), which satisfy the following commutation relations (see (7.8) and (7.22)):

$$[T_a, T_b] = i f_{abc} T_c, \tag{7.40}$$

where a summation is taken over the repeating indexes. The real structure constants of the group f_{abc}, given in (7.18), are antisymmetric with respect to all three indexes.

2. The irreducible fundamental representation $D^{(1,0)}$ (three-dimensional) of the SU_3^c group is related to the color quark triplet. The generators $T_a^{(\text{fund})}$ in the fundamental representation are usually given by a set of eight Gell-Mann λ-matrices (7.16)

$$T_a^{(\text{fund})} = \frac{1}{2}\lambda_a,\ T_a^{(\text{fund})} T_a^{(\text{fund})} = \frac{4}{3},\ Sp(T_a^{(\text{fund})} T_b^{(\text{fund})}) = \frac{1}{2}\delta_{ab}. \tag{7.41}$$

3. It is postulated that the irreducible added regular representation $D^{(1,1)}$ (eight-dimensional) of the SU_3^c group is related to the color gluon octet. Matrix elements of the generators $T_a^{(\text{con})}$ in the added representation are expressed in terms of the structure group constants (7.18) as follows

$$(T_a^{(\text{con})})_{bc} = -i f_{abc},\ T_a^{(\text{con})} T_a^{(\text{con})} = 3,\ Sp(T_a^{(\text{con})} T_b^{(\text{con})}) = 3\delta_{ab}. \tag{7.42}$$

4. By analogy with quantum electrodynamics (QED) where one electron bispinor field $\Psi(x)$ and one electromagnetic 4-vector field $A_\mu(x)$ are introduced, three color bispinor quark fields $q^\alpha(x)$ and eight 4-vector gluon fields $B_\mu^a(x)$ depending on a space-time coordinate $x = (\mathbf{r}, t)$ or x^μ are inserted into the QCD.

It is known that the equations of motion of an arbitrary field $\Phi(x)$ can be obtained from the general *Euler-Lagrange equation*

$$\frac{\partial}{\partial x_\mu}\left(\frac{\partial \mathcal{L}}{\partial(\frac{\partial \Phi}{\partial x_\mu})}\right) - \frac{\partial \mathcal{L}}{\partial \Phi} = 0, \tag{7.43}$$

where $\mathcal{L} \equiv \mathcal{L}(\Phi, \frac{\partial \Phi}{\partial x_\mu}, x_\mu)$ is the Lagrangian $\int d^3x\mathcal{L}$ density. A quantity $\mathcal{L}$ is usually called the *Lagrangian*. To write down the Lagrangian $\mathcal{L}_c$ in the QCD, we use the fact that chromodynamics, as well as electrodynamics, is gauge theory. Recall that a physical theory is called the *gauge theory*, if it is invariant with respect to local symmetry transformations (in the case of QCD we deal with transformations of the SU_3^c group). Basically, we require this invariance for the Lagrangian and then obtain $\mathcal{L}_c$ and the corresponding QCD equations of motion.

Let us consider, firstly, in brief rather simple and more familiar gauge theory of the QED. After this, by use of some analogies, we can make generalizations and write down the main relations of the more complex QCD gauge theory . The *Lagrangian of the free electron complex field* ψ takes the known form

$$\mathcal{L}_{EM}^{(0)} = i\bar{\psi}\gamma^\mu \partial_\mu \psi - m_e \bar{\psi}\psi, \ \bar{\psi} = \psi^+ \gamma^0, \ \partial_\mu \equiv \frac{\partial}{\partial x^\mu}, \tag{7.44}$$

where γ^μ are the Dirac matrices, m_e is the electron mass. Obviously, this Lagrangian is invariant with respect to the phase transformations $\psi \to \psi' = e^{i\xi}\psi$ with a constant phase ξ. As it was noted before, this is the global transformation of the ψ-field. In electrodynamics, we can also provide a local gauge invariance of the Lagrangian $\mathcal{L}_{EM}$ treating the phase as a coordinate function $\xi(x)$. At the same time, Lagrangian (7.44) is not invariant with respect to substitution $\psi \to e^{i\xi(x)}\psi$, because the derivatives of $\xi(x)$ appear: $\partial_\mu \psi \to \partial_\mu \psi' = \partial_\mu e^{i\xi(x)}\psi = e^{i\xi(x)}\partial_\mu \psi + i(\partial_\mu \xi(x))e^{i\xi(x)}\psi$. We introduce now the electromagnetic field $A_\mu(x)$ interacting with the field of charged particles $\psi(x)$, which can be determined within the accuracy of the gradient (gauge) transformation $A_\mu(x) \to A'_\mu = A_\mu(x) - \frac{1}{e}\partial_\mu(x)$, where e is the electric charge of a particle (electron), and $\xi(x)$ is an arbitrary function of x. Due to this gradient invariance of the 4- vector potential of the electromagnetic field $A_\mu(x)$, the total Lagrangian of the system in the form

$$\mathcal{L}_{EM} = i\bar{\psi}\gamma^\mu D_\mu \psi - m_e \bar{\psi}\psi - \frac{1}{4}F_{\mu\nu}F^{\mu\nu}, \ F_{\mu\nu} = \partial_\mu A_\nu - \partial_\nu A_\mu, \tag{7.45}$$

where $D_\mu = \partial_\mu + ieA_\mu$ is locally invariant at simultaneous transformation $\psi \to e^{i\xi(x)}\psi$ and $A_\mu \to A_\mu - \frac{1}{e}\partial_\mu \xi(x)$. In (7.45), we added the electromagnetic field energy containing the gauge invariant field tensor $F_{\mu\nu}$. With use of the Eqs. (7.43) and (7.45), we can obtain the Dirac and Maxwell equations. From the formal view point, the transformation $\psi(x) \to \psi'(x) = e^{i\xi(x)}\psi(x)$ is nothing more than the local transformation connected with the simplest Abelian (commutative) unitary U_1 group (its unique generator reduces to unity). Thus, one can say that the Abelian local U_1-invariance of the Lagrangian takes place in the QED.

We shall see that the *gauge-invariant Lagrangian* $\mathcal{L}_c$ *of quantum chromodynamics* can be written in the same manner, by changing the gauge group U_1 in the QED to the group SU_3^c of phase transformations of the quark color fields in the QCD. However, if the Lagrangian in the QED is the Abelian locally U_1-invariant, in the QCD we deal with the non-abelian locally gauge invariant Lagrangians, since the operations of the group SU_3^c, in a general case, do not commute. The theory possessing the non-abelian gauge invariance are called the *Yang-Mills theories* (Yang,Mills,1954).

Denoting, as earlier, the quark state vectors by the same letters as quarks themselves, we can write down the quark Lagrangian of the quark free fields q^α (analogously to (7.44)) in the following form

$$\mathcal{L}_c^{(0)} = \sum_q \sum_\alpha \bar{q}_\alpha (i\gamma^\mu \partial_\mu - m_q) q^\alpha \tag{7.46}$$

(here m_q are the current quark masses). The sum symbols over two repeating quark flavors $q = u, d, s, c, b, t$ and quark colors α=1,2,3 will be omitted latter to avoid the encumbering of formulas. The requirement of the QCD Lagrangian invariance with respect to local phase transformations of $q(x)$, namely (see (7.32) and (7.41)),

$$q(x) \to q'(x) = Uq(x) = \exp(i \sum_{a=1}^{8} \xi_a(x) T_a^{(\mathrm{fund})}) q(x), \tag{7.47}$$

requires to introduce eight-gauge fields of the strong interactions B_μ^a (a=1, 2, ... , 8; μ=0,1,2,3) in the QCD. To construct the SU_3^c-invariant Lagrangian in the QCD, it is enough to consider even the infinitesimal group parameters $\xi_a(x)$.

Because of the fact that not all generators $T_a^{(\mathrm{fund})}$ of the non-abelian group SU_3^c commute with each other, it is not enough to require that the fields B_μ^a are transformed like fields $A_\mu(x)$ in the QED, accordingly $B_\mu^a \to B_\mu^{'a} = B_\mu^a - \frac{1}{g}\partial_\mu \xi_a$, where g is the coupling constant (charge) of the strong interactions. Introducing, as well as in the QED, the "extended" *covariant derivative*

$$D_\mu = \partial_\mu + igT_a B_\mu^a, \tag{7.48}$$

or more precisely, $D_\mu q^\alpha(x) = \sum_{\beta=1}^3 (\delta_{\alpha\beta}\partial_\mu + ig\sum_{a=1}^8 T_q^{\alpha\beta} B_\mu^a) q^\beta(x)$, changing ∂_μ in (7.46) to D_μ and introducing more complex transformations for the QCD fields $B_\mu^a(x)$:

$$B^a_\mu \to B'^a_\mu = B^a_\mu - \frac{1}{g}\partial_\mu \xi_a - f_{abc}\xi_b B^c_\mu, \tag{7.49}$$

we can obtain the *gauge-invariant Lagrangian in the QCD*

$$\begin{aligned}\mathcal{L}_c &= \bar{q}(i\gamma^\mu\partial_\mu - m_q)q - g(\bar{q}\gamma^\mu T_a q)B^a_\mu - \frac{1}{4}G^a_{\mu\nu}G^{\mu\nu}_a \\ &\equiv i\bar{q}(\gamma^\mu D_\mu + im_q)q - \frac{1}{4}G^a_{\mu\nu}G^{\mu\nu}_a,\end{aligned} \tag{7.50}$$

$$G^a_{\mu\nu} = \partial_\mu B^a_\nu - \partial_\nu B^a_\mu - gf_{abc}B^b_\mu B^c_\nu, \tag{7.51}$$

where we have introduced the total field energy of the strong interaction $-\frac{1}{4}G^a_{\mu\nu}G^{\mu\nu}_a$ (in (7.50)). (The color indexes a, b, c=1,2,...,8, as well as α, β=1,2,3, are written arbitrarily in the form of upper or lower indexes.) The QCD field tensor $G^a_{\mu\nu}$ is nonlinear in the 4-potential B^a_μ, contrary to the QED. Owing to the non-ableian gauge symmetry group SU^c_3 in the QCD Lagrangian (7.50), the terms proportional to the third and fourth power of the gluon field B^a_μ appear. This reflects the fact that gluons, as well as quarks, carry color charges and interact with each other. This leads to the conclusion that, in addition to the Feynman diagrams, describing the quark-gluon interaction (Fig. 7.1), the Feynman diagrams with three-gluon (Fig. 7.2) and four-gluons vertexes appear. However, the local gauge invariance, as well as in the photon case, requires gluons to be massless.

Substituting the QCD Lagrangian (7.50) in the Euler-Lagrange equation (7.43), and considering in turn the quark $q^\alpha(x)$, $\bar{q}_\alpha(x)$ and gluon $B^a_\mu(x)$ fields as the field $\Phi(x)$, we obtain the following *equations of motion of the fields* $q(x)$ *and* $B(x)$ in the QCD

$$\bar{q}(x)(i\gamma_\mu \overleftarrow{D}^\mu + m_q) = 0, \ (i\gamma_\mu D^\mu - m_q)q(x) = 0, \tag{7.52}$$

$$D_\mu G^{\mu\nu}_a(x) \equiv \partial_\mu G^{\mu\nu}_a(x) - gf_{abc}B^b_\mu(x)G^{\mu\nu}_c(x) = 0. \tag{7.53}$$

The arrow directed in the left side over the operator D^μ in (7.52) means that it acts on $\bar{q}(x)$. We can see that equations (7.53) are nonlinear with respect to the gluon field. When obtaining the equations of motion related to Lagrangian (7.50), we have to keep in mind that $\partial_\mu \bar{q}$ does not appear in $\mathcal{L}_c$ explicitly. Because of this, $\dfrac{\partial \mathcal{L}_c}{\partial(\partial_\mu \bar{q})} = 0$ and with use of the Euler-Lagrange equation, we obtain $\dfrac{\partial \mathcal{L}_c}{\partial \bar{q}} = 0$. This in turn gives rise to the equation of motion $(i\gamma_m D^\mu - m_q)q(x) = 0$, presented in (7.52). Equations of QCD (7.52) and (7.53) are the analogous to the Dirac and Maxwell equations in the QED.

7.2. Quark-Hadron Systems

Systems of two color objects. Mesons. In this section we consider the intrinsic hadron states of the quark and antiquark systems, discuss some general properties of the wave functions of the intrinsic motion of quarks (antiquarks) in hadrons, as well as the problems of quark-quark and quark-antiquark interaction in hadrons. Since quarks (antiquarks), as well as nucleons, are fermions, then in studies of the quark systems, we can use the known facts and approaches, which have been established and successfully applied to the study of the nuclear systems. In particular, we can use the developed group methods in studies of the quark systems, for example, of the nuclear shell structure. Now we have to take into account the appearance of new degrees of freedom in the quark systems connected with color. Because of this, the total wave functions of nucleons, as well as complex nuclei treated as the quark systems, can be presented as antisymmetric superpositions of products of spatial, spin, flavor (here isospin), and color wave functions. In this approximation u- and d-quarks do not differ (i.e., the u- and d-quarks are identical with introduction of the isotopic spin). It is assumed that any of the above-mentioned wave functions satisfies the orthonormalization condition.

Since the flavor SU_3^f-symmetry of hadrons is considerably broken, the total wave function of the barion intrinsic state is not, strictly speaking, antisymmetric with respect to permutations of quarks. Recall that a barion consists of u- and d-quarks, as well as more heavy s-quarks. But in the rough approximation of the SU_3^f-symmetry, which does not distinguish the lightest three quarks u, d, s, the intrinsic total wave function of a strange barion can be treated as symmetric with respect to permutation of any two quarks (for example, u- and s-quarks).

We consider, firstly, more closely the simplest colorless bound system of two subparticles consisting of quark 1 and antiquark 2: $q(1)\bar{q}(2)$, i.e., the physically observed meson. Let us dwell on the mesons, which contain the lightest strangeless quarks u, d and antiquarks $\bar{u}, \bar{d}$, and restrict ourselves to the flavor symmetry of the isospin SU_2^T group. Let us construct, above all, the isospin functions of a system of two subparticles. For the system of two quarks $q(1)q(2)$ (this system has a nonzero color) the isotopic wave functions $Z_{T,M_T}(1,2)$ with definite total isospin $T = 0$ or $T = 1$ and its projections M_T have the same form as for the system of two nucleons (see (1.151), (1.152))

$$\begin{cases} Z_{1,+1}(1,2) &= u(1)u(2), \\ Z_{1,0}(1,2) &= \frac{1}{\sqrt{2}}\{u(1)d(2)+d(1)u(2)\}, \\ Z_{1,-1}(1,2) &= d(1)d(2); \end{cases} \tag{7.54}$$

$$Z_{0,0}(1,2) = \frac{1}{\sqrt{2}}\{u(1)d(2) - d(1)u(2)\}. \tag{7.55}$$

Here, the one-quark isospin functions are denoted by the same letters as the quarks themselves u and d.

When considering the system of a quark and antiquark $q(1)\bar{q}(2)$, quark 2 is formally substituted by antiquark 2 in formulas (7.54) and (7.55). In doing so we do not obtain, in a general case, the state with definite values of T and M_T. For example, the function (7.55), changed in this way, consists of two terms $\frac{1}{\sqrt{2}}u(1)\bar{d}(2)$ and $-\frac{1}{\sqrt{2}}d(1)\bar{u}(2)$, which have different projections of the isotopic spin $M_T = +1$ and $M_T = -1$. The latter cannot be related to the state with $T = 0$ as function (7.55), because it is necessary to have $|M_T \leq T$. It turns out to be that for the system quark-antiquark $q(1)\bar{q}(2)$, it is possible to construct the isotopic wave functions of the isospin triplet ($T = 1$) and isospin singlet ($T = 0$), corresponding to the quark-quark functions (7.54) and (7.55), i.e., with definite values of T and M_T. This is associated with the specific properties of the SU_2^T group.

Using the properties of the Pauli matrices, we present (7.27) in the following more convenient form

$$\begin{aligned} U &= \exp\{\frac{i}{2}\boldsymbol{\xi\tau}\} = \exp\{\frac{i}{2}\xi\mathbf{n}\boldsymbol{\tau}\} = \cos\frac{\xi}{2} \\ &+ i(\mathbf{n}\boldsymbol{\tau})\sin\frac{\xi}{2}, \ \mathbf{n}^2 = 1. \end{aligned} \tag{7.56}$$

In particular, a current element of the SU_2^T group is thus seen to be a rotation through the angle ξ about $\mathbf{n}$ axis in the isospin (isotopic) space. If the unit vector $\mathbf{n}$ is directed along the second axis (y-axis) in the isotopic space, the isotopic wave function-spinor $\xi = \begin{pmatrix} u \\ d \end{pmatrix}$ is transformed by operation (7.56) into the following isotopic spinor $\xi' = U\begin{pmatrix} u \\ d \end{pmatrix} = \begin{pmatrix} u' \\ d' \end{pmatrix}$ related to the isotopic *quark doublet* $u'd'$, whose components are superpositions of the components u, d of the initial isotopic quark doublet:

$$\xi' = \begin{pmatrix} u' \\ d' \end{pmatrix} = \begin{pmatrix} \cos\frac{\xi}{2} & \sin\frac{\xi}{2} \\ -\sin\frac{\xi}{2} & \cos\frac{\xi}{2} \end{pmatrix}\begin{pmatrix} u \\ d \end{pmatrix}, \tag{7.57}$$

whence

$$u' = u\cos\frac{\xi}{2} + d\sin\frac{\xi}{2},$$
$$d' = -u\sin\frac{\xi}{2} + d\cos\frac{\xi}{2}. \tag{7.58}$$

Let us act by operation of charge conjunction $\hat{C}$ on the left and right parts of the latter two equalities. Then quarks u, d are transformed into antiquarks $\bar{u}, \bar{d}$, in this case $\bar{d}$ has $M_T = +\frac{1}{2}$ and $\bar{u}$ has $M_T = -\frac{1}{2}$. If we would like the antiquark doublet $\bar{u}, \bar{d}$ to be transformed by operations of the SU_2^T group in the same manner as the quark doublet u, d, and if we could unite the quark and antiquark states by use of the same coefficients of the vector addition of isotopic moments, then it would be necessary to put a subparticle with a positive isospin projections, i.e., with $M_T = +\frac{1}{2}$, in the upper part of the isotopic antiquark doublet, and to put a subparticle with $M_T = -\frac{1}{2}$ in the lower part of this doublet, as well as in the quark doublet. Then, using the antiquark transformations analogous to (7.58)

$$\bar{d}' = +\bar{d}\cos\frac{\xi}{2} - \bar{u}\sin\frac{\xi}{2},$$
$$\bar{u}' = \bar{d}\sin\frac{\xi}{2} + \bar{u}\cos\frac{\xi}{2}, \tag{7.59}$$

we obtain the transformation in the matrix form

$$\begin{pmatrix} \bar{d}' \\ \bar{u}' \end{pmatrix} = \begin{pmatrix} \cos\frac{\xi}{2} & -\sin\frac{\xi}{2} \\ \sin\frac{\xi}{2} & \cos\frac{\xi}{2} \end{pmatrix} \begin{pmatrix} \bar{d} \\ \bar{u} \end{pmatrix}. \tag{7.60}$$

This transformation still differs from (7.57). But, if we determine the antiquark doublet as $\begin{pmatrix} \bar{d} \\ -\bar{u} \end{pmatrix}$, i.e., rearrange minus sign in (7.60) from the transformation matrix in the *antiquark doublet*, we obtain the transformation analogous to (7.57)

$$\begin{pmatrix} \bar{d}' \\ -\bar{u}' \end{pmatrix} = \begin{pmatrix} \cos\frac{\xi}{2} & \sin\frac{\xi}{2} \\ -\sin\frac{\xi}{2} & \cos\frac{\xi}{2} \end{pmatrix} \begin{pmatrix} \bar{d} \\ -\bar{u} \end{pmatrix}. \tag{7.61}$$

Thus, the lightest antiquarks with the isotopic spin are transformed accordingly the same fundamental representation of the SU_2^T group as the corresponding quarks. More exactly, the antiquark doublet $\bar{d}, -\bar{u}$ are transformed in the same manner as the quark doublet u, d. This follows from the fact that the irreducible representation $\overset{*}{D}{}^{1/2}(U)$ conjugated to the fundamental representation $D^{1/2}(U)$ of the SU_2^T group is equivalent to the representation $D^{1/2}(U)$ owing to the reality of characters of the irreducible representations of the SU_2 group. (Recall that the representations of a group are equivalent if they have the same characters, i.e., spurs, or the traces of representation matrices for every element of the group.) However, for the unimodular groups SU_n with $n > 2$ there is no such property. Because of this, it is impossible to construct the antiquark SU_3^f-triplet so that it will be transformed as the triplet of the corresponding quarks: triplet $\bar{u}, \bar{d}, \bar{s}$ is transformed according to the representation $\overset{*}{D}{}^{(1,0)} = D^{(0,1)}$ conjugated to the fundamental representation $D^{(1,0)}$ of the group SU_3^f. The representations $\overset{*}{D}{}^{(1,0)}$ and $D^{(1,0)}$ are nonequivalent in this case.

In particular, we obtain at $\xi = \pm\pi$ from (7.61)

$$\begin{pmatrix} \bar{d}' \\ -\bar{u}' \end{pmatrix} = \begin{pmatrix} 0 & \pm 1 \\ \mp 1 & 0 \end{pmatrix} \begin{pmatrix} \bar{d} \\ -\bar{u} \end{pmatrix} = \begin{pmatrix} \mp\bar{u} \\ \mp\bar{d} \end{pmatrix}, \tag{7.62}$$

i.e., $\bar{d}$ is transformed in $\mp\bar{u}$ and $\bar{u}$ – in $\pm\bar{d}$. Thus, when going from the two-quark system $q(1)q(2)$ to the *quark-antiquark system* $q(1)\bar{q}(2)$, it is necessary to substitute in (7.54) and (7.55) $u(2)$ with $-\bar{d}(2)$, and $d(2)$ with $\bar{u}(2)$, or substitute $u(2)$ with $\bar{d}(2)$, and $d(2)$ with $-\bar{u}(2)$. This means that compound system of a pair quark-antiquark (meson) have, within accuracy of an unessential common phase (sign), the following isospin wave functions $Z'_{TM_T}(1,2)$:

$$\begin{cases} Z'_{1,+1}(1,2) &= -u(1)\bar{d}(2), \\ Z'_{1,0}(1,2) &= \frac{1}{\sqrt{2}}\{u(1)\bar{u}(2) - d(1)\bar{d}(2)\}, \\ Z'_{1,-1}(1,2) &= d(1)\bar{u}(2); \end{cases} \tag{7.63}$$

$$Z'_{0,0}(1,2) = \frac{1}{\sqrt{2}}\{u(1)\bar{u}(2) + d(1)\bar{d}(2)\}. \tag{7.64}$$

The *isospin wave functions of the nucleon-antinucleon system* take an analogous form. They can be obtained from (7.63) and (7.64) by changing the denotations $u \to p$, $d \to n$, $\bar{u} \to \bar{p}$, $\bar{d} \to \bar{n}$.

In the previous section we did not indicated in Fig. 7.5 and 7.6 how the genuine neutral mesons could be constructed from the combinations $u\bar{u}$, $d\bar{d}$, $s\bar{s}$. We can say now that, according to (7.63), the isotopic wave function of π^0- and ρ^0-mesons of the SU_3^f-octets is the triplet in isospin wave function $Z'_{1,0}(1,2)$ (7.63), i.e., the quark-antiquark composition of these mesons is determined by a combination $\frac{1}{\sqrt{2}}(u\bar{u} - d\bar{d})$. The isotopic wave function of the vector SU_3^f-singlet ω-meson (Fig.7.6) is the singlet with respect to the isospin wave function $Z'_{0,0}(1,2)$ (7.64), i.e., the composition of ω-meson is determined by a superposition $\frac{1}{\sqrt{2}}(u\bar{u}+d\bar{d})$, which is orthogonal to the combination $\frac{1}{\sqrt{2}}(u\bar{u}-d\bar{d})$ of ρ^0-meson. In a general case, the product of *quark-antiquark isotopic wave functions* $q\bar{q}$ is transformed according to the direct product of the fundamental representation of the SU_2^T group, which is conjugated to its representation. The flavor combination $\bar{u}u+\bar{d}d \equiv \sum_{q=u,d}\bar{q}q$, in fact, is the SU_2^T-singlet, i.e., a scalar (invariant). Acting on the quark wave function q by operation (7.27) or (7.56) and on the antiquark function $\bar{q}$ by the operation, complex-conjugated to (7.27) , we obtain

$$\sum_{q=u,d}\bar{q}q \to \sum_{q,d}\bar{q}'q' = \sum_{q=u,d}(\overset{*}{U}q)(Uq) = \sum_{q=u,d}\bar{q}U^{+}Uq = \sum_{q=u,d}\bar{q}q, \qquad (7.65)$$

because $U^+U = 1$. (Here and further $\bar{q}q \equiv \bar{q}(k)q(j)$, where k, j=1,2.)

In a similar manner the flavor combination $u\bar{u}+d\bar{d}+s\bar{s} \equiv \sum_{q=u,d,s}\bar{q}q$ is the SU_3^f-invariant or the flavor SU_3^f-singlet combination, that can be proved exactly as (7.65). This combination presents the SU_3^f-singlet in Fig. (7.5), i.e., the pseudoscalar η'-meson

$$\eta' = \frac{1}{\sqrt{3}}(u\bar{u}+d\bar{d}+s\bar{s}) \equiv \frac{1}{\sqrt{3}}\sum_{q=u,d,s} q\bar{q}. \qquad (7.66)$$

The eighth-meson η^0 of the pseudoscalar meson octet is described by a normalized (by unity) superposition

$$\eta^0 = \frac{1}{\sqrt{6}}(u\bar{u}+d\bar{d}-2s\bar{s}), \qquad (7.67)$$

which is orthogonal to both functions $\pi^0 = \frac{1}{\sqrt{2}}(u\bar{u}-d\bar{d})$ and (7.66). All three given mesons (π^0, η^0, η') appear in the same nonet of pseudoscalar mesons. The eighth meson φ of the vector meson octet consists of only one combination of a strange quark and antiquark

$$\varphi = s\bar{s}, \qquad (7.68)$$

which is orthogonal to the flavor wave functions of the rest genuine-neutral mesons of the vector octet $\rho^0 = \frac{1}{\sqrt{2}}(u\bar{u}-d\bar{d})$ and $\omega = \frac{1}{\sqrt{2}}(u\bar{u}+d\bar{d})$. The flavor wave functions of the rest mesons of both nonets in Fig. 7.5 and 7.6, determined up to an inessential phase (sign), consist of only one term, for example, $\pi^+ = u\bar{d}$, $K^0 = d\bar{s}$, and so on. Thus, due to a strong violation of the flavor SU_3^f-symmetry in the nature, the eighth (genuine-neutral) mesons of the SU_3^f-octets and SU_3^f-singlet mesons are partly mixed. This phenomena is known as the "mixing". Obviously, the mixing of the vector mesons is much stronger than the pseudoscalar mesons. It is clear from the above-considered that the mixing of neutral combinations $u\bar{u}, d\bar{d}, s\bar{s}$ for genuine-neutral mesons cannot be established uniquely theoretically in spite of the fact that the meson wave functions within each of two meson nonets in Fig. 7.5 and 7.6 are orthogonal. Experiment can help here, and we can use, first of all, the experimentally measured meson masses. It is worth noting that , the mixing of combinations $u\bar{u}$, $d\bar{d}$, and $s\bar{s}$ in the nonet of vector mesons occurrs in such a way that the observed genuine-neutral mesons are basically constructed either only from strange ($s\bar{s}$) or only from strangeless ($u\bar{u}, d\bar{d}$) quarks or antiquarks.

We consider now the *color meson states.* The color dependence of the total meson wave function $|M\rangle$ is usually united with flavor. While on the subject of the color meson wave function ψ_c^M, it will be considered to be the flavor (in particular, isospin) wave function as well. To express the color meson wave function ψ_c^M in terms of the corresponding quark and antiquark wave functions, we have to keep in mind that a meson is the color SU_3^c-singlet, and that it is colorless, as well as all hadrons. Because of this, the color meson wave function ψ_c^M is an invariant (scalar) when the operations U_c from the SU_3^c group act on it.

We show now that such a singlet meson wave function, normalized by unity, presents the following simple combination of color (and flavor) wave functions of a quark p^α and antiquark $\bar{q}_\alpha$ (q and p, in a general case, are different flavors):

$$\psi_c^M \equiv \psi_c^M(1,2) = \frac{1}{\sqrt{3}} \sum_{\alpha=1}^{3} \bar{q}_\alpha(1) p^\alpha(2) \equiv \frac{1}{\sqrt{3}} \bar{q}_\alpha p^\alpha, \tag{7.69}$$

where the summation is carried out over all three different color quantum numbers or color charges α. Since the color quark p^α and antiquark $\bar{q}^\alpha$ are transformed according to the fundamental $D^{(1,0)}$ and conjugated to it $D^{(0,1)} = \overset{*}{D}{}^{(1,0)}$ irreducible representations of the color group SU_3^c, we obtain (analogously (7.65)):

$$\begin{aligned} \bar{q}_a p^\alpha \to \bar{q}'_\alpha p'^\alpha &= \sum_\alpha (\overset{*}{U}_c\, \bar{q}_\alpha)(U_c p^\alpha) = \sum_\alpha \sum_\beta D^{(0,1)}_{\beta\alpha} \bar{q}_\beta \sum_\gamma D^{(1,0)}_{\gamma\alpha} p^\gamma \\ &= \sum_{\beta,\gamma} \Big(\sum_\alpha D^{+(1,0)}_{\alpha\beta} D^{(1,0)}_{\gamma\alpha} \Big) \bar{q}_\beta p^\gamma = \bar{q}_\beta p^\beta. \end{aligned} \tag{7.70}$$

We used here that $D^{(1,0)} D^{+(1,0)} = 1$. It is clear that in a general case, we do not need necessary to single out the color-flavor meson wave function ψ_c^M from the total function $|M>$, which depends also on space and spin coordinates. We can just speak about the function $|M>$ as the color singlet wave function. This isolation has been done only from the procedure considerations.

Note that color singlet (7.69) has the same structure as flavor singlet (7.66). There is no surprise since both (7.66) and (7.64) are invariants of the same mathematical group SU_3. In this connection recall that a gluon, as the system of color and anticolor, cannot have the SU_3^c-singlet color combination analogous to (7.69), i.e., the combination (in denotations of Fig. 7.1 and 7.2)

$$\frac{1}{\sqrt{3}}(r\bar{r} + b\bar{b} + y\bar{y}). \tag{7.71}$$

Because of this, there are eight possible independent orthonormal *color-anticolor combinations for gluons* :

$$\begin{gathered} r\bar{y}, r\bar{b}, y\bar{r}, y\bar{b}, b\bar{r}, b\bar{y}, \\ \frac{1}{\sqrt{2}}(r\bar{r} - y\bar{y}), \frac{1}{\sqrt{6}}(r\bar{r} + y\bar{y} - 2b\bar{b}), \end{gathered} \tag{7.72}$$

where two latter color superpositions are not the color singlets. They are analogous to the observed flavor combinations, $\pi^0 = \frac{1}{\sqrt{2}}(u\bar{u} - d\bar{d})$ and $\eta^0 = \frac{1}{\sqrt{6}}(u\bar{u} + d\bar{d} - 2s\bar{s})$, respectively, which are not the SU_3^f-singlets.

While obtaining the spatial wave function $\varphi(\mathbf{r})$ of the relative quark-antiquark motion in a meson ($\mathbf{r}$ is a radius-vector connecting the meson quark and antiquark), we deal with the problem analogous to the two-nucleon problem of a deuteron. But here it is necessary to know (specify) the quark-antiquark potential $V_{q\bar{q}}(\mathbf{r})$. In the nonrelativistic approximation, the wave functions $\varphi(\mathbf{r})$ and energy eigenvalues (i.e., the meson mass M_M spectrum) of the relative motion can be obtained from the Schrödinger equation. The latter for a quark and antiquark of the same mass M_q takes a simple form

$$\{-\frac{1}{M_q}\Delta + V_{q\bar{q}}(\mathbf{r})\}\varphi(\mathbf{r}) = (M_M - 2M_q)\varphi(\mathbf{r}), \tag{7.73}$$

where M_q is the block quark mass. The possible model expressions of the quark-quark and quark-antiquark potentials are given at the end of this section. The nonrelativistic approximation is more applicable for heavy quarks c, b, t and corresponding antiquarks. The relativistic corrections are important for light quarks u, d, s and antiquarks $\bar{u}, \bar{d}, \bar{s}$.

If the angular momentum of the relative quark-antiquark motion L in a meson can be treated as a good quantum number, then the momentum L

can be treated as one of the quantum characteristics of the meson spatial wave function $\varphi(\mathbf{r})$. Since the intrinsic parities of a quark and antiquark are opposite, the *meson parity* $\Pi_M = -(-1)^L$. On increasing L by unity, the meson excitation energy increases approximately by 600 MeV due to the orbital motion. The radial excitations are also possible, so that we can obtain new nonets of mesons with increased masses compared with the meson masses in Fig. 7.5 and 7.6 even at $L = 0$.

Since mesons consist of only two subparticles with spin 1/2, the admissible meson spin wave functions have quite simple structure, as well as for a system of two nucleons. They can be either symmetric when the total spin of quark and antiquark $S = 1$ or antisymmetric when $S = 0$ with respect to permutations of spin variables.

As an example, we give the general form of the *total wave function of π^+-meson* , which includes the spatial, flavor, color, and spin components

$$|\pi^+ >\equiv \pi^+ = \varphi(\mathbf{r})\frac{1}{\sqrt{3}}\sum_{\alpha=1}^{3}\bar{d}_\alpha(1)u^\alpha(2)\frac{1}{\sqrt{2}}$$
$$\times\left(\chi_{\frac{1}{2},+\frac{1}{2}}(1)\chi_{\frac{1}{2},-\frac{1}{2}}(2) - \chi_{\frac{1}{2},-\frac{1}{2}}(1)\chi_{\frac{1}{2},+\frac{1}{2}}(2)\right), \tag{7.74}$$

where $\chi_{\frac{1}{2},\pm\frac{1}{2}}$ are the quark and antiquark spin wave functions. The color and flavor (in this case – the isotopic or isospin) states are united in one normalized combination $\psi_c^{\pi^+} = \frac{1}{\sqrt{3}}\sum_{\alpha=1}^{3}\bar{d}_\alpha(1)u^\alpha(2)$.

Completing the general description of the states of a system of two color objects, in particular mesons, we mention the so-called G-parity which is inherent to some mesons. The operator of G-parity $\hat{G}$ is a product of the charge conjugation operator $\hat{C}$ and the operation of rotation through 180° about the second axis (y-axis) in the isospin space

$$\hat{G} = \hat{C}e^{i\pi\hat{T}_2}, \tag{7.75}$$

where $\hat{T}_2$ is one of the generators of the SU_2^T group being the projection of the isotopic spin operator $\hat{\mathbf{T}}$ on the second axis of the isotopic space. From the definition of operator (7.75) follows that any Hamiltonian of hadron system, which is invariant with respect to the charge conjugations and rotations in the isotopic space, commutes with the operator of G-parity. The G-parity is available in all process conditioned by the strong interactions.

The operator (7.75) for the quark isospinor $\begin{pmatrix} u \\ d \end{pmatrix}$, where $\hat{\mathbf{T}} = \frac{1}{2}\boldsymbol{\tau}$, by means of (7.56) can be presented in the form

$$\hat{G} = \hat{C}e^{i\frac{\pi}{2}\tau_2} = \hat{C}i\tau_2 = \hat{C}\begin{pmatrix} 0 & 1 \\ -1 & 0 \end{pmatrix}.$$

Acting on the isospin wave functions (7.25) by this operator and taking account of $\hat{C}u \to \bar{u}, \hat{C}d \to \bar{d}$, we obtain

$$\hat{G}\xi_u = \hat{C}\begin{pmatrix} 0 & 1 \\ -1 & 0 \end{pmatrix}\begin{pmatrix} 1 \\ 0 \end{pmatrix} = \hat{C}\begin{pmatrix} 0 \\ -1 \end{pmatrix} = -\hat{C}\xi_d \to -\xi_{\bar{d}},$$

$$\hat{G}\xi_d = \hat{C}\begin{pmatrix} 0 & 1 \\ -1 & 0 \end{pmatrix}\begin{pmatrix} 0 \\ 1 \end{pmatrix} = \hat{C}\begin{pmatrix} 1 \\ 0 \end{pmatrix} = \hat{C}\xi_u \to \xi_{\bar{u}},$$

Thus, the operator $\hat{G}$ transforms the states u in $-\bar{d}$, and d in $\bar{u}$ ($u \to -\bar{d}, d \to \bar{u}$). Since $\hat{C}\bar{u} \to u$, $\hat{C}\bar{d} \to d$, then acting in the same manner, we can see that $\bar{u} \to -d$ and $\bar{d} \to u$ when the G-parity operator acts on the antiquark states $\bar{u}$ and $\bar{d}$.

Let us act now on the pion isotriplet by the G-parity operator, taking into account the meson quark-antiquark structure and using the results of the $\hat{G}$ operator action on the quark and antiquark states

$$\begin{aligned} \hat{G}\pi^+ &= \hat{G}u\bar{d} \to -\bar{d}u = -\pi^+; \\ \hat{G}\pi^- &= \hat{G}\bar{u}d \to -d\bar{u} = -\pi^-; \\ \hat{G}\pi^0 &= \hat{G}\frac{u\bar{u} - d\bar{d}}{\sqrt{2}} \to \frac{\bar{d}d - \bar{u}u}{\sqrt{2}} = -\pi^0. \end{aligned} \tag{7.76}$$

Note that the result does not change when substituting $u \to \bar{d}$, $d \to -\bar{u}$, $\bar{u}tod$, $\bar{d} \to -u$, since the common phase (in this case, common sign) may be arbitrary. Therefore, all three pions are eigenstates of the G-parity operator $\hat{G}$ with the same eigenvalue -1. The G-parity is presented as multiplicative quantum number following from the definition (7.75). Because of this, for example, a system of ν pions must have the G-parity equal to $(-1)^\nu$. From this, in particular, follows that an even number of pions cannot be transformed in odd number of pions by virtue of the strong interaction.

The ω- and φ-mesons have the negative G-parity, as well as pions, and η- and ρ-mesons have the positive G-parity. Whereas a definite value of the charge parity (a proper value of $\hat{G}$ operator) is inherent only to genuine neutral particles, the G-parity can take a definite value in both neutral and charged particles. Because of this, the eigenvalues of the $\hat{G}$ operator (G-parity) is the convenient generalization of the charge parity idea on the state with nonzero electric charge. Therefore, the idea of the G-parity is useful, if a particle or a system of particles are transformed into themselves by operation $\hat{G}$. At the same time, the corresponding state does not change or changes its sign. Apparently, it is necessary that the barion charge of a particle B be equal to zero (that is why barions do not have a definite value of the G-parity) and, moreover, the particle hypercharge Y be equal to zero, as for some other mesons. The important property of the G-parity is that, as we have seen from the pion example, it is the same for all particles of the same isotopic multiplet. This can be understood from the following reasoning. Both operators $\hat{C}$ and $\mathrm{e}^{i\pi\hat{T}_2}$, appearing in $\hat{G}$, change the electric charge of a particle or system of particles on the opposite (in sign) charge. But when they act simultaneously these changes are mutually compensated. Whereas both operators $\hat{C}$ and $\mathrm{e}^{i\pi\hat{T}_2}$ are related to the electric charge, only the operator $\hat{C}$ in $\hat{G}$ is related to the barion charge and hypercharge.

One further general comment should be made about eigenvalues of the operators $\hat{G}$ and $\hat{C}$. Basically, it was already shown that mesons with zero hypercharge have a definite value of the G-parity. But it is impossible to determine theoretically which value +1 or -1 is realized for particular mesons, since in the operation of charge conjuction $\hat{C}$ a common arbitrary phase (sign) factor appears. (Because of this, we have written $\hat{C}u \to \bar{u}$, and so on.) It is conventional to choose phase so that all members of the meson isospin multiplet have the same G-parity as the electrically neutral meson appearing in the multiplet. The G-parity is usually determined from experiment. Thus, for example, the state π^0 must be an eigenvalue of the operator of charge conjugation $\hat{C}$ related to the eigenvalue +1: $\hat{C}\pi^0 = +\pi^0$, since π^0-meson decays mainly into two γ-quanta, i.e., in an even number of charge-odd particles – photons. This leads to $\hat{G}\pi^0 = -\pi^0$ as well as $\hat{G}\pi^\pm = -\pi^\pm$, if $\hat{C}\pi^\pm = -\pi^\mp$ is assumed. Sometimes, a rotation through 180° around the second axis is substituted with rotation around the first axis, i.e., the G-parity operator is specified as $\hat{G} = \hat{C}\mathrm{e}^{i\pi\hat{T}_1}$. In this case the above-reported results would remain the same provided that $\hat{C}\pi^\pm = +\pi^\mp$.

Barions as three-quark systems. Further we consider the neutral in color systems consisting of only quarks. These systems (barions, complex atomic nuclei) contain no less than three quarks, and we have the many-body problem unlike the meson problem. We discuss first the inner barion states as systems of three color objects (quarks).

Barions, as well as all observed strongly interacting particles, are the color SU_3^c-singlets, i.e., colorless particles. This means that the color part of the total barion wave function is invariant with respect to the color group SU_3^c operations. We prove now that the *barion normalized color wave function* ψ_c^B, consisting of various quark flavors $q(1), q(2), q(3)$, can be presented in the form of the following determinant

$$\psi_c^B \equiv \psi_c^B(1,2,3) = \frac{1}{\sqrt{6}}\varepsilon_{\alpha\beta\gamma}q^\alpha(1)q^\beta(2)q^\gamma(3), \qquad (7.77)$$

where the sum is taken over repetitive color indexes $\alpha, \beta, \gamma = 1, 2, 3$ and $\varepsilon_{\alpha\beta\gamma}$ is the totally antisymmetric unit tensor of the third rank. In fact, acting on function (7.77) by operations of general form (7.32) of the SU_3^c group and using the fact that the color quark triplet q^α is the basis of the fundamental representation $D^{(1,0)}$ of the SU_3^c group, we obtain

$$\psi_c^B \to \psi_c^{B'} = U_c \psi_c^B = \frac{1}{\sqrt{6}} \sum_{\bar{\alpha},\bar{\beta},\bar{\gamma}=1}^{3}$$
$$\times \left\{ \sum_{\alpha,\beta,\gamma=1}^{3} \varepsilon_{\alpha\beta\gamma} D_{\bar{\alpha}\alpha}^{(1,0)}(U_c) D_{\bar{\beta}\beta}^{(1,0)}(U_c) D_{\bar{\gamma}\gamma}^{(1,0)}(U_c) \right\} q^{\bar{\alpha}}(1) q^{\bar{\beta}}(2) q^{\bar{\gamma}}(3)$$
$$= \frac{1}{\sqrt{6}} \sum_{\bar{\alpha},\bar{\beta},\bar{\gamma}=1}^{3} \varepsilon_{\bar{\alpha}\bar{\beta}\bar{\gamma}} q^{\bar{\alpha}}(1) q^{\bar{\beta}}(2) q^{\bar{\gamma}}(3) = \psi_c^B, \tag{7.78}$$

where we have used the known relation

$$\sum_{\alpha,\beta,\gamma=1}^{3} \varepsilon_{\alpha\beta\gamma} D_{\bar{\alpha}\alpha}^{(1,0)}(U_c) D_{\bar{\beta}\beta}^{(1,0)}(U_c) D_{\bar{\gamma}\gamma}^{(1,0)}(U_c) = \varepsilon_{\bar{\alpha}\bar{\beta}\bar{\gamma}} \mathrm{Det} D^{(1,0)}(U_c),$$

and the fact that $\mathrm{Det} D^{(1,0)}(U_c) = 1$.

The total barion wave function $|B>$ depending on color, spatial, spin and flavor variables must be completely antisymmetric in the approximation of flavor symmetries SU_2^T, SU_3^f, SU_4^f, and so on, where quarks of different flavor are indistinguishable. Since the color barion wave function (7.77) is totally symmetric with respect to permutations of quarks 1,2,3, the part of the wave function depending on the spatial, spin, and flavor variables must be totally antisymmetrical in the flavor symmetry approximations.

The structure of the *color three-quark function* (7.77) is determined by the barion colorless condition (the color charge of a barion $C = 0$). For colorless quark systems, unlike mesons, the color part of the total wave function is not usually united with its flavor part. Taking into account antisymmetry of the barion color wave function ψ_c^B, we present it as the corresponding Young tableaux in the form of a columns with three boxes (accordingly the number of quarks) $\psi_c^B([111]_c) \equiv \psi_c^B([1^3]_c)$ or simply $[1^3]_c$. (One-column Young tableaux with k boxes is usually denoted as $[1^k]$.) Further, we put a sign on the lower right of the Young scheme, which specifies the nature of variables appearing in the corresponding part of the wave function with respect to the permutation symmetry.

Note that a length of the color Young tableaux column of a barion

$[1^3] \equiv$  is maximum, since all three possible color states are occupied

by three quarks. Because of this, the corresponding color wave function of a barion u is an invariant with respect to the operations of the color SU_3^c group. There is some analogy with the antisymmetric spatial (orbital) wave function of three identical nucleons in p-states (with angular momenta $l = 1$), which is also described by the one-column Young tableaux with three boxes. Its length is maximum as well. Since all three states with three possible projections of the angular momentum are occupied in this case by nucleons, the spatial (orbital) wave function of the three-nucleon system is invariant with respect to operations of the three-dimensional rotations R_3 corresponding to zero value of the total angular momentum L.

We dwell now on the spin wave functions of the *three-quark system* with a particular value of the total spin. The spin wave function of the k-th quark is denoted as $\chi_{\frac{1}{2}m_k}(k) \equiv \chi_{m_k}(k)$, where $m_k = \pm\frac{1}{2}$. The total spin function of the three-quark system, for example, according to (4.199) can be written in the form

$$\left[\left(\frac{1}{2}\frac{1}{2}\right)^{J_{12}}\frac{1}{2}\right]^{J}_{M_J} = \sum_{m_1,m_2}\left(J_{12}m_1 + m_2\frac{1}{2}M_J - m_1 - m_2|JM_J\right)$$
$$\times\left(\frac{1}{2}m_1\;\frac{1}{2}m_2|J_{12}m_1 + m_2\right)\chi_{m_1}(1)\chi_{m_2}(2)\chi_{M_J-m_1-m_2}(3), \qquad (7.79)$$

where J_{12} is the total spin of the 1-st and 2-d quarks, and J and M_J are the barion spin and its projection. The spin wave function of the barion decinet in Fig. 7.4 with spin $\frac{3}{2}$ is totally symmetric with respect to permutations of three quarks and equal, in agreement with (7.79) to the following expression

$$\left[\left(\frac{1}{2}\frac{1}{2}\right)^{1}\frac{1}{2}\right]^{\frac{3}{2}}_{M_J} = \sum_{m_1,m_2}\left(1m_1 + m_2\frac{1}{2}M_J - m_1 - m_2|\frac{3}{2}M_J\right)$$
$$\times\left(\frac{1}{2}m_1\;\frac{1}{2}m_2|1m_1 + m_2\right)\chi_{m_1}(1)\chi_{m_2}(2)\chi_{M_J-m_1-m_2}(3). \qquad (7.80)$$

It is transformed according to the irreducible representation $D^{3/2}$ of a dimension 4 ($M_J = \pm\frac{3}{2}, \pm\frac{1}{2}$) of the spin group SU_2^S. Apparently, the corresponding Young tableaux, which characterizes the permutation symmetry of function (7.80), is represented as a line with three boxes (according to the quark number) and denoted as $[3]_s$. The spin barion wave function of the octet in Fig. 7.3 with spin $J = \frac{1}{2}$ can be symmetric with respect to permutations of only two first of three quarks and, accordingly the general expression (7.79), is equal to

$$\Big[\Big(\frac{1}{2}\frac{1}{2}\Big)^{1}\frac{1}{2}\Big]^{\frac{1}{2}}_{M_J} = \sum_{m_1,m_2}\Big(1m_1+m_2\frac{1}{2}M_J-m_1-m_2\Big|\frac{1}{2}M_J\Big)$$
$$\times\Big(\frac{1}{2}m_1\,\frac{1}{2}m_2\Big|1m_1+m_2\Big)\chi_{m_1}(1)\chi_{m_2}(2)\chi_{M_J-m_1-m_2}(3). \qquad (7.81)$$

It is precisely the wave function that describes a proton and neutron as three-particles systems where two quarks of the same flavor have the total spin equal to 1. The corresponding Young tableau of the spin nucleon state $[21]_s$ determines the mixed symmetry with respect to permutation of three quarks. Since the function (7.81) is symmetric with respect to the permutation of quarks with numbers 1 and 2 ($J_{12} = 1$), the corresponding Yamanouchi symbol of the spin nucleon state takes the form

(2, 1, 1) =

1	2
3	

.

With (7.81) in mind, the theoretical value for the ratio between the neutron μ_n and proton μ_p magnetic moments in the approximation of the flavor symmetry of the SU_6^{sf} turns out to be close to the experimental ratio μ_n/μ_p. We assume that the magnetic moment μ_q of the quark q is connected with its charge Q_q (in the proton charge units e) and a mass M_q by the relation

$$\mu_q = \frac{eQ_q}{2M_q}, \qquad (7.82)$$

as well as for the pointlike Dirac particle ($\hbar = c = 1$). Then the operator of the nucleon magnetic moment $\hat{\boldsymbol{\mu}}_N$ can be presented as a sum over three quarks, which are supposed to be in the S-state,

$$\hat{\boldsymbol{\mu}}_N = \sum_{j=1}^{3}\mu_{q(j)}\boldsymbol{\sigma}(j), \ N = p, n. \qquad (7.83)$$

The observed nucleon magnetic moment μ_N is determined as a mean value of the projection of $\hat{\boldsymbol{\mu}}_N$ on z-axis in the state with the nucleon spin projection $M_J = +\frac{1}{2}$

$$\mu_N = \Big\langle\Big[\Big(\frac{1}{2}\frac{1}{2}\Big)^{1}\frac{1}{2}\Big]^{\frac{1}{2}}_{\frac{1}{2}}\Big|\hat{\mu}_{Nz}\Big|\Big[\Big(\frac{1}{2}\frac{1}{2}\Big)^{1}\frac{1}{2}\Big]^{\frac{1}{2}}_{\frac{1}{2}}\Big\rangle. \qquad (7.84)$$

Since the operator in (7.83) acts only on quark spin variables, we can perform the average in (7.84), using only the spin wave function

$$\left[\left(\frac{1}{2}\frac{1}{2}\right)^1\frac{1}{2}\right]^{\frac{1}{2}}_{\frac{1}{2}} = \sqrt{\frac{2}{3}}\chi_{\frac{1}{2}}(1)\chi_{\frac{1}{2}}(2)\chi_{-\frac{1}{2}}(3)$$
$$-\frac{1}{\sqrt{6}}\chi_{\frac{1}{2}}(1)\chi_{-\frac{1}{2}}(2)\chi_{\frac{1}{2}}(3) - \frac{1}{\sqrt{6}}\chi_{-\frac{1}{2}}(1)\chi_{\frac{1}{2}}(2)\chi_{\frac{1}{2}}(3), \qquad (7.85)$$

which is obtained from (7.81) by substitution of numerical values of the corresponding Clebsh-Gordan coefficients. Taking into account that $\sigma_z(j)\chi_m(j) = 2m\chi_m(j)$ and $Q_u = +\frac{2}{3}$, $Q_d = -\frac{1}{3}$, we obtain from (7.84) by use of (7.83) and (7.85) expressions for the magnetic moments of proton and neutron, respectively,

$$\mu_p = \frac{2}{3}\mu_u + \frac{2}{3}\mu_u - \frac{1}{3}\mu_d = \frac{4}{3}\mu_u - \frac{1}{3}\mu_d, \; \mu_n = \frac{4}{3}\mu_d - \frac{1}{3}\mu_u.$$

Since the quarks d and u are equal in mass in the approximation of flavor symmetries SU_2^T, SU_3^f, and SU_6^{sf}, we obtain the interrelation $\mu_u = -2\mu_d$, which can be used to find $\mu_p = -3\mu_d$ and $\mu_n = 2\mu_d$. Hence, we obtain the desirable *ratio of a neutron and proton magnetic moments*

$$\frac{\mu_n}{\mu_p} = -\frac{2}{3}, \qquad (7.86)$$

that is rather close to the experimental ratio of these quantities

$$\frac{\mu_n}{\mu_p} = \frac{-1,91}{2,79} = -0,685. \qquad (7.87)$$

Thus, the nucleon quark model gives approximately correct value of a ratio of the magnetic moments of a neutron and proton.

The isotopic wave functions of barions, which consist of only u- and d-quarks, have the same mathematical structure as the spin wave functions of the three-quark system (7.79). The isotopic wave functions $[(\frac{1}{2}\frac{1}{2})^{T_{12}}\,\frac{1}{2}]^T_{M_T}$ are transformed, accordingly the irreducible representation D^T of the isotopic SU_2^T group. The isotopic quartet of the Δ-resonances or Δ-isobars with $T = \frac{3}{2}$ is characterized by the totally symmetric isotopic wave function $[(\frac{1}{2}\frac{1}{2})^1\,\frac{1}{2}]^{3/2}_{M_T}$ with respect to quark permutations, i.e., the permutation symmetry is described by the Young tableaux $[3]_T$. Barions, i.e., isobars of the quartet $\Delta^{++}, \Delta^+, \Delta^0, \Delta^-$ have the isospin projections $M_T = +\frac{3}{2}, +\frac{1}{2}, -\frac{1}{2}, -\frac{3}{2}$. The nucleon isotopic doublet is characterized by the isotopic wave function $[(\frac{1}{2}\frac{1}{2})^1\,\frac{1}{2}]^{1/2}_{M_T}$ with $M_T = +\frac{1}{2}$ for a proton and $M_T = -\frac{1}{2}$ for a neutron. This function is symmetric ($T_{21} = 1$) with respect to permutations of only two quarks with the same flavor, so that it corresponds to the Young tableaux $[21]_T$ for the mixed permutation symmetry. The isotopic wave functions of barions, consisting of one or two quarks with the isotopic spin (i.e., u- or d-quark), are rather simple. If two such quarks appear in a barion, then the triplet barion wave function $Z_{1,M_T}(1,2) \equiv (\frac{1}{2}\frac{1}{2})^1_{M_T}$ with $T = 1$ (see (7.64)) corresponds to the Young tableaux $[2]_T$ and is symmetric, the singlet function $Z_{0,0}(1,2) \equiv (\frac{1}{2}\frac{1}{2})^0_0$ with $T = M_T = 0$ (see (7.55)) corresponds to the Young tableaux $[11]_T$ and is antisymmetric. (This is, for example, the sigma-hyperon with $T = 1$ and the lambda-hyperon with $T = 0$.)

Let us consider the *symmetry properties of the barion flavor wave functions* in the more general form. For this purpose, we expand the quark composition of barions, introducing quarks of different flavor within the framework of the flavor group SU_n^f approximation, i.e., assuming quarks of different flavor to be indistinguishable particles for the strong interactions. The barion flavor wave function with quark flavors q_1, q_2 and q_3 is a product of the flavor wave functions of individual quarks and is denoted as $\psi_{q_1q_2q_3} \equiv \psi_{q_1q_2q_3}(1,2,3)$. At the same time, three unitary (flavor) indexes q_1, q_2, q_3 can run independently n different values (flavors): $q = 1, 2, \ldots, n$ so that a general number of function $\psi_{q_1q_2q_3}$ is equal to n^3. The operations U from the flavor group SU_n^f, acting in the n^3-dimensional linear space of functions $\psi_{q_1q_2q_3}$, transform these functions in the following way

$$\psi_{q_1q_2q_3} \to \psi'_{q_1q_2q_3} = U\psi_{q_1q_2q_3} = \sum_{q'_1q'_2q'_3} U_{q'_1q_1}U_{q'_2q_2}U_{q'_3q_3}\psi_{q'_1q'_2q'_3}, \tag{7.88}$$

where $U_{q'_kq_k}$ is the matrix of the fundamental representation of the SU_n^f group for the k-th quark. Thus, the function $\psi_{q_1q_2q_3}$ is transformed according to the irreducible representation $U \otimes U \otimes U \equiv n \otimes n \otimes n$ of the group SU_n^f.

We determine now the commutation relations of basis functions of the irreducible representation of the group SU_n^f which appear in the expansion of the direct product of representations $U \otimes U \otimes U$. Since transformations (7.88) are continuous, the permutation properties of the functions $\psi_{q_1q_2q_3} \equiv \psi_{q_1q_2q_3}(1,2,3)$ do not change. In particular, if the wave function $\psi_{q_1q_2q_3}(1,2,3)$ is symmetric or antisymmetric with respect to the permutations of any pair of quarks, then the transformed function $\psi'_{q_1q_2q_3}(1,2,3)$ preserves this property, i.e., functions with a given symmetry form the closed space of functions and are transformed through themselves.

We find first the total number N_a of the antisymmetric wave functions $\psi^{(a)}_{q_1q_2q_3}$, which are, of course, particular linear combinations of the initial functions $\psi_{q_1q_2q_3}$. It is clear that $\psi^{(a)}_{q_1q_2q_3} \neq 0$ only if all three flavor indexes q_1, q_2 and q_3 are different. Because of this, the number N_a of independent components $\psi^a_{q_1q_2q_3}$ is equal to a number of combinations from n by 3, i.e.,

$$N_a = \frac{1}{3!} n(n-1)(n-2). \tag{7.89}$$

This number determines the dimension of the antisymmetric irreducible representation of the SU_n^f group.

We determine now a number N_s of the totally symmetric wave functions $\psi^{(s)}_{q_1 q_2 q_3}(1,2,3)$ with respect to the permutations of two any quarks. Apparently, the number of symmetric functions $\psi^{(s)}_{q_1 q_2 q_3}$ with different flavor indexes is equal to N_a, the number of functions $\psi^{(s)}_{q_1 q_2 q_3}$ with two equal indexes and one different index is $n(n-1)$, and, at least, the number of $\psi^{(s)}_{q_1 q_2 q_3}$ with three equal indexes is equal to n. Summing up these three numbers, we obtain the number of all independent totally symmetric wave functions $\psi^{(s)}_{q_1 q_2 q_3}$:

$$N_s = \frac{1}{3!} n(n+1)(n+2). \tag{7.90}$$

The rest numbers $n^3 - N_a - N_s = \frac{2}{3} n(n^2 - 1)$ of functions $\psi_{q_1 q_2 q_3}(1,2,3)$ have the mixed permutation symmetry, in this case, a half of these functions are symmetric and the other half are antisymmetric with respect to the permutations of only first two quarks. Thus, in a general case, for n flavors, we have the following expansion of the direct product of three fundamental representations of the SU_n^f group in four irreducible representations of this group (with indication of their dimension)

$$\begin{aligned} n \bigotimes n \bigotimes n &= \frac{1}{6} n(n-1)(n-2) \\ &\oplus \frac{1}{6} n(n+1)(n+2) \bigoplus \frac{1}{3} n(n^2-1) \bigoplus \frac{1}{3} n(n^2-1) \end{aligned} \tag{7.91}$$

with four sets of the corresponding basis functions, which have the above-established permutation properties. For example, we have $4 \bigotimes 4 \bigotimes 4 = 4 \bigoplus 20 \bigoplus 20 \bigoplus 20$ and $6 \bigotimes 6 \bigotimes 6 = 20 \bigoplus 56 \bigoplus 70 \bigoplus 70$ for SU_4^f and SU_6^f groups.

In the important particular case $n = 3$, we have $N_a = 1$, $N_s = 10$ for the flavor group SU_3^f based on three quarks u, d, s. Therefore, the one-dimensional irreducible representation $D^{(0,0)}$ of the SU_3^f group, appearing in the expansion (7.36) $3 \bigotimes 3 \bigotimes 3 = 1 \bigoplus 10 \bigoplus 8 \bigoplus 8$, has one totally antisymmetric normalized flavor wave function with respect to permutations of three quarks

$$\psi^{(a)}_{uds}(1,2,3) = \frac{1}{\sqrt{6}}(uds - usd + sud - sdu + dsu - dus)$$

$$\equiv \frac{1}{\sqrt{6}} \begin{vmatrix} u(1) & d(1) & s(1) \\ u(2) & d(2) & s(2) \\ u(3) & d(3) & s(3) \end{vmatrix}, \tag{7.92}$$

where $uds \equiv u(1)d(2)s(3)$ and so on. This function has the same structure as color singlet barion wave function (7.77). The ten-dimensional irreducible representation $D^{(3,0)}$, appearing in expansion (7.36), for the basis flavor wave functions has ten totally symmetric functions $\psi^{(s)}_{q_1q_2q_3}(1,2,3)$ with respect to the possible permutations of three quarks. The remaining two eight-dimensional irreducible representations $D^{(1,1)}$ of the group SU_3^f, appearing in expansion (7.36), have the basis functions of mixed symmetry. In this case, all eight functions of one of the basis sets are symmetric, and all eight functions of the other set are antisymmetric with respect to permutations of only two quarks.

As to the spatial (orbital) quark states in the lightest barions (nucleons and Δ-isobars consisting of only u- and d-quarks with almost the same block masses), it can be assumed from the same positions of quarks in the SU_2^T-, SU_3^f-, and SU_3^c-symmetries and zero value of the total relative angular momentum L, that each of three quarks in a barion is in the s-state with the angular moment $l = 0$, and the spatial wave function of the relative quark motion $\varphi(\mathbf{r}_1, \mathbf{r}_2, \mathbf{r}_3)$ is totally symmetric in respect of quark permutations. (Here $\mathbf{r}_1, \mathbf{r}_2, \mathbf{r}_3$ are the quark radius-vectors counted from the barion mass center $\mathbf{r}_1 + \mathbf{r}_2 + \mathbf{r}_3 = 0$.) Because of this, the spatial part of the barion wave function, as well as the system of three quarks, is described by the Young tableaux $[3]_x$, where x indicates the type of symmetry related to the quark spatial state.

The function $\varphi(\mathbf{r}_1, \mathbf{r}_2, \mathbf{r}_3)$ itself must be determined from a solution of the Schrödinger equation with given qq-interactions. We discuss this problem at the end of this section. If attractive forces act between quarks, then the state with minimum energy is described by the symmetric spatial wave function, as it was assumed above.

The total wave function of a barion $|B\rangle$, as a system of identical quarks (fermions) u and d, is totally symmetric with respect to permutations of any pair of quarks, i.e., is characterized by the Young tableaux $[1]^3_{XCST}$, where spatial, color, spin nd flavor (isospin) quark degrees of freedom are indicated. The three-quark wave function of barion is factorized with isolation of a flavor factor only if the flavor and spin parts of the wave function are separately either totally symmetric or antisymetric. If they have the mixed symmetry in respect of quark permutations, then we construct a symmetrical superposition of products of the flavor and spin wave functions. Taking into account the symmetry of the spatial and asymmetry of the flavor wave functions of a barion, we obtain the total wave function antisymmetric with respect to quark permutations. In particular, the *total antisymmetric nucleon wave function* $|N\rangle$ written like the wave function of a system of identical nucleons in the nuclear shell model, accordingly the above-reported, is described by the following main quantum characteristics, i.e., by symmetries

$$|N\rangle = |s^3[3]_X[1^3]_C[21]_S[21]_T||[1^3]_{XCST}\rangle. \tag{7.93}$$

The total wave function of the Δ-isobar in the analogous denotations takes the form

$$|\Delta\rangle = |s^3[3]_X[1^3]_C[3]_S[3]_T||[1^3]_{XCST}\rangle. \tag{7.94}$$

We consider more fully some structure details of the total wave function (7.93) and (7.94). In particular, it is necessary to show the consistency of the total Young tableaux $[1^3]_{XCST}$ with the partial Young tableaux, appearing in these functions and determining symmetries of the separate parts of the total wave functions associated with different quark degrees of freedom. In other words, we have to show that the representation $[1^3]_{XCST}$ appears when expanding the inner (Kronecker) products of the irreducible representations of the permutation groups $[3]_X \times [1^3]_C \times [21]_S \times [21]_T$ and $[3]_X \times [1^3]_C \times [3]_S \times [3]_T$. (We will use the convention denotation of the irreducible representation $D^{(\lambda)} \equiv [\lambda]$.) For this purpose we recall the definition of the inner (Kronecker) product of two irreducible representations $[\lambda] \times [\lambda']$ of the permutation group given in the section 4.4. In particular, we have already used the result (see (4.94) or (4.104)) that the one-dimensional antisymmetrical representation $[11\dots1] \equiv [1^k]$ for k particles with a coefficient equal to unity appears in the expansion (in the Clebsh-Gordan series) of the inner product $[\lambda] \times [\tilde{\lambda}]$ (where $[\lambda]$ and $[\tilde{\lambda}]$ are the conjugated representations) in the irreducible representations of the permutation group. It can be also shown that the one-dimensional symmetric representation of the permutation group appears in the expansion of $[\lambda] \times [\lambda]$.

Both the inner (Kronecker) and external *product of the permutation* group representations do not depend on the order of representations being multiplied. From this fact and the obvious equalities

$$([\lambda] \times [\mu]) \times [\nu] = [\lambda] \times ([\mu] \times [\nu]) = ([\lambda] \times [\nu]) \times [\mu]$$

and using that the mutually conjugated representations $[\lambda]$ and $[\tilde{\lambda}]$ are connected (by definition)

$$[\lambda] \times [1^k] = [\tilde{\lambda}], \tag{7.95}$$

we obtain the relation

$$[\lambda] \times [\tilde{\mu}] = [\tilde{\lambda}] \times [\mu] \tag{7.96}$$

at $[\nu] = [1^k]$. Substituting $[\tilde{\lambda}] \leftrightarrow [\lambda]$ we obtain one more useful relation

$$[\tilde{\lambda}] \times [\tilde{\mu}] = [\lambda] \times [\mu]. \tag{7.97}$$

In particular, for the three-quarks systems ($k = 3$), since $[\tilde{3}] = [1^3]$, $[\tilde{1^3}] = [3]$, $[\tilde{21}] = [21]$, we get from (7.95) – (7.97)

$$\begin{aligned} [3] \times [3] &= \left[1^3\right] \times \left[1^3\right] = \left[\tilde{1}^3\right] = [3], \\ [3] \times [21] &= \left[1^3\right] \times [21] = \left[\tilde{21}\right] = [21], \\ [3] \times \left[1^3\right] &= \left[\tilde{3}\right] = \left[1^3\right]. \end{aligned} \tag{7.98}$$

Returning to functions (7.93) and (7.94), we can see that by use of (7.98) it is possible to introduce the generalizing Young tableaux (in some cases uniquely):

$$\begin{aligned} \left[1^3\right]_C \times [21]_S &= [21]_{CS}, \ \left[1^3\right]_C \times [3]_S = \left[1^3\right]_{CS}, \\ [3]_X \times [21]_{CS} &= [21]_{XCS}, \ [3]_X \times \left[1^3\right]_{CS} = \left[1^3\right]_{XCS} \end{aligned} \tag{7.99}$$

and, finally,

$$[21]_{XCS} \times [21]_T = [3]_{XCST} \bigoplus [21]_{XCST} \bigoplus [1^3]_{XCST}, \ [1^3]_{XCS} \times [3]_T = [1^3]_{XCST}. \tag{7.100}$$

This proves that it is possible to construct the antisymmetric functions (7.93) and (7.94) on the basis of the partial Young tableaux for the spatial, color, spin, and isospin states. From (7.99) is also seen that nucleon wave function (7.93) can be uniquely characterized by the Young tableaux $[21]_{CS}$ and $[21]_{XCS}$, as the isobar wave function (7.94) by the Young tableaux $[1^3]_{CS}$ and $[1^3]_{XCS}$.

The barion and the barion system wave functions can be written in the form of sums of products of the one-quark wave functions with corresponding coefficients like the many-nucleon wave functions in the nuclear shell model can be expressed in terms of the one-nucleon wave functions by use of the vector addition and relationship (parentage) coefficients.

The wave functions of antibarions treated as systems of three antiquarks have analogous structure and symmetry properties as the barion wave functions.

The quark-quark and quark-antiquark interaction potentials. The discovery in 1974 of the comparatively long-lived (with a small width $\Gamma \equiv \frac{1}{\tau} \approx 70$ KeV) heavy *vector resonance* J/ψ with a mass about 3.1 GeV in the collision experiments (electron-positrons and proton-beryllium nucleus collisions) was an important event in the elementary particle physics. This meson consisted of a heavy charmed quark c and related antiquark $\bar{c}$: $J/\psi = c\bar{c}$. Thus, the J/ψ meson is a genuine neutral particle with the hidden charm. Later, new mesons of the same composition $c\bar{c}$ but with larger mass have been discovered. They can be interpreted as the excited states (energy spectrum) of the J/ψ meson. Such a neutral system containing two subparticles is analogous to the positronium e^-e^+. It is naturally, that the system $c\bar{c}$ is known as *charmonium.*

The study of charmonium is comparatively easy because it is a system of only two bodies with large masses. The spatial wave function $\varphi(\mathbf{r})$ of the relative motion of the c-quark and $\bar{c}$-antiquark as well as the charmonium spectrum can be found by solving the nonrelativistic Schrödinger equation (for example, (7.73)). The quark-antiquark potential $V_{c\bar{c}}(\mathbf{r})$ must be chosen to correctly obtain the observed spectrum of the charmonium excited states.

In rather rough approximation, specifically in the central field approximation, neglecting the spin and the relativistic corrections, the phenomenological potential of the $c\bar{c}$-interaction $V_{c\bar{c}}(\mathbf{r})$ can be simulated by the formula (r is measured in fm)

$$V_{q\bar{q}}(\mathbf{r}) = -\frac{a}{r} + Ar + V_0, \tag{7.101}$$

where $a = \frac{4}{3}\alpha_c$, A and V_0 are the positive constants. At small distances Eq. (7.101) is Coulomb potential with the additive constant, i.e., the one-gluon exchange potential. It corresponds to the domain of the so-called asymptotic freedom. The term Ar in (7.101) with $A \approx 1\,\mathrm{GeV}\cdot\mathrm{fm}^{-1}$ approximately takes into acount the confinement phenomena. This is the cutoff long-range part of the potential. The quark-quark interaction potential can be also approximated by (7.101) but with $a = \frac{2}{3}\alpha_c$.

At present, the question of behavior of the qq- and $q\bar{q}$-potentials is still open. Therefore, at $r \geq 1$ fm, we can consider a few different versions of the potential long-range part, which is responsible for the confinement and differs from the linear dependence on r in (7.101). For example, it is possible to propose the potential of a harmonic oscillator type, the potential growing with r as r^δ $(1 < \delta < 2)$, and the logarithmically growing potential. The latter gives better agreement between the calculated charmonium energy spectrum and the experimental one in comparison with two other potentials. The following *phenomenological $q\bar{q}$-interaction potential* seems to be more realistic than (7.101):

$$V_{q\bar{q}}(\mathbf{r}) = -\frac{a}{r} + k(lnr) + V_0', \; k \approx 0,7\,\text{GeV}. \tag{7.102}$$

We consider now other candidates for the *qq- and $q\bar{q}$-potentials*, which are more complex in comparison with (7.101) and (7.102), in the colorless three-quark (barion) and quark-antiquark (meson) states. The total model Hamiltonian of the hadron, taking into account the strong and electromagnetic interactions with comparatively simple coordinate dependence between particles (the *Breit-Wigner quasistatic interaction*), can be presented in the form

$$\hat{H} = \sum_j \left(M_j + \frac{\hat{\mathbf{P}}_j^2}{2M_j} \right) + \sum_{j>k} (\alpha Q_j Q_k + \xi \alpha_c) \hat{f}_{jk}(\mathbf{r}) + W_c, \tag{7.103}$$

$$\hat{f}_{jk}(\mathbf{r}) = \frac{1}{r} - \frac{1}{2M_j M_k} \Big[\frac{(\hat{\mathbf{P}}_j \hat{\mathbf{P}}_k)}{r} + \frac{(\mathbf{r}\hat{\mathbf{p}}_j)(\mathbf{r}\hat{\mathbf{p}}_k)}{r^3} \Big]$$

$$- \frac{8\pi}{3M_j M_k} (\mathbf{s}_j \mathbf{s}_k) \delta(\mathbf{r}) - \frac{1}{2M_j M_k}$$

$$\times \Big[\frac{6(\mathbf{s}_j \mathbf{r})(\mathbf{s}_k \mathbf{r})}{r^2} - 2(\mathbf{s}_j \mathbf{s}_k) + (\hat{\mathbf{l}}_j \mathbf{s}_k) - (\hat{\mathbf{l}}_k \mathbf{s}_j) \Big]$$

$$- \frac{1}{2r^3} \Big[\frac{(\hat{\mathbf{l}}_j \mathbf{s}_j)}{M_j^2} - \frac{(\hat{\mathbf{l}}_k \mathbf{s}_k)}{M_k^2} \Big] - \frac{\pi}{2} \Big(\frac{1}{M_j^2} + \frac{1}{M_k^2} \Big) \delta(\mathbf{r}) + \ldots, \tag{7.104}$$

where $\mathbf{r} \equiv \mathbf{r}_{jk} = \mathbf{r}_j - \mathbf{r}_k$, Q_k is an electric charge, M_k is the block mass, $\mathbf{s}_k = \frac{1}{2}\boldsymbol{\sigma}_k$ is a spin of the k-th quark, α is the fine structure constant, α_c is the strong interaction effective constant, $\xi = -\frac{4}{3}$ for the $\bar{q}q$ system and $\xi = -\frac{2}{3}$ for qqq system, W_c is the cutoff long-range potential part responsible for the confinement. The first term in the right-hand side of (7.104) is the static Coloumb potential both for the electromagnetic and strong interactions. In the latter case, this dependence corresponds to one-gluon exchange between subparticles at small distances. Further, there are different corrections including the relativistic ones, in particular, the spin-spin contact ($\sim \delta(\mathbf{r})$), tensor and spin-orbital interactions.

The model Hamiltonian (7.103) – (7.104) with rather simple dependence of spatial coordinate $\mathbf{r}$ was proposed, first of all, to describe qualitatively the charmonium $c\bar{c}$ energy spectrum. There are only two terms in sums (7.103) and $\xi = -\frac{4}{3}$ in this case. The model Hamiltonian can be also used for barion systems, where $\xi = -\frac{2}{3}$, although we deal with the three-body (quarks) problem. A quantitative description of the $c\bar{c}$ system comes across definite problems, and that is why some complicated modifications of potential (7.103) – (7.104), which gave the more satisfactory charmonium spectrum, were suggested. The best evaluation of the charmonium effective coupling constant gives $\alpha_c \approx 0.4$. The model interaction potential can pretend only on the approximate description of the systems $q\bar{q}$ and qqq, because of the fact, that, at the moment, there is some arbitrariness when choosing the potentials of $q\bar{q}$- and qq-interactions and, particularly, the long-range part of potentials.

In conclusion of this section, we present one more frequently used potential $\hat{V}_{qq}(jk)$ of interaction between light quarks j and k in nucleons and Δ-isobars (Δ-resonances):

$$\hat{V}_{qq}(jk) = \sum_{a=1}^{8} \lambda_a(j)\lambda_a(k) f_{jk}(\mathbf{r}_{jk}, \mathbf{s}_j, \mathbf{s}_k), \tag{7.105}$$

$$\begin{aligned} f_{jk}(\mathbf{r}_{jk}, \mathbf{s}_j, \mathbf{s}_k), = {} & f_{jk}^{c}(\mathbf{r}_{jk}) + \frac{1}{4M_q}\Big(1 + \frac{8}{3}\mathbf{s}_j\mathbf{s}_k\Big)\Delta f_{jk}^{ss}(\mathbf{r}_{jk}) \\ & + \frac{3}{2M_q^2}(\mathbf{l}_j - \mathbf{l}_k)(\mathbf{s}_j + \mathbf{s}_k)\frac{1}{r_{jk}}\frac{df^{\text{sor}}(r_{jk})}{dr_{jk}} + \frac{1}{2M_q^2} \\ & \times\Big[6(\mathbf{r}_{jk}\mathbf{s}_j)(\mathbf{r}_{jk}\mathbf{s}_k) - 2\mathbf{s}_j\mathbf{s}_k\Big]\Big[\frac{d^2 f^t(r_{jk})}{dr_{jk}^2} - \frac{1}{r_{jk}}\frac{df^t(r_{jk})}{dr_{jk}}\Big], \end{aligned}$$

where $\lambda_a(j)$ are the Gell-Mann matrices of j-th quark. A central part $f^c(r)$ takes into account the one-gluon exchange at small distances and the confinement phenomena at large distances ($r \sim r_0$). It is usually chosen in the form

$$f^c(r) = -\frac{2}{3}\frac{\alpha_c}{r}[1 - (\frac{r}{r_0})^{n+1}]\vartheta(r_0 - r),\ n = 1, 2, \tag{7.106}$$

where r_0 is the confinement radius, $\vartheta(r_0 - r) = \begin{cases} 1, & r < r_0 \\ 0, & r > r_0 \end{cases}$ is the Heavyside function, α_c is the strong interaction coupling constant. The spin-spin $f^{ss}(r)$, spin-orbit $f^{\text{sor}}(r)$, and tensor $f^t(r)$ interaction parts are customarily chosen in two versions: 1) it is assumed that they emerge due to the gluon exchange and can be written in the form

$$f^{ss}(r) = f^{\rm sor}(r) = f^{t}(r) = -\frac{2}{3}\frac{\alpha_c}{r}, \qquad (7.107)$$

2) it is assumed that these parts of the interaction give a contribution to the confinement

$$f^{ss}(r) = f^{\rm sor}(r) = f^{t}(r) = f^{c}(r). \qquad (7.108)$$

The parameters in (7.106) – (7.108) are chosen in such a way to get the observed spectra of nucleons (and Δ-isobars) and their general characteristics.

In the next section, we consider the quark aspects of the nuclear forces between the nucleons (it is the six-quark problem) and return to the interaction between lightest quarks which compose nucleons and Δ-isobars.

7.3. Quark in a few-nucleon systems

Deuteron as a six-quark system. A complex atomic nucleus can be treated as a system of interacting colorless three-quark clusters of nucleons. In bound, as well as in unbound nuclear systems, i.e., in nuclei and under collisions of different particles and nucleus-nucleus collisions in various nuclear processes the nontrivial *many-quark systems* can appear. In the first turn these are $6q$-, $9q$-, $12q$-systems, which are overlappings of two, three and more nucleons. It is naturally to expect that in the ground and weakly excited nuclear states, the overlappings result in small corrections to the ground (nonoverlapping) state (channel) $\Psi(AN)$ in the total nuclear function

$$\Psi(AN) = \alpha\Psi(AN) + \beta\Psi(3Aq), \qquad (7.109)$$

i.e., $|\beta|^2 \ll 1$, $|\alpha|^2 \approx 1$ in this case. If the collision energy is large enough and exceeds masses of the colliding particles and nuclei, then the *nucleon overlapping* , determined by the second term $\beta\Psi(3Aq)$ in (7.109), can be dominant. In this case $|\beta|^2 \approx 1$ and $|\alpha|^2 \ll 1$. At the same time, we assume that the wave functions Ψ, $\Psi(AN)$ and $\Psi(3Aq)$ are individually normalized. The functions $\Psi(AN)$ and $\Psi(3Aq)$ are not orthogonal in general. But their mutual overlapping, as a rule, is small enough and can be often neglected. Thus, in this case, we can assume that $|\alpha|^2 + |\beta|^2 \approx 1$. For example, the contribution of these *overlappings in a deuteron* is only of the order of 0.1%.

We consider, as an example, the simplest two-nucleon nucleus, known as a deuteron. The deuteron ground state presents, first of all, a weakly bounded state of two three-quark clusters: a proton p and a neutron n with a weight $|\alpha|^2$ close to unity, and a small addition of the *six-quark state* with a small weight $|\beta|^2 \ll 1$. Thus, the corresponding deuteron wave function Ψ_d, as in a general case (7.109), takes the form

$$|d\rangle \equiv \Psi_d = \alpha\Psi(pn) + \beta\Psi(6q). \qquad (7.110)$$

Two nucleons, as the colorless three-quark clusters, are weakly bound in a deuteron and can be separated by large distances. This is due to the fact that two quarks from different clusters interact very weakly, since their fields are screened by the fields of other quarks. But in the *six-quark state* described by the wave function $\Psi(6q)$ there is no separation of colorless clusters. The whole system of six quarks can be treated as colorless, where all six quarks strongly interact. Hence, the system has a comparatively small volume. The *six-quark wave function* takes a large magnitude in the deuteron center and decreases fast even inside the deuteron, being practically zero on the deuteron border unlike the two-cluster wave function $\Psi(pn)$ describing the relative motion of a proton and neutron. Because of this, the first term in (7.110), namely $\alpha\Psi(pn)$, is called the *external part of the total wave function of the deuteron* Ψ_d, and the second term $\beta\Psi(6q)$ in (7.110) is called the *inner part of the deuteron wave function* Ψ_d.

The deuteron spatial wave function, on taking account of the *quark degrees of freedom*, can be constructed within the framework of the simplest model, where each quark moves in the field of an isotropic harmonic oscillator with a frequency ω. There is a complete analogy with the nucleon motion in nuclei in the oscillator potential (see sections 4.2, 4.6, and problem 4.8). A similar nonrelativistic oscillator model of a quark motion in nucleons and complex nuclei takes into account qualitatively both the *quark asymptotic freedom* and the *confinement phenomena.* These circumstances dictate the terminology of the related phenomenological models describing the bound states of the color objects introducing the so-called *bag models for quarks.* Quarks interact weakly inside the bag, but they cannot escape from the bag with sharp borders which envelopes the whole colorless system consisting of color objects.

Let us denote radius vectors of six deuteron quarks as $\mathbf{r}_j$ (j=1,2,...,6) and, assuming the block masses of u- and d-quarks being the same and equal to M_q, introduce the following vector of the Jacobi coordinates, which contain the vector $\mathbf{r}$ uniting both nucleons in the deuteron:

$$\begin{aligned}
\boldsymbol{\xi}_1 &= \frac{1}{6}(\mathbf{r}_2 + \mathbf{r}_3 - 2\mathbf{r}_1), \quad \boldsymbol{\xi}_2 = \frac{1}{2\sqrt{3}}(\mathbf{r}_3 - \mathbf{r}_2), \\
\boldsymbol{\eta}_1 &= \frac{1}{6}(\mathbf{r}_5 + \mathbf{r}_6 - 2\mathbf{r}_4), \quad \boldsymbol{\eta}_2 = \frac{1}{2\sqrt{3}}(\mathbf{r}_6 - \mathbf{r}_5),
\end{aligned}$$

$$\mathbf{r} \equiv 2\boldsymbol{\rho} = \frac{1}{3}\Big(\sum_{j=1}^{3}\mathbf{r}_j - \sum_{j=4}^{6}\mathbf{r}_j\Big), \tag{7.111}$$

where the first three quarks are related to one nucleon (for example, to the neutron), and the rest three quarks – to the other nucleon (proton). Here we introduce the radius-vector of the deuteron center-mass $\mathbf{R} = \frac{1}{6}\sum_{j=1}^{6}\mathbf{r}_j$. The spatial wave functions of the inner states of nucleons φ_N, where $N = n, p$, and of the relative inner motion of all six quarks in the deuteron φ_{6q} and nucleon relative motion φ_0, normalized by unity in the oscillator quark model, take the form

$$\begin{aligned}
\varphi_n(\boldsymbol{\xi}_1, \boldsymbol{\xi}_2) &= \left(\frac{\Omega}{\pi}\right)^{3/2} \mathrm{e}^{-\frac{\Omega}{2}(\boldsymbol{\xi}_1^2+\boldsymbol{\xi}_2^2)},\\
\varphi_p(\boldsymbol{\eta}_1, \boldsymbol{\eta}_2) &= (\frac{\Omega}{\pi})^{3/2}\mathrm{e}^{-\frac{\Omega}{2}(\boldsymbol{\eta}_1^2+\boldsymbol{\eta}_2^2)},\\
\varphi_{6q}(\boldsymbol{\rho}, \boldsymbol{\xi}_1, \boldsymbol{\xi}_2, \boldsymbol{\eta}_1, \boldsymbol{\eta}_2) &= (\frac{2\Omega}{\pi})^{15/4}\exp\{-\Omega(\boldsymbol{\rho}^2 + \boldsymbol{\xi}_1^2 + \boldsymbol{\xi}_2^2 + \boldsymbol{\eta}_1^2 + \boldsymbol{\eta}_2^2)\},\\
\varphi_0(\mathbf{r}) &= (\frac{\Omega}{2\pi})^{3/4}\mathrm{e}^{-\frac{\Omega \mathbf{r}^2}{4}},
\end{aligned} \tag{7.112}$$

where $\Omega = 3M_q\omega/\hbar$. A numerical value of the structure parameter Ω is approximately equal to $1\,\mathrm{GeV}^2/c^2\hbar^2$. From (7.112), in particular, it follows that the $6q$-system has mean squared radius in $\sqrt{2}$ times smaller than $3q$-system. According to (7.11) and (7.112) the spatial wave functions φ_n and φ_p are symmetric with respect to quark permutations inside each nucleon. The spatial wave function φ_{6q} is symmetric with respect to permutations of all six quarks, i.e., it is described by the Young tableaux $[6]_X$ (recall that, accordingly the rules of expansion of the external product of two irreducible representations of the permutation group, the following relation can be written: $[3]_X \bigotimes [3]_X = [6]_X \bigoplus [42]_X \bigoplus [51]_X \bigoplus [3^2]_X$), and the function $\varphi_0(P\mathbf{r})$ is symmetric with respect to the quark permutations only within the individual clusters-nucleons. It follows from (7.112) that the state of six quarks corresponds to the spatial s^6 configuration.

The external part of $\Psi(pn)$ of the total deuteron wave function (7.110) in the nonrelativistic approximation can be written in the form of the antisymmetric product of functions of different variables

$$\Psi(pn) = \mathrm{e}^{\frac{i}{\hbar}\mathbf{PR}}\hat{a}\Big\{\varphi_n(\boldsymbol{\xi}_1, \boldsymbol{\xi}_2)\psi_c^n\varphi_p(\boldsymbol{\eta}_1, \boldsymbol{\eta}_2)\psi_c^p\varphi_0(\mathbf{r})\chi_{1M_s}Z_{00}\Big\}, \tag{7.113}$$

where $\mathbf{P}$ is the total deuteron momentum, ψ_c^n and ψ_c^p are the antisymmetric singlet color wave functions of a neutron and proton, χ_{1M_s} is the symmetric (with respect to permutations of two nucleons) spin wave function of the deuteron (the total spin is $S = 1$); Z_{00} is the antisymmetric (with respect to permutations of nucleons) isospin deuteron wave function (according to Table 1.2 the deuteron state ${}^3S_1 + {}^3D_1$ corresponds to the isospin $T = 1$). The wave function $\varphi_0(\mathbf{r})$ in (7.112) does not take into account the contribution of the D-wave in the deuteron. To include this contribution we have to change a product $\varphi_0(\mathbf{r})\chi_{1M_s}$ in (7.113) with function (1.111). Appearing in (7.113) the antisymmetrization operator $\hat{a}$ is equal to $\hat{a} = \frac{1}{\sqrt{10}}(1 - \sum_{1\leq i\leq 3}\sum_{4\leq j\leq 6}\hat{P}_{ij})$, where $\hat{P}_{ij}$ is the exchange operator of quarks i and j which belong to different nucleons.

We construct now the antisymmetric wave function of the inner six-quark state in the deuteron $\Psi(6q)$, which appears in (7.110). It can be presented in the factorized form

$$\Psi(6q) = \mathrm{e}^{\frac{i}{\hbar}\mathbf{PR}}\varphi_{6q}(\boldsymbol{\rho}, \boldsymbol{\xi}_1, \boldsymbol{\xi}_2, \boldsymbol{\eta}_1, \boldsymbol{\eta}_2)\Psi_{CST}, \tag{7.114}$$

where Ψ_{CST} is the totally antisymmetric spin-isospin part of the deuteron inner wave function. The function Ψ_{CST} is a superposition of products of the six-quark color wave function $\psi_c([\lambda]_c, r_c)$ (where $[\lambda]_c$ is, accordingly a quark number, the six-box color Young tableaux of the deuteron, r_c is one of the possible corresponding Yamagouchi symbols) and the spin-isospin wave function of the system of six quarks $\psi_{s=1,T=0}([\lambda]_c, r_c)$ with the Young tableaux $[\tilde{\lambda}]_c$ and the Yamagouchi symbol $\tilde{r}_c$, which is conjugated to $[\lambda]_c$ and r_c, respectively, so that the function Ψ_{CST} be antisymmetric

$$\Psi_{CST} = \frac{1}{\sqrt{n_\lambda}}\sum_r \Lambda_r^{[\lambda]_c}\psi_c([\lambda]_c r_c)\psi_{s=1,T=0}([\tilde{\lambda}]_c\tilde{r}_c). \tag{7.115}$$

Here n_λ is the dimension of the irreducible representation $D^{[\lambda]_c} \equiv [\lambda]_c$ of the permutation group of six quarks, and $\Lambda_r^{\lambda]_c}$ is a sign factor, which is included in one of the functions ψ_c or $\psi_{s=1,T=0}$. (A quantity $\frac{1}{\sqrt{n_\lambda}}\Lambda_r^{[\lambda]_c}$ presents the simplest *Clebsch-Gordan coefficient of the permutation group* , in the given particular case $[\Lambda_r^{[\lambda]_c}]^2 = 1$.)

The *six-quark color wave function* $\psi_c([\lambda]_c r)$ is the invariant (singlet) with respect to operations from the color group SU_3^c. This is automatically true for the *Young color tableaux*

$$[\lambda]_c = [222]_c \equiv [2^3]_c = \tag{7.116}$$
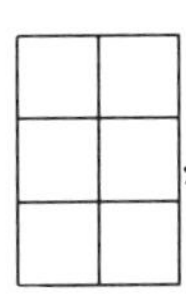

consisting of two columns of a maximum length in 3 boxes, accordingly a number of the possible color charges, in each column. Note that for the six-quark system Young tableaux (7.116), characterizing the colorless (singlet) state, is unique. The corresponding spin-isospin wave function $\psi_{S=1,T=0}([\tilde{\lambda}]_c\tilde{r}_c)$ is characterized by the conjugated Young tableaux

$$[\lambda]_{ST} = [\tilde{\lambda}]_c = [33]_{ST} \equiv [3^2]_{ST} =$$ 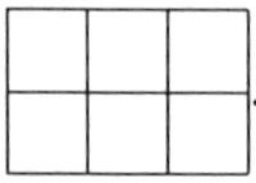 . (7.117)

The dimension of the representation $D^{[2^3]}$, as well as the representation $D^{[3^2]}$, is equal to $n_\lambda = 5$, and five Yomanouchi symbols r_c related to the Young tableaux $[2^3]_c$ are the following

$$(332211), (323121), (323211), (332121), (321321). \qquad (7.118)$$

The quantity $\Lambda_{r_c}^{[\lambda]_c}$ is equal to +1 or -1 depending on the manner in which the Yomanouchi symbol with numbers arranged in an increasing order from right to left (in this particular case from r_c^0=(332211)) is obtained from the symbol r_c^0 by even or odd number of permutations. Thus, $\Lambda_{r_c}^{[\lambda]_c} = +1$ for two first Yomanouchi symbols in (7.118) and $\Lambda_{r_c}^{[\lambda]_c} = -1$ for the rest three symbols.

The *six-quark spin-isospin wave function* $\psi_{S=1,T=0}([3^2]_{S,T}r)$, in its turn, can be presented in the form of a superposition of products of spin $\chi_{S=2}$ and isospin $Z_{T=0}$ wave functions

$$\psi_{S=1,T=0}([3^2]_{S,T}r) = \sum_{r_s,r_T} C^{[3^2]_{ST}r}_{[42]_s r_s [3^2]_T r_T} \chi_{S=1}([42]_s r_s) Z_{T=0}([3^2]_T r_T), \qquad (7.119)$$

where the superposition coefficients are the *Clebsh-Gordan coefficients of the permutation group*. They can be found from the recurrent formulas. In view of the fact that spin and isospin of u- and d-quarks, as well as nucleons, are equal to $\frac{1}{2}$, the Young tableaux for the spin and isospin state of six-quarks are two-line tableaux

$$[42]_S =$$ 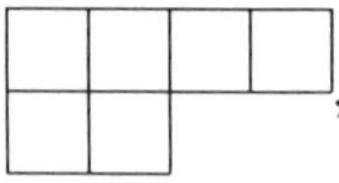 , $$[3^2]_T =$$ 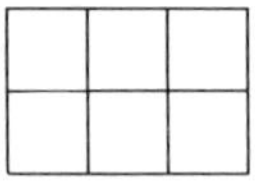 (7.120)

and provide the necessary values of spin $S = 1$ and isospin $T = 0$ of a deuteron, which are equal to a half difference between the box numbers of two lines of the related Young tableaux (see (4.116)).

Thus, the inner deuteron wave function in the s^6-configuration for the spatial state of six quarks, as well as the wave functions of three-quark barion systems (7.93) and (7.94), can be briefly written with notation of the symmetry type in terms of the nuclear shell model as follows

$$\Psi(6q) = |s^6[6]_X[2^3]_C[42]_S[3^2]_T||[1^6]_{XCST}\rangle. \qquad (7.121)$$

As we have seen before, this wave function is characterized by the Young tableaux $[3^2]_{ST}$ and, certainly, by the Young tableaux $[1^6]_{CST} = [\tilde{6}]_X$ that follows directly from the Pauli principle: $[6]_X \times [1^6]_{CST} = [\tilde{6}]_{XCST} = [1^6]_{XCST}$ (see (7.95)).

The considered six-quark model contains two empirical parameters: the structure parameter Ω, associated with the oscillator parameter (frequency) ω, and a probability of realization of the six-quark state in the deuteron $|\beta|^2$. The parameter Ω, determined from the comparison of the calculated electromagnetic nucleon form-factors with the experimental ones, turns out to be equal to $\Omega \approx 25\,\text{fm}^{-2}$. To determine $|\beta|^2$ we compared the expression of the deuteron magnetic moment obtained by means of its total wave function (7.110) in the *oscillator quark model* with expression (1.118) obtained neglecting the six-quark state in a deuteron, and with the experimental deuteron magnetic moment. From a difference between the experimental value and (1.118) we found the *six-quark state contribution* in a deuteron. It happened to be that a value $|\beta|^2$ is as much as 2%, i.e., it is small enough within the framework of the oscillator quark model. Note that the calculated value of $|\beta|^2$ can strongly depend on the used deuteron spatial wave functions.

In the framework of the oscillator quark model, we have calculated the cross-section of the elastic electron-nucleus scattering. Since $|\beta|^2 \ll 1$, we can take into account only external (nucleon) part of the deuteron wave function at small transferred momenta $q \lesssim 3\,\text{fm}^{-1}$. In this approximation, we can satisfactorily describe the relevant experiments. In the latter experiments at large transferred momenta $q \gtrsim 5\,\text{fm}^{-1}$, an electron can penetrate into very small space volume of the order of $10^{-2}\,\text{fm}^3$, where, according to the current state of research, the quark structure of the nuclear matter has to manifest itself. In particular, at such a large q, the deuteron six-quark state contribution becomes important in different processes. A consideration of this contribution enables one to describe satisfactorily experiments on the electron-deuteron scattering through large angles at high energies.

We have to note that the behaviour of the scattering cross-section at large q cannot be explained by taking into account only relativistic effects and neglecting the many-quark states contribution. All the above-reported concerning the elastic electron-deuteron scattering is illustrated in Fig. 7.7 and 7.8, where the calculated dependencies of the *deuteron form-factors* $A(q^2)$ *and* $B(q^2)$ presented at large transferred momenta q on taking account of the six-quark state contribution in a deuteron (solid

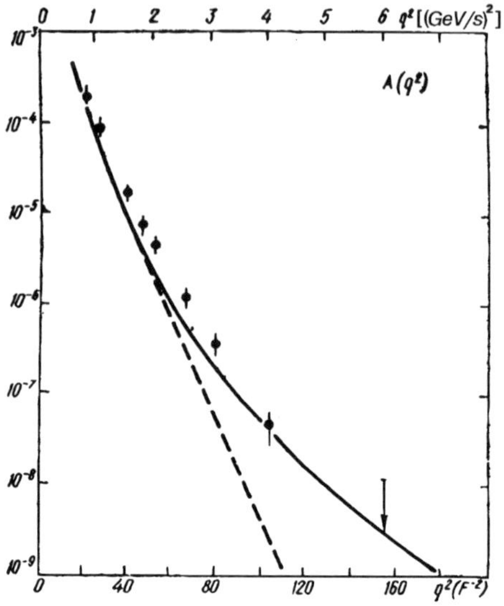

Figure 7.7. Dependencies of the deuteron form-factor $A(q^2)$ on a squared transferred electron momenta q^2 calculated on taking account of the six-quark state contribution in a deuteron (solid line) and neglecting this contribution (dotted line) and the experimental values of $A(q^2)$.

lines) and neglecting this contribution and the experimental values of the form-factors, as well (Kobushkin, 1978). We recall that the form-factors $A(q^2)$ and $B(q^2)$ associated with the deuteron structure appear in the differential cross-section of the elastic electron-deuteron scattering (in unit solid angle)

$$\frac{d\sigma}{d\Omega} = \frac{\sigma_M(\theta)}{1 + \frac{2k}{M_d}\sin^2\frac{\theta}{2}}\Big\{A(q^2) + B(q^2)\mathrm{tg}^2\frac{\theta}{2}\Big\},$$

where θ and k are the electron scattering angle and momentum, respectively, M_d is the deuteron mass ($\hbar = c = 1$), $\sigma_M(\theta)$ is the Mott cross-section of the scattering by a point charge.

In regard to the quantum state of many quark systems in nuclei and, in particular, the six-quark system in a deuteron, we may note that many questions are still unanswered , in particular, related to the quark-quark interactions. Up to the present, we assume that all six quarks are in the s-state in the deuteron inner state. It is highly probable but not necessary, as it can be seen further. The point is that we can assume such quark-quark interactions so the simplest quark configuration s^6 may correspond to the higher energy than the spatial configuration s^4p^2. This means that two quarks out of six can have an angular momentum $l = 1$. In this case, the spatial and color-spin permutation symmetries of the corresponding wave

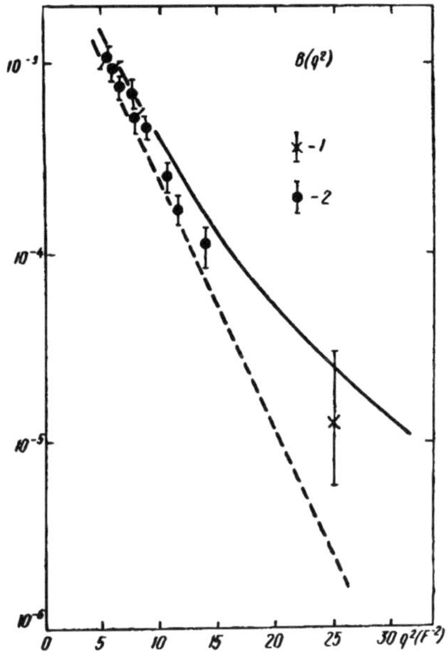

Figure 7.8. The same dependencies and experimental data as in Fig. 7.7 but for the deuteron form-factor $B(q^2)$.

functions are described, as we will see later, by the same Young tableaux $[42]_X$ and $[42]_{CS}$, but not $[6]_X$ and $[2^3]_{CS}$, as for the s^6-configuration. However, the spatial configuration s^6 is used more frequently.

The interpolation model of many-quark systems. Analyzing the inner six-quark part of the deuteron wave-function $\Psi(6q)$ in (7.110), we have found a value of the parameter β, which characterizes this contribution, from experiment. However, the inner (many-quark) $\beta\Psi(3Aq)$ and external (nucleon) $\alpha\Psi(AN)$ parts of the total wave function Φ of a complex nucleus (7.109) must be related to each other. This relation is automatically established, if we start with a given q–q interactions when obtaining the nuclear total wave function (7.109) from the general equations of motion. Here we report the approximate, but rather efficient, method of obtaining the total wave function, which interrelates the inner and external parts of the function in the nonrelativistic approximation. This method has been successfully used earlier in the nonrelativistic nuclear physics for a few nucleon systems without regard for the quark structure; it is usually called the interpolation model. This model is enough general and flexible and allows one to study both bound and unbound nucleon systems, i.e., to study the structure and dynamics of complex nuclei, as well as different nuclear processes.

Usually the *interpolation model* [5] is illustrated on an example of two interacting colorless three-quark clusters of nucleons or a system of six quarks. This can be a deuteron, but the six-quark systems may be formed inside of the more complex many-nucleon nuclei, as well. The model can be easily generalized on the case of nine, twelve, and so on quarks. We consider now the colorless two-cluster system of $\nu = \nu_1 + \nu_2$ quarks with ν_1 and ν_2 quarks in the first and second clusters respectively. We are interesting, first of all, in the spatial part of the wave functions omitting color, spin, and isospin coordinates and summing over them.

We introduce now a $3\nu - 3$-dimensional space according to a number of the relative spatial coordinate of ν-quarks. Let us pass to spherical (hyperspherical) coordinates in the $(3\nu - 3)$-dimensional space, introducing $3\nu - 4$ angular variables $\Omega_{3\nu-4}$ and the so-called global or collective coordinate ρ with dimension of length, accordingly (see Chapters 2 and 4),

$$\rho^2 = \frac{1}{\nu}\sum_{k>j}^{\nu}(\mathbf{r}_k - \mathbf{r}_j)^2 = \sum_{k=1}^{\nu}(\mathbf{r}_k - \mathbf{R})^2 = \nu\mathbf{r}_q^2,\ \mathbf{R} = \frac{1}{\nu}\sum_{k=1}^{\nu}\mathbf{r}_k, \tag{7.122}$$

where $\mathbf{r}_k$ is a radius-vector of the k-th quarks, $\sqrt{\mathbf{r}_q^2}$ is the mean distance between quarks in the system. It will be convenient to introduce the global coordinates ρ_1 and ρ_2 for each of two clusters:

$$\rho_1^2 = \frac{1}{2\nu_1}\sum_{k,j=1}^{\nu_1}(\mathbf{r}_k - \mathbf{r}_j)^2,\ \rho_2^2 = \frac{1}{2\nu_2}\sum_{k,j=\nu_1+1}^{\nu}(\mathbf{r}_k - \mathbf{r}_j)^2, \tag{7.123}$$

and the radius-vector connecting the cluster centers of mass

$$\mathbf{r} = \frac{1}{\nu_1}\sum_{k=1}^{\nu_1}\mathbf{r}_k - \frac{1}{\nu_2}\sum_{k=\nu_1+1}^{\nu}\mathbf{r}_k \equiv (\frac{\nu}{\nu_1\nu_2})^{1/2}\mathbf{z}_{12}. \tag{7.124}$$

There is a simple constraint

$$\rho^2 = \rho_1^2 + \rho_2^2 + z_{12}^2, \tag{7.125}$$

which allows us to deduce that $z_{12} \approx \rho$ at large distances between clusters.

The total wave function of colorless quark (nuclear) system dependent on variables $\boldsymbol{\rho} = (\rho, \Omega_{3\nu-4})$ we present as a sum of the external $\Psi_{\rm ext}$ and inner $\Psi_{\rm in}$ parts without specifying the weight factors α and β (as in (7.109)) in the form:

[5]The interpolation model of many-quark systems is described in the review by Burov, Luk'yanov and Titov (1984).

$$\Psi = \Psi_{\text{ext}} + \Psi_{\text{in}}. \tag{7.126}$$

The total Hamiltonian of a nuclear system with the quark structure is $\hat{H} = \hat{T} + \hat{V}$, where $\hat{T}$ is the operator of kinetic energy of quarks and $\hat{V} = \sum_{k<j}^{\nu} \hat{V}_{qq}(kj)$ is the operator of the quark-quark interaction in the system. Let us present $\hat{H}$ in the form $\hat{H} = \hat{H}_0 + \hat{V} - \hat{V}_0$, where the "truncated" part of the $\hat{H}_0 = \hat{T} + \hat{V}_0$ being the part of the total Hamiltonian acts in the inner domain ($\rho < \rho_0$) confined by the ν-quark bag. We assume that $\hat{V}$ coincides with the "truncated" interaction $\hat{V}_0$ in the inner domain.

The eigenfunctions of the "truncated" Hamiltonian $\psi_\lambda(\boldsymbol{\rho})$, satisfying the Schrödinger equation $\hat{H}\psi_\lambda = E_\lambda\psi_\lambda$, are nonzero only within the quark bag. Sometimes, they are called the bag functions. Apparently, it would be worthwhile to expand the desirable inner part Ψ_{in} of the total wave function (7.126) in a series of the functions $\psi_\lambda(\boldsymbol{\rho})$:

$$\Psi_{\text{in}} = \sum_\lambda C_\lambda \psi_\lambda. \tag{7.127}$$

Within the framework of the interpolation model, the functions $\psi_\lambda(\boldsymbol{\rho})$ are assumed to be known. Thus, we have to find the coefficients C_λ in (7.127). The inner domain restricted by the quark bag of radius r_0 gives a dominant contribution in different collision processes at sufficiently large transferred momenta $q > q_0 \approx \hbar/r_0$. In the case of one nucleon, the domain of a radius r_0 is called the *nucleon quark kern* and usually $r_0 \approx 0,4\,\text{fm}$.

The external part Ψ_{ext} of the wave function (7.126) has to describe, first of all, the motion of colorless clusters (mainly nucleons) separated by sufficiently large ($\rho > \rho_0$) distances (external domain) and can be presented in the form

$$\begin{aligned} \Psi_{\text{ext}} &= \Phi(\rho) W(\boldsymbol{\rho}), \\ W(\boldsymbol{\rho}) &= \Big(\frac{\nu_1!\nu_2!}{\nu!}\Big)^{1/2} \hat{a}_{12}\Big(z_{12}^L Y_{LM}(\hat{\mathbf{z}}_{12})\varphi_1(\boldsymbol{\rho}_1)\varphi(\boldsymbol{\rho}_2)\Big). \end{aligned} \tag{7.128}$$

Here L is the relative angular momentum of two clusters, $\hat{\mathbf{z}}_{12} = \mathbf{z}_{12}/z_{12}$, $\boldsymbol{\rho}_1 = (\rho_1, \Omega_{3\nu_1-4})$, $\boldsymbol{\rho}_2 = (\rho_2, \Omega_{3\nu_2-4})$; the cluster wave functions $\varphi_1(\boldsymbol{\rho}_1)$ and $\varphi(\boldsymbol{\rho}_2)$ are antisymmetric with respect to permutations of quark coordinates inside every cluster. Because of this, the operator of quark permutations $\hat{a}_{12}$ contains only permutations transferring quarks from one cluster into another (as well in (7.113)). The unknown function $\Phi(\rho)$, appearing in (7.128), at large distances between clusters ($\rho \gg \rho_1$, $\rho \gg \rho_2$, $r \approx z_{12}$) can be treated as the wave function of the cluster relative motion. The cluster wave functions $\varphi_1(\boldsymbol{\rho}_1)$ and $\varphi(\boldsymbol{\rho}_2)$ (and the function $W(\boldsymbol{\rho})$), unlike the function $\Phi(\rho)$, are given in the interpolation model.

The coefficients C_λ and function $\Phi(\rho)$ can be obtained as solutions of the system of the coupled integro-differential equations. This system can be obtained by multiplying the left-hand side of equation of the total wave function

$$(\hat{H}_0 + \hat{V} - \hat{V}_0 - E)(\Psi_{\text{ext}} + \Psi_{\text{in}}) = 0 \tag{7.129}$$

firstly by $\psi^*_\lambda(\boldsymbol{\rho})$ through integrating over $d\boldsymbol{r} = \rho^{3\nu-4} d\rho d\Omega_{3\nu-4}$, then multiplying (7.129) by $W^*(\boldsymbol{\rho})$ and integrating over the angular variables $d\Omega_{3\nu-4}$. Using the fact that functions (7.127) and (7.128) do not practically overlap, we obtain the following system of equations for determination of C_λ and $\varphi(\rho) = \Phi(\rho)\sqrt{\mu(\rho)}$:

$$-(E_\lambda - E)C_\lambda = \int\limits_0^\infty d\rho \varphi(\rho) F^*_\lambda(\rho), \tag{7.130}$$

$$\Big(-\frac{\hbar^2}{2M_q}\frac{d^2}{d\rho^2} + U - \varepsilon\Big)\varphi(\rho) = -\sum_\lambda C_\lambda F_\lambda(\rho). \tag{7.131}$$

Here the following denotations are introduced:

$$\begin{aligned}
F_\lambda(\rho) &= \frac{\rho^{3\nu-4}}{\sqrt{\mu(\rho)}} \int d\Omega_{3\nu-4} W^*(\boldsymbol{\rho})\{(E_\lambda - E) + (\hat{V} - \hat{V}_0)\}\psi_\lambda(\boldsymbol{\rho}), \\
\mu(\rho) &= \rho^{3\nu-4} \int d\Omega_{3\nu-4} W^*(\boldsymbol{\rho}) W(\boldsymbol{\rho}), \\
U &= U_{NN} + V(\rho),\ V(\rho) = \frac{\hbar^2}{2M_q}\frac{1}{2\sqrt{\mu(\rho)}}\frac{d}{d\rho}\Big(\frac{1}{\sqrt{\mu(\rho)}}\frac{d\mu(\rho)}{d\rho}\Big), \\
U_{NN} &= \int d\Omega_{3\nu-4} W^*(\boldsymbol{\rho})(\hat{H} - M_1 - M_2)W(\boldsymbol{\rho}) \Big/ \int d\Omega_{3\nu-4} W^*(\boldsymbol{\rho}) W(\boldsymbol{\rho}) \\
&= \frac{\rho^{3\nu-4}}{\mu(\rho)} \int d\Omega_{3\nu-4} W^*(\boldsymbol{\rho})(\hat{H} - M_1 - M_2) W(\boldsymbol{\rho}), \\
\varepsilon &= E - M_1 - M_2,
\end{aligned} \tag{7.132}$$

where M_1 and M_2 are the cluster masses (at $\nu = 6$, for the six-quark system, these are the nucleon masses M), ε is the relative energy of the cluster motion.

The system of equations (7.130) – (7.131) for the *interpolation wave function* can be solved by use of the modern computers. In the interpolation model, the potentials $\hat{V}_{qq}$ are assumed to be specified (see, for example, the potential of V_{qq} interaction (7.105)), but they are largely model and obtained on the basis of the general principles of the elementary particle physics, quantum chromodynamics and experimental data. Since the potentials $\hat{V}_{qq}$ are still largely undetermined, then many conclusions based on calculations with these potentials must be treated as qualitative. Moreover, in many cases, they are questionable. For this reason, the contribution of the many-quark states, which is dictated by the inner part of the wave function (7.127), determined directly from experiments is more relevant, as it has been shown on the deuteron example.

The effective cluster (nucleon) interaction potential U in equation (7.131) is represented by two parts: the average potential of the nucleon-nucleon (cluster-cluster) interaction U_{NN}, having the repulsive character at small distances, which is caused, first of all, by the peculiarities of the qq-interaction, and cluster repulsion potential $V(\rho)$, originated from the Pauli principle. At $\rho > \rho_0$, when the cluster wave functions begin to overlap, $V(\rho)$ increases sharply and at a very small ρ becomes equal to the kinetic energy of ν quarks localized in a small domain of space. Thus, the quark-quark interaction is concentrated mainly in U_{NN}, and the quark kinetic energy, associated with the Pauli principle, turns out to be concentrated in $V(\rho)$. At $\rho > \rho_0$, i.e., out of the quark bag, this part of the kinetic energy is completely taken into account in the cluster wave functions $\varphi_1(\boldsymbol{\rho}_1)$ and $\varphi_2(\boldsymbol{\rho}_2)$. The potential of NN-interaction U_{NN} is borrowed mainly from experiments on the nucleon-nucleon and nucleon-nucleus interactions and from the nuclear properties. However, the potential U_{NN} is also connected with the potentials $\hat{V}_{qq}$, that can be seen from (7.132). Thus, it can be derived from the given $q-q$ interactions. We return to this issue later. Note that at present, the quark-nucleon interaction potential $\hat{V}_{qN}$ is the subject of much investigation.

Equations (7.130) and (7.131), in spite of many used approximations, have rather general character. In particular, one can obtain the equation of a nucleon motion in the averaged nuclear field. As it was mentioned, the contribution of the inner many-quark part Ψ_{in} in the total wave function (7.109) or (7.126) of the ground and excited nuclear states is very small. Because of this, the expansion coefficients in (7.127) are also small, $|C_\lambda|^2 \ll 1$. This enables one to neglect the right-hand side of equation (7.131). The latter is separated from the system and becomes independent. Introducing the effective nucleon mass $M_{\text{eff}} = 3M_q$ and a distance between nucleon and the rest part of the system (nucleus) $r \approx \frac{1}{\sqrt{3}}\rho$, we obtain the usual Schrödinger equation for a nucleon in the nuclear self-consistent field (instead of (7.131))

$$\left(-\frac{\hbar^2}{2M_{\text{eff}}}\frac{d^2}{dr^2}+U(r)-\varepsilon\right)\varphi(r\sqrt{3})=0. \qquad (7.133)$$

The self-consistent potential $U(r)$ in (7.133) is usually given phenomenologically in the nuclear physics.

The *resonant group method* , developed previously in the theoretical nuclear physics and considering the whole system of particles as consisting of subsystems – clusters, is close to the interpolation model using hyperspherical coordinates. In this method the Jacobi vector coordinates are usually used. The both approaches are efficiently used, at present, for calculations of the nuclear system properties and description of nuclear processes on the base of the given NN-potentials, as well as for calculations of the barion and few-nucleon nuclei on the base of the primary model quark-quark interactions.

Using the above-described approaches with some simplifying assumptions, we carried out the calculations of the inner many-quark impurities in the lightest nuclei with a quark number $\nu \leqslant 12$. In the $6q$-system all six quarks can be arranged, as mentioned above, within the s-shell, which is not prohibited by the Pauli principle (quarks are characterized by color). The spatial s^6-configuration is assumed to be the most energy profitable, although it differs (in energy) not very much in comparison with other configurations, for example, configurations $s^{\nu_1}p^{\nu_2}$ with $\nu_1+\nu_2=6$. These states are called *dibarions*. Calculations show that the $6q$-systems with color clusters have close energies. The calculated total energy of the six-quark system in the s^6-configuration exceeds the deuteron mass approximately by 270 MeV. This also indicates that a deuteron is mainly the two-nucleon system rather than $6q$-system. The calculated dibarion spectra is always by 200-300 MeV higher than the total mass of two nucleons, i.e., high enough. This energy range (~300 MeV) determines an order of the barrier which separates two phases in a nucleus: the nucleon and quark ones.

The six-quark state contribution in a deuteron has a profound effect on the scattering of high energy electrons at large transferred momenta q up to almost $q \approx 15\,\text{fm}^{-1}$. A consideration of this contribution allows one to describe satisfactorily by means of the above-reported methods the relevant experimental data. However, it increases a contribution of $6q$-systems in a deuteron (~8%) in comparison with the results obtained for the deuteron magnetic moment in the quark oscillator model. In this case, the radius r_0 of the quark s^6-configuration was taken equal to 0.77 fm with respect to the relativistic effects. However, at small and average values of the transferred momentum, the contribution of the $6q$-system in the electron-deuteron scattering is negligible.

On taking account of the six-quark component in the deuteron wave function, one can better describe some particular deuteron properties and different processes with its participation. In particular, the consideration of the $6q$-system helps to describe the deuteron photosplitting by γ-quanta $\sim$1 GeV of energy, as well as the momenta distribution of protons obtained in the collision reactions of deuterons with nuclei with the escaping protons at small angles. Figure 7.9 shows the experimental data and the calculated dependence of a product of the differential cross-section $\frac{d^3\sigma}{d\mathbf{p}_p}$ (where $d\mathbf{p}_p = p_p^2 dp_p d\Omega_p$) and energy of the formed proton E_p on the proton momentum p_p for the splitting process of nonpolarized deuterons with a momentum 8,9 GeV/c by carbon nuclei when protons escape at practically zero angle[6]. The used neutron-proton wave function was calculated with the Paris NN-potential. The solid curve in the figure obtained on taking account of the six-quark state in a deuteron, and dashed curve obtained when this state was neglected. It can be seen from the figure that to describe satisfactorily the corresponding experiment in the wide range of the proton momenta (200 MeV/c$\lesssim p_p \lesssim$450 MeV/c), we have to consider the $6q$-component in the total deuteron wave function. Calculations have shown that the processes with participation of polarized deuterons are very sensitive to admixture of the $6q$-state in a deuteron.

The elastic scattering of high energy electrons by the ^3He nuclei on taking into account of admixture of the $6q$-states have been studied in many papers. By and large it improves the agreement between the calculations and experiment within a wide range of the transferred momenta $q \lesssim 8\,\text{fm}^{-1}$ in comparison with the purely nuclear three-nucleon calculations for the given NN-interactions. It is shown that interference of the 3-nucleon and 6-quark states is of crucial importance. Figure 7.10 presents the calculated dependence of a modulus of the charge form-factor $F(q)$ of the ^3He nucleus on a transferred momentum q obtained by means of the Gauss (oscillator) nucleon and quark wave functions and the corresponding experimental data. The solid line is obtained on taking account of the $6q$-state admixture in the ^3He nucleus. The dashed line is calculated neglecting the many-quark states in nuclei, i.e., by use of the conventional nuclear three-nucleon wave functions. Whereas a consideration of the $6q$-states improves a correspondence between the calculated form-factor and experiment, the addition of the $9q$-quark states spoils immediately the results at large transferred momenta q. There are the magnetic form-factor calculations of the ^3He nucleus, where the $9q$-states are occurring. On taking account of these states, as well as the three-nucleon states, one can satisfactorily describe the data on the ^3He nucleus magnetic form-factor in a range of transferred momenta

[6]The figure 7.9 is taked out the review by Kobushkin and Shelest (1983).

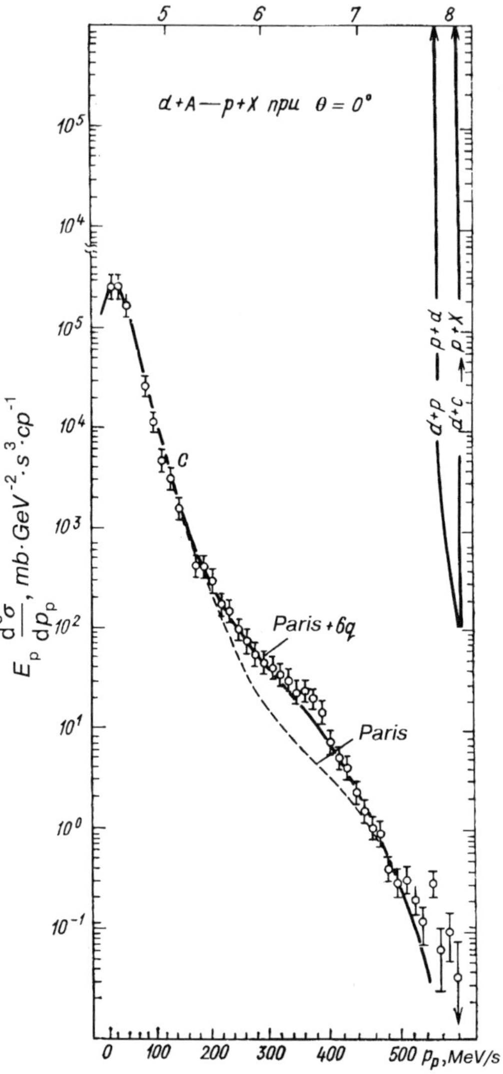

Figure 7.9. Dependencies of the deuteron splitting cross-section (on carbon nuclei) multiplied by an energy Ep of originating protons on a proton momentum Pp taking into account the six-quark state in a deuteron (solid line) and neglecting it (dotted line) compared with experiment.

$q \lesssim 5,7\,\mathrm{fm}^{-1}$. There are some theoretical evidence and experimental data pointing to the possible existence of the *many-quark states in the lightest nucleus* . (For example, it is very impressive phenomenon when the dependence of the product $(q^2)^{\nu-1}F(q^2)$ on q^2 flattens out at asymptotically large q for hadrons and some lightest nuclei, i.e., becomes a constant. Here ν is a

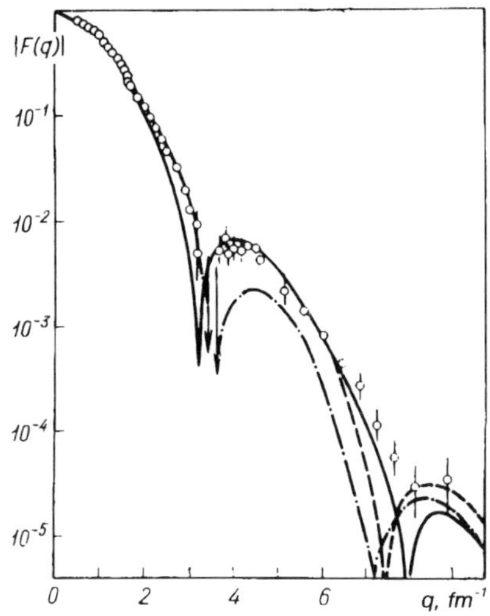

Figure 7.10. A comparison with experiment of the change form-factor modulus of the ^3He nucleus as a function of a transferred momentum taking into account $6q$-states in the ^3He nucleus (solid line), $6q$- and $9q$-states (dotted line) and neglecting the many-quark states (dashed line).

quark number in hadron or complex nucleus, $F(q^2)$ are the corresponding form-factors.)

Quark-quark interactions and nuclear forces. One of the most important problems of the modern nuclear physics and elementary particle physics is the establishment of interrelations between the given *quark-quark interactions* and nucleon-nucleon (barion-barion) *interactions.* The nuclear forces, in this formulation of the problem, i.e., the forces between colorless nucleons within the framework of the hadron quark model, must be the second in importance. Basically, they can be obtained from the given quark-quark forces, which must be primary, like the forces between the electrically neutral atoms are the second in importance, and can be obtained from the primary interactions of moving electrically charged particles constituting atoms.

However, the *reconstruction of the NN-forces* from the qq-forces at the moment is the extremely complex problem. Within the framework of quantum chromodynamics the relevant calculations can be performed, as it was mentioned above, only at small distances between quarks, i.e., in the quark asymptotic freedom domain. This domain corresponds to the repulsive core of the nucleon-nucleon (barion-barion) interactions. Outside of the core in the confinement domain, quantum chromodynamics does not have, at the moment, the rigorous and reliable calculation methods. Because of this, at average and large distances $\gtrsim 1$ fm between quarks, one has to use either the model approaches or approaches of the hadron quantum field theory where the forces between nucleons (barions) emerge as a result of interchange by different mesons: pions at comparatively large distances ($\gtrsim 2$ fm) between nucleons, ρ- and ω-mesons and other heavy nucleons. The so-called "tail" of the NN-meson interchange potential is significant at distances between nucleons more than $0.8 \div 1$ fm.

The *meson theory of nuclear forces* can be applied to the description of the nucleon-nucleon interaction at energies which do not exceed 400 MeV, when comprehansive data on the NN-potential behaviour with a distance is not important, and the quark structures of interacting nucleons are hidden from each other. It is interesting to note that the meson theory, based on the conception of the NN-potential, can explain the data on NN-scattering for considerably higher energies $\approx$1 GeV and sometimes up to energies $\approx$3 GeV. In particular, it is shown that the ω-meson interchange can result in the appearance of the repulsive core in the NN-potential. Thus, the Yukawa fundamental idea (Yukawa, 1935) that the nucleon-nucleon forces can be explained by interchange of massive hadron quanta, i.e., mesons, providing a small radius of the NN-forces, is justified up to such high energies of the NN-collisions.

The *quark bag model for a nucleon* (barion) is one of the possibilities of unification of two above-mentioned approaches to describe the quark and meson NN-interaction. By virtue of this model a nucleon (barion) is treated as a bag with three-valence quarks and gluons surrounded by a meson (mainly pion) cloud, interacting with nucleon (barion) quarks through the bag surface. The nuclear forces problem on the quark level is reduced to the study of the six-quark systems which are unbound, in a general case, and have a whole spectrum of possible different quantum characteristics. At the same time, the applicability of the above-developed phenomenological bag models to many-quark systems, in particular, to the six-quark systems, which appear in the process of two-nucleon collisions, needs an additional study. There are some uncertainities when a radius r_0 of the three-quark bag is choosing. If r_0 is taken of the order of that of the nucleon or the average distances in a nucleus, then the attractive part of the NN-potential within a range of $r \sim 1$ fm (r is a distance between nucleons) cannot be obtained. However, at such large distances we can simulate the periphery of the NN-interaction whose nature is apparently associated with a meson cloud. If we take into account the meson interchange in the explicit form by using the experimental phenomenological NN-forces at large distances r, the r_0 is naturally to be connected with dimensions of the quark nucleon kern r_0^N=0.4 fm. This idea corresponds to the model of the so-called "small bags' surrounded by a meson cloud. For numerical evaluations, the *six-quark bag radius* r_0 in the case of the NN-interactions is taken to be equal to a *double radius of the nucleon quark kern* $r_0 \approx 2r_0^N = 0.8\,\text{fm}$.

At present we can follow some basic properties of the NN-interaction potential at small and intermediate distances in the calculations with the model qq-interactions $\hat{V}_{qq}$ (see (7.105)) and in the calculations, using, first of all, some general principles of the nuclear interactions symmetry on the quark level, i.e., the largely modelless, but using the experimental data. We dwell briefly on the model approach. There are some hints that the NN-interaction can be treated as nonlocal. However, this nonlocality is usually neglected in the qualitative analysis. In this approach, the NN-potential $U_{NN}(r)$ as a function of a distance between nucleon centers, obtained by use of (7.105) has the soft kern, which appears, mainly, from the spin-spin part of the qq-interaction in (7.105). The kern size increases, if we take into account the confinement phenomena in the spin-spin part of the qq-interaction (see (7.108)) in comparison with the case when the gluon part is included only as in (7.107).

The nucleon-nucleon effective potential $U_{NN}(r)$, emerging as a result of averaging of the interaction between quarks from different nucleons with respect to the quark motion inside of nucleons, essentially depends on the total spin S and isospin T of two nucleons. However, the qualitative features of the r dependence at different S and T remain approximately the same. In Fig. 7.11 and 7.12 the typical $U_{NN}(r)$ dependence obtained by using (7.105) for all possible pairs of the spin-isospin quantum numbers (S, T) at $n = 2$ in (7.106) are shown (Kopaileshvili, 1981). The magnitude of the U_{NN}-potential repulsive part is of the order of 600 MeV and depends weakly on a radius of the three-nucleon bag r_0. If $r_0 = 0.3\,\text{fm}$ then the attractive part of the NN-potential is formed in a range $0.5 \lesssim r \leqslant 1\,\text{fm}$ with a depth $U_{NN} \lesssim 30\,\text{MeV}$, i.e., the radius of the attractive core in this case is close to $0.5\,\text{fm}$. At $r_0 \approx 0.9\,\text{fm}$, when quarks occupy the whole nucleon volume, there is no the attractive part of the U_{NN}-potential at $r \lesssim 2\,\text{fm}$, i.e., $U_{NN}(r) > 0$ everywhere. Therefore, a radius of the repulsive core of the NN-potential seems to be strongly dependent on the three-quark bag dimensions.

These and other qualitative results can be obtained even for the more simple explicit form of the qq-interaction in comparison with (7.105). It happened to be conveniently to use the following model operator of the NN-interaction potential in the numerical calculations

$$U_{NN}(r) = \theta(r_0 - r) \sum_{k>j=1}^{6} \hat{V}_{qq}(kj) + \theta(r_0 - r) \times \Big[\sum_{k>j=1}^{3} \hat{V}_{qq}(kj) + \sum_{k>j=4}^{6} \hat{V}_{qq}(kj) + V^M(r) \Big], \qquad (7.134)$$

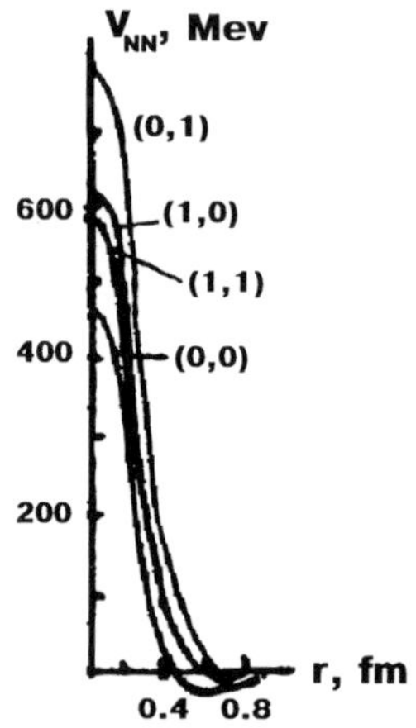

Figure 7.11. Nucleon-nucleon potentials $U_{NN}(r)$ for possible pairs of the total spin and isospin (S,T) for the quark bag radius r_0=0,3 fm.

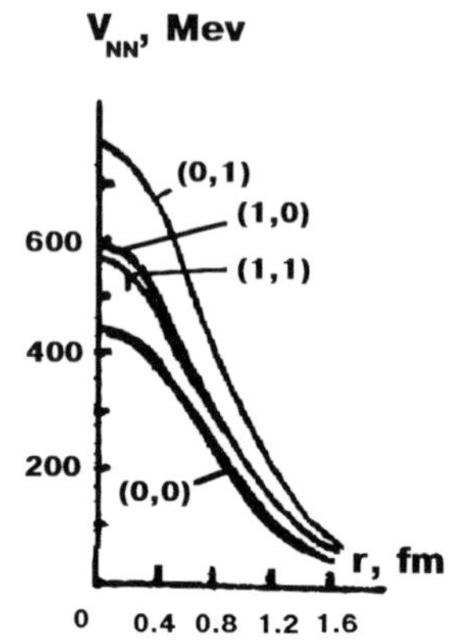

Figure 7.12. The same dependencies as in Fig. 7.11 but for r_0=0,9 fm.

where the qq-interaction operator is chosen in one of the forms:

$$\hat{V}_{qq}(kj) = -\frac{\kappa_c}{2}\sum_{a=1}^{8}\lambda_a(k)\lambda_a(j)(\mathbf{r}_k-\mathbf{r}_j)^2 - \mu\sum_{a=1}^{8}\lambda_a(k)\lambda_a(j)\boldsymbol{\sigma}(k)\boldsymbol{\sigma}(j), \qquad (7.135)$$

$$\hat{V}_{qq}(kj) = -\Big[\frac{\alpha_c}{4}\Big(\frac{1}{r_{kj}} - \frac{\pi}{M_q^2}\delta(\mathbf{r}_{kj})\Big) - Ar_{kj}\Big]\sum_{a=1}^{8}\lambda_a(k)\lambda_a(j) - \frac{\alpha_c}{4}\frac{\pi}{M_q^2}\delta(\mathbf{r}_{kj})\sum_{a=1}^{8}\lambda_a(k)\lambda_a(j)\boldsymbol{\sigma}(k)\boldsymbol{\sigma}(j). \qquad (7.136)$$

Here $\theta(x) = \begin{cases} 0, & x<0 \\ 1, & x \geqslant 0 \end{cases}$ is the Heavyside function, $V^M(r)$ is a "tail" of the meson interchange potential. The potential (7.134) acts in the entire space coinciding with the six-quark potential at $r < r_0$ and determining the interaction of isolated nucleons in the external region $r > r_0$. The parameter κ_c is interrelated with parameter Ω in (7.112) by $\kappa_c = \dfrac{\Omega^2}{288M_q}$

and at M_q=0.3 GeV and $\Omega = 1\,\text{GeV}^2$ is equal to $\kappa_c = 1,16 \cdot 10^{-3}$ GeV (we use the system of units where $\hbar = c = 1$). The parameter μ, appearing in the so-called *color-magnetic term of the qq-interaction* , is evaluated from the dimensionality considerations of the Δ-isobar and nucleon masses in the modelless treatment of the NN- and $\Delta\Delta$-interactions and turns out to be 19 MeV.

It is interesting to compare the interaction between two nucleons with that of two Δ-isobars, which consist of the same u- and d-quarks as the nucleons. Two Δ-isobars in the state with total spin $S = 3$ and isospin $T = 3$ repulse at small distances. The repulsive core existence is a result of the Pauli principle. In fact, in every Δ-isobar all three quarks are in the same spin and isospin states (quark angular momenta $l = 0$, and their spins and isospins are parallel in the corresponding spaces) but in different color states. Because of this, the Pauli principle prevents the confluence of two Δ-isobars in one six-quark system with $S = T = 3$ without exciting the orbital states. And vice versa, the Pauli principle permits to arrange six quarks (when two nucleons interact) in different quantum states corresponding to the spatial s^6-configuration at $S = 0$ and $T = 0$ or $S = 1$ and $T = 1$. Because of this, the observed repulsive core in the NN-potential origins due to the specific character of the qq-interaction. If the system of two Δ-isobars is in the state with $S = 3$ and $T = 0$, then an attraction appears at small distances, which can result in the bound state origin, i.e., a *dibarion resonance.* In this case, the six-quark bag with the spatial (orbital) s^6-configuration is formed by confluence. Note that the NN-interaction is caused not only by states of the nucleon three-quark clusters but, apparently, by isobar two-cluster states and states with color clusters as well.

We pass now to the description of the NN-interactions, proceeding from the nucleon quark structure, reducing a number of the possible model assumptions and by using the general symmetry properties of the interactions[7]. Since the form of the spatial dependence of the qq-interactions, in a way, is arbitrary (see (7.103) – (7.105), (7.134) – (7.136)), we can write down the total interaction potential of a colorless system $\hat{U}$ of ν quarks in the form similar to (7.134) or (7.135) but not assuming any explicit dependence on space coordinates for pair qq-interactions

$$\hat{U} = \sum_{k>j=1}^{\nu} \sum_{a=1}^{8} \lambda_a(k)\lambda_a(j)V^c(\mathbf{r}_{kj})$$

[7]The general symmetry properties of the NN-interactions is used to the description of the NN-interactions, proceeding from the nucleon quark structure, in the review by Neudachin, Obukhovskii and Smirnov (1984).

$$- \sum_{k>j=1}^{\nu} \sum_{a=1}^{8} \lambda_a(k)\lambda_a(j)\boldsymbol{\sigma}(k)\boldsymbol{\sigma}(j)V(\mathbf{r}_{kj}). \tag{7.137}$$

Here we preserved the upper limit of summation and put it equal to ν. The spatial dependencies of potentials $V^c(\mathbf{r}_{kj})$, $V(\mathbf{r}_{kj})$ on coordinates are not specified. In (7.137), we did not single out the long range part of the meson exchange potential. The first part of potential (7.137), consisting of $V^c(\mathbf{r}_{kj})$, is not much of importance. Because of this, we have to evaluate numerically the intensity of rather powerful *color-magnetic quark interaction* which is described by the spatial potentials $V(\mathbf{r}_{kj})$ and contains both the Gell-Mann matrices λ_a and the Pauli matrices $\boldsymbol{\sigma}$. The evaluation of an average value of $\langle V \rangle \equiv \langle V(\mathbf{r}_{kj}) \rangle$ on the quark orbital wave functions will be carried out by using the general symmetry properties interaction (7.137).

The interaction operator (7.137), as a part of the total quark Hamiltonian, must be invariant (scalar) with respect to a whole set of the symmetry transformations. In particular, operators

$$\hat{O}_\lambda(\nu) = \sum_{k>j=1}^{\nu} \sum_{a=1}^{8} \lambda_a(k)\lambda_a(j),$$

$$\hat{O}_{\lambda\sigma} = - \sum_{k>j=1}^{\nu} \sum_{a=1}^{8} \lambda_a(k)\lambda_a(j)\boldsymbol{\sigma}(k)\boldsymbol{\sigma}(j), \tag{7.138}$$

whose average values appear in eigenvalues of the operator (7.137), are invariant with respect to the color SU_3^C group transformations, spin group SU_2^S, direct product of these groups $SU_3^C \times SU_2^S$, as well as group SU_6^{CS} containing the group $SU_3^C \times SU_2^S$ as a subgroup of the $SU_3^C \times SU_2^S$. (Operators (7.138) are invariant with respect to permutations of other symmetry groups, for example, SU_2^T.) To emphasize this invariance the sum $\sum_{a=1}^{8} \lambda_a(k)\lambda_a(j)$ is written in the form of scalar product in 8-dimensional space $\boldsymbol{\lambda}(k)\boldsymbol{\lambda}(j)$, as well as $\sum_{\alpha=1}^{3} \sigma_\alpha(k)\sigma_\alpha(j) \equiv \boldsymbol{\sigma}(k)\boldsymbol{\sigma}(j)$ in 3-dimensional space. This invariance means that operators (7.138) can be expressed in terms of the Casimir operators of the above-mentioned groups (see section 7.1). Since operators (7.138) are quadratic in the Gell-Mann and Pauli matrices, they can be expressed in terms of the Casimir operators $\hat{C}_2^{(n)}$ of SU_n groups with n=2,3, and 6.

For an arbitrary unimodular group SU_n, the matrix h_{ij} in (7.12) will be diagonal and proportional to a unit matrix:

$$h_{ab} = \sum_{s,t=1}^{n^2-1} C_{at}^{s} C_{bs}^{t} = -n\delta_{ab}. \tag{7.139}$$

Using (7.139) for the Casimir operator $C_2^{(n)}$, we obtain from (7.11) the following expression

$$C_2^{(n)} = \sum_{a=1}^{n^2-1}\left(\frac{1}{2}X_a\frac{1}{2}X_a\right) \equiv \frac{1}{4}(\mathbf{XX}), \tag{7.140}$$

where

$$X_a = \sum_{j=1}^{\nu} X_a(j), \; SpX_a(k)X_a(k) = 2, \; , a = 1, 2, \ldots, n^2 - 1 \tag{7.141}$$

are generators of the SU_n group for a system of ν particles (quarks). The generators $X_a(k)$, related to a k-th particle, are normalized in the same manner as the Pauli matrices $Sp\sigma_\alpha(k)\sigma_\alpha(k) = 2$ (there is no summation over a and k, as well as on a and k in (7.141)). Formula (7.140), presenting the Casimir operator $C_2^{(3)}$ of the SU_3 group, is a generalization of the Eq.(7.19). The operator $\hat{O}_\lambda(\nu)$ in (7.138) can be straight forward interrelated with the Casimir operator $C_2^{(3)}$

$$\hat{O}_\lambda = 2C_2^{(3)} - \frac{8\nu}{3}, \tag{7.142}$$

where we have used (7.140) (or (7.19)) and the normalization condition of the Gell-Mann λ-matrices $\sum_{a=1}^{8} \frac{1}{2}\lambda_a(k)\frac{1}{2}\lambda_a(k) = \frac{4}{3}$ (there is summation over k), which can be obtained directly from (7.16). To express the operator $\hat{O}_{\lambda\sigma}(\nu)$ in (7.138) in terms of the Casimir operators of the unimodular groups, we write down, first of all, generators of the SU_6 group. It has $6^2 - 1 = 35$ generators X_s, which can be connected with generators of the SU_3 and SU_2 groups:

$$X_s = \lambda_a, \sigma_\alpha, \lambda_a \times \sigma_\alpha, \; 1 \leqslant s \leqslant 35, \; 1 \leqslant a \leqslant 8, \; 1 \leqslant \alpha \leqslant 3. \tag{7.143}$$

Using (7.140) and (7.143), we can present the operator $\hat{O}_{\lambda\sigma}(\nu)$, determined in (7.138), in the form

$$\hat{O}_{\lambda\sigma}(\nu) = -4C_2^{(6)} + 2C_2^{(3)} + \frac{4}{3}C_2^{(2)} + 8\nu. \tag{7.144}$$

Eigenvalues of the interaction operator (7.137) or average values of this operator with respect to the states of the ν-quark system, which are characterized by the definite symmetry properties (we relate them to the specific Young tableaux $[\lambda]$ determined the irreducible representations of SU_n

group), are expressed, according to (7.142) and (7.144) in terms of the Casimir operator eigenvalues. At the same time, the diagonal matrix element $\langle V\rangle$ for the pair potential $V(\mathbf{r}_{kj})$ (analogously, $\langle V^c\rangle$ for $V^c(\mathbf{r}_{kj})$) in (7.137) is the same for any quark pair in the same orbital state (for example, three quarks in the configuration s^3 and six quarks in the configuration s^6). Eigenvalues of the Casimir operator $C_2^{(n)}$, i.e., the average values $\langle[\lambda]C_2^{(n)}|[\lambda]\rangle$ of this operator on the wave functions $|[\lambda]\rangle$ of a system of ν quarks, which are the basis functions of the irreducible representations of SU_n group, can be expressed in terms of line lengths λ_i of the Young tableaux $[\lambda]=[\lambda_1,\lambda_2,\ldots,\lambda_n]$, or, more precisely, in terms of line lengths $\lambda'_i=\lambda_i-\lambda_n$ of the corresponding truncated Young tableaux in the following way

$$\langle[\lambda]|C_2^{(n)}|[\lambda]\rangle=\frac{1}{2}[\lambda'_1(\lambda'_1+n-1)+\lambda'_2(\lambda'_2+n-3)$$

$$+\ldots+\lambda'_{n-1}(\lambda'_{n-1}-n+3)]-\frac{1}{2n}\Big(\sum_{i=1}^{n-1}\lambda'_i\Big)^2. \qquad (7.145)$$

In particular, at $n=3$ in the Elliott denotations (4.124), we have

$$\langle[\lambda]|C_2^{(3)}|[\lambda]\rangle=\frac{1}{3}(\bar{\lambda}^2+\bar{\mu}^2+\bar{\lambda}\bar{\mu})+\bar{\lambda}+\bar{\mu}. \qquad (7.146)$$

We evaluate now the *color-magnetic contribution* of the three-quark interaction in barions consisting of the lightest u- and d-quarks. From (7.137) and (7.138) and using the wave functions (7.93) and (7.94), as well as (7.99), we can write the average values of the interaction of quarks in a nucleon and in Δ-isobar ($\nu=3$) :

$$\langle N|\hat{U}|N\rangle = \langle V^c\rangle\langle[1^3]_c|\hat{O}_\lambda(3)|[1^3]_c\rangle+\langle V\rangle\langle[21]_{CS}|\hat{O}_{\lambda\sigma}(3)|[21]_{CS}\rangle,$$
$$\langle\Delta|\hat{U}|\Delta\rangle = \langle V^c\rangle\langle[1^3]_c|\hat{O}_\lambda(3)|[1^3]_c\rangle+\langle V\rangle\langle[1^3]_{CS}|\hat{O}_{\lambda\sigma}(3)|[1^3]_{CS}\rangle. \qquad (7.147)$$

On taking account of (7.142), (7.144) – (7.146) these average values can be presented in the form

$$\langle N|\hat{U}|N\rangle=-8\langle V^c\rangle-8\langle V\rangle, \qquad (7.148)$$

$$\langle\Delta|\hat{U}|\Delta\rangle=-8\langle V^c\rangle+8\langle V\rangle. \qquad (7.149)$$

We can compare the quark interaction contributions (7.137) into the nucleon mass and Δ-isobar from difference between (7.149) and (7.148)

$$\langle\Delta|\hat{U}|\Delta\rangle - \langle N|\hat{U}|N\rangle = 16\langle V\rangle. \tag{7.150}$$

Equating this quantity to the known difference between masses of the Δ-isobar and nucleon $M_\Delta - M_N \simeq 300\,\mathrm{MeV}$, we can estimate the diagonal orbital matrix element $\langle V\rangle$ of the pair *quark-quark color-magnetic interaction* in the s^3-configuration

$$\langle V\rangle \approx \frac{300}{16}\,\mathrm{MeV} \approx 19\,\mathrm{MeV}. \tag{7.151}$$

The obtaining of a contribution of the color-magnetic interaction in the nuclear forces (i.e., the NN-forces) is a considerably more complex problem, because we deal with the six-quark system ($\nu = 6$). Consider, firstly, the colorless six-quark system with the total spin $S = 1$ and isospin $T = 0$ (as in a deuteron). According to (7.116) and (7.120), the color and spin Young tableaux, in this case, take the form $[2^3]_C$ and $[42]_S$. Unlike the simple relations (7.99) for three-quark systems, an expansion of the inner product of two irreducible representations of the permutations group with the color $[2^3]_C$ and spin $[42]_S$ Young tableaux for the six-quark system becomes more complex:

$$[2^3]_C \times [42]_S = [42]_{CS} \bigoplus [321]_{CS} \bigoplus [2^3]_{CS} \bigoplus [31^3]_{CS} \bigoplus [21^4]_{CS}. \tag{7.152}$$

The color Young tableaux $[2^3]_C$ for the six-quark system with $S = 0$ and $T = 1$ remains as before and the spin one $[3^2]_S$ becomes different (see (4.116)). The expansion of the inner product for these two Young tableaux takes the following form

$$[2^3]_C \times [3^2]_S = [3^2]_{CS} \bigoplus [41^2]_{CS} \bigoplus [2^2 1^2]_{CS} \bigoplus [1^6]_{CS}. \tag{7.153}$$

In the same manner, as we have calculated the average values of operators (7.138) of the three-nucleon systems, we can find the average values of the $\hat{O}_{\lambda\sigma}(6)$ operators of the six-quark systems

$$\langle O(6)\rangle_{CS} \equiv \langle [\lambda]_{CS}|\hat{O}_{\lambda\sigma}(6)|[\lambda]_{CS}\rangle$$

for all *color-spin Young tableaux* $[\lambda]_{CS}$ appearing in the right-hand side of equalities (7.152) and (7.153). It turns out that for the most symmetric color-spin states, described by the Young tableaux with longest lines and shortest columns (they are called the most symmetric tableaux), as well as $[42]_{CS}$, $[321]_{CS}$, $[3^2]_{CS}$, and $[41^2]_{CS}$, the average value $\langle O(6)\rangle_{CS}$ is negative, i.e., here we deal with the *color-magnetic quark attraction* . For the rest color-spin states the average value $\langle O(6)\rangle_{CS}$ is positive, i.e., the *color-magnetic repulsion* occurs.

The totally symmetric orbital state, described by the Young tableaux $[\lambda]_X = [6]_X$, corresponds to the orbital configuration of six quarks s^6. The space-isospin state at $[\lambda]_X = [6]_X$ is uniquely determined according to (7.95) – (7.97), by an isospin value: at $T = 0$ we have $[\lambda]_{XT} = [3^2]_{XT}$, at $T = 1$ $[\lambda]_{XT} = [42]_{XT}$, as far as $[6]_X \times [\lambda]_T = [\lambda]_{XT}$. Since the total wave function of the six-quark system is antisymmetric, i.e., it corresponds to the generalized Young tableaux $[1^6]_{XCST}$, the corresponding color-spin Young tableaux $[\lambda]_{CS} = [\tilde{\lambda}]_{XT}$. Then for the spatial quark configuration s^6, the Pauli principle admits only one color-spin state with the Young tableaux $[\lambda]_{CS} = [\tilde{3^2}]_{XT} = [2^3]_{CS}$ at $S = 1$, $T = 0$, and one color-spin state with the Young tableaux $[\lambda]_{CS} = [\tilde{42}]_{XT} = [2^2 1^2]_{CS}$ at $S = 0$, $T = 1$, which are available in expansions (7.152) and (7.153), respectively. Both these Young tableaux, as it was above-noted, correspond to the color-magnetic repulsion (but relatively small repulsion). Thus, we can treat the core between nucleons at small distances as the many-particle (many-quark) phenomena.

For another spatial configuration, namely, s^4p^2 with the orbital Young tableaux $[42]_X$, all color-spin states in the right-hand sides of equalities (7.152) and (7.153) are possible. Due to this, in some cases, the strong color-magnetic attraction at small distances between the three-quark nucleon clusters in the six-quark configuration s^4p^2 can take place, as well as at their partial or total overlapping. Since in the state $[42]_{CS}$ with the maximum color-magnetic attraction for $S = 1$, $T = 0$ there is the comparatively large benefit in energy ($\sim 600\,\text{MeV}$), in comparison with the state $[2^3]_{CS}$ for the same S and T, the six-quark configuration s^4p^2 in the state $[42]_X \times [42]_{CS}$ can dominate in the NN-system (for example, in the nucleon-nucleon scattering, as well as in a deuteron) over the s^6-configuration in the state $[6]_X[2^3]_{CS}$. With account of the preceding, we can write down the inner six-quark deuteron wave function specifying its symmetry in the form different from (7.121):

$$\Psi'(6q) = |s^4p^2[42]_X[2^3]_C[42]_S[3^2]_T||[1^6]_{XCST}\rangle. \qquad (7.154)$$

At the same time, it is possible that the color-magnetic state $[42]_{CS}$ to be realized. Apparently, the Pauli principle determines the Young tableaux $[\lambda]_{CST} = [\tilde{42}]_X = [2^2 1^2]_{CST}$. This effect of the color-magnetic attraction with the rearrangement of an orbital quark configuration can be more pronounced when passing to the three-and four-nucleon systems. In view of the above-reported, the inner wave function of the colorless nine-quark system with $S = T = \frac{1}{2}$ (as well as for the three-nucleon nuclei ^3H) and specified symmetry properties in the case of the s^9-configuration written in the same form as the inner deuteron wave function (7.121)

$$\Psi(9q) = |s^9[9]_X[3^3]_C[54]_S[54]_T|[1^9]_{XCST}\rangle \tag{7.155}$$

is related, apparently, to higher energy than the wave function for the configuration s^4p^2

$$\Psi'(9q) = |s^5p^4[52^2]_X[3^3]_C[54]_S[54]_T||[1^9]_{XCST}\rangle. \tag{7.156}$$

The point is that the latter state, accordingly the Pauli principle, is compatible with the most symmetric color-spin Young tableaux $[621]_{CS}$, which appears in an expansion

$$\begin{aligned} & [3^3]_C \times [54]_S \\ = \ & [621]_{CS} \bigoplus [54]_{CS} \bigoplus [531]_{CS} \bigoplus [521^2]_{CS} \bigoplus [432]_{CS} \bigoplus [431^2]_{CS} \\ \oplus \ & [42^21]_{CS} \bigoplus [421^3]_{CS} \bigoplus [3^321]_{CS} \bigoplus [32^21^2]_{CS} \bigoplus [321^4]_{CS} \bigoplus [2^41]_{CS}. \end{aligned} \tag{7.157}$$

The maximum color-magnetic attraction between quarks is associated with the Young tableaux $[621]_{CS}$ in this expansion. Note that the orbital state s^5p^4 has a minimum quark number with $l = 1$, i.e., four, which permits to satisfy the Pauli principle, if we demand the color-magnetic (color-spin) Young tableaux for the state of nine quarks to be $[621]_{CS}$. At the same time, the Young tableaux $[3^21^3]_{CST}$, appearing in the inner product $[621]_{CS} \times [54]_T$, is determined uniquely, as well as the spatial Young tableaux $[52^2]_X$, which is conjugated to the Young tableaux $[3^21^3]_{CST}$. As this takes place, according to the Pauli principle, the wave function (7.155) with the configuration s^9 corresponds to the color-magnetic Young tableaux $[2^41]_{CS}$. It is the least symmetric term in expansion (7.157), i.e., the energy unprofitable one.

Considering the colorless *twelve-quark system* with $S = T = 0$ (as in the ^4He nucleus) and acting in the same manner, we can show that the color-magnetic attraction in the state with s^6p^6-configuration at $[\lambda]_X = [62^3]_X$ and $[\lambda]_{CS} = [82^2]_{CS}$ is the most efficient. Thus, the state with the wave function

$$\Psi'(12q) = |s^6p^6[62^3]_X[4^3]_C[6^2]_S[6^2]_T||[1^{12}]_{XCST}\rangle \tag{7.158}$$

is the energy profitable one. In a state with the symmetric configuration s^{12}, namely,

$$\Psi(12q) = |s^{12}[12]_X[4^3]_C[6^2]_S[6^2]_T||[1^{12}]_{XCST}\rangle \tag{7.159}$$

for a possible, accordingly the Pauli principle, the Young color-magnetic (color-spin) tableaux $[2^6]_{CS}$, which is the lowest symmetric in an expansion $[4^3]_C \times [6^2]_S$, the magnetic repulsion dominates. Thus, for the ^4He nucleus the most probable $12q$-state, apparently, is (7.158), but not (7.159).

In conclusion we have to note that the above-described approaches for the reconstruction of some properties of the nucleon-nucleon potentials on the basis of the assumed quark-quark interactions are naive and rough to some extent for the moment. Even using the most general physical principles, some unsufficiently justified assumptions would have been made. In the many-quark system theory we use the mathematical apparatus developed in the nonrelativistic nuclear models: shell model, interpolation model, resonant group method. But the relativistic effects are very essential for the lightest u- and d-quarks constituting nucleons. It is not easy even to determine how the used potential models of the quark systems can be justified in general, and, particularly, in relation to the most difficult problem of the QCD, i.e., the confinement problem. In spite of some uncertainties, the model approach to the quark aspect of the nuclear forces is useful, since it allows one to clarify qualitatively in combination with experiment, at least, some features of the nuclear complex structures and nuclear interactions.

References

Akhiezer, A.I., Beresterskii, V.B. (1981) *Quantum Electrodynamics*. Nauka, Moscow.

Akhiezer, A.I., Sitenko, A.G., and Tartakovskii, V.K. (1994) *Nuclear Electrodynamics*. Springer, Berlin.

Bogolyubov, N.N. (1967) *Elementary Particle Symmetry theory (In: High energy physics and elementary particle theory)*. Naukova Dumka, Kiev.

Bogolyubov, N.N., Shirkov, D.V. (1984) *Introduction to the Quantum Field Theory*. Nauka, Moscow.

Bogolyubov, N.N., Struminskii, B.V., Tawkhelidze, A.N. (1965) *Preprint JINR, D-1968*, Dubna.

Burov, V.V., Luk'yanov, V.K., and Titov, A.I. (1984) *Particles and Nuclei*, **15**, 1249–1295.

Fermi, E., Yang, C.N. (1949) *Phys. Rev.*, **76**, N 4, 1739.

Gell-Mann, M. (1961) *Caltech Preprint CTSL-20.*

Gell-Mann, M. (1953) *Phys. Rev.*, **93**, 833.

Gell-Mann, M. (1964) *Phys. Lett.*, **8**, 214.

Gell-Mann, M., Zweig, G. (1964) *Phys. Rev. Lett.*, **12**, 237.

Han, M.Y., Nambu, Y. (1965) *Phys. Rev.*, **139**, N5, 1006.

Heisenberg, W. (1932) *Zs. Phys.*, **77**, 1.

Kobushkin, A.P. (1978) *Journal of Nuclear Physics*, **28**, N2(8), 495.

Kobushkin, A.P., Shelest, V.P. (1983) *Particles and Nuclei*, **14**, 1146-1192.

Kopaileshvili, T.I. (1981) *Quark Aspects of Nuclear Forces.* (In: Multiquark interactions and quantum chromodynamics, VI International seminar on problems of the high energy physics, JINR) Dubna).

Markov, M.A. (1940) *J. Phys. USSR*, **3**, 452.

Markov, M.A. (1968) *Hyperons and K-mesons.* Fizmatgiz, Moscow.

Ne'eman, Y. (1961) *Nucl. Phys.*, **26**, 222.

Neudachin, V.G., Obukhovskii, I.T., and Smirnov, Yu.F. (1984) *Particles and Nuclei*, **15**, 1165–1207.

N'guen Van H'eu (1967) *Theory of Elementary Particle Unitary Symmetry.* Atomizdat, Moscow.

Nishijima, Y. (1955) *Prog. Theor. Phys.*, **13**, 285.

Okun', L.B. (1990) *Leptons and Quarks.* Nauka, Moscow.

Sakata, S. (1956) *Prog. Theor. Phys.*, **16**, 686.

Wigner, E.P., Feenberg, E. (1941) *Rep. Prog. Phys.*, **8**, 274.

Yang, C.N., Mills, R.L. (1954) *Phys. Rev.*, **96**, 191.

Yukawa, H. (1935) *Prog. Math. Soc. Jap.*, **17**, 48.

Zweig, G. (1964) *CERN Preprints Th. 401 and 412.*

AUTHOR INDEX

SUBJECT INDEX

Fundamental Theories of Physics

Series Editor: Alwyn van der Merwe, *University of Denver, USA*

1. M. Sachs: *General Relativity and Matter.* A Spinor Field Theory from Fermis to Light-Years. With a Foreword by C. Kilmister. 1982 ISBN 90-277-1381-2
2. G.H. Duffey: *A Development of Quantum Mechanics.* Based on Symmetry Considerations. 1985 ISBN 90-277-1587-4
3. S. Diner, D. Fargue, G. Lochak and F. Selleri (eds.): *The Wave-Particle Dualism.* A Tribute to Louis de Broglie on his 90th Birthday. 1984 ISBN 90-277-1664-1
4. E. Prugovečki: *Stochastic Quantum Mechanics and Quantum Spacetime.* A Consistent Unification of Relativity and Quantum Theory based on Stochastic Spaces. 1984; 2nd printing 1986 ISBN 90-277-1617-X
5. D. Hestenes and G. Sobczyk: *Clifford Algebra to Geometric Calculus.* A Unified Language for Mathematics and Physics. 1984 ISBN 90-277-1673-0; Pb (1987) 90-277-2561-6
6. P. Exner: *Open Quantum Systems and Feynman Integrals.* 1985 ISBN 90-277-1678-1
7. L. Mayants: *The Enigma of Probability and Physics.* 1984 ISBN 90-277-1674-9
8. E. Tocaci: *Relativistic Mechanics, Time and Inertia.* Translated from Romanian. Edited and with a Foreword by C.W. Kilmister. 1985 ISBN 90-277-1769-9
9. B. Bertotti, F. de Felice and A. Pascolini (eds.): *General Relativity and Gravitation.* Proceedings of the 10th International Conference (Padova, Italy, 1983). 1984 ISBN 90-277-1819-9
10. G. Tarozzi and A. van der Merwe (eds.): *Open Questions in Quantum Physics.* 1985 ISBN 90-277-1853-9
11. J.V. Narlikar and T. Padmanabhan: *Gravity, Gauge Theories and Quantum Cosmology.* 1986 ISBN 90-277-1948-9
12. G.S. Asanov: *Finsler Geometry, Relativity and Gauge Theories.* 1985 ISBN 90-277-1960-8
13. K. Namsrai: *Nonlocal Quantum Field Theory and Stochastic Quantum Mechanics.* 1986 ISBN 90-277-2001-0
14. C. Ray Smith and W.T. Grandy, Jr. (eds.): *Maximum-Entropy and Bayesian Methods in Inverse Problems.* Proceedings of the 1st and 2nd International Workshop (Laramie, Wyoming, USA). 1985 ISBN 90-277-2074-6
15. D. Hestenes: *New Foundations for Classical Mechanics.* 1986 ISBN 90-277-2090-8; Pb (1987) 90-277-2526-8
16. S.J. Prokhovnik: *Light in Einstein's Universe.* The Role of Energy in Cosmology and Relativity. 1985 ISBN 90-277-2093-2
17. Y.S. Kim and M.E. Noz: *Theory and Applications of the Poincaré Group.* 1986 ISBN 90-277-2141-6
18. M. Sachs: *Quantum Mechanics from General Relativity.* An Approximation for a Theory of Inertia. 1986 ISBN 90-277-2247-1
19. W.T. Grandy, Jr.: *Foundations of Statistical Mechanics.* Vol. I: *Equilibrium Theory.* 1987 ISBN 90-277-2489-X
20. H.-H von Borzeszkowski and H.-J. Treder: *The Meaning of Quantum Gravity.* 1988 ISBN 90-277-2518-7
21. C. Ray Smith and G.J. Erickson (eds.): *Maximum-Entropy and Bayesian Spectral Analysis and Estimation Problems.* Proceedings of the 3rd International Workshop (Laramie, Wyoming, USA, 1983). 1987 ISBN 90-277-2579-9

Fundamental Theories of Physics

22. A.O. Barut and A. van der Merwe (eds.): *Selected Scientific Papers of Alfred Landé. [1888-1975]*. 1988 ISBN 90-277-2594-2
23. W.T. Grandy, Jr.: *Foundations of Statistical Mechanics.* Vol. II: *Nonequilibrium Phenomena.* 1988 ISBN 90-277-2649-3
24. E.I. Bitsakis and C.A. Nicolaides (eds.): *The Concept of Probability.* Proceedings of the Delphi Conference (Delphi, Greece, 1987). 1989 ISBN 90-277-2679-5
25. A. van der Merwe, F. Selleri and G. Tarozzi (eds.): *Microphysical Reality and Quantum Formalism, Vol. 1.* Proceedings of the International Conference (Urbino, Italy, 1985). 1988 ISBN 90-277-2683-3
26. A. van der Merwe, F. Selleri and G. Tarozzi (eds.): *Microphysical Reality and Quantum Formalism, Vol. 2.* Proceedings of the International Conference (Urbino, Italy, 1985). 1988 ISBN 90-277-2684-1
27. I.D. Novikov and V.P. Frolov: *Physics of Black Holes.* 1989 ISBN 90-277-2685-X
28. G. Tarozzi and A. van der Merwe (eds.): *The Nature of Quantum Paradoxes.* Italian Studies in the Foundations and Philosophy of Modern Physics. 1988 ISBN 90-277-2703-1
29. B.R. Iyer, N. Mukunda and C.V. Vishveshwara (eds.): *Gravitation, Gauge Theories and the Early Universe.* 1989 ISBN 90-277-2710-4
30. H. Mark and L. Wood (eds.): *Energy in Physics, War and Peace.* A Festschrift celebrating Edward Teller's 80th Birthday. 1988 ISBN 90-277-2775-9
31. G.J. Erickson and C.R. Smith (eds.): *Maximum-Entropy and Bayesian Methods in Science and Engineering.* Vol. I: *Foundations.* 1988 ISBN 90-277-2793-7
32. G.J. Erickson and C.R. Smith (eds.): *Maximum-Entropy and Bayesian Methods in Science and Engineering.* Vol. II: *Applications.* 1988 ISBN 90-277-2794-5
33. M.E. Noz and Y.S. Kim (eds.): *Special Relativity and Quantum Theory.* A Collection of Papers on the Poincaré Group. 1988 ISBN 90-277-2799-6
34. I.Yu. Kobzarev and Yu.I. Manin: *Elementary Particles. Mathematics, Physics and Philosophy.* 1989 ISBN 0-7923-0098-X
35. F. Selleri: *Quantum Paradoxes and Physical Reality.* 1990 ISBN 0-7923-0253-2
36. J. Skilling (ed.): *Maximum-Entropy and Bayesian Methods.* Proceedings of the 8th International Workshop (Cambridge, UK, 1988). 1989 ISBN 0-7923-0224-9
37. M. Kafatos (ed.): *Bell's Theorem, Quantum Theory and Conceptions of the Universe.* 1989 ISBN 0-7923-0496-9
38. Yu.A. Izyumov and V.N. Syromyatnikov: *Phase Transitions and Crystal Symmetry.* 1990 ISBN 0-7923-0542-6
39. P.F. Fougère (ed.): *Maximum-Entropy and Bayesian Methods.* Proceedings of the 9th International Workshop (Dartmouth, Massachusetts, USA, 1989). 1990 ISBN 0-7923-0928-6
40. L. de Broglie: *Heisenberg's Uncertainties and the Probabilistic Interpretation of Wave Mechanics.* With Critical Notes of the Author. 1990 ISBN 0-7923-0929-4
41. W.T. Grandy, Jr.: *Relativistic Quantum Mechanics of Leptons and Fields.* 1991 ISBN 0-7923-1049-7
42. Yu.L. Klimontovich: *Turbulent Motion and the Structure of Chaos.* A New Approach to the Statistical Theory of Open Systems. 1991 ISBN 0-7923-1114-0

Fundamental Theories of Physics

43. W.T. Grandy, Jr. and L.H. Schick (eds.): *Maximum-Entropy and Bayesian Methods.* Proceedings of the 10th International Workshop (Laramie, Wyoming, USA, 1990). 1991 ISBN 0-7923-1140-X
44. P.Pták and S. Pulmannová: *Orthomodular Structures as Quantum Logics.* Intrinsic Properties, State Space and Probabilistic Topics. 1991 ISBN 0-7923-1207-4
45. D. Hestenes and A. Weingartshofer (eds.): *The Electron.* New Theory and Experiment. 1991 ISBN 0-7923-1356-9
46. P.P.J.M. Schram: *Kinetic Theory of Gases and Plasmas.* 1991 ISBN 0-7923-1392-5
47. A. Micali, R. Boudet and J. Helmstetter (eds.): *Clifford Algebras and their Applications in Mathematical Physics.* 1992 ISBN 0-7923-1623-1
48. E. Prugovečki: *Quantum Geometry.* A Framework for Quantum General Relativity. 1992 ISBN 0-7923-1640-1
49. M.H. Mac Gregor: *The Enigmatic Electron.* 1992 ISBN 0-7923-1982-6
50. C.R. Smith, G.J. Erickson and P.O. Neudorfer (eds.): *Maximum Entropy and Bayesian Methods.* Proceedings of the 11th International Workshop (Seattle, 1991). 1993 ISBN 0-7923-2031-X
51. D.J. Hoekzema: *The Quantum Labyrinth.* 1993 ISBN 0-7923-2066-2
52. Z. Oziewicz, B. Jancewicz and A. Borowiec (eds.): *Spinors, Twistors, Clifford Algebras and Quantum Deformations.* Proceedings of the Second Max Born Symposium (Wrocław, Poland, 1992). 1993 ISBN 0-7923-2251-7
53. A. Mohammad-Djafari and G. Demoment (eds.): *Maximum Entropy and Bayesian Methods.* Proceedings of the 12th International Workshop (Paris, France, 1992). 1993 ISBN 0-7923-2280-0
54. M. Riesz: *Clifford Numbers and Spinors* with Riesz' Private Lectures to E. Folke Bolinder and a Historical Review by Pertti Lounesto. E.F. Bolinder and P. Lounesto (eds.). 1993 ISBN 0-7923-2299-1
55. F. Brackx, R. Delanghe and H. Serras (eds.): *Clifford Algebras and their Applications in Mathematical Physics.* Proceedings of the Third Conference (Deinze, 1993) 1993 ISBN 0-7923-2347-5
56. J.R. Fanchi: *Parametrized Relativistic Quantum Theory.* 1993 ISBN 0-7923-2376-9
57. A. Peres: *Quantum Theory: Concepts and Methods.* 1993 ISBN 0-7923-2549-4
58. P.L. Antonelli, R.S. Ingarden and M. Matsumoto: *The Theory of Sprays and Finsler Spaces with Applications in Physics and Biology.* 1993 ISBN 0-7923-2577-X
59. R. Miron and M. Anastasiei: *The Geometry of Lagrange Spaces: Theory and Applications.* 1994 ISBN 0-7923-2591-5
60. G. Adomian: *Solving Frontier Problems of Physics: The Decomposition Method.* 1994 ISBN 0-7923-2644-X
61 B.S. Kerner and V.V. Osipov: *Autosolitons.* A New Approach to Problems of Self-Organization and Turbulence. 1994 ISBN 0-7923-2816-7
62. G.R. Heidbreder (ed.): *Maximum Entropy and Bayesian Methods.* Proceedings of the 13th International Workshop (Santa Barbara, USA, 1993) 1996 ISBN 0-7923-2851-5
63. J. Peřina, Z. Hradil and B. Jurčo: *Quantum Optics and Fundamentals of Physics.* 1994 ISBN 0-7923-3000-5
64. M. Evans and J.-P. Vigier: *The Enigmatic Photon.* Volume 1: The Field $\boldsymbol{B}^{(3)}$. 1994 ISBN 0-7923-3049-8
65. C.K. Raju: *Time: Towards a Constistent Theory.* 1994 ISBN 0-7923-3103-6
66. A.K.T. Assis: *Weber's Electrodynamics.* 1994 ISBN 0-7923-3137-0

Fundamental Theories of Physics

67. Yu. L. Klimontovich: *Statistical Theory of Open Systems.* Volume 1: A Unified Approach to Kinetic Description of Processes in Active Systems. 1995
ISBN 0-7923-3199-0; Pb: ISBN 0-7923-3242-3
68. M. Evans and J.-P. Vigier: *The Enigmatic Photon.* Volume 2: Non-Abelian Electrodynamics. 1995 ISBN 0-7923-3288-1
69. G. Esposito: *Complex General Relativity.* 1995 ISBN 0-7923-3340-3
70. J. Skilling and S. Sibisi (eds.): *Maximum Entropy and Bayesian Methods.* Proceedings of the Fourteenth International Workshop on Maximum Entropy and Bayesian Methods. 1996 ISBN 0-7923-3452-3
71. C. Garola and A. Rossi (eds.): *The Foundations of Quantum Mechanics – Historical Analysis and Open Questions.* 1995 ISBN 0-7923-3480-9
72. A. Peres: *Quantum Theory: Concepts and Methods.* 1995 (see for hardback edition, Vol. 57) ISBN Pb 0-7923-3632-1
73. M. Ferrero and A. van der Merwe (eds.): *Fundamental Problems in Quantum Physics.* 1995 ISBN 0-7923-3670-4
74. F.E. Schroeck, Jr.: *Quantum Mechanics on Phase Space.* 1996 ISBN 0-7923-3794-8
75. L. de la Peña and A.M. Cetto: *The Quantum Dice.* An Introduction to Stochastic Electrodynamics. 1996 ISBN 0-7923-3818-9
76. P.L. Antonelli and R. Miron (eds.): *Lagrange and Finsler Geometry.* Applications to Physics and Biology. 1996 ISBN 0-7923-3873-1
77. M.W. Evans, J.-P. Vigier, S. Roy and S. Jeffers: *The Enigmatic Photon.* Volume 3: Theory and Practice of the $\boldsymbol{B}^{(3)}$ Field. 1996 ISBN 0-7923-4044-2
78. W.G.V. Rosser: *Interpretation of Classical Electromagnetism.* 1996
ISBN 0-7923-4187-2
79. K.M. Hanson and R.N. Silver (eds.): *Maximum Entropy and Bayesian Methods.* 1996
ISBN 0-7923-4311-5
80. S. Jeffers, S. Roy, J.-P. Vigier and G. Hunter (eds.): *The Present Status of the Quantum Theory of Light.* Proceedings of a Symposium in Honour of Jean-Pierre Vigier. 1997
ISBN 0-7923-4337-9
81. *Still to be published*
82. R. Miron: *The Geometry of Higher-Order Lagrange Spaces.* Applications to Mechanics and Physics. 1997 ISBN 0-7923-4393-X
83. T. Hakioğlu and A.S. Shumovsky (eds.): *Quantum Optics and the Spectroscopy of Solids.* Concepts and Advances. 1997 ISBN 0-7923-4414-6
84. A. Sitenko and V. Tartakovskii: *Theory of Nucleus.* Nuclear Structure and Nuclear Interaction. 1997 ISBN 0-7923-4423-5
85. G. Esposito, A.Yu. Kamenshchik and G. Pollifrone: *Euclidean Quantum Gravity on Manifolds with Boundary.* 1997 ISBN 0-7923-4472-3
86. R.S. Ingarden, A. Kossakowski and M. Ohya: *Information Dynamics and Open Systems.* Classical and Quantum Approach. 1997 ISBN 0-7923-4473-1
87. K. Nakamura: *Quantum versus Chaos.* Questions Emerging from Mesoscopic Cosmos. 1997 ISBN 0-7923-4557-6

KLUWER ACADEMIC PUBLISHERS – DORDRECHT / BOSTON / LONDON